AF734670

79
77
Portobelo
El Porvenir
Panama Canal
Madden Lake
Chepo
Cañita
SAN BLAS
Lago Bayano
PANAMÁ
Río Chepo
Panama
Balboa
Bay of Panama
Río Chucunaque
ARCHIPIÉLAGO
DE LAS PERLAS
La Palma
Río Tuira
Yaviza
DARIÉN
Gulf of Panama
Río Balsas
Río Atrato
COLOMBIA
PACIFIC OCEAN
79

Contents

Not shown on our maps but in his journal Pine described the site as a mining camp operated by Texasgulf, Inc., on the slopes of Cerro Colorado [ca. 0832/8148] (USNM). According to specimen labels, the camp was located 1.25 km N of Escopeta Camp [ca. 0830/8148] and ca. 24 km NNE of San Félix [0817/8152 (USBGN)]. Pine and Izor, GL = San Félix, at 1,175-1,400 m, 23-24.5 km NNE San Félix, plain grass, brush, some forest, 1980 May 29, Jun 1, 16, 19, 23-25, 27, 29, Jul 1, 5-6, specimens at 1,425-1,500 m are labeled 25 km NNE of San Félix, Bocas del Toro province, although it is unclear if Pine and Izor crossed the divide, see San Félix for further discussion, 1980 May 21, Jun 8, 13, 22 (USNM, as "Colorado Camp," and as "San Félix").

COLUMBUS; see Colón, Isla

COLUMBUS ISLAND; see Colón, Isla

COMARCA DE BARÚ; see Barú, Comarca de

COMARCA DEL DARIÉN; see Darién, Provincia del

COMARCA NGÖBE BUGLÉ; see Buenos Aires

CONCEPCIÓN; see La Concepción

CONCEPCIÓN, LA; see La Concepción

CONCEPCIÓN, RÍO; Veraguas (Atlantic)
0850/8059 (USBGN)
Originates in the mountains of NE Veraguas province and flows N into Golfo de los Mosquitos [0900/8115 (USBGN)] 27 km E of Calovébora [0848/8113 (USBGN)] and 12 km W of the Veraguas/Colón provincial boundary (JOG). See also: Cucuyos de Veragua.

CONCEPCIÓN LA VIEJA; Panamá (E, Pacific)
ca. 0902/7927
Not shown on our maps but located in E Panamá province in Corregimiento Juan Díaz [0902/7927 (USBGN)] near the town Juan Díaz [0903/7927 (USBGN)] (q. v.). Pérez, Madrid, and Sanguillen, 1990 Apr 8 (MVUP).

CONDADO DEL REY; Panamá (E, Pacific)
ca. 0902/7932
60 m, section of greater Panama City [0858/7932 (USBGN)] located in Corregimiento Betania [0900/7932 (USBGN)]. Pérez and Wong, 1989 Dec 2 (MVUP).

CONGAL; Veraguas (Pacific)
ca. 0803/8117
Described by Wetmore as an area of hills covered with natural forest on the Río San Pablo [0751/8110 (USBGN)] ca. 3 mi (5 km) above the mouth of the Río Cobre [0801/8118 (USBGN)] (USNM). On JOG this region is ca. 6 km NNE of Soná [0801/8119 (USBGN)] just E of Los Bajos del Cobre [0803/8119 (USBGN)]. USBGN also indicates a town El Congal [0802/8118] in the same region. Wetmore, GL = Soná, Río San Pablo, small creeks and trails through the forest, ivory nut palm and considerable climbing bamboo in the lower areas, cut-up woodlands bordering the forest, 1953 June 13 (USNM). Not shown on our maps.

CONGO, RÍO; Darién

0825/7822 (USBGN)

Originates on the E slopes of Cerro Chucantí [ca. 0850/7828] and flows S forming an inlet on the N coast of Golfo de San Miguel [0822/7817 (USBGN)] 25 km W of La Palma [0825/7809 (USBGN)] (JOG; MGRP). Angehr, Christian, Castillo, and Jordán, GL for Cerro Chucantí, 1996 Aug 27-Sep 3 (MVUP; Angehr and Christian, 2000). (Angehr, 2003:312, as "Estuarios del Congo y Cucunatí").

CONTADORA, ISLA; Panamá, Archipiélago de las Perlas

0838/7902 (USBGN)

Island with a land area of ca. 2 km^2 (200 ha), located 20 km NNW of Isla del Rey [0822/7855 (USBGN)] (JOG), and according to a map section in Wetmore's journal, the elevation is 150 feet (45 m) (USNM). According to Murphy (1944b), the woods were "thin and dry…gulches were dusty; cactus (*Holocereus*) was growing on the rocks." Bovallius, no fresh water, "totally blasted and showed but a very poor animal life," see discussion of Bovallius under Pueblo Nuevo [ca. 0820/8013 ?], 1882 (NHRM; Rendahl, 1920); Smith, R. W., 1937 Feb 2-3 (ANSP); Wetmore, abrupt rocky headlands with sandy beaches between them, ridge covered with small trees, open gallery forest, flowering mangrove swamp, 1960 Jan 17 (USNM); (MacArthur et al., 1972—species list; see comments in Bibliography). (N. Smith, 1971).

CONTRACTORS HILL; Panamá (Canal Zone) (W, Atlantic, Pacific)

0902/7939 (USBGN)

100 m, hill on the continental divide located on the W side of the Panama Canal [0920/7955 (USBGN)] just N of the mouth of the Río Grande [0902/7938 (USBGN)] (E761B; E8660). Morton, 1 mi (1.5 km) N, 1966 Jul 26, 1969 Nov 20 (AMNH; FMNH); Hespenheide, observations just N of, 1969 (Hespenheide, 1975); Morton and Hespenheide, N of, 1970 Jul 20 (CM); Morton, studies, 1973-1975 (Morton, 1980).

CONTRERAS, ISLAS; Veraguas (Pacific)

0750/8147 (USBGN)

Group of ca. 10 islands in the Pacific Ocean 20 km N of Isla de Coiba [0728/8147 (USBGN)] and 20 km SW of Isla de la Corocita [0758/8139 (USBGN)] (JOG, as "Las Islas de Contreras"). The only collection from the group appears to be that of Wetmore on Isla Brincanco [0751/8147 (USBGN)], although he made distant observations of seabirds on Isla Uva [0748/8146 (USBGN)] (q. v.). Wetmore, GL for Isla Brincanco, 1962 Mar 23-24 (USNM; Olson, 1980b, as "Islas Contreras, Isla Brincanco").

COORPORACIÓN BAYANO; see Bayano, Lago

COPÉ; see El Copé

COPÉ, EL; see El Copé

COPETE; see Copete, Cerro

COPETE, CERRO; Chiriquí

ca. 0848/8231

10,000 feet (3,050 m), described by Blake as a peak, that is also called La Campana, located E of Volcán Barú [0848/8233 (USBGN)] and W of Casita Alta [ca. 0849/8229] and Peña Blanca [ca. 0848/8229] (Blake, 1958:502 (map)). Mönniche, at 9,500-10,500 feet (2,900-3,200 m), slopes, summit, 1933 Jul-Aug, 1934 Jul, 1937, 1939 (FMNH, as "Copete"; MCN, as "Copete"); Hartman, at 6,000-7,000 feet (1,850-2,150 m), above Boquete, 1949 Feb 9-10, 12, 14, 1953 Jan 29-31, Feb 1-4, 1956 Feb 9-10, 14, 16, and as a GL for Casita Alta, 1949 Feb 8, 11-12 (USNM, as "Copete Mountain," as "Mount Copete," and as "Cerro Copete"; Hartman, 1955, as "Mount Copete"; Hartman, 1957, as "Mount Copete").

COPEY, QUEBRADA EL; see El Copey, Quebrada

COPUNATI, RÍO; see Cucunatí, Río

CORDILLERA; see Los Andes, Cordillera

CORDILLERA, POTRERO; Chiriquí

ca. 0848/8235

ca. 1,900 m, described by Wetmore as an area in the upper end of Llanos del Volcán [ca. 0847/8238] S of Potrero Verde [ca. 0848/8235] (q. v.) in a valley that cuts into the W face of Volcán Barú [0848/8233 (USBGN)]. Located in a box canyon so broken above Potrero Cordillera as to be practically impassible, Wetmore, GL = Volcán, Volcán Barú, at 6,000 feet (1,850 m), from Potrero Verde S along a ridge to Aguacate [0848/8235 (USBGN)], trees with twisted limbs able to withstand strong winds, undergrowth of shrubs with some bamboo, followed a cattle trail through woods that led steeply down to Potrero Cordillera, here springs broke out of the base of the precipitous slope creating a narrow stream lasting only 300-400 yards (270-370 m) before submerging, dry woodland, *Yucca* amid rough stones on the slopes, 1955 Mar 13 (USNM).

CORDILLERA ANDINO; see Alto Carvallo

CORDILLERAS; see Los Andes, Cordillera

CORDILLERA VERAGUA; see Central, Cordillera

COROBÁ, LOMA; Darién

ca. 0730/7802

Alt. ?, described by Wetmore as a hill covered with heavy, undisturbed forest on the Río Jaqué [0731/7810 (USBGN)] between the Arroyo Buradó [ca. 0731/7804] and the Arroyo Tortadó [ca. 0730/7801], Wetmore and Perrygo, GL = Río Jaqué, near the base, 1947 Mar 27 (USNM). Not shown on our maps.

COROCITA, ISLA; see Corocita, Isla de la

COROCITA, ISLA DE LA; Veraguas (Pacific)

0758/8139 (USBGN)

Sea level, an island 8 km SSW of Puerto Vidal [0803/8136 (USBGN)]. The thin, 14 km long island is parallel to the coast and separated from it by a maze of rivers and mangrove swamps between the mouth of the Río Tabasará [0802/8142 (USBGN)] and Boca de Pajaroncito [0756/8139 (USBGN)] (JOG). Wetmore and Perrygo, GL = Puerto Vidal, Río Vidal [ca. 0801/8138], landed opposite the mouth of the Río Vidal, swampy woodland, high forest beyond a cleared area, 1953 June 8 (USNM, as "Isla Corocita"; Olson, 1980b, as "Puerto Vidal, Río Vidal, Isla Corocita").

CORONADO, PLAYA; see Playa Coronado

CОROTÚ; Chiriquí
0817/8254 (USBGN)
50 m, town 3 km W of Puerto Armuelles [0817/8252 (USBGN)] on the Río Corutú [0816/8254 (USBGN)] 3 km above its mouth (JOG). Wetmore, GL = Puerto Armuelles, collected at two locations on a logging road that led W along the Río Corutú to the Costa Rican border. The first one was ca. 3-4 mi (5-6 km) from Puerto Armuelles (according to JOG, this area is 2 km W of Corotú) where the valley walls were steep and irregular with forest on the slopes and to a lesser extent in the level areas, collected in recent clearings, in cultivated areas and in older, abandoned clearings with *rastrojo*. The second location was 7.5 mi (12 km) from Puerto Armuelles (according to JOG, this area is 7 km W of Corotú) at the logging camp near the Costa Rican border, collected in open pasture near the forest being logged, on hill slopes covered with forest, along a bulldozed trail that followed the ridge just inside the border, 1966 Feb 9-10, 12, 15, 22, Mar 8 (USNM).

COROTÚ, RÍO; Panamá (E, Pacific)
0851/7838 (USBGN)
Originates on the W end of Serranía de Majé [0855/7830 (USBGN)] E of Pico Columna [0857/7838 (USBGN)], flows SW then turns sharply to the SSE flowing parallel to the Río Chimán [0843/7838 (USBGN)] for 3 km before joining it 11 km N of Chimán [0842/7837 (USBGN)] (JOG). According to Wetmore, the Río Corotú was 40-50 feet (10-15 m) wide at its mouth and marked the head of tidewater on the Río Chimán, Wetmore and Perrygo, GL = Chimán, Río Chimán, they collected both up- and downstream of their camp, which was ca. 1 mi (1.5 km) above the river mouth near a series of 10-12 foot (3-4 m) deep pools, in dense forest, in areas of creepers and heavy undergrowth, on hills covered with gallery forest, inland to tributaries, on an open beach, in old clearings and along dry creeks on the Río Chimán near the mouth of the Río Corotú, 1950 Feb 18, 20, 23-28 (USNM; Wetmore, 1965:73, as "Río Curutú").

COROTÚ, RÍO; see Corutú, Río

COROZAL; Panamá (Canal Zone) (E, Pacific)
0859/7934 (USBGN)
Sea level, town 4 km NNW of Balboa [0857/7934 (USBGN)] and ca. 1 km E of the Panama Canal [0920/7955 (USBGN)], located on a section of the Panama Railroad [ca. 0922/7954] that was not relocated during the construction of the Panama Canal (E661; PR; RM1). Hallinan, shrubbery near a clearing, trees in mangrove swamp, muddy tidal flats, in a house, 1907 Jul 28, Sep 8, 1914 Jun 12, 1915 Mar 10, 12, Aug 29, Nov 12, Dec 2, 25, 1916 Jan 30, May 6, 21, and as a GL for Las Cruces Trail, 1915 Sep 5, 19, Dec 24 (AMNH; Hallinan, 1924; Griscom, 1926; Eisenmann, 1962a); Goldman, 1911 Jun 15, 19-20 (USNM; Todd, 1913);

Kelsall and Cullingford, 1924 Jun, Nov (BMNH; Douglas and Johnson, 1926); Benson, at 100 feet (30 m), 1931 Jul 2 (USNM); Miller, near, 1937 Feb 10 (USNM); Meyer, 1941 Jun 15 (USNM); Gale, 1960 Jan 9, Sep 12-13, 1963 May 11-12 (USNM; Wetmore, 1962; Loftin and Olson, 1963); GMLP, 1960 Jun 12 (GML); Alvarez, 1962 Oct (GML); Van Horn, 1965 Mar 29 (GML); Yunker, 1961 Jun 26 (USNM); Loftin and Gale, ditch near the canal, 1962 Oct-Nov (Loftin, 1963a); Olson, 1963 May 12 (USNM); Loftin, 1967 Oct 18 (GML); Keenan, 1968 Feb 20 (GML). (Wetmore, 1952:18; Wetmore, 1970; Olson, 1980b; Olson, 1981d). See also: Corozal, Cerro; Corozales Adendro; and Las Cruces Trail.

COROZAL, CERRO; Panamá (Canal Zone) (E, Pacific)
0859/7934 (USBGN)
120 m, hill in Corozal [0859/7934 (USBGN)] 4 km N of Balboa [0857/7934 (USBGN)] (E962). Miller, 1937 Feb 2, 4-5, 10, 15-16, 19, 22, Mar 1 (USNM, as "Corozal Hill").

COROZALES ADENTRO; Panamá (W, Pacific)
0855/7955 (USBGN)
200 m, town located 15 km NW of La Chorrera [0853/7947 (USBGN)] (MGRP, as "Corozales Adentro o El Saino"). The town Corozales Afuera [0855/7952 (USBGN)] is 4 km SE of Corozales Adentro and 11 km NW of La Chorrera. Carrasco, GL = La Chorrera, presumably collected 10-15 km NW of La Chorrera, 1976 Jun 15-16 (GML, as "Corozal, near Chorrera").

COROZALES AFUERA; see Corozales Adentro

COROZAL HILL; see Corozal, Cerro

COROZO; Chiriquí
ca. 0819/8237
Not shown on our maps but located in SW Chiriquí province in Corregimiento Palo Grande [0819/8237 (USBGN)] near Palo Grande [0820/8236 (USBGN)] (q. v.). Miranda, 1983 Sep 25 (MVUP); González, 1990 Aug 7 (MVUP).

CORPUS CHRISTI, RÍO; Panamá (E, Pacific)
0915/7910 (USBGN)
Originates in the hills NW of Chepo [0910/7906 (USBGN)] and flows E entering the Río Mamoní [0908/7904 (USBGN)] 9 km NNW of Chepo and 7 km NNW of Las Margaritas [0911/7905 (USBGN)] (MGRP, as "Río Corpus Cristi"). Guerrero, Mendieta, and Rodríguez, GL = Las Margaritas, 1985 Mar 16 (MVUP).

CORPUS CRISTI, RÍO; see Corpus Christi, Río

CORRIENTE GRANDE; Bocas del Toro
0918/8232 (USBGN)
ca. 25 m, town on the Río Changuinola [0928/8227 (USBGN)] 15 km W of Almirante [0918/8224 (USBGN)] and 15 km S of Changuinola [0926/8231 (USBGN)] (JOG). According to E762, the town is located along a 500 m section of rapids. Delgado and GMLP, is also labeled Río Changuinola, 1980 Mar 21-22 (GML).

CORTATITA, PLAYA; Darién

ca. 0754/7729

Described by Wetmore in 1959 as a point on the E bank of the Río Paya [0757/7730 (USBGN)] where the river banks were still steep and abrupt, dropping 12-18 feet (4-5 m) down to the water. The area was reached by canoe and by a trail cut to camp (USNM). See also: Tuira, Río. Not shown on our maps.

CORTATITA, QUEBRADA; Darién

ca. 0756/7732

Described by Wetmore in 1959 as a 50-75 feet (20-25 m) wide, almost stagnant, bywater of the Río Tuira [0821/7803 (USBGN)] located downstream from the mouth of the Río Paya [0757/7730 (USBGN)]. The water was 2 inches (5 cm) to 2 feet (0.6 m) deep and the banks rose 10-30 foot (3-9 m) above the water (USNM). See also: Tuira, Río. Not shown on our maps.

CORTE CULEBRA; see Gaillard Cut

CORTE DE CULEBRA; see Gaillard Cut

CORTEZA; see El Cortezo

CORTEZO; see El Cortezo

CORUTÚ, RÍO; Chiriquí

0816/8254 (USBGN)

Originates W of Puerto Armuelles [0817/8252 (USBGN)] near the Costa Rican border and flows ESE entering Bahía de Charco Azul [0812/8245 (USBGN)] 2 km S of Puerto Armuelles (JOG). Wetmore, GL = Puerto Armuelles, ca. 3-4 mi (5-6 km) from Puerto Armuelles on the Corotú [0817/8254 (USBGN)] (q. v.) logging road, 1966 Mar 11 (USNM, as "Río Corotú").

COSTA DEL ESTE; Panamá (E, Pacific)

0901/7928 (Buehler et al., 2004)

Sea level, Buehler et al. (2004) gives the coordinates 09°00'40"/79°27'31" for the site, which is located 3 km E of Old Panama [0900/7929 (USBGN)] and "comprises broad mudflats, flanked to the west by the Río Matías Hernández [0901/7929 (USBGN)] and to the east by mangroves." Buehler et al. (2004), observations, 2003 Sep 7-27 (Buehler et al., 2004). Not shown on our maps.

COSTA RICAN BORDER; see Santa Clara

CÔTE FERME

A Verreaux specimen so labeled was regarded by Deignan (1961:393) as a syntype of *Thryophilus rufalbus castanonotus* Ridgway and for some reason he equated the locality with Panama. The term actually appears to have applied to Venezuela. There does not seem to have been any reason whatever to relate this specimen to Panama.

COTITO, RÍO; Chiriquí

0848/8244 (USBGN)

Originates on the continental divide NNE of Santa Clara [0850/8246 (USBGN)] and flows SSW entering the Río Chiriquí Viejo [0820/8241 (USBGN)] 3 km S of

Santa Clara (E762). See also: Santa Clara, Quebrada.

COTÓN, RÍO; Panamá (W, Pacific)
0825/8002 (USBGN)
Originates SE of El Valle [0836/8008 (USBGN)] and flows SSE entering the Gulf of Panama [0800/7930 (USBGN)] just E of the Coclé/Panamá provincial border ca. 16 km ENE of Río Hato [0823/8010 (USBGN)] and ca. 9 km ENE of Santa Clara [0823/8007 (USBGN)] (E761A). Smith, N., observations above Santa Clara in the valley of, 1980 Jul (Ridgely and Gwynne, 1989:322).

CRAPECHE ISLAND; see Trapiche, Isleta

CREEK, CAMP; see Camp Creek

CREO, PLAYA DEL; Panamá (E, Pacific)
ca. 0841/7837
Sea level, described by Wetmore as a small bay opening to the Gulf of Panama [0800/7930 (USBGN)] ca. 2 mi (3 km) S of Chimán [0842/7837 (USBGN)], Wetmore and Perrygo, GL = Chimán, on a small, sandy beach with rocks and gravel inland, in forest and thickets behind the beach, inland along a small stream, along a small stream just S of Playa del Creo, along a trail leading N across the point to La Aguadilla [ca. 0842/7838], 1950 Feb 21-22 (USNM). Not shown on our maps.

CRESTA, LA; see La Cresta

CRICAMOLA; see Boca de Cricamola

CRICAMOLA, BOCA DE; see Boca de Cricamola

CRICAMOLA, RÍO; Bocas del Toro
0859/8154 (USBGN)
Originates at the merger of several streams that drain the Cordillera Central [0838/8150 (USBGN)]. The river flows NNW through grass and marsh entering the S coast of Laguna de Chiriquí [0903/8200 (USBGN)] 15 km W of the base of Península Valiente [0905/8151 (USBGN)] at Boca de Cricamola [0859/8155 (USBGN)] (E762; JOG). Wedel, according to his specimen labels, Wedel traveled 24 mi (39 km) upstream, collected in a dry stream bed near the river ca. 15 mi (24 km) S of its mouth, over a pasture near a farmhouse on the river ca. 16 mi (26 km) S of its mouth, apparently his early collecting was entirely from the settlement, Boca de Cricamola (q. v.), 1932 Dec, 1934, 1937 Apr-Jun, Sep-Nov, 1938 Jan-Mar, 1939 Sep (CMNH, as "Cricamola River"; FMNH, as "Cricamola," and as "Cricamola River"; MVZ, as "Cricamola R.," and as "Cricamola River"; SDSNH, as "Cricamola River," UCLA, as "Cricamola R," and as "Cricamola River"; Rogers, 1939, as "Cricamola River"; Lincoln, 1944, as "Cricamola"); Miller and Richardson, GL = Laguna de Chiriquí, collected in Laguna de Chiriquí along the 12 mi (19 km) boat trip S from Laguna de Bluefield [0909/8155 (USBGN)] and along the 7 mi (11 km) they traveled upstream. According to Miller, the jungle came right down to the shore of the 12-14 foot (4 m) deep stream and the only variety was a few areas of tall grass and a few bare sandbars upstream. The area was relatively uninhabited as

only 3-4 huts were noted along the river, 1936 Mar 2 (UCLA, as "Cricamola River"; Miller, 1937, as "Cricamola River").

CRICAMOLA RIVER; see Cricamola, Río

CRISTO, EL; see El Cristo

CRISTÓBAL; Colón (Canal Zone) (E)
0921/7955 (USBGN)
Sea level, town immediately SW of Colón [0922/7954 (USBGN)] on the E coast of Puerto Cristóbal [0921/7955 (USBGN)], which opens to Bahía Limón [0920/7956 (USBGN)] (E8660). Smith, R. W., 50 mi (80 km) N of, 1937 Jul 5 (ANSP); Wetmore, observations, 60 mi (100 km) N of, on board a ship to Costa Rica, 1940 Oct 11 (Wetmore, 1965:256); Ambrose, observations, 1955 (Wetmore, 1972:417); Barnard, observations, 1955 Dec 14 (Wetmore, 1965:63); Collector ?, 1962 Oct 20 (FMNH); Tejera, 1992 Nov 13 (MVUP). See also: Portobelo.

CRISTÓBAL, ISLA; see San Cristóbal, Isla

CRISTÓBAL, PUERTO; see Cristóbal

CRUCES, CAMINO DE; see Las Cruces Trail

CRUCES, LAS; see Las Cruces

CRUCES TRAIL; see Las Cruces Trail

CRUCES TRAIL, LAS; see Las Cruces Trail

CRUZ, PUNTA; see Cruz, Punta de

CRUZ, PUNTA DE; Panamá, Archipiélago de las Perlas
0813/7908 (USBGN)
Not shown on our maps, but USBGN indicates a point on the E coast of Isla San José [0815/7907 (USBGN)], and according to Wetmore (1946b:2) the point is at the terminus of a 1.5 mi (2.5 km) extension. Pacific Project, GL = Gulf of Panama, Isla San José, off of island, 1969 Aug 29 (USNM, as "Punta Cruz").

CUAN BAY; Bocas del Toro
0912/8218 (USBGN)
Small bay enclosed on the E and on the SW by 2 mainland peninsulas that extend 2 km N into Ensenada Grande [0914/8219 (USBGN)], the bay is W of Laguna de Tierra Oscura [0912/8216 (USBGN)] and E of Ensenada Shepherd [0913/8221 (USBGN)] (JOG). See also: Cuan Creek.

CUAN CREEK; Bocas del Toro
0912/8217 (USBGN)
Short mainland river that enters the central S coast of Cuan Bay [0912/8218 (USBGN)] (E762). SBE, 1989 Feb 27, Mar 4 (USNM).

CUATRO, QUEBRADA; see Chucunaque, Río

CUATRO CALLES; Colón (Canal Zone) (E)
0858/8033
A Western Watershed study site (MVUP; Louis Berger Group et al., 2003)

located near the Quebrada Cuatro Callitas on the west bank of Río Coclé del Norte [0905/8035 (USBGN)], opposite the mouth of the Río Cuatro Calles [0908/7943 (USBGN)] on the east nank of Río Coclé del Norte, (G. Angehr, pers. comm.). Garcés, Jiménez, López, Ramos, and Rodríguez, 2001 Jun 2 (MVUP).

CUATRO CALLITAS, QUEBRADA, see Cuatro Calles

CUATRO MIL; see Cuatro Mil, Potrero

CUATRO MIL, POTRERO; Chiriquí

ca. 0850/8233

ca. 7,000 feet (2,150 m), described by Wetmore in 1955 as a great pasture, cut 2 years previously, but not yet producing grass, reached by a logging road leading E from the main highway near Cerro Punta [0852/8235 (USBGN)], Wetmore, GL = Cerro Punta, at 6,800-7,000 feet (2,050-2,150 m), clearing bordered by low growth of bracken, pokeweed and other scrubby brush, trails through forest with much bamboo, 1-2 day old clearings, 1955 Mar 8, 10 (USNM, as "Cuatro Mil." and as "Potrero Cuatro Mil"). Not shown on our maps.

CUCARACHA HILL; Panamá (Canal Zone) (E, Pacific)

ca. 0902/7938

Not shown on our maps but USBGN and E8660 both indicate Cucaracha Slide [0902/7938 (USBGN)], which is the unstable side of a hill, adjacent to Cucaracha Reach [0902/7938 (USBGN)]. The hill shown on E8660 above Cucaracha Slide is 150 m and located 2 km NW of Paraiso [0902/7938 (USBGN)]. MCU, GL = Paraiso, 1955 Jul 11 (GML).

CUCHERBO, ISLA; see Cuchirvo

CUCHIRBO; see Cuchirvo

CUCHIRO; see Cuchirvo

CUCHIRVO; Panamá (Canal Zone) (W, Atlantic)

0904/7957 (USBGN)

Sea level, town located on the SW coast of Lago Gatún (Gatun Lake) [0912/7955 (USBGN)] 29 km NW of La Chorrera [0853/7947 (USBGN)]. The town is located on the SE coast of Bahía Trinidad [0905/8000 (USBGN)] ca. 10 km N of the mouth of the Río Trinidad [0901/7958 (USBGN)] (E8660, as "Cuchiro"; JOG, as "Cuchiro"). E8660 also indicates Isla Cucherbo [0904/7958 (USBGN)], a small island in the bay, 2 km W of Cuchirvo. Based on the proximity of Cuchirvo to the Río Trinidad, we presumed that Cuchirvo and Maxon Ranch were the same locality. Shiras Expedition, GL = Río Trinidad, near, 1914 Mar 7-17 (AMNH, as "Cuchirbo"; Anthony, 1916, as "Maxon Ranch, Río Trinidad").

CUCULLO; see Cucuyos de Veragua

CUCUNATÍ, RÍO; Darién

0827/7815 (USBGN)

Originates in the SE foothills of Serranía de Cañazas [0852/7822 (USBGN)] and flows S widening into a narrow estuary that enters Golfo de San Miguel [0822/

7817 (USBGN)] 11 km WNW of La Palma [0825/7809 (USBGN)] (MGRP; MHA, as "Río Cupanuti"). Festa, according to his travel account and map, he traveled through the lower part of the estuary before heading W through channels that led into the bay below the Río Chanatí [0827/7816 (USBGN)], 1895 Jun 14, 22 (Salvadori and Festa, 1899, as "Río Coconati"; Festa, 1909, as "Río Cocunati"). (Delgado, 1983; Angehr, 2003:312, as "Estuarios del Congo y Cucunatí").

CUCUYOS DE VERAGUA; Veraguas (Atlantic)
ca. 0845/8056
100 m, apparently an old mining locality near the Río Concepción [0850/8059 (USBGN)] about 10 km above its mouth. A town named "Cucullo" appears on the index to E762 in Veraguas province NNE of Santa Fé [0831/8105 (USBGN)] and SSW of Belén [0853/8052 (USBGN)] at roughly the spot marked "Cucuyos de Veragua" by Wetmore on one of our other maps (USNM). Mitchell, 18?? (ANSP; Cassin, 1857, as "Cocuyos de Veragua"; Salvin and Godman, 1888-1904, vol. 2:558, as "Cocuyos de Veragua").

CUESTRA DE PIEDRA; Chiriquí
ca. 0845/8238
Described by Sousa and Herman as a farm on the slopes of Volcán Barú [0848/8233 (USBGN)] below the town Volcán [0846/8238 (USBGN)], Sousa and Herman, blood smears, c. 1980 (Sousa and Herman, 1982). Not shown on our maps.

CUIPO; Colón (W)
0905/8003 (USBGN)
100 feet (30 m), town 12 km SW of Escobal [0909/7958 (USBGN)] on the SW coast of Bahía Trinidad [0905/8000 (USBGN)] of Lago Gatún (Gatun Lake) [0912/7955 (USBGN)] (E661, as "Quipo"; E761A). Chapin, Rogers and Mathews, 1923 Aug 14-15, 19-23, and as a GL for La Ullama, 1923 Aug 15-16, 18, 20-23 (FMNH, as "Quipo").

CULEBRA; Panamá (Canal Zone) (W, Atlantic)
ca. 0903/7940
200 feet (60 m), former station on the original Panama Railroad [ca. 0922/7954] now located W of the Panama Canal [0920/7955 (USBGN)] section called Gaillard Cut [0903/7940 (USBGN)] just N of the continental divide (E761A; PR). Presumably the town moved slightly W during construction of the canal as Hallinan labeled his specimens "New Culebra." Bovallius, 1883 Mar 12 (NHRM; Rendahl, 1919:2, 11, 25); Hallinan, power lines, trails, cultivated areas, vegetation at the edge of a sandy beach, 1915 Mar 21, Oct 23-24, Nov 3, 21, 25 (AMNH; Hallinan, 1924; Bond and Meyer de Schauensee, 1944:41; and Zimmer, 1950:8, all as "New Culebra").

CULEBRA, CORTE DE; see Gaillard Cut

CULEBRA, ISLA; Panamá (Canal Zone) (E, Pacific)
0855/7932 (USBGN)
20 m, peninsula that extends 300 m from the S coast of Isla Naos [0855/7932 (USBGN)] (E962). Seutin, 1994 Oct 24 (MVUP). See also: Fortified Islands.

CULEBRA, RÍO; Bocas del Toro

A local name for the upper reaches of the Río Changuinola (see Holcomb Trail).

CULEBRA-ARRAIJAN TRAIL; Panamá (Canal Zone) (W, Atlantic, Pacific)
ca. 0903/7940
Apparently a trail that extended 10 km N-S from Culebra [ca. 0903/7940] to Arraijan [0857/7939 (USBGN)]. E761A indicates a trail named the "Río Grande Trail" that extends N from Arraijan to the Río Grande [0902/7938 (USBGN)], where it connects with trails that extend along the Panama Canal [0920/7955 (USBGN)] to Culebra. None of our maps show a more direct route. Hallinan, trees in the jungle, palms, 1914 Nov 26, 1915 Jan 1 (AMNH; Hallinan, 1924).

CULEBRA CUT; see Gaillard Cut

CULEBRA TRAIL; see Holcomb Trailt

CULUBRE, RÍO; Bocas del Toro
0908/8230 (USBGN)
Originates on the continental divide between Cerro Fabrega [0907/8252 (USBGN)] and Cerro Pando [0855/8243 (USBGN)] and flows NE and E entering the Río Changuinola [0928/8227 (USBGN)] 34 km S of Changuinola [0926/8231 (USBGN)] (CARP; JOG; MGRP). This river appears to be the one called the Río Changuena (q. v.) by collectors.

CUNNINGHAM, CAMP; see Camp Cunningham

CUPANUTI, RÍO; see Cucunatí, Río

CUPE, BOCA DE; see Boca de Cupe

CUPÉ, RÍO; Darién
0804/7735 (USBGN)
Originates on the NE slopes of Serranía de Pirre [ca. 0749/7744] and flows NE entering the Río Tuira near Boca de Cupe [0803/7735 (USBGN)] (MGRP; SMN2B). Anthony, Richardson, and Ball, GL = Cituro (q. v.), 1915 May 12-13, 18, 20 (AMNH; Anthony, 1916; Griscom, 1932a, as "Cupé River"; Chapman, 1926, as "eastern Panama"); Benson, specimens are also labeled El Tigre, a locality that we could not find, but which must be located on or near the Río Cupé. According to Griscom (1929a), Benson collected at only three localities, El Real, Cupe, and Cana, therefore it is possible that El Tigre refers to a point along the route from Boca de Cupe to Cana, 1928 Feb 1-4, 10, 12 (ANSP, as "El Tigre"; MCZ, as "El Tigre, Río Cupe"; Griscom, 1929a, as "El Tigre, Río Cupé"; Wetmore, 1953, as "El Tigre"; Haffer, 1967b); Wetmore, GL = Boca de Cupé, mouth of, 1964 Mar 21 (USNM).

CUPÉ RIVER; see Cupé, Río

CURUNDU; Panamá (Canal Zone) (E, Pacific)
0859/7933 (USBGN)
25 m, section of greater Panama City [0858/7932 (USBGN)] located N of Ancón [0858/7933 (USBGN)] and W of Calidonia [0858/7933 (USBGN)] (E962) that was originally part of Fort Clayton Army Reservation [0901/7934 (USBGN)].

USBGN treats Curundu Army Reservation as an alternate name for the town. Gale, 1961 Oct (Loftin, 1963a); Loftin, 1962 May 26-27, Oct-Nov (USNM; Loftin, 1963a, as "Curundu," and as "Churundu"); Collector ? [Loftin ?], 1962 May 26-27, Sep 9 (GML; MCN); Loftin and Gale, 1962 Nov (Loftin, 1963a); Loftin et al., 1963-1964 (Loftin et al., 1966; Eisenmann, 1966); Olson et al., jungle behind, woods near, 1963 Feb 19-21, Mar 3-9, 14, 17, 20, 27, Apr 19-20, 25, 27-28, May 11, 1966 Apr 30, May 1-2, 24-25, Jun 22-23, 27, 29-30, Jul 8, 26, 31, Aug 1-3 (AMNH; FMNH; UF; USNM; Loftin and Olson, 1963; Olson, 1968; Olson, 1965; Olson, 1981a; Olson, 1981d; Olson, 1983); Chapman and Tyson, 1 mi (1.5 km) E, 1964 Feb 11-13 (FSU); Powell, 1966 Sep 23, 1969 Jan 4 (AMNH); Buskirk, 1968 Dec 11 (AMNH); Morton, studies, 1973-1975 (Morton, 1980); Grundor, PAD area, 1976 Nov 10 (GML); Santamaría, 1986 Apr 9 (MVUP). (Loftin, 1975). See also: Fort Clayton Army Reservation.

CURUTÚ, RÍO; see Corotú, Río

DAMANÍ, LAGUNA DE; see Cañaveral, Río

DAMAS, BAHÍA; Veraguas, Isla de Coiba
0727/8141 (USBGN)
Bay on the central E coast of Isla de Coiba [0728/8147 (USBGN)], extending 9 km N-S from Punta Damas [0730/8141 (USBGN)] to Punta Felipa [0724/8142 (USBGN)] (JOG). Described by Wetmore (1957:9) as extending S as far as Estero Boca Grande [0724/8139 (USBGN)], although both USBGN and JOG identify two bays on the SE coast, the more S being Ensenada María [0724/8141 (USBGN)] (q. v.). (Routledge, 1920).

DAMAS, PUNTA; Veraguas, Isla de Coiba
0730/8141 (USBGN)
Sea level, point on the central E coast of Isla de Coiba [0728/8147 (USBGN)] on the SE tip of the 4 km wide peninsula that extends 4 km E into the Pacific Ocean (E762, as "Punta Damasi"; JOG, as "Punta Damasi"). See also: Punta Damas, Campamento de.

DARIÉN; see Darién Railroad Stop

DARIÉN, EL; see El Darién

DARIÉN, GULF OF; see Urabá, Golfo de

DARIÉN, ISTHMUS OF; Darién
ca. 0810/7745
According to Selfridge (1874:7-8) in his account of the U. S. Naval surveys conducted in search of a route for the canal, "the terms Darien and Panama are indiscriminately applied to the narrow neck of land between latitudes 8° and 10° north, connecting North and South America. Properly speaking, the Isthmus of Panama comprises all the territory watered by the Chagres and its tributaries across the divide to the Pacific. The Isthmus of Darien extends from the San Blas Mountains, which separate the head-waters of the Mandinga [0928/7904 (USBGN)] and Marmoni [probably refers to the Río Mamoní [0908/7904 (USBGN)]] from those of the Chagres [0907/7942 (USBGN)], to the boundaries of the State of Choco, or to the mountain-

range from which the Tuyra [sic] or Darien River takes its rise." Maack, who traveled as a naturalist with the surveys, apparently used the terms "properly" as specimens are labeled both "Isthmus of Panama" and "Isthmus of Darien." Western Colombia has also been considered the Isthmus of Darién as seen in Hellmayr (1910a) who refers to *Calliste emiliae* from "Isthme de Darien, Río Truando," although Dalmas (1900) did not include "Isthme de Darien" in his original description. Wood and Michler, according to Wetmore et al. (1984:431) the type specimen *Calliste Lavinia* Cassin is now lost, specimens from the Ríos Truando and Atrato were collected in Colombia, 18?? (Cassin, 1858; Sclater, 1863b; Hellmayr, 1910a; Wetmore et al., 1984:431); Viguier, according to Wyse (1886), Viguier worked in the valley of the Río Tuira near Paya, Pinogana, and Yaviza during French surveys in Darién province, Griscom indicates (1932 and 1935b) that he worked in the "Gulf of Darien," but the source for this information is unclear, 1876 Nov-1877 Apr (MNHN; Wyse, 1886:47, 134-161; Hellmayr, 1911a:17; Hellmayr, 1934:492; Griscom, 1932a:303, as "Gulf of Darien"; Griscom, 1935b:362, as "Gulf of Darién"; Wetmore et al., 1984:182); Maack, G. A., the surveys were conducted from Jan-Jun 1871, several specimens are labeled Mar, which appears to correspond with the time that the teams worked along the Río Tuira [0821/7803 (USBGN)] in Darien province, although many specimens are not dated. Maack's narrative (1872:155) indicates that he collected on both the Atlantic and the Pacific drainages. On the Atlantic side he explored "the Gulf of Darien del Norte [Golfo de Urabá [0825/7653 (USBGN)] in Colombia], the eastern Antioquian Mountains [Colombia], the Atrato Valley [Colombia] with its western mountainous boundary, the Manzanillo Island at Aspinwall [see Colón [0922/7954 (USBGN)]], and the whole region between Aspinwall and Panama," and on the Pacific side he explored "the different islands in the Bay of Panama, the Bay of Saint Miguel, or Gulf of Darien del Sur, the country up the river Tuyra [sic], and the Cupica Bay [Colombia], with the River Napipi [Colombia]." He later explains (1872:170) that he "brought home from the interior of the Isthmus a large collection of...some hundred stuffed and skinned birds and mammals, besides many alcoholic specimens," 1871 Jan-Jun (MCZ; USNM). (Sclater, 1861b; Anthony, 1923). See also: Panama, Isthmus of.

DARIÉN, PARQUE NACIONAL; Darién

0749/7744 (Angehr, 2003:303)

National park that extends N-S over a roughly 20-40 km strip along the Colombian border. The S boundary of the park includes much of the Pacific Coast on either side of the town Jaqué [0731/7810 (USBGN)], although the town is outside of its limits. The towns Yaviza [0811/7741 (USBGN)] and El Real de Santa María [0808/7743 (USBGN)] are just outside of the park's W boundary. The park includes the headwaters for the Ríos Balsas [0815/7759 (USBGN)], Tuira [0821/7803 (USBGN)], Sambú [0805/7818 (USBGN)], and Jaqué [0731/7810 (USBGN)] in addition to Cana [ca. 0745/7742] and Cerros Pirre [0746/7744 (GA)], Malí [0807/7714 (FH)] and Tacarcuna [0811/7717 (USBGN)] (RDP2). According to Angehr (2003:303), the park's 579,000 ha include areas of semideciduous lowland, evergreen lowland, submontane and montane forests. The park was created in 1980 shortly after which it was designated a World Heritage Site (1981) and a Biosphere Reserve (1983).

Alvarez and Ellis, observations of nests in the buffer zone of, 1991 (Alvarez and Ellis, 1994). See also: Pirre Station.

DARIÉN, PROVINCIA DEL; (Pacific)
0810/7750 (USBGN)
The most eastern province in Panamá extending from the low mountains on the Panamá provincial border to the Colombian border. The continental divide forms the N and E border and extends roughly NW-SE along the Comarca de San Blas border and roughly N-S along the Colombian border. The S coast is divided into two distinct regions. Golfo de San Miguel [0822/7817 (USBGN)], which receives the effluent of most Darién rivers, forms an irregular and lowland coastline on the NW, whereas low hills extend along the SW coast. In spite of its isolation and its reputation for pirates and bandits - or perhaps even on account of these factors - numerous collectors and expeditions have visited Darién province making an exhaustive list impractical here. The earliest ornithological specimens from the region were collected in the late 1800s. G. A. Maack, who worked in 1871 as the naturalist with U. S. Navy surveys of the region brought back both birds and mammals, but much of the locality data was not specific. Heyde and Lux collected a few specimens near El Real de Santa María [0808/7743 (USBGN)] in 1889, and later in 1895 Festa made extensive collections throughout the Golfo de San Miguel. During the 1900s the coastal regions and waterways were explored by expeditions such as the Barbour, Brooks and Underwood, the Griscom Crosby, the Askoy, the Vanderbilt and the British Trans-Americas. Mountainous areas, specifically the region near Cana, and the divide near Cerros Malí and Tacarcuna, were explored by Goldman in 1912, Benson in 1928, Anthony, Richardson and Ball in 1915, Wetmore in 1964, and researchers at GML in the 1950s-1960s to name a few. Wetmore and Perrygo worked along the length of the Río Jaqué, and later Wetmore explored the valleys of and the tributaries of the Ríos Chucunaque and Tuira.

DARIÉN, SERRANÍA DEL; Darién
0830/7730 (USBGN)
Mountain range forming the continental divide in E Panamá. The W portion of the range extends NW-SE along the Comarca de San Blas/Darién provincial border and the E portion of the range extends roughly N-S along the Colombian border (RDPF). Chapman (1915b:420) calls this range the Santa Espiritu Mountains of eastern Panamá and Robbins et al. (1985) refer to it as the Serranía de Tacarcuna. The Serranía del Darién is often mentioned in conjunction with specimens collected near Cerro Malí [0807/7714 (FH)] (q. v.) and Cerro Tacarcuna [0811/7717 (USBGN)] (q. v.) (Anthony, 1916, as "Darién Cordillera"; Wetmore, 1963b; Wetmore, 1967). Heyde and Lux, GL for El Real de Santa María, 1889 Jun (WFVZ). (Myers, 1969, as "Camp Summit (Seranía del Darién)"; Wege, 1996, as "Tacarcuna Ridge").

DARIÉN CORDILLERA; see Darién, Serranía del

DARIÉN NATIONAL PARK; see Darién, Parque Nacional

DARIÉN NATIONAL PARK STATION; see Pirre Station

DARIÉN RADIO STATION; see Darién Railroad Stop

DARIÉN RAILROAD STOP; Panamá (Canal Zone) (E, Atlantic)
0907/7946 (USBGN)
Sea level, a stop on the Panama Railroad [ca. 0922/7954] located on the E coast of Lago Gatún (Gatun Lake) [0912/7955 (USBGN)] 7 km SSE of Frijoles [0910/7948 (USBGN)]. E8661 indicates a radio tower 500 m S of the station on a small island in Lago Gatún (E661; E866). Hallinan (1924:304) describes the Darien Radio Station as being "on the Canal, about 22 miles from the Atlantic entrance, on the Atlantic Slope." Hallinan, dying trees and floating vegetation in Lago Gatún, 1915 Dec 12 (AMNH; Hallinan, 1924, as "Darien Radio Station"); Weber and Wonder, near, between Frijoles and, 1928 Dec 26 (FMNH, as "Darien"). See also: El Darién.

DARKLAND; see Tierra Oscura

DAVID; Chiriquí
0826/8226 (USBGN)
40 m, capital of Chiriquí province located on the Carretera Interamericana [0832/8242 (USBGN)] 1 km W of the Río David [0822/8224 (USBGN)] and ca. 8 km N of the river deltas, islands and mangroves that form at the mouth of the Río Chiriquí [0820/8226 (USBGN)] (E762; JOG). This area of mangroves is considered an Important Bird Area (Angehr, 2003). David was one of the early settlements and was accessible from Panama City [0858/7932 (USBGN)] by sea, and less easily overland, and therefore visited by early naturalists who worked in Panamá. Collector ?, date ?, GL for Boquete, date ? (FMNH; USNM); Whitely, date ? (OC); Hicks, 18??, 1855 Feb, 1861 Feb, 1865 Feb 13-14, Mar, 1895 Feb 7, W coast of Chiriquí, 1865 Feb 4 (BMNH; FMNH; USNM; Lawrence, 1867d; Salvin, 1867b; Sclater and Salvin, 1869a; Salvin and Godman, 1879-1904, vol. 1; Davidson, 1938); Warszewicz, near, c. 1850 (Gould, 1850b; Griscom, 1924b); Bridges, T., in town and vicinity of, in plains, thickets, tall trees, forests, and dense jungle, on river banks, near orange trees, *Tamarindus indicus*, *Erythrinae*, and *Ficus*, 1856 Jan-Mar (BMNH; Sclater, 1856c; Gould, 1856; Gould, 1861b:65; Salvin, 1867b; Sclater and Salvin, 1867:123, as "Veragua," and as "David in Southern Veragua"; Salvin and Godman, 1888-1904, vol. 2:416); Arcé, reached David and worked in the area at "Mina de Chorcha," "Volcán de Chiriquí," and "Bugaba" (Salvin, 1870b:177); Brown, "heavily forested lowlands," "cool tropical forest or on the plantations at about 200 feet (60 m) above sea level," 1900 Oct 15-22, 1901 July 25 (ANSP; MCZ; USNM; Bangs, 1901d; Bangs, 1902a; Bangs and Penard, 1921:391; Griscom, 1933); Watson, date ? (BMNH; Salvin and Godman, 1897-1904, vol. 3:106); Davidson, 1931 Nov 23 (CAS); Martín and Saldaña, 1985 Apr 11, and as a GL for Río Risacua, 1985 Apr 11 (MVUP); Madrid, 1985 Mar 24 (MVUP). (Berlepsch, 1911; Todd, 1913; Chapman, 1917; Hellmayr, 1918; Davidson, 1938; Olson, 1981d; Angehr, 2000).

DAVID, RÍO; Chiriquí
0822/8224 (USBGN)
Originates on the S slopes of Volcán Barú [0848/8233 (USBGN)] and flows SSE and S entering the Río Chiriquí [0820/8226 (USBGN)] 8 km SSE of David [0826/8226 (USBGN)] (E762). Bridges, T., banks of, thick bush on the margins of, in a

thicket on the flat land between the rivers David and Chiriquí, 1856 Jan-Mar (BMNH; Sclater, 1856c).

DAVIS, FORT; see Fort William D. Davis Army Reservation

DEATH BEACH; see Muerto, Playa

DEER, ISLA; Bocas del Toro, Archipiélago de Bocas del Toro
0914/8208 (USBGN)
10 m, small island with a land area of ca. 0.5 km^2 (50 ha) just N of Isla Popa [0911/8207 (USBGN)]. Its entire S coast is enclosed by a portion of the N coast of Isla Popa where the two islands are separated from one another by a 100-300 m wide channel (E762). SBE, GL = Laguna de Chiriquí [0903/8200 (USBGN)], used to further modify specimens from Isla Popa (q. v.), 1987 Mar 9, 30-Apr 13 (USNM, "Deer Island Channel").

DEER ISLAND CHANNEL; see Deer, Isla

DEJAL BEACH; see Fort Kobbe Army Reservation

DELORES, LOMA; Darién
ca. 0732/7809
Alt. ?, described by Wetmore as a hill on the Río Jaqué [0731/7810 (USBGN)], less than 1 mi (1.5 km) upstream from Jaqué [0731/7810 (USBGN)], reached by a trail that begins in mangroves and leads up a steep incline to a narrow ridge covered with gallery forest with an undergrowth of palm, Wetmore and Perrygo, GL = Jaqué, 1946 Mar 20 (USNM). Not shown on our maps.

DIABLO; see Diablo Heights

DIABLO, RÍO; Panamá (E, Pacific)
0907/7834 (see text)
Originates in eastern Panamá province on the Panamá/San Blas provincial border and flows S entering the NE coast of Lago Bayano [0908/7835 (USBGN)] ca. 5 km S of the town Río Diablo [0911/7832 (USBGN)] (q. v.) and ca. 60 km E of Chepo [0910/7906 (USBGN)] (MGRP; RDP2). USBGN treats Río Diablo as an alternate name for Río Capandí, yet treats Capandí as an alternate name for the town Río Diablo. As none of our maps use the name Río Capandí for the river, we used the name Río Diablo here.

DIABLO HEIGHTS; Panamá (Canal Zone) (E, Pacific)
0858/7934 (USBGN)
10 m, section of greater Panama City [0858/7932 (USBGN)] adjacent to the E side of the Panama Canal [0920/7955 (USBGN)] 1 km NW of Ancón [0858/7933 (USBGN)] (E962, as "Diablo"). Hunt and Seutin, at 10 m, 1994 Apr 27, 1995 Mar 9, 17 (LSUMZ, as "Diablo").

DIRGADO, PUNTA; Darién
0720/7801 (USBGN)
Not shown on our maps but USBGN indicates a point on the S coast of Darién province ca. 20 km NNW of the Colombian border, ca. 11 km NNW of Punta Cocalito [0719/7759 (USBGN)] and ca. 5 km SSE of Punta Guayabo Chiquito

[0724/7802 (USBGN)]. Murphy, S of, 1937 Sep 11 (AMNH, as "Point Dirgado").

DIRGADO POINT; see Dirgado, Punta

DIVALÁ; Chiriquí

0825/8243 (USBGN)

ca. 25 m, town on the W bank of the Río Divalá [0821/8243 (USBGN)], 16 km SW of La Concepción [0831/8237 (USBGN)] and 9 km N of the coast (JOG). According to Wetmore's journal, in 1960 the former forests had been cut except for limited tracts, so that numerous species taken here earlier were rare or had disappeared. In Wetmore's gazetteer notes he indicated that "some of Arcé's early collections for Salvin were made here," although his supporting evidence was not documented. Angehr (in litt., Oct 2003) suggested that Bibalaz is an older spelling for Divalá, and the above statement suggests that Wetmore may also have believed this. If these two names are synonymous, then Arcé's collections from Bibalaz were from Divalá Arcé, 18?? (BMNH; Salvin and Godman, 1888-1904, vol. 2, as "Bibalaz," and as "Vibalá"); Brown, "heavily forested lowlands," "cool tropical forest or on the plantations at about 200 feet (60 m) above sea level," 1900 Oct 27-Dec 19 (ANSP; MCZ; OC; USNM; Bangs, 1901c; Bangs, 1901d; Bangs, 1901e; Bangs, 1903a; Bangs, 1905a; Bangs, 1905b; Bangs, 1908; Bangs and Penard, 1921:380, 391; Thayer and Bangs, 1905:146, 149; Thayer and Bangs, 1906; Todd, 1913; Griscom, 1929b; Griscom, 1933; Zimmer, 1929; Peters, 1931a:313, 318; Peters and Griscom, 1929; Davidson, 1938; Moore and Peters, 1939; Eisenmann, 1950; Howell, 1956; Aldrich and Duvall, 1958; Wetmore, 1959b:18; Wetmore, 1970; Monroe and Howell, 1966; Olson, 1981d; Olson, 1983); Underwood, 1900 Nov 9 (ANSP); Miranda, GL for Finca El Jobito, 1983 Sep 25 (MVUP); Atencio, Ayarza, Gómez, Ponce, and Serrano, 1990 Jan 21 (MVUP); DaCosta, Guitton-Mayerma, Klicka, Lelevier, Miller, and Spellman, N of Bahía de Charco Azul, at 27 m, at 08°23'37"/82°43'46", 3 km W of, and at 29 m, at 08°23'58"/82°39'50", 4 km E of, 2006 Mar 3 (MBM; STRI). (Chapman, 1915a). See also: Bibalaz; and Rico, Estero.

DIVISA; Herrera

0808/8041 (USBGN)

ca. 25 m, town in N Herrera province 33 km E of Santiago [0806/8059 (USBGN)] on the Carretera Interamericana [0832/8242 (USBGN)] just S of the Río Santa María [0806/8029 (USBGN)], where it forms the Herrera/Coclé provincial boundary (JOG). Loftin, 5 mi (8 km) S of, observations, 1961 Oct 21 (Loftin, 1963a); Dutary, both specimens are labeled Coclé, 1965 Feb 5 (GML).

DOCK 45 ROAD; see Fort William D. Davis Army Reservation

DOLEGA; Chiriquí

0834/8225 (USBGN)

220 m, town 15 km N of David [0826/8226 (USBGN)] on the Río David [0822/8224 (USBGN)] (E762; JOG; MHA). According to Bangs (1902a:17), the town was located on the route taken by Brown from David to the peak of Volcán Barú [0848/8233 (USBGN)], and his account indicates that "the elevation rose through

a sparsely wooded region to the pueblo de Dolega [which had] great plantations of cocoa, coffee, sugarcane and bananas, at an elevation of about 700 feet (210 m)." Pérez, 1981 Jan 1 (MVUP).

DOÑA JUANA; Los Santos
ca. 0755/8027
Not shown on our maps but located in N Los Santos province in Corregimiento Los Santos [0755/8027 (USBGN)] near the town Los Santos [0756/8025] (q. v.). Hernández, 1985 Mar 29 (MVUP).

DONATO, PUNTA; Bocas del Toro
0923/8221 (USBGN)
Point on the mainland 11 km NNE of Almirante [0918/8224 (USBGN)] and 10 km N of Punta Pondsock [0918/8221 (USBGN)] at the S end of Boca del Drago [0925/8220 (USBGN)] channel (E762). Wetmore, GL = Almirante, in the bay near, 1958 Feb 20 (USNM, as "Donato Point").

DONATO POINT; see Donato, Punta

DORADO, EL; see El Dorado

DOS RIOS HOTEL; Chiriquí
ca. 0846/8238
Described by Engleman et al. (1989:498) as a hotel near the town Volcán [0846/8238 (USBGN)]. Ridgely, observations, 1976 Feb 3, Mar 6, 1979 Mar 20 (Wetmore et al., 1984:305, 393, 409).

DRAGO, BOCA DEL; Bocas del Toro
0925/8220 (USBGN)
Marine channel that extends 5 km N-S between the mainland and Isla Colón [0924/8217 (USBGN)] connecting Bahía Almirante [0920/8218 (USBGN)] on the S with the Caribbean Sea [1500/7500 (USBGN)] on the N. There is also a town called Boca del Drago [0925/8220 (USBGN)] on Isla Colón (JOG). Wetmore GL = Almirante [0918/8224 (USBGN)], at Changuinola Canal (q. v.), 1958 Feb 20 (USNM; Wetmore, 1959b:13); GMLP, presumably these specimens were collected along the mainland coast of Boca del Drago channel, although they may have also come from Changuinola Canal [ca. 0919/8221] (q. v.), or another canal just S of it as is also labeled Channel 2 [ca. 0922/8222 ?] (q. v.), 1960 Aug 25, 1962 Jun 3 (GML, as "Boca del Drago," as Boca del Drago Canal," and as "Bocas del Drago, Channel 2"); Loftin, observations, the bay near Boca del Drago, 1963 May 18-21 (Loftin and Olson, 1963, as "Boca del Drago Bay").

DRAGO, BOCA DEL;
0920/8218 (USBGN)
Bocas del Toro, Archipiélago de Bocas del Toro
Community [0925/8220 (USBGN)] on the NW end of Isla Colón (JOG, E762). Billo, GL = Isla Colón, 2007 Apr 6-7, "1/2 km NE of Boca del Drago Rd, 2 km from NW side of island, on trail leading from little community" and "1/2 km NE of Boca del Drago Rd. 4 km from STRI station, on hill across cow pasture," (USNM).

DROP ZONE; see Gatún Drop Zone

DUABLO, RÍO; Chiriquí
0820/8241 (USBGN)
Originates a short distance NW of La Concepción [0831/8237 (USBGN)] and flows S between the Río Escarrea [0820/8241 (USBGN)] and the Río Gariché [0821/8243 (USBGN)] entering the Río Chiriquí Viejo [0820/8241 (USBGN)] very near its mouth (MHA). See also: Rico, Estero; and Santo Domingo, Quebrada.

DUCK CLUB; see La Jagua Hunting Club

DULCE, GULF OF; see Chiriquí, Golfo de

DUPÍ, RÍO; Chiriquí
0816/8156 (USBGN)
Originates in the hills 10 km NW of San Félix [0817/8152 (USBGN)] and flows S and SW entering the Estero El Cabuyal [0813/8157 (USBGN)] 8 km W of Las Lajas [0815/8152 (USBGN)] (JOG). Wetmore, GL = San Félix, collected near Pozos de Salud de Galíque [ca. 0818/8155] (q. v.), behind the buildings above a creek on slopes covered with partly cut forest, 1956 Feb 18 (USNM).

DUQUE, QUEBRADA; see Duque, Río

DUQUE, RÍO; Panamá (E, Atlantic)
0916/7941 (USBGN)
Originates NW of Represa Madden (Madden Dam) [0913/7937 (USBGN)] and flows S crossing the Carretera Boyd-Roosevelt [0917/7943 (USBGN) 16 km N of Gamboa [0907/7942 (USBGN)]. The river then turns W and enters the Rio Agua Sucia [0917/7946 (USBGN)], which continues W and enters the NE coast of Lago Gatún (Gatun Lake) [0912/7955 (USBGN)] (E661, as "Quebrada Duque"; RM2). See also: Frijolito.

EAST DIVERSION; Colón (Canal Zone) (E)
0920/7953 (USBGN)
Waterway that extends NNE-SSW from Río Folk [0921/7954 (USBGN)] to the entrance of the Panama Canal [0920/7955 (USBGN)] just S of Bahía Limón [0920/7956 (USBGN)] (E8660). See also: Gatún.

EAST HARBOR; Panamá, Archipiélago de las Perlas
ca. 0815/7904
Apparently a harbor on the E coast of Isla San José [0815/7907 (USBGN)], located E of the mouth of the Río Marina [ca. 0815/7906], presumably on the E end of Ensenada Playa Grande [0815/7906 (USBGN)] near the airport. The term East Harbor appears in Wetmore (1946b), but specimens collected by Wetmore and Morrison in 1944 all appear to be labeled San Jose Island. See also: San José, Isla.

EDWARDS CREEK; see Bryan Creek

EL AGALLITO; see El Agallito, Playa

EL AGALLITO, LA PLAYA; see El Agallito, Playa

EL AGALLITO, PLAYA; Herrera
ca. 0800/8024
Sea level, beach on the coast of Bahía de Parita [0808/8024 (USBGN)] located just

NNE of the Chitré Airport [0759/8025 (USBGN)] near mangroves and salt beds (Delgado and Butler, 1993 (map), as "El Agallito Beach"). Delgado, 1980 Feb 27 (Wetmore et al., 1984:634), observations, 1984 Jan 10-Feb 3 (Ridgely and Gwynne, 1989:63); Wiedenfeld, observations in medium-height black mangroves, 1986-1988 (Wiedenfeld, 1992); Egeler and Williams, 1995 Dec, 1996 May (Egeler and Williams, 2000); Backwell et al., observations, "large, unvegetated, intertidal mudflat," 1995 Oct-1996 Jan (Backwell et al., 1998, as "La Playa El Agallito"). Mendez (1979:51, as "El Agallito").

EL AGALLITO BEACH; see El Agallito, Playa

EL AGUACATE; see El Aguacate Número Uno

EL AGUACATE NÚMERO UNO; Panamá (W, Atlantic)
0845/7958 (USBGN)
300 m, town on the N slopes of Cerro Trinidad [0844/7958 (USBGN)] ca. 12 km W of Capira [0845/7953 (USBGN)] (JOG; MGRP, as "El Aguacate"). JOG indicates that a road, which becomes a footpath, extends roughly E-W between the town and Capira. Our maps also indicate another town named Aguacate [0857/7959 (USBGN)], at 100 m, and located 25 km NNW of Capira and ca. 5 km S of Bahía Trinidad [0905/8000 (USBGN)] (JOG; MGRP, as "El Aguacate"), but specimens probably came from the town 12 km W of Capira. Montenegro, GL = Capira, some specimens are also labeled La Coca [0853/7953 (USBGN)] (q. v.), 1970 Aug 21, 1971 Nov 21, 1972 Jun 28, 1973 Jul, 1976 Mar 6, 1978-1981, 1982 Jun 14, 1983 Aug 15-16, 1984 May 16, 25, and as a GL for Trinidad Arriba, 1978 Mar 12 (GML).

EL ÁGUILA; Coclé (Atlantic)
0857/8011 (USBGN)
ca. 50 m, town in NE Coclé province on the Río Uracillo [0858/8011 (USBGN)] 3 km SW of Boca de Uracillo [0858/8011 (USBGN)] (JOG, as "Aguila"). Wetmore and Perrygo, GL = Boca de Uracillo, 1952 Feb 24 (USNM). See also: Uracillo, Río

EL ATILLO; Panamá (E, Pacific) ?
ca. 0901/7931 ?
Possibly refers to the section of greater Panama City [0858/7932 (USBGN)] named El Hatillo that is located SE of Betania [0901/7932 (USBGN)] and NW of Parque Lefevre [0901/7929 (USBGN)], although Hatillo is a common name for towns throughout Panamá. Terriente, 1977 Oct 25 (GML).

EL BAISAL; see Baizal

EL BAJO CHITRA; see Chitra

EL BANCO; Chiriquí
0841/8231 (USBGN)
According to a letter written by Mönniche on July 6, 1955, El Banco was a large, green pasture "sloping down towards the southeast." It was located "north and slightly west of Potrerillos, [and from] the highway to Boquete and looking west this green pasture dominat[ed] the view." Currently, the name refers to a town, at 1,000 m, located 14 km S of Volcán Barú [0848/8233 (USBGN)], 15 km SSW of

Boquete [0847/8226 (USBGN)] and 3 km WNW of Potrerillos Abajo [0839/8229 (USBGN)], near the peak Cerro El Banco [0842/8232 (USBGN)] (JOG). Collector ? [Watson ?], at 3,700 feet (1,150 m), date ? (ANSP); Watson, at 3,000-3,700 feet (900-1,150 m), 1903 Mar (ANSP; CU; FMNH; LSUMZ; MCZ; OC; USNM; Oberholser, 1904; Zimmer, 1941b:14; Wetmore, 1952:16).

EL BANCO; Chiriquí

ca. 0825/8152

900 feet (270 m), described by Davidson (1938:255) as a locality within Llanos de Cacicón [ca. 0826/8152] (q. v.) 15 km N of San Félix [0817/8152 (USBGN)], that is presumably located S of the current town of Cacicón [0826/8153 (USBGN) and W of the Río San Félix [0810/8151 (USBGN)]. According to JOG and MGRP, the areas N of San Félix that lie below 400 m are close to either the Río San Félix or the Río San Juan [0816/8157 (USBGN)]. Davidson, at 900 feet (270 m), 1930 Dec 15-26 (CAS; CM; Blake, 1958; Davidson, 1938). Not shown on our maps.

EL BARRERO; Herrera

0756/8031 (USBGN)

ca. 25 m, small town 7 km S of Parita [0800/8031 (USBGN)] (JOG). Wetmore and Perrygo, GL for: Quebrada Agua Caliente, Quebrada de los Pajaros, and Quebrada Limón, 1948 Feb 28, Mar 18, 22-23 (USNM; Wetmore, 1950a; Wetmore, 1952:17; Wetmore, 1953; Wetmore, 1970); Atencio, 1983 Jul 23 (MVUP).

EL BARÚ; see Barú, Volcán

EL BRAZO; Darién

ca. 0732/7808

Sea level, described by Wetmore as a river bend section of the Río Jaqué [0731/7810 (USBGN)] ca. 3 river mi (5 km) upstream from Jaqué [0731/7810 (USBGN)], Wetmore and Perrygo, GL = Jaqué, Río Jaqué, mangrove swamp with areas of cultivation, banks grown with *Tilia* and cane, small streams, heavy forest, S in hills covered with low, dry forest near Loma González [ca. 0729/7806] and near Quebrada del Tigre (not located), 1946 Mar 22, 25, Apr 5, 7-9 (USNM; WFVZ). Not shown on our maps.

EL BRUJO; Coclé (Pacific)

ca. 0814/8032

The region S of Aguadulce [0815/8033 (USBGN)] and N of the Río Estero Salado [0811/8029 (USBGN)] (JOG, as "Finca El Brujo"). Tejera, GL = Aguadulce, *1.6 km de* (1.6 km from) Aguadulce, 1975 Mar 29, Sep 3 (MVUP, as "Ciénaga El Brujo," and as "El Brujo").

EL BRUJO, CIÉNAGA; see El Brujo

EL BRUJO, FINCA; see El Brujo

EL CABIMO; Chiriquí

ca. 0821/8242

Sea level, town 7 km S of Divalá [0825/8243 (USBGN)] on the Río Chiriquí Viejo [0820/8241 (USBGN)] (q. v.) in an area of tropical grass just below the mouth of the Río Divalá [0822/8243 (USBGN)]. One km S of town the mouth of the Río

Chiriquí Viejo widens in a mangrove swamp and flows parallel to the coast for 2 km, separated from the ocean by a thin, sandy strip of land (JOG). See also: Rico, Estero.

EL CABUYAL, ESTERO; Chiriquí
0813/8157 (USBGN)
An estuary at the lower end of the Río Dupí [0816/8156 (USBGN)] extending through mangrove swamp for 6-10 km and opening to the Golfo de Chiriquí [0800/8220 (USBGN)] 12 km SW of Las Lajas [0815/8152 (USBGN)] (JOG, as "Estero Cabuyal"). See also: Dupí, Río; and Santa Cruz.

EL CACAL; see Piña

EL CALVARIO; Chiriquí
0834/8237 (USBGN)
300 m, a town in W Chiriquí province 3 km N of La Concepción [0831/8237 (USBGN)] (JOG, as "Calvario"; MGRP). Morales, 1968 Jan 28, 31 (GML, as "Calvario," and as "El Calvario").

EL CAMINO A CHIVA CHIVA; see Chiva Chiva, Camino

EL CANDELARIA, ESTACIÓN HIDRO; see Candelaria, Estación Hidro El

EL CAÑO; Coclé (Pacific)
0824/8031 (USBGN)
ca. 50 m, town on the Carretera Interamericana [0832/8242 (USBGN)] 18 km N of Aguadulce [0815/8033 (USBGN)] (JOG). Cooke, observations, 1970 Feb-Mar (Wetmore, 1972:374).

EL CARMEN; Panamá (E, Pacific)
0900/7932 (USBGN)
30 m, section of greater Panama City [0858/7932 (USBGN)] NNE of Calidonia [0858/7933 (USBGN)] and S of Betania [0901/7932 (USBGN)] (E962). Collector ?, 1967 Nov 8 (GML); Mendez, some specimens are labeled GL = Panama City, 1969 Apr 20, 1972 Oct 23, 1973 Dec 27, 1976 Oct 4 (GML).

EL CAUCHAL; Panamá ? (W, Atlantic) ?
ca. 0846/7959 ?
No province was indicated on the specimen but in 1979 Montenegro worked in Panamá province near Capira [0845/7953 (USBGN)] (q. v.). Therefore, El Cauchal may refer to the town Cauchales [0846/7959 (USBGN)], which is shown on JOG at 300 m, 11 km W of Capira and 2 km NNW of El Aguacate Número Uno [0845/7958 (USBGN)]. Montenegro, 1979 Apr 4 (GML).

EL CAVIMO; see Finca El Jobito

EL CHILAR; Colón (W)
0909/8012 (USBGN)
Sea level, town on the W bank of the Río Indio [0912/8011 (USBGN)] 4 km above its mouth [0910/8012 (USBGN)] (E761B, as "Chilar"; MHA, as "Río Indio o Chila"). Described by Wetmore as a settlement of ca. 15 houses on open ground 10-15 feet (3-5 m) above the river and surrounded by lower swampy areas, Wetmore

and Perrygo, GL for: Quebrada Sereña, Quebrada Torno Rompido, Río El Chilar, and Río Indio, 1952 Feb 17-19, Mar 7-11 (USNM, as "Chilar"; Wetmore, 1959b:10, 19; Eisenmann and Howell, 1962, as "Quebrada Torno Rompío, Chilar"; Huff and Wetmore, 1967, as "Chilar"; Parkes, 1976, as "Chilar"; Olson, 1980a, as "Río Indio, Chilar"; Olson, 1981d, as "Chilar, Río Indio"; Olson, 1983, as "Chilar").

EL CHILAR, RÍO; Colón (W)
0910/8012 (USBGN)
Described by Wetmore in 1952 as a stream that enters the Río Indio [0912/8011 (USBGN)] a short distance N of El Chilar [0910/8012 (USBGN)], Wetmore and Perrygo, GL = El Chilar, swampy forest, 1952 Mar 9 (USNM, as "Río Chilar"). Not shown on our maps.

EL CHOROGO; see Bartolo Arriba

EL CHORRO; Panamá (W, Pacific)
ca. 0857/7951
Not shown on our maps but located in W Panamá province in Distrito de La Chorrera [0857/7951 (USBGN)] near La Chorrera [0853/7947 (USBGN)] (q. v.). Pagan, 1985 Mar 15 (MVUP); Candanedo, 1985 Apr 16 (MVUP, as "El Chorro de La Chorrera"); Centeño, 1985 Apr 18 (MVUP, as "El Chorro de La Chorrera").

EL CHORRO DE LA CHORRERA; see El Chorro

EL COCAL; Panamá, Archipiélago de las Perlas
0825/7906 (USBGN)
Sea level, town on the N coast of Isla Pedro González [0824/7906 (USBGN)] (RDPF). Shown as Pedro González on RDPP and Atlas (1988:205) and as Pedro de Cocal on JOG. Described by Wetmore in 1944 as a small settlement of 40-50 inhabitants, Wetmore, GL = Isla Pedro González, along the beach, inland on a volcanic dyke, in open forest and near banana plantations and cane fields, 1944 Mar 9, 11 (USNM, as "Cocal").

EL COCO; Coclé (Pacific)
0824/8021 (USBGN)
ca. 50 m, a town 13 km S of Penonomé [0831/8022 (USBGN)] and 10 km W of Antón [0824/8016 (USBGN)] (MGRP). Wetmore, collected N of El Coco on the road to Penonomé [0831/8022 (USBGN)] in level savannas with occasional stretches of brush and slightly larger scrub, 1962 Mar 10 (USNM; Olson, 1980c). See also: El Rosario.

EL CONGAL; see Congal

EL COPÉ; Coclé (Pacific)
0837/8036 (USBGN)
800 m, city located in NW Coclé province just S of the continental divide (JOG; MGRP; MHA, as "Copé"; SMN2B, as "El Cupé"). Our more recent map RDP2 indicates a much lower elevation of 200 m. Described by Wetmore as a town on a high ridge, at ca. 475 m that overlooked the valley of the Río Barrigón [ca. 0837/8034], Wetmore, the secondary localities attributed to El Copé were located within 3 km of town along the Río Barrigón, GL for: Calvarito, Quebrada El Copey, Río

Barrigón, and Sabaneta, 1962 Feb 21-25 (USNM; Wetmore, 1967, as "Alto Carvallo, N of El Copé"; Wetmore, 1970; Olson, 1983). See also: Omar Torrijos Parque Nacional.

EL COPÉ NATIONAL PARK; see Omar Torrijos Parque Nacional

EL COPEY, QUEBRADA; Coclé (Pacific)
ca. 0837/8036
Described by Wetmore as a small stream near El Copé [0837/8036 (USBGN)], Wetmore, GL = El Copé, 1962 Feb 24 (USNM). Not shown on our maps.

EL CORTEZ; see El Cortezo

EL CORTEZO; Los Santos
0726/8038 (USBGN)
ca. 50 m, town 22 km WNW of Tonosí [0724/8027 (USBGN)] on the Río Tonosí [0721/8022 (USBGN)] (JOG, as "El Cortez"; MGRP; MHA, as "Cortezo"; SMN2A, as "Corteza"). Delgado, from El Cortezo worked NW of town along Filo de Marroquin [0729/8042 (USBGN)] (q. v.) and perhaps E of town near Bijaguales [0726/8032 (USBGN)] (q. v.), date ?, 1979 Feb 26, Jul 3, 1980 Jan 8 (AMNH, as "Los Piraguales, El Cortezo de Tonosi"; GML, as "Filo Marroquin, El Cortezo, Distrito de Tonosi," as "Filo Marroquin, Tonosi," and as "Los Piraguales, El Cortezo, Tonosí"; CRUV; Wetmore et al., 1984:635, as "El Cortozo de Tonosi"; Delgado, 1985, as "Los Piraguales, El Cortezo de Tonosí"; Joseph, 2000, as "Las Piraguales, El Cortezo de Tondsi" [sic]). See also: Azuero, Península de.

EL CORTEZO DE TONDSÍ; see El Cortezo

EL CORTEZO DE TONOSÍ; see El Cortezo

EL CRISTO; Coclé (Pacific)
0815/8037 (USBGN)
ca. 25 m, town 8 km W of Aguadulce [0815/8033 (USBGN)] and 4 km N of the Carretera Interamericana [0832/8242 (USBGN)] (JOG). Wetmore, 1 mi (1.5 km) N of town on a horse trail through rolling hills covered sparingly in *chumico* and thorn scrub, occasional creeks, 1963 Jan 24 (USNM); Tejera, 1975 Mar 29 (MVUP).

EL CUPÉ; see El Copé

EL DARIÉN; Panamá ?
ca. 0907/7946 ?
Unclear if this name refers to the province of Darién, to the train station named Darién Railroad Stop [0907/7946 (USBGN)] (q. v.), or to some other locality, but because Zetek was based in the Canal Zone it seems likely that these specimens came from the railroad station. Collector ? (from Zetek), date ? (rec'd 1915) (USNM); Zetek, date ? (rec'd at USNM in 1915) (UF; USNM).

EL DORADO; Panamá
0901/7933 (USBGN)
Section of greater Panama City [0858/7932 (USBGN)] located in NW Corregimiento Betania [0900/7932 (USBGN)] (Atlas, 1988:183). Rovi, 1985 Feb 8 (MVUP); Pérez and Sánchez, Urbanización Las Mercedes, shown in W Betania on Atlas (1988:208,

as "Las Mercedes"), 1990 Sep 5 (MVUP).

EL EMPALME; Bocas del Toro
0925/8231 (USBGN)
ca. 25 m, town located 3 km S of Changuinola [0926/8231 (USBGN)] on the Río Changuinola [0928/8227 (USBGN)] 10 km SW of its mouth (JOG, as "Empalme"). See also: Changuinola, Río.

ELENA; Bocas del Toro
0934/8246 (USBGN)
Sea level, town in the northernmost tip of Bocas del Toro province on the S banks of the Río Sixaola [0934/8234 (USBGN)] in a cultivated area at the end of a railroad line 30 km NW of Changuinola [0926/8231 (USBGN)] (JOG, as "Elena (Finca Elena)"). Wedel, GL = Río Sixaola, 1927 Jan 15 (MCZ; Peters, 1931a:302, as "Sixaola, Elena").

EL ESPINO; Panamá (W, Pacific)
0829/8001 (USBGN)
ca. 75 m, town 18 km SW of Chame [0835/7953 (USBGN)] and 18 km SE of El Valle [0836/8008 (USBGN)] and on the road labeled "Panama 71," which extends ca. 23 km between El Valle and the Carretera Interamericana [0832/8242 (USBGN)] (JOG; MGRP). Wetmore and Perrygo, GL for Río Calabazo, 1951 Mar 24, Apr 1 (USNM; Aldrich and Duvall, 1958); Wetmore, collected W of El Espino near the Carretera Interamericana [0832/8242 (USBGN)] and near the Coclé/Panamá provincial boundary (according to JOG, this area is ca. 7 km SW of El Espino), from an old road to a small stream with little water surrounded by barren terrain and dry vegetation, a few straggling blossoms, trees filled with berries, 1958 Mar 14, 17 (USNM; Wetmore, 1956a).

EL FARALLÓN DEL CHIRÚ; Coclé (Pacific)
0821/8008 (USBGN)
Sea level, coastal town 3.5 km SSE of Río Hato [0823/8010 (USBGN)] (E761B, as "Port Farallón"; JOG). See also: Farallón del Chirú, Isla.

EL FICAL; see Los Voladores

EL FILO DE TALLO, SERRANÍA; Darién
0827/7757 (USBGN)
200 m, range of low hills in central Darién province 19 mi (31 km) E of La Palma [0825/7809 (USBGN)], extending NW to SE for ca. 40 mi (64 km), E of the lower end of the Río Tuira [0821/7803 (USBGN)] and W of the lower end of the Río Chucunaque [0809/7744 (USBGN)] (DP; RDPF).

EL FRANCÉS; see Francés

EL GAGO; Coclé (Pacific)
0822/8026 (USBGN)
20 feet (6 m), town 18 km SSW of Penonomé [0831/8022 (USBGN)] and 9 km SE of Río Grande [0825/8029 (USBGN)] (E761A; JOG). Described by Wetmore as a small settlement with scattered houses amidst open, level savannas, Wetmore, savannas above, 1963 Jan 23, and as a GL for: Cerro Pan de Azucar, and Gaucho,

1963 Jan 20, 23 (USNM, as "Gago"; Wetmore, 1970, as "Gago"; Olson, 1980c, as "Gago, Gaucho"; Olson, 1981d, as "Gago").

EL GALLITO; Herrera
ca. 0759/8025
Not shown on our maps but located in N Herrera province in Corregimiento Chitré [0759/8025 (USBGN)] near Chitré [0758/8026 (USBGN)] (q. v.). López, 1990 Feb 29 (MVUP).

EL GALLO, ISLETA; see Gallo

EL GUABAL; see Guaval

EL GUABO; Bocas del Toro
0853/8211 (USBGN)
60 m, town on the Río Guabo [0854/8211 (USBGN)] 10.5 km SW of Chiriquí Grande [0857/8207 (USBGN)] on the "Camino Tres de Noviembre," which is presumed to be the old Boquete Trail [ca. 0857/8207] (q. v.) (E762). Peters (1931:295) could not locate the town on his maps, but presumed that the birds had been collected on the banks of the Río Guabo, and Wetmore believed that the locality was an outlying farm near the river (USNM). Presumably the current town El Guabo has developed near the original locality called Guabo. Wedel, 1928 Apr 2-16 (ANSP; MCZ; UCLA; and Peters, 1931a, all as "Guabo"). See also: Guabo, Río.

EL HATILLO; Herrera
0755/8038 (USBGN)
ca. 75 m, town 2 km W of Pesé [0754/8037 (USBGN)] on the road to La Trinidad [0755/8043 (USBGN)] (JOG; MGRP). Gazin, 1951 (USNM; Wetmore, 1956b; Wetmore, 1965:134).

EL HATILLO; see El Atillo

EL HATO; see Volcán

EL HATO DEL VOLCÁN; see Volcán

EL JOBITO; see Finca El Jobito

EL JOBITO, FINCA; see Finca El Jobito

EL JOBO; Coclé (Pacific)
0829/8031 (USBGN)
Not shown on our maps but USBGN indicates a town, at ca. 100 m, 4 km S of El Potrero [0831/8032 (USBGN)] near the Río Grande [0818/8024 (USBGN)]. See also: Grande, Río.

EL LIMÓN; Coclé (Pacific)
0817/8032 (USBGN)
ca. 25 m, town 5 km N of Aguadulce [0815/8033 (USBGN)] (MGRP). Tejera, GL = Aguadulce, 1978 Jun 20 (MVUP).

EL LIMÓN; Herrera
0803/8046 (USBGN)

ca. 25 m, town in N Herrera province located 6 km S of the Carretera Interamericana [0832/8242 (USBGN)] and 33 km SW of Aguadulce [0815/8033 (USBGN)] (JOG). Miller and Weir, at less than 100 m, 2003 Apr 18 (CU; STRI; UAM).

EL LLANO; Panamá (E, Pacific)

0913/7859 (USBGN)

ca. 50 m, town on the Río Chepo [0900/7906 (USBGN)] and on the Carretera Interamericana [0832/8242 (USBGN)] 14 km NE of Chepo [0910/7906 (USBGN)] (JOG). In 1962 Wetmore described El Llano as an old agricultural area that had been cleared of forest to create pastures. Wetmore, the secondary localities attributed to El Llano were located in the region extending 8 km W of town along the Río Chepo and its tributaries, GL for: Platanales, Río Chepo, Río Terable, and Río Uní, 1962 Feb 3-5, 9 (USNM; Wetmore, 1970); Smith, N., observations in the hills N of, 1973 Sep 3 (Wetmore et al., 1984:388); Eisenmann, Ridgely, Pujals, and N. Smith, observations near town, and N of El Llano in lowlands of the Río Chepo basin forest along the El Llano-Cartí Road [ca. 0928/7858], 1972 Aug 2-4, 1973-1975 (Wetmore et al., 1984:27, 34, 121). See also: El Llano-Cartí Road.

EL LLANO; see Volcán, Llanos del

EL LLANO-CARTÍ ROAD; Panamá/San Blas (Atlantic, Pacific)

ca. 0928/7858

Name used by Ridgely for the road named Panama 87 on MGRP, which extends ca. 30 km N-S between the Cartí Tupile [0929/7858 (USBGN)] airport on the W coast of Comarca de San Blas and El Llano [0913/7859 (USBGN)] on the Carretera Interamericana [0832/8242 (USBGN)] in E Panamá province. According to Angehr (in litt., Aug 2005) "there is no settlement to speak of" at the N terminus of the road, only the airport; Cartí Tupile is located on an island ca. 4 km offshore. The road crosses the continental divide and the provincial border at 520 m ca. 10 km NNW of El Llano (MGRP) and in Comarca de San Blas the road passes through Nusagandí [ca. 0919/7859]. Agro and Bevier, 0916/7858, the coordinates given by ANSP indicate that they collected in Panamá province at 200 m, ca. 6 km NNE of El Llano, 1996 Feb 29 (ANSP, as "Nusugandi Road"). See also: El Llano; and Nusagandí.

EL MACANO; Chiriquí

0819/8152 (USBGN)

Not shown on our maps but USBGN indicates a town ca. 3 km N of San Félix [0817/8152 (USBGN)], 1 km W of the Río San Félix [0810/8151 (USBGN)], and 2.5 km E of the Río Dupí [0816/8156 (USBGN)]. Wetmore, GL = San Félix, ca. 500 feet (150 m), in forest on the level edge of a steep slope above the Río San Félix, some larger trees removed, with areas of heavy undergrowth, elsewhere open undergrowth, the area was bordered by *rastrojo*, small fields and pastures, in a small tract of timber with tall *pava* tree loaded with berries, 1956 Feb 16, 19, 27 (USNM). The reference in Wetmore et al. (1984:162) to "the area called Macano" that is "near the coast" is erroneous as the specimens in question are of an upland species and all are labeled "El Macano." See also: San Félix, Río.

EL MANGLE, CIÉNAGA; Herrera
0804/8029 (USBGN)
Sea level, coastal marsh extending 6 km S from banks of the Río Santa María [0806/8029 (USBGN)] and 3 km inland from the SW coast of Bahía de Parita [0808/8024 (USBGN)] (JOG). Described by Angehr (2003, as "Cenegón del Mangle") as an area of "mangroves (300 ha) and associated brackish wetlands (350 ha) and dry scrub at the mouth of the Santa Maria River." Delgado, 1972 (?) Aug 23, 30, 1977 Jun 20, Aug 20, 23, Oct 10, (GML, as "Cenegón del Mangle"; Delgado, 1983; Wetmore et al., 1984:634, as "Cenegón del Mangle," Ridgely and Gwynne, 1989:128, as "Cenegón del Mangle"). See also: Alvina.

EL MINERAL DE VERAGUAS; Veraguas (Atlantic, Pacific)
ca. 0830/8100
Described by Salvin (1870:177) and Lawrence (1862a:4) as a district adjacent to Belén district and near Santiago [0806/8059 (USBGN)] and Calobre [0819/8051 (USBGN)]. This district included Arcé's localities Calovébora, Calobre, Castillo, Chitra, and Cordillera del Chucú. The district is not shown on our maps, but it must have surrounded the town Mineral [ca. 0845/8058] (q. v.) and possibly spanned both the Atlantic and the Pacific drainages. Wetmore (1968:296) indicated that the district was located "inland from the N coast at the foothills along the Río Santiago [0848/8058 (USBGN)]," which corresponds with the map location that we found for Mineral. Described by Merritt as a district on the Caribbean slope of the divide SE of Laguna de Chiriquí [0903/8200 (USBGN)] that was "exceedingly humid" and "entirely clothed with the densest tropical forest" (Salvin, 1874:328). Merritt, 18??, (Lawrence, 1862a, as "El Mineral"; Salvin and Godman, 1888-1904, vol. 2:343-344, as "El Mineral"; Wetmore, 1968:49, 296).

EL MINERAL; see El Mineral de Veraguas

EL MIRADOR; see Mirador

EL MONTUOSO, LOMA; Herrera
0743/8050 (USBGN)
A peak on the W slope of Península de Azuero [0740/8035 (USBGN)] between the headwaters of the Río Suay [0744/8102 (USBGN)] and the Río Mariato [0737/8059 (USBGN)] (Aldrich and Bole, 1937:4 (map), as "Cerro Montuosa"). The peak is not shown on our maps, but the USBGN coordinates indicate that it is in a region where the elevation ranges from 500 m at the town La Loma de la Montuoso [0745/8051 (USBGN)] to 953 m on Cerro Alto Higo [0743/8051 (USBGN)]. Angehr (in litt., Oct 2003) indicates that the peak is shown on map 4039-III [E762 series] with an elevation of 939 m, and although it is within the Reserva Forestal El Montuoso [0743/8050 (Angehr, 2003:203)], it has been extensively deforested. According to Angehr (2003:203), the surrounding forest reserve covers 10,375 ha, contains semideciduous forest and on the higher peaks some submontane elements including *Podocarpus* trees can be found. The peak is not quite 3,200 feet (1,000 m) as indicated by Aldrich and Bole (1937), nor is it in the center of Península de Azuero as described by Griscom (1927a:4, as "Cerro Montosa"). Benson, at 2,500 feet (750 m), heavy forest, 1925 Jul 26-Aug 8 (AMNH, as "Cerro Montosa, Cape

Mala"; Griscom, 1927a, as "Cerro Montosa"; Peters, 1931a:299, as "Cape Mala"; Aldrich and Bole, 1937:6, as "Cerro Montuosa"; Zimmer, 1939, as "Cerro Montosa"; Zimmer, 1941b:7, as "Cerro Montosa"; Zimmer, 1950:21, 26, as "Cerro Montosa"); Delgado, GL for Chepo, 1977 Aug 29 (GML, as "Cerro Montuoso"). See also: Cerro Largo.

EL MONTUOSO, RESERVA FORESTAL; see El Montuoso, Loma

EL MORRO; see Taboga, Morro de

EL MORRO, ISLA; see Taboga, Morro de

EL PELADO, ISLOTE; see Pelado, Isla

EL PELUCA, ESTACIÓN HIDRO; see Peluca, Estación Hidro El

EL PORVENIR; San Blas
0934/7857 (USBGN)
Sea level, town on an island 2 km E of the easternmost land of Peninsula de San Blas [0933/7903 (USBGN)] in the northern waters of Golfo de San Blas [0930/7900 (USBGN)] (JOG). Jolly, observations on the coast near, 1985 Oct 14 (Ridgely and Gwynne, 1989:376).

EL POTRERO; Coclé (Pacific)
0831/8032 (USBGN)
ca. 75 m, town on the Río El Potrero [0829/8032 (USBGN)] 19 km W of Penonomé [0831/8022 (USBGN)] and 10 km NNW of Río Grande [0825/8029 (USBGN)] (JOG). Wetmore, at 50 m, 1962 Mar 8-12, and as a GL for: Río El Potrero, and Río Grande, 1962 Mar 6-9 (USNM; Wetmore, 1970; Parkes, 1976).

EL POTRERO, RÍO; Coclé (Pacific)
0829/8032 (USBGN)
Originates E of El Copé [0837/8036 (USBGN)], flows S and enters the Río Grande [0818/8024 (USBGN)] 2 km S of El Potrero [0831/8032 (USBGN)] (JOG). Described by Wetmore as a clear, shallow, 30-40 foot (9-10 m) wide stream, Wetmore, GL = El Potrero, worked downstream to the junction with the Río Grande where the forest was largely uncut, huge shade trees, gravel bars, brush, an extensive tomato planting, sugar cane, 1962 Mar 6-7 (USNM, as "Río Potrero").

EL PUEBLO, QUEBRADA; see Pueblo, Quebrada del

EL PUERTO; see Puerto Aguaducle

EL PUERTO DE PIÑA; Los Santos
0722/8024 (USBGN)
Sea level, not shown on our maps, but USBGN indicates a town ca. 6 km SE of Tonosí [0724/8027 (USBGN)] on the Río Tonosí [0721/8022 (USBGN)]. Described by Wetmore as a small port above the mouth of the Río Tonosí, near the head of tidewater, Wetmore, GL = Tonosí, along the river in mangrove swamp NW of town, 1957 Mar 27 (USNM, as "Puerto Piña"; Olson, 1980b, as "Tonosí, Puerto Piña").

EL QUIEL; see Quiel

EL REAL; see El Real de Santa María

EL REAL DE SANTA MARÍA; Darién

0808/7743 (USBGN)

Sea level, city in S central Darién province on the Río Tuira [0821/7803 (USBGN)] 2 mi (3 km) SE of the mouth of the Río Chucunaque [0809/7744 (USBGN)] (SMN2B). Heyde and Lux, GL = Serranía del Darién, open forest, 1889 Jun, and as a GL for Río Pirre, 1889 Sep 3 (USNM; WFVZ, as "El Real"; Ridgway, 1911a, as "El Real"; Deignan, 1961:221); Richardson, GL for Río Tuira, 1914 Dec 24, 29, 1915 Jan 9, 23 (AMNH, as "El Real"; Anthony, 1916, as "El Real"; Chapman, 1917, as "El Real"; Hellmayr, 1929:84, as "El Real, Río Tuyra"; Zimmer, 1934:14, as "El Real, Río Tuyra"; Zimmer, 1939, as "El Real, Río Tuyra"; Zimmer, 1941b:3, 11, 14, as "El Real"; Zimmer, 1950:8, 26, 50 as "El Real, Río Tuyra," and as "El Real"; Zimmer, 1955:5, as "El Real, Río Tuyra"; Olson, 1981d, as "El Real, Río Tuira"); Baer, 1924 Feb 7 (USNM, as "Real de Santa María"); Benson, 1928 Jan 17-19 (MCZ, as "El Real"; Chapman, 1929b, as "El Real"; Griscom, 1929a, as "El Real"); Wetmore, 1964 Jan 11, the secondary localities attributed to El Real de Santa María in 1964 were located within 6 mi (10 km) of town, GL for: Aeropuerto, Río Pirre, Río Tuira, Uruseca, and Urutí, 1964 Jan 8-9, 11, 22-27, Feb 16-18, Mar 21 (USNM, as "El Real"); Fish, 1966 Jan 26 (MCN, as "El Real"); Slattery, GL for Río Tuira, 1966 May 31 (GML, as "El Real"); Collector ? [Slattery], GL for Río Tuira, 1966 May 31 (GML); Cavalier, GL for Rio Pirre, 1967 Sept 14 (MVZ, as "El Real"); Slattery, 1967 Aug 28, Oct 9 (GML, as "El Real, Air Strip"; MVZ, as "El Real"); Collector ?, 1968 Mar 20, 1972 Mar 1, Río Tuira, 1966 May 31 (GML); Burton, British Trans-Americas Expedition, 1972 Feb-Mar (BMNH, as "El Real"; Burton, 1973b, as "El Real"; Burton, 1975, as "El Real"); Othon, 1975 Aug 16 (GML, as "El Real"); Wolf and Braun, observations, photographs, 1985 Jan 27, Feb 3 (ANSP, as "El Real"; Braun and Wolf, 1987, as "El Real"). (Wetmore, 1953, as "El Real"; Haffer, 1962, as "El Real"). See also: Yaviza.

EL REY, ISLA; see Rey, Isla del

EL RINCÓN; Herrera

0807/8037 (USBGN)

ca. 25 m, town in N Herrera province 2 km S of the Coclé provincial boundary and 5 km E of Santa María [0807/8040 (USBGN)] (JOG). Wetmore and Perrygo, GL for Ciénaga de la Macana, 1948 Mar 1, 3, 6, 17, 25 (USNM; Wetmore, 1950a; Wetmore, 1950c; Parkes, 1976); local hunter, near, 1962 Oct (Loftin, 1963a); Olson and Loftin, near, 1963 Feb 23 (GML; USNM).

EL ROBLE; Coclé (Pacific)

0810/8039 (USBGN)

ca. 25 m, town 15 km SW of Aguadulce [0815/8033 (USBGN)] on the Carretera Interamericana [0832/8242 (USBGN)] (JOG). Wetmore, GL for Río Santa María, 1963 Jan 24 (USNM); Dutary, 1977 Nov 13 (GML, as "Robles").

EL ROMPIO; Coclé (Pacific)

ca. 0811/8029

Not shown on our maps but USBGN treats Estero El Rompio as an alternate name for Boca de Estero Salado [0811/8029 (USBGN)] (q. v.), therefore the locality is presumed to be near the river mouth in the region ca. 6 km SE of Aguadulce [0815/8033 (USBGN)]. Angehr (in litt., Aug 2005) points out that there is a better known Playa El Rompio near Chitré [0758/8026 (USBGN)], but the database indicates that the specimens are from Coclé province near Aguadulce. Tribaldos, 1986 Apr 12 (MVUP).

EL ROMPÍO; see Las Palmitas

EL ROSARIO; Coclé (Pacific)

0825/8020 (USBGN)

ca. 25 m, town located 10.5 km S of Penonomé [0831/8022 (USBGN)] and 4 km NE of El Coco [0824/8021 (USBGN)]. The town is on the Río Chorrera [0818/8023 (USBGN)] 9 km N of the coast (JOG, as "Finca El Rosario"; MGRP). Laguna del Rosario refers to a coastal region near areas of rice and shrimp cultivation (Adames et al., 1993). According to Angehr (in litt., Aug 2005), Laguna del Rosario "is located at UTM 921500 N 572000 (0820/8021), 8 km S of the town of El Rosario." González, GL = Penonomé, 1977 Nov 26 (GML); Herrera, 1984 Jul 25 (MVUP, as "Laguna del Rosario"); De Leon, the specimen at GML indicates GL = Penonomé, 1984 Jul 25, Aug 8, Sep 11 (GML, as "Rosario"; MVUP, as "Laguna del Rosario"); Tejera, 1984 Sep 27 (MVUP, as "Laguna del Rosario"); Tejera and Araúz, 1985 Aug 6 (MVUP); Tejera and Guardia, 1985 Oct 29 (MVUP); Weir, at less than 10 m, at 0820/8020, near Anton, 2003 Apr 14, 22 (CU, as "Rosario"; LSUMZ as "Rosario").

EL SAINO; see Corozales Adentro

EL SALADO; Coclé (Pacific)

0811/8029 (USBGN)

Not shown on our maps but USBGN indicates a town at the mouth of the Río Estero Salado [0811/8029 (USBGN)]. Described by Wetmore as a small fishing village beyond Gallo [ca. 0812/8030] near the mouth of the Río Estero Salado at the coastal terminus of the road extending 9 km SE from Aguadulce [0815/8033 (USBGN)]. Wetmore, GL = Aguadulce, near the salt ponds, 1963 Jan 17, 21 (USNM, as "Salado"); Wetmore, observations, on mudflats, 1963 Jan 25 (Wetmore, 1965:125, as "Puerto Salado"). See also: Gallo.

EL SALTO; Chiriquí

0848/8229 (USBGN)

1,600 m, town 5 km WNW of Boquete [0847/8226 (USBGN)] (E762). Presumably this town, or the region between El Salto and the peak Cerro El Salto [0847/8227 (USBGN)], which is located 1 km NW of Boquete (E762), is the locality Davidson called "Salta." Davidson, GL = Boquete, at 5,000-5,400 feet (1,500-1,650 m), 1934 Jan 9-19 (CAS, as "Salta"; CM, as "Salta"; Davidson, 1938, as "Salta"); Wetmore, GL = Boquete, at 4,000-4,500 feet (1,200-1,350 m), sloping plateau that leads to lower levels of Volcán Barú, trails along stands of bracken, 1960 Mar 15-18 (USNM; Wetmore, 1962); Guitton-Mayerma, Miller, and Winker, at Finca Bermingham (which is Eldredge Bermingham's finca near Alto Quiel [0848/8228 (USBGN)]), 2004 Feb 17-18 (STRI; UAM).

EL SALTO; Darién

ca. 0815/7745

According to Wetmore, El Salto is a section of rapids on the Río Chucunaque [0809/7744 (USBGN)] above the mouth of the Río Tupisa [0813/7742 (USBGN)] and below the mouth of the Río Canglón [0820/7746 (USBGN)] (USNM). According to SMN2B, the Río Tupisa enters the Río Chucunaque from the E ca. 3 mi (5 km) N of Yaviza [0811/7741 (USBGN)]. Wetmore, observations, 1959 Mar 19 (Wetmore, 1965:93). Not shown on our maps.

EL SERENO; see Río Sereno

EL SILENCIO; Coclé (Atlantic)

ca. 0857/8011

Described by Wetmore as a region along the Río Uracillo [0858/8011 (USBGN)] between El Águila [0857/8011 (USBGN)] and Boca de Uracillo [0858/8011 (USBGN)] (USNM), presumably near the mouth of the Quebrada El Silencio [0857/8011 (USBGN)] (q. v.). See also: Uracillo, Río. Not shown on our maps.

EL SITIO; Panamá (E, Pacific)

0902/7928 (USBGN)

ca. 50 m, town ca. 3 km W of Juan Díaz [0903/7927 (USBGN)] (JOG). Villareal, 1992 Aug 16 (MVUP).

EL TIGRE; Darién

0826/7803 (USBGN)

Pearson, collected at Cana [ca. 0745/7742] and at Cerro Pirre [0749/7744 (GA)] in the interior of Darién in 1938 Feb-Apr and must have taken a specimen en route. We have no notes about this locality and can only assume that it comes from some point along the Río Tuira [0821/7803 (USBGN)]. Benson used the name El Tigre in conjunction with the Río Cupé [0804/7735 (USBGN)] (q. v.), and according to Angehr (in litt., Aug 2005), Pearson must have referred to Benson's locality. "The normal route up to Cana was up the valley of the Cupé, and Pearson surely would have passed that way." Pearson, 1938 Jan 30 (ANSP).

EL TIGRE; see Tigre

EL TIGRE, PUERTO; see Puerto El Tigre

EL TIGRE RÍO CUPE; see Cupé, Río

EL URACILLO; see Boca de Uracillo

EL UVITO; Veraguas (Pacific)

0807/8059 (USBGN)

ca. 50 m, town 2 km NW of Santiago [0806/8059 (USBGN)] (MGRP). Bradvica, 1983 Feb 19 (MVUP); Chandler, 1983 Mar 5 (MVUP).

EL VALLE; Coclé (Pacific)

0836/8008 (USBGN)

600 m, city in E Coclé province near the Panamá provincial border, located just S of Cerro Gaital [0837/8007 (USBGN)] in a large valley that extends 5 km E-W and 4 km N-S. The town is connected to the Carretera Interamericana [0832/8242

(USBGN)] by the N-S road labeled "Panama 71" on MGRP (E761A, as "El Valle de Antón"; JOG; MGRP). USBGN also identifies Valle de Antón [0837/8008 (USBGN)], designated as a crater or a central depression of a volcanic cone, as a distinct locality. The region N of town is considered an Important Bird Area (Angehr, 2003, as "El Valle de Antón"). Wetmore and Perrygo, the secondary localities attributed to El Valle in 1951 were located within 7 km of town, GL for: Cerro Agua, Cerro Gaital, Cerro La India Dormida, Río Antón, and Río Mata Ahogado, 1951 Mar 23-24, 26-31, at 2,800 feet (850 m), familiar trails, 1953 June 22 (USNM; Wetmore, 1952:17, 18, as "El Valle de Antón"; Wetmore, 1962); Eisenmann, ca. 2,000 feet (600 m), observations, "eroded badlands near the road to El Valle," c. 1948-1956 Jun-Jul (Eisenmann, 1962a, as "El Valle de Antón," and as "El Valle"); Hartman, 1955 Aug 14 (USNM); Olson et al., 1966 Jun 18 (UF); Ridgely, observations, 1969 Apr 27 (Ridgely and Gwynne, 1989); Collector ?, 1978 Apr 14 (GML, as "Valle de Anton"); Garcés, 1990 Mar 24 (MVUP).

EL VALLE DE ANTÓN; see El Valle

EL VELO; see Velo

EL VELO, QUEBRADA; Chiriquí
0849/8228 (USBGN)
Originates on the E slopes of Cerro Copete [ca. 0848/8231] W of Finca Lérida [0848/8229 (USBGN)] and flows W through Casita Alta [ca. 0849/8229] entering the Río Caldera [0837/8221 (USBGN)] near Bajo Mono [0848/8228 (USBGN)] (E762; Blake, 1958:502 (map, as "Q. Velo," and as "Q. Velos")). Mönniche, at 5,400-5,500 feet (1,650-1,700 m), 1933, 1938, 1939 (FMNH, MCN, as "Quebrada de Velo," and as "Quebrada Velo"; Blake, 1958, as "Quebrada Velo").

EL VIGÍA; Colón ? (Canal Zone) (E, Atlantic)
ca. 0914/7935
140 feet (43 m), former town on the Río Chagres [0907/7942 (USBGN)] 5.5 km NE of Alhajuela [ca. 0913/7937] (E761A) that is submerged in Lago Alajuela (Madden Lake) [0915/7935 (USBGN)] near what is now Isla Vigía [0913/7935 (USBGN)]. This locality is near what is now the Colón/Panamá provincial boundary, and it is estimated to lie within Colón province. See also: La Puente, Río.

EL VILLANO; Veraguas (Pacific)
0757/8059 (USBGN)
ca. 50 m, town 17 km S of Santiago [0806/8059 (USBGN)] on a tributary of the Río San Pedro [ca. 0750/8105], which flows into the Golfo de Montijo [0740/8107 (USBGN)] (JOG). Griscom (1927a) states that in 1924 "Mr. Benson worked in the vicinity of Santiago, working southward to the coastal forest at the head of Montijo Gulf," thereby indicating that both El Villano and La Colorada were located between Santiago and the N coast of Golfo de Montijo. Specimens, however, are labeled 15 miles SE of Santiago, which led Wetmore (1950a) to believe that El Villano was located in the Río Santa María [0806/8029 (USBGN)] drainage, a river that empties E of Golfo de Montijo in the Bahía de Parita [0808/8024 (USBGN)]. Older maps (MHA; MRP) show Santiago due N of Puerto Mutis, which is at the head of Golfo

de Montijo, whereas newer maps (JOG; SMN2B) indicate that Santiago is NNE of Puerto Mutis. This slight change in the map location of Santiago could explain how El Villano could have been located on older maps both near the head of Golfo de Montijo and SE of Santiago. As JOG shows both El Villano and La Colorada in the region S of Santiago near the head of the Golfo de Montijo, these towns are presumed to be the 1924 collecting localities of Benson and Seaman despite the fact that they are not SE of Santiago. Although Wetmore described the locality in 1950, later (1953:8) when reviewing Zimmer (1941), he indicated that he could not locate El Villano. Benson and Seaman, GL = Santiago, specimens labeled ca. 15 mi (24 km) SE of Santiago, 1924 Jun 19-Sep 24 (AMNH; Griscom, 1924b, as "El Villano, Santiago"; Griscom, 1927a, as "15 mi SE of Santiago"; Zimmer, 1939; Zimmer, 1941a:10; Zimmer, 1941b:3, 7, 14; Zimmer, 1950:50; Wetmore, 1950a; Wetmore, 1953; Parkes, 1976, as "15 mi SE of Santiago").

EL VOLCÁN; see Volcán

EL VOLCÁN BARÚ; see Barú, Volcán

EL ZAPOTILLO; see Zapotillo

EMPALME; see El Empalme

EMPIRE; Panamá (Canal Zone) (Atlantic)
ca. 0904/7941
Former station on the original Panama Railroad [ca. 0922/7954] on the W bank of the Río Camacho [0906/7941 (USBGN)] ca. 4 mi (6 km) SSE of Gamboa [0907/7942 (USBGN)] (PR, as "Emperador") near the present location of Empire Known Distance Range [0904/7940 (USBGN)] (E8660). Collector ?, Panama RR, 1878 Jan 28 (MVZ); Goldman, vicinity of town, across the Panama Canal [0920/7955 (USBGN)] in old fields, 1912 Feb 6, 8-9, 12, 14, 17-19 (USNM; Wetmore, 1963a). See also: Empire Known Distance Range.

EMPIRE FIRING RANGE; see Empire Known Distance Range

EMPIRE KNOWN DISTANCE RANGE;
0904/7940 (USBGN)
Panamá (Canal Zone) (W, Atlantic)
75 m, former United States Government establishment that was located 7 km SSE of Gamboa [0907/7942 (USBGN)] on the W side of the Panama Canal [0920/7955 (USBGN)] near the Río Camacho [0906/7941 (USBGN)]. Several specimens bear the labels "Empire Trail" and "Empire Road," although our maps indicate that several roads passed through the range. Carretera Borinquén [0901/7938 (USBGN)] (q. v.), along the W side of the canal, changed from an all weather road to a dry weather road ca. 1 km S of the range and continued N for 7 km beyond the range before terminating at the mouth of the Río Mandinga [0906/7942 (USBGN)]. Road K19 [ca. 0853/7947] (q. v.) extended from La Chorrera [0853/7947 (USBGN)] through Nuevo Emperador [0900/7944 (USBGN)] to the range and was labeled "Chorrera Empire Road" on E761A. In addition, several other short, dry weather roads extended through the area (E8660; RMM2, as "Camp Empire"). Miller, 1937 Feb 20-21, 24 (USNM, as "Empire Trail"); Scholes and Scholes, observations,

1951 Mar 4 (Scholes and Scholes, 1954, as "Camp Empire"); Chapelle, observations, 1954 Oct 16 (Eisenmann, 1955c, as "Empire Rifle Range"); MCU, 1955 Oct 31, and as a GL for: Carretera Borinquén, and Road K10, 1955 Aug 29, Oct 3 (GML, as "Empire Range"; MCN, as "Empire Range"; USNM, as "Empire Range," and as "Empire Road"); Morton, studies, 1973-1975 (Morton, 1980, as "Empire Range"); Pujals et al., observations on K-2 road [Carretera Borinquén], 1977 Jun (Pujals et al., 1977, as "Empire (firing) Range"); Lane, *2 km OSO de las esclusas de Miraflores* (2 km WSW of Miraflores Locks), which according to our maps is S of Empire Range on the Pacific drainage, 1995 May 16 (MVUP, as "Empire Range"); Collector ? [Lane ?], 1995 May 16, Bayonet [Range ?], 1995 Nov 29, and as a GL for Cerro Santa Cruz, 1995 Dec 3 (LSUMZ, as "Empire Range"); Christian, GL for Cerro Santa Cruz, 1997 Feb 28 (LSUMZ, as "Empire Range"); Cardiff, Christian, Dittmann, Lovette, and Naoki, GL = Paraiso, 110 m, 7 km W of Paraiso, 1997 Mar 23 (LSUMZ, as "Empire Range"; MVUP). See also: Empire.

EMPIRE RANGE; see Empire Known Distance Range

EMPIRE RIFLE RANGE; see Empire Known Distance Range

EMPIRE ROAD; see Empire Known Distance Range

EMPIRE TRAIL; see Empire Known Distance Range

ENEA, LA; see Guararé

ENSENADA; see La Ensenada

ENSENADA, LA; see La Ensenada

ENSENADA ÁLVAREZ; see Alberrys Creek

ENSENADA DE BODEGA; see Bodega, Ensenada de

ENSENADA PLAYA GRANDE; see Playa Grande, Ensenada

ESCAPETA; see Escopeta Camp

ESCARREA, RÍO; Chiriquí

0820/8241 (USBGN)

Originates SE of Las Lagunas [0846/8241 (USBGN)] and flows S crossing the Carretera Interamericana [0832/8242 (USBGN)] 5 km WNW of La Concepción [0831/8237 (USBGN)]. The river enters Bahía de Charco Azul [0812/8245 (USBGN)] 21 km ENE of Puerto Armuelles [0817/8252 (USBGN)] near the mouth of the Río Chiriquí Viejo [0820/8241 (USBGN)] (JOG). Wetmore described the lower section of the river near Canta Gallo Abajo [0822/8238 (USBGN)] (q. v.) as shallow, placid and ca. 100 feet (30 m) wide, Wetmore, GL = Buena Vista [0839/8238 (USBGN)], at 2,300 feet (700 m), near the headwaters, drove just over 1 mi (1.5 km) N of Buena Vista through a coffee plantation and into a stand of original forest, 1960 Mar 9 (USNM). See also: Buena Vista.

ESCARTÍN, QUEBRADA; Panamá (E, Pacific)

ca. 0848/7838

Described by Wetmore as a stream immediately N of Juan Avele [ca. 0848/7838], which is an area ca. 10 km N of Chimán [0842/7837 (USBGN)]. Presumably the stream is a tributary of the Río Chimán [0843/7838 (USBGN)], Wetmore and Perrygo, GL = Chimán, Río Chimán, upstream to large stones and an open stand of trees, the higher slopes were covered with black palms and were cut by small streams, 1950 Feb 18 (USNM). Not shown on our maps.

ESCOBAL; Colón (Canal Zone) (W)
0909/7958 (USBGN)
Sea level, town on the former Colón/Canal Zone border located on the NW coast of Bahía Trinidad [0905/8000 (USBGN)] of Lago Gatún (Gatun Lake) [0912/7955 (USBGN)] (E661; E761A; RRM2). RMM1 indicates Camp Escobal ca. 1 mi (1.5 km) N of where Escobal is located on the other maps. Chapin, Rogers and Mathews, GL for La Ullama, 1923 Aug 15-16, 18, 20-23 (FMNH); Ridgely, observations (Wetmore et al., 1984:378). See also: Gatún-Escobal Road.

ESCOBAL, CARRETERA; see Gatún-Escobal Road

ESCOBAL ROAD; see Gatún-Escobal Road

ESCOPETA; see Escopeta Camp

ESCOPETA CAMP; Chiriquí
ca. 0830/8148
Mining camp operated by Texasgulf, Inc. located N of Hato Chamí [0827/8146 (USBGN)], S of the continental divide, and E of the upper reaches of the Río San Félix [0810/8151 (USBGN)] (Gjording, 1991:12). Pine and Izor's specimen labels indicate that the camp was located at 900 m and 23 km NNE of San Félix [0817/8152 (USBGN)] (q. v.). Pine and Izor, GL = San Félix, at 900 m, ca. 23 km NNE of San Félix, area heavily grazed, brush and trees along the streams, 1980 May 15-17, and used to further modify specimens collected at Cerro Bollo [ca. 0830/8146] and at Colorado Camp [ca. 0831/8148] (USNM, as "Escapeta," as "Escopeta," as "Escopeta Camp," and as "San Félix").

ESCOTÁ, QUEBRADA; Los Santos
0733/8003 (USBGN)
Not shown on our maps, but the USBGN coordinates indicate a stream that enters the Río Pedasí [0734/8002 (USBGN)] near its mouth, and according to Wetmore (USNM), the river originates S of Pedasí [0732/8002 (USBGN)] and flows N. Wetmore, GL = Pedasí, 2 mi (3 km) W of Limón [0730/8002 (USBGN)] where the stream banks were covered with virgin forest, the undergrowth was open with occasional thickets of thorny, corozo palm, 1957 Mar 17 (USNM).

ESCOTÁ, RÍO; Herrera
0806/8033 (USBGN)
Originates in NW Herrera province and flows N then E joining the Río Santa María [0806/8029 (USBGN)] 6 km N of Paris [0803/8033 (USBGN)] and 12 km E of Santa María [0807/8040 (USBGN)] (JOG). Wetmore and Perrygo, GL = Paris, 4 mi (6 km) N of Paris, shallow water, largely standing in pools, 1948 Feb 24, GL = Santa María, 3 mi (5 km) SW of Santa María, bridge over the shallow, 8-15 foot (2-5 m) wide stream, tract of gallery forest and savannas beyond, 1948 Mar 8, 19 (USNM). See also: La Concepción.

ESCUDO DE VERAGUAS, ISLA;

0906/8133 (USBGN)

Bocas del Toro, Archipiélago de Bocas del Toro

13 m, island with a land area of 4 km² (400 ha) (Angehr, 2003:105) in the Golfo de los Mosquitos [0900/8115 (USBGN)] 17 km NE of Cocoplum Point [0902/8143 (USBGN)] (JOG, as "Escudo de Veraguas"; MGRP; SMN2B). According to Wetmore, in 1958 the island was low in elevation without any prominent hills and the coastline was comprised of sandy beaches and 40 foot (10 m), steep-faced, sandy-clay bluffs. Inland behind the bluffs the land was broken into ridges that were steep, rough and wooded with wild pineapple, whereas the central region was somewhat depressed and swampy with a few darkly colored streams that ran out to the ocean. The vegetation consisted of coconut palms, low jungle, 20-30 foot (6-9 m) scrub that was very tangled with creepers and 40-60 foot (10-20 m) trees with more open undergrowth. The island was not inhabited by humans and access was somewhat limited. Miller and Richardson, GL = Laguna de Chiriquí, although the GL is Laguna Chiriquí, the island is outside of the lagoon in the NW waters of Golfo de los Mosquitos, 1936 Mar 3 (MVZ; UCLA, as "Escudo Is."; Wetmore, 1959b:3). Wetmore, on the SE coast along a long sandy beach and a projecting sandy point and inland primarily in the more level areas, 1958 Mar 1-2 (USNM, as "Escudo de Veraguas"; Wetmore, 1959b); Handley and Greenwell, "in a shore camp," lagoon, 1962 Mar 20-24 (USNM; Wetmore, 1963b); SBE, collected on the E side, SE shore, SE side, SE swamp, W side, W point, 1990 Mar 29-Apr 2, 1991 Mar 19, 21-30 (MVUP; USNM; Olson, 1993a); Billo, GL = "Isla Escudo", 2007 Apr 12, "West Beach" (USNM). .

ESCUDO, ISLA; see Escudo de Veraguas, Isla

ESMERALDA; see La Esmeralda

ESMERALDA, LA; see La Esmeralda

ESNAPE; Darién

ca. 0804/7812

ca. 25 m, a camp on the Río Esnape [0805/7813 (FH)] that was used by the Barbour, Brooks, Underwood Expedition. According to Barbour and Brooks (1923), they "made [their] first camp on the Quebrada Esnápe [sic], a tributary of the Río Taimatí [0808/7815 (USBGN)], several hours' march northeast of Boca de Sábalo" where they spent several days. According to their map, the camp was probably 4-8 km N of Boca de Sábalo [0802/7813 (USBGN)]. All specimens in Bangs and Barbour are listed as Río Esnape, and apparently Río Esnape and Esnape both refer to the camp, Barbour, Brooks, Underwood Panama Expedition, 1922 April 3-4, 22 ? (ANSP; MCZ; Bangs and Barbour, 1922).

ESNAPE, QUEBRADA; see Esnape, Río

ESNAPE, RÍO; Darién

0805/7813 (Fairchild and Handley, 1966)

Originates between the Río Taimatí [0808/7815 (USBGN)] and the Río Sambú [0805/7818 (USBGN)] and flows NNW entering the Río Taimatí 8 km SE of its mouth and 8 km N of Boca de Sábalo [0802/7813 (USBGN)] (Barbour and Brooks, 1923 (map, as "Quebrada Esnápe")). Barbour, Brooks, Underwood Panama Expedition, 1922 April 3-4 (MCZ; Bangs and Barbour, 1922; Hellmayr, 1929:84; Wetmore, 1953).

ESPARTAL ISLAND; see Toro, Isla

ESPAVÉ, RÍO; see Punorgandí, Río

ESPERANZA; see La Esperanza

ESPERANZA, LA; see La Esperanza

ESPERANZA, RÍO; Panamá (E, Atlantic)
0920/7923 (USBGN)
Originates on the Colón/Panamá provincial border in the mountains NW of Cerro Brewster [0921/7915 (USBGN)] and flows SW entering the Río Chagres [0907/7942 (USBGN)] ca. 40 km NNE of Panama City [0858/7932 (USBGN)] (JOG; MGRP). Birks, Davis, Faucett, and Rohwer, GL for Playa Grande, 2002 Mar 5-11 (UWBM).

ESPIGADILLA, ESTERO; Los Santos
ca. 0753/8019
The name does not appear on our maps, but it must refer to the lower section of the river with the same name. According to E761A, the Quebrada La Espigadilla [ca. 0753/8019] originates S of Los Santos [0756/8025 (USBGN)] and flows E, and after crossing the main highway, it widens and enters a 4 km long section of mangroves before discharging into the Gulf of Panama [0800/7930 (USBGN)]. See also: La Honda.

ESPIGADILLA, QUEBRADA LA; see Espigadilla, Estero

ESPINAR; see Fort Gulick

ESPINAR, FORT; see Fort Gulick

ESPINO; see El Espino

ESPINO, EL; see El Espino

ESPÍRITU SANTO, ISLA DEL; Panamá, Archipiélago de las Perlas
0826/7851 (USBGN)
Sea level, small island with a land area of ca. 5 ha located 100 m off the NE coast of Isla del Rey [0822/7855 (USBGN)] (MGRP). (MacArthur et al., 1972—species list; see comments in Bibliography).

ESQUIGUITA, RÍO; Herrera
0751/8035 (USBGN)
Originates in the hills of W central Herrera province and flows E and NE joining the Río La Villa [0759/8023 (USBGN)] 8 km SSE of Pesé [0754/8037 (USBGN)] (JOG). Wetmore and Perrygo, GL = Pesé, according to Wetmore's journal they collected 5 mi (8 km) W of Pesé where the highway crossed the river, but JOG shows the river coming from the SW and crossing the main highway 7 km S of Pesé. Collected on a shallow section of river ca. 35 feet (11 m) wide with a rocky

bed, bordered by thickets and shaded by good-sized trees, 1948 Mar 24 (USNM).

ESTACIÓN EUSEÑAT; see Aguadulce

ESTACIÓN PIRRE; see Pirre Station

ESTANCIA, RÍO; Coclé (Pacific)
0818/8019 (USBGN)
Originates W of El Valle [0836/8008 (USBGN)] on the SW slopes of Cerro Gaital [0837/8007 (USBGN)] and flows SSW entering the N coast of Bahía de Parita [0808/8024 (USBGN)] 12 km SSW of Antón [0824/8016 (USBGN)] (JOG). Wetmore, GL = Antón, left the Carretera Interamericana [0832/8242 (USBGN)] ca. 1 mi (1.5 km) NW of Antón and walked N along the river following cattle trails and a fence line for 1 mi (1.5 km), 1958 Mar 16 (USNM).

ESTERO BOCA GRANDE; see Boca Grande, Estero

ESTERO EL CABUYAL; see El Cabuyal, Estero

ESTERO FUERTES; see Fuertes Estero

ESTERO RICO; see Rico, Estero

ESTERO SALADO, BOCA DE; Coclé (Pacific)
0811/8029 (USBGN)
Mouth of the Río Estero Salado [0811/8029 (USBGN)], which discharges into the Bahía de Parita [0808/8024 (USBGN)] 6 km SE of Aguadulce [0815/8033 (USBGN)] in an area of mangroves (E761A, as "Estero Salado"). See also: El Rompio.

ESTERO SALADO, RÍO; Coclé (Pacific)
0811/8029 (USBGN)
Originates NW of Aguadulce [0815/8033 (USBGN)] and flows SE entering the W central shore of Bahía de Parita [0808/8024 (USBGN)] 9 km SE of Aguadulce (JOG). Wetmore, observations, mangroves bordering the river, 1963 Jan 25 (Wetmore, 1965 236). See also: El Salado; and Estero Salado, Boca de.

ESTRELLA; see La Estrella

ESTRELLA, LA; see La Estrella

FABREGA, CERRO; Bocas del Toro
0907/8252 (USBGN)
3,335 m, peak in extreme W Bocas del Toro province just N of the continental divide, located 7 km E of the Costa Rican border and 3 km NNE of the Chiriquí provincial border (JOG). See also: Changuena, Río.

FAIRCHILD PLANT; see Gatún, Lago

FARALLÓN, ISLA Panamá (W, Pacific)
0848/7931 (USBGN)
Sea level, rock located in Bahía de Panamá [0850/7915 (USBGN)] 500 m S of Isla Taboguilla [0848/7931 (USBGN)] (E761A, as "Farallon Island"). Meyer, GL = Isla Taboguilla, 1942 Dec 13, 1944 Apr 9 (USNM, as "Farallon Rock").

FARALLÓN DEL CHIRÚ; see Farallón del Chirú, Isla

FARALLÓN DEL CHIRÚ, ISLA; Coclé (Pacific)

0821/8007 (USBGN)

129 feet (39 m), rock in the Gulf of Panama [0800/7930 (USBGN)] 3 km E of El Farallón del Chirú [0821/8008 (USBGN)] (E761B, as "Farallón del Chirú I.") that is considered an Important Bird Area (Angehr, 2003, as "Farallón del Chirú"). Described by Wetmore as a rocky islet, a few low bushes and stands of cactus covered the crest and long grass covered the level sections, Wetmore, 1957 Feb 28 (USNM, as "Farallón del Chirú"; Olson, 1997a, as "Farallón del Chirú").

FARALLON ISLAND; see Farallón, Isla

FARALLÓN ROCK; see Farallón, Isla

FARALLÓN SUCIO; Colón (E)

ca. 0939/7938

Group of rocks ca. 3 km NNW of Punta Cacique [0937/7937 (USBGN)], which is a point located on the E coast of Colón province ca. 42 km NE of Colón [0922/7954 (USBGN)] (E761A; MHA). (Wetmore, 1965:58).

FAR FAN; see Farfan

FARFAN; Panamá (Canal Zone) (W, Pacific)

0856/7935 (USBGN)

10 m, town on a tributary of the Río Farfan [0856/7934 (USBGN)] 2 km W of Farfan Beach [0856/7934 (USBGN)] and 7 km SW of Panama City [0858/7932 (USBGN)] (E962). The name occasionally appears in the literature as Far Fan. Hallinan, clearings, mangroves, cacti and shrubbery, trees near the shore, jungle, 1915 Feb 28, Mar 21, 26, May 2, 9, 16, 23, Jun 20, Aug 22, Dec 19, 1916 Jan 9, 23, Feb 6, May 11, 14 (AMNH; Hallinan, 1924; Zimmer, 1955:5); Benson, 0-100 feet (30 m), 1931 Jun 26-27, 29, Jul 21 Aug 15, 1932 Jan 27 (USNM); Blanton, 1953 Sep 28, Oct 5-6, 19, 27, Nov 9, and as a GL for Road K7, 1955 Oct 17 (USNM); MCU, 1955 Jul 28, Oct 17 (USNM); Morton, 1970 Oct 29, Dec 20, 1971 Jan 3, 5, observations 1970s (AMNH; USNM; Morton, 1982, as "Far Fan"); Trainer, observations, 1981-1982 (Trainer, 1989); Guardia and Suazo, 1983 Aug 19 (MVUP); Lane and Seutin, *al oeste de la salida sur del canal de Panamá* (S of the W side of the Panama Canal [0920/7955 (USBGN)] entrance), 1994 Oct 9 (MVUP). (Wetmore, 1970). See also: Farfan Beach.

FARFAN, CERRO; Panamá (Canal Zone) (W, Pacific)

0856/7934 (USBGN)

101 m, hill on the W side of the Pacific entrance of the Panama Canal [0920/7955 (USBGN)] just S of the Carretera Interamericana [0832/8242 (USBGN)] 1.5 km W of Balboa [0857/7934 (USBGN)]. The hill is just N of Farfan Beach [0856/7934 (USBGN)] and ca. 1 km N of the mouth of the Río Farfan [0856/7934 (USBGN)] (E762, as "Farfan Hill"; E962). Cardiff, GL for Río Farfan, 1997 Mar 11 (LSUMZ); Cardiff and Lovette, at ca. 30 m, 1.25 mi (2 km) SW of, 1997 Mar 11 (LSUMZ). See also: Farfan, Río.

FARFAN, PLAYA; see Farfan Beach

FARFAN, RÍO; Panamá (Canal Zone) (W, Pacific)

0856/7934 (USBGN)

Short river that originates on the E slopes of Cerro Galera [0856/7937 (USBGN)] and flows E widening in mangrove swamps as it discharges into Bahía de Panamá [0850/7915 (USBGN)] 2 km WSW of Panama City [0858/7932 (USBGN)] (E761A). The "Río Farfán Diversion and Spillway" is located at the mouth of the river where a road extends along the coast across the river mouth (E661). Farfan Marsh presumably refers to the area of tropical grass and mangroves that extend 2-3 km inland along the river's mouth. Blanton, Farfan Spillway, 1953 Sep 25 (USNM); Overton, GL = Panama City, 5.5 km SW of Panama City, 1994 Sep 28 (LSUMZ, as "Farfan Marsh"); Cardiff, Dittmann, Lovette, and Renner, at 10-50 m, at the mouth, 0.5 km SSW of mouth, and 1 mi (1.5 km) S Cerro Farfan, 1997 Mar 11, 13 (LSUMZ, as "Farfan," and as "Farfan Marsh"; MVUP).

FAR FAN BEACH; see Farfan Beach

FARFAN BEACH; Panamá (Canal Zone) (W, Pacific)

0856/7934 (USBGN)

Coastal beach 1 km N of the mouth of the Río Farfan [0856/7934 (USBGN)] and 4 km W of Panama City [0858/7932 (USBGN)] (E661; E962A, as "Playa Farfán"). The name occasionally appears in the literature as Far Fan Beach. Buckalew and Low, 1941 Jun 15 (Eisenmann, 1951:184); Scholes and Scholes, observations, 1951 Jan 14 (Scholes and Scholes, 1954); Blanton, 1953 Sep 27-28, Oct 1, 5, 19, 26-27 (USNM); Olson, 1963 Mar 1 (USNM); Smith, N., 1970 Oct 20 (USNM); Schneider, observations, "a protected beach of cobble, sand, and silt...south of the Thatcher Ferry Bridge (Bridge of the Americas) [0856/7933 (USBGN)]," 1978 Jan-May (Schneider and Mallory, 1982, as "Farfan"). (Olson, 1980b).

FARFAN HILL; see Farfan, Cerro

FARFAN MARSH; see Farfan, Río

FARFAN ROAD; Panamá (Canal Zone) (W, Pacific)

ca. 0857/7934

Short road that extends ca. 1 km N from Farfan Beach [0856/7934 (USBGN)] to the Carretera Interamericana [0832/8242 (USBGN)] (E661). According to RM1, Farfan Road refers to the N section of Road K1 (RM1, as "Road K1"). Morton, 1970 Feb 13, 1971 Mar 11 (USNM, as "Farfan Road"); Warkentin and Morton, observations, 1973, 1974, 1991 (Warkentin and Morton, 1995).

FARFAN SPILLWAY; see Farfan, Río

FARM 6; Bocas del Toro

ca. 0928/8231

10 m, town in a cultivated area 3 km N of Changuinola [0926/8231 (USBGN)] on the Río Changuinola [0928/8227 (USBGN)] (E762). Kennard et al., GL = Almirante, 1926 Feb 14 (MCZ; Kennard and Peters, 1928:443-444, 447).

FARM 8; see Finca Ocho

FEA, PUNTA; Veraguas, Isla de Coiba

ca. 0724/8139

Described by Wetmore (1957:9) as a point on the SE coast of Isla de Coiba [0728/8147 (USBGN)] at the entrance of Estero Boca Grande [0724/8139 (USBGN)] (MHA, as "Punta Fea o del Cedro"). The locality does not appear on specimen labels, but was used to define the southernmost limit of his collecting on the island. See also: Coiba, Isla de.

FELIPA, PUNTA; Veraguas, Isla de Coiba

0724/8142 (USBGN)

Sea level, point located on the SE coast of Isla de Coiba [0728/8147 (USBGN)] that separates Bahía Damas [0727/8141 (USBGN)] (q. v.) on the N from Ensenada María [0724/8141 (USBGN)] on the S (JOG; MHA, as "Punta Observatorio").

FELIPILLO; Panamá (E, Pacific)

0906/7918 (USBGN)

Not shown on our maps but USBGN indicates a town ca. 3 km N of Pacora [0905/7917 (USBGN)]. Saucedo and Boyd, 1991 Oct 17 (MVUP); Arosemena and Jipson, 1992 Jun 5 (MVUP, as "Ingenio Felipillo"); Guitton-Mayerma, Lelevier, Maley, Miller, and Winker, at 09°03'27"/79°18'08", old airstrip, the coordinates indicate a region on JOG below 100 m ca. 3 km SW of Pacora very near the Río Pacora [0901/7917 (USBGN)], 2004 Feb 10, 2005 Dec 5, 19-20, 2006 Jan 22 (STRI, as "Filipillo"; UAM, as "Filipillo"); DaCosta, Klicka, and Spellman, at 0-25 m, at 09°04'55"/79°18'32", 22 km ENE of Panama City, Felipillo marsh (old airstrip), 2006 Feb 19 (MBM; STRI).

FELIPILLO MARSH; see Felipillo

FÉLIX, QUEBRADA; Darién

0822/7747 (USBGN)

Originates on the S end of Serranía el Filo de Tallo [0827/7757 (USBGN)] and flows E entering the Río Chucunaque [0809/7744 (USBGN)] 1 km SW of the Río Tuquesa [0823/7747 (USBGN)] (DP). Wetmore described the river as 20-35 feet (6-11 m) wide with water only at its mouth, and most of its course was a 15 foot (5 m) deep ravine with occasional pools of water, Wetmore, GL for Río Chucunaque, 1959 Mar 23 (USNM).

FERRY; see West Ferry Ramp

FILA LÉRIDA; see Lérida, Fila

FILIPILLO; see Felipillo

FILIPILLO MARSH; see Felipillo

FILO DE TALLO, SERRANÍA EL; see El Filo de Tallo, Serranía

FILO MARROQUIN; see Marroquin, Filo de

FINCA 4; see Finca Cuatro

FINCA 8; see Finca Ocho

FINCA 32; see Changuinola

FINCA 34; see Changuinola

FINCA BERMINGHAM; see El Salto

FINCA CUATRO; Bocas del Toro

0924/8230 (USBGN)

ca. 25 m, town located 4.5 km SSE of Changuinola [0926/8231 (USBGN)] and 17 km NW of Almirante [0918/8224 (USBGN)] on the railroad that connects the two towns (JOG). Hinds, GL = Almirante, 1960 Mar 21 (USNM, as "Finca 4").

FINCA EL BRUJO; see El Brujo

FINCA EL JOBITO; Chiriquí

ca. 0821/8242

Not shown on our maps but USBGN indicates that El Jobito is an alternate name for the town El Cavimo [0821/8242 (USBGN)], which is shown on JOG (as "El Cabimo") ca. 5 km S of Divalá [0825/8243 (USBGN)] near coastal mangroves. Presumably Finca El Jobito is also in the region 5 km S of Divalá. Miranda, GL = Divalá, 1983 Sep 25 (MVUP).

FINCA EL ROSARIO; see El Rosario

FINCA FLORENTINA; see Volcán

FINCA HARTMANN; Chiriquí

ca. 0850/8246

A coffee plantation near Santa Clara [0850/8246 (USBGN)] (q. v.) that was operated by Alois (Luis) Hartmann and his family. Wilson and Kalar, 1999 Jan 1, 11, 14, 28, Feb 24, 27, Mar 1, 2000 Mar 2 (MVUP).

FINCA LA LOMA; see Bahía Honda, Bocas del Toro

FINCA LÉRIDA; Chiriquí

0848/8229 (USBGN)

1,700 m, settlement 6.5 km NW of Boquete [0847/8226 (USBGN)] and 2 km W of Bajo Mono [0848/8228 (USBGN)] that grew up around the original location of Mönniche's house and plantation (E762; Blake, 1958:502 (map), 503, as "Lerida"). Mönniche, at 5,300-5,400, 6,800 feet (1,600-1,650, 2,050 m), 1928, 1931-1940, 1946-1947, 1952 (ANSP, as "Lerida"; CAS, as "Lerida," and as "Lerida River"; FMNH, as "Lerida"; GML, as "Lerida"; MCN, as "Lerida"; Eisenmann, 1962c; Monroe, 1963b, as "Lerida"); Worth, GL = Boquete, specimens are labeled Boquete, although his publications indicate that he lodged at the Mönniche plantation and specifically mentions Velo, 1937 Jul 16-18, Aug 2-5 (ANSP, as "Boquete"; Worth, 1939a; Worth, 1939b). See also: Lérida, Fila.

FINCA MARTÍNEZ; Chiriquí

ca. 0852/8235

ca. 1,900 m, settlement located ca. 1 km NNW of Cerro Punta [0852/8235 (USBGN)] (VCP). Ambrose (MARU), GL = Cerro Punta, at 6,250 feet (1,900 m), 1961 Apr 27 (AMNH).

FINCA MORENA; see Molejon

FINCA OCHO; Bocas del Toro

0926/8231 (USBGN)

Sea level, town ca. 1 km SSW of Changuinola [0926/8231 (USBGN)] on the railroad that connects Changuinola and Almirante [0918/8224 (USBGN)] (JOG, as "Finca 8"). Kennard et al., GL = Almirante, 1926 (MCZ, as "Farm 8"; Kennard and Peters, 1928:443-444, as "Farm 8"); Wedel, 1927 Dec 2 (MCZ, as "Farm 8"). See also: Changuinola.

FINCA OJO DE AGUA; see Ojo de Agua

FINCA PALO SANTO; see Palo Santo

FINCA PASO BLANCO; Panamá (E, Pacific)

ca. 0908/7914

Presumably located near the town Paso Blanco [0908/7914 (USBGN)], which is at ca. 50 m on the Carretera Interamericana [0832/8242 (USBGN)] 18 km WSW of Chepo [0910/7906 (USBGN)] (MGRP). Braun, M., Ponce, and Sandoval, 1985 Mar 30 (MVUP).

FINCA SAN JOSÉ; see La Jagua, Río

FINCA SANTA MÓNICA; Coclé (Pacific)

0822/8012 (USBGN)

ca. 25 m, town located 4 km WSW of Río Hato [0823/8010 (USBGN)] and ca. 4 km N of the coast (JOG). Eisenmann, observations, year ? Sep 26 (Wetmore et al., 1984:378).

FINCA WILLIAMS; Bocas del Toro

ca. 0918/8225

10 m, apparently on the railroad near Milla 2 [ca. 0918/8225], which is ca. 3 km NW of Almirante [0918/8224 (USBGN)] and 1 km NE of Fruitdale [0918/8225 (USBGN)]. GMLP, GL = Almirante (q. v.), Milla 2, 1962 Nov 10 (USNM). Not shown on our maps.

FLAMENCO, ISLA; Panamá (Canal Zone) (E, Pacific)

0854/7931 (USBGN)

280 feet (85 m), island with a land area of ca. 0.1 km^2 (10 ha) located in Bahía de Panamá [0850/7915 (USBGN)] 1 km SE of Isla Naos [0855/7932 (USBGN)] (E761A, as "Flamenco Island"). Loftin, 1962 Mar 5 (Loftin and Olson, 1963, as "Flamenco Island"); Pennycuick, observations, "small, steep island," 1980 May 21-Jun 6 (Pennycuick, 1983). See also: Fortified Islands; and Panamá, Bahía de.

FLAMENCO ISLAND; see Flamenco, Isla

FLAT ROCK BEACH; Bocas del Toro, Archipiélago de Bocas del Toro

0925/8215 (USBGN)

Sea level, beach on the NE coast of Isla Colón [0924/8217 (USBGN)], extending 4 km NNW from Paunch [0923/8214 (USBGN)] (JOG). SBE, GL = Isla Colón, 1989 Jan 30 (USNM).

FLORENTINA, FINCA; see Volcán

FLORES, CERRO; Chiriquí

0829/8144 (USBGN)

According to Griscom (1924b:1), Cerro Flores was located in extreme eastern

Chiriquí province closer to the mountains of Veraguas than to Volcán de Chiriquí [= Volcán Barú [0848/8233 (USBGN)]]. He later described Cerro Flores as two or three minor peaks reaching elevations of ca. 7,000 feet (2,150 m) along the continental divide, which itself was rather uniform and lacking prominent peaks (Griscom, 1935b:269). Correspondence between Myers (1969:32) and Benson suggests that "after leaving San Felix, Chiriquí, and arriving in high country, [Griscom et al.] never quite knew where they were. However, they managed to reach cloud forest that evidently [was] of the montane thicket or elfin woodland type." Myers and Duellman (1982:3) indicate that Cerro Santiago [0833/8144 (USBGN)] is not at an elevation of 2,826 m as indicated on many maps, but at a much lower elevation of ca. 2,100 m. They further point out that when Griscom (1924:516) noted "to the west about ten miles away...a cone-shaped peak about 1,000 feet [300 m] higher than the crest on which [they] were standing," that he believed was "the real Cerro Santiago," he was likely looking at Cerro Sagui (C. Ratón), the highest peak in the area at ca. 2,400 m. Therefore, it appears that what Griscom called Cerro Flores — "two or three minor peaks...along the continental divide" — may in fact refer to Cerro Santiago and the nearby peak named Cerro Flores. E562 indicates that Cerro Flores is at 1,605 m, 4 km SSW of Cerro Santiago and 4 km NNE of Hato Chamí [0827/8146 (USBGN)]. Griscom, specimens are attributed to Griscom, Benson, Boulton, Valiente, and others, at 3,400-6,000 feet (1,050-1,850 m), camped on the slope of Cerro Flores at 3,700 feet (1,150 m), dense forest, "heart of the subtropical zone," above 5,000 feet (1,500 m) the ground was mossy and very wet, reached the continental divide at 6,000 feet (1,850 m), 1924 Mar 3-18 (AMNH; CM; Griscom, 1924a; Griscom, 1924b; Griscom, 1927a; Davidson, 1932; Davidson, 1938; Zimmer, 1939; Zimmer, 1941b:14; Zimmer, 1944; Zimmer, 1955:25; Chapman, 1940; Olson et al., 1968; Parkes, 1976); Davidson, 1932 Jan 7 (CAS). See also: Hato Chamí.

FOCOUME; see Tocumen

FOLK, RÍO; see Río Folk

FOLK RIVER; see Río Folk

FOREST PRESERVE; see Madden Road Timber Preserve

FOREST RESERVE; see Madden Road Timber Preserve

FORT AMADOR; see Fort Amador Army Reservation

FORT AMADOR ARMY RESERVATION;
0856/7933 (USBGN)
Panamá (Canal Zone) (E, Pacific)
Sea level, former United States Government military base that was located in the Canal Zone 1 km S of Balboa [0857/7934 (USBGN)]. The base was located on a peninsula immediately SW of Panama City [0858/7932 (USBGN)] that is connected to the Fortified Islands [ca. 0855/7931] by a 3 km long causeway. The city Fuerte Amador [0856/7933 (USBGN)] is now located at the site (E761A, as "Fort Amador"; E962, as "Amador"). Bunker, date ? (USNM); Imhof, observations, 1942 Jun 15, Sep 16, 19-Oct 1, 14 (Imhof, 1950, as "Fort Amador"); Loftin, 1964 Oct 4 (GML,

as "Amador Flats"); McCafferty, 1994 Nov 2 (LSUMZ, as "Fort Amador"); Miller, 1972 Apr 18 (GML, as "Fort Amador"); Collector ?, 1995 Feb (LSUMZ, as "Fort Amador"). (N. Smith, 1982). See also: Fortified Islands.

FORT AMADOR CAUSEWAY; see Fortified Islands

FORT CLAYTON; see Fort Clayton Army Reservation

FORT CLAYTON ARMY RESERVATION;
0901/7934 (USBGN)
Panamá (Canal Zone) (E, Pacific)
10 m, former United States Government military base that was located 5 km N and NNW of Balboa [0857/7934 (USBGN)] on the Panama Canal [0920/7955 (USBGN)] E of Lago de Miraflores [0901/7936 (USBGN)] (E762, as "Fort Clayton," and as "Fort Clayton Army"). The boundaries are not clearly indicated on our maps, but the base also included Curundu [0859/7933 (USBGN)] to the S and E. The city Fuerte Clayton [ca. 0900/7935, USBGN mistakenly indicates it is at 7945] is now located at the site (E962, as "Fuerte Clayton"). The locality has always been written on specimen labels and mentioned in the literature as simply "Fort Clayton." Fischer, 1929 May 19, 25, 26 (USNM); Imhof, observations, 1942 Oct 4, 14-18 (Imhof, 1950); Meyer, 1943 May 2 (USNM); Blanton, 1953 Sep 17, 22, and as a GL for Road C12E, 1953 Sep 17, 22, Oct 13 (USNM); MCU, 1955 Sep 8 (GML); Loftin, shrubs and lawn, near Curundu, 1961 Oct, 1962 Mar, Oct-Nov (Loftin, 1963a; Loftin and Olson, 1963), Curundu area of Fort Clayton, 1969 Jan 4 (MCN); Keenan, 1963 Mar 19 (Loftin and Olson, 1963); Loftin and Olson, 1963 May 3, 11-12 (Loftin and Olson, 1963); Tyson, 1966 Jul 26 (UF); Willis, 1966 Aug 3 (Willis, 1972:36). (Wetmore, 1970; Parkes, 1976). See also: Metropolitano, Parque Natural.

FORT DAVIS; see Fort William D. Davis Army Reservation

FORT GULICK; Colón (Canal Zone) (E)
0919/7952 (USBGN)
25 m, former United States Government military base that existed in 2 sections. One section was located at sea level on a strip of land between the NE coast of Río Folk [0921/7954 (USBGN)] and the SE coast of Bahía Manzanillo [0922/7953 (USBGN)]. The other, somewhat larger section was located ca. 2 km SSE of the first on the NE coast of Lago Gatún (Gatun Lake) [0912/7955 (USBGN)] ca. 1 km SE of Rainbow City [0921/7953 (USBGN)] (E8660, as "Fort Gulick Army Reservation"). According to Angehr (in litt., Aug 2005), "the name was later changed to Ft. Espinar, and the area today is called just 'Espinar.'" Keenan, "bldg 400 on Gatun Lake," 1963 Sep 17, 20 (GML); Collector ?, 1965 Mar (GML); Gardner, 1986 May 3, 13 (WFVZ); Akers, 1987 Nov (WFVZ).

FORT GULICK ARMY RESERVATION; see Fort Gulick

FORT HOWARD AIR FORCE BASE; see Howard Air Force Base

FORTIFIED ISLANDS; Panamá (Canal Zone) (E, Pacific)
ca. 0855/7931
Refers to the four islands — Isla Naos [0855/7932 (USBGN)], Isla Culebra [0855/7932 (USBGN)], Isla Perico [0855/7931 (USBGN)], and Isla Flamenco [0854/

7931 (USBGN)] — that are located in Bahía de Panamá [0850/7915 (USBGN)] ca. 4 km SSE of Panama City [0858/7932 (USBGN)]. The islands are connected to the mainland at Fuerte Amador [0856/7933 (USBGN)], as well as to each other, by a 3 km causeway. The name Fortified Islands does not appear in USBGN or on our maps. Loftin, observations, 1962 Mar 5, Aug-Nov (Loftin, 1963a, as "Amador Causeway"; Loftin and Olson, 1963, as "Fort Amador Causeway"); Pacific Project, GL = Gulf of Panama, near the Fortified Islands, 1969 Jan 27 (GML; USNM; Loftin, 1991). (Pennycuick, 1983). See also: Panamá, Bahía de.

FORT KOBBE; see Fort Kobbe Army Reservation

FORT KOBBE ARMY RESERVATION;
0854/7935 (USBGN)
Panamá (Canal Zone) (W, Pacific)

10-90 m, former United States Government military base that was located 6 km SSW of Balboa [0857/7934 (USBGN)] and 0.5 mi (1 km) inland from the coast (E762, as "Fort Kobbe"; E962, as "Fuerte Kobbe"). Kobbe Beach is on the coast at the terminus of a road between Punta Bruja [0853/7935 (USBGN)] and Punta Batele [0854/7934 (USBGN)] (E761B, as "Dejal Beach"; E962, as "Balneario Kobbe"). Blanton, 1953 Jul 6 (USNM, as "Fort Kobbe"); Todd, 1961 Jul 20, 24 (USNM, as "Fort Kobbe"); Kuns, 1962 Aug (MCN, as "Fort Kobbe"); Morton, some specimens are also labeled "Bayonet Range," 1970 Dec 24, 1971 Jan 13-15, 19, 21, 26, Feb 2-3, 6, 9, 18, 27-28, Mar 2-3, 14, 17, 26, Apr 10, 18, May 14, studies 1973-1975 (AMNH, as "Fort Kobbe"; USNM, as "Fort Kobbe"; YPM, as "Fort Kobbe"; Morton, 1980, as "Fort Kobbe"; Morton, 1982, as "Fort Kobbe," and as "Fort Kobbe Mangrove Island"); Barrett, 1973 Jul 20, Dec 20 (GML, as "Kobbe"); Trainer and Will, observations, 1981-1983 (Trainer and Will, 1984, as "Kobbe Beach"; Trainer, 1989, as "Kobbe Beach"); Tejera, 1993 May 3 (MVUP, as "Kobbe").

FORT KOBBE MANGROVE ISLAND; see Fort Kobbe Army Reservation

FORT LORENZO; see Fuerte San Lorenzo

FORT RANDOLPH; see Fort Randolph Army Reservation

FORT RANDOLPH ARMY RESERVATION; Colón (Canal Zone) (E)
0923/7953 (USBGN)
Sea level, former United States Government military base that was located on Margarita Island [0923/7953 (USBGN)] (q. v.) (E8660). Tomas, 1939 May 8 (SDSNH, as "Fort Randolph").

FORT RODMAN; see Rodman United States Naval Reservation

FORT SAN LORENZO; see Fuerte San Lorenzo

FORT SHERMAN; see Fort Sherman Army Reservation

FORT SHERMAN ARMY RESERVATION; Colón (Canal Zone) (W)
0920/7958 (USBGN)
100 m, former United States Government military base that extended from the E coast of Bahía Limón [0920/7956 (USBGN)] to an area 5 km SW of the mouth of

the Río Chagres [0919/8000 (USBGN)] covering a land area of ca. 40 km^2 (4,000 ha) (E8660; JOG). Several specimens are labeled "Fort Sherman Road," which according to our maps, refers to either Road S2 [ca. 0920/7958] (q. v.) or Road S8 [ca. 0912/8008] (q. v.). All specimens, unless otherwise noted, were labeled simply Fort Sherman. Meyer, 1943 May 15, 1944 Mar 16 (USNM); Willis, 1961 May 23 (Willis, 1972, as "Fort Sherman Road"); Powell, 1966 Sep 25-26, Oct 2, 18-19 (AMNH); Wallis, 1966 Sep 30 (AMNH); Fleming, 1967 Feb 2 (USNM); Gardner, 1970, 1973 Nov 12, 1974 Jul 4, Feb 27, near, Post 85-86 on Fort Sherman Road, 1985 Nov 18, post 75 on Fort Sherman Road, 1986 Jan 13 (WFVZ, as "Ft. Sherman," and as "Fort Sherman"); MARU, 1961 Aug 11 (USNM); Yunker, 1961 Mar 17 (USNM); Handley, airstrip, 1962 Mar 19 (Wetmore, 1963b); Morton, studies, 1970s (Morton, 1980, as "Fort Sherman"; Morton, 1982, as "Fort Sherman"); Martínez, GL for Gatún Drop Zone, 1992 Sep 13 (MVUP, as "Fuerte Sherman"); Seutin, 1995 Mar 7 (MVUP, as "Fuerte Sherman"); Cardiff and Lovette, GL = Esclusas Gatún (Gatún Locks), 1.5-3 mi (2.5-5 km) NW of Gatun Locks, 1997 Mar 19 (LSUMZ, as "Fort Sherman"). (Caldwell, 1986). See also: Galeta, Isla; and Gatún, Esclusas.

FORT SHERMAN ROAD; see Fort Sherman Army Reservation

FORTUNA; Chiriquí
0843/8216 (USBGN)
1,100 m, town 19 km ESE of Boquete [0847/8226 (USBGN)], 5.5 km S of the continental divide, and 1 km S of Cerro Fortuna [0844/8216 (USBGN)] (E762). The town is reached by the Gualaca-Chiriquí Grande Road [ca. 0857/8207]. The Fortuna Dam [ca. 0843/8216] on the upper reaches of the Río Chiriquí [0820/8226 (USBGN)] creates Lago Fortuna [ca. 0842/8216], a reservoir that extends E of town covering a surface area of 1.03 km^2. Fortuna Reserva Forestal [0842/8212 (Angehr, 2003:131)] surrounds the reservoir (RDP2), and according to Engleman et al. (1989:497), "the forest of the reservoir's watershed is rigidly protected and includes cloud forest and very humid montane forest." Angehr (2003) notes that Cerro Chorcha [0839/8207 (USBGN)] is the reserve's high point and that small amounts of submontane and high montane forest can be found in the reserve, which is considered an Important Bird Area. Ridgely and Hinds, at 3,100-3,500 feet (950-1,050 m), some specimens are also labeled N side, and *sitio de presa* (at the dam), 1976 Feb 10-Mar 26, and as a GL for: Cerro Hornito, and Río Hornito, 1976 Aug 14 (GML; Wetmore, 1984 as "Fortuna," as "Fortuna area," as "Fortuna Dam" and as "Fortuna Dam Site"; Ridgely and Gwynne, 1989 as "Fortuna," as "Fortuna area," as "Fortuna Dam," and as "Fortuna Dam Site"; Olson, 1993b); Sousa and Herman, blood smears, c. 1980 (Sousa and Herman, 1982, as "La Fortuna"); Tejera, observations at sites between Los Planes and the continental divide, 1984 Apr, Nov (Tejera, 2001); Arosemena and Olmos, *sitio de presa* (at the dam), 1987 Nov 2 (MVUP, as "Fortuna," and as "Proyecto de Fortuna"); Mobley and Robbins, 1995 Apr 2, 4-6, 8-9, 14, 16, 21 (MVUP); Cardiff, Dittmann, and Lovette, GL = Cordillera Central, at 1,300 m, Distrito Gualaca [0837/8215 (USBGN)], 4.3 km by road S of Lago Fortuna Dam (at the STRI Fortuna Field Station), 1996 Mar 4-8 (LSUMZ; MVUP, as "Hornito"); Christian, at ca. 1,100 m, observations, cloud forest, 1996 Aug

(Christian, 2001, as "Fortuna Forest Reserve"); Braun, M., and Tejera, 1997 Mar 1 (MVUP); Conos and Rowlet, km 59, according to our maps, 59 km N of the Carretera Interamericana [0832/8242 (USBGN)] appears to be near the continental divide if not on the Atlantic drainage, 2001 Mar 22 (MVUP); DaCosta, Guitton-Mayerma, Klicka, Lelevier, Miller, and Spellman, GL = Gualaca, at 1,215 m, at 08°43'03"/82°14'08", 22 km NNE of Gualaca, and at 1,245m, 08°42'27"/82°13'57", 20 km NNE of Gualaca, 2006 Feb 25-28 (MBM, as "Reserva Forestal Fortuna"; STRI, as "Reserva Forestal Fortuna"). Smithsonian field parties collected near Fortuna on the Gualaca-Chiriquí Grande Road (q. v.) (Olson, 1993b). (Delgado, 1983).

FORTUNA, LAGO; see Fortuna

FORTUNA, PROYECTO DE; see Fortuna

FORTUNA AREA; see Fortuna

FORTUNA DAM; see Fortuna

FORTUNA DAM SITE; see Fortuna

FORTUNA RESERVA FORESTAL; see Fortuna

FORTUNA RESERVOIR; see Fortuna

FORTUNA ROAD; see Gualaca-Chiriquí Grande Road

FORT WILLIAM D. DAVIS ARMY RESERVATION;
0917/7953 (USBGN)
Colón (Canal Zone) (E)
50 m, former United States Government military base that was located 2 km NE of Gatún [0916/7955 (USBGN)] adjacent to the S border of Coco Solo Naval Reservation [0923/7952 (USBGN)] (E8660). Weber, 1928 Jul 26, 28 (FMNH, as "Fort Davis"; USNM, as "Ft. Davis," and as "Fort Davis"; Eisenmann, 1962a, as "Fort Davis"); Imhof, observations, 1942 Oct 20-28 (Imhof, 1950, as "Fort Davis (Caribbean Side)"); Ambrose, Dock 45 Road, according to E8660, Dock 45 is on Lago Gatún (Gatun Lake) [0912/7955 (USBGN)], and a road extends here E-W along the lakeshore, 1960 Jun 14-15, Sep 12 (LSUMZ, as "Fort Davis"); Willis, 1961 May 22 (Willis, 1972, as "Fort Davis"); Simpson, 1966 Oct 15 (GML, as "Fort Davis"); Gardner, 1973, 1985 Oct 18, near, 1988 Jan 3 (WFVZ, as "Ft. Davis," and as "Fort Davis"). (Eisenmann, 1961, as "Fort Davis").

FRAILES DEL NORTE; Los Santos
0722/8009 (USBGN)
Sea level, very small island 6 km SSE of Venado [0725/8011 (USBGN)] (JOG). Described by Wetmore in Mar 1962 as a low, barren rock, swept by storms and without vegetation (USNM). GMLP, observations, 1949 May 6 (Wetmore, 1965:462).

FRAILES DEL SUR; Los Santos
0720/8008 (USBGN)
10 m, two rocky islets (Angehr, 2003:213, as "Frailes del Sur Islands") that are located 10 km SSE of Venado [0725/8011 (USBGN)] and 16 km SW of Punta Mala [0728/8000 (USBGN)] (JOG). According to Angehr (2003:213) the larger

islet is ca. 200 m long, neither has much vegetation, and landing is difficult due to sea conditions and the steep cliffs. Wetmore also uses the term "Los Frailes" for the islets. Observer ?, photographs, 1949 May 6 (Ridgely and Gwynne, 1989:161, as "Los Frailes del Sur"); Wetmore, GL = Punta Mala, off Punta Mala, observations, 1956 Feb 6 (Wetmore, 1957:33-34, as "Frailes del Sur Rocks," and as "Frailes"), observations, 1957 Feb 26, collected near the shore from a boat, 1962 Mar 18 (USNM, as "Islas Frailes del Sur"; Olson, 1997a); Willis, observations, 1961 Feb (Ridgely and Gwynne, 1989:161, as "Los Frailes del Sur"); Delgado, observations, c. 2000 (Angehr, 2003:213).

FRAILES DEL SUR, ISLAS; see Frailes del Sur

FRAILES DEL SUR ISLANDS; see Frailes del Sur

FRAILES DEL SUR ROCKS; see Frailes del Sur

FRANCE FIELD; Colón (Canal Zone) (E)
0922/7952 (USBGN)
Sea level, an airfield located on the former Coco Solo Naval Reservation [0923/7952 (USBGN)] ca. 2 km E of the E coast of Bahía Manzanillo [0922/7953 (USBGN)](E8660). Keenan, 1961 Jun 16, Sep 8 (USNM); Yunker, Horse Corral, 1962 Apr 10 (USNM); Engleman, D. and L. Engleman, 1984 Oct 20 (WFVZ).

FRANCÉS; Chiriquí
0837/8225 (USBGN)
400 m, town 5 km N of Dolega [0834/8225 (USBGN)] (E762). According to correspondence between Mrs. Terry and Wetmore, this was a station on the railroad between Boquete [0847/8226 (USBGN)] and David [0826/8226 (USBGN)] (USNM, as "Villa Francés"). Watson, 1905 Sep-Nov (AMNH; BMNH; CM; CU; FMNH, as "El Frances," and as "Frances"; MCZ; Hellmayr, 1906; Hellmayr, 1910b; Eisenmann, 1962a; Wetmore, 1962; Olson, 1980c).

FRANCISCO CREEK; Bocas del Toro
ca. 0915/8223
Described by Wetmore as a very small inlet among mangroves on the S coast of Ambrosia Bight [0916/8222 (USBGN)] SE of Río Oeste [0916/8224 (USBGN)] and WNW of Punta de Gallinazo [0915/8222 (USBGN)], Wetmore, GL = Almirante [0918/8224 (USBGN)], from the shore entered tall, white mangroves and followed a trail along the foot of a hill that was largely cleared, but with some brush on it, on the shore of Punta de Gallinazo, 1958 Jan 18 (USNM). Not shown on our maps.

FRENCH CANAL; Colón (Canal Zone) (E, W, Atlantic)
0920/7955 (USBGN)
The abandoned cut of the canal that extends 9 km SW from Cristóbal [0921/7955 (USBGN)] just SW of Colón [0922/7954 (USBGN)], crosses the Panama Canal [0920/7955 (USBGN)] and terminates in a basin 1 km W of Gatún [0916/7955 (USBGN)] (E761A). E8660 indicates that the central section of the cut, on the E side of the Panama Canal, has been filled creating two waterways that are named French Canal. This abandoned cut was dug during the late 1800s during the French attempts to build a sea level canal. The name appears in USBGN and on some

maps as Canal French, but according to Angehr (in litt., Aug 2005) the name has never been used as such. Presumably the name Canal French is a typographical error — perhaps arising from the Spanish to English translation — and presumably should be listed as French Canal. Jewel, at the mouth, near Colón, 1911 Dec 10, 13 (ANSP; Jewel, 1913, as "Old French Canal"). See also: Gatún; and Gatún, Esclusas.

FRIJOL, RÍO; see Frijoles, Río

FRIJOLES; Colón (Canal Zone) (E)
0910/7948 (USBGN)
Sea level, a railroad station that was located in the valley of the continuous Río Chagres (see Río Chagres [0919/8000 (USBGN)]) on the original Panama Railroad [ca. 0922/7954], and therefore relocated 1-2 km NE of its original site c. 1912 when Lago Gatún (Gatun Lake) [0912/7955 (USBGN)] was created. The station is now on the present Panama Railroad [ca. 0922/7954] 23 km SSE of Colon [0922/7954 (USBGN)] on the E coast of Bahía Frijoles [0910/7949 (USBGN)] (E661; MIP, as "Frijol Station"). McLeannan, apparently he served as stationmaster here when the railway first opened before working at Lion Hill [ca. 0913/7954] (see Panama Railroad), most specimens are also labeled "Panama Railroad" or "Line of Panama Railroad," 18?? (ANSP; FMNH; USNM; Salvin and Godman, 1879-1904, vol. 1:26, as "Frijole Station"; Ridgway, 1901, as "Frijole Station, Panama R. R."); Goldman, 1911 May 9, 11 (USNM; Wetmore, 1959b); Van Tyne, 1926 May 17 (Eisenmann, 1952:19, 34); Gross, 1927 Nov 5 (Eisenmann, 1952:18); Weber, 1928 Jul 23-24, Aug 8-10 (FMNH; USNM); Wonder, between Darién Railroad Stop and, 1928 Dec 26 (FMNH); McCoy, 1929 Feb 25-26, Mar 12, 14, 19 (LACM); Howells, 1929 Aug 13 (UCLA); Gilliard, GL = Lago Gatún, 1937 Feb 7, 13, 17 (AMNH; Eisenmann, 1952:14); Imhof, observations, between Isla Barro Colorado [0909/7951 (USBGN)] and, 1942 Apr 17 (Imhof, 1950); Eisenmann, 1951 Jul 3 (Eisenmann, 1952:46); Willis, 2.4 km N, 4.4 km W, 1962 Feb 5-7 (MVZ; Willis, 1972); Morton, 1964 Jul 1, 16, 20, 2 mi (3 km) W, 1966 Jul 7, and as a GL for Río Agua Salud, 1966 Jul 5 (CU; FMNH); Miller, 1965 Feb 25, and as a GL for Punta Buena Vista, 1965 Feb 21, 24 (MVZ); Kilham, observations 1965 Feb, 1970 Nov-1971 Feb (Kilham, 1972); Smith, N., GL for Río Agua Salud, 1965 Jun 1 (USNM); White, 1 mi (1.5 km) SE, 1972 Mar 13, and as a GL for Península Bohío, 1972 May 3 (ANSP). (Wetmore, 1970; Olson, 1981d; Greenberg, 1981a; Moynihan, 1963).

FRIJOLES; Veraguas (Pacific)
0739/8143 (USBGN)
Group of rocks 1 km NW of Isla Ranchería [0738/8142 (USBGN)] (JOG; Wetmore, 1957:5, as "Aaron Rocks"). See also: Coibita, Isla.

FRIJOLES, BAHÍA; Colón (Canal Zone) (E)
0910/7949 (USBGN)
Bay that extends ca. 4 km N-S along the E coast of Lago Gatún (Gatun Lake) [0912/7955 (USBGN)]. It opens on the W to the Panama Canal [0920/7955 (USBGN)] and Isla Barro Colorado [0909/7951 (USBGN)] (E762).

FRIJOLES, RÍO; Colón (Canal Zone) (E)
0909/7946 (USBGN)

Originates W of Palenque [0913/7941 (USBGN)], flows S and SW entering Lago Gatún (Gatun Lake) [0912/7955 (USBGN)] 6 km WNW of Gamboa [0907/7942 (USBGN)] and 5 km SSE of Frijoles [0910/7948 (USBGN)] (JOG). Wetmore, GL = Gamboa, trails through tall, dense grass and low forest above the river, upstream on the river through a swampy woodland, some of the specimens are also labeled Smithsonian Tract, which presumably refers to a tract of forest along the Pipeline Road [ca. 0915/7948] ca. 5 km NW of Gamboa, where the Ríos Frijoles and Frijolito [0909/7944 (USBGN)], a tributary of the Río Frijoles, both cross the Pipeline Road, 1960 Jan 9-10, 13 (USNM); Chapman and Tyson, 5 mi (8 km) N of Gamboa, 1964 Jan 15, 27-31, Feb 1-4 (FSU, as "Río Frijol"). (Poulin and Lefebvre, 1996).

FRIJOLE STATION; see Frijoles

FRIJOLITA, RÍO; Colón (Canal Zone) (E)
0910/7947 (USBGN)
Originates N of Gamboa [0907/7942 (USBGN)], flows S then SW entering Bahía Frijoles [0910/7949 (USBGN)] S of the Río Agua Salud [0911/7947 (USBGN)] and N of the Río Frijoles [0909/7946 (USBGN)]. Pipeline Road [ca. 0915/7948] crosses the river 8 km NW of Gamboa [0907/7942 (USBGN)] (RM1, as "Río Frijolito"; E866, as "Río Mendosa"). Wetmore, GL = Frijolito, at 800 feet (240 m), head of, forest near logged and cleared area, along a high ridge where some forest remained, 1962 Jan 8-9 (USNM, as "Río Frijolito"); Hespenheide and Karr, GL = Gamboa, 7 mi (11 km) NW of Gamboa, 1962 Jul 20 (CM, as "Río Mendoza"); Karr, 1969 Jul (Karr, 1971a, as "Río Mendoza"). (Karr, 1990, as "Río Mendoza").

FRIJOLITO; Colón (E)
ca. 0913/7944
Described by Wetmore as an area at the headwaters of the Río Frijolita [0910/7947 (USBGN)], reached by traveling on the Carretera Boyd-Roosevelt [0917/7943 (USBGN)] 5 mi (8 km) NW of Alhajuela [ca. 0913/7937], then walking 5 mi (8 km) SSW to the headwaters (USNM). He also says that he left the highway where the Río Duque [0916/7941 (USBGN)] crosses it (Wetmore, 1968:188). Not named on our maps, but the region described by Wetmore appears on RM1 ca. 6 mi (10 km) NNW of Gamboa [0907/7942 (USBGN)]. Frijolito appears on Angehr's (in litt., Aug 2005) copy of "4243-I (Buena Vista) [E762] at UTM 1018500 N 642100 E (0913/7942), about 2 km east of [our] coordinates," although the same does not appear on our version of 4243-I. Wetmore, GL for Río Frijolita, 1962 Jan 8-9 (USNM).

FRIJOLITO, RÍO; see Frijoles, Río

FRIJOLITO, RÍO; see Frijolita, Río

FRIJOL STATION; see Frijoles

FRUITDALE; Bocas del Toro
0918/8225 (USBGN)
Sea level, town 3 km WNW of Almirante [0918/8224 (USBGN)] on the Quebrada Nigua [0917/8224 (USBGN)] (JOG). Wedel, 1927 May-1929 Jan (ANSP; MCZ; USNM; YPM; Peters, 1931a; Rogers, 1939; Olson, 1981c). See also: Nigua,

Quebrada.

FT. DAVIS; see Fort William D. Davis Army Reservation

FT. LORENZO; see Fuerte San Lorenzo

FT. SAN LORENZO; see Fuerte San Lorenzo

FT. SHERMAN; see Fort Sherman Army Reservation

FUERTE AMADOR; see Fort Amador Army Reservation

FUERTE SAN LORENZO; Colón (Canal Zone) (W)
0919/8000 (USBGN)
Sea level, ruins of the old fortified entrance to the transisthmian route located at the mouth of the Río Chagres [0919/8000 (USBGN)]. The fort was rebuilt in 1751-1752 and served as the entry point across the isthmus until c. 1851 when the Panama Railroad [ca. 0922/7954] (q. v.) began taking passengers from Colón [0922/7954 (USBGN)] (HMCZ, as "Ft. San Lorenzo"; IP, as "Fort San Lorenzo"). Fuerte San Lorenzo is included in the Important Bird Area also named Fuerte San Lorenzo [0916/7958 (Angehr, 2003:217)], a partially protected area that includes evergreen lowland forest and "deciduous forest on limestone outcrops." Jewel, palm tree, low bushes along the river bank, sea shore, 1911 Jan 7 (?), Feb 26, Jun 2, 11, 21, Jul 11, 1912 Jan 7 (ANSP, as "Ft Lorenzo," as "Ft San Lorenzo," and as "Fort San Lorenzo"; Stone, 1918, as "Fort Lorenzo"); Goldman, mouth of the Río Chagres, 1911 Jun 21 (USNM, as "Fort Lorenzo"; Olson, 1981d, as "Fort Lorenzo"); Boulton, 1924 Apr 5, 8-10, 12 (CM, as "San Lorenzo"); Bull and Eisenmann, 1952 Jun 28 (Eisenmann, 1957, as "Fort San Lorenzo"); Loftin, GL for Río Chagres, 1964 Sep (GML, as "San Lorenzo"). (Eisenmann, 1955c). See also: Mojinga Swamp.

FUERTES COVE; see Fuertes Estero

FUERTES ESTERO; Panamá (Canal Zone) (W, Atlantic)
ca. 0910/7951
Eisenmann (1952:41, 44, 50) mentions observations at Fuertes Estero and at Fuertes House, both of which are on the N coast of Isla Barro Colorado [0909/7951 (USBGN)] (Willis and Eisenmann, 1979:3, as "Fuertes Cove," and as "Fuertes House"). See also: Salud, Punta.

FUERTE SHERMAN; see Fort Sherman Army Reservation

FUERTES HOUSE; see Fuertes Estero

FUNDEADERO, PUNTA; Darién
ca. 0729/7808
Described by Wetmore as a rocky projection ca. 4 mi (6 km) SE of Jaqué [0731/7810 (USBGN)] that blocked farther travel S along the beach (USNM). The point is not named on our maps, but a bay in the same region is name "El Fondeadero" on MHA. See also: Fundeadero, Quebrada; and Playita del Medio, Loma.

FUNDEADERO, QUEBRADA; Darién
ca. 0729/7808
Described by Wetmore as a fair-sized brook on Punta Fundeadero [ca. 0729/7808]

ca. 4 mi (6 km) SE of Jaqué [0731/7810 (USBGN)], which flowed down over a stony, sandy beach, with two 4 foot (1 m) falls, Wetmore and Perrygo, GL = Jaqué, 1946 Apr 4 (USNM). Not shown on our maps.

GAGO; see El Gago

GAGO, EL; see El Gago

GAILLARD, CARRETERA;
0904/7940 (USBGN)
Colón/Panamá (Canal Zone) (E, Atlantic, Pacific)
Highway that extends ca. 15 km N-S between Gamboa [0907/7942 (USBGN)] and Balboa [0857/7934 (USBGN)]. Summit Park [0904/7939 (USBGN)] is located on the road 8.5 km SE of Gamboa at the junction with Camino Madden (Madden Road) [0907/7937 (USBGN)], and the road extends through Parque Nacional Soberanía [0909/7941 (Angehr, 2003:239)]. The road is also known as Road C2 and is often called Gamboa Highway by collectors (E866, as "Gaillard Highway"; E8660, as "Gaillard Highway"; RM1, as "Gaillard Highway," and as "Road C2"). Meyer, "Gamboa Road Junction" may refer to the junction of Carretera Gaillard with Camino Madden, 1941 May 11, 15, 17, Jul 7, 30 (USNM, as "Gamboa Road," and as "Gamboa Road Junction"); MCU, 1955 Oct 18 (USNM, as "Gamboa Road"); Yunker, GL = Gamboa, 1962 Feb 8 (USNM, as "Gamboa Road"); Olson and Wiese, GL = Gamboa, Summit Park, collected at Mile 12, between Summit Park and Gamboa in shaded second growth jungle with old cacao trees, GH-M12 appears on many labels and refers to a sign post on the road at mile 12 and near the trail head for Cacao Plantation Road q.v. [ca. 0907/7938], 1966 May 8-10, 13, 15, 25, Jun 13, 15 (USNM, as "Gamboa Highway," and as "GH-M12"); White, GL = Summit Park, 2 mi (3 km) E, 1972 Apr 6 (ANSP, as "Graillard Highway"); Morton, GL for Summit Park, 1974 Mar 12 (CAS, as "GH Road"); Braun, M., GL for Summit Park, 1994 Apr 13, May 6 (USNM, as "Gamboa Road," and as "Old Gamboa Road"); Agro, 1996 Feb 27 (MVUP, as "Carretera de Gamboa"); Collector ?, near Renacer Jail, (located just S of Gamboa), 1997 Mar 29, at mile 14, 1998 Jul 1, GL = Summit Park, 1998 Oct, GL for Cacao Plantation Road, 1999 Mar 17 (LSUMZ, as "Gaillard Highway"). See also: Cacao Plantation Road; and Old Gamboa Road.

GAILLARD CUT; Panamá (Canal Zone) (Atlantic, Pacific)
0903/7940 (USBGN)
Section of the Panama Canal [0920/7955 (USBGN)] just W of Paraiso [0902/7938 (USBGN)] that extends through the continental divide (E762, as "Corte Culebra"). Tejera, Duran, and Garcés, 40-160 m, "*áreas boscosas, áreas abiertas y el área acuática*" (forested, open and aquatic areas), "*el bosque es secundarío*" (secondary growth forest), 1993 Feb 8-9 (MVUP, as "Corte de Culebra"; Tejera, 1995:7-8, as "Corte Culebra"); Collector ?, GL = Panama Canal, 1998 Jun 24 (LSUMZ, as "Culebra Cut").

GAILLARD HIGHWAY; see Gaillard, Carretera

GAITAL, CERRO; Coclé (Atlantic, Pacific)
0837/8007 (USBGN)
1,185 m, isolated peak near the E end of the Cordillera Central [0838/8150

(USBGN)], on the continental divide 3 km N of El Valle [0836/8008 (USBGN)] (E761B; JOG). Cerro Gaital is included in the Important Bird Area located N of El Valle (Angehr, 2003:190-191). Wetmore and Perrygo, GL = El Valle, at 2,500 feet (760 m), worked both sides of the continental divide, the growth on the Atlantic drainage was dense and wet, near the head of the Río Antón [0818/8016 (USBGN)] (Pacific drainage) with slopes above denuded and overgrown farms below, near clearings for potatoes, overgrown orange groves and fields with forest on the higher slopes, 1951 Mar 26-27 (USNM).

GALERA, CERRO; Panamá (Canal Zone) (W, Pacific)
0856/7937 (USBGN)
360 m, peak located 11 km SW of Panama City [0858/7932 (USBGN)] and 5.5 km N of the coast (E762). According to E661, this was the highest peak in the former Canal Zone. Wetmore, GL for Road K6, 1960 Jan 14, at 1,000 feet (300 m), drove up Road K6 [ca. 0900/7939], hunted along the narrow roadway on steep, forested slopes, 1961 Jan 8 (USNM).

GALERA, ISLA; Panamá, Archipiélago de las Perlas
0812/7847 (USBGN)
42 m, small island with a land area of 21 ha (Angehr, 2003:279) located 13 km ESE of Punta Cocos [0813/7854 (USBGN)] on Isla del Rey [0822/7855 (USBGN)] (JOG). Angehr also indicates that the island is steep sided and uninhabited. Chapin, Rogers and Mathews, 1923 Aug 11 (FMNH); Griscom and Crosby, 1927 Feb 23 (AMNH, as "Galera Island"); Robins, observations, 1957 Jul 15-26 (Robins, 1958); Wetmore, hunted from the boat without landing, but noted an extensive sandy beach behind which the island was rocky, a good stand of fairly tall trees and an extensive stand of coarse, heavy grass, 1960 Jan 28 (USNM); Pacific Project, GL = Gulf of Panama, off Isla Galera, 1968 Mar, 1969 Aug 30 (GML, as "Galera Island"; Loftin, 1991).

GALERA ISLAND; see Galera, Isla

GALETA, ISLA; Colón (Canal Zone) (E)
0924/7952 (USBGN)
Sea level, an island with a land area of ca. 1.5 km^2 (150 ha) located 5.5 km NE of Colón [0922/7954 (USBGN)] and separated from the mainland by a narrow stream channel. The island was included in the limits of the former Coco Solo Naval Reservation [0923/7952 (USBGN)] (E8660). According to Wiedenfeld (1992) the area includes "low red mangroves, with some strips of low black mangroves along the road from Coco Solo." Morton, studies, 1970s (Morton, 1980, as "Galeta Island"; Morton, 1982, as "Galeta Mangroves"); Caldwell, observations along the Atlantic coast between Fort Sherman Army Reservation [0920/7958 (USBGN)] and, 1975 Dec, 1977 Jan-1978 Jan, 1982 Nov-1983 Nov (Caldwell, 1986, as "Galeta Island"); Schneider, observations, the longitude [09°24'18"/79°57'48"] presented appears to be erroneous and presumably should read 79°52', 1978 Mar-Apr (Schneider, 1985); Ridgely, observations, 1980 May 1 (Wetmore et al., 1984:305, as "Galeta Island"); Cubit, 1980 Nov 25 (WFVZ, as "Galeta I."); Gardner, 1985 Nov 8 (WFVZ, as "Galeta I."); Wiedenfeld, observations, 1986-1988 (Wiedenfeld, 1992); Lefebvre

and Poulin, observations and studies, the coordinates given [09°20' N, 79°09' W] appear to be erroneous, 1990s (Poulin and Lefebvre, 1995; Lefebvre and Poulin, 1996; Lefebvre and Poulin, 1997; Lefebvre and Poulin, 2000).

GALETA, PUNTA; Colón (Canal Zone) (E)
0924/7952 (USBGN)
Sea level, point on the N coast of Isla Galeta [0924/7952 (USBGN)], near coral reefs (E8660, as "Galeta Point"). Noyes, 1947 Aug 18 (MVZ, as "Galeta Point"); Eisenmann, observations, 1958 Aug 28 (Wetmore, 1972:382, as "Galeta Point"); Guevara, 1991 Sep 19 (MVUP).

GALETA ISLAND; see Galeta, Isla

GALETA MANGROVES; see Galeta, Isla

GALETA POINT; see Galeta, Punta

GALETA ROAD; Colón (Canal Zone) (E)
ca. 0924/7952
Presumably refers to the section of Road R12 that extends 4 km E and N from Coco Solo [0922/7953 (USBGN)] to Punta Galeta [0924/7952 (USBGN)]. The road also extends S from Coco Solo to Mount Hope [0920/7954 (USBGN)] and is located within the former limits of Coco Solo Naval Reservation [0923/7952 (USBGN)] (E8660, as "Road R12"). Gardner, GL = Coco Solo, 1984 Jul 6 (WFVZ); Engleman, D., GL = Coco Solo, 1984 Nov 6 (WFVZ).

GALINAZO, PUNTA; see Gallinazo, Punta de

GALLINAZO, PUNTA DE; Bocas del Toro
0915/8222 (USBGN)
50 m, point extending ca. 1 km E from the mainland on the S coast of Ambrosia Bight [0916/8222 (USBGN)] and separated from the NW shore of Isla Pastores [0914/8220 (USBGN)] by a 1 km wide channel. The point is located ca. 6 km SE of Almirante [0918/8224 (USBGN)] (E762, as "Punta de Gallinazo (Jungle Point)"; JOG, as "Punta Galinazo"). Punta de Gallinazo is the Spanish name, whereas the English name is John Crow Point. Interestingly, in Panamá *gallinazo* refers to the Black Vulture *Coragyps atratus*, whereas in Jamaica, whence many of the English names in Bocas del Toro are derived, "John Crow" refers to the Turkey Vulture *Cathartes aura.* Wetmore, GL = Almirante, the point extends E-W with a fairly high ridge in the center, walked along the base of the slope in a wooded swamp, up through cacao and patches of bananas, to the level summit in a pasture and in cultivated areas, on the shore at the edge of mangroves, 1958 Jan 27, Feb 7 (UF, as "John Crow Point"; USNM, as "John Crow Point"). See also: Francisco Creek; and Valle de Agua.

GALLINERO; Panamá (E, Pacific)
ca. 0902/7911
Apparently a pond ca. 2 mi (3 km) SE of the clubhouse at La Jagua Hunting Club [ca. 0903/7911], Wetmore, GL = Río Chico [0902/7910 (USBGN)], La Jagua Hunting Club, in trees and grass along the pond, 1958 Mar 21 (USNM). Not shown on our maps.

Introduction

It was always intended that there be a gazetteer to accompany Alexander Wetmore's *Birds of the Republic of Panamá* (Wetmore, 1965, 1968, 1972; Wetmore et al., 1984) but after the species accounts were finished, following Wetmore's death (7 December 1978—see Ripley and Steed, 1987), the enterprise was repeatedly postponed. Wetmore left a card file of place names and a bibliography and using this as a basis I attempted to expand the information desultorily over the years, with considerable assistance from Mrs. Wetmore (d. 1 March 1997—see Olson, 1997b). Because it was nearly impossible to carry out such labors on a part-time basis, it was decided to use some of the funds that the Wetmores bequeathed to the Division of Birds to hire someone to bring the work to completion. Fortunately, Deborah Siegel proved the perfect choice and has brought understanding and painstaking care to the project. The manuscript was largely completed by May 2006 but some additions and alterations were made up until June 2008. Further corrections and additions will be gratefully received <olsons@si.edu>.

Acknowledgments.—George Angehr, Smithsonian Tropical Research Institute (STRI), provided many helpful comments on the manuscript and much assistance to DCS in Panamá. Bird Division (USNM) volunteer Frederick H. May verified all coordinates against those given in USBGN. Persons supplying collection data are acknowledged in the section on museum acronyms and their contributions are deeply appreciated.

Storrs L. Olson, Washington, D.C., 10 June 2008

PORTRAIT OF ALEXANDER WETMORE

This large (30 X 35 inches) oil portrait of Alexander Wetmore, believed never to have been previously published, is shrouded in mystery. On 22 March 1999, it was discovered by J. Phillip Angle, then collections manager in the Division of Birds, National Museum of Natural History, in an auction house a block from the museum. Listed only as a "Portrait of a Gentleman," it was to be sold for its frame. Angle bid successfully on it and carried it off for $70.00. It was painted in 1941, four years before Wetmore became Secretary of the Smithsonian, by Bjorn Egeli (1900-1984), a well-known Norwegian/American portraitist who painted official portraits of presidents Eisenhower and Nixon. On the back was an inscription saying that it had been "given to Samuel B. Pratt by the artist." Pratt (1903-1999) was a local landscape painter who had connections with the Corcoran Gallery in Washington. The seller in 1999 was identified only as a "woman from Potomac [Maryland]" who had purchased the painting from the same auction house four or five years previously "for quite a bit of money," there being no record of who the seller was at that time. Angle contacted Egeli's son, Cedric B. Egeli, who had no knowledge of the portrait. Wetmore's daughter, Margaret Harlan, who would have been 25 years of age in 1941, had no recollection whatever of this portrait and in fact was astounded by it, as she related that her father would have detested sitting for a portrait. This raises the possibility that the portrait may have been painted from a photograph, but whoever commissioned it evidently did not pay for or claim the picture, as it remained with the artist. It is doubtful that the painting was rejected as inadequate because it is a very good likeness. It is cataloged in the National Portrait Gallery and currently hangs in the Division of Birds.

Alexander Wetmore
(1886 - 1978)

ALEXANDER WETMORE (1886-1978)
AND THE ORNITHOLOGY OF PANAMÁ

Alexander Wetmore was one of the predominant figures in New World ornithology in the 20th century. He was born 18 June 1886 in North Freedom, Wisconsin, and knew from a very early age that he wanted to devote himself to the study of birds. At the age of 18 he moved with his mother to Independence, Kansas, where he attended high school and also began collecting and preparing specimens of birds. One of his early specimen catalogs contains entries for "Indian Territory," as Oklahoma was not then a state. He received a bachelor's degree from the University of Kansas in 1912 but prior to that held the position of Agent for the Biological Survey of the U. S. Department of Agriculture, in which capacity he conducted field investigations in Wyoming and Alaska and in late 1911 began collecting birds for nearly a year in Puerto Rico, which resulted in the baseline monograph of the avifauna of that island.

A promotion to Assistant Biologist took Wetmore to Washington in 1913, where he associated with many of the luminaries in natural history associated with the Survey and the Smithsonian Institution. The next few years found him engaged in field work in various parts of the United States. In 1920 and 1921, he conducted extensive fieldwork in Argentina, Uruguay, Paraguay, and Chile, resulting in another important monograph. Along the way he picked up MS (1916) and PhD (1920) degrees from George Washington University. In 1923, he was a major participant in the *Tanager* expedition to the northwestern Hawaiian Islands, and Johnston and Wake islands (Olson, 1996). Afterwards he was promoted in 1924, seemingly rather belatedly, to Biologist, but that was not sufficient to head off his departure from the Survey to the Smithsonian Institution. A patient and circumspect person by nature, Wetmore was also strong-willed and I think always knew exactly what he wanted for himself, which probably did not square with whatever the equally strong-willed Chief of the Survey, E. W. Nelson, may have had in mind.

The unexpected death of the Superintendent of the National Zoological Park, part of the Smithsonian, created a position to which Wetmore was assigned in 1924 by Smithsonian Secretary Charles D. Walcott, of Burgess Shale fame. Only a matter of months passed before Walcott appointed Wetmore to the post of Assistant Secretary in charge of the U. S. National Museum, as it was then known, a position that Wetmore must have found agreeable, as he occupied it for nearly 20 years, through the entire tenure (1928-1944) of the 5th Secretary, astronomer C. G. Abbott, who had other fish to fry and evidently was content to let Wetmore run the museum end of things.

As Assistant Secretary, Wetmore maintained a rigorous program of research and fieldwork. He explored and collected in Hispaniola in 1927, which resulted in another seminal monograph on the avifauna of that island co-authored with B. H. Swales, who I think was more of a sponsor than a contributor to the work. I once innocently asked Wetmore about Swales, not knowing anything about his participation in the endeavor, and Wetmore became slightly defensive, I thought, stating that Swales had earned co-authorship on the volume, as though others may have questioned his inclusion in the past. In any case, with his contributions to the ornithology of Puerto Rico and Hispaniola, Wetmore became a stalwart of Antillean ornithology. During his tenure as Assistant Secretary, Wetmore also conducted fieldwork in Guatemala, Costa Rica, Colombia, and Venezuela.

Wetmore was appointed 6th Secretary of the Institution in 1945 from which position he retired in 1952, making sure that his emeritus research was well-endowed with funds that had been left to the Institution by Abbott and others. Tall, handsome, stately in bearing and demeanor, Wetmore was the personification of authority, dignity, and institutional memory, so his emeritus research arrangements would never have been questioned. By then, of course, Wetmore had occupied important offices in dozens of scientific societies, received awards and diplomas galore, and had a publication record second to none, which included over 150 titles on fossil birds alone (Olson, 1976).

Shortly before he assumed the Secretary's mantle, Wetmore participated in a collecting expedition to the Pearl Islands of Panamá in 1944 as part of what I understand to have been a wartime effort to locate remote areas to make tests for gas warfare. Wetmore must have found Panamá to be exceedingly agreeable to his disposition, as after the war, at the age of 60, when most men might be thinking of retirement and a life of leisure, he began what became his *magnum opus* on the birds of Panamá, a country in which he collected tirelessly for 20 consecutive years, from 1946 to 1966. From 1946 through 1953 he was assisted by his protégé Watson M. Perrygo (1906-1984), but in subsequent years he made other arrangements.

In 1953, Wetmore's first wife, Fay Holloway, died after a period of illness. In the same year, Wetmore married Annie Beatrice Thielen, whom he had met in 1944 (Olson, 1997b). This struck some in Washington society as precipitate and Wetmore once told me that it sparked comment from Washington columnist Drew Pearson, whereafter Wetmore lost no love on the Washington press. Bea accompanied Alec in the field in Panamá on most of his subsequent expeditions (Olson, 1997). Wetmore's collecting was planned to cover the entire country sampling birds to determine their distribution and geographic variation. His trips always took place during the northern winters, which in most of Panamá coincides with the annual dry season, so that there is a decided bias to his samples with all specimens being taken in the period from December to April (with the exception of 1953 in which he was mainly in Coclé and Veraguas in May and June)

The gentlemanly Wetmore was affectionately and respectfully known throughout Panamá as "Don Alejandro" and his energy in the field was legendary. He was known, for example, to bound up the very long, steep stairway to the facilities on Barro Colorado Island far ahead of the much younger folks who trailed him. Once a visitor to one of his camps, on seeing a row of specimens laid out all apparently alike, asked him "How many specimens of each species do you need?", to which he responded "Just one more."

Wetmore ceased field work in Panamá after 1966 and concentrated on writing up the results of his years of effort for his *Birds of the Republic of Panamá*, the first three volumes of which appeared from 1965 to 1972. As he worked up his Panamá collections he also identified very extensive collections from Colombia that he had commissioned M. A. Carriker, Jr., to collect (Olson 2007) with money from the Abbott Fund. I have a vivid recollection in the early 1970's of Wetmore repeatedly poring over specimens of Purple Martins (*Progne subis*) trying to determine whether he could recognize a subspecies that W. H. Behle had described from Utah in 1968 and whether any of the specimens from Panamá might belong to that form. He did not make much more progress on volume 4 before his advanced age made claims on his abilities. At the instigation of Dillon Ripley, Roger Pasquier and I were able to finish the last volume of *Birds of the Republic of Panamá* for publication in 1984. The present gazetteer brings Wetmore's efforts in Panamá to a long overdue conclusion. For more details of Wetmore's biography see Oehser (1980), Ripley and Steed (1987; and the Smithsonian Archives website at http://siarchives.si.edu/findingaids/faru7006.htm#FARU7006h.

Storrs L. Olson, Washington, D. C., 12 May 2008

Using the Gazetteer

For numerous reasons, we have chosen to employ the format for the ornithological gazetteers of South America produced by the late Raymond A. Paynter, Jr. et al. (e.g. Paynter 1997). We have relied most heavily on computer inventories of museum collections and are greatly indebted to all those who assisted in providing this information. Nowhere was this more important than with the Smithsonian's collections, which are the most comprehensive for Panamanian birds anywhere, thanks largely to Alexander Wetmore's long efforts. Wetmore's first collecting trip to Panamá was in 1944, when he was 58 years of age. He returned in 1946 and then every year after that through 1966, accumulating a total of over 14,000 specimens. His detailed journals, maintained in the Division of Birds, have enabled us to supply ecological notes for most of Wetmore collecting localities. Whenever we have cited "Wetmore" without further attribution, the information may be assumed to have been taken from Wetmore's journals. Also at the Smithsonian we have had access to the journals of E. A. Goldman and C. O. Handley, Jr.

Alphabetizing.—-The entries were alphabetized using the English alphabet; therefore *ch*, *ll*, *ñ*, and *rr* are not treated as separate letters as they are in Spanish.

Coordinates.—Coordinates for rivers indicate the mouth of the river. Coordinates for roads and trails include either the northernmost or westernmost terminus. Coordinates for parks, airfields, etc. are estimates of the center of the locality.

Dates.—We have attempted to be as exact as possible when giving dates of collection or observation, but in the case of residents, such as Mönniche or Watson, who collected year round, we generally give only the years that a collector was active.

Distances.—Distances within locality descriptions are airline distances unless specifically stated to be river miles, road miles, etc. Distances taken from collectors' notes may be of various means.

Literature citations.—The page numbers cited for *Birds of the Republic of Panamá* and Ridgely and Gwynne (1989) indicate the beginning of the species account, not necessarily the page where a given locality is mentioned. All other page numbers indicate the actual page.

Not shown on our maps.—This indicates that a locality was not *named* on any of our maps, even if the physical entity (river, hill, etc.) is indicated.

Observers.—We have not necessarily included every reference to observations and collections for a given locality, especially in heavily studied regions, such as Barro Colorado Island, although most or all known observations will be included for less well known localities.

Provinces.—We have treated the Comarca (Territory) de San Blas as being equivalent to a province but we do not refer to any of the other Native American *comarcas* that have been designated subsequently. The Canal Zone no longer exists as a political entity but because so many specimens are so labeled, we have indicated when a locality was in the former Canal Zone as well as the province to which it now belongs. Because the Canal marks a biogeographically significant area, we have also indicated whether a given locality in Panamá or Colón provinces lies east (E) or west (W) of the Canal.

Versants.—All localities in the provinces of Chiriquí, Herrera, Los Santos, and Darién lie on the Pacific slope, and all localities in Bocas del Toro, Colón, and San Blas are on the Atlantic slope. For Veraguas, Coclé, and Panamá, which straddle the continental divide, we have indicated in the heading whether a locality is on the Pacific or Atlantic versant. This has been omitted for Archipíelago de las Perlas (Panamá) and Isla de Coiba (Veraguas), as these islands are obviously in the Pacific Ocean.

Map Abbreviations

Unless otherwise indicated, maps are single sheets.

AC — Aviation Chart V14-136, Barranquilla-Cali. 1945. The legend and publishing information are missing from the chart, along with the lower portion of the map. The map title and date were written on the chart by Wetmore.

Atlas, 1988 — Atlas Nacional de la República de Panamá, Tercera Edición (3rd Edition). 1988. x + 222 pages. Instituto Geográfico Nacional "Tommy Guardia," Ministerio de Obras Públicas, Panamá.

CARP — Carta Aeronáutica de la República de Panamá (1:500,000), edition 1. 1976. One of two sheets covering Panamá. Instituto Geográfico Nacional "Tommy Guardia," Panamá.

DB — República de Panamá, Provincia de Bocas del Toro, Distrito de Bastimentos (1:100,000). 1968. Sección de Cartografía de la Dirección de Estadística y Censo, Panamá.

DP — República de Panamá, Provincia de Darién, Distrito de Pinogana (1:125,000). 1957. Sección de Cartografía de la Dirección de Estadística y Censo, Panamá.

E501 — Central America, Panamá (1:250,000), series E501, editions: prov, 1, 2. 1929-1950. Maps sections covering Panamá. We have only the following seven sheets: NC 17S VI (Panamá), NC 17S II (Donoso), NB 17N II (Chitre), NC 17S IV (David), NC 17S V (Santiago), NC 17S I (Cartago, Costa Rica; Panamá), and Prov C 17S III (Colón). The Colón map is an unedited provisional edition, dated 1941 and printed by the Engineer Production Plant, U. S. Army, The Army War College, Washington, D.C. Edition 1 maps (David, Cartago) are dated 1929 or 1930, and edition 2 maps (Panama, Donoso, Chitre, Santiago) are dated 1941, 1949, or 1950. The edition 1 and 2 maps were printed by the Army Map Service, Corps of Engineers, Washington, D.C.

E562 — David (1:250,000), edición 11, series E562. 2002. We have only sheet 2, David, but the complete set covers the entire country and includes 12 sheets. Instituto Geográfico Nacional "Tommy Guardia," Ministerio de Obras Públicas, Panamá.

E661 The Panama Canal, Map of the Canal Zone and Vicinity (1:100,000), edition 4. 1957. The map contains an inset of Madden Lake Watershed (1:335,000) and a separate index sheet. Army Map Service, Corps of Engineers, Washington, D.C. We also have edition 3 (1949) and have noted few, if any, differences between the two.

E761A Republic of Panama and Canal Zone (1:62,500), series E671. 1928-1947. According to the index sheet (1946), the complete set includes 37 quadrangles that cover central Panamá from approximately longitude 81° to 79°. Not all quadrangles within the limits were mapped and some maps are incomplete. Our set includes the following 31 quadrangles: Aguadulce, Antón, Balboa, Calobre, Chame, Chepillo, Chepo, Chitre, Chorrera, Ciri, Colón, Estibana, Gamboa, Indio, Lake, Macaracas, Majagual, Mensabé, Natá, Ocu, Otoque, Pacora, Palenque, Parita, Penonomé, Pese, Picacho, Piña, Porto Bello, Santiago, and Upper Chagres. Army Map Service, Corps of Engineers, Washington, D.C.

E761B Panama (1:50,000), series E761, editions 1-4. 1950, 1955. Based on the E761A maps, and as with that set not all quadrangles were mapped and some maps are incomplete. Our set of 31 quadrangles includes the same as listed for E761A with the exception of Lake, Natá, and Pacora, and the addition of Cañazas, Sipurio, and Sixaola, the latter three being W of longitude 81°. Edition 1-3 maps are dated 1950, and edition 4 maps are dated either 1950 or 1955. Army Map Service, Corps of Engineers, Washington, D.C.

E762 Panama (1:50,000), series E762, editions 1-3. 1960-1995. Quadrangles, each labeled with a regional name and a number, that cover Panama. We have 33 edition 1 maps that were printed from 1960 to 1995 by either the Army Map Service, Washington, D.C., or the Instituto Geográfico Nacional "Tommy Guardia," Ministerio de Obras Públicas, Panamá. Our set is incomplete and includes: Almirante (3743 IV), Bocas del Toro (3744 II), Boquete (3742 III), Canquintú (3842 IV), Capira (4242 III), Cauchero (3743 III), Cerro Chorcha (3741 I), Cerro Peña Blanca (3643 II), Cerro Pittier (3643 III), Cerro Punta (3642 I), Changuinola (3644 II), Chiriquí Grande (3742 I), Corriente Grande (3643 I), Cusapín (3843 IV), David (3741 III), El Valle (4141 I),

Escudo de Veraguas (3843 II), Gran Galera de Chorcha (3741 II), Gualaca (3741 IV), Horconcitos (3740 I), Isla Colón (3744 III), Isla de Coiba North (3838 IV), Isla de Coiba SE (3838 II), Isla de Coiba SW (3838 III), Isla Popa (3743 I), Muturicoto (3842 I), Península Valiente (3843 III), Punta Róbalo (3743 II), Quebrada de Yuca (3742 II), Río Changuinola (3742 IV), Río Quebro (4038 III), Río Sucut (3644 III), and Río Teribe (3643 IV). We have eight edition 2 maps including: Buena Vista (4243 I), Chame (4241 IV), David (3741 III), Escobal (4243 III), Gatún (4243 IV), Gualaca (3741 IV), Panamá (4242 I), and Volcán (3642 II). These maps were printed in 1977-2001 by either the Defense Mapping Agency or the Instituto Geográfico Nacional "Tommy Guardia," Ministerio de Obras Públicas, Panamá. We have a single edition 3 map (Alcalde Díaz (4243 II)) that was printed in 1993 by the Instituto Geográfico Nacional "Tommy Guardia," Ministerio de Obras Públicas, Panamá. Three of our maps (La Unión (3642 IV), Puerto Viejo (3644 IV), and Río Sixaola (3644 I)) include large sections of Costa Rica and technically belong to the series E763, but because these maps were used in conjunction with the adjoining sheets in the E762 series, they were referenced as E762. In addition, George Angehr provided us with information from map sheets that we did not have. The names of those maps are cited directly in the text.

E866 Panama (1:25,000), series E866, edition 1. 1958-1960. Maps covering sections of the Canal Zone. Of the entire series we have only the following eight sheets: Barro Colorado (4243 III NE), Colon (4244 III SW), Escobal (4243 III NW), Frijoles (4243 IV SE), Gamboa (4243 II NW), Gatun (4243 IV NW), Gatun Lake (4243 IV SW), and Nuevo San Juan (4243 I SW). Army Map Service, Corps of Engineers, U. S. Army, Washington, D.C.

E8660 Special Training Pictomap, Panama (1:25,000), series E8660, edition 1. 1968. One of fourteen maps covering the Canal Zone. Our set is missing sheets 4 and 14. Army Map Service, Corps of Engineers, U. S. Army, Washington, D.C.

E962 Ciudad de Panamá (1:12,500), series E962, edición 1. 1985. One of nine sheets covering Panama City. Our set includes only sheets 1-3, 5. Instituto Geográfico Nacional "Tommy Guardia," Ministerio de Obras, Panamá.

HMCZ — Historical Map of Canal Zone and Vicinity, Showing various Routes, Settlements, etc. between 1502 to 1855. 1939. Compiled by Kirkpatrick, Pearl and Orsini, made by Pearl and Orsini, The Panama Canal, Section of Surveys.

IP — Carte Topographique de la Partie de L'Isthme de Panama, Comprise entre Panama et Chagrés, avec le Tracé du Canal Maritime (1:200,000). 1845. Napolléon Garella, Ingenieur en Chef au Corps Royal des Mines.

JOG — Joint Operation Graphics (Ground) (1:250,000), series 1501, edition 1. 1968, 1970. One of 13 sheets covering Panama. Our set includes: NB 17-2 (Isla de Coiba), NB 17-3 (Chitré), NB 17-4 (Jaqué), NC 17-10 (Almirante), NC 17-11 (Colón), NC 17-12 (Pedregal), NC 17-13 (Golfito, Costa Rica), NC 17-14 (David), NC 17-15 (Panamá), NC 17-16 (La Palma), and NC 18-9 (Ustupu). We also have an edition 2 copy of Map NC-17-10 (Almirante), 1970, which is significantly different from the edition 1 copy. We are missing the two easternmost sheets (NC-18-13 and NB-18-01). Army Map Service, Corps of Engineers, U. S. Army, Washington, D.C.

LZP — Provisional Life Zones of Panama. 1918. E. A. Goldman, Bureau of Biological Survey, U. S. Department of Agriculture, Washington, D.C. Printed as plate 1 in Goldman (1920).

ME — Mapa Esso, La República de Panamá. Date not given, probably c. 1950. Fold-up map that includes a map of Panamá on the recto and a map of Central America on the verso. General Drafting Co., Inc., Convent Station, New Jersey. Olson recalls that in 1963 the inaccuracy of this map was a joke among Panamanians, who assumed that any lost motorist must have been following the Esso Map.

MGP — Mapa-Guía de Panamá. c. 2003. Fold-up map that includes a map of Panamá on the recto and a map of greater Panama City (Ciudad de Panamá) on the recto. Purchased in Panamá from National Car Rental in Jan 2003.

MGRP — Mapa General de la República de Panamá (1:250,000), edición 10. 1985. One of 12 sheets covering Panamá. Instituto

Geográfico Nacional "Tommy Guardia," Ministerio de Obras Públicas, Panamá.

MHA Map of Hispanic America (1:1,000,000), sheet NC-17 (Panama), series 1301, edition 1-AMS. 1954. Copyright 1927 by the American Geographical Society, New York. Printed by Army Map Service, Corps of Engineers. The map does not include the region E of 78° longitude.

MIP Map of the Isthmus of Panamá, Representing the Line of the Panama Rail Road as constructed under the direction of George M. Totten, Chief Engineer &c. Reduced and compiled from the original surveys by Thos. Harrison Crown Surveyor, Jamaica, 1857. Lith. of Endicott & Co., New York.

MRP Mapa de la República de Panamá (1:750,000). Date not given, but the map was created after the 1930s as indicated by the presence of Represa Madden. Instituto Geografico de Agostini, Novara, Italia.

PC Map of the United States Canal Zone showing the completed Panama Canal. No date given, but it was received at USNM in 1932. C. S. Hammond and Co., New York.

PDV Provincia de Veraguas, Distrito de Santa Fé, Corregimiento Calovébora (1:50,000). 1968. Sección de Cartografía de la Dirección de Estadística y Censo, Controlaría General de la República, Panamá.

PR Panama Route, Map Showing Location of Proposed Canal (1:100,000). 1901. We do not have an original map, but a tracing done by Wetmore. Isthmian Canal Commission.

PTC Map of Pacific Terminal Cities (1:12,000). 1952. Engineer Base Reproduction Plant, Panama Canal Department, Office Department Engineer, Corozal, Canal Zone.

RDP República de Panamá (1:250,000), Carta Preliminar. 1954. We have only sheet 3, which covers the region between longitude 78°30' and 79°45'. From the legend the entire set includes 4 sheets. Cartografía, Ministerio de Obras Públicas, Panamá.

RDP2 República de Panamá (1:1,000,000). 2002. A single sheet that contains a *Mapa Físico* and a *Mapa Político* on the recto and *Síntesis Geográfico, Edición 3* (a collection of 8 small maps and data tables) on the verso. Ministerio de Obras Públicas, Instituto Geográfico Nacional "Tommy Guardia," Panamá.

RDPF República de Panamá, Mapa Físico (1:500,000), edición 5. 1973. One of two sheets covering Panama. Instituto Geográfico Nacional "Tommy Guardia," Ministerio de Obras Públicas, Panamá.

RDPP República de Panamá, Mapa Político (1:500,000), edición 14. 1984. One of two sheets covering Panama. Instituto Geográfico Nacional "Tommy Guardia," Ministerio de Obras Públicas, Panamá. We also have edición 5 (1973) and found only a few differences between the editions.

RM1 Road Map of the Canal Zone and Vicinity (1:100,000). 1952. The verso consists of three small maps – Atlantic Terminal Cities, Pacific Terminal Cities, and a Highway Map of Panama (1:1,250,000). Engineer, U.S. Army Caribbean, Corozal, Canal Zone.

RM2 Road Map of the Canal Zone and Vicinity (1:100,000). This version has no publication data, but appears to be a more recent version of RM1.

RMM1 Road and Mileage Map of the Canal Zone and Vicinity (scale not given). 1943. The sheet contains three small insets – Atlantic Terminal Cities, Pacific Terminal Cities, and Road Map, Republic of Panama, which covers only a small section of the country W of the Canal Zone. Engineer Reproduction Plant, Panama Canal Department, U. S. Army, Corozal, Canal Zone.

RMM2 Road and Mileage Map of the Canal Zone and Vicinity (1:100,000). 1947. The verso consists of three small maps — Atlantic Terminal Cities, Pacific Terminal Cities, and a Highway Map of Panama (1:1,250,000). Engineer Reproduction Plant, Panama Canal Department, U. S. Army, Corozal, Canal Zone.

SMN2A Republic of Panama, Special Map No. 2 (1:500,000). 1943, 1946. One of three sheets covering Panama. We are missing the western section sheet. Engineer Reproduction Plant, Panama Canal Department, U. S. Army, Corozal, Canal Zone.

SMN2B Republic of Panama, Special Map No. 2 (1:500,000). No date given, but assumed to be an updated version of SMN2A. One of three sheets covering Panama. Army Map Service, Corps of Engineers, USARCARIB, Fort Clayton, Canal Zone.

TM Republic of Panama, Terrain Map, Military Survey of Panama (1:20,000). 1938-1947. A full set includes nine sheets for each quadrangle in series E761, but we have only sheets 82-83, 85-88 for the Chepo Quadrangle, and sheets 63-66, 69 for the Pacora Quadrangle. Sheet 66 was printed in 1938 at Engineering Reproduction Plant, U. S. Army, Fort Humphreys, Washington, D.C. All others were printed in 1942-1947 by the Panama Canal Department, U. S. Army, Corozal, Canal Zone.

VCP El Volcán-Cerro Punta Region. Date ? Map given to Wetmore in 1961 by V. J. Tipton of the Environmental Health Bureau, Office of the Surgeon General, USARCARIB.

Museum Acronyms

Whenever possible, museum acronyms follow Leviton et al. (1985). Those who provided information or assisted in other ways are acknowledged in parentheses.

AMNH	American Museum of Natural History, New York, New York (Allison Andors, Mary LeCroy).
ANSP	Academy of Natural Sciences, Philadelphia, Pennsylvania (Leo Joseph).
BMNH	British Museum (Natural History), Bird Section, Tring, England (Mark Adams, Michael Walters).
BYU	Brigham Young University, Provo, Utah. Monte L. Bean Life Science Museum (Wetmore et al., 1984:304).
CAS	California Academy of Sciences, San Francisco, CA (online database).
CM	Carnegie Museum of Natural History, Pittsburgh, Pennsylvania (Robin Panza, Kenneth C. Parkes).
CMNH	Cincinnati Museum of Natural History, Cincinnati, Ohio (Robert Kennedy).
CRUV	Central Regional University of Veraguas, Panama (Delgado, 1985).
CU	Cornell University, Ithaca, New York (Kevin McGowan).
FMNH	Field Museum, Chicago, Illinois (David Willard). Includes collections formerly housed at Princeton University and University of Miami.
FSU	Florida State University, Tallahassee, Florida (Frances James).
GML	Gorgas Memorial Laboratories. Instituto Commemorativo Gorgas de Estudios de la Salud, Panama City, Ancón, Panamá (Evelia Quiroz, Miguel Morales).
LACM	Natural History Museum of Los Angeles County, Los Angeles, California (Kimball Garrett).

LIV	Liverpool Museum, Liverpool, England (Clemency Fisher).
LSUMZ	Louisiana State University Museum of Zoology, Baton Rouge, Louisiana (Robb Brumfield, J. V. Remsen).
MBM	Marjorie Barrick Museum of Natural History, University of Nevada, Las Vegas (John Klicka, Matthew J. Miller).
MCN	Museo de Ciéncias Naturales, Panama City, Panamá (Nuria Esquivel de Barrillas, Martín Martinez).
MCZ	Museum of Comparative Zoology, Harvard University, Cambridge, Massachusetts (Raymond A. Paynter, Jr., Alison Pirie).
MNHN	Museum National d'Histoire Naturelle, Paris (Griscom, 1932a:303).
MVUP	Museo de Vertebrados, Universidad de Panamá (Víctor H. Tejera N.).
MVZ	Museum of Vertebrate Zoology, University of California, Berkeley, California (Carla Cicero, Ned K. Johnson).
NHRM	Naturhistoriska Rijksmuseum, Stockholm, Sweden (Per Ericson).
OC	Moore Laboratory of Zoology, Occidental College, Los Angeles, California (James Northern).
SDSNH	San Diego Museum of Natural History, San Diego, California (Philip Unitt).
STRI	Smithsonian Tropical Research Institute, Panama City, Panamá (George Angehr, Matthew J. Miller).
UAM	University of Alaska Museum, Fairbanks (Matthew J. Miller).
UCLA	University of California at Los Angeles, Los Angeles, California (Kathy Molina).
UF	Florida Museum of Natural History, Gainesville, Florida (David W. Steadman, Tom Webber).

UMMZ	University of Michigan Museum of Zoology (Strauch, 1975a, 1977).
USNM	National Museum of Natural History, Smithsonian Institution, Washington, D.C. (James Dean, Craig Ludwig).
UWBM	University of Washington Burke Museum, Seattle, Washington (Robert C. Faucett, Sievert Rohwer, Chris Wood).
WFVZ	Western Foundation of Vertebrate Zoology, Camarilla, California (Lloyd Kiff)
YPM	Peabody Museum of Natural History, Yale University, New Haven, Connecticut (Eleanor Stickney, online database).
ZSM	Zoologisches Sammlung des Bayerischen Staates, Munich, Germany, (Hellmayr, 1910b, 1911, 1921).

Other Abbreviations

ADL	Alexandria Digital Library. 2004. Board of Regents of the University of California. Accessed via the World Wide Web at <http://middleware.alexandria.ucsb.edu/client/gaz/adl/index.jsp>
FH	Fairchild and Handley, 1966
GA	George Angehr, from correspondence during 2003-2005.
GG	Global Gazetteer. 1996-2004. Falling Rain Genomics, Inc. Accessed via the World Wide Web at <http://www.fallingrain.com/world/PM/>
GL	General Locator. A locality that was used as a reference point for one or more secondary localities. Used especially on Wetmore's labels where the general locator may be rather far removed from the actual site of collection. When collectors used more than one locality on a label, we designated one as the GL.
GMLP	Gorgas Memorial Laboratories Personnel. See GML under museum acronyms.
MARU	Personnel of the Middle America Research Unit of the National Institutes of Health. Some specimens also have C. E. Yunker's or James E. Ambrose's name on them, although they were not necessarily the collectors. All of the specimens that we know of with this label were collected in 1961. See also: NIID.
MCU	Malaria Control Unit Personnel. Wetmore (1957:9) refers to the Malaria Control Force of the U. S. Army. 25th Medical Detachment (Preventive Medicine Survey). Mosquito Control Force.
NIID	National Institute of Infectious Diseases Personnel. The only NIID that we know of is located in Japan, although NIID may have been associated with MARU as some specimens also have C. E. Yunker's name on them. All of the specimens that we know of with this label were collected in 1961-1962. See also: MARU.

SBE	Smithsonian Bocas Expeditions. Applies to collections, mainly of terrestrial vertebrates, made by staff of the National Museum of Natural History, on the islands and lowlands of Bocas del Toro from 1987 through 1992 (see Olson 1993a).
USBGN	Gazetteer of Panama. Names Approved by the United States Board on Geographic Names. 1990. 3rd edition. xvii + 509 pages. Defense Mapping Agency, Washington, D.C.
USBGN2	Gazetteer of Colombia. Names Approved by the United States Board on Geographic Names. 1988. 3rd edition. xviii + 859 pages. Defense Mapping Agency, Washington, D.C.

Botanical Terms

Information on plant names is mainly from Duke (1972, 1986).

agave—-*Agave* spp., century plant

almacigo—-*Bursera simaruba*, gumbo-limbo tree

bamboo, climbing or creeping—-*Chusquea* spp.

bejuco—thin pliable vine, as used for cordage

black palm—-*Astrocaryum standleyanum*

bracken—*Pteridium aquilinum*

caña blanca—-*Gynerium sagittatum*, giant cane

cativo—-*Prioria copaifera* (Angehr, 2003:217)

chumico—-*Curatella americana*, sandpaper tree

corozo—-*Elaeis oleifera*, American oil palm

cuipo—*Cavanellisia platanifolia*

emajagua—the sea hibiscus *Hibiscus tiliaceus* (now = *Talipariti tiliaceum* var. *pernambucense*); several places in his journal Wetmore refers to "*Tilia*," which from the context is certainly this species.

Erythrina—*Erythrina* spp., coral trees

espavé—-*Anacardium excelsum*, wild cashew (not the true cashew *A. occidentale*)

Ficus—-*Ficus* spp., fig trees

guarumo—*Cecropia peltata*, trumpet tree

guinea grass—-*Panicum maximum*

Heliconia—-*Heliconia* spp., wild plantain

Ipomea—-*Ipomea pes-caprae*, beach morning-glory

mahogany—-*Swietenia* spp.

mangrove, black—-*Avicennia germinans*

mangrove, red—-*Rhizophora mangle*

mangrove, white—-*Laguncularia racemosa*

morning glory—see *Ipomea*

nance—-*Byrsonima crassifolia*, golden spoon, wild cherry

pava—-*Didymopanax morototoni*, morototo or yagrumo, macho tree

pineapple, wild—-*Aechmea magdalenae*

rastrojo—brushy second-growth, especially successional stages following cultivation

sea grape—-*Coccoloba uvifera*

tagua—-*Phytelephas seemannii*, ivory palm

Tamarindus indicus—-tamarind tree

tecas, tecales—-*Tectona grandis,* teak tree

Tilia—-see *emajagua*

water hyacinth—-*Eichornia crassipes*

yuca—-*Manihot esculenta*, cassava

Yucca—-*Yucca elephantipes*, Spanish bayonet

Collecting Itinerary of Alexander Wetmore in Panamá 1944, 1946-1966

The localities listed reflect only Wetmore's general locator (i.e. the locality on the front of specimen labels) which often subsumes a variety of localities, some rather distant from one another. The number in parentheses is the total of specimens cataloged from a given locality. The annual total is in parentheses below the year. The total for all 22 years of collecting in Panama is 14,493 specimens, for an average of 659 per year and a range from 305 to 970.

1944
(340)

Isla San José, Perlas	Feb 9-Mar 8, 10, 12-13	(291)
Isla Pedro González, Perlas	Mar 9, 11	(49)

1946
(638)

Jaqué, Darien	Mar 14-Apr 15	(638)

1947
(462)

Río Jaqué, Darién	Mar 27-Apr 14	(462)

1948
(601)

Paris, Herrera	Feb 24; Mar 1, 4-5, 11	(82)
Monagrillo, Herrera	Feb 25	(20)
Portobellilo, Herrera	Feb 26	(15)
Parita, Herrera	Feb 26-29; Mar 2-9, 12-13, 15, 22-25, 27-30	(52)
El Barrero, Herrera	Feb 28; Mar 18, 22-23	(82)
El Rincón, Herrera	Mar 1, 3, 6, 17, 25	(78)
Potuga, Herrera	Mar 2	(21)
Santa María, Herrera	Mar 8-10, 19	(88)
La Cabuya, Herrera	Mar 12-13	(35)
Los Santos, Los Santos	Mar 15-16, 20	(50)
Pesé, Herrera	Mar 23-24, 29-30	(63)
Punta Mala, Los Santos	Mar 27	(15)

1949
(927)

Chico, Panamá	Mar 13-16, 18, 20, 22, 24-30; Apr 1-2	(220)
Pacora, Panamá	Mar 15, 19, 21, 26-27; Apr 2-3, 21	(126)
Utivé, Panamá	Mar 17, 19, 25	(59)
Cerro Azul, Panamá	Mar 23, 29, 31; Apr 12, 19, 23, 25	(170)
Chepo, Panamá	Apr 6-11, 13-18, 20-22, 24, 26	(352)

1950
(970)

Chimán, Panamá	Feb 16-28; Mar 31-Apr 1	(339)
Majé, Panamá	Mar 2-3, 30	(31)
Charco del Toro, Panamá	Mar 3, 20-28	(233)
Cerro Chucantí, Panamá	Mar 5-18	(367)

1951
(534)

Cerro Campana, Panamá	Mar 2-3, 5-8, 10, 12-14, 16, 19-21	(300)
Bejuco, Panamá	Mar 15	(27)
Nueva Gorgona, Panamá	Mar 17	(29)
El Valle, Coclé	Mar 23-24, 26-31	(160)
El Espino, Panamá	Mar 24; Apr 1	(18)

1952
(692)

Río Indio, Colón/Panamá/Coclé	Feb 14-17, 20; Mar 8	(97)
Chilar, Colón	Feb 17-19; Mar 7-11	(161)
El Uracillo, Coclé	Feb 22-Mar 5	(307)
Isla Taboga, Panamá	Mar 15-17, 21-23	(74)
Isla Taboguilla, Panamá	Mar 18, 20	(46)
Isla Uravá, Panamá	Mar 22	(7)

1953
(655)

Antón, Coclé	May 10, 13; Jun 20	(13)
Río Hato, Coclé	May 13	(2)

Soná, Veraguas	May 14-21, 23, 25-28, 30-Jun 7, 9-13, 15	(447)
Río de Jesús, Veraguas	May 22, 27	(44)
Puerto Vidal, Veraguas	May 29-30; Jun 6, 8	(59)
Quebrada Piedra, Chiriquí	Jun 6	(16)
Puerto Mutís, Veraguas	Jun 11	(7)
Ancón, Canal Zone	Jun 18	(1)
Penonomé, Coclé	Jun 20	(5)
Cerro Campana, Panamá	Jun 21	(3)
Chame, Panamá	Jun 21-22	(3)
San Carlos, Panamá	Jun 21-22	(5)
El Valle, Coclé	Jun 22	(1)
Nueva Gorgona, Panamá	Jun 23	(7)
Gamboa, Canal Zone	Jun 24	(8)
Chico, Panamá	Jun 26-28	(34)

1954
(305)

El Volcán, Chiriquí	Feb 24-Mar 13, 21-22, 24-29	(190)
Santa Clara, Chiriquí	Mar 2, 15-20	(115)

1955
(916)

Juan Mina, Canal Zone	Jan 6-22	(320)
El Volcán, Chiriquí	Jan 31-Feb 7, 9-12, 26; Mar 13-16	(264)
Santa Clara, Chiriquí	Feb 15-19, 21-22, 24	(23)
Sereno, Chiriquí	Feb 21-23	(60)
Cerro Punta, Chiriquí	Feb 28-Mar 8, 10-11	(185)
Cerro Chame, Panamá	Mar 26-28	(35)
Cerro Azul, Panamá	Mar 29	(11)
Chico, Panamá	Mar 30	(16)

1955-1956
(904)

Juan Mina, Canal Zone	Dec 9-16	(53)
Gamboa, Canal Zone	Dec 10	(2)

Cerro Campana, Panamá	Dec 21	(2)
Isla Taboguilla, Panamá	Dec 24	(7)
Isla Taboga, Panamá	Dec 25, 26	(6)
Isla Uravá, Panamá	Dec 26	(8)
Chiva Chiva, Canal Zone	Dec 29; Jan 1	(6)
Isla Coiba, Veraguas	Jan 7-Feb 3	(541)
Isla Ranchería, Veraguas	Feb 4	(11)
San Félix, Chiriquí	Feb 13, 16-19, 21, 23, 27	(140)
Las Lajas, Chiriquí	Feb 14-15, 20, 22, 24-26	(123)
Chico, Panamá	Mar 5-6	(5)

1957

(828)

Chiva Chiva, Canal Zone	Jan 13	(1)
Mandinga, San Blas	Jan 22-Feb 17	(426)
Ensenada Venado, Los Santos	Feb 25-26	(12)
Isla Iguana, Los Santos	Feb 26-27	(22)
Isla Villa, Los Santos	Feb 28	(1)
Farallón del Chirú, Coclé	Feb 28	(1)
Cerro Azul, Panamá	Mar 2	(4)
Río Boquerón, Colón	Mar 3	(6)
Pedasí, Los Santos	Mar 8-18, 20	(210)
Los Asientos, Los Santos	Mar 18-19	(39)
Tonosí, Los Santos	Mar 22-28	(106)

1958

(693)

Almirante, Bocas del Toro	Jan 15-18, 20-23, 25, 27-29, 31 Feb 1, 4-8, 10-24, 26, 28; Mar 3, 5-7, 9	(568)
Changuinola, Bocas del Toro	Jan 24, 30; Feb 13, 27; Mar 4	(52)
Swan Key, Bocas del Toro	Jan 26	(2)
Tiger Cays, Bocas del Toro	Feb 25	(9)
Isla Escudo de Veraguas, Bocas	Mar 1-2	(28)
El Espino, Panama	Mar 14, 17	(12)
Antón, Coclé	Mar 15-17	(4)
San Carlos, Panamá	Mar 15	(5)
Pacora, Panama	Mar 19, 22	(2)
Chico, Panamá	Mar 19-21	(11)

1959

(746)

Juan Mina, Canal Zone	Jan 27-Feb 1	(31)
Gamboa, Canal Zone	Jan 29	(6)
Río Paya, Darién	Feb 13-Mar 17	(500)
Río Chucunaque, Darién	Mar 20-Apr 3	(209)

1960

(760)

Gamboa, Canal Zone	Jan 9-10, 13	(11)
Chiva Chiva, Canal Zone	Jan 12	(4)
Cerro Galera, Canal Zone	Jan 14	(5)
Isla Pachequilla, Perlas	Jan 16	(2)
Isla Contadora, Perlas	Jan 17	(12)
Isla Chapera, Perlas	Jan 18	(9)
Isla Saboga, Perlas	Jan 19	(13)
Isla Pacheca, Perlas	Jan 20	(1)
Isla de Cañas, Perlas	Jan 20-23	(47)
Isla del Rey, Perlas	Jan 24-25, 27	(51)
Isla Santelmo, Perlas	Jan 26	(17)
Isla Galera, Perlas	Jan 28	(3)
Isleta Malaga, Perlas	Jan 29	(15)
Isla Bayoneta, Perlas	Jan 30	(21)
El Volcán, Chiriquí	Feb 6, 8-13, 15-20, 22-25; Mar 20, 22	(275)
Cerro Punta, Chiriquí	Feb 25-26; Mar 19, 21	(54)
Buena Vista, Chiriquí	Feb 29; Mar 1, 4, 6-7, 9, 11, 13, 18	(71)
Concepción, Chiriquí	Mar 2	(15)
Alanje, Chiriquí	Mar 3-5, 8, 10, 12-13	(106)
Boquete, Chiriquí	Mar 15-18	(28)

1961

(597)

Chiva Chiva, Canal Zone	Jan 7; Mar 19	(1)
Juan Mina, Canal Zone	Jan 10-18	(91)
Cerro Pirre, Darién	Jan 27-Feb 6	(147)
Ancón, Canal Zone	Feb 15	(1)
Pacora, Panamá	Feb 15; Mar 23, 25	(16)
Peluca Hydrographic Station, Panamá	Feb 17-27	(162)

Candelaria Hydrographic Station, Panamá	Mar 2-11	(147)
La Jagua, Panamá	Mar 18, 20-24	(25)
Chico, Panamá	Mar 24	(7)

1962
(672)

Chiva Chiva, Canal Zone	Jan 7	(3)
Frijolito, Colón	Jan 8-9	(8)
La Jagua, Panamá	Jan 12-14	(30)
Las Palmitas, Los Santos	Jan 21-28	(102)
El Llano, Panamá	Feb 3-5, 9	(56)
Cañita, Panamá	Feb 6-8	(55)
El Copé, Coclé	Feb 21-25	(71)
Barrigon, Coclé	Feb 26	(1)
Río Guabal, Coclé	Feb 26-Mar 3	(101)
El Potrero, Coclé	Mar 6-12	(80)
Río Grande, Coclé	Mar 9	(7)
El Coco, Coclé	Mar 10	(11)
Puerto Aguadulce, Coclé	Mar 11-12	(14)
Islas Frailes del Sur, Los Santos	Mar 18	(4)
Isla Canal de Afuera, Veraguas	Mar 20-21, 26	(12)
Isla Afuerita, Veraguas	Mar 22, 25	(19)
Isla Brincanco, Veraguas	Mar 23-24	(26)
Isla Cébaco, Veraguas	Mar 27-28	(27)

1963
(577)

Chiva Chiva, Canal Zone	Jan 12	(3)
Aguadulce, Coclé	Jan 16-19, 21-25	(71)
Gago, Coclé	Jan 20, 23	(13)
Penonomé, Coclé	Jan 23	(6)
El Cristo, Coclé	Jan 24	(6)
El Roble, Coclé	Jan 24	(1)
Isla Bolaños, Chiriquí	Feb 1, 6	(14)
Isla Parida, Chiriquí	Feb 2-5, 7	(37)
Puerto Obaldía, San Blas	Feb 16-18; Mar 14-18	(104)
Armila, San Blas	Feb 21-Mar 13	(322)

1964
(769)

El Real, Darién	Jan 8-9, 11, 22-27; Feb 16-18; Mar 21	(146)
Pucro, Darién	Jan 29-Feb 11	(167)
Cerro Malí, Darién	Feb 19-Mar 3	(252)
Tacarcuna Village, Darién	Mar 4-17	(200)
La Laguna, Darién	Mar 19	(1)
Boca de Cupé, Darién	Mar 21	(1)
La Jagua, Panamá	Mar 24	(2)

1965
(457)

Isla Gobernadora, Veraguas	Jan 8-11	(52)
Isla Cébaco, Veraguas	Jan 11,13-20	(118)
Volcán Barú, Chiriquí	Feb 24; Mar 1-5, 8-12, 17	(111)
El Volcán, Chiriquí	Feb 26-28; Mar 2, 7, 9, 14-18, 20, 22-24	(166)
Cerro Punta, Chiriquí	Mar 19	(10)

1966
(450)

Puerto Armuelles, Chiriquí	Jan 29-Feb 12, 14-19, 21-24, 26, 28; Mar 1-4, 7-13	(442)
Punta Balsa, Chiriquí	Feb 25	(8)

Ornithological Gazetteer of Panamá

AARON ROCKS; see Frijoles

ABAJO, RÍO; Panamá (E, Pacific)
0901/7929 (USBGN)
Originates S of Santa Rita [0905/7930 (USBGN)] and flows S and SE entering Bahía de Panamá [0850/7915 (USBGN)] at Old Panama [0900/7929 (USBGN)] (E661; E761A). Both Rendahl (1919:4) and Hallinan (1924) refer to a "Río Algarrobo" that enters Bahía de Panamá 7 mi (11 km) NE of the Pacific entrance of the Panama Canal [0920/7955 (USBGN)] On modern maps the Río Algarrobo (which USBGN now calls the Río Tesorera [0903/7932 (USBGN)]) originates S of Santa Rita near the former Canal Zone border and joins the Río Abajo ca. 5 km N of the coast, and from here the Río Abajo continues S entering Bahía de Panamá at Old Panama, therefore it appears that Rendahl and Hallinan were both referring to what is now called the Río Abajo. Hallinan, the valley of, clearings, "jungle along the river," "deep jungle," 1915 May 30, Jun 6, Jul 4, 1916 Jun 6 (?) (AMNH; Hallinan, 1924, as "Río Algarrobo"); Benson, at 200 feet (60 m), unclear if specimens were collected on the river or in the town Río Abajo [0901/7930 (USBGN)] (q. v.), 1931 Jun 20 (MCN; USNM; Wetmore, 1970).

ABAJO, RÍO; see Río Abajo

ABOGADO, ISLA; Colón (Canal Zone) (E)
0912/7952 (USBGN)
86 m, small island with a land area of 13.6 ha (Wright, 1985) located in Lago Gatún (Gatun Lake) [0912/7955 (USBGN)] 300 m off the NW coast of Península Bohío [0912/7950 (USBGN)] (E762, as "Abogado Island"). Wright, censuses, 1976-1978 (Wright, 1979, Wright, 1985, both as "Abogado Island").

ABOGADO ISLAND; see Abogado, Isla

ACHIOTE; Colón (W)
0913/8001 (USBGN)
100 m, town located 20 km SW of Colón [0922/7954 (USBGN)] and 7 km S of the coast (JOG). Collector ?, date ? (GML); Van Horn, 1963 Apr 9, 1964 Feb 20, Mar 6, Nov 8, 1967 Apr 18, 1969 Apr 11, 16, 18, 1974 Jan (GML; USNM; Wetmore, 1967); Quinones, 1967 Oct 25 (GML); Telford, 1968 Jul 9 (GML); Thatcher, 1964 Oct 20, Nov 9, 1965 Aug 18, Dec 7, 1967 Jan 11, 13 (GML); GMLP, 1969 Apr 11, 1973 Feb 21 (GML); Montenegro, 1970 Apr 11 (GML); Ridgely and Gwynne, 1970 Jul 4 (GML); Mendoza, 1972 Feb 20, 23 (GML); Araúz, Coronas, Garcés, and Holness, 1986 Aug 9 (MVUP); Araúz, Garcés, and Holness, 1987 Mar 29, May 14 (MVUP); Seutin, GL for Achiote Road, 1994 Sep 23, 25 (LSUMZ). See also: Achiote Road.

ACHIOTE, CARRETERA; see Achiote Road

ACHIOTE ROAD; Colón (Canal Zone) (W)

ca. 0917/8003

According to Engleman et al. (1989:487-488), the name refers to a 7 km stretch of gravel road between the Gatún-Escobal Road [ca. 0916/7955] and the town of Achiote [0913/8001 (USBGN)]. According to Angehr (in litt., Aug 2005), the road is now paved to the coast. According to JOG, the road extends 15 km roughly NW-SE between Piña [0917/8003 (USBGN)] on the Caribbean coast and La Treinticinco [0911/7958 (USBGN)] on the NW coast of Lago Gatún (Gatun Lake) [0912/7955 (USBGN)]. This road is named Road S11 on RM1 and RMM1, although Achiote itself does not appear on those two maps. Eisenmann, observations, 1967 Mar 23 (Wetmore et al., 1984:62); Smith, N., observations, SE of Achiote, 1969 Jan, 1977 May 27 (Wetmore et al., 1984:269, 336); Collector ?, 1970 Jul 5 (GML); Morton, studies, 1973-1975 (Morton, 1980); Seutin, GL = Achiote, 3.5 km SE Achiote, 1994 Sep 23, 25 (LSUMZ); Cardiff, Christian, Dittmann, Naoki, and Lovette, GL for: Quebrada Treinticinco, and Río Providencia, 1997 Mar 16-22 (LSUMZ; MVUP, as "Achiote"). See also: Gatún-Escobal Road.

ACKER STATION; see John W. Acker Field Station

AEROPUERTO; Darién

ca. 0808/7743

Sea level, SMN2B indicates that the airport is just NE of El Real de Santa María [0808/7743 (USBGN)] near both the Río Tuira [0821/7803 (USBGN)] and the mouth of the Quebrada Urutí [ca. 0808/7743]. Wetmore, GL = El Real de Santa María, open savannas, 1964 Jan 22 (USNM).

AFUERA; see Afuera, Isla de

AFUERA, ISLA; see Canal de Afuera, Isla del

AFUERA, ISLA DE; Panamá, Archipiélago de las Perlas

0828/7857 (USBGN)

Sea level ?, described by Rendahl (1920:24, as "Afuera") as "a holm off the town of San Miguel." Not shown on our maps, but USBGN indicates an island just NW of San Miguel [0827/7856 (USBGN)], a town on the N end of Isla del Rey [0822/7855 (USBGN)]. Bovallius, 1882 May 4 (NHRM; Rendahl, 1920, as "Afuera").

AFUERA, ISLA DEL CANAL DE; see Canal de Afuera, Isla del

AFUERA ISLAND; see Canal de Afuera, Isla del

AFUERITA, ISLA; Veraguas (Pacific)

0742/8138 (USBGN)

Sea level, small island with a land area of ca. 0.2 km^2 (20 ha), located in the Pacific Ocean off the S coast of Veraguas province just NW of Isla del Canal de Afuera [0741/8137 (USBGN)] (JOG). Wetmore, landed at a small, sandy beach, up a steep slope through open woodland, trees in flower and with fruit, interior valley cut and cleared for planting, open *rastrojo*, 1962 Mar 22, 25 (USNM; Olson, 1997a).

AGLAITIGUAR, RÍO; see Aila Tiwar

AGUA, CAYO; Bocas del Toro, Archipiélago de Bocas del Toro

0909/8202 (see text)
35 m, island in the Archipiélago de Bocas del Toro [0920/8210 (USBGN)] with a land area of 14 km^2 (1,400 ha) (Angehr, 2003:102) located in the N waters of Laguna de Chiriquí [0903/8200 (USBGN)] 3 km E of Isla Popa [0911/8207 (USBGN)] and 6 km W of Península Valiente [0905/8151 (USBGN)] (JOG; MHA, as "Cayo de Agua"; SMN2B, as "Isla Cayo Agua"). USBGN is unclear about the official name of the island and lists the following names: Agua, Cayo; Agua, Cayo de; Agua Key; Cayo Agua, Isla; Cayo de Agua, Isla; and Water Cay. Most of these names are cross-referenced to one another and all have the same coordinates. As ornithological specimens from the island all appear to be labeled Cayo Agua, we have used that name here. One of the boatmen employed by SBE indicated that Isla Popa used to be, or maybe sometimes still is, called Cayo Agua. Miller and Richardson, GL = Laguna de Chiriquí, on the water near Cayo Agua, on the coast in mangroves and a short distance into the jungle, 1936 Feb 28 (UCLA); Handley, 1962 Mar 28, and as a GL for Punta Norte, 1963 Feb 12-14 (USNM); SBE, collected on the E and the N ends of the island, 1990 Apr 10, 12-13, and as a GL for Punta Limón, 1990 Apr 5-13 (MVUP; USNM; Brumfield and Braun, 2001; Olson, 1993a); Billo, "12/km NW of S tip, Douglas Property," and as GL for Punta Limon and Punta Norte, 2007 May 4-5 (USNM).

AGUA, CAYO DE; see Agua Cayo

AGUA, CERRO; Panamá (W, Atlantic, Pacific)
ca. 0838/8005
ca. 1,200 m, described by Wetmore as a peak on the continental divide NE of Cerro Gaital [0837/8007 (USBGN)] between the headwaters of the Río Indio [0912/8011 (USBGN)] (Atlantic drainage) and the headwaters of the Río Mata Ahogado [0828/7957 (USBGN)] (Pacific drainage) (USNM). Wetmore and Perrygo, GL = El Valle [0836/8008 (USBGN)], at 2,200 feet (670 m), slopes of Cerro Agua, up a steep trail to the summit of the ridge connecting Cerro Gaital [0837/8007 (USBGN)] and Cerro Agua, above the valley of the Río Mata Ahogado, 1951 Mar 30-31 (USNM). Not shown on our maps.

AGUA, VALLE DE; see Valle de Agua

AGUA BUENA; see Aguas Buenas

AGUA CALIENTE; Chiriquí
ca. 0850/8229
6,000 feet (1,850 m), according to Blake (1958:502 (map)), the town was located SW of Cerro Horqueta [0850/8226 (USBGN)] on the Quebrada Cerro Punta [= Río Caldera [0837/8221 (USBGN)]] where the Cerro Punta-Boquete Trail [ca. 0852/8235] splits from the Holcomb Trail [ca. 0901/8215]. The town is not shown on E762, but the corresponding region is located ca. 9.5 km NNW of Boquete [0847/8226 (USBGN)] and ca. 10 km E of Cerro Punta [0852/8235 (USBGN)]. See also: Boquete Trail; Cerro-Punta Boquete Trail; and Holcomb Trail.

AGUA CALIENTE, QUEBRADA; Herrera
ca. 0759/8034

Not shown on our maps, but Wetmore described the creek as a small tributary of the Río Parita [0802/8027 (USBGN)] ca. 3 mi (5 km) W of El Barrero [0756/8031 (USBGN)], Wetmore and Perrygo, GL = El Barrero, running water bordered by brush, pastures and occasional trees, crossed over a shallow valley with low hills, thin soil and numerous rock outcroppings, to the settlement of Los Hatillos [0756/8033 (USBGN)], 1948 Mar 18 (USNM).

AGUA CATA; see Aguacate

AGUA CLARA; Panamá ? (Canal Zone) (W, Atlantic)
ca. 0911/7957
Former town 6 mi (10 km) S of Gatún [0916/7955 (USBGN)] on the Río Trinidad [0901/7958 (USBGN)] 2.5 mi (4 km) above its mouth and on the W end of Agua Clara Swamp [ca. 0901/7956] (PR). The locality is now submerged in Lago Gatún (Gatun Lake) [0912/7955 (USBGN)] NW of Bahía Agua Clara [0910/7954 (USBGN)] near the Colón/Panamá provincial boundary (E8660; MGRP). Goldman, GL = Río Trinidad, by canoe on the Río Trinidad to a point several miles above Agua Clara, 1911 Jan 23, 26 (USNM); Jewel, GL = Río Trinidad, near Agua Clara, scrub overgrown with vines, 1911 Mar 12, Apr 30, 1912 Mar 24, May 19 (ANSP; Stone, 1918; Meyer de Schauensee, 1950a). (Olson, 1981d, as "Río Trinidad, Agua Clara").

AGUA CLARA; see Aguas Claras

AGUA CLARA, BAHÍA; Panamá (Canal Zone) (W, Atlantic)
0910/7954 (USBGN)
Bay centrally located in Lago Gatún (Gatun Lake) [0912/7955 (USBGN)] ca. 3 km W of Isla Barro Colorado [0909/7951 (USBGN)]. The bay extends 3 km S into the mainland coast and it opens on the N to the reservoir just S of the Panama Canal [0920/7955 (USBGN)] (E8660, as "Agua Clara Bay"). See also: Agua Clara.

AGUA CLARA, QUEBRADA; Colón (Canal Zone) (E)
0910/7941 (USBGN)
Originates in hills NW of Aguas Claras [0911/7941 (USBGN)] and flows SE entering the Río Aguardiente [0910/7941 (USBGN)] (q. v.) ca. 1 km above its mouth (E866; E8660). Wetmore, GL = Juan Mina [0910/7939 (USBGN)], Wetmore used the name "Quebrada Agua Clara" for the Río Aguardiente below the mouth of the Quebrada Agua Clara as well as for the Quebrada Agua Clara itself, therefore specimens with this label may have been collected on either stream. Landed in the bay near the mouth of the Río Aguardiente, followed a 20-25 foot (6-8 m) wide trail that extends 3 mi (5 km) to Aguas Claras through heavy, original forest in the former Canal Zone and through cultivated areas in Colón province, then walked E from Aguas Claras to Santa Rosa [0911/7939 (USBGN)] past corn, yuca, *rastrojo* and farm land, 1955 Jan 20, 1961 Jan 13 (USNM; Brumfield et al., 2001).

AGUA CLARA SWAMP; Panamá (Canal Zone) (W, Atlantic)
ca. 0911/7955
Former wetland along the S banks of the Río Trinidad [0901/7958 (USBGN)] between its mouth and Agua Clara [ca. 0901/7957] (IP, as "Marais d'Agua Clara";

PR). The locality is now submerged in Lago Gatún (Gatun Lake) [0912/7955 (USBGN)] in what is now Bahía Agua Clara [0910/7954 (USBGN)] (E8660; MGRP). See also: Agua Clara; and Trinidad, Río.

AGUA DULCE; see Aguadulce

AGUA KEY; see Agua, Cayo

AGUA SALUD, RÍO; Colón (Canal Zone) (E)
0911/7947 (USBGN)
Originates SE of Colón [0922/7954 (USBGN)] and E of Península Bohío [0912/7950 (USBGN)], flows SW crossing the Pipeline Road [ca. 0915/7943], and enters a small bay of Lago Gatún (Gatun Lake) [0912/7955 (USBGN)] ca. 1 km NE of Frijoles [0910/7948 (USBGN)] (E762; RM1). Willis, observations, 1960-1970 (Willis, 1966a; Willis, 1972:36); Smith, N., GL = Frijoles, 1.4 mi (2.3 km) ENE of Frijoles, S bank of the river, 0.7 mi (1.1 km) E of mouth, 1965 Jun 1 (USNM); Morton, GL = Frijoles, near Frijoles, 1966 Jul 5 (FMNH); Tejera, 1984 Mar 15 (MVUP); Cardiff, Dittmann, and Lovette, GL = Gamboa, at 40-120 m, 17 km by road NW of Gamboa, 1996 Mar 15-18 (LSUMZ; MVUP).

AGUA SUCIA, RÍO; see Duque, Río

AGUACATAL; see Aguacate

AGUACATE; Chiriquí
0848/8235 (USBGN)
1,900 m, town located 7 km NE of Volcán [0846/8238 (USBGN)], 4 km W of Volcán Barú [0848/8233 (USBGN)] and 2 km NNE of Cerro Aguacate [0848/8236 (USBGN)] (JOG). The region near Aguacate is presumably the wild area "Aguacatal" mentioned in Wetmore's journals in 1955 and again in 1960. Wetmore, GL = Volcán, at 6,400 feet (2,000 m), drove across Llanos del Volcán [ca. 0847/8238] and left the jeep near private property, hunted on an abandoned logging road with occasional tracts of trees, 1960 Feb 24 (USNM, as "Aguacatal"); Tyson, he was working near Volcán Barú (q. v.) at this time, but specimen is not labeled as such, 1965 Mar 18 (MCN, as "Agua Cata"). See also: Barú, Volcán.

AGUACATE; see El Aguacate Número Uno

AGUACATE, CERRO; Chiriquí
0848/8236 (USBGN)
2,200 m, peak located SW of Volcán Barú [0848/8233 (USBGN)] and 6 km ENE of Volcán [0846/8238 (USBGN)] (JOG). See also: Aguacate; and Barú, Volcán.

AGUACATÓN; Chiriquí
ca. 0820/8248
Sea level, described by Wetmore as a forested area NNE of Puerto Armuelles [0817/8252 (USBGN)] between the main highway and Olivo [ca. 0819/8247], Wetmore, GL = Puerto Armuelles, forest, cornfield, 1966 Feb 1-2, 5, 7 (USNM). Not shown on our maps.

AGUADILLA; Coclé (Pacific)
ca. 0822/8034

Not shown on our maps but located in S Coclé province in Distrito Natá [0822/8034 (USBGN)] near Natá [0820/8031 (USBGN)] (q. v.). Alcedo, 1985 Aug 16 (MVUP); Quezada, 1985 Aug 29 (MVUP).

AGUADILLA, LA; see La Aguadilla

AGUADULCE; Coclé (Pacific)

0815/8033 (USBGN)

ca. 25 m, city in SW Coclé province on the Carretera Interamericana [0832/8242 (USBGN)] 9 km inland from the W coast of Bahía de Parita [0808/8024 (USBGN)] (JOG). In early times the town was an important terminal for travel by sea from Panama City [0858/7932 (USBGN)] and therefore visited by persons interested in natural history, though seldom mentioned. The dealers Hübner and Schlesinger, as well as Boucard, sold skins from Aguadulce, and the early collectors on record include Arcé and Herrera. The specimens were well prepared with firm, filled bodies and shortened necks so that they presented a rounded form, an appearance that was sometimes called the "Aguadulce make" by Hellmayr, von Berlepsch, and others of their period. Originally the town's name was written as two words, Agua Dulce, and translates to "fresh water" in English, an interesting name as there were natural salt flats just E and SE of town. Angehr (in litt., Aug 2005) distinguishes between *albinas*, which are natural salt flats and "*salinas*, [which] are artificial ponds from which salt is extracted by evaporation." Cooke and Bush (1989) indicate the salt flats are being replaced by shrimp farms and in danger of being eliminated altogether. According to Angehr (in litt., Aug 2005), many *albinas* have already been converted to *salinas* or shrimp farms. Angehr (2003) lists the wetlands including the *salinas* as an Important Bird Area. Observations and collections made near the *salinas* were made a few kilometers SE of town at Gallo [ca. 0812/8030] (q. v.), near El Salado [0811/8029 (USBGN)] (q. v.), at Playa Caleta [0810/8030 (Cooke and Bush, 1989)], or anywhere along the coast between Aguadulce and Chitré [0758/8026 (USBGN)]. Arcé, 1876 (USNM, as "Agua Dulce"); Collector ? [Herrera ?], 1877 Nov, Dec (FMNH); Herrera, 18??, 1877 (FMNH, as "Agua Dulce"; USNM, as "Agua Dulce"); Schlesinger and Hübner [Herrera ?], 18??, 1877 Dec (USNM, as "Agua Dulce"); Collector ?, 1905 Jun 1 (GML); Benson, arid plains and nearby coastal swamps, 1925 Sep 3-26 (AMNH; UCLA; Griscom, 1927a, as "Agua Dulce"; Griscom, 1927b, as "Agua Dulce"; Dickey and van Rossem, 1930; Wetmore, 1962); Wetmore, country road outside of town, 1963 Jan 22-23, and as a GL for: El Salado, Gallo, La Isleta, and Río Pocrí, 1963 Jan 16-19, 21-22, 24-25 (USNM); Dutary, 1963 Feb 2, 1964 Oct, 1965 Nov 7 (GML); Olson and Loftin, 1963 Feb 23 (USNM); Ordoñez, 1968 (MVUP); Cooke, observations, c. 1970 (Wetmore, 1972:374, as "salinas near Aguadulce"); Eisenmann and Pujals, observations, 1971 Jun 26 (Ridgely and Gwynne, 1989:156; Wetmore et al., 1984:633, as "Aguadulce Salinas"); Hinds and Ridgely, 1974 Jan 6 (GML); Stiles and Ridgely, observations, on rocks on coast S of, 1974 Jan 6 (Ridgely and Gwynne, 1976:116; Wetmore et al., 1984:633); Tejera, some specimens are also labeled "Caleta," "La Albina," and "La Albina La Caleta," as well as "Las Salinas," and "Salinas," 1975 Mar 29, 1976, 1977 Jun 20, Aug 15, Dec 26, 1978 Feb 2, Apr 21, May 27, Jun 20-21, 26, 1979 Jun 2,

Nov 25, 1980 Apr 25, Mar 18, Aug 2, and as a GL for: El Brujo, El Limón, Gallo, Las Mineras, Panchillo, Puerto Aguadulce, Río Simón Gómez, 1975 Mar 29, Sep 3, 1977 Dec 26, 1978 Apr 10, Jun 20-21, Dec 30, 1979 Nov 2, 1980 Apr 25 (MVUP); Quezada, 1983 Aug 21 (MVUP); Quezada and Vargas, 1983 Oct 30 (MVUP); Tuñun, at Estación Euseñat (not located), 1984 Apr 23 (MVUP); Andrión, "Salinas de Aguadulce," 1985 Apr 15 (MVUP); Justavino, 1985 Aug 31 (MVUP); Tribaldos, GL for: El Rompio, and El Salado, 1986 Apr 12 (MVUP); Brown et al., observations at a shallow lagoon, 1987 Sep 24, Oct 1, 18 (Brown and Engleman, 1988); Casas, GL for La Laguna de Pocrí, 1989 Nov 22 (MVUP); Tejera, GL for Gallo, 1990 Feb 18 (MVUP); Weir, at less than 10 m, at 0813/8029, 2003 Apr 14 (CU; LSUMZ). (Berlepsch, 1880, as "Agua dulce"; Boucard, 1892, as "Agua Dulce"; Zimmer, 1941a:7, as "Agua Dulce"; Wetmore, 1950a; Wetmore, 1953; Wetmore, 1970; Monroe and Howell, 1966; Parkes, 1976; Olson, 1980b).

AGUADULCE, PUERTO; see Puerto Aguadulce

AGUADULCE SALINAS; see Aguadulce

AGUARDIENTE, RÍO; Colón (Canal Zone) (E)
0910/7941 (USBGN)
Originates in hills S of Aguas Claras [0911/7941 (USBGN)] and flows E and SE entering a 1 km wide bay on the W bank of the Río Chagres [0907/7942 (USBGN)]. The river mouth is 2 km W of Juan Mina [0910/7939 (USBGN)] and 200 m NW of the mouth of the Río Palenque [0910/7940 (USBGN)] (E866). Chapin, Rogers and Mathews, GL = Río Chagres, mouth of the Río Aguardiente, 1923 Jul 17 (FMNH); Wetmore, GL = Juan Mina, landed in the bay near the mouth of the Río Aguardiente and followed a 20-25 foot (6-8 m) wide trail inland through heavy, original forest, 1955 Jan 20, 1961 Jan 13 (USNM, as "Quebrada Agua Clara"). See also: Agua Clara, Quebrada; and Chagres, Río.

AGUAS BUENAS; Panamá (E, Atlantic)
0907/7936 (USBGN)
100 m, town on the Carretera Boyd-Roosevelt [0917/7943 (USBGN)] 11 km E of Gamboa [0907/7942 (USBGN)] and 4 km SSE of Chilibre [0909/7937 (USBGN)] (E761C, as "Agua Buena"; E8660, as "Agua Buena"; MGRP). Pinto, 1985 Mar 11 (MVUP, as "Agua Buena").

AGUAS CLARAS; Colón (E)
0911/7941 (USBGN)
70 m, small town 4 km NW of Juan Mina [0910/7939 (USBGN)] just N of the former Canal Zone border and near Quebrada Agua Clara [0910/7941 (USBGN)] (E762). See also: Agua Clara, Quebrada; and Palenque, Río.

AGUAS CLARAS, RÍO; Panamá (E, Pacific)
0913/7843 (see text)
Originates in eastern Panamá province on the Panamá/San Blas provincial border and flows S entering the N central coast of Lago Bayano [0908/7835 (USBGN)] ca. 45 km E of Chepo [0910/7906 (USBGN)] (JOG, as "Río Icantí"; MGRP, as "Río Icantí o Aguas Claras"). According to USBGN, Río Icantí [0913/7843

(USBGN)] is the river's official name. (Bennett, 1962, as "Río Icandí"). See also: Aguas Claras Arriba.

AGUAS CLARAS ARRIBA; Panamá (E, Pacific)

ca. 0918/7842

Not shown on our maps, but presumably refers to a town or simply the region near the upper reaches of the Río Aguas Claras [0913/7843] (q. v.). Río Aguas Claras flows S into Lago Bayano [0908/7835 (USBGN)] ca. 45 km E of Chepo [0910/7906 (USBGN)]. USBGN indicates that Aguas Claras is an alternate name for the town Icantí [0913/7843 (USBGN)], although Icantí is near the river's mouth not "arriba." Hinds, 1974 Feb 22 (GML).

ÁGUILA, EL; see El Águila

AILA, RÍO; see Aila Tiwar

AILA TIWAR; San Blas

0851/7741 (USBGN)

A stream that originates on the continental divide and flows N entering the Caribbean Sea [1500/7500 (USBGN)] 5 km E of Cerro Sasardí [ca. 0856/7751] and ca. 38 km NW of Puerto Obaldía [0840/7725 (USBGN)] (MGRP, as "Río Aglaitiguar"). According to USBGN, Tiwar is the Cuna word for stream. Hudson and Sugden, observations, 1979 Feb 15-Mar 12 (Hudson, 1984, as "Río Aila"; Hudson and Sugden, 1984, as "Río Aila").

AILIGANDÍ; San Blas

0914/7801 (USBGN)

Sea level, a town on the island Ailigandí [0914/7801 (USBGN)], located 500 m N of the central San Blas coast due N of the N end of the Panamá/Darién provincial border (MGRP; SMN2B). According to Wetmore, a U. S. Fish and Wildlife Service banded bird was found here, 1955 Nov 12 (Wetmore, 1963b, as "Aligandí").

ALAJUELA; see Alhajuela

ALAJUELA, LAGO; Colón/Panamá (Canal Zone) (E, Atlantic)

0915/7935 (USBGN)

100 m, reservoir on the Río Chagres [0907/7942 (USBGN)] formed by Represa Madden (Madden Dam) [0913/7937 (USBGN)]. The reservoir has a surface area of 57 km^2 (RDP2) with the Río Boquerón [0921/7933 (USBGN)] and the Río Pequení [0921/7933 (USBGN)] entering from the N and the Río Chagres [0907/7942 (USBGN)] flowing through the reservoir. The former Canal Zone boundary extended around the reservoir with Colón province to the W of it and Panamá province to the E. The reservoir is commonly called Madden Lake (Atlas, 1988:204; E661, as "Madden Lake"; JOG, as "Lago Madden") and the Boy Scout Camp is located 2 km E of Represa Madden (Madden Dam) (q. v.). Miller, 1937 Mar 6 (USNM, as "Madden Lake"); Meyer, 1941 May 13 (USNM, as "Madden Lake"); Morton, 1971 Feb 26, Mar 25, observations, 1970s (YPM, as "Madden Lake"; Morton, 1971, as "Madden Lake"; Morton, 1982, as "Madden Lake"); Tejera, GL = San Juan, 1978 May 6, Jun 3, Aug 19, Sep 21, Oct 31, Nov 30, Dec 30, 1979 Jan 11, Feb 11, 1980 Nov 30 (MVUP); Hernández, GL = San Juan, 1978 Aug 19, Dec

30 (MVUP); Pineda, GL = San Juan, 1978 Nov 7 (MVUP); Collector ?, 1995 Mar (LSUMZ, as "Madden Lake"); Dittmann, Río Chagres, Scout Camp, 1996 Dec 6 (LSUMZ, as "Madden Lake"); Bacon and Miller, 2002 Sep 16 (STRI, as "Madden Lake"; UAM, as "Madden Lake"). (Wetmore, 1962:8, as "Madden Lake"; Delgado, 1983).

ALANJE; Chiriquí
0824/8233 (USBGN)
ca. 40 m, 15 km SSE of La Concepción [0831/8237 (USBGN)] (JOG). Wetmore, the secondary localities attributed to Alanje were located in the low coastal region extending 11 km S and 17 km SW of town, GL for: Canta Gallo Abajo, Estero Rico, Guásimo, La Barqueta, La Martina, and Paja Blanca, 1960 Mar 3-5, 8, 10, 12-13 (USNM; Wetmore, 1962; Parkes, 1976; Olson, 1980b, as "Alanje, Estero Rico").

ALBERRYS CREEK; Bocas del Toro, Archipiélago de Bocas del Toro
0917/8207 (USBGN)
Sea level, town on the S coast of Isla Bastimentos [0918/8208 (USBGN)] located 5 km SW of Old Point [0917/8205 (USBGN)] and 0.5 km SW of Cedar Creek [0917/8207 (USBGN)] in an area of coastal mangrove swamp (E762, as "Alberry's Creek"). Although the official name is given as Alberrys, this is surely a corruption of the local English pronunciation of Alvarez. According to Handley's journal, this was one of several saltwater creeks on the S coast of Isla Bastimentos, SBE, GL = Isla Bastimentos, Laguna de Chiriquí [0903/8200 (USBGN)], 1987 Mar 21, 26, 28-29 (MVUP, as "Ensenada Alvarez"; USNM, as "Alvarez Creek").

ALBINA DE GALLO; see Gallo

ALBINAS; see Aguadulce

ALBROOK AIR FORCE BASE; Panamá (Canal Zone) (E, Pacific)
0859/7933 (USBGN)
Sea level, a former United States Government military base N of Balboa [0857/7934 (USBGN)] and SSW of Curundu [0859/7933 (USBGN)]. The base was bordered on the S and W by the Panama Railroad [ca. 0922/7954] and included a landing field, which is now used for domestic flights (E761A, as "Albrook Field"). Blanton, some specimens are also labeled Landing Field or Road C15 [ca. 0902/7932], 1953 Sep 20, Oct 6, 21-22 (USNM, as "Albrook," as "Albrook AFB," and as "Albrook Field"); Loftin, observations, grassy field at, jungly area behind, 1961 Sep-Nov, 1962 Sep 9 (Loftin, 1963a, as "Albrook," and as "Albrook Air Force Base"); Hunt, observations in a field behind, Tropical Survival School (not located), 1962 Aug 29 (Loftin, 1963a, as "Albrook Field"); Olson and Fischer, 1-4 mi (1.5-6 km) E, "dense secondary jungle," "hill outside of Panama City," 1963 Feb 1-3, 13, 28, Mar 5, 9, 12, 14-15, 17, May 4, 6, 10-13, 17-19, 21, 23 (USNM, as "Albrook Field"; Loftin and Olson, 1963, as "Albrook Field"; Olson, 1965, as "Albrook Field"; Olson and Eisenmann, 1966, as "Albrook Field"); Gale, 2 mi (3 km) E, 1963 Feb 3, May 25-26 (GML, as "Albrook Field"; USNM, as "Albrook Field"; Loftin and Olson, 1963); Wiese, 1966 Jul 28 (USNM, as "Albrook Field"); Renner, Albrook Tower Road, E762 indicates a tower on the E side of the field, 1967 May

4 (GML); Morton, 1971 Feb 19 (AMNH; USNM; YPM; Morton, 1971); González, 1994 Aug 10 (MVUP, as "Albrook Field"); González, 2001 Nov 30 (MVUP, as "Albrook Field").

ALBROOK FIELD; see Albrook Air Force Base

ALBROOK TOWER ROAD; see Albrook Air Force Base

ALCALDE DÍAZ; Panamá (E, Atlantic)

0907/7933 (USBGN)

150 m, town 17 km N of Panama City [0858/7932 (USBGN)] just N of the continental divide (E762; E8660, as "Alcalde Díaz (Peñoncito)"). The name appears in USBGN as Alcaldedíaz, which is presumably a typographical error. Pérez, 1985 Apr 17 (MVUP); Guerra, 1990 Apr 5 (MVUP).

ALCALDEDÍAZ; see Alcalde Díaz

ALEGRE, PUNTA; Bocas del Toro

ca. 0909/8154

Sea level, point on the N coast of Laguna de Bluefield [0909/8155 (USBGN)] 3.5 km SW of Punta Valiente [0911/8155 (USBGN)] (Olson, 1993a:101 (map)). SBE, GL = Península Valiente, 1990 Apr 17-29 (MVUP; USNM; Olson, 1993a).

ALGARROBO, RÍO; see Abajo, Río

ALGARROBOS, QUEBRADA; Chiriquí

0842/8218 (USBGN)

Originates as intermittent streams on the continental divide, flows S crossing La Caldera Road [ca. 0839/8226] 10 km NE of Caldera [0839/8223 (USBGN)], and enters the Río Chiriquí [0820/8226 (USBGN)] 1 km S of the road (E762). Ridgely and Gwynne, observations, 1982 Jun 30 (Ridgely and Gwynne, 1989:295, 376, as "Río Algarrobos").

ALGARROBOS, RÍO; see Algarrobos, Quebrada

ALHAJUELA; Colón (Canal Zone) (E)

ca. 0913/7937

120 feet (37 m), town on the Río Chagres [0907/7942 (USBGN)] 13 km NE of Gamboa [0907/7942 (USBGN)] (E761A) on the Carretera Boyd-Roosevelt [0917/7943 (USBGN)] just N of Represa Madden (Madden Dam) [0913/7937 (USBGN)] (RM1). The area was not submerged in Lago Alajuela (Madden Lake) [0915/7935 (USBGN)], although the town does not appear on our more recent maps. Goldman, in 1912 the savannas near Alhajuela extended intermittently to the Pacific coast, traveled by canoe on the Río Chagres (see both Río Chagres [0907/7942 (USBGN)] and Río Chagres [0919/8000 (USBGN)]) from Matachin [ca. 0907/7943] to Alhajuela, 1912 Jan 29 (USNM); Benson, at 100 feet (30 m), 1931 Jul 7, 12, Sep 3, 27 (MCN, as "Alajuela"; USNM, as "Alajuela," and as "Alhajuela"); Sackett, 1931 Sep 27 (USNM, as "Alajuela"); Candanedo, Gallardo, and Jiménez, the Corregimiento is listed as Lago Cantera (not located), 1990 Oct 8 (MVUP, as "Alajuela"). (Olson, 1980a, as "Alajuela"). See also: Chilibrillo, Río; Frijolito; and La Puente, Río.

ALIGANDÍ, see Ailigandí

ALISA, QUEBRADA; Darién

ca. 0730/7808 ?

Described by Wetmore as a stream near Loma González [ca. 0729/7806], unclear where the stream begins or empties and unclear whether Quebrada Lisa (q. v.) and Quebrada Alisa are 2 distinct localities, Wetmore and Perrygo, GL = Jaqué, 1946 Mar 23, 31, Wetmore and Perrygo (from Vaughan), GL = Jaqué, 1946 Apr 8 (USNM). Not shown on our maps.

ALMEJAS, ISLA; see Sabaneta, Isla

ALMIJAS, ISLA; see Sabaneta, Isla

ALMIRANTE; Bocas del Toro

0918/8224 (USBGN)

Sea level, coastal town in NW Bocas del Toro province. Despite the shared name, Almirante is not on the coast of Bahía Almirante [0920/8218 (USBGN)], but S of it on the W coast of Ambrosia Bight [0916/8222 (USBGN)] (JOG). Almirante was founded as one of the original headquarters of the United Fruit Company and the company installed a railroad connecting the town to Changuinola [0926/8231 (USBGN)] and the surrounding plantations. The mile markers (indicated on E762 as "Milla 2," as "Milla 3," etc.) indicated on many specimen labels refer to the distance from Almirante along this line, but according to E762, these stations are not spaced exactly one mile from each other with variations from 0.5-1.5 mi (1-2.5 km). The railroad extends 26 km between the 2 towns and passes E of Fruitdale [0918/8225 (USBGN)] at Milla 2 [ca. 0918/8225], through hills rising above 200 m W of the tracks, and near Finca Cuatro [0924/8230 (USBGN)], El Empalme [0925/8231 (USBGN)], and Finca Ocho [0926/8231 (USBGN)] before reaching Changuinola. In the area W of Changuinola the railroad line branches into a network of railroad lines connecting numerous small settlements in the cultivated areas of Bocas del Toro province and Costa Rica (E762; JOG). According to Kennard and Peters (1928:443), the railroad extended through the "jungle-clad foothills and adjacent swamps, through miles of cacao, cocoanut, and banana plantations" as well as crossing the Río Banano [920/8223 (USBGN)], the Río Changuinola [0928/8227 (USBGN)], and beyond the town of Changuinola it crossed the Río San San [0931/8231 (USBGN)]. In addition to the railroad, MGRP indicates that Almirante and Changuinola are now connected by a road, which extends roughly parallel to the railway and crosses it near Finca Cuatro. Blair, 1923-1926 (Peters, 1931a); J. D. S. [= Smith], 1926 Feb 17 (ANSP); Collector ? [Smith ?], 1926 Mar 20 (YPM); Kennard, accompanied by J. D. Smith and H. Wedel, arrived in Almirante on or just after Feb 8, 1926 and remained there until Mar 6, collecting in the vicinity of town, through the foothills near town, at "the golf links" maintained by the United Fruit Company, and near banana clearings, in coconut groves and on brushy hillsides. In addition Almirante was a GL for: Changuinola Canal, Farm 6, Finca Ocho, Río Banano, Río Oeste, Rubber Tree, Valle de Agua, and Zegla. Although Kennard and Peters (1928) state that the outfit departed Almirante on Mar 6 to begin 3 weeks of collecting along the Boquete Trail [ca. 0857/8207], the publication describes specimens from Almirante as late as Mar 20. Smith was forced to turn back from

Buena Vista about Mar 10, and whether he collected for Kennard in Almirante during this period is unclear. Specimens may have also been collected by local hunters, although the text states that "two *Galbulas* were seen in the brush beside the railroad track on Mar 16" indicating that the author or a trusted colleague had been in Almirante on Mar 16 (MCZ; Kennard and Peters, 1928; Peters, 1931a; Peters and Griscom, 1929; Eisenmann, 1950); Smith, 1927 Apr-Aug (YPM; Peters, 1931a; Griscom, 1933); Wedel, 1927 Apr-1928 Dec, 1938 Oct-Dec, 1939 Feb-Mar, and as a GL for: Milla 7, Quebrada Nigua, and Río Oeste (ANSP; CMNH; FMNH; MCZ; MVZ; YPM; Peters, 1927; Peters, 1931a; Eisenmann, 1950; Wetmore, 1959b:9); Benson, 1927 May 3, 16, 20, Aug 22, 24, Sep 30, Oct 1, and as a GL for: Cayos Zapatilla, and Milla 7, 1927 Jul 2, Sep 11 (AMNH; USNM; Griscom, 1928; Peters, 1931a; Eisenmann and Howell, 1962; Phillips and Dickerman, 1965:298; Wetmore, 1968:7); Skutch, observations at Almirante and in western Panamá, c. 1935 (Skutch, 1945a); Eisenmann, observations, 1956 Jun-Jul (Eisenmann, 1957); Wetmore, 1958 Feb 18, Mar 5, 9. The secondary localities attributed to Almirante were located in the coastal areas within 13 km of town on both the mainland and the neighboring islands. According to Wetmore, the coastline was generally low in elevation, muddy and swampy and the higher ground, which was largely converted to banana and cacao plantations, was reached by crossing slippery, makeshift walkways or tree roots. The river mouths were typically enclosed by mud bars and mangroves that separated their more silty waters from the coral in the bays. GL for: Bahía Almirante, Boca del Drago, Cayo Roldán, Changuinola Canal, Ensenada Shepherd, Francisco Creek, Isla Pastores, Isla San Cristóbal, Punta de Gallinazo, Punta Donato, Punta Pondsock, Punta Rodríguez, Quebrada Garay, Quebrada Jones, Quebrada Nigua, Quebrada Pastores, Río Banano, Río Oeste, Tirbi Bight, and Valle de Agua, 1958 Jan 15-Mar 9 (MVUP; USNM; Wetmore, 1959b:13; Payne, 1974); Munch, GL for Río Oeste, 1958 Feb 25 (USNM); Handley and Rhymer, collected near Milla 1, 1960 Jan 15-Mar 7 (USNM); GMLP, Gorgas Memorial Laboratory personnel set up mist nets for bird banding at Milla 1, Milla 2, Milla 5, and ca. 2 mi (3 km) NW of Almirante S of the railroad line on what Galindo calls the "Bocas-Chiriqui road," the entire banding area "enclos[ed] a variety of habitats including freshwater marshes, swampy lowland tropical rain-forest, second-growth, banana groves, cacao orchards and grassy pastures," most specimens are labeled only Almirante, the exact dates are unclear on several specimens, 1960-1964, 1965 Feb, Mar, Jun, Nov, 1967 Jan, Mar, Apr, Jul, Sep, 1970 Jun 28, 1971 Mar 11, 1973 Jan 10 (GML; USNM; Galindo et al., 1963; Galindo and Méndez, 1965; Loftin and Olson, 1963; Olson, 1980c); Galindo (GMLP), 1960 Nov 11, 1961 Sep 10, 14, 1962 Apr 22, 1964 Oct 16, 1965 Jun 10 (LSUMZ; USNM); Loftin, mist netting, observations near, 1961 Aug 1-Nov 30, 1962 Apr 17-21, Aug 1-Nov 30 (Loftin and Olson, 1963; Loftin, 1963a; Loftin, 1963b); Hinds (GMLP), 1960-1963, 1964 Dec, 1965 Mar, Jul, Oct, 1967 Sep, 1981 Oct 9, and as a GL for: Finca Cuatro, Milla 2, and Milla 5, 1960 Mar 21, 1961 Jan, 1962 Mar 28 (GML; LSUMZ; MCN; USNM; Wetmore, 1963a); Mendez (GMLP), 1960 May 14, 1963 Apr 20-25, and as a GL for Milla 2, 1963 Aug 23 (USNM; Loftin and Olson, 1963; Wetmore, 1963b); Quintero, 1961 Apr 10, 1963 May 4, Oct-Nov (GML); Quinones,

1962 Oct 24, Nov 2 (GML); Rogers, data from rainstorm kills, 1963 Oct 13-15 (USNM; Rogers, 1965; Child, 1969; Hicks et al., 1967); Barrett, 1962 Nov 19 (GML); Olson, GL for Milla 2, 1963 Apr 6-12 (MCN; USNM; Loftin and Olson, 1963); Hicks, 1964 Mar 21-22, 29, Apr 2, Sep 30, Oct 4, 10, 12-13, 16-17, 19-21, 24-26 (AMNH; USNM; Loftin et al., 1966; Eisenmann, 1966; Hicks, 1967; Hicks et al., 1967); Larson, 1964 Jun (GML); Morton and Loftin, 1965 Oct 17-18 (AMNH); Heatley, 1967 Oct 6, 9 (GML; MCN); Kirmse and Heatley, 1967 Oct 13-14 (USNM; Wetmore, 1970); Kleen and Heatley, 1967 Oct 23, 31 (AMNH); Brumfield, at 50 m, 2 km W on the road to Río Oeste [0916/8224 (USBGN)], 1994 Jan 20 (USNM); Rodríguez and Vanegas, 1997 Jun 8 (MVUP). (Rogers, 1939; Rogers and Odum, 1966; Zimmer, 1941b:14; Zimmer, 1950:26, 50; Zimmer, 1955:5; Howell, 1956; Parkes, 1976; Olson, 1981a; Olson, 1981c; Olson, 1981d; Olson, 1983; Westwater, 1988; Escalante-Pliego and Peterson, 1992; McDonald et al., 2001). See also: Finca Williams; Fruitdale; Milla 1; Milla 2; Milla 5; Milla 7; Nigua, Quebrada; Oeste, Río; and Tierra Oscura.

ALMIRANTE, BAHÍA; Bocas del Toro

0920/8218 (USBGN)

Bay on the NW coast of Bocas del Toro province that extends 10 km N-S along the mainland between Punta Donato [0923/8221 (USBGN)] and Punta Pondsock [0918/8221 (USBGN)]. The bay opens on the S to Ambrosia Bight [0916/8222 (USBGN)] and Isla San Cristóbal [0915/8216 (USBGN)] and on the NE to Isla Colón [0924/8217 (USBGN)] with access to the Caribbean Sea [1500/7500 (USBGN)] only by narrow channels on either side of Isla Colón. Despite the shared name, the town Almirante [0918/8224 (USBGN)] is not located in Bahía Almirante, but S of it on the coast of Ambrosia Bight [0916/8222 (USBGN)] (JOG; MHA, as "Bahía de Almirante"). Collectors working in the region used the name Bahía Almirante to cover a broad area including the waters of Ambrosia Bight (q. v.) and Ensenada Grande [0914/8219 (USBGN)] (q. v.), although JOG and USBGN treat these bays as separate localities. Kennard et al., along small streams flowing into, shores of, 1926 Feb 18, 26 (Kennard and Peters, 1928); Wedel, 1938 Oct-Dec, 1939 Feb-Apr (FMNH, as "Almirante Bay"); Wetmore, GL = Almirante, 1958 Jan 25 (USNM, as "Almirante Bay"); Collector ?, 1958 Feb 7 (Hofslund, 1959, as "Bay at Bocas del Toro"); SBE, between Isla San Cristóbal [0915/8216 (USBGN)] and Isla Pastores [0914/8220 (USBGN)], 1989 Feb 28, GL = Bocas del Toro [0920/8215 (USBGN)], 1 mi (1.5 km) S of Bocas del Toro, 1989 Mar 6 (USNM; Olson, 1993a). See also: Oeste, Río; Swan Cay; and Tierra Oscura.

ALMIRANTE BAY; see Almirante, Bahía

ALTA PEÑA; Chiriquí

ca. 0848/8228

Not shown on our maps, but apparently a region NW of Boquete [0847/8226 (USBGN)] and near Quiel [ca. 0848/8228]. As the term is not used by Mönniche, who lived in the area at Finca Lérida [0848/8229 (USBGN)], it seems likely that Alta Peña is a mix of the terms Peña Blanca [ca. 0848/8229] and Casita Alta [ca. 0849/8229] – both of which were also visited in late Jan 1933 and modified by

Quiel in the database. Davidson, GL = Boquete, Quiel, at 6,800-7,000 feet (2,050-2,150 m), 1933 Jan 26 (CAS).

ALTO, PALO; see Palo Alto

ALTO CARVALLO; Coclé (Atlantic, Pacific)

ca. 0839/8036

Described by Wetmore as a summit on the continental divide on the ridge known as Cordillera Andino (not located), the summit was located 6.5 km N of El Copé [0837/8036 (USBGN)] and 2.5 km S of Tigre [ca. 0842/8036], Wetmore, GL = Río Guabal [ca. 0846/8036], at 900 m, above the head of the Río Guabal, small, green opening in the forest, flowers, 1962 Mar 3 (USNM; Wetmore, 1967). Not shown on our maps.

ALTO CHIQUERO; see Alto de Chiquero

ALTO DE CHIQUERO; Chiriquí

ca. 0850/8229

ca. 5,400-6,000 feet (1,650-1,850 m), described by Blake (1958:502 (map)) as the local name for a plateau N of the Río Chiquero [ca. 0849/8229], NW of Bajo Mono [0848/8228 (USBGN)] and SW of Agua Caliente [ca. 0851/8229]. Blake's (1958:564) single use of Alto de Chiriquí, a locality not in his gazetteer, is presumably a lapsus for Alto de Chiquero. Mönniche, at 5,300-6,800 feet (1,600-2,050 m), 1932-1933, 1936-1937 (FMNH; GML; MCN; Blake, 1958); Guitton-Mayerma, Miller, and Winker, 2004 Feb 13-17 (STRI, as "Alto Chiquero"; UAM, as "Alto Chiquero").

ALTO DE CHIRIQUI; see Alto de Chiquero

ALTO DE PIEDRA; Veraguas (Pacific)

ca. 0831/8106

1,000 m, town 3 km W of Santa Fé [0831/8105 (USBGN)] and ca. 3 km E of the continental divide, which is at 1,400-1,500 m W and N of town (MGRP). Hinds, Ridgely, and Stiles, GL = Santa Fé, at 2,800 feet (850 m), between Alto de Piedra and the continental divide, on the Pacific slope, 1974 Jan 3-5 (GML).

ALTO DE QUIEL; see Quiel

ALTO JARAMILLO; see Jaramilla

ALTO QUIEL; Chiriquí

0848/8228 (USBGN)

1,500 m, town 5 km NW of Boquete [0847/8226 (USBGN)] just S of Quebrada El Velo [0849/8228 (USBGN)] (E762). See also: El Salto; and Quiel.

ALTOS CACAO; Herrera

ca. 0739/8049

1,400-1,500 feet (450 m), described by Aldrich and Bole (1937:4 (map), 8) as an abandoned rubber camp in a small clearing ca. 16 mi (26 km) E of Golfo de Montijo [0740/8107 (USBGN)] on a steep-sided ridge between the valleys of the Río Negro [0737/8059 (USBGN)] and the Río Mariato [0737/8059 (USBGN)]. Presumably the camp was located near the current town El Cacao [0739/8051 (USBGN)].

Although the authors attribute the locality to Veraguas province, recent maps indicate that it is in Herrera province. For further discussion of the provincial borders see Península de Azuero [0740/8035 (USBGN)]. Aldrich and Bole, GL = Golfo de Montijo, Península de Azuero, collected on top of the ridge, in "small draws that drain into the 2 rivers on both sides," heavy tropical rainforest, and a cultivated, 2 acre clearing, 1932 Feb 28-Mar 5 (CM; Aldrich and Bole, 1937).

ALTOS DE BAYANO; see Majé, Serranía de

ALTOS DEL GOLF; Panamá (E, Pacific)

0900/7930 (USBGN)

Not shown on our maps but USBGN indicates a section of greater Panama City [0858/7932 (USBGN)] ca. 1 km SW of Old Panama [0900/7929 (USBGN)]. Quintero and Sánchez, 1990 Sep 18 (MVUP).

ALTOS DE MAJÉ; see Majé, Serranía de

ALTOS DE PACORA; see Cerro Azul

ÁLVAREZ, ENSENADA; see Alberrys Creek

ALVAREZ CREEK; see Alberrys Creek

ALVINA; Herrera

ca. 0805/8032

Sea level, described by Wetmore as a strip of open land 3-4 mi (5-6 km) N of Parita [0800/8031 (USBGN)] (USNM). According to JOG, this region is bordered on the N by the Ríos Santa María [0806/8029 (USBGN)] and Escotá [0806/8033 (USBGN)] and on the E by the Ciénaga el Mangle [0802/8029 (USBGN)], Wetmore and Perrygo, GL = Parita, 4 mi (6 km) N of Parita in extensive open flats with 6-8 foot (2 m) rocky hills and knolls covered with dry scrub, cacti, low trees, wild pineapple and other thorny growth, 1948 Feb 27, GL = Paris [0803/8033 (USBGN)], 4 mi (6 km) ENE of Paris, shallow, salt lagoons bordered by a broad open beach, an extensive mangrove swamp, slightly higher ground with *Cereus* and spiny shrubs, 1948 Mar 11 (USNM; Olson, 1980b, as "4 miles NE of Alvina," but the specimen was taken at Alvina, 4 mi (6 km) NE of Paris).

AMADOR, FORT; see Fort Amador Army Reservation

AMADOR CAUSEWAY; see Fortified Islands

AMADOR FLATS; see Fort Amador Army Reservation

AMBROSIA, ENSENADA; see Ambrosia Bight

AMBROSIA BIGHT; Bocas del Toro

0916/8222 (USBGN)

Bay on the NW coast of Bocas del Toro province that extends 6 km N from Punta de Gallinazo [0915/8222 to (USBGN)] to Punta Pondsock [0918/8221 (USBGN)] with its mainland coast well-protected 6 km W of these two points. The bay opens on the E to Ensenada Grande [0914/8219 (USBGN)] and on the NE to Bahía Almirante [0920/8218 (USBGN)] (JOG, as "Ensenada Ambrosia"). Early collectors did not distinguish this bay from Bahía Almirante as indicated by Peters (1931:294), who describes Almirante as a coastal town on the W coast of Bahía Almirante. See also: Almirante, Bahía; and Oeste, Río.

ANACHUCUNA; see Anachukuna

ANACHUCUNA, BAHÍA; see Anachucuna Bay

ANACHUCUNA, CERRO; Darién
0836/7735 (USBGN)
700 m, peak in central Serranía del Darién [0830/7730 (USBGN)] 12 mi (19 km) SW of Puerto Obaldía [0840/7725 (USBGN)] (RDPF; SMN2B). See also: Pito, Río; Tuquesa, Río; and Ucurgantí, Río.

ANACHUCUNA BAY; San Blas
ca. 0842/7729
Bay on the SE San Blas coast that extends 15 mi (24 km) W from the Colombian border. Puerto Obaldía [0840/7725 (USBGN)] (q. v.) is on the E coast of the bay, and Anachukuna [0843/7733 (USBGN)] (q. v.) is on its W coast (SMN2B, as "Bahía Anachucuna").

ANACHUKUNA; San Blas
0843/7733 (USBGN)
Sea level, town located 21 km W of the Colombian border on the W coast of Anachucuna Bay [ca. 0842/7729] (MGRP, as "Anachucuna"). Used by Griscom (1932:303 as "Anachucuna") to locate Permé [ca. 0845/7734] (q. v.), which was 5 mi (8 km) W of Anachukuna.

ANA LUZ; Panamá (E, Pacific)
ca. 0908/7905
Sea level, town 6 km SSE of Chepo [0910/7906 (USBGN)] on the N bank of the Río Chepo [0900/7906 (USBGN)] 1.5 km below the mouth of the Río Mamoní [0908/7904 (USBGN)] (E761A, as "Ana Luz Landing"). Described by Wetmore as a region of swampy woods and savannas along the N bank of the 250 foot (75 m) wide Río Chepo, Wetmore and Perrygo, GL = Chepo, Río Chepo, within 1.5 mi (2.5 km) of the Río Chepo, 1949 Apr 15-17, 20 (USNM).

ANAM HEADQUARTERS; see Coiba, Isla de

ANAM STATION; see Azul, Cerro

ANCÓN; Panamá (Canal Zone) (E, Pacific)
0858/7933 (USBGN)
Sea level, town in the former Canal Zone that was adjacent to the SW edge of Panama City [0858/7932 (USBGN)]. Balboa [0857/7934 (USBGN)] is just WSW of town, and Cerro Ancón (Ancón Hill) [0857/7933 (USBGN)] is within Ancón. The town is now part of greater Panama City (E962), and is the location of the Smithsonian Tropical Research Institute's main offices (STRI). Hallinan, 1916 Jan 16 (AMNH; Hallinan, 1924); Benson, 1925 May 8 (AMNH); Chapman, 1932 Mar 26 (AMNH); Greenway, at the hospital piazza, 1932 Jul (Greenway, 1934; Eisenmann, 1952:25); Griswold, 1936 Apr 20, 23, 27 (MCZ); Earhart, observations, 1939 Feb (Chapman, 1941); Scholes and Scholes, observations, 1951 Jan 25, Feb 3

(Scholes and Scholes, 1954); Wetmore, 1953 June 18, 1961 Feb 15 (USNM); Yunker (MARU), 1961 (USNM); Loftin and Belton, banding, 1962-1964 (Loftin, 1975); GMLP, 1962 Dec 9 (GML); Hespenheide, 1969 Jul (CM); Morton, 1970 Sep 13-14, Dec 3, 7, 9-11, 1971 Jan 26, Feb 1, Mar 28 (USNM); Cooke, 1977 Oct (USNM); Portillo, 2001 Mar 27 (GML); Collector ?, 2002 Jan 28 (GML); Santos, STRI, 2002 Aug 22 (MVUP).

ANCÓN, CERRO; Panamá (Canal Zone) (E, Pacific)
0857/7933 (USBGN)
199 m, a hill located in Ancón [0858/7933 (USBGN)] in the former Canal Zone. Balboa [0857/7934 (USBGN)] is located near the SW base of the hill (E962). Bovallius, the date May 22 in Rendahl (1919:15) is erroneous, 1883 Mar 22 (NHRM; Rendahl, 1919:15, 23, as "Cerro di Ancon"); Hallinan, dense vegetation on the slopes, 1916 Jan 29 (AMNH; Hallinan, 1924, as "Ancon Hill"); Barbour, Brooks, Underwood Panama Expedition, 1922 Mar 26 (MCZ, as "Ancon Hill"); Chapman, observations, 1920s (Chapman, 1929a, as "Ancon Hill"); Barnard, observations, 1955 May (Barnard, 1956, as "Ancon Hill"); Wetmore (from Mrs. Paul Barnard), GL = Balboa, 1955 Dec (USNM, as "Ancon Hill"); Loftin, 1961 Nov (Loftin, 1963a, as "Ancon Hill"); Gale, 1962 Dec 23 (USNM, as "Ancon Hill"); Loftin et al., 1963-1964 (Loftin et al., 1966, as "Ancon Hill"; Eisenmann, 1966); Belton, 1964 Apr 24 (MCN, as "Ancon Hill"); Morton, 1969 Dec 13 (AMNH, as "Ancon Hill"); Christian, GL = Panama City, Balboa, base of, 1997 Mar 26 (LSUMZ, as "Ancon Hill").

ANCÓN, CERRO DE; see Ancón, Cerro

ANCÓN, CERRO DI; see Ancón, Cerro

ANCÓN HILL; see Ancón, Cerro

ANDINO, CORDILLERA; see Alto Carvallo

ÁNGULO, RÍO; Veraguas/Herrera (Pacific)
0742/8102 (USBGN)
Originates on the slopes of Cerro Alto Higo [0743/8051 (USBGN)] in W Herrera province and flows W entering Veraguas province, the river discharges into mangroves on the E coast of Golfo de Montijo [0740/8107 (USBGN)] 14 km ESE of Isla Leones [0744/8107 (USBGN)] (JOG). Aldrich and Bole, used to describe Paracoté [ca. 0740/8101] (q. v.), 1932 Feb 5-19, Mar 20-Apr 2 (CM, as "Angulo River"; Aldrich and Bole, 1937, as "Angulo River"; Olson, 1981d, as "Angulo River").

ANGULO RIVER; see Ángulo, Río

ANTARAL, RÍO; Darién
0743/7801 (Angehr et al., 2004)
Not shown on our maps, but refers to a stream in the S hills of the Sierra de Jungurudó [0735/7754 (USBGN)] that originates on the Cerro Antaral [0737/7800 (Angehr et al., 2004)] and empties into the Río Sambú [0805/7818 (USBGN)] (Angehr et al., 2004). Angehr, Christian, and Aparicio, 195-1,000 m, trails and streams, cloud forest c. 900 m, base camp at 840 m, mouth of, *cabecera de* (head of),

1997 Aug 9-22 (AMNH; CU; MVUP; Angehr et al., 2004). See also: Jungurudó, Sierra de; and Sambú, Río.

ANTENNA; see Howard Air Force Base

ANTÓN; Coclé (Pacific)
0824/8016 (USBGN)
ca. 25 m, city on the Carretera Interamericana [0832/8242 (USBGN)] 17 km SE of Penonomé [0831/8022 (USBGN)] and 11 km N of the coast (JOG). Wetmore and Perrygo, 3 mi (5 km) NW of Antón on the S side of the Carretera Interamericana in an area of recently burned grass, 1953 May 10, 13; Wetmore, 2 mi (3 km) E of, and 3 mi (5 km) NW of Antón near the Carretera Interamericana, 1953 June 20, from the Carretera Interamericana to an old road leading to a small stream, 1958 Mar 15, and as a GL for Río Estancia, 1958 Mar 16 (CM; USNM; Wetmore, 1953; Wetmore, 1956a; Aldrich and Duvall, 1958; Olson, 1980c, as "3 miles NW of Antón"); Chapman and Tyson, 1964 Jan 9 (USNM, as "Antone"); Correa, *laguna*, 1965 Feb 7 (MCN); Araúz, 1983 Aug 23 (MVUP); Torres, 1985 Apr 10 (MVUP); Lamela, 1992 Nov 25 (MCN).

ANTÓN, RÍO; Coclé (Pacific)
0818/8016 (USBGN)
Originates on the E slopes of Cerro Gaital [0837/8007 (USBGN)] and flows S then W draining the large valley surrounding El Valle [0836/8008 (USBGN)], and from here it flows SW to the N end of Bahía de Parita [0808/8024 (USBGN)] (JOG). Wetmore and Perrygo, GL = El Valle, at 1,900 feet (600 m), SW end of the valley where the stream was 20-25 feet (6-8 m) wide and heavily shaded by trees, river dropped ca. 40 feet (10 m) through a series of waterfalls, 1951 Mar 28 (USNM; Wetmore, 1952:17).

ANTONE; see Antón

ANTONIO, PORT; see San Antonio

ANTONIO, PUERTO; see San Antonio

ANTONIO, RÍO; see San Antonio, Río

APOSENTRO, ISLA; see Morena, Isla

ARCHIPELAGO, VERAGUA; see Veragua Archipelago

ARCHIPIÉLAGO DE BOCAS DEL TORO; Bocas del Toro, Archipiélago de

ARCHIPIÉLAGO DE LAS PERLAS; see Perlas, Archipiélago de las

ÁREA SILVESTRE DE NARGANÁ; see Nusagandí

ARENA, LA; see La Arena

ARENA, QUEBRADA DE; Chiriquí
ca. 0845/8212
Described by Ridgely and Gwynne (1989:323, 348) as a tributary of the Río Chiriquí [0820/8226 (USBGN)] located above Fortuna [0843/8216 (USBGN)]. According to Angehr (in litt., Aug 2005), the stream appears on 3742-III (Quebrada de Yuca) [E762] where it is shown joining "the Rio Chiriqui at about UTM 967300 N 368400

E (0845/8212). It is the last drainage crossed by the Oleoducto Road [Gualaca-Chiriquí Grande Road, ca. 0857/8207] on the Pacific slope before it crosses the continental divide into Bocas."

ARENAS, ENSENADA; Veraguas, Isla de Coiba
0733/8142 (USBGN)
A cove on the NE coast of Isla de Coiba [0728/8147 (USBGN)] located S of Campamento de Juncal [0734/8144 (USBGN)] and 5 km NNW of Punta Damas [0730/8141 (USBGN)] (JOG). Mentioned by Wetmore (1957:9, as "Ensenada Arenosa") in defining the northernmost limit of his work on Isla de Coiba (q. v.), although the name does not appear on specimen labels. However, it appears that Wetmore worked just N of this bay when he worked at Campamento de Juncal on Bahía de Juncal [0734/8143 (USBGN)], a bay that is not labeled on JOG.

ARENOSA; Panamá (Canal Zone) (W, Atlantic)
0902/7957 (USBGN)
ca. 40 m, town on the SW coast of Lago Gatún (Gatun Lake) [0912/7955 (USBGN)] on the SE coast of Bahía Trinidad [0905/8000 (USBGN)] 26 km NW of La Chorrera [0853/7947 (USBGN)] (E762; JOG). Fernández, 1992 Aug 25 (MVUP).

ARENOSA, ENSENADA; see Arenas, Ensenada

ARMILA; San Blas
0840/7728 (USBGN)
Sea level, coastal town 2 mi (3 km) W of Puerto Obaldía [0840/7725 (USBGN)] at the mouth of the Río Armila [0840/7728 (USBGN)] (SMN2A). Wedel, 1932 Jul 4 (FMNH, as "Puerto Obaldia"; Rogers, 1939); Handley, GL for: Quebrada Venado, and Río Pito, 1962 Mar 21, 23, 27, Apr 1, 1963 Mar 17, 22-23, 29, 31 (USNM); Wetmore, down to the old cacao plantation, 1963 Mar 8, and as a GL for: Quebrada Venado, and Río Pito, 1963 Feb 21-Mar 13 (LSUMZ; MVUP; USNM); Feinstein, GL for Quebrada Venado, 1963 Mar 8 (USNM). (Olson, 1980a; Olson, 1983).

ARMILA, RÍO; San Blas
0840/7728 (USBGN)
Originates on the continental divide E of Cerro Anachucuna [0836/7735 (USBGN)] and flows N entering the Caribbean Sea [1500/7500 (USBGN)] at Armila [0840/7728 (USBGN)] (RDPF). CARP shows the river originating S of Puerto Obaldía [0840/7725 (USBGN)] near the Colombian border and flowing NW to the Caribbean Sea at Armila. See also: Armila.

ARMUELLES, PUERTO; see Puerto Armuelles

ARRAIJAN; Panamá (W, Pacific)
0857/7939 (USBGN)
100 m, town on the Carretera Interamericana [0832/8242 (USBGN)] 13 km W of Panama City [0858/7932 (USBGN)] (JOG). Blanton, either collected on Carretera Thatcher [0858/7938 (USBGN)], which extends through town, or on Road K10 [ca. 0901/7942], which extends N from town, 1953 Oct 5 (USNM, as "Arraijan Road"); MCU, GL for Road K6, 1955 Aug 22 (GML); Mendez, 1973 Apr 9 (GML); Barrett, 2.8 km NE of, 1974 Jan 30 (GML); Vega, GL for Cerro Silvestre, 1978 Jan

17 (GML); Petit et al., surveys in pastures between Río Hato [0823/8010 (USBGN)] and, Dec 1993-Feb 1994 (Petit et al., 1999).

ARRAIJAN ROAD; see Arraijan

ARRAIJAN TANK FARM; Panamá (Canal Zone) (W, Pacific)
0857/7937 (USBGN)
100 m, former United States Military installation that housed underground oil tanks located on Carretera Thatcher [0858/7938 (USBGN)] 4.5 km E of Arraijan [0857/7939 (USBGN)] (E762) and 2 km W of Rodman United States Naval Reservation [0857/7935 (USBGN)], observations, 1981-1983 (Trainer, 1989, as "Tank Farm"); Fleischer and Smith, 1986 (Fleischer and Smith, 1992).

ARRIBA DE TRINIDAD; see Trinidad Arriba

ARROYO TORTADÓ; see Tortadó, Arroyo

ARUSA; see Aruza

ARUZA; Darién
0803/7737 (USBGN)
ca. 100 m, town 5 km SW of Boca de Cupe [0803/7735 (USBGN)] on the Río Aruza [0807/7736 (USBGN)] 7 km above its mouth (DP; SMN2A, as "Arusa"). Described by Goldman as a rubber plantation ca. 6 mi (10 km) SW of Boca de Cupe, Goldman, at 500 feet (150 m), his journal indicates that he was at Aruza from Jun 15-16, but the 3 specimens are dated 1912 Jun 19 (USNM).

ARUZA, RÍO; Darién
0807/7736 (USBGN)
Originates on the N slopes of Cerro Setetule [0747/7736 (USBGN)] and flows NNE entering the Río Tuira [0821/7803 (USBGN)] 6 km NNW of Boca de Cupe [0803/7735 (USBGN)] (DP). See also: Aruza.

ASIENTOS, LOS; see Los Asientos

ASPINWALL; see Colón

AZUERO, PENÍNSULA DE; Veraguas/Herrera/Los Santos (Pacific)
0740/8035 (USBGN)
1,500 m, very large peninsula, 100 km wide extending 100 km S off the coast of SE Veraguas province, its NW coast drains into the Golfo de Montijo [0740/8107 (USBGN)], its E coast drains into the Gulf of Panama [0800/7930 (USBGN)], and its S and SW coasts drain into the Pacific Ocean (RDPF). The peninsula is occasionally referred to as Mala Peninsula, presumably after Punta Mala [0728/8000 (USBGN)], a point on SE coast of the peninsula. Los Santos and Herrera provinces are entirely on the peninsula, but the provincial boundaries vary between maps. Older maps (MHA; MRP) indicate that the Veraguas/Herrera and the Veraguas/Los Santos provincial borders extended along the ridge separating the E and the W watersheds of the peninsula, but on more recent maps (JOG; MGRP) the border has shifted to the W indicating that the ridge and therefore the headwaters of rivers that discharge into Golfo de Montijo [0740/8107 (USBGN)], on the W, are now located in Herrera province. In addition, more recent maps indicate that

Herrera province now extends farther S. Older maps indicated that the S limit of the province was roughly at latitude 0738, but recent maps place this limit near 0730. Therefore, the region that was previously located on the Veraguas/Los Santos provincial border is now located either in Herrera province or on the Veraguas/ Herrera provincial border. In particular, Aldrich and Bole (1937:7) indicate that they used MHA to locate and describe their localities, which explains the change of province noted herein for their collecting localities. Kelsall and Cullingford, at sea, off Azuero Peninsula, 100 mi (160 km) S of Panama, 1924 Aug (BMNH; Douglas and Johnson, 1926); Benson, interior of the peninsula, mountainous region covered with heavy tropical forest, faint tinge of the subtropical zone, and according to Aldrich and Bole (1937:6), Benson worked at Loma El Montuoso [0743/8050 (USBGN)] (q. v.) and at Cerro Largo [0750/8050 (USBGN)] (q. v.), 1925 summer, (Griscom, 1927a, as "Cape Mala Peninsula," and as "Cape Mala"; Aldrich and Bole, 1937, as "Cape Mala [= Azuero] Península"); Aldrich and Bole, the secondary localities are all located on the W slope of the peninsula, GL for: Altos Cacao, Cavulla Camp, Cerro Viejo Camp, Mariato River Camp, and Paracoté, 1937 (Aldrich, 1937, as "Azuero Peninsula"; Aldrich and Bole, 1937, as "Azuero Peninsula"; Rogers, 1939, as "Azuero Peninsula"); Murphy, in the Gulf of Panama, "toward Cape Mala," on the open ocean ca. "90 mi (145 km) S of Cape Mala," 1956 Nov 26 (Murphy, 1958); Delgado, SW end of the peninsula, humid forests, forest edge, near Cerro Hoya and El Cortezo [0726/8038 (USBGN)] (q. v.), 1979 Jan-Feb, 1980 Jan (AMNH; CRUV; Delgado, 1985; Angehr, 2003:172); Agro et al., collected on the SW end of the peninsula, GL for: Cambutal, Cascajilloso, and Cobachón, 1996 Feb 14-17, 19-23 (ANSP). (Wetmore, 1952:18, 25). See also: Palo Seco, Río.

AZUL, BAHÍA; see Bluefield, Laguna de

AZUL, CERRO; Chiriquí
0848/8223 (USBGN)
2,310 m, peak located 6 km NE of Boquete [0847/8226 (USBGN)] (E762). Davidson, GL = Boquete, at 5,500 feet (1,700 m), a few specimens are labeled Work Finca or Finca Cerro Azul, 1934 Jan 24-26, 29 (CAS; Davidson, 1938:256).

AZUL, CERRO; Panamá (E, Pacific)
0910/7924 (USBGN)
ca. 600 m, peak on the continental divide 10 km NNW of Tocumen [0905/7923 (USBGN)] (E661), the peak is not shown on our more recent maps, which do, however, indicate the town Cerro Azul [0910/7925 (USBGN)] at the same location (JOG; MGRP). Many collectors (Goldman, Karr, Wetmore, and presumably others) apply the name Cerro Azul to the ridge that includes Cerro Azul and other peaks. Goldman applies the name Cerro Azul to a peak ca. 15 mi (24 km) WNW of Chepo [0910/7906 (USBGN)] and 15 mi (24 km) N of the coast (LZP; Nelson, 1911a) indicating that his work at Cerro Azul was N of the actual peak itself. Given Goldman's descriptions of Cerro Azul as a peak near the head of the Río Chagres [0907/7942 (USBGN)] and the Río Pacora [0901/7917 (USBGN)] with a summit at 3,000 feet (900 m) (at 3,200 feet (1,000 m) in Nelson, 1911a), as well as the locations and descriptions of his camps, he probably worked ca. 14 km NE of

Cerro Azul on Cerro Prominente [0914/7917 (USBGN)] (q. v.). Goldman, most descriptions come from his journal entries, which are similar to the account given in Goldman (1920), although slight variations are noted, at 300 feet (90 m) on the S base at a camp in the savannas called Llano de la Mesa (see La Mesa [0911/7916 (USBGN)]), ca. 15 mi (24 km) NW of Chepo, 1911 Mar 20, at 300-3,000 feet (90-900 m), departed Llano de la Mesa on Mar 21 and moved the camp ca. 3 mi (5 km) [NW ?], from the savannas followed stream beds with large, water worn boulders into heavy, unbroken forest and made camp at 800 feet (240 m) at Cabobré (not found, but see Río Cabobré [0906/7917 (USBGN)] and La Mesa) on a tributary of the Río Pacora. From camp cut a trail to the peak, which had been cleared of timber by engineers ca. 2 years previously, growth at the higher elevations was lower and denser with thick undergrowth, above 2,000 feet (600 m) the forest was more humid, and at 3,000 feet (900 m) on the summit and down the N slopes to ca. 500 feet (150 m) the land was covered with a dense growth of low trees, moss, orchids, and bromeliads. On Mar 27 the group returned to Llano de la Mesa at 300 feet (90 m). Goldman (1920) indicates that the group returned to Chepo on Mar 27, although the journal and specimen labels indicate otherwise. The majority of Goldman's specimens probably came from the S slopes, as a few specimens were labeled "S base," "S slopes," and "S face," whereas none of the specimens appear to have been labeled "N slopes." In either case, Cerro Prominente is SE of the continental divide and both its N and its S slopes are on the Pacific drainage, 1911 Mar 21-28 (USNM; Nelson, 1911a; Nelson, 1912a; Goldman, 1920); Wetmore applied the name Cerro Azul to the entire range that marks the continental divide and extends NE from Cerro Azul to Cerro Prominente [0914/7917 (USBGN)] and beyond as indicated by his journal entry: "the range should properly be known as Serranía Azul." Therefore, specimens labeled Cerro Azul were collected over a large area and on both sides of the continental divide. Wetmore and Perrygo, the secondary localities attributed to Cerro Azul were located in the area extending 13 km NNE of the peak and 20 km ENE of the peak, GL for: Bajo Grande, Cerro Ultima, Quebrada Carriaso, Río Pacora, and San Miguel, 1949 Mar 23, 29, 31, Apr 12, 19, 23, 25; Wetmore, at 2,800 feet (850 m), walked several miles along the ridge, at 2,600-2,700 feet (790-820 m) there was a 30-40 acre (12-16 ha) open, bald spot with grass clumps and a few bushes in the center of the ridge, recently enlarged by forest cutting and regular burning, E along the ridge through low, dense forest with short trees and brush able to withstand strong winds, to the Atlantic drainage into wetter forest, below the crest to more open gallery forest with larger trees (see also: Pilón, Cerro), 1955 Mar 29, and as a GL for La Zumbadora, 1961 Jan 31, GL for Cerro Pilón, 1957 Mar 2 (ANSP; LSUMZ; USNM); Blanton, 1953 Sep 1, 18, 24, 27, 30, Oct 2, 8 (USNM); MCU, 1955 Oct 11 (USNM, as "Blue Mt. (= Cerro Azul)"); Handley, apparently used the term Cerro Azul to refer to the entire range, GL for: Cerro Jefe, and Río Pacora, 1958 Jan 25, 27, Feb 7 (USNM); Vicente, 1958 Jan 27 (USNM); Davis, 1960 Apr 10 (Eisenmann, 1962a); Hinds, Mendez, Renner, Rojas, Kaster, and GMLP, additional labels include "Camp 4, Mandinga Trail," and "Altos de Pacora," 1960 Sep 12, 26, 1965 May 15, Jun 12, Jul 20, Sep 4, Nov 8, 23, Dec 11, 1966 Oct 26, 1967 Mar 29, Apr 17, May 11, 25, 1968 Mar 12, 1969 Aug

3 (GML); Willis and Eisenmann, at 800 m, observations, 1964 Jun 23 (Olson and Eisenmann, 1966; Willis, 1972); Morton, at 2,500-7,500 feet (750-2,300 m), 4-7 mi (6-11 km) E, N, and NE, 1965 Jul, Dec 10, 1966 Jan 6, Feb 15, Mar 10, Apr 23 and as a GL for Goofy Lake, 1971 Feb 7, Mar 5 (AMNH; FMNH; YPM); Rojas, 1965 Jul 17, Sep 5 (GML, as "Cerro Azul," and as "Altos de Pacora, Cerro Azul"); Karr, mist nets, "under this name I include the elevated areas referred to as Cerro Azul, Cerro Jefe, and other peaks on the same ridge (maximum elevation about 1000 m). Mature, humid forest is still present in some areas although much of the area consists of cut over forest and forest-edge," 1969 Apr 9 (Karr, 1971a); Tejera, 1978 Apr 29, May 20, Jul 12, Sep 19, Oct 26, Dec 21, 1979 Jan 14 (MVUP); Bacon, Guitton-Mayerma, Lelevier, Maley, Miller, and Winker, ANAM Station [Autoridad Nacional del Ambiente, 0912/7930], 2002 Oct 19, Nov 24, Dec 7, 2004 Feb 6-8, 2006 Feb 3 (STRI; UAM). Guitton-Mayerma, Miller, and Winker, Rio Mono (not located), at 0913/7925, 2004 Feb 6-7 (STRI; UAM); DaCosta, Klicka, and Spellman, at 650-700 m, at 0913/7925, 24 km NNE of Panama City, Rio Mono, 2006 Feb 20 (MBM; STRI). (Wetmore, 1952:19; Wetmore, 1962; Wetmore, 1970; Olson, 1980a; Olson, 1981c; Olson, 1983; Petit et al., 1999).

BAHÍA, AZUL; see Bluefield, Laguna de

BAHÍA BIQUE; see Bique, Bahía

BAHÍA HONDA; see Honda, Bahía

BAHIA PENINSULA; see Bohío, Península

BAIZAL; Darién

ca. 0748/7731

According to Angehr (in litt., Aug 2005), Baizal appears to refer to a town shown on E562, sheet 12 (1:250:000 series) (1997), named El Baisal [ca. 0748/7731], which is located on the Rio Tuira [0821/7803 (USBGN)] ca. 6 km SW of Púcuro [0801/7730 (USBGN)]. He further mentions that a town named "Balsal" is indicated on the older 1:250,000 Special Map/Mapa Especial of Darién, which contains numerous errors, near where El Baisal appears on E562, but that "Baisal" is indicated farther away on the Río Púcuro [0801/7733 (USBGN)] near Púcuro. Emanuel and Ridgely, observations, on the trail between Púcuro and, 1981 Mar 7 (Wetmore et al., 1984:633, as "Baizal").

BAJADERO; Los Santos

ca. 0733/8001

Sea level, described by Wetmore as a low, swampy area 2 mi (3 km) NNE of Pedasí [0732/8002 (USBGN)], located between the beach and 70-100 foot (20-30 m) bluffs covered with dry scrub, Wetmore, GL = Pedasí, 40-70 foot (10-20 m) white mangroves, some in bloom, near the mouth of the Río Pedasí [0734/8002 (USBGN)] where sand dunes dropped down to the mangroves, 1957 Mar 8, 15 (USNM; Olson, 1980b, as "Pedasí, Bajadero"). Not shown on our maps.

BAJO BOQUETE; see Boquete

BAJO BOYARENA, ISLA; Panamá, Archipiélago de las Perlas

0833/7902 (USBGN)

According to Angehr (in litt., Aug 2005), the island appears on map 4341-I (Saboga) [E762]. "It is a rocky reef mostly submerged at high tide [and] entirely so during spring high tides" The island is not shown on our maps but the USBGN coordinates indicate that it is located ca. 12 km NW of Isla del Rey [0822/7855 (USBGN)], N of Isla Bolaños [0832/7902 (USBGN)] and S of Isla Mogo Mogo [0834/7902 (USBGN)]. Levings et al., observations, 1982-1983 (Levings et al., 1986).

BAJO CHITRA; see Chitra

BAJO CHUMBAGO; Chiriquí

ca. 0852/8234

ca. 2,000 m, described by Wetmore as an area N of Cerro Punta [0852/8235 (USBGN)] near the continental divide, presumably near the current town Chumbago [0852/8234 (USBGN)] (q. v.). The name is a corruption of Schönberger, an early German settler in the region, and Wetmore originally entered "Bajo Schönberger" in his field catalog, but later changed it to "Bajo Chumbago." Wetmore, GL = Cerro Punta, at 6,500 feet (2,000 m), followed a foot trail through rough pastures and bamboo to 6,500 feet (2,000 m) ending in a rough, unkempt pasture with short, bright green grass, muddy cattle tracks and many thickets, bracken, berry bushes and other brush grew around the large, fallen logs, the area was cleared of forest except on the high, steep slopes where conditions were still wild, 1955 Mar 3 (USNM). Not shown on our maps.

BAJO ESPAÑA; Veraguas, Isla de Coiba

ca. 0730/8142

Described by Wetmore as a coastal area on Isla de Coiba [0728/8147 (USBGN)] between Colonia Penal [ca. 0730/8142] and Campamento de Catival [0729/8144 (USBGN)] (USNM). Kelsall and Cullingford, GL = Isla de Coiba, near the mouth of the Río Catival, small stream, mangroves, low bluff, 1924 Aug 31 (BMNH, as "Coiba Island"; Douglas and Johnson, 1926, as "Coiba Island"; Wetmore, 1957:8); Wetmore, GL = Isla de Coiba, tract of mangrove swamp, coconut palms along the beach, brush on the shoreline, broad pasture with scattered trees, small streams, 1956 Jan 8, 12, 14, Feb 2 (USNM; Wetmore, 1957). Not shown on our maps.

BAJO GRANDE; Chiriquí

0852/8233 (USBGN)

2,100 m, town 2 km ENE of Cerro Punta [0852/8235 (USBGN)] (JOG). Described by Wetmore as a level valley drained by the headwaters of the Río Chiriquí Viejo [0820/8241 (USBGN)]. The forest had been cut by 1955, but many living trees remained and short, green grass as well as garden plots with lettuce, cabbage, bushes, and flowers covered the open ground. Wetmore, GL = Cerro Punta, at 6,700 feet (2,050 m), probably the E end of Bajo Grande, as specimens are also labeled Cerro Respingo [0851/8232 (USBGN)] (q. v.), 1955 Feb 28, at 7,000 feet (2,150 m), above, on, and near the Cerro Punta-Boquete Trail [ca. 0852/8235] (q. v.), old trails, in woodland with heavy undergrowth, through excellent forest to a recent clearing and down to Bajo Grande, 1955 Mar 1-2, 4, 7 (USNM).

BAJO GRANDE; Panamá (E, Pacific)

ca. 0912/7915

ca. 100-200 m, described by Wetmore in 1949 as a great level stretch a short distance S of San Miguel [0914/7915 (USBGN)] where the Río Pacora [0901/7917 (USBGN)] passes through a low divide between Cerro Carbunco [0911/7914 (USBGN)] and Cerro Azul [0910/7924 (USBGN)], Wetmore and Perrygo, GL = Cerro Azul, Río Pacora, at 500 feet (150 m), gallery forest with heavy undergrowth, forest cut at lower end, 1949 Mar 23 (USNM). Not shown on our maps.

BAJO MONO; Chiriquí

0848/8228 (USBGN)

1,500 m, town 5 km NW of Boquete [0847/8226 (USBGN)] on the Río Caldera [0837/8221 (USBGN)] near the mouth of the Quebrada El Velo [0849/8228 (USBGN)] (E762). Mönniche, at 5,200-6,500 feet (1,600-2,000 m), 1932-1933, 1936-1939 (CAS, as "Bajo Mono River"; FMNH; MCN; Blake, 1958; Monroe, 1963b); Davidson, GL = Boquete, a few specimens are labeled Leandro Trail or Bajo Mono-Leandro Trail (not found, but perhaps an error for Lérida Trail), 1933 Dec 6-8, 15-19, 22, 25, 28-29 (CAS).

BAJO MONO RIVER; see Bajo Mono

BAJO MONO-LEANDRO TRAIL; see Bajo Mono

BAJO PALENQUE; see Palenque, Río

BAJO QUIEL; Chiriquí

0848/8227 (USBGN)

1,400 m, town 4 km NW of Boquete [0847/8226 (USBGN)] just N of Callejón Seco [0848/8228 (USBGN)] (E762). See also: Quiel.

BAJO SCHÖNBERGER; see Bajo Chumbago

BAJOS DEL MUÑOZ; Los Santos

ca. 0737/8006

Probably refers to the lower reaches of the Río Muñoz [0737/8005 (USBGN)], a river that originates in SE Los Santos province and flows NE entering the Gulf of Panama [0800/7930 (USBGN)] 14 km NNW of Pedasí [0732/8002 (USBGN)]. The lower reaches pass through mangroves and widen forming a 1-2 km long coastal inlet (JOG). May also refer to the town named Río Muñoz [0737/8006 (USBGN)] that is shown on JOG 1 km N of the Río Muñoz [0737/8005 (USBGN)] and ca. 5 km above its mouth. Gómez and Lasso, 1984 Sep 1 (MVUP).

BALBOA; Panamá (Canal Zone) (E, Pacific)

0857/7934 (USBGN)

Sea level, town in the former Canal Zone just SW of Ancón [0858/7933 (USBGN)] and Panama City [0858/7932 (USBGN)] on the Pacific entrance of the Panama Canal [0920/7955 (USBGN)] (E962; JOG). Osgood and Anderson, 1912 Feb 8, 10-12 (FMNH); Van Name, 1924 Apr 15 (AMNH), Hallinan, power lines, tall grass in a clearing, mango tree, over the water, mangrove swamp, mudbanks of a tidal stream, tall grass, sewer mouth on the mud beach, 1915 Feb 26, Apr 25, Jun 20, 24-25, 27, 30, Aug 17, Sep 3, Dec 25, 1916 Feb 13, May 6, 13, 15, 18, 21 (AMNH; Hallinan, 1924; Eisenmann, 1962a, "May 13, 1913," which should read 1916); Fagan, at sea 180 mi (290 km) S of Balboa, which is S of the Panama/Colombia land boundary, 1922 Dec 22 (USNM; Wetmore, 1923; Wetmore, 1965:41, as "290 kilometers to the southwest"); Kelsall and Cullingford, 1924 Aug, Nov (BMNH; Douglas and Johnson, 1926); Weber, mangrove swamp, 1928 Dec 12 (FMNH); Deignan, observations, 1932 Jul (Deignan, 1933, as "Balboa," and as "Balboa Station"); Chapman, observations, 1938 Feb (Chapman, 1941, as "Balboa Railway Station"); Meyer, 1941 Jun 12, 14, 21, Jul 6, 9, 13, 16 (USNM; Wetmore, 1946a); Eisenmann, GL for Bahía de Panamá, 1948 Jun 28 (Eisenmann, 1951); Wetmore, observations between Isla Taboga [0847/7933 (USBGN)] and, 1952 (Wetmore, 1952:7, 13); Wetmore (from Mrs. Paul Barnard), GL for Cerro Ancón (Ancón Hill), Dec 1955 (USNM); Murphy, observations, 100 mi (160 km) S of, 1956 Dec 1 (Murphy, 1958); Robins, observations, 1957 Jul 15-26 (Robins, 1958); Loftin, Balboa Train Station, grassy lawn, 1961 Sep, 1962 Aug, Oct (Loftin, 1963a); Gale, J. 1962 Apr 1 (USNM); Renner, 1967 Sep 5 (GML); Sharp, Amador Road, 1971 Oct 17 (GML); Beason, observations, 1972 Nov 2-3 (Beason, 1976); Barrett, 4 km SW of, 1973 Aug 29 (GML); Dawson, 6 km W of, 1973 Sep 10 (GML); Justines, 1977 May 31 (GML); Morton, observations, 1970s (Morton, 1982). (Zimmer, 1939; Zimmer, 1955:5; Odum, 1947, as "Balboa Heights"; Olson, 1981d). See also: Ancón, Cerro.

BALBOA HEIGHTS; see Balboa

BALBOA RAILWAY STATION; see Balboa

BALBOA TRAIN STATION; see Balboa

BALD HILL; Panamá, Archipiélago de las Perlas

ca. 0818/7906

Described by Wetmore as a small savanna area in the N of Isla San José [0815/7907 (USBGN)]. According to Angehr (in litt., Aug 2005), this is certainly Cerro Pelado, a small grass-covered peak, 129 m, at UTM 918000 N 708600 E (0818/7906) on map 4340-I [E762], near the N end of Isla San José. Specimens collected by Wetmore and Morrison are labeled only Isla San José, although Wetmore (1946b) repeatedly refers to Bald Hill.

BALNEARIO KOBBE; see Fort Kobbe Army Reservation

BALSA, PUNTA; Chiriquí

ca. 0809/8251

Sea level, point on the E coast of Península Burica [0810/8255 (USBGN)] 5 km N of Punta Burica [0802/8255 (USBGN)] (MHA). Wetmore, GL = Punta Burica, 5 km N of Punta Burica, through brush to a small ravine that was dry except for a line of small pools, forest along a ravine and on the slopes above, the level land was farmed and partly grown in *rastrojo* with scattered palms and other trees, 1966 Feb 25 (USNM).

BALSAL; see Baizal

BALSAS, RÍO; Darién

0815/7759 (USBGN)

Originates in SE Darién province on the Colombian border and flows N entering the Río Tuira [0821/7803 (USBGN)] 27 km SE of La Palma [0825/7809 (USBGN)].

This is the main river that drains the W slopes of Serranía de Pirre [ca. 0749/7744] (MGRP, as "Río Balsas o Tucutí"). See also: Manené.

BAMBITO; Chiriquí
0850/8237 (USBGN)
1,600 m, town on the Río Chiriquí Viejo [0820/8241 (USBGN)] 5 km SW of Cerro Punta [0852/8235 (USBGN)] (JOG). Yunker, Casa Winkloski (not located), 1961 Apr 28 (USNM); Morton, GL = Volcán Barú, at 5,300 feet (1,600 m), 1966 Apr 22-23, Jul 18-20 (FMNH); Olson et al., at 5,200 feet (1,600 m), 1966 Jul 3-4 (UF; USNM); Cardiff and Dittmann, at 1,720 m, 2 km NE Bambito, 1996 Mar 9 (LSUMZ); Weir, at 1,700 m, 2003 Apr 9 (CU). See also: Chiriquí Viejo, Río; and Llanos del Volcán.

BANANA RIVER; see Banano, Río

BANANO, RÍO; Bocas del Toro
0920/8223 (USBGN)
A river that enters the W shore of Bahía Almirante [0920/8218 (USBGN)] ca. 5 km N of Almirante [0918/8224 (USBGN)] (MHA, as "R. Banana"). Kennard et al., GL = Almirante, collected near the river, although it is not clear whether the name appears on specimen labels, 1926 (Kennard and Peters, 1928:443-444, as "Banana River"); Wedel 1927 Jul, Dec, 1928 Jan, Apr, Jul (MCZ, as "Banana River"); Wetmore, GL = Almirante, inland above its mouth the 4-6 foot (1-2 m) deep river narrowed quickly to 10-15 feet (3-5 m) and the banks offered no landing or open ground on which to walk through the surrounding mangroves that extended N to Changuinola Canal [ca. 0919/8221], traveled upriver ca. 1 mi (1.5 km), 1958 Feb 6 (USNM, as "Banana River").

BANCO HANNIBAL; see Hannibal, Banco

BANDERA, LOMA DE LA; see La Bandera, Loma de

BARCA QUEBRADA, ISLA; see Barco Quebrada, Isla

BARCA QUEBRADO; see Barco Quebrada, Isla

BARCO QUEBRADA; see Barco Quebrada, Isla

BARCO QUEBRADA, ISLA; Veraguas (Pacific)
0719/8141 (USBGN)
The spelling varies between Isla Barco Quebrad*a* and Isla Barco Quebrad*o* on maps and specimens. USBGN and JOG both use Isla Barco Quebrada for the small islet located 1 km off the S coast of Isla de Coiba [0728/8147 (USBGN)] in the E waters of Canal de Jicarón [0720/8146 (USBGN)] (JOG). Angehr (in litt., Aug 2005) points out that Isla Barco Quebrado would be correct Spanish, and that the name appears as such on map 3837-I (Isla Jicaron) [E762]. At the same time, his map 3838 II shows a "Playa Barco Quebrada" on Isla de Coiba just to its N. The island is now located within Isla de Coiba Parque Nacional [ca. 0730/8140] (RDP2). Ridgely, 1976 Apr 10 (Ridgely and Gwynne, 1989:135, as "Isla Barca Quebrada"); Garcés and Mena, 1996 Feb 16-14, Mar 2 (MVUP, as "Barca Quebrada," and as "Barco Quebrada").

BARCO QUEBRADA, PLAYA; see Barco Quebrada, Isla

BARQUETA, LA; see La Barqueta

BARRERO, EL; see El Barrero

BARRIADA CHANIS; see Chanis, Barriada

BARRIADA SAN ANTONIO; see San Antonio

BARRIADO OBARRIO; see Obarrio, Barriado

BARRIGÓN; Coclé (Pacific)
0838/8037 (USBGN)
1,000 m, town 2 km N of El Copé [0837/8036 (USBGN)] (MGRP) and according to Wetmore, the town is on the Río Barrigón [ca. 0837/8034], Wetmore, GL for Río Barrigón, 1962 Feb 26 (USNM).

BARRIGÓN, RÍO; Coclé (Pacific)
ca. 0837/8034
According to Wetmore, El Copé [0837/8036 (USBGN)] was built on a ridge with views to the N and E of the river valley. Presumably the river originated N or NW of El Copé and flowed SE entering the Río Harino [0833/8034 (USBGN)], a tributary of the Río Grande [0818/8024 (USBGN)], a few km E of El Copé. Wetmore, GL = El Copé, coffee area with bits of forest near town, the area upriver was completely cultivated in coffee and oranges, 1962 Feb 21-23, GL = Barrigón, near the settlement, 1962 Feb 26 (USNM). Not shown on our maps.

BARRILES; Chiriquí
0848/8242 (USBGN)
1,300 m, described by Wetmore as an abandoned town S of Palo Santo [0849/8241 (USBGN)] on the W end of Llanos del Volcán [ca. 0847/8238] (USNM). Davidson, GL = Volcán, at 4,200-5,000 feet (1,300-1,500 m), "*llanos*," "*lagunas*," 1929 Dec 14-24, 26-27, 1931 Jan 12-15, 17-31, Feb 2-6, 8 (CAS; CM; Davidson, 1932; Davidson, 1938; Blake, 1958:531); Barrows, 1929 Dec 24 (CAS); Wetmore, GL = Volcán, at 4,400 feet (1,350 m), pastures, cultivated land, and *rastrojo*, 1954 Mar 25, GL = Volcán, forest cover gone since his visit in 1954, open pasture, to Quebrada Barriles [0848/8241 (USBGN)], small tract of forest, tall *rastrojo*, 1960 Feb 17 (USNM). Not shown on our maps.

BARRILES, QUEBRADA; Chiriquí
0848/8241 (USBGN)
Originates W of Volcán [0846/8238 (USBGN)] and flows NW entering the Río Chiriquí Viejo [0820/8241 (USBGN)] 6 km NW of Volcán (E762). Described in Wetmore's journal entries for 1960 as a clear, swift stream, ca. 20 feet (6 m) wide. See also: Barriles; and Las Lagunas.

BARRIO COLÓN, CORREGIMIENTO; Panamá (W, Pacific)
0854/7945 (USBGN)
Not shown on our maps but USBGN indicates a third-order administrative division (below both province (comarca) and district) located on the SW coast of Panamá province ca. 35 km SW of Panama City [0858/7932 (USBGN)]. Badillo (or Bandillo),

1985 Mar 18-19 (MVUP).

BARRO COLORADO; Panamá (Canal Zone) (W, Atlantic)
0910/7950 (USBGN)
50 m, settlement on the NE coast of Isla Barro Colorado [0909/7951 (USBGN)] in the cove formed between Punta Salud [0910/7950 (USBGN)] and Punta Colorado [0910/7949 (USBGN)]. This is the location of the Smithsonian Tropical Research Institute's laboratory as well as the starting point for many of the trails (E866; Willis and Eisenmann, 1979:3 (map)). Howell, 1929 Jul 31, Aug 4-6, 9-19 (UCLA). See also: Barro Colorado, Isla.

BARRO COLORADO, ISLA; Panamá (Canal Zone) (W, Atlantic)
0909/7951 (USBGN)
171 m, island in Lago Gatún (Gatun Lake) [0912/7955 (USBGN)] located 15 km WNW of Gamboa [0907/7942 (USBGN)] and 22 km SSE of Colón [0922/7954 (USBGN)] just W of the Panama Canal [0920/7955 (USBGN)]. The island was clearly a hilltop before Lago Gatún was formed ca. 1912 during the construction of the Panama Canal. Numerous trails, which are named in Eisenmann (1952) and depicted on a map in Willis and Eisenmann (1979), dissect the ca. 14.8 km^2 (1,480 ha) island providing access to all regions (E866; JOG; Willis and Eisenmann, 1979). According to Eisenmann (1952), the island had 25 mi (40 km) of shoreline and its mainland was wooded aside from a few cleared acres. The laboratory is located at the settlement Barro Colorado [0910/7950 (USBGN)] on the NE coast of the island in Bahía Frijoles [0910/7949 (USBGN)] (Willis and Eisenmann, 1979:3 (map)). Eisenmann (1952) lists numerous collections and observations from the island not all of which are included herein. The name often appears as Barro Colorado Island in the literature and on specimen labels cited here unless otherwise specified. Van Tyne, 1925 Jul 28, Aug 11, 1926 Mar 3, 11-13, 15, 22-23, Apr 9, 22-23, 25, May 1, 4, 12, 18, 1927 Mar 8, Apr 11, 14, 17-18, 27, 29, May 14, 28, Jun 18, Jul 27, Aug 5, 17-18 (Eisenmann, 1952; Van Tyne, 1926; Van Tyne, 1937; Van Tyne, 1944; Van Tyne, 1950); Chapman and Potter, 1925 Dec 28-1926 Feb 20 (AMNH, as "Barro Colorado"; Eisenmann, 1952); Chapman, 1927 Jan 16, 29, Feb 19, Apr 1, 1929 Jan 30, Feb 5, 25, 1932 Mar 20 (AMNH, as "Barro Colorado"; Chapman, 1929a; Chapman, 1931b; Chapman, 1933; Chapman, 1935; Chapman, 1938; Zimmer, 1939; Eisenmann, 1952); Gross, 1927 Oct 30, Nov 6, 12 (USNM,; Eisenmann, 1952); Weber 1928 Dec 13-19, 23 (FMNH, as "Barro Colorado Id."); Wonder 1928 Dec 14-16, 18 (FMNH); Howell, 1929 Aug 15-17, 20-21, 23, 25 (UCLA); Arbib and Loetscher, 1934 Aug 1 (Arbib and Loetscher, 1935, as "Barro Colorado," and as "Barro Colorado Island"); Niedrach and Rogers, 1936 Mar 30 (Eisenmann, 1952:15); Griswold, 1936 Apr 16 (MCZ, as "Barro Colorado I."); Gilliard, 1937 Jan 23 (AMNH); Imhof, observations, between Frijoles and, 1942 Apr 17 (Imhof, 1950); Johnson, observations in 1948 (Johnson, 1953; Johnson, 1954); Eisenmann, GL for Lago Gatún, 1949 Jun 28 (Eisenmann, 1951); Koford, near the laboratory, Thomas Barbour Trail near marker 9, and Van Tyne Trail, 1951 Feb 25-26 (MVZ; Lundy, 1957); Murphy, GL = Lago Gatún, observations "in front of the Barro Colorado Island Laboratory," 1956 Nov 28 (Murphy, 1958);

Willis, 1961 Oct 4, 17, 19, 27, observations, banding, 1960-1982 (MVZ, as "Barro Colorado Id."; Willis, 1966a; Willis, 1966b; Willis, 1972, as "Barro Colorado"; Willis, 1973; Willis, 1974; Willis, 1980; Willis, 1982; Willis, 1983a; Willis, 1983b; Willis, 1983c; Willis, 1984; Willis, 1985; Willis, 1992; Willis and Oniki, 1992); Hunt, 1962 Jun 18, 25, Jul 1 (YPM); Loftin, 1962 Aug 7 (Loftin, 1963a, as "Barro Colorado"); Olson, captive bird, 1963 Feb 16 (USNM); Leck, 1969 Mar 30 (AMNH; Leck, 1972a); Thorington, R., 1972 Mar 7 (USNM); Thorington, E., 1972 Mar 21 (USNM); White, clearing, 1972 Mar 21 (ANSP, as "Barro Colorado Isl."); Gardner, 1977 Dec 27 (USNM); Howe and Vande Kerckhove, observations, 1979 May 1-Sep 7 (Howe and Vande Kerckhove, 1981); Acosta, GL for Lago Gatún, 1981 Jan 15 (MVUP, as "Isla de Barro Colorado"); Handley, lab clearing, 1982 Oct 27 (USNM). (Chapman, 1928a; Chapman, 1928b; Griscom, 1929a; Gross and Van Tyne, 1929; Shurcliff, 1930; Skutch, 1934; Skutch, 1945b; Skutch, 1946a; Skutch, 1948a; Skutch, 1949a; Jacques, 1942; Hobson, 1948; Moore, 1948; Ellis, 1952; Gross, 1952; Gross, 1964; Laughlin, 1952; Zimmer, 1955:5; Eisenmann, 1961; Moynihan, 1962b; Moynihan, 1963; Eisenmann, 1968; Ricklefs, 1971; Ricklefs, 1980; Kilham, 1972; Oniki, 1972; Jones, 1977; McHargue, 1977; McHargue, 1981; Wright, 1979; Fontaine, 1980; Greenberg, 1981a; Greenberg, 1981b; Greenberg and Gradwohl, 1980a; Greenberg and Gradwohl, 1980b; Greenberg and Gradwohl, 1983; Greenberg and Gradwohl, 1985; Greenberg and Gradwohl, 1986; Greenberg and Gradwohl, 1997; Gradwohl and Greenberg, 1980; Gradwohl and Greenberg, 1983; Gradwohl and Greenberg, 1984; Angehr, 1981; Angehr, 2003; Howe, 1981; Howe and De Steven, 1979; Howe and Kerckhove, 1981; Olson, 1981d; Vleck, 1981; Karr, 1982a; Karr, 1982b; Karr, 1990; Karr et al., 1990b; Willson et al., 1982; Young, 1982; Loiselle and Hoppes, 1983; Dyrcz, 1984; Dyrcz, 1999; Dyrcz, 2000; Dyrcz, 2002; Mindell and Black, 1984; Roubik et al., 1985; Houston, 1986; Houston, 1987; Wright, 1988; Wiedenfeld, 1991; Mabey and Morton, 1992; Sieving, 1992; Moore et al., 1999; Poulin et al., 1999; Robinson, 1999; Robinson, 2001). See also: Barro Colorado.

BARRO COLORADO, ISLA DE; see Barro Colorado, Isla

BARRO COLORADO ISLAND; see Barro Colorado, Isla

BARTHOLOMEW, ISLA; see Bartolomé, Isla

BARTOLO ARRIBA; Chiriquí

ca. 0818/8256

According to Jason Weir (personal communication), this refers to a settlement on the N end of Península Burica [0810/8255 (USBGN)], on the Río San Bartolo [0817/8251 (USBGN)] (q. v.), along the road to El Chorogo [0818/8301 (Angehr, 2003:139)]. El Chorogo is an Important Bird Area along the Costa Rican border that includes the headwaters of the Río San Bartolo (Angehr, 2003:139). Bartolo Arriba is not shown on our maps but two towns with similar names are shown along the upper reaches of the Río San Bartolo. San Bartolo Arriba [ca. 0819/8255] appears on MGRP at 100 m, 9 km WNW of Puerto Armuelles [0817/8252 (USBGN)]. Cabecera de San Bartolo [ca. 0817/8257] appears on JOG at 100 m, ca. 11 km WNW of Puerto Armuelles. Weir, GL = Península Burica, 157 m, at

08°18'42"/82°57'15", according to Weir, he collected beyond the settlement along the first 2 km of the trail to El Chorogo, "along the side of a ravine off the trail in open cow pasture with scattered trees," and "along a wooded ravine with stream and gallery forest," 2003 Apr 7-8 (CU; LSUMZ).

BARTOLOMÉ, ISLA; Panamá, Archipiélago de las Perlas

0839/7902 (USBGN)

Sea level, very small island with a land area of 1.2 ha (Angehr, 2003:267) located in Bahía de Panamá [0850/7915 (USBGN)] 23 km NNW of Isla del Rey [0822/7855 (USBGN)] between Islas Pacheca [0840/7903 (USBGN)] and Contadora [0838/7902 (USBGN)] (E501). According to Angehr (2003), the island is uninhabited, "covered with palms and a few other trees on the eastern end [and with] herbaceous vegetation on the west," and an Important Bird Area. Smith, R. W., 1937 Feb 3 (ANSP, as "Isla Bartholomew").

BARÚ, COMARCA DE; Chiriquí

Probably a lapsus for Corregimiento or Distrito de (Wetmore 1959b:19).

BARÚ, VOLCÁN; Chiriquí

0848/8233 (USBGN)

3,474 m, the highest point in Panamá, a volcano 12 km NE of Volcán [0846/8238 (USBGN)] and 6 km SE of Cerro Punta [0852/8235 (USBGN)] (JOG, as "Volcán de Chiriquí"; MGRP). Authors and collectors often referred to the peak as "Volcán de Chiriquí," as "Mount Chiriquí," as "Volcán," or simply as "the volcano," although one should note that "Volcán" now specifies the nearby town. The name was applied to localities extending tens of kilometers from the peak itself, and those marked the "Caribbean Slope of Volcán de Chiriquí" were probably collected in Bocas del Toro province along the Boquete Trail [ca. 0857/8207] (q. v.) (Davidson, 1936:232). The volcano is now included in Parque Nacional Volcán Barú (La Bastille, 1973; La Bastille, 1974), which is considered an Important Bird Area (Angehr, 2003:126). Warszewicz, at 9,000-10,000 feet (2,750-3,050 m), 1849 (Gould, 1850b, as "Volcano of Chiriquí"; Gould, 1860, as "Volcano of Chiriquí in Costa Rica"; Gould, 1861b:141, as "Volcano of Chiriquí"; Salvin, 1867b, as "Volcano of Chiriquí"); Arcé, according to Salvin (1870:177) specimens from "Mina de Chorcha, Bugaba, and Volcan de Chiriqui, were made since Arcé reached David...these collections, too, were formed on the southern slope of the Volcano, the highest point reached being about 6,500 feet (2,000 m)," 18?? (BMNH, as "Volcán de Chiriquí"; FMNH, as "Volcán de Chiriquí"; USNM, as "Bolcan M," and as "Volcán de Chiriquí"; Salvin, 1870bb, as "Volcán de Chiriquí"; Salvin, 1872, as "Volcán de Chiriquí"; Salvin, 1891, as "Volcan de Chiriquí"; Salvin and Godman, 1879-1904, vol. 1:305, vol. 2:94-95, as "Volcán de Chiriquí"; Salvadori, 1895, as "Volcán de Chiriquí"; Griscom, 1935a, as "Volcan de Chiriquí"); Coale, H. K. (?), 1900 Sep 10 (UCLA, as "Volcán de Chiriquí"); Brown, at 7,000-7,500, and at 11,000 feet (2,150-2,300, and 3,350 m), these specimens belonged to the collection of E. A. and O. Bangs, and occasionally they are credited with collecting them, but all were collected by Brown. *Venilornis oleaginus sanguinolentus* (Bangs, 1902a:34) was apparently collected at 7,000 feet, but on the "Caribbean Slope," which suggests the Boquete Trail, but the elevation corresponds with specimens credited to the volcano itself, 1901 Feb 15-27, May 21, 26, 28-29, Jun 6 (ANSP, as "Volcán de Chiriquí"; MCZ, as "Volcán de Chiriquí"; USNM, as "Volcán Chiriqui," as "Mount Chiriqui," and as "Volcán de Chiriquí"; Bangs, 1902a, as "Volcán de Chiriquí," and as "Caribbean Slope of Volcan de Chiriquí"; Bangs, 1902c, as "Volcán de Chiriquí"; Bangs, 1906b, as "Volcán de Chiriquí"; Peters, 1931a:320, as "Volcan de Chiriquí"; Griscom, 1935a, as "Volcan de Chiriquí"; Davidson, 1938, as "Volcan de Chiriquí," and as "Caribbean Slope of Volcan de Chiriquí"; Olson, 1981c, as "Mount Chiriquí"); Collector ? [Brown ?], at 11,000 feet (3,350 m), 1901 Jun (ANSP, as "Volcán de Chiriquí"); Watson, at 7,000-9,000 feet (2,150-2,750 m), 1904 Oct (CM, as "Volcán de Chiriquí"; CU, as "Volcán Chiriqui"; OC, as "Volcán Chiriqui"); Mönniche, at 5,400-10,100 feet (1,650-3,100 m), 1932-1933, 1950 (MCN; Blake, 1956:386-388, as "Volcán de Chiriquí"; Blake, 1958:575, as "Volcán"), Wetmore, GL = Volcán, at 6,500 feet (2,000 m), collected on the W slopes in the area 5-7 km NE of Volcán, S along the mountain ridge toward Aguacate [0848/8235 (USBGN)], cattle trail through woodland, at Potrero Cordillera (q. v.), at Potrero Verde (q. v.) 1955 Mar 13, at Potrero Seco (q. v.), 1955 Feb 26, collections on the W side of the volcano, at 6,000-6,300 feet (1,800-1,900 m) on the W base where *Yucca* plants were blooming, at 6,200-10,400 feet (1,900-3,200 m) on the W face, at 11,000-11,200 (3,400 m) near the summit, 1965 Feb 24, Mar 1-5, 8-12, 17 (MVUP, as "El Volcán Barú," and as "Volcán Barú"; USNM); Moynihan, 4,500-7,500 feet (1,350-2,300 m), observations, on the W slopes from Volcán to above Cerro Punta, at the edges of forest and scrub, 1958 Sep 17-21, 1959 Mar 2-9, 1960 Mar 19-30, Oct 3-10 (Moynihan, 1962a:71, as "Volcán de Chiriquí (El Barú)"); Means, Koslovsky, and Sawyer, GL for: Nueva Suiza, and Cerro Punta, 1967 Feb 25-26, 28, Mar 1-2, 4-6, 8-13, 15-20, 23-25, 30, Apr 1, 14 (AMNH, as "Volcán de Chiriquí"; FSU, as "Volcán de Chiriquí"; GML, as "Volcán de Chiriquí"; USNM, as "Volcán de Chiriquí"); Tyson, E., at 6,300-7,600, and at 11,000 ? feet (1,900-2,300, and 3,350 m), on the W face, dry canyon, 1965 Feb 16-19, 23, 26, Mar 2-5, 15-16, 18-19 (AMNH, as "El Barú"; GML, as "El Baru"; MCN, as "El Baru; MVUP, as "El Volcán Barú"; USNM, as "El Baru"); Tyson, A., at 7,600 feet (2,300 m), W face, 1965 Mar 3 (FSU; GML); Morton, GL for: Bambito, and Cerro Punta, 1965 Dec 30, 1966 Apr 22-24, Jul 18-20 (FMNH, as "Volcán de Chiriquí"; YPM, as "Volcán de Chiriquí"); Stevenson, GL for Cerro Punta, 1968 Aug 31, Sep 2-5, 7-8 (FSU, as "Volcán de Chiriquí"; USNM, as "Volcán de Chiriquí"); Tejera and Ponce, 1981 Feb 26, Mar 1-2 (MVUP, as "El Volcán Barú"); Tejera, Ponce, and Ramos, 1983 Apr 19-21 (MVUP, as "El Volcán Barú," and as "Volcán Barú"); Weir, at 2,400-2,500 m, at 0851/8232, 2003 Apr 9-12 (CU, as "Volcan Baru Parque National"; LSUMZ, as "Volcan Baru Parque National"); Guitton-Mayerma, Miller, and Winker, 2004 Feb 15 (STRI; UAM). (Boucard, 1892, as "Volcán de Chiriquí"; Griscom, 1924b, as "Volcan de Chiriquí"; Griscom, 1927a, as "Volcan de Chiriquí"; Griscom, 1929a, as "Volcan de Chiriquí"; Rogers, 1939, as "Volcan de Chiriquí"; Pitelka, 1951, as "Volcán de Chiriquí"; Parkes, 1954a, as "Mt. Chiriquí"; Wetmore, 1956a, as "Volcan de Chiriquí"; Wetmore, 1963a, "The Volcano"; Wetmore, 1967, as

"Volcán de Chiriquí"). See also: Boquete; Francés; and Nueva Suiza.

BAS OBISPO; Panamá (Canal Zone) (Atlantic)

ca. 0907/7942

Former station on the original Panama Railroad [ca. 0922/7954] and on the continuous Río Chagres (see Río Chagres [0919/8000 (USBGN)]) at the mouth of the Río Camacho [0906/7941 (USBGN)] (PR). According to E8660, the site is now submerged in Lago Gatún (Gatun Lake) [0912/7955 (USBGN)] just SE of Gamboa [0907/7942 (USBGN)]. Maack, 1871 (MCZ, as "Obispo, Isthmus of Panama"; Maack, 1872); Hassler Expedition, 1872 Jul (USNM, as "Obispo"); Salvin, 1873 Apr-May (Salvin and Godman, 1879-1904, vols. 1-3, as "Obispo," and as "Obispo Station"); Goldman, crossed the Gamboa bridge to the hills (at 100-150 m according to E8660) on the E side of the river and returned to Matachin [ca. 0907/7943], 1911 May 5 (USNM; Wetmore, 1970; Olson, 1981d).

BASTAMIENTOS, ISLA; see Bastimentos, Isla

BASTIMENTOS, ISLA;

0918/8208 (USBGN)

Bocas del Toro, Archipiélago de Bocas del Toro

73 m, irregularly shaped island with a land area of 51 km^2 (5,100 ha) (Angehr, 2003:99) located 3 km SE of Isla Colón [0924/8217 (USBGN)] and 2 km N of Isla Popa [0911/8207 (USBGN)] (JOG). USBGN indicates that Isla Provisión is an older name for the island, and Angehr (2003) includes it as an Important Bird Area. Handley, 1962 Feb 8, and as a GL for Old Point, 1963 Feb 7-8, 10 (USNM, as "Isla Bastamientos," and as "Isla Bastimentos"); SBE, GL for: Alberrys Creek, Bryan Creek, Cayos Zapatilla, Long Beach, Old Point, and Sal Creek, 1987 Mar 13-29 (MVUP; USNM, as "Bastimentos Island"; Olson, 1993a, as "Bastimentos Island"); Billo, GL for Playa Larga, ½ km SW, 2007 Apr 18 (USNM as "Bastimentos"); Billo, GL for "Finca La Loma on N side of Bahia Honda near east end," 2007 Mar 30-Apr 2 (USNM) (Collar and Dunn, 1989).

BASTIMENTOS ISLAND; see Bastimentos, Isla

BATELE, PUNTA; Panamá (Canal Zone) (W, Pacific)

0854/7934 (USBGN)

Sea level, point on the coast of Bahía de Panamá [0850/7915 (USBGN)] 7 km SW of Panama City [0858/7932 (USBGN)] and 2 km NE of Punta Bruja [0853/7935 (USBGN)] (E761A, as "Batele Point"). See also: Changamé, Isla; and Palo Seco Leprosarium.

BATELE POINT; see Batele, Punta

BATIPA, CERRO; Chiriquí

0820/8215 (USBGN)

325 m, hill located 23 km SE of David [0826/8226 (USBGN)] on a peninsula that extends S into Golfo de Chiriquí [0800/8220 (USBGN)] that is largely surrounded by estuaries and mangroves (JOG). Angehr (2000) indicates that the Cerro Batipa is " a private reserve owned and protected by Luis Ríos and that it is a "forested hill at the eastern edge of the mangroves." The Important Bird Area includes the

"largest patch of lowland forest in Chiriquí province east of El Chorogo [0818/ 8301 (Angehr, 2003:139)] (Angehr, 2003:150). Observer ?, 1997 (Angehr, 2000).

BAVARENA, ISLA; Panamá, Archipiélago de las Perlas
0831/7901 ?
The name presumably refers to Isla Boyarena, a very small island in Bahía de Panamá [0850/7915 (USBGN)] 9 km NNW of Isla del Rey [0822/7855 (USBGN)] and 3 km E of Isla Bolaños [0832/7902 (USBGN)] (E501, as "Isla Boyarena"). USBGN indicates that Isla Boyarena (MHA and E501) is an older name for Isla Casayeta [0831/7901 (USBGN)], the name that appears on our more recent maps (MGRP; RDP2). Described by Rendahl (1920:4) as a "small, low-lying," but wooded island.

BAYANO; Herrera
0755/8035 (USBGN)
50 m, town in central Herrera province 4 km NE of Pesé [0754/8037 (USBGN)] (JOG). See also: Los Pájaros, Quebrada de.

BAYANO; see Chepo, Río

BAYANO; see Majé, Isla

BAYANO, COORPORACIÓN; see Bayano, Lago

BAYANO, LAGO; Panamá (E, Pacific)
0908/7835 (USBGN)
A reservoir with a surface area of 185 km^2 (RDP2) in eastern Panamá province created in 1978 by damming the Río Chepo [0900/7906 (USBGN)] for a hydroelectric project (Delgado, 1983). The dam is located 32 km NE of the river mouth and 24 km E of Chepo [0910/7906 (USBGN)] and the reservoir extends E of the dam through the Río Chepo valley. The Carretera Interamericana [0832/ 8242 (USBGN)] crosses the W section of the reservoir (Atlas, 1988:205; MGRP). According to Angehr (2003:280), the lake is ca. 37 km in length with a very irregular shoreline that contains many islands, the largest of which is Isla Majé [ca. 0909/ 7851], located near the dam. GMLP, GL for Isla Majé, 1972 Oct 4, 1977 Mar 17-18, Jul 22, 1978 Jul 30 (GML, as "Bayano"); Trejos, *instalaciones de la subestación hidroeléctrica del Bayano* (installations of the hydroelectric substation), 1983 Feb 27 (MVUP, as "Subestación Hidroeléctrica del Bayano"); Young, 1983 Nov 15 (MVUP, as "Coorporación Bayano"); Samudio, 1987 Mar 19 (GML, as "Hidroeléctrica del Bayano"); Polshek and Hinds, GL for Isla Majé, 1989 Jan 14, Feb 13, 20, 22, Mar 5, 17, 20, 28, Apr 15, 19-20 (GML; UF, as "Bayano Lake"); DaCosta, Guitton-Mayerma, Klicka, Miller, and Spellman, at 125 m, at 0909/7842, 20 km ESE of Cañita, 2006 Feb 11-13 (MBM; STRI). See also: Calle Larga; Chepo, Río; Majé, Isla; and Majé, Serranía de.

BAYANO, RÍO; see Chepo, Río

BAYANO LAKE; see Bayano, Lago

BAYANO RIVER; see Chepo, Río

BAYANO RIVER RESEARCH STATION; see Majé, Isla

BAY AT BOCAS DEL TORO; see Almirante, Bahía

BAY OF PANAMA; see Panamá, Bahía de

BAY OF SAINT MIGUEL; see San Miguel, Golfo de

BAYONET; see Empire Known Distance Range

BAYONET RANGE; see Fort Kobbe Army Reservation

BAYONETA, ISLA; Panamá, Archipiélago de las Perlas
0829/7904 (USBGN)
Island with a land area of 2.26 km^2 (226 ha) (Wright et al., 1985) located 11 km WNW of Isla del Rey [0822/7855 (USBGN)] (JOG), and according to a map section in Wetmore's journal, the elevation is 150 feet (45 m) (USNM). Bovallius, rocks on the coast were worn, inland mountain ridges were forested and separated by valleys and mangroves, "bird-life was eminently rich," see discussion of Bovallius under Pueblo Nuevo [ca. 0820/8013 ?], 1882 Apr 14, 16, 18-19 (NHRM; Rendahl, 1920; Wetmore, 1946b:28); Wetmore, small sandy beach, rocky point, inland through a stand of low trees with a few winding creepers, to a low ridge, to an extensive mangrove swamp that was completely open beneath tall trees, 1960 Jan 30 (USNM; Olson, 1980b); Tyson, off the island, 1968 Sep 18 (GML, as "Bayoneta Island"); (MacArthur et al., 1972—species list; see comments in Bibliography); Wright and Faaborg, mist nets, observations, 1976 Jun-Aug (Wright et al., 1985); Levings et al., observations, 1982-1983 (Levings et al., 1986). (N. Smith, 1971).

BAYONETA ISLAND; see Bayoneta, Isla

BEJUCO; Panamá (W, Pacific)
0836/7953 (USBGN)
20 m, town 3 km N of Chame [0835/7953 (USBGN)] on the Carretera Interamericana [0832/8242 (USBGN)] (E762; JOG). Wetmore and Perrygo, GL for Puerto El Tigre, 1951 Mar 15 (USNM; Wetmore, 1962; Olson, 1980b, as "Bejuco, Puerto El Tigre"); Quinones, 1962 Jan 10, 1964 Aug 9 (GML).

BELÉM; see Belén

BELÉN; Veraguas/Colón ? (Atlantic)
0853/8052 (USBGN)
Sea level, coastal town in extreme NE Veraguas province or in extreme W Colón province at the mouth of the Río Belén [0853/8052 (USBGN)] 25 km SW of Coclé del Norte [0905/8035 (USBGN)] (PDV; RDP2; Salvin, 1870bb (map)). Merritt indicates that he was "'stationed in the mountainous district of Belén, province of Veraguas, New Grenada'" (Lawrence, 1858:140), which presumably extended from the coast to the continental divide. This district was adjacent to El Mineral de Veraguas [ca. 0830/8100] (Lawrence, 1862a:4). Merritt, collected near camp, near an aquatic environment, 1852 Sep (Lawrence, 1858; Salvin, 1867b:152, as "District of Belén, Veragua"; Salvin and Godman, 1888-1904, vol. 2:268-269). Ridgely and Gwynne's (1989:218) mention of Belém in W Colón province presumably refers to Belén.

BELÉN, RÍO; Veraguas/Colón (Atlantic)
0853/8052 (USBGN)
Originates near the continental divide and flows NNW along the Veraguas/Colón

provincial boundary (JOG; PDV). Handley, at sea 15 km from the mouth of, 1962 Mar 20 (Wetmore, 1963b).

BELÉN DISTRICT; see Belén

BELLA VISTA; Panamá (E, Pacific)

0858/7932 (USBGN)

20 m, section of greater Panama City [0858/7932 (USBGN)] located 1.5 km NE of Calidonia [0858/7933 (USBGN)] (E962). The *Universidad de Panamá* is located in Bella Vista, and unless otherwise noted, specimens were collected on the campus. Tejera et al. (2001) provide details of the campus. Wong, 1977 Oct 18 (MVUP); Tejera, 1978 Aug 7, 1982 Nov 11, 1999 Apr 20 (MVUP); Achurra, Parque Urracá, according to Atlas (1988:183) Parque Urracá is located on the coast, 1979 Oct 14 (MVUP); Chong, 1979 Oct 17 (MVUP); Miguelena, 1980 Apr 20-1 (MVUP); Jackson, 1980 Sep 16 (MVUP); Conte, 1980 Oct 20 (MVUP); Quiros, 1983 Mar 19 (MVUP); Cedeño, 1983 Aug 22 (MVUP); Hernández, 1985 Apr 13 (MVUP); Acosta, 1985 Apr 15 (MVUP); Guerrero, 1985 Apr 16 (MVUP); Cross, 1987 Aug 4 (MVUP); Fabrega, 1990 Jun 1 (MVUP); Bonilla, 1990 Jul 18 (MVUP); Araúz, 1990 Oct 26 (MVUP); Cardenas, 1990 Dec 3 (MVUP); Jiménez, 1991 Aug 22 (MVUP); Tejera, 1994 Mar 22 (MVUP); Jiménez, 1994 Jul 9 (MVUP); Gómez and Sánchez, 1999 Mar 24 (MVUP); Tejera, Dominguez, Pacheco, and Vega, 1996 May 10 (MVUP); Dominguez and Vega, 1996 Jul 21 (MVUP); Pérez, Campines, Cordoba, and Jiménez, 2000 Nov 15 (MVUP); Becarra, 2001 Jul 4 (MVUP); Lara, 2001 Oct 17 (MVUP). See also: Obarrio, Barriado.

BERBÁ; Chiriquí

ca. 0823/8249

ca. 25 m, town 13 km NNE of Puerto Armuelles [0817/8252 (USBGN)] and 8 km SSE of Progreso [0827/8250 (USBGN)] (JOG). Sánchez, *entrada de* (entrance of) Berbá, 1983 Oct 22 (MVUP).

BERMEJO; Coclé (Pacific)

0840/8033 (USBGN)

600 m, town in NW Coclé province 8 km NE of El Copé [0837/8036 (USBGN)] at the headwaters of the Río Bermejo [0838/8034 (USBGN)] (JOG), although SMN2A shows the town 5 mi (8 km) N of El Copé at 2,000 feet (600 m) on the Atlantic drainage at the headwaters of the Río Cascajal [0850/8029 (USBGN)], a tributary of Río Coclé del Norte [0905/8035 (USBGN)]. See also: Vermejo, Montaña del.

BERMEJO, RÍO; Coclé (Pacific)

0838/8034 (USBGN)

Short river that originates on the continental divide and enters the Río Harino [0833/8034 (USBGN)] ca. 3 km E of El Copé [0837/8036 (USBGN)] (JOG). See also: Vermejo, Montaña del.

BERMINGHAM, FINCA; see El Salto

BETANIA; Panamá (E, Pacific)

0901/7932 (USBGN)

80 m, section of NW greater Panama City [0858/7932 (USBGN)] located NNE of El Carmen [0900/7932 (USBGN)] (E762; E8660, as "Bethania"). Nortiz (or N. Ortiz), 1968 Dec 14 (GML). See also: Condado del Rey; El Dorado; La Locería; and Santa María.

BETHANIA; see Betania

BIBALAZ; Chiriquí
ca. 0825/8243 ?
ca. 25 m ?, not shown on our maps, but described by Salvin and Godman (1888-1897, vol. 2:158) as a collecting site of Enrique Arcé on the S slopes of Volcán de Chiriquí [= Volcán Barú [0848/8233 (USBGN)]]. The collection of *Automolus fumosus* from Bibalaz led Wetmore to believe that Bibalaz was N of La Concepción [0831/8237 (USBGN)] within the subtropical zone, at 1,000-1,200 m (USNM), although elsewhere (e.g. Wetmore, 1968:301, 535) he describes Bibalaz as a lowland locality. Arcé probably applied the name to a substantial area W of David and S of the continental divide. The name "Vibalá" is presumed to be an alternate spelling of Bibalaz, and according to Angehr (in litt., Oct 2003), both probably refer to Divalá [0825/8243 (USBGN)] (q. v.). Arcé, 18?? (BMNH; Salvin and Godman, 1888-1904, vol. 2, as "Bibalaz," and as "Vibalá").

BIG CREEK; Bocas del Toro, Archipiélago de Bocas del Toro
0922/8215 (USBGN)
Sea level, settlement on the S coast of Isla Colón [0924/8217 (USBGN)] 3 km N of Bocas del Toro [0920/8215 (USBGN)] (JOG). SBE, GL = Isla Colón, Bocas del Toro, 1 mi (1.5 km) N of Bocas del Toro, 1988 Feb 6, and at La Represa [ca. 0922/8215], 1989 Jan 28 (USNM).

BIG PLANTAIN CAY; Bocas del Toro, Archipiélago de Bocas del Toro
0908/8148 (USBGN)
USBGN treats Plantain Cay as an alternate name for Big Plantain Cay, which is shown on our maps as the larger of two small islands located 2 km off the NE coast of Península Valiente [0905/8151 (USBGN)] (JOG). Although Punta Chiriquí [0911/8153 (USBGN)] is not shown on our maps, the USBGN coordinates indicate that it is located on the N coast of Península Valiente. Wetmore, observations "near the mainland coast from the vicinity of Plantain Cay to Chiriqui Point," 1958 Mar 2 (Wetmore, 1959b:13, as "Plantain Cay").

BIG ZAPATILLA KEY; see Zapatilla, Cayos

BIJAGUALES; Los Santos
0726/8032 (USBGN)
ca. 50 m, town on the Río Tonosí [0721/8022 (USBGN)] 10 km W of Tonosí [0724/8027 (USBGN)] and 11 km E of El Cortezo [0726/8038 (USBGN)] (JOG). See also: El Cortezo.

BIQUE, BAHÍA; Panamá (W, Pacific)
0853/7940 (USBGN)
Bay on the Pacific coast within Bahía de Panamá [0850/7915 (USBGN)] 15 km SW of Panama City [0858/7932 (USBGN)] and 7 km S of Arraijan [0857/7939

(USBGN)]. The bay extends 3 km E-W between Punta Bique [0853/7939 (USBGN)] and Punta Vaca Monte [0852/7940 (USBGN)] (E762). On older maps Bique is spelled Vique (E761A, as "Vique Cove"). Castrellón and Murrillo, 1983 Oct 23 (MVUP, as "Playa Bique").

BIQUE, PLAYA; see Bique, Bahía

BIQUE, PUNTA; see Bique, Bahía

BLANCA, PAJA; see Paja Blanca

BLANCA, PLAYA; Veraguas, Isla de Coiba
0723/8139 (USBGN)
Sea level, sandy beach on the E shore of Ensenada María [0724/8141 (USBGN)] of Isla de Coiba [0728/8147 (USBGN)] (E762). Wetmore, GL = Isla de Coiba, rocky shore NW of Campamento de Playa Blanca [0723/8140 (USBGN)], E along the beach past the camp, trail into woods and pasture, tall mangroves and other trees along the beach, to the E end of the beach near Estero Boca Grande [0724/8139 (USBGN)], 1955 Jan 28 (USNM; Wetmore, 1957); Garcés and Mena, 1996 Feb 28-29 (MVUP). See also: Damas, Bahía.

BLUE MOUNTAIN; see Azul, Cerro

BLUEFIELD, LAGUNA DE; Bocas del Toro
0909/8155 (USBGN)
Bay that extends 7 km E into the NW tip of Península Valiente [0905/8151 (USBGN)] separating the tip into 2 small peninsulas. The bay opens on the WNW to Canal del Tigre [0915/8158 (USBGN)] between Punta Valiente [0911/8155 (USBGN)] on the N and Punta Bluefield [0908/8157 (USBGN)] on the S (E762, as "Laguna de Bluefield (Bahía Azul)"; JOG). Miller and Richardson, GL = Laguna de Chiriquí [0903/8200 (USBGN)], Península Valiente, the house boat that served as their base was anchored in or near Laguna de Bluefield. According to Miller, the lagoon (probably Laguna de Bluefield) was "really a bay more than a lagoon" and was enclosed by sharply elevated hills that were wooded to the water's edge. The tidal fluctuations in the bay created "little in the way of mudflats." They collected on the water from boats and went ashore at the NW end of Península Valiente following trails that had been cut by the U. S. Naval Crew surveying the area, collected in the jungle among banana clearings on a steep hillside, above the landing on a ridge with great trees, in mangroves on the shore, and above a jungle choked ravine near hills on Punta Valiente, traveled "a long way up" a winding creek that was completely enclosed by overhanging mangroves, and collected along the 12 mi (19 km) boat trip S to the mouth of the Río Cricamola [0859/8154 (USBGN)], 1936 Feb 28 (?), 29, Mar 2, 7-8, 10, 16, 18-19, 21, 23, 25, Apr 1-2 (MVZ, as "Chiriquí Lagoon"; UCLA, as "Bluefields," as "Bluefield Creek," as "Bluefield Lagoon," and as "Bluefields Lagoon"); SBE, GL = Península Valiente, collected at the head of, and on the S side of, 1990 Apr 23, 25 (USNM, as "Bahía Azul"). Delgado (1983, as "Bahía Azul").

BLUEFIELD, PUNTA; Bocas del Toro
0908/8157 (USBGN)

Point on the NW coast of Península Valiente [0905/8151 (USBGN)] located 5 km SW of Punta Valiente [0911/8155 (USBGN)] on the peninsula that forms the S coast of Laguna de Bluefield [0909/8155 (USBGN)] (MHA). On many maps Punta Valiente, which is on the northern coast of Laguna de Bluefield, is labeled Punta de Bluefield or Punta Bluefield. See also: Toro, Cayo.

BLUEFIELD, PUNTA DE; see Valiente, Punta

BLUEFIELD CREEK; see Bluefield, Laguna de

BLUEFIELD LAGOON; see Bluefield, Laguna de

BLUEFIELD POINT; see Bluefield, Punta

BLUEFIELDS; see Bluefield, Laguna de

BLUEFIELDS LAGOON; see Bluefield, Laguna de

BOCA, LAGUNA DE LA; see La Boca, Laguna de

BOCA BRAVA, ISLA; Chiriquí

0813/8216 (USBGN)

177 m, irregularly shaped island with a land area of 28 km^2 (2,800 ha) (RDP2) located in the Golfo de Chiriquí [0800/8220 (USBGN)] 30 km SSE of David [0826/8226 (USBGN)] (JOG; MGRP; MHA; MRP, as "I. Brava"). Batty, specimens at AMNH and at BMNH are dated 1902 Jan 26-30, but Batty's data were completely fabricated and his specimens labeled "Brava Island" were actually collected in 1901 near his mainland bases at Boquete [0847/8226 (USBGN)] and Boquerón [0830/8234 (USBGN)] (AMNH, as "Brava Island"; BMNH, as "Brava Island"; Eisenmann, 1950, as "Brava I."; Zimmer, 1950:8, 50, as "Brava Is."; Zimmer, 1955:5, as "Brava Island"; Wetmore, 1957:6, as "Brava Island"; Olson,2008); Duval, 2000 Mar 15, Apr 29 (MVUP).

BOCA CHICA; Chiriquí

0813/8213 (USBGN)

Sea level, coastal town on the mainland 34 km SE of David [0826/8226 (USBGN)] less than 1 km E of Isla Boca Brava [0813/8216 (USBGN)] (JOG; MGRP; MHA). Sclater (1856c:138-139) described Boca Chica as a port at the mouth of the Río David [0822/8224 (USBGN)], but, the Río David enters the Río Chiriquí [0820/8226 (USBGN)] 8 km SSE of David and 26 km NW of Boca Chica.

BOCA DE CAÑITA; Panamá (E, Pacific)

0912/7855 (USBGN)

ca. 50 m, town 10 km ESE of El Llano [0913/7859 (USBGN)] on the Río Chepo [0900/7906 (USBGN)] at the mouth of the Río Cañita [0912/7855 (USBGN)] (RDP; SMN2B). Wetmore, the secondary localities attributed to Boca de Cañita were located in the area extending 5 km SE of town along the Río Chepo, GL for: Canchigua, Juan Pelón, and Río Chepo, 1962 Feb 6-8 (USNM, as "Cañita"; Wetmore, 1970, as "Cañita"). See also: Las Cañitas.

BOCA DE CRICAMOLA; Bocas del Toro

0859/8155 (USBGN)

Sea level, settlement on an island in the mouth of the Río Cricamola [0859/8154

(USBGN)] (E762). The locality has always been written on specimen labels and mentioned in the literature as simply "Cricamola." Wedel's use of "Cricamola" appears to encompass considerable distance, presumably along the Río Cricamola (q. v.), as specimens of both *Ramphocelus passerinii* and *R. icteronotus* are thus labeled. SBE never found the contact between these two species, which lies somewhere between Chiriquí Grande [0857/8207 (USBGN)] and the base of the Península Valiente [0905/8151 (USBGN)]. Wedel, 1927 Aug 13-Sep 22, 1928 Feb 10-Aug 28, 1936 Jun, Aug-Dec, 1937 Feb-Oct, 1938 Jan-Mar, Jul-Aug, Oct-Dec (ANSP; CM; CMNH; FMNH; MCZ; MVZ; UCLA; YPM; Peters, 1931a, as "Cricamola"; Peters and Griscom, 1929; Wetmore, 1959b:8, 10, 19).

BOCA DE CUPE; Darién
0803/7735 (USBGN)
ca. 25 m, town 12 mi (19 km) SE of El Real de Santa María [0808/7743 (USBGN)] on the Río Tuira [0821/7803 (USBGN)] near the mouth of the Río Cupe [0804/7735 (USBGN)] (SMN2B). Goldman, at 250 feet (75 m), 1912 June 14, 18-19 (USNM; Nelson, 1912b:5; Parkes, 1976); Richardson, GL for Río Tuira, 1915 Apr 30, May 2 (AMNH; Anthony, 1916; Chapman, 1917; Chapman, 1929b; Friedmann, 1927, as "Boca de Cupe, Río Tuyra"; Zimmer, 1950:8; Olson, 1981d, as "Boca de Cupe, Río Tuira"); Wetmore, GL for Río Cupé, 1964 Mar 21 (USNM). Specimens examined, collector ?, date ? (Zimmer, 1934:14; Wetmore, 1953; Wetmore, 1952:19). See also: Cana.

BOCA DE ESTERO SALADO; see El Rompio.

BOCA DE GATUNCILLO; Colón (Canal Zone) (E)
0912/7939 (USBGN)
30 m, town on the Río Chagres [0907/7942 (USBGN)] 4.5 km N of Juan Mina [0910/7939 (USBGN)] at the mouth of the Río Gatuncillo [0913/7939 (USBGN)] (E866). Wetmore, GL = Juan Mina, on the E bank of the Río Chagres opposite Boca de Gatuncillo, 1955 Dec 12, GL = Juan Mina, Río Chagres, brush grown bar, side channel, 1959 Jan 28, 30, GL = Juan Mina, to an area known as Bayo de Gustavo (not located), series of sloughs opposite Boca de Gatuncillo, 1961 Jan 12, 17 (USNM, as "Gatuncillo").

BOCA DEL DRAGO; Bocas del Toro, Archipiélago de Bocas del Toro
0925/8220 (USBGN)
Sea level, town on the NW coast of Isla Colón [0924/8217 (USBGN)] on Boca del Drago [0925/8220 (USBGN)] channel (JOG). Handley, Hinds and Rhymer, 1960 Feb 16, 21, 23 (USNM). See also: Drago, Boca del.

BOCA DEL DRAGO BAY; see Drago, Boca del

BOCA DEL RÍO INDIO; Colón (W)
0912/8011 (USBGN)
Sea level, coastal town in W Colón province 7 km W of Salud [0912/8008 (USBGN)] at the mouth of the Río Indio [0912/8011 (USBGN)] (E761B). Described by Wetmore as a small settlement that floods even during the dry season (USNM). Pujals, observations, 1947 Feb 14 (Wetmore et al., 1984:81, as "Rio Indio Village");

Wetmore and Perrygo, GL = Río Indio, mouth of the Río Indio, woodland adjacent to town, 300 yards (270 m) upriver to a trail through forest and old clearings, to the beach and W over a 50-75 foot (15-25 m) rocky headland known as El Morro (not located), its peak covered with green grass and bushes, ca. 1 mi (1.5 km) W to a sandy beach and small streams amid mangroves, 1952 Feb 14-17, 20 (USNM; Parkes, 1969; Olson, 1980a, as "Río Indio, Boca del Río Indio").

BOCA DEL RÍO RISCO; see Risco, Río

BOCA DEL TORO; Panamá (E, Pacific)
ca. 0913/7904
Not shown on our maps, but probably located along the Río Mamoní [0908/7904 (USBGN)] between Charare [ca. 0914/7904] and Chepo [0910/7906 (USBGN)], Wetmore and Perrygo, GL = Chepo, Río Mamoní, 1949 Apr 22 (USNM).

BOCA DEL TORO; see Bocas del Toro

BOCA DE PAJARONCITO; see Pajaroncito, Boca de

BOCA DE PAVARANDÓ; see Sambú, Río

BOCA DE PAYA; see Paya, Río

BOCA DE RÍO PAYA; see Paya, Río

BOCA DE RÍO TERIBE; see Teribe, Río

BOCA DE SÁBALO; Darién
0802/7813 (USBGN)
ca. 25 m, town 17 km ESE of Garachiné [0804/7822 (USBGN)] on the Río Sambú [0805/7818 (USBGN)] 3 km NW of the mouth of the Río Sábalo [0802/7812 (USBGN)] (E501, as "Sabalo"; MGRP). The Vanderbilt Expedition collected on the Río Sambú (q. v.) near Boca de Sábalo.

BOCA DE TOABRÉ; see Toabré

BOCA DE URACILLO; Coclé (Atlantic)
0858/8011 (USBGN)
ca. 50 m, town on the Río Indio [0912/8011 (USBGN)] (q. v.) 25 km S of Boca del Río Indio [0912/8011 (USBGN)] at the mouth of the Río Uracillo [0858/8011 (USBGN)] (JOG, as "La Boca de Uracillo"). Wetmore and Perrygo, GL for: Río Indio, Río Negro, and Río Uracillo, collected within 6 km of town near the three rivers, worked along small creeks and on forested hills, on rolling hills cleared of forest, in cultivated areas and pastures, along cattle trails through saw-grass, on forested slopes above the Río Negro and above the Río Indio, near the Río Negro on the Camino Real (not located) leading through bare hill slopes, *rastrojo*, thickets, stands of second growth forest and blooming orange groves, 1952 Feb 22-Mar 5 (USNM, as "El Uracillo"; Wetmore, 1952:31, as "El Uracillo"; Wetmore, 1953, as "El Uracillo"; Wetmore, 1959b:10, as "El Uracillo"; Wetmore, 1962, as "Río Uracillo, near the town of that name"; Wetmore, 1970, as "El Uracillo"; Huff and Wetmore, 1967, as "El Uracillo"; Olson, 1980a, as "El Uracillo, Río Indio"; Olson, 1981d, as "El Uracillo, Río Negro"; Olson, 1983, as "El Uracillo").

BOCA GRANDE; see Boca Grande, Estero

BOCA GRANDE, ESTERO; Veraguas, Isla de Coiba
0724/8139 (USBGN)
Tidal creek on the SE coast of Isla de Coiba [0728/8147 (USBGN)]. Originates 1 km S of Campamento de Playa Blanca [0723/8140 (USBGN)], flows 4-5 km NE widening to 0.5 km at its mouth E of Playa Blanca [0723/8139 (USBGN)] (q. v.) (E762). Wetmore (1957:9, as "Boca Grande"). See also: Blanca, Playa.

BOCA GRANDE, ISLA; Darién
0826/7809 (USBGN)
Sea level, island located 4 km N of La Palma [0825/7809 (USBGN)] in the mouth of the Río Tuira [0821/7803 (USBGN)] (MGRP). Eisenmann, observations, 1967 Mar 29 (Wetmore et al., 1984:269).

BOCA LA CAJA; Panamá (E, Pacific)
0859/7930 (USBGN)
Sea level, section of greater Panama City [0858/7932 (USBGN)] immediately S of San Francisco de la Caleta [0859/7930 (USBGN)] on the coast 1.5 km NE of Punta Paitilla [0858/7931 (USBGN)] (E762). Garrido, 1991 Aug 4 (MVUP); Guerra and Pérez, 1991 Sep 10 (MVUP).

BOCA RÍO CATIVÁ; see Catival, Río

BOCA RÍO CATIVAL; see Catival, Río

BOCAS, ISLA; see Colón, Isla

BOCAS DEL DRAGO; see Boca del Drago

BOCAS DEL TORO; Bocas del Toro, Archipiélago de Bocas del Toro
0920/8215 (USBGN)
Sea level, the capital of Bocas del Toro province, located on the S coast of Isla Colón [0924/8217 (USBGN)] (JOG). The town or perhaps the entire province was attributed to Costa Rica by early collectors (Gould, 1861a, as "Bocca del Toro, Costa Rica"). The use of "Bocas del Toro" on specimen labels is frequently ambiguous, as at times it is uncertain whether the town or the province is intended. Compounding the uncertainties is the fact that some collectors used "Bocas del Toro" to mean Isla Colon. Verreaux and Verreaux, c. 1853 (BMNH; Verreaux and Verreaux, 1853:194, as "Bocos del Toro (Nouvelle-Grenade)"; Gould, 1861a, as "Bocca del Toro in Costa Rica"; Griscom, 1928, as "Bocca del Toro, Costa Rica (now western Panama)"; Eisenmann and Howell, 1962, as "The Bocca del Toro in Costa Rica [= Bocas del Toro, Panama]"); Abbott, 1887 May 5 (ANSP); Collector ? [Abbott ?], 1887 May 1, 5 (ANSP); Smith, these dates overlap those listed under Chiriquí Grande [0857/8207 (USBGN)]) and include many species not found on Isla Colón, 1927 May 3-5 (YPM; Griscom, 1933); Benson, 1927 (Griscom, 1928, as "Boca del Toro," the text also refers to the region as the "Caribbean Slope of Western Panama"; Peters, 1931a); Wedel, 1927 Aug, Oct, Nov, 1936 Jun-Sep, 1937 Apr, Jun, Aug, Oct, 1938 Mar, May-Dec, 1939 Feb (CMNH; FMNH; MCZ; MVZ; Peters, 1931a); Eisenmann, observations, "spent a couple of hours in the

open scrubby second growth immediately back of town," 1956 Jul 4 (Eisenmann, 1957); Loftin, observations, the bay near, 1963 May 18-21 (Loftin and Olson, 1963); Collector ?, 1980 May (UF, as "Bocas del Toro Airport"); SBE, GL = Isla Colón, unmodified and 1 mi (1.5 km) N of, 1988 Feb 7, 1989 Feb 27, and as a GL for: Bahía Almirante, Big Creek, Botel Thomas, Paunch, and Sandfly Bay, 1987 Mar 12, 1988 Feb 6-7, 17, 1989 Mar 6, 1990 Apr 15-16 (USNM; Hekstra, 1982; Olson, 1993a).

BOCAS DEL TORO, ARCHIPIÉLAGO DE; Bocas del Toro
0920/8210 (USBGN)
Group of more than 100 islands and islets located very close to the NW coast of Bocas del Toro province. The islands are located between Punta Terraba [0926/8220 (USBGN)] and Península Valiente [0905/8151 (USBGN)] N and NW of Laguna de Chiriquí [0903/8200 (USBGN)] and they form numerous bays and channels between the mainland coast and the Caribbean Sea [1500/7500 (USBGN)]. The larger islands in the group include Cayo Agua [0909/8202], Isla Bastimentos [0918/8208 (USBGN)], Isla Colón [0924/8217 (USBGN)], Isla Popa [0911/8207 (USBGN)], and Isla San Cristóbal [0915/8216 (USBGN)] (MGRP). See also: García, Cayo; Pastores, Isla; Roldán, Cayo; Tiger, Cayos; and Zapatilla, Cayos.

BOCAS DEL TORO, PROVINCIA DE; (Atlantic)
0850/8210 (USBGN)
The most northwestern province in the country, extending along the Caribbean coast from the Río Calovébora [0848/8113 (USBGN)] on the Veraguas provincial boundary in the E to the Costa Rican border on the W. The northern, lowland coastal regions rise to elevations above 2,000 m along the continental divide, which marks the border with both Veraguas and Chiriquí provinces in the S. Its major physiographic feature is a large embayment, the Bahía Almirante [0920/8218 (USBGN)] and the Laguna de Chiriquí [0903/8200 (USBGN)], replete with islands, the Archipiélago de Bocas del Toro [0920/8210 (USBGN)], and the only real settlements – as well as scientific study for that matter – have been in this area. Despite a visit by Christopher Columbus in 1502 and official control by the Spaniards, many early settlers, merchants and traders seeking turtles, mahogany and other local treasures, were from Britain, the Caribbean islands and the United States. Bocas del Toro [0920/8215 (USBGN)], the capital of the province, was founded on Isla Colón [0924/8217 (USBGN)] in 1826 by immigrants from Jamaica, Providencia, and San Andres and the economic importance of the port grew with the development of new economic sectors. In the 1890s The United Fruit Company, among other agricultural companies, developed large plantations at Changuinola [0926/8231 (USBGN)] and following the construction of the Panama Canal [0920/7955 (USBGN)], took on the recently unemployed Caribbean laborers, including many Jamaicans. The apparent Spanish uninterest in the region encouraged a variety of British, Caribbean, and U. S. activities in Bocas del Toro, with the result that nowhere else in the country is English so prevalent, either in everyday speech or in geographical names. The province scarcely contains an anthill without both a Spanish and an English name, each of which may be equally well used and understood. There is

practically no exception to this. As an example, Tierra Oscura, where SBE collected in 1989 appeared to have no antecedents in ornithology, until a specimen was found in the Cincinnati Museum collected by H. von Wedel that was labeled, almost illegibly, "Darkland," the English equivalent. Because both languages are well understood and commonly used in the province, many maps note the English name along with the Spanish name, and the labels on ornithological specimens reflect this blend of the languages. However, to maintain a consistent format throughout the gazetteer, official names have been used whenever they were available and all other variations of the names have been cross-referenced accordingly. See also: Bocas del Toro.

BOCATORITO; Bocas del Toro, Archipiélago de Bocas del Toro
0914/8216 (USBGN)
Sea level, settlement on the SE coast of Isla San Cristóbal [0915/8216 (USBGN)] on the NW coast of Laguna Bocatorito [0914/8215 (USBGN)] 1 km NE of Cerro de Bocatorito [0913/8216 (USBGN)] (JOG). SBE, GL = Isla San Cristóbal, camp was located ca. 1 km W of the settlement in a lowland area inland from the coast between cacao plantations to the S and second growth forest to the N, collected near camp, 1 mi (1.5 km) N and NE of camp, 1989 Jan 30-31, Feb 3-16, 20, 24, Mar 3, 15 (MVUP; USNM).

BOCATORITO, CERRO; see Bocatorito, Cerro de

BOCATORITO, CERRO DE;
0913/8216 (USBGN)
Bocas del Toro, Archipiélago de Bocas del Toro
95 m, peak on the S coast of Isla San Cristóbal [0915/8216 (USBGN)] just N of Laguna de Tierra Oscura [0912/8216 (USBGN)] (E762). SBE, GL = Isla San Cristóbal, on the S shore of Laguna Bocatorito [0914/8215 (USBGN)] at the N foot of the peak, 1989 Feb 7-8, 11 (USNM, as "Cerro Bocatorito").

BOCATORITO, LAGUNA; Bocas del Toro
0914/8215 (USBGN)
The more eastern of two well-protected bays between the S coast of Isla San Cristóbal [0915/8216 (USBGN)] and the mainland coast. The S shores of Laguna Bocatorito are formed by the mainland where the bay extends 2 km from E-W, the N boundary of the bay is formed by the SE coast of Isla San Cristóbal and several small mangrove islands off its E coast (E762; JOG, as "Laguna Porras"; MGRP, as "Laguna Porras"). See also: Bocatorito; and Green, Bahía de.

BOCATORITO CAMP; see Bocatorito

BOCA TORO, ISLA, see Toro, Isla

BOCCA DEL TORO; see Bocas del Toro

BOCOS DEL TORO; see Bocas del Toro

BODEGA, ENSENADA DE; Panamá, Archipiélago de las Perlas
0816/7908 (USBGN)
Cove centrally located on the W coast of Isla San José [0815/7907 (USBGN)] (JOG), and described by Wetmore (1946b:2, as "Ensenada de la Bodega") as the

principal bay on the W side of the island. See also: Mata Puerco, Río.

BODEGA, ENSENADA DE LA; see Bodega, Ensenada de

BOFUVA; see Bugaba

BOGABA; see Bugaba

BOGAVA; see Bugaba

BOGAVE; see Bugaba

BOGOBA; see Bugaba

BOHÍO; Colón (Canal Zone)

ca. 0912/7952

Former station on the original Panama Railroad [ca. 0922/7954] that was also known as Bohío Soldado or the "soldier's hut." The station, now submerged in Lago Gatún (Gatun Lake) [0912/7955 (USBGN)], was located 7.5 mi (12 km) SE of Gatún [0916/7955 (USBGN)] on the continuous Río Chagres (see Río Chagres [0919/8000 (USBGN)]) near what is now Punta Bohío [0911/7952 (USBGN)] (PR; IP, as "Bujio Soldado"). Goldman, across the Río Chagres at a pond with a dense growth of white pond lilies, 1911 Apr 29, 1912 Feb 2 (USNM; Olson, 1981a; Olson, 1981d).

BOHÍO, PENÍNSULA; Colón (Canal Zone) (E)

0912/7950 (USBGN)

121 m, 1 km wide peninsula extending 5 km WSW from the mainland into Lago Gatún (Gatun Lake) [0912/7955 (USBGN)] S of Islas Juan Gallegos [0913/7952 (USBGN)] and N of Isla Barro Colorado [0909/7951 (USBGN)] (E661). Willis, observations, "tall secondary forest," 1960-1970 (Karr, 1971a; Willis, 1966a; Willis, 1967; Willis, 1972:36); White, GL = Frijoles, 6 mi (10 km) NNW Frijoles, 1972 May 3 (ANSP, as "Bahia Peninsula").

BOHÍO, PUNTA; Colón (Canal Zone) (E)

0911/7952 (USBGN)

50 m, point on the W coast of Península Bohío [0912/7950 (USBGN)] in Lago Gatún (Gatun Lake) [0912/7955 (USBGN)] 6 km WNW of Frijoles [0910/7948 (USBGN)] (E866, as "Bohío Point"). See also: Bohío.

BOHÍO PENÍNSULA; see Bohío, Península

BOJAVA; see Bugaba

BOLAÑOS, ISLA; Chiriquí

0805/8216 (USBGN)

55 m, small island with a land area of 0.7 km^2 (70 ha) located in the Golfo de Chiriquí [0800/8220 (USBGN)] 7 km E of Isla Parida [0807/8220 (USBGN)] (JOG; Olson, 1997a). Wetmore, sandy beach on the NE coast where the land rose steeply to 100-110 feet (30-33 m), up through a broad valley with low, open brush between large trees with vines and tangled undergrowth at higher elevation, landed at a rock shelf on the central E coast, tall, gallery forest with open undergrowth, 1963 Feb 1, 6 (USNM; Olson, 1980b; Olson, 1997a).

BOLAÑOS, ISLA; Panamá, Archipiélago de las Perlas

0832/7902 (USBGN)

Sea level, small island with a land area of 21 ha (Wright et al., 1985), located 11 km NW of Isla del Rey [0822/7855 (USBGN)] (E501; JOG, as "Isla Bolaño"). Described by Rendahl (1920:4) as a "small, low-lying," but wooded island. Wright and Faaborg, mist nets, observations, 1976 Jun-Aug (Wright et al., 1985); Brawn et al., mist nets, 199? (Brawn et al., 1996).

BOLCAN; see Barú, Volcán

BOLIVAR, CARRETERA; Colón (Canal Zone) (E)

0918/7953 (USBGN)

Road that extends NNE-SSW from Colón [0922/7954 (USBGN)] along the East Diversion [0920/7953 (USBGN)], through Fort William D. Davis Army Reservation [0917/7953 (USBGN)] and New Gatún [ca. 0917/7955] and terminates at Gatún [0916/7955 (USBGN)] (E761A). See also: Mindi.

BOLIVAR HIGHWAY; see Bolivar, Carretera

BOLLO, CERRO; Chiriquí

ca. 0830/8146

Not shown on our maps but Pine and Izor's specimen labels indicate that Cerro Bollo is 3.5 km E of Escopeta Camp [ca. 0830/8148], and ca. 23 km NNE of San Félix [0817/8152 (USBGN)]. This description suggests that they worked just W to SW of Cerro Santiago [0833/8144 (USBGN)]. Pine and Izor, GL = San Félix, at 1,800-1,850 m, 3.5 km E of Escopeta Camp, specimens are labeled Bocas del Toro/Chiriquí provincial border, but it is it not clear whether they worked on the divide, see San Félix for further discussion, 1980 Jun 16, 26, 29, Jul 3 (USNM). See also: Flores, Cerro; and Santiago, Cerro.

BONÁ, ISLA; Panamá (Pacific)

0834/7936 (USBGN)

205 m, uninhabited island with a land area of 0.83 km^2 (83 ha) located in Bahía de Panamá [0850/7915 (USBGN)] 42 km S of Panama City [0858/7932 (USBGN)] and 2 km S of Isla Otoque [0836/7936 (USBGN)] (JOG; Angehr, 2003:233). The island is an Important Bird Area (Angehr 2003). Angehr (in litt., Aug 2005) states that "based on my recent visit there, Boná is well forested, with scrub on the steeper cliffs, rather than 'covered with herbaceous vegetation, except for scrub on the steeper cliffs'as I have it in the IBA Directory (Angehr 2003:233)." Wetmore, observations at sea near the island, 1956 Feb 6 (Wetmore, 1957:32), 1962 Mar 31 (Wetmore, 1965:63); Aid, Montgomery, and Mock, 1983 Apr 3, 24, 1984 Jan 18 (MVUP; Aid et al., 1985). See also: Iguana, Isla; Mala, Punta; and Otoque, Isla.

BONITA, QUEBRADA; Colón (Canal Zone) (E)

0920/7936 (USBGN)

Originates N of Lago Alajuela (Madden Lake) [0915/7935 (USBGN)] and flows S entering the NW coast of the reservoir ca. 1 km W of Salamanca [0920/7935 (USBGN)] (E8660). Van Horn, 1966 Nov 16 (GML, as "Q. Bonito").

BONITO, QUEBRADA; see Bonita, Quebrada

BOQUERÓN; Chiriquí
0830/8234 (USBGN)
200 m, town 5 km E of La Concepción [0831/8237 (USBGN)] and 3 km N of the Carretera Interamericana [0832/8242 (USBGN)] (JOG; MHA). Batty, 1901 Sep 27-Dec 29, although Batty may have left before Dec 29 as there are few specimens with dates after Dec 17, (AMNH; FMNH; USNM; Miller and Griscom, 1921; Griscom, 1927a; Griscom, 1933; Zimmer, 1941b:14; Zimmer, 1950:50; Eisenmann, 1950; Wetmore, 1952:16; Wetmore, 1957:6; Parkes, 1976; Olson, 1981d; Olson, 2008).

BOQUERÓN, RÍO; Colón/Panamá (Canal Zone) (E, Atlantic)
0921/7933 (USBGN)
Originates SW of Nombre de Dios [0935/7928 (USBGN)] flowing S along the Colón/Panamá provincial boundary, enters the former Canal Zone just below Estación Hidro el Peluca [0923/7934 (USBGN)] and discharges into the N end of Lago Alajuela (Madden Lake) [0915/7935 (USBGN)] 2 km W of the mouth of the Río Pequení [0921/7933 (USBGN)] (E661). Wetmore, GL for Manganese Mine, 1957 Mar 3, GL = Estación Hidro el Peluca, collected up- and downstream on the river within 4 km of Estación Hidro el Peluca, specimens were labeled either Panamá or Colón, although some of those collected below the station must have come from what was then the Canal Zone. The Colón side of the river had been logged 2-3 years previously. Collected upriver on gravel beaches, in areas with steep banks that were largely forested, near clearings for corn, along an old, overgrown road, along small streams, in logging areas, upriver to Charco Número Uno [ca. 0923/7933] (q. v.), waterfalls and beaches with water worn rock, along trails above Charco Número Dos [ca. 0924/7933] (q. v.), a dry creek, border of the rocky river with occasional short beaches, upriver to Campamento Número Uno [ca. 0924/7933] (q. v.) where the river banks were lower. Collected downriver in small streams, on gravel beaches partly grown with willows and partly open, on the banks in undergrowth and in high trees, along an old bulldozer cut and a large deforested area with a few tall trees and second growth, along a deep area of the river with rock exposures, 1961 Feb 18-27 (USNM; Wetmore, 1962; Wetmore, 1963a; Olson, 1980a).

BOQUETE; Chiriquí
0847/8226 (USBGN)
1,200 m, city in NW Chiriquí province, 12 km ESE of Volcán Barú [0848/8233 (USBGN)] (JOG). Described by Sclater (1856c:138, as "Boqueti") as an "elevated savannah of about 4,000 feet (1,200 m) above the sea-level, lying on the western slope of the volcano of Chiriquí [= Volcán Barú]." In early writings the name was often spelled "Boqueti," and more recently the name "Bajo Boquete" appears on both maps and databases, although Bajo Boquete [0846/8226 (USBGN)] is a town just S of Boquete. Collector ?, at 1,000-4,000 feet (300-1,200 m), date ?, 1875, 1883 May 20, 1900 Feb 1, Dec 11, 1903 May 2, 1905 Mar 4, Apr 23, Oct 7, 1913 Apr 20, Dec 16 (ANSP, as "Boquete," and as "Boquite"; CU; FMNH; OC; YPM); Collector ?, [Watson ?, prepared by Arcé ?], at 3,500 feet (1,050 m), date ? (USNM); Arcé, date ? (BMNH; Salvin and Godman, 1879-1904, vol. 1:91, as "Boqueti"); Coale, H. K., date ? (MVZ); Fijereta, date ? (FMNH, as "Bajo Boquete");

Underwood, date ? (FMNH); Bridges, T., at ca. 4,000 feet (1,200 m), savanna, dense forest, tree tops, at the base of Volcán Barú, malvaceous plants, *Salvia*, and *Quercus*, 1856 Jan-Mar (BMNH; Sclater, 1856c, as "Boqueti"), Brown, at 3,500-7,000 feet (1,050-2,150 m), described by Bangs (1902a:16) as a "small village on the southern slope of the volcano [Volcán Barú], in the mountain forest zone," although Boquete is E of the volcano, Olson (1981c:365) suggests that Bangs used the term "south" to indicate that the town was on the Pacific drainage vs. the northern Caribbean drainage, 1901 Jan 15-Apr 28 (ANSP; CM, as "Bajo Boquete"; MCZ; UCLA; USNM; Bangs, 1902a; Bangs, 1903c, as "Boquete, Volcán de Chiriquí"; Bangs, 1906b; Bangs, 1908; Ridgway, 1902a; Ridgway, 1903b; Nelson, 1904; Nelson, 1912b:22; Thayer and Bangs, 1906; Bangs and Penard, 1921:380, 391; Griscom, 1930; Griscom, 1934; Davidson, 1938); Batty, is labeled April 14, but Batty's correspondence puts him in David on that date, Batty probably ranged considerable distances from his base in Boquete, 1901 Apr 19-Sep 24 (AMNH; CM, as "Bajo Boquete"; CU; FMNH; MCN; UCLA; USNM; Chapman, 1915a; Van Rossem, 1934a; Wetmore, 1957:6; Olson, 2008); Allen, at 4,500 feet (1,350 m), 1902 Sep 21 (MCZ); Smith, A., at 3,500 feet (1,050 m), 1903 (OC); Watson, at 3,500-4,500 feet (1,050-1,350 m), 1903 Jan, Mar-May, Oct-Nov, 1904 May, Nov-Dec, 1905 Jan-Apr, Nov-Dec (AMNH; ANSP; CM, as "Bajo Boquete"; CU; FMNH; MCZ; MVZ; OC; UCLA; USNM; YPM; ZSM; Hellmayr, 1910b; Hellmayr, 1911b; Phillips, 1969:615); Rosenberg (collection) [Watson ?], at 4,000-6,000 feet (1,200-1,850 m), date ?, 1903 May 13, 1905 Mar 24, May 6 (ANSP; NHRM; OC; YPM); Palmer, at 3,500 feet (1,050 m), 1904 Dec 6 (ANSP); Watson [in the database as Wedel], 1905 Oct 4 (YPM); Maxon, at 1,300 m, above, 1911 Mar 6, 19 (USNM); Kennard, 1926 Mar 14 (YPM); Smith, 1927 Aug 16 (YPM); Benson, GL for: Río Caldera, and Quiel, 1931 Oct 24-Dec 20 (ANSP; USNM; WFVZ; Griscom, 1934; Davidson, 1938; Olson, 1981c, as "Boquete, Palo Alto," as "Boquete, Quiel," and as "Boquete, Río Caldera"); Weber, 1932 Mar 17 (YPM); Mönniche, at 5,000-9,500 feet (1,500-2,900 m), as "Boquete R." [= Río Caldera [0837/8221 (USBGN)] ?], 1932-1934, 1936-1937, GL for: Jaramilla, and Cordillera Los Andes, 1930, 1933, 1938-39 (ANSP; FMNH; GML; Blake, 1958); Davidson, GL for: Alta Peña, Bajo Mono, Casita Alta, Cerro Azul, Chiquero, Horqueta, Peña Blanca, Quiel, Río Caldera, and Salta, 1933 Jan 9, 16, 19-22, 24-27, Dec 6-8, 11, 14-19, 22, 24-25, 28-29, 1934 Jan 9-19, 24-26, 29, Feb 1, 4-6, 8-10, 12-13, 15 (CAS; CM; Davidson, 1938); Hartmann, at 4,100-11,200 feet (1,250-3,400 m), 1937 Jun 23, 30, Jul 17, 19-22, 27-Aug 3, 8-9 (ANSP); Worth, at 5,400-10,200 feet (1,650-3,400 m), GL for: Finca Lérida and Velo, 1937 Jul 16-18, Aug 2-5 (ANSP; Worth, 1939a; Worth, 1939b); Enders, at 10,000 feet (3,050 m), 1937 Jul 23 (ANSP); Hartman, GL for Casita Alta, 1949 Feb 9, 14 (USNM); Gallo, 1949 Apr 12 (LSUMZ); Morton, 7 mi (11 km) NW, 1965 Dec 26, 28 (AMNH); Olson, at 6,000 feet (1,850 m), above, 1966 Jul 23 (UF; USNM); Wetmore, 3 mi (5 km) S on the road, 1960 Mar 18. The secondary localities attributed to Boquete were located within 10 km of town, GL for: El Salto, Palo Alto, and La Estrella, 1960 Mar 15-18 (LSUMZ; USNM); Larson, 1964 Jun (GML); Hartmann, R., 1970 Feb 5 (GML); Collector ?, 1971 Nov 1 (GML). (Oberholser, 1904; Chapman, 1915b; Chapman, 1917; Chapman, 1940; Hellmayr, 1918; Griscom, 1924b; Griscom, 1927a; Peters and

Griscom, 1929; Moore and Peters, 1939; Zimmer, 1939; Zimmer, 1941b:10, 14; Zimmer, 1950:15, 26, 50; Loetscher, 1952; Parkes, 1954a; Wetmore, 1952:16; Wetmore, 1967; Wetmore, 1970; Parkes, 1976; Olson, 1981c). See also: Boquete Trail.

BOQUETE, CAMINO A; see Cerro Punta-Boquete Trail

BOQUETE DE CHITRA; see Chitra

BOQUETE RIVER; see Caldera, Río

BOQUETE TRAIL; Bocas del Toro

ca. 0857/8207

The term Boquete Trail appears to refer to three different routes that terminate in Boquete [0847/8226 (USBGN)]. One route is now called the Cerro Punta-Boquete Trail and is listed as a separate entry. The entire length of this route is at relatively high elevations within Chiriquí province, and as its name implies, it connects Cerro Punta [0852/8235 (USBGN)] and Boquete. The Boquete Trail described later in this entry connects Chiriquí Grande [0857/8207 (USBGN)] with Boquete and rises from sea level at Laguna de Chiriquí [0903/8200 (USBGN)] in Bocas del Toro province to ca. 6,000 feet (1,850 m) where it crosses both the continental divide and the provincial border. The third route was described by Blake (1958), who identifies the Holcomb Trail as an alternate name for the Boquete Trail. However the small portion of this route identified on his map does not match the Boquete Trail as described by Kennard and Peters (1928) or by Peters (1931). In addition Blake indicates that the Holcomb Trail was constructed c. 1910, although Davidson (1936:232) indicates that Brown collected along a Boquete Trail that terminated in Chiriquí Grande as early as 1901, further suggesting two distinct routes. Therefore, because it appears to be a second route from Boquete across the continental divide into Bocas del Toro province, the Holcomb Trail is also listed as a separate entry. Peters (1931:294) described the Boquete Trail as a footpath that connected Chiriquí Grande [0857/8207 (USBGN)] in Bocas del Toro province with Boquete [0847/8226 (USBGN)] in Chiriquí province. According to Kennard and Peters (1928:444-445) and to Peters (1931:294), the trail extended inland from Chiriquí Grande to Chiriquicito [ca. 0857/8211] and from here continued first along the Río Guarumo [0900/8211 (USBGN)], then to Buena Vista [0850/8212 (USBGN)], to Camp Feliz [ca. 0848/8215] and finally crossed the continental divide and the Bocas del Toro/Chiriquí provincial border at ca. 6,000 feet (1,850 m) 12-15 mi (19-24 km) E of Boquete [0847/8226 (USBGN)]. Although not shown on our maps, E762 indicates that the trail probably extended ca. 11 km W and SW from Chiriquí Grande to Punta de Peña [0855/8211 (USBGN)] on the Río Guarumo where it met a foot trail now labeled "Camino Tres de Noviembre." According to E762, Camino Tres de Noviembre continues S first along the Río Guabo [0854/8211 (USBGN)] then along its tributaries to Buena Vista. From here it leaves the river banks and continues SSW crossing the continental divide 20 km E of Boquete at 1,900 m, consistent with the descriptions given by Kennard and Peters. From the continental divide the trail continues S and SSW along the upper reaches of the Río Chiriquí [0820/8226 (USBGN)], where several foot trails appear, one of which extends NW to Boquete.

Tim Billo (in litt to Olson 21 June 2007), who hiked the Holcomb Trail (q.v.) in 2007 made the following observations on the Boquete trail (accents added): "The [present gazetteer] descriptions . . . of the Río Guabo and Río Guarumo suggest that at least the first part of the C[hiriquí] G[rande]-Boquete trail paralleled the location of the road today that goes up to the divide. . . . .It must have then left this river valley and cut across on a ridge system above Coclecito [0849/8033 (USBGN)] to the headwaters of the [Río] Romero [not in USBGN] and [Río] Changuinola [0928/8227 (USBGN)] . . . . . There is another trail from Coclecito that heads uphill to the south and joins a higher elevation trail. This must be the trail coming from Chiriqui Grande. I did not see where this trail enters Boquete. Supposedly this trail crosses the headwaters of the Río Romero, but is dangerous in the wet season. It must also cross the headwaters of the Changuinola before reaching Boquete. It is also supposedly more forested, following ridges and crossing steeper terrain less suitable for ranching." Specimens that were collected on the Caribbean Slope of Volcán de Chiriquí [= Volcán Barú] were probably collected in Bocas del Toro province on the Boquete Trail (Davidson, 1936:232). Brown, at 700-4,000 (?) feet (210-1,200 m), 1901 Jun 10-16 (USNM, as "Caribbean Slope"; MCZ, as "Caribbean Slope"; Bangs, 1902a, as "Caribbean Slope of Volcán de Chiriquí"; Davidson, 1936:232, as "Caribbean Slope of Volcán de Chiriquí"; Olson, 1981c, as "Caribbean Slope"); Kennard et al., at 1,000-4,500 feet (300-1,350 m), 1926 Mar 8-23, 27-28, 30 (MCZ; YPM, as "Boquete"; Kennard and Peters, 1928; Griscom, 1935b:300, as "Caribbean Slope of Volcán de Chiriquí"); Wedel, at 1,300-1,900 feet (400-600 m), 1928 Mar 12, 14, 16-17, 19, 20-25, Apr 1, at 3,000-3,500 feet (900-1,050 m), 1928 Mar 27-30 (ANSP; MCZ; Peters, 1931a). (Friedmann, 1935; Eisenmann and Howell, 1962; Olson et al., 1968; Rabenold, 1980). See also: Gualaca-Chiriquí Grande Road.

BOQUETI; see Boquete

BOQUETI DE CHITRA; see Chitra

BOQUISCO; Bocas del Toro

0920/8231 (USBGN)

20 m, town located 11 km S of Changuinola [0926/8231 (USBGN)] on a bend of the Río Changuinola [0928/8227 (USBGN)] in the foothills S of the coastal lowlands (E762). Braun, M., and McDonald, GL for Río Changuinola, 1994 Apr 20-23, 25-28 (USNM).

BÓQUITA, ISLA; Chiriquí

0817/8221 (USBGN)

109 m, eastern section of a single island that contains Isla Bóquita, Isla Sabaneta [0817/8223 (USBGN)] and Isla Sabino [0816/8224 (USBGN)], although the narrow isthmuses connecting the island sections are omitted on many maps. Isla Bóquita has a land area of ca. 12 km^2 (1,200 ha) and is located 20 km SSE of David [0826/8226 (USBGN)] in Bahía de Charco Azul [0812/8245 (USBGN)] at the mouth of the Río Chiriquí [0820/8226 (USBGN)] (JOG; MGRP; MHA, as "Isla Palenka"; MRP, as "I. Bóquita"). Batty, specimens at AMNH are dated 1902 Jan 5-6, but Batty's data were completely fabricated and his specimens labeled "Palenque Island"

were actually collected in 1901 near his mainland bases at Boquete [0847/8226 (USBGN)] and Boquerón [0830/8234 (USBGN)] (AMNH, as "Palenque Island"; Olson, 2008).

BORACHO, LOMA; Colón (Canal Zone) (W)
ca. 0917/7955
Wetmore indicated this locality on our maps, but it is uncertain that it was ever used on specimen labels or in publication. Located 3 km NNW of Gatún [0916/7955 (USBGN)] and ca. 1 km S of Bahía Limón [0920/7956 (USBGN)].

BORINQUÉN, CARRETERA;
0901/7938 (USBGN)
Panamá (Canal Zone) (W, Atlantic, Pacific)
Road that extends along the W side of the Panama Canal [0920/7955 (USBGN)] between Fort Kobbe Army Reservation [0854/7935 (USBGN)] and the mouth of the Río Mandinga [0906/7942 (USBGN)]. The USBGN coordinates indicate that the road terminates ca. 6 km SE of Empire Known Distance Range [0904/7940 (USBGN)], although E8660 indicates that this is where the road changes from an all weather road to a dry season road that continues N through Empire Known Distance Range terminating ca. 4 km NW of the range. The road is also called Road K2 (E8660; RM1, as "Borinquén Road," and as "Road K2"). MCU, GL = Empire Known Distance Range, 1955 Aug 29 (GML, as "K-2 Road"; USNM, as "K-2 Road"). See also: Empire Known Distance Range.

BORINQUEN ROAD; see Borinquén, Carretera

BORRELA; see Borrola

BORROLA; Herrera
0749/8038 (USBGN)
100 m, town in Distrito Pesé [0754/8037 (USBGN)] 10 km SSW of Pesé [0754/8037 (USBGN)] (JOG, as "Borrela"; MGRP). Rodríguez, date ? (MCN, as "Borrola de Pesé").

BORROLA DE PESÉ; see Borrola

BOSQUE COMÚN; see Macaracas

BOSQUE DE CHONGO; Chiriquí
ca. 0818/8300
ca. 400 m, located near the Costa Rican border in the region ca. 16 km W of Puerto Armuelles [0817/8252 (USBGN)], N of the head of the Río San Bartolo [0817/8251 (USBGN)] and S of the town Quebrada Sábalo [ca. 0820/8300]. Araúz and Ponce, *entre la cabecera del Río San Bartolo y Quebrada Sábalo* (between the head of the Río San Bartolo and Quebrada Sábalo), 1996 May 3 (MVUP).

BOSQUE PROTECTOR DE PALO SECO; see Palo Seco Bosque de Protección

BOTEL THOMAS; Bocas del Toro, Archipiélago de Bocas del Toro
ca. 0920/8215
Sea level, waterfront hotel in Bocas del Toro [0920/8215 (USBGN)] town, SBE, GL = Bocas del Toro, 1990 Apr 15 (USNM).

BÓVEDAS, LAS; see Las Bóvedas

BOYARENA, ISLA; see Bavarena, Isla

BOYD-ROOSEVELT, CARRETERA;
0917/7943 (USBGN)
Colón/Panamá (Canal Zone) (E, Atlantic, Pacific)
Road that connects Colón [0922/7954 (USBGN)] and Panama City [0858/7932 (USBGN)] and extends parallel to, but 10-20 km E of the Panama Canal [0920/7955 (USBGN)]. The road crosses the Río Chagres [0907/7942 (USBGN)] 0.5 mi (1 km) S of Represa Madden (Madden Dam) [0913/7937 (USBGN)]. The road is also labeled Road R20 in Colón province and Road C20 in Panamá province (E8660; JOG; RM1, as "Boyd Roosevelt Highway"). Referred to as the "Transisthmian Highway" by Wetmore and on E8660. See also: Chagres, Río.

BOYD ROOSEVELT HIGHWAY; see Boyd-Roosevelt, Carretera

BOY SCOUT CAMP; see Madden, Represa

BRAVA, ISLA; see Boca Brava, Isla

BRAVA, PUNTA; Veraguas (Pacific)
0736/8114 (USBGN)
Sea level, a point on the southernmost mainland of the W shore of Golfo de Montijo [0740/8107 (USBGN)] (JOG; Wetmore, 1957:13).

BRAVA ISLAND; see Boca Brava, Isla

BRAZO, EL; see El Brazo

BREWSTER, CERRO; Panamá/San Blas (Atlantic, Pacific)
0921/7915 (USBGN)
Peak located on the continental divide ca. 35 km NW of El Llano [0913/7859 (USBGN)] and ca. 50 km NE of Panama City [0858/7932 (USBGN)] (MHA, as "C. Brewster"; SMN2A, as "C. Brewster"). The elevation is given as 920 m on MHA and SMN2A, but our more recent maps (JOG; MGRP) indicate that the elevation in this area is ca. 600 m. Blake (1989) indicates that the elevation is 850 m. Blake, mist nets, premontane rain forest, 1985 Apr 22-26 (Blake, 1989; Ridgely and Gwynne, 1989:215).

BRIDGE OF THE AMERICAS; Panamá (Canal Zone) (Pacific)
ca. 0856/7933
Bridge on the Carretera Interamericana [0832/8242 (USBGN)] that crosses the Pacific entrance of the Panama Canal [0920/7955 (USBGN)] just W of Panama City [0858/7932 (USBGN)] (Atlas, 1988:182, 208, as "Puente de las Americas"; MGP, as "Puente de las Americas"). Boyden, observations at flowering bushes on a woods trail 2 km W of, 1974 Jul 3 (Boyden, 1978). See also: Farfan Beach; Mangrove; and West Ferry Ramp.

BRINCANCO, ISLA; Veraguas, Islas Contreras (Pacific)
0751/8147 (USBGN)
181 m, small, irregularly shaped island with a land area of 2.9 km^2 (290 ha), located in Islas Contreras [0750/8147 (USBGN)] 5 km NNW of Isla Uva [0748/8146

(USBGN)] (JOG; Olson, 1997a). The island is now within Isla de Coiba Parque Nacional [ca. 0730/8140]. Beaudette and Northern, 1959 Mar 26 (LACM, as "Brincanco Is"; Olson 2007, as "Isla Brincanco"); Wetmore, GL = Islas Contreras, creek behind a gravel beach, up through tall, green trees, open undergrowth with few creepers, 1962 Mar 23-24 (USNM; Olson, 1980b, as "Islas Contreras, Isla Brincanco"; Olson, 1997a); Garcés and Mena, 1996 Feb 25-26 (MVUP).

BRUJA, CERRO; Colón (E)
0929/7934 (USBGN)
979 m, peak 40 km NE of Colón [0922/7954 (USBGN)] and 13 km SE of Portobelo [0933/7939 (USBGN)] (JOG). The peak is now included in Parque Nacional Portobelo [0931/7954 (Angehr, 2003:221)]. Goldman, at 100-2,000 feet (30-600 m), camp on the Río Cascajal [0934/7939 (USBGN)] near Cerro Bruja, cut a trail to ca. 2,000 feet (600 m) which was described in his journal as along "a narrow ridge between two good-sized streams," 1911 Jun 4-9 (USNM; Nelson, 1911b, as "Cerro Brujo"; Goldman, 1920; Olson et al., 1968; Olson, 1980a; Wetmore, 1970); Observer ?, below the summit to ca. 800 m, 1996 May 24-27 (Adsett and Wege, 1998). (Myers, 1969).

BRUJA, PUNTA; Panamá (Canal Zone) (W, Pacific)
0853/7935 (USBGN)
Sea level, mainland point 11 km SW of Panama City [0858/7932 (USBGN)] between the mouths of the Río Venado [0854/7936 (USBGN)] and the Río Farfan [0856/7934 (USBGN)] (E761A, as "Bruja Pt."; MGRP, as "Punta Brujas"), and the nearest mainland to Isla Taboga [0847/7933 (USBGN)] (Wetmore, 1952:1). Weber and Wonder, near Balboa [0857/7934 (USBGN)], 1928 Dec 27 (FMNH, as "Bruja Point"; Wetmore, 1945, as "Brujas Point").

BRUJA POINT; see Bruja, Punta

BRUJO, CERRO; Bocas del Toro
0911/8213 (USBGN)
140 m, peak on the mainland peninsula that forms the NW coast of Laguna de Chiriquí [0903/8200 (USBGN)]. The peak is located on the N central section of the peninsula 2 km E of Bahía de Green [0912/8214 (USBGN)] (E762; JOG; MGRP, as "Cerro Bruja"). SBE, GL = Bahía de Green, E of Bahía de Green, 1989 Feb 27 (USNM).

BRUJO, CERRO; see Bruja, Cerro

BRUJO, EL; see El Brujo

BRYAN CREEK; Bocas del Toro, Archipiélago de Bocas del Toro
ca. 0907/8207
According to Handley's journal, this is one of several saltwater creeks on Isla Bastimentos [0918/8208 (USBGN)] SW of Old Point [0917/8205 (USBGN)]. Located near Cedar Creek [0917/8207 (USBGN)] and Alberrys Creek [0917/8207 (USBGN)], the creek was named after Edward Bryan who farmed the area, and therefore locally known as both Edward's creek and Bryan's creek, SBE, GL = Isla Bastimentos, 1987 Mar 27 (USNM). Not shown on our maps.

BUBÍ, RÍO; Veraguas (Pacific)

0759/8138 (USBGN)

Originates in the hills SW of La Arena [0800/8126 (USBGN)], flows NW and W entering Boca de Pajaroncito [0756/8139 (USBGN)] opposite the SE coast of Isla de la Corocita [0758/8139 (USBGN)] (JOG). Wetmore, GL = Zapotillo, 4 mi (6 km) W of Zapotillo, near the road in a pasture with standing water, 1955 Mar 19 (USNM). See also: Tuancle, Río.

BUENAVENTURA, BAHÍA DE; Colón (E)

0932/7941 (USBGN)

Small bay on the N coast of Colón province 32 km NE of Colón [0922/7954 (USBGN)] (E761A, as "Buenaventura Bay"). See also: Los Mogotes, Isla.

BUENAVENTURA BAY; see Buenaventura, Bahía de

BUENAVENTURA ISLAND; see Los Mogotes, Isla

BUENA VISTA; Bocas del Toro

0850/8212 (USBGN)

300 m, settlement 15 km SW of Chiriquí Grande [0857/8207 (USBGN)] (E762). According to Kennard and Peters (1928:445) and to Peters (1931:294), Buena Vista was a camp at 1,200 feet (350 m) on the Boquete Trail [ca. 0857/8207] (q. v.) and was estimated to be ca. 18 mi (29 km) from Chiriquí Grande [0857/8207 (USBGN)] and ca. 15 mi (24 km) from the continental divide via the Boquete Trail. The settlement appears to have developed at the former camp site. Kennard et al., GL = Boquete Trail, 1926 Mar (MCZ; Kennard and Peters, 1928:445, 450); Smith, at 1,000 feet (300 m), 1926 Mar 9 (ANSP); Wedel, 1928 Mar 11-25 (Peters, 1931a).

BUENA VISTA; Chiriquí

?

Unclear where Galindo collected. He may have worked at Buena Vista [0839/8238 (USBGN)] (q. v.), although USBGN also lists Buena Vista [0837/8244 (USBGN)] in W Chiriquí, and Buena Vista [0823/8137 (USBGN)] in E Chiriquí near the Río Tabasará [0823/8137 (USBGN)]. Galindo, 1971 May 20 (GML).

BUENA VISTA; Chiriquí

0839/8238 (USBGN)

Town on the road between La Concepción [0831/8237 (USBGN)] and Volcán [0846/8238 (USBGN)] E of the Río Escarrea [0820/8241 (USBGN)] (MRP). The USBGN coordinates correspond with a region on JOG at ca. 800 m, ca. 15 km N of La Concepción and ca. 1 km E of the river. Wetmore, at 2,300-2,500 feet (700-760 m), walked N following a steep, rocky trail to Quebrada Grande [0840/8239 (USBGN)], crossed it to an area of pastures with brushy and weedy growth along the fences and trails, scattered trees and tracts of low *rastrojo*, ca. 1 mi (1.5 km) N on the road to a stand of original forest near the head of the Río Escarrea, 1960 Feb 29, Mar 1, 4, 6-7, 11, 13, 18, and as a GL for Río Escarrea, 1960 Mar 9 (USNM; Wetmore, 1967; Olson, 1981d).

BUENA VISTA; Colón (E)

0916/7942 (USBGN)

ca. 50 m, town on the Carretera Boyd-Roosevelt [0917/7943 (USBGN)] ca. 25 km ESE of Colón [0922/7954 (USBGN)] (E8660; JOG; RDP2). Van Horn, 1971 Dec 24, 1972 Jan 29 (GML); Pitman, 1985 Dec 9 (LACM); Castillo, Johnson, Ortega, and Walker, 1991 Aug 20 (MVUP).

BUENA VISTA, PUNTA; Panamá (Canal Zone) (E, Atlantic)
0911/7950 (USBGN)
50 m, point located 4 km WNW of Frijoles [0910/7948 (USBGN)] on a peninsula that extends ca. 3 km W from the mainland forming the N coast of Bahía Frijoles [0910/7949 (USBGN)]. The point is the nearest mainland to the NE coast of Isla Barro Colorado [0909/7951 (USBGN)] (E866). Willis, observations, medium second growth woodland, "tall secondary forest," 1960-1970 (Willis, 1966a; Willis, 1967; and Willis, 1972, all as "Buenavista Point"); Miller, GL = Frijoles, 1965 Feb 21, 24 (MVZ, as "Buena Vista Point").

BUENA VISTA POINT; see Buena Vista, Punta

BUENAVISTA POINT; see Buena Vista, Punta

BUENOS AIRES; Panamá (W, Pacific)
0838/7957 (USBGN)
200 m, town located on a road leading NW of the Carretera Interamericana [0832/8242 (USBGN)] from Bejuco [0836/7953 (USBGN)], ca. 7 km NW of Bejuco, 19 km E of Cerro Gaital [0837/8007 (USBGN)] and ca. 7 km S of Cerro Campana [0841/7955 (USBGN)] (JOG). Bacon and Miller, 2002 Sept 14 (STRI; UAM).

BUENOS AIRES; Veraguas (Pacific)
0824/8129 (USBGN)
ca. 300 m, a district capital within Comarca Ngöbe Buglé [ca. 0835/8145], which extends over land that was formerly included in western Veraguas, eastern Chiriquí and eastern Bocas del Toro provinces. The town is located on the Río Cobre [0801/8118 (USBGN)] ca. 30 km NE of Tolé [0814/8141 (USBGN)] and ca. 30 km WNW of Cañazas [0819/8113 (USBGN)] (RDP2). The town is not shown on MGRP, but the coordinates indicate that the town is located in Distrito Cañazas [0823/8120 (USBGN)]. Delgado, observations, 1982 May (Ridgely and Gwynne, 1989:423, as "Buenos Aires de Cañazas").

BUENOS AIRES DE CAÑAZAS; see Buenos Aires

BUGABA; Chiriquí
0829/8237 (USBGN)
ca. 150 m, town 3 km S of La Concepción [0831/8237 (USBGN)] (JOG; MHA). Collector ?, date ?, 1903 Oct 22, 1912 Jun or Jul (FMNH, as "Bojava"; OC, as "Bogava," and as "Bogave"; USNM, as "Bugara"); Arcé, date ?, 186?, 1869 (BMNH; USNM; Salvin, 1870bb; Salvin, 1871; Salvin and Godman, 1879-1904, vol. 1:91, 305; Sharpe, 1882:230; Salvin and Godman, 1888-1904, vol. 2:58; Salvadori, 1891); Heyde, 1886 Dec 10 (USNM); Brown, at 600 feet (180 m), "in the foothills [of the volcano]," 1901 Jul 2-16 (MCZ, as "Bogaba"; USNM, as "Bogaba," and as "Bugaba"; Bangs, 1902a, as "Bogaba"; Bangs and Penard, 1921:380, 391, as "Bogaba"; Peters and Griscom, 1929, as "Bogaba"); Frazar, at 600 feet (180 m), 1901 Jul 6 (ANSP, as

"Bogoba"); Watson, at 800 feet (240 m), 1903 Sep-Dec, 1904 Oct, 1905 Oct, 1907 Nov (AMNH, as "Bogava"; ANSP, as "Bogava"; CM, as "Bofuva"; FMNH; MCZ; OC, as "Bogava"; Zimmer, 1941b:7, as "Bogava"; Zimmer, 1950:8, 26, 50, as "Bogava"; Eisenmann, 1950, as "Bogava"; Howell, 1956, as "Bogava"). (Hellmayr, 1918; Davidson, 1938; Wetmore, 1952:16; Wetmore, 1959b:18).

BUGABITA ABAJO; see Bugabito

BUGABITA ARRIBA; see Bugabito

BUGABITO; Chiriquí

ca. 0831/8238

ca. 200 m, presumably refers to the area near the towns Bugabita Arriba [0831/8238 (USBGN)] and Bugabita Abajo [0830/8238 (USBGN)]. According to E562, Bugabita Arriba is located 3 km NW of La Concepción [0831/8237 (USBGN)] and Bugabita Abajo is located 3 km SW of La Concepción. Quintero, *ruina de* (ruins of) Bugabito, 1980 Jan 2 (MVUP).

BUGARA; see Bugaba

BUHO, CIÉNAGA DE; Herrera

ca. 0804/8043

Described by Wetmore as an open lagoon, ca. 125 yards (115 m) wide, ca. 2 mi (3 km) W of La Concepción [0804/8042 (USBGN)] and W of the Río Escotá [0806/8033 (USBGN)] (USNM). See also: La Concepción. Not shown on our maps.

BURADÓ, ARROYO; Darién

ca. 0731/7804

Described by Wetmore as a tributary of the Río Jaqué [0731/7810 (USBGN)] that entered from the N upstream from the mouth of the Río Pavarandó [0749/7804 (USBGN)], Wetmore and Perrygo, GL = Río Jaqué, 1947 Mar 27 (USNM). Not shown on our maps.

BURICA, ISLA; see Burica, Isla de

BURICA, ISLA DE; Chiriquí

0801/8253 (USBGN)

Sea level, island with a land area of ca. 0.2 km^2 (20 ha) located in the Golfo de Chiriquí [0800/8220 (USBGN)] 1 km S of Punta Burica [0802/8255 (USBGN)] (JOG; MGRP; MHA, as "Isla Burica"). Batty, specimens at AMNH are dated 1902 Feb 4, but Batty's data were completely fabricated and his specimens labeled "Burica Island" were actually collected in 1901 near his mainland bases at Boquete [0847/8226 (USBGN)] and Boquerón [0830/8234 (USBGN)] (AMNH, as "Burica Island"; Eisenmann, 1950, as "Burica I."; Olson, 2008).

BURICA, PENÍNSULA; Chiriquí

0810/8255 (USBGN)

Peninsula on the extreme W coast of Chiriquí province that extends 27 km S from Puerto Armuelles [0817/8252 (USBGN)] and varies in width from 4 to 11 km. The Costa Rican border extends S through the peninsula to Punta Burica [0802/8255 (USBGN)] (JOG). Ridgely, observations, 1982 Jun 26 (Ridgely and Gwynne,

1989); Weir, GL for: Bartolo Arriba, Limones, and Yerbazales, 2003 Feb 26-28, Apr 4-8 (CU; LSUMZ); DaCosta, Guitton-Mayerma, Klicka, Lelevier, Miller, and Spellman, GL = Puerto Armuelles, at 50 m, at 08°12'18"/82°52'39", 8 km SSW of Puerto Armuelles, [Bahía de] Charco Azul, and at 345 m, 08°19'27"/82°58'20", 14 km NW of Puerto Armuelles, 2006 Mar 1-2 (MBM; STRI). (Angehr, 2000). See also: Burica, Punta.

BURICA, PUNTA; Chiriquí
0802/8255 (USBGN)
Sea level, point on the S tip of Península Burica [0810/8255 (USBGN)] 27 km S of Puerto Armuelles [0817/8252 (USBGN)] (JOG, as "Punta de Burica"). Wetmore, GL for Punta Balsa, 1966 Feb 25 (USNM).

BURICA, PUNTA DE; Burica, Punta

BURICA ISLAND; see Burica, Isla de

BURUNGA; Panamá (W, Pacific)
0857/7941 (USBGN)
100 m, town 3 km NW of Arraijan [0857/7939 (USBGN)] (MGRP, as "Burunga o Pozo Hondo"; RDP2). Vergara, *las tecas* (teak trees), 1990 Sep 10 (MVUP); Tejada, 1994 Aug 10 (MVUP).

C-12-E ROAD; see Road C12E

C-15 ROAD; see Road C15

C-29 ROAD; see Road C29

C. I. M.; Panamá (E, Pacific)
0908/7915 (Burton, 1973b)
According to Burton (1973b), this was a military instruction center located near Pacora [0905/7917 (USBGN)], Burton, 1972 Jan (BMNH; Burton, 1973b; Burton, 1975). Not shown on our maps.

CABIANCHO, RÍO; Colón (E)
ca. 0915/7955 ?
Possibly refers to the Rio Quebrancha, which originated E of Bahía Limón [0920/7956 (USBGN)] and flowed SW entering the Río Gatún [0916/7947 (USBGN)] near its mouth (MIP; PR). The river is now submerged in the NE waters of Lago Gatún (Gatun Lake) [0912/7955 (USBGN)], and Isla Quebrancha [0917/7952 (USBGN)] remains above water near its former mouth. Observer ?, "a stream called the Cabiancho, a tributary of the Gatún," 186? (Lawrence, 1867a; Wetmore, 1968:25).

CABIMA; Panamá (E, Pacific)
ca. 0903/7930
Not shown on our maps but located in NE greater Panama City [0858/7932 (USBGN)] in Distrito San Miguelito [0903/7930 (USBGN)]. Batista, *entrada de* (entrance of) Cabima, 1990 Sep 25 (MVUP).

CABIMA; see La Cabima

CABIMO, EL; see El Cabimo

CABOBRÉ, RÍO; Panamá (E, Pacific)

0906/7917 (USBGN)

Originates on the SE slopes of Cerro Prominente [0914/7917 (USBGN)] and flows S joining the Río Pacora [0901/7917 (USBGN)] 3 km N of Pacora [0905/7917 (USBGN)] (MGRP). Wetmore and Perrygo, GL = Pacora, 2-3 mi (3-5 km) N of Pacora, valley of the Río Cabobré, woodland, gallery forest and secondary growth, scrub along the stream banks, 1949 Mar 19, 21, specimens labeled "Utivé" were collected near the Río Utivé, not the Río Cabobré (USNM). See also: Azul, Cerro; and Utivé, Río.

CABRA; see Cabra, Río

CABRA, CERRO; Panamá (W, Atlantic)

0854/7938 (USBGN)

507 m, peak located just W of the former Canal Zone border ca. 3 km SW of Cerro Galera [0856/7937 (USBGN)], 13 km SW of Panama City [0858/7932 (USBGN)], and 2 km N of the coast (E761A; E8660; JOG). MCU, specimens are also labeled Canal Zone, 1955 Oct 6 (USNM, as "Cabra Hill"); Collector ?, at 20 m, 3 km WNW, 1995 Dec 31 (LSUMZ). See also: Taboga, Isla.

CABRA, RÍO; Panamá (E, Atlantic)

0903/7924 (USBGN)

Originates NE of Goofy Lake [ca. 0910/7924] flowing S and SE, crosses the Carretera Interamericana [0832/8242 (USBGN)] at Cabra [0907/7919 (USBGN)] and enters mangroves on the coast of Bahía de Panamá [0850/7915 (USBGN)] SW of Pacora [0905/7917 (USBGN)] and NW of the mouth of the Río Pacora [0901/7917 (USBGN)]. The coordinates given are for the region where the river enters the mangroves, not for the open ocean itself. The coastline is at ca. N 09°00'30" in this area (JOG). Miller and Winker, at 09°01'24"/79°17'30", 2004 Feb 9-10 (STRI; UAM).

CABRA HILL; see Cerro, Cabra

CABUYA; Herrera

0802/8038 (USBGN)

ca. 50 m, town 10 km SSE of Santa María [0807/8040 (USBGN)] (JOG; MHA, as "Cabuyao"). Wetmore and Perrygo, stream on the edge of town, pasture and low thorny growth, 1948 Mar 12, and as a GL for Santo Domingo, 1948 Mar 13 (USNM, as "La Cabuya"; Wetmore, 1970, as "La Cabuya").

CABUYA, LA; see Cabuya

CABUYAL, ESTERO EL; see El Cabuyal, Estero

CACAL, EL; see Piña

CACAO PLANTATION; see Cacao Plantation Road

CACAO PLANTATION ROAD; Panamá (Canal Zone) (E, Atlantic)

ca. 0907/7938

Foot trail within Parque Nacional Soberanía [0909/7941 (Angehr, 2003:239)] that was originally a road that led to a cacao plantation. The trail extends roughly SW-NE between Carretera Gaillard and Las Cruces Trail [ca. 0904/7936]. The trail head (S terminus) is located on Carretera Gaillard [0904/7940 (USBGN)] 6 km SSE of Gamboa (near the highway marker for mile 12) at ca. 100 m, and is marked with a sign labeled "Camino Plantation." Also known as Plantation Road (Ridgely and Gwynne, 1989:480-1, as "Plantation Road"). Yunker (from NIID), GL = Gamboa, 1961 Oct 26, Nov 20, 1962 Jan 23, 29, Feb 2, 5, 8, Mar 28 (USNM, as "Cacao Plantation," and as "Cacao Plantation Road"); Collector ?, GL = Carretera Gaillard, 1999 Mar 17 (LSUMZ, as "Plantation Loop"). (Karr, 1990, as "Plantation Road").

CACICÓN; Chiriquí
0826/8153 (USBGN)
500 m, town 15 km NNW of San Félix [0817/8152 (USBGN)] E of the Río San Juan [0816/8157 (USBGN)] (JOG). See also: El Banco.

CACICÓN, LLANOS DE; Chiriquí
ca. 0826/8152
200-500 m, an area in E Chiriquí province 15 km N of San Félix [0817/8152 (USBGN)] extending ca. 10 km from the W bank of the Río San Félix [0810/8151 (USBGN)] (MHA). See also: El Banco.

CACIQUE, PUNTA; see Farallón Sucio

CACIQUE, RÍO; Panamá, Archipiélago de las Perlas
0818/7854 (USBGN)
Described by Wetmore as a river on Isla del Rey [0822/7855 (USBGN)] that flows S into the N end of Bahía San Telmo [0816/7853 (USBGN)] (USNM). Beaudette, GL = Isla del Rey, 1959 Mar 22 (LACM, as "Cacique R."); Wetmore, GL = Isla del Rey, the mouth was a shallow channel winding through muddy shores and mangroves, inland to coarse grass, *rastrojo* and small stands of forest, 1960 Jan 27 (USNM; Wetmore, 1970; Wetmore et al., 1984:71). (Murphy, 1944b). Not shown on our maps.

CAIMÁN, CALETA; see Naranjo, Ensenada del

CAIMITILLO, RÍO; Panamá (Canal Zone) (E, Pacific)
0902/7936 (USBGN)
Originates N of Chiva Chiva [0902/7935 (USBGN)] near the former Panamá province/Canal Zone boundary, flows S entering the N end of Caimitillo Arm [0901/7936 (USBGN)] (q. v.) (E661, as "Río Caimito"; E8660). Hallinan, trees in the valley of, 1915 Oct 20 (AMNH; Hallinan, 1924); Wetmore, GL = Chiva Chiva, from Camino Chiva Chiva (Chiva Chiva Road) [0902/7935 (USBGN)] followed an abandoned road near the former Canal Zone/Panamá provincial boundary through tall grass and shrubby growth along the upper course of the river, 1960 Jan 12 (USNM). See also: Caimito Arm.

CAIMITILLO ARM; Panamá (Canal Zone) (E, Pacific)
0901/7936 (USBGN)
A 200 m wide inlet that extends 1 km N from the E shore of Lago de Miraflores

[0901/7936 (USBGN)]. Camino Chiva Chiva (Chiva Chiva Road) [0902/7935 (USBGN)] extends along its E shore. Its N end near the mouth of the Río Caimitillo [0902/7936 (USBGN)] is swampy (E8660). According to Angehr (in litt., Aug 2005), "the lake just W of the start of Chiva Chiva Road on the Caimitillo Arm is generally called Chiva Chiva Lake. However, all three small lakes, including the one W of the highway and the Brazo Camarón [0901/7936 (USBGN)], are sometimes collectively referred to as the 'Chiva Chiva Lakes.'" According to Wetmore, in 1955 Chiva Chiva Lake was covered with floating masses of water hyacinths and was bordered by and filled with islands of tall marsh grass (USNM, as "Chiva Chiva Lake"). Scholes and Scholes, "lake formed by damming of the Río Caimito [sic], E of Red Tank [ca. 0901/7936]," 1951 Feb 25 (Scholes and Scholes, 1954); MCU, 1955 Sep 12 (USNM, as "Chiva Chiva Lake"). See also: Chiva Chiva.

CAIMITO, RÍO; Panamá (W, Pacific)
0852/7943 (USBGN)
Originates N of Cerro Trinidad [0844/7958 (USBGN)] and flows N, then it turns E and passes N of La Chorrera [0853/7947 (USBGN)], from here it crosses the Carretera Interamericana [0832/8242 (USBGN)] and turns S widening as it enters mangrove swamps before discharging into Bahía Chorrera [0850/7943 (USBGN)] at Puerto Caimito [0852/7943 (USBGN)] (E761A; JOG). Bovallius, the mouth was blocked by a sand bar at low tide, the tide and mangroves extended ca. 3 km inland, tall trees with many palms grew behind the mangroves, cultivated areas near La Chorrera, see discussion of Bovallius under Pueblo Nuevo [ca. 0820/8013 ?], 1883 Jun 8 (NHRM; Bovallius, 1877, as "Caymito River"; Rendahl, 1919:2, 15, 16, as "Caymito River"). See also: Paja, Río.

CAIMITO, RÍO; see Caimitillo, Río

CAIMITO, RÍO; see Caimitillo Arm

CAJA, BOCA LA; see Boca La Caja

CALABASO; Panamá (E, Pacific)
ca. 0903/7911
Described by Wetmore as a section of the Río La Jagua [0902/7909 (USBGN)] near La Jagua Hunting Club [ca. 0903/7911], Wetmore, GL = La Jagua Hunting Club, belt of timber that borders the marsh where the coarse, 6-10 foot (2-3 m) grass was still green, above Calabaso in forest, 1962 Jan 12, 14 (USNM). See also: La Jagua, Río. Not shown on our maps.

CALABAZO, RÍO; Panamá (W, Pacific)
0827/7958 (USBGN)
Originates in the hills SE of El Valle [0836/8008 (USBGN)], flows SE parallel to the Río Mata Ahogado [0828/7957 (USBGN)] and enters the Gulf of Panama [0800/7930 (USBGN)] 3 km SE of San Carlos [0829/7957 (USBGN)]. The river flows E of and within 2 km of "Panama 71," the road that connects El Espino [0829/8001 (USBGN)] to El Valle [0836/8008 (USBGN)] (E761B; MGRP). Wetmore and Perrygo, GL = El Espino, at 1,000 feet (300 m), on the road to El Valle at road marker "km 12" located ca. 5 mi (8 km) N of El Espino, steep dry

slopes, grassland, brush, small plantations, 1951 Mar 24, near El Espino, 1951 Apr 1 (USNM).

CALAFATE, PUNTA; see San Felipe

CALDERA; Chiriquí
0839/8223 (USBGN)
300 m, town 15 km SSE of Boquete [0847/8226 (USBGN)] E of the Río Caldera [0837/8221 (USBGN)] (E762; JOG). Mentioned in Ridgely and Gwynne (1989, as "La Caldera"). See also: La Caldera Road.

CALDERA, RÍO; Chiriquí
0837/8221 (USBGN)
Originates on the SW slopes of Cerro Horqueta [0850/8226 (USBGN)], flows E to Agua Caliente [ca. 0851/8229], then flows SE to Boquete [0847/8226 (USBGN)] and continues S and SE entering the Río Chiriquí [0820/8226 (USBGN)] 19 km SSE of Boquete (E762; Blake, 1958:502 (map), as "Quebrada Cerro Punta"). Benson, GL = Boquete, at 5,100-5,900 feet (1,550-1,800 m), 1931 Nov 24-25, Dec 2-3, 12-15 (MCZ; USNM; Olson, 1981c, as "Boquete, Río Caldera"); Davidson, GL = Boquete, at 3,800 feet (1,150 m), observations, near Boquete, c. 1933-1934 (Davidson, 1938:256, as "Caldera River"); Mönniche, at 5,500-6,000 feet (1,700-1,850 m), 1933, 1938-1939, ANSP specimen at 6,400 feet (1,950 m), listed as Boquete R., presumably refers to the Río Caldera or one of its tributaries, 1934 (ANSP; FMNH; MCN; Blake, 1958). See also: Holcomb Trail.

CALDERA, RÍO; Los Santos
0728/8000 (USBGN)
Originates near Los Asientos [0731/8008 (USBGN)] in E Los Santos province, flows SE and enters the Gulf of Panama [0800/7930 (USBGN)] just N of Punta Mala [0728/8000 (USBGN)] (JOG). Wetmore, GL = Pedasí [0732/8002 (USBGN)], Wetmore reached the river by two different roads leading S from Pedasí and therefore collected near the mouth and farther inland. Near the mouth he came to the river at the crossing of the road to Punta Mala [0728/8000 (USBGN)] in the area shown on JOG as 1.5 km WNW of Punta Mala; collected in mangrove swamps, in an area of original forest with an undergrowth of corozo palm and open shrubs bound with creepers, a few *Cereus* cactus, and at the edge of a canefield. The upstream collecting area was near a *finca* behind Punta Mala at the head of tidewaters where the river was 30-40 feet (9-10 m) wide and bordered by fair-sized trees mixed with smaller growth, collected in pastures, in savannas, near a small creek and in an area of uncut forest, 1957 Mar 9, 11, 13, 16, 20 (USNM).

CALEDONIA; see Calidonia

CALEDONIA, BAHÍA DE; San Blas
0855/7742 (USBGN)
Bay on the central San Blas coast SE of Ailigandí [0914/7801 (USBGN)] (SMN2A, as "Calidonia Bay"). Ashcraft, 1939 Apr 3-5, 14, 21, 26-27 (LACM, as "Caledonia Bay").

CALEDONIA BAY; see Caledonia, Bahía de

CALETA; see Aguadulce

CALETA, PLAYA; Coclé (Pacific)
0810/8030 (Cooke and Bush, 1989)
Salt flats located 1 km from the coast of Bahía de Parita [0808/8024 (USBGN)] and ca. 9 km SSE of Aguadulce [0815/8033 (USBGN)], Cooke and Bush, photographs, 1987 Jun 24, Aug 29 (ANSP; Cooke and Bush, 1989). Not shown on our maps.

CALETA CAYMAN; see Naranjo, Ensenada del

CALIDONIA; Panamá (E, Pacific)
0858/7933 (USBGN)
Sea level, a section of greater Panama City [0858/7932 (USBGN)] located on the coast just NE of Ancón [0858/7933 (USBGN)] (E962). While collecting on the Savanna of Panama [ca. 0859/7931] (q. v.) near Panama City in May 1904, Brown was based "at Calidonia at the edge of the swamp of Panama, about a mile from the seashore and about seventy-five yards from the beginning of the mangroves." Specimens apparently were labeled "near Panama" and "Savanna of Panama" (Bangs, 1906a). Granados, 1980 Oct 27 (MVUP); González, 1996 May 8 (MVUP). Wetmore (1972:578, as "Caledonia").

CALIDONIA BAY; see Caledonia, Bahía de

CALLE LARGA; Panamá (E, Pacific)
0911/7849 (USBGN)
ca. 50 m, not shown on our maps, but USBGN indicates a town on the Río Chepo [0900/7906 (USBGN)] ca. 30 km E of Chepo [0910/7906 (USBGN)]. The town is on the Carretera Interamericana [0832/8242 (USBGN)] on the W end of Lago Bayano [0908/7835 (USBGN)]. Galindo and Hinds, GL = Río Chepo, 1973 Jan 26, Feb 4, 9, Apr 2, May 10, Sep 2 (GML; LSUMZ).

CALLE LARGA, LOMA; Darién
ca. 0727/7757
ca. 350 m, described by Wetmore as a summit ca. 750 feet (230 m) above the surrounding area that dropped down to the E bank of the Río Jaqué [0731/7810 (USBGN)] ca. 1 mi (1.5 km) above the mouth of the Río Imamadó [ca. 0727/7758]. Its S slopes lead back down to Chicao [ca. 0727/7758], Wetmore and Perrygo, GL = Río Jaqué, 1947 Apr 2 (USNM). Not shown on our maps.

CALLEJÓN DE PALMAS; Chiriquí
ca. 0849/8231
ca. 2,000 m, described by Blake (1958:502 (map)) as a canyon on Casita Alta [ca. 0849/8229] just NW of Fila Lérida [ca. 0847/8230], sloping eastward from the foot of Cerro Copete [ca. 0848/8231]. Mönniche, at 5,200-6,700 feet (1,600-2,050 m), 1932, 1937 (FMNH; MCN; Blake, 1958).

CALLEJÓN SECO; Chiriquí
0848/8228 (USBGN)
1,400 m, town 4 km NW of Boquete [0847/8226 (USBGN)] (E762). Described by Blake (1958:502) as a canyon whose N rim forms the southern boundary of Quiel

[ca. 0848/8228] (q. v.).

CALOBÉBORA; see Calovébora

CALOBRE; Veraguas (Pacific)

0819/8051 (USBGN)

100 m, town 28 km NE of Santiago [0806/8059 (USBGN)] and 17 km S of Laguna de la Yeguada [0828/8052 (USBGN)] on the Río San Juan [0812/8049 (USBGN)], which is a tributary of the Río Santa María [0806/8029 (USBGN)] (JOG; MHA, as "Colobre"; Salvin, 1870bb (map)). Collector ? [Arcé ?], 18?? (USNM); Arcé, 18??, 1869 (BMNH; USNM; Salvin, 1870bb; Salvin, 1874; Salvin, 1876, as "district around Calobre"; Todd, 1913; Griscom, 1924b; Griscom, 1927a; Peters, 1931a:299; Kelso and Kelso, 1935; Moore and Peters, 1939; Zimmer, 1951:13; Wetmore, 1956a; Olson et al., 1968). See also: Veraguas, Provincia de.

CALOVÉBORA; Veraguas (Atlantic)

0848/8113 (USBGN)

Sea level, town centrally located on the coast of Golfo de los Mosquitos [0900/8115 (USBGN)] at the mouth of the Río Calovébora [0848/8113 (USBGN)] in extreme NW Veraguas province (JOG). Arcé, specimens labeled "Calovévora" are from the Atlantic drainage and at least some are from the coastal region as Sclater and Salvin (1867:135) indicates that six specimens were labeled "near the village of Calovevora." That said, we must point out that a specimen of *Sporophila americana* (BMNH 1886.12.14.464) labeled "Calovevora" and taken by Enrique Arcé in 1869 is not the expected all black Caribbean subspecies *corvina* and must have been taken on the Pacific slope. Either it is simply mislabeled or Arcé used Calovébora in a broad sense to include specimens taken in transit to or from Calovébora, 18??, 1867, 1868, 1869 (FMNH; USNM,; Sclater and Salvin, 1867 [and as "Veragua"]; Salvin, 1870bb; Berlepsch, 1911; Hellmayr, 1918; Griscom, 1927a; Moore and Peters, 1939; Zimmer, 1941b:5; and Zimmer, 1950:15, all as "Calovevora"). See also: Veraguas, Provincia de.

CALOVÉBORA, RÍO; Bocas del Toro/Veraguas (Atlantic)

0848/8113 (USBGN)

Originates on the continental divide WNW of Santa Fé [0831/8105 (USBGN)] and flows N forming the Bocas del Toro/Veraguas provincial boundary entering Golfo de los Mosquitos [0900/8115 (USBGN)] at Calovébora [0848/8113 (USBGN)] (JOG; MGRP). Benson and Gaffney, collected near the river at two different camps after crossing into the Caribbean drainage above Santa Fé. The first camp was named Guabal [ca. 0835/8114] (q. v.) and was located at 1,500 feet (450 m), and while the second camp was not named it was located at 800 feet (240 m) ca. 10 km S of the coast and the river's mouth. Griscom (1927a; 1928) indicates that collections were mdae in Veraguas province. However, according to a map that accompanied a letter from Benson to Griscom dated Nov. 7, 1926 (AMNH), both camps were located on the W banks of the river in Bocas del Toro province, 1926 Aug 5-Sep 14 (AMNH, as "Río Calovévora"; Griscom, 1927a, as "Caribbean rainforest in W. Panamá," as "Guabal, Río Calovévora," and as "Río Calovévora"; Griscom, 1928, as "Río Calovévora, Veraguas"; Zimmer, 1934:14; Zimmer, 1941b:10;

and Zimmer, 1950:16, 26, 50, all as "Río Calovevora "; Peters, 1931a:297, 324, as "Calovevora"; Wetmore, 1959b:19; Eisenmann and Howell, 1962, as "Guaval, Río Calovévora"; Olson, 1981a, as "Guabal, Río Calovévora"; Olson, 1981d, as "Río Calovévora, Caribbean slope"; Escalante-Pliego and Peterson, 1992).

CALOVÉVORA; see Calovébora

CALOVÉVORA, PICO; Veraguas (Atlantic, Pacific)

ca. 0832/8106

ca. 1,400 m, according to Salvin's map (1870), the peak is located on the continental divide N of Santa Fé [0831/8105 (USBGN)] and W of the trail connecting that town and Mineral [ca. 0845/8058]. The Río Calovébora [0848/8113 (USBGN)] originates on its N and W slopes. Our maps indicate that the peak is near Cerro Tute [0832/8106 (USBGN)] if it is not Cerro Tute itself. Mentioned by Griscom (1924b:1) in regards to work planned for and later executed by Benson on the Río Calovébora [0848/8113 (USBGN)].

CALOVÉVORA, RÍO; see Calovébora, Río

CALVARIO; see El Calvario

CALVARITO; Coclé (Pacific)

ca. 0838/8037

Described by Wetmore as a wooded cove NW of El Copé [0837/8036 (USBGN)] (USNM), a region that is shown on JOG at 800 m, Wetmore, GL = El Copé, at 400 m, 1962 Feb 23-25 (USNM). Not shown on our maps.

CALZADA LARGA; Panamá (E, Atlantic)

0910/7934 (USBGN)

400 feet (120 m), town 10.5 km E of Juan Mina [0910/7939 (USBGN)] (E761A, as "Casa Larga"). Described by Goldman in 1912 as a group of scattered Indian huts covering several hundred acres on an open rolling savanna (USNM, as "Casa Larga"). Hallinan, forest, open savannas, 1914 May 31 (AMNH; Hallinan, 1924, as "Casa Largo"). See also: La Cabima; and La Puente, Río.

CAMACHO, RÍO; Panamá (Canal Zone) (W, Atlantic)

0906/7941 (USBGN)

Originates on the continental divide and flows N entering the Panama Canal [0920/7955 (USBGN)] at Las Cascadas Reach [0905/7941 (USBGN)] 6 km SSE of Gamboa [0907/7942 (USBGN)] (E8660). Originally the river continued N in the area now occupied by the Panama Canal [0920/7955 (USBGN)] and entered the continuous Río Chagres (see Río Chagres [0919/8000 (USBGN)]) at Bas Obispo [ca. 0907/7942] (PR). See also: Bas Obispo; Empire Known Distance Range; Empire; Panama Canal; and Panama Railroad.

CAMARÓN; Panamá (E, Pacific)

ca. 0912/7905

ca. 100 feet (30 m), described by Wetmore in 1949 as an area ca. 3 mi (5 km) N of Chepo [0910/7906 (USBGN)] near the Río Mamoní [0908/7904 (USBGN)], Wetmore and Perrygo, GL = Chepo, Río Mamoní, along the river near rice and yuca plantations and in the surrounding hills covered with tall, dark forest, 1949

Apr 7, 22 (USNM). Not shown on our maps.

CAMARÓN, BRAZO; see Camarón, Río

CAMARÓN, QUEBRADA; Darién
ca. 0801/7730
Described by Wetmore as a tributary of the Río Púcuro [0801/7733 (USBGN)] downstream from Púcuro [0801/7730 (USBGN)], Wetmore, GL = Púcuro, followed the dry river bed through forest to an area of small pools and more open forest, 1964 Feb 7 (USNM). Not shown on our maps.

CAMARÓN, RÍO; Panamá (Canal Zone) (E, Pacific)
0902/7936 (USBGN)
Short river that originates just N of Chiva Chiva [0902/7935 (USBGN)], flows SSW and crosses the Camino Chiva Chiva (Chiva Chiva Road) [0902/7935 (USBGN)] before entering Brazo Camarón [0901/7936 (USBGN)], which is an inlet on the E shore of Lago de Miraflores [0901/7936 (USBGN)] (E8660). MCU, GL = Chiva Chiva, 1955 Jun 13 (GML); Wetmore, GL = Chiva Chiva, along the river near the Camarón bridge (presumably where the Camino Chiva Chiva crosses the river), dry bed that had occasional water holes, 1961 Jan 7, Mar 19 (USNM).

CAMARÓN, RÍO; Panamá (W, Pacific)
0842/7950 (USBGN)
Short river that originates in the E foothills of Cerro Campana [0841/7955 (USBGN)], flows E crossing the Carretera Interamericana [0832/8242 (USBGN)] 3 km SE of Campana [0843/7954 (USBGN)] and discharges into the Río Capira Salado [0838/7947 (USBGN)] (E762). Wetmore and Perrygo, GL = Campana, entered the stream from the Carretera Interamericana, walked 0.5 mi (1 km) upstream along intermittent pools of the sand and gravel bed in an area that was largely forested, but with many large trees removed, downstream ca. 1 mi (1.5 km) to an area of forest that floods in the rainy season, 1951 Mar 6, 9 (USNM). See also: Sajalices, Río.

CAMBUTAL; Los Santos
0716/8029 (USBGN)
ca. 50 m, town in S Los Santos province 2 km N of the S coast of Península de Azuero [0740/8035 (USBGN)] on a tributary of the Río Cambutal [0715/8029 (USBGN)] (JOG). MGRP indicates that the town is located on a river that enters the Pacific Ocean E of the Río Cambutal. Agro, Ridgely, Pederson, Peet, and Conyne, GL = Península de Azuero, beach at, 1996 Feb 14-15, 23 (ANSP; MVUP).

CAMBUTAL, RÍO; Los Santos
0715/8029 (USBGN)
Short river that originates on the S coast of Península de Azuero [0740/8035 (USBGN)] and flows SE entering the Pacific Ocean 10 km S of Las Palmitas [0720/8030 (USBGN)] (JOG). See also: Cambutal.

CAMERÓN; see Veracruz

CAMINO A BOQUETE; see Cerro Punta-Boquete Trail

CAMINO DE CRUCES; see Las Cruces Trail

CAMINO PLANTATION; see Cacao Plantation Road

CAMINO REAL; see Boca de Uracillo

CAMINO TRES DE NOVIEMBRE; see Boquete Trail

CAMINO VIEJO DE GAMBOA; see Old Gamboa Road

CAMISETA; Chiriquí
ca. 0847/8231
ca. 2,000 m, described by Blake (1958:502 (map)) as the local name for a ridge S of Cerro Copete [ca. 0848/8231]. Mönniche, 1932, 1936 (FMNH; Blake, 1958).

CAMOTE, ISLA; Panamá, Archipiélago de las Perlas
0817/7848 (USBGN)
23 m, very small island with a land area of 0.7 ha (Angehr, 2003:278) located in Bahía de Panamá [0850/7915 (USBGN)] 7 km off the SE coast of Isla del Rey [0822/7855 (USBGN)] and 7 km E of Isla San Telmo [0817/7851 (USBGN)] (MHA). Robins, observations, 1957 Jul 15-26 (Robins, 1958, as "rocky isle of Camote"); Pacific Project, GL = Gulf of Panama, Loftin described the island as treeless, 1969 Mar (Loftin, 1991).

CAMPAMENTO DE CATIVAL; see Catival, Campamento de

CAMPAMENTO DE HATO; see Hato, Campamento de

CAMPAMENTO DE JUNCAL; see Juncal, Campamento de

CAMPAMENTO DE PLAYA BLANCA; see Playa Blanca, Campamento de

CAMPAMENTO DE PUNTA DAMAS; see Punta Damas, Campamento de

CAMPAMENTO DE SAN JUAN; see San Juan, Campamento de

CAMPAMENTO NÚMERO UNO; see Número Uno, Campamento

CAMPANA; Panamá (W, Pacific)
0843/7954 (USBGN)
100 m, town in the E foothills of Cordillera del Llorón [0843/7957 (USBGN)] 6 km NE of Cerro Campana [0841/7955 (USBGN)] on the Carretera Interamericana [0832/8242 (USBGN)] (JOG; MHA, as "La Campana"). Wetmore and Perrygo, at 200 feet (60 m), ca. 1 mi (1.5 km) W of Campana, along a small creek with pooled water amid boulders, trees and thickets on the slopes above, 1951 Mar 4, and as a GL for Río Camarón, 1951 Mar 6, 9 (USNM; Wetmore, 1952:6, 18; Wetmore, 1953; and Wetmore, 1959b:10, 12, all as "La Campana"). Collector ? [Loftin ?], 1968 Sep 10 (MCN).

CAMPANA, CERRO; Panamá (W, Pacific)
0841/7955 (USBGN)
956 m, peak 3 km SE of the continental divide at the S end of Cordillera del Llorón [0843/7957 (USBGN)] 14 km NNW of Chame [0835/7953 (USBGN)] (E761B; JOG). The peak is now included in Parque Nacional Altos de Campana [0842/7957 (Angehr, 2003:227)]. Wetmore and Perrygo, at 2,000-2,800 feet (600-850 m), S and SE slopes, worked in tracts of forest and grassland, at 2,000 feet (600 m) bunch grass and low sedge with volcanic rock boulders, forested area with many

waterfalls, at 2,800 feet (850 m) on the lower edge of the subtropical zone in an area of 80 foot (25 m) trees with heavy undergrowth, moss and epiphytes, in open, rolling areas covered in grass or bare rock, forest tracts of moderate height with 40-80 foot (10-25 m) trees and a dense undergrowth of creeping bamboo and other vines, 1951 Mar 2-3, 5-8, 10, 12-14, 16, at 3,000-3,300 feet (900-1,000 m), all specimens labeled S face, although Wetmore's journal describes the collecting habitat on the N and NE sides of the mountain, at 3,000 feet (900 m), steep slopes covered with heavy forest and dense undergrowth, the area was wetter than the opposite, sheltered side, 1951 Mar 19-20, at 3,000-3,300 feet (900-1,000 m), S face, along the rocky ridge that gives the mountain its name, treeless top covered with low brush and dense creeping bamboo, 1951 Mar 21, at 2,800 feet (850 m), sloping valleys along the road, forest trails, 1953 June 21, 1955 Dec 21 (LSUMZ; USNM; Wetmore, 1951a; Wetmore, 1962; Parkes, 1954a); Eisenmann, observations, 1955 Jul 10, 1956 Jun 24-26 (Eisenmann, 1962a; Parkes, 1969); MCU, 1955 Oct 13 (USNM); Ryan and Boyajian, observations, 1960 Jun 14 (Eisenmann, 1962a); MARU, at 2,700 feet (800 m), 1961 May 7 (AMNH); Eisenmann and Ambrose, 1961 May 6-7 (AMNH; Eisenmann, 1962a); Willis, at 800-900 m, observations, "patchy forests," 1961 Aug 27-29, 1968 Jun 24-27 (Parkes, 1969; Willis, 1967; Willis, 1972); Loftin and Belton, banding, 1962-1964 (Loftin, 1975); Olson, Wiese, Blum, Rassmussen, Fischer, at 2,000-3,000 feet (600-900 m), at ca. 1,025 m, 1963 Feb 17, May 11, 1966 Jun 19, Jul 9 (FMNH; UF; USNM, as "Cerro Campana," and as "La Campana Mountain"; Olson et al., 1968); Powell, at 1,000 feet (300 m), vicinity of the John W. Acker Field Station, 1966 Oct 29-Nov 13 (AMNH; Olson et al., 1968); Hespenheide and Karr, 1969 Jul 20 (Karr, 1971a); Morton, 1970 Aug 3, 1971 Mar 6, May 27 (AMNH; USNM); Strauch, 1971-1973 (UMMZ; Strauch, 1977); Cornejo, *las cumbres de* (the peaks of), 1978 Nov 19 (MVUP); Weir, at 800-900 m, 2003 Mar 28, 30-31 (CU; LSUMZ). (Myers, 1969; Olson, 1980a; Olson, 1981c).

CAMPANA, CIÉNAGA; Panamá (E, Pacific)

ca. 0902/7916

Described by Wetmore as a swamp between Sabana de San José [ca. 0905/7915] and the coast near Río Chico [0902/7911 (USBGN)], Wetmore and Perrygo, GL = Río Chico, largely dry with water in the center only, surrounded by brush and forest, 1949 Mar 18, Apr 1 (USNM; Wetmore, 1951b). Not shown on our maps.

CAMPANA, LA; see Campana

CAMP CILINDRO; see Camp Cylindro

CAMP COLORADO; see Colorado Camp

CAMP CREEK; Darién

ca. 0813/7735

ca. 50 m, Baer, GL = Yaviza [0811/7741 (USBGN)], Río Chucunaque, Camp Townsend, according to Marsh's (1934:76-92) account, Baer traveled up the Río Chico [0811/7741 (USBGN)] and spent Feb 10-14 in an Amerindian village that was located "a distance by airline from Yaviza of only about seven miles, and perhaps twice that amount by the river's course." According to our maps, the village was 7 mi (11 km) ENE of Yaviza, and presumably Camp Creek was Baer's designation for the village. In addition, Baer's specimens from Feb 11 are labeled "Río Chicaque," which is either a misspelling for the Río Chico or a small stream near the village, 1924 Feb 11, 14 (USNM, as "Camp Creek," and as "Río Chicaque"). For more on Baer see comments under Marsh (1934). Not shown on our maps.

CAMP CUNNINGHAM; Panamá ? (Canal Zone) ?

?

Not located on our maps. Presumably a former military site located in Panamá province between the mouth of the Río Chagres [0907/7942 (USBGN)], where Chapin et al. were collecting on Jul 17, and Tapia Número Dos [0903/7924 (USBGN)], their location on Jul 23. Chapin and Rogers, 1923 Jul 19 (FMNH).

CAMP CYLINDRO; Bocas del Toro

ca. 0853/8228

5,200 feet (1,600 m), described by Blake (1958:502) as one of three collecting sites used by Mönniche along the Atlantic drainage of the Holcomb Trail [ca. 0901/8215] (q. v.). Presumably located ca. 11 km NNW of Boquete [0847/8226 (USBGN)] along the upper reaches of the Río Changuinola [0928/8227 (USBGN)]. Mönniche, at 4,500-6,000 feet (1,350-1,850 m), 1933 (FMNH; MCN; Blake, 1958). Not shown on our maps.

CAMP EMPIRE; see Empire Known Distance Range

CAMP ESCOBAL; see Escobal

CAMP FELIZ; Bocas del Toro

ca. 0848/8215

4,100 feet (1,250 m), described by Kennard and Peters (1928:444-445) as a permanent camp on the Boquete Trail [ca. 0857/8207] 12 mi (19 km) farther inland [= S] than Buena Vista [0850/8212 (USBGN)] and ca. 3 mi (5 km) from the summit of the pass. The camp is not shown on our maps, but the corresponding region is shown on JOG ca. 23 km SW of Chiriquí Grande [0857/8207 (USBGN)], ca. 8 km SW of Buena Vista, and ca. 1 km N of the continental divide. From our maps, it is difficult to reconcile all the elevations and distances given by Kennard and Peters, and the best estimate places Camp Feliz much closer to Buena Vista than their description. The authors may have over-estimated the distance traveled due to the increased time required to traverse the steep and narrow trail, or the trail may have meandered greatly as it ascended to the continental divide. Kennard et al., GL = Boquete Trail, arrived from Buena Vista on 1926 Mar 12 and collected for 3 weeks along the Boquete Trail, all specimens came from the wet, N side of the continental divide, which they described as a "jungle-clad mountain side," unclear if this name appears on specimen labels, 1926 Mar 12-early Apr (MCZ; Kennard and Peters, 1928:444-445).

CAMP HOLCOMB; Bocas del Toro

ca. 0853/8228

5,000 feet (1,500 m), described by Blake (1958:502) as one of three collecting sites used by Mönniche along the Atlantic drainage of the Holcomb Trail [ca. 0901/8215] (q. v.). Presumably located ca. 12 km NNW of Boquete [0847/8226 (USBGN)] on the upper reaches of the Río Changuinola [0928/8227 (USBGN)]. Mönniche, at 5,000 feet (1,500 m), 1933 (GML; FMNH; Blake, 1958). Not shown on our maps.

CAMP JESUCITO; see Jesusito

CAMP JESUSITO; see Jesusito

CAMP LAGUNA; see Laguna Camp

CAMPO CHAGRES; see Madden, Represa

CAMPO DE PARACAIDISMO; see Gatún Drop Zone

CAMPO PRODUCCIÓN; Veraguas (Pacific)

ca. 0733/8142

According to Cardiel et al. (1997:22-23), this locality is represented by UTM coordinates MU 2233, which correspond to an area on the NE side of Isla de Coiba [0728/8147 (USBGN)] on the coast of Ensenada Arenas [0733/8142 (USBGN)] ca. 5 km SE of Campamento de Juncal [0734/8144 (USBGN)]. Garcés and Mena, 1996 Feb 16-17, 26, Mar 2 (MVUP). Not shown on our maps.

CAMP PARAISO; see Paraiso

CAMP PIÑA; Colón (Canal Zone) (W)

ca. 0917/7959

50 m, located on Road S1 [ca. 0918/8001] between the W bank of the Río Chagres [0919/8000 (USBGN)] and the E bank of the Río Piña [0917/8003 (USBGN)] near the headwaters of the Río Indio [0917/7958 (USBGN)] and the S border of Fort Sherman Army Reservation [0920/7958 (USBGN)] (E8660). Imhof, observations, 1942 Dec 17 (Imhof, 1950); Noyes, 1947 Sep 5, Oct 13 (MVZ). See also: Piña.

CAMP SUMMIT; see Darién, Serranía del

CAMP TOWNSEND; Darién

ca. 0810/7741

ca. 25 m, located on the Río Chucunaque [0809/7744 (USBGN)] ca. 0.25 mi (0.5 km) below Yaviza [0811/7741 (USBGN)]. The camp was near a coconut grove and on a level bench that extended back to low hills (Marsh, 1934:63-64). Baer, GL = Yaviza, Río Chucunaque, 1924 Feb 22, and at Camp Creek (q. v.), 1924 Feb 11, 14 (USNM). For more on Baer see comments under Marsh (1934).

CAMP WILCOX; see Wilcox Camp

CANA; Darién

ca. 0745/7742

1,800 feet (550 m) (Nelson, 1913a); 600 m (Robbins et al., 1985); an old Spanish gold mining site that operated from 1680-1727 (Chibas, 1898), and intermittently through the early 1900s, located 24 mi (39 km) S of El Real de Santa María [0808/7743 (USBGN)] on the Río Seteganti [0746/7738 (USBGN)] ca. 5-6 km SE of the peak collectors call Cerro Pirre [0749/7744 (GA)] (HMCZ; LZP, as "Santa Cruz de Cana"; SMN2A; Goldman, 1920; Barbour and Brooks, 1923). Although described as just SE of Cerro Pirre (q. v.) (Woakes, 1900; Anthony, 1916; Goldman, 1920), Cana is actually located on the SE end of the Serranía de Pirre [ca. 0749/7744] much closer to Alturas de Nique [0740/7745 (USBGN)] than to the peak officially named Cerro Pirre [0756/7743 (USBGN)]. In the early 1900s Cana was reached by a small railway that led into the mountains from Boca de Cupe [0803/7735 (USBGN)] via Paca [0759/7735 (USBGN)]. According to Goldman's journal, Cana was located 30 mi (48 km) by train from Boca de Cupé, but Barbour and Brooks (1923) indicated that the railroad extended 40 mi (64 km) between the two towns. On LZP this distance is only 20 direct miles (32 km). According to Goldman's journal, beyond Paca the railroad entered the foothills of Cerro Pirre and followed the canyon of a small stream (the Río Paca [0759/7732 (USBGN)] (Anthony, 1916 (map)) until a few miles below Cana where it came out onto the region that Goldman called the Cana Plateau or the Cana Valley. The valley's relatively level land extended for several miles, had once been planted in sugar cane, and received both the Río Cana [ca. 0745/7742] and the Río Seteganti in marshy areas. The Plateau de Cana with these rivers is pictured on the map in Woakes (1900). Several of the plates in Woakes and in Goldman (1920) indicate that in the early 1900s there were many buildings and extensive clearings at Cana, but by Jan 1960 Wetmore found the area abandoned and completely reforested (USNM). The mines were re-opened in the 1920s, which made Benson's work in the region possible (Griscom, 1929a:149). Robbins et al. (1985:200) indicate that in 1982 the settlement at Cana was only 5 buildings and that clearing in the area was limited to these buildings and a trail to the airstrip. From Cana "the initial 4-5 km of the Boca de Cupe trail passed through a relatively young second-growth forest with an average canopy height of 10 to 15 m. The understory was nearly impenetrable without the aid of a machete...5 km (by trail) northeast of Cana, shortly after crossing a major stream, the forest became less dense with many emergent trees...[and] an extensive marsh fed by several small streams abut[ted] the base of the mountains at the eastern end of the airstrip." According to Angehr (in litt., Aug 2005), "the site was still under the control of a mining company until the early 1990s, when it was turned over to the conservation organization ANCON. It is now operated as a small ecotourism lodge by ANCON Expeditions, a private firm." Williams, 1908 Apr, Jun 3, 1912 Mar 22 (USNM); Goldman, at 1,800-4,000 feet (550-1,200 m), specimens are labeled both Cana and Santa Cruz de Cana, and it does not appear that these two names represent separate collecting localities. His field notes refer only to Cana, the dates of collection for Cana and for Santa Cruz de Cana overlap, and the elevational range is the same for both names. SMN2A shows Santa Cruz de Cana as a city 2 mi (3 km) SE of Cana, but other maps show only one or the other name at this location, and Woakes (1900:249) indicates that Santa Cruz de Cana is an older name for Cana. According to his journal, he collected in the vicinity of Cana and above, across the Río Seteganti to virgin forest and level ground near the Cana Valley, along the Río Seteganti as far as the power plant at Chuyua (or Chuyuca ?) (not located). Goldman (1920:12) adds that "heavily forested mountain slopes cut by numerous streams were of easy access behind the town, while open fields, old clearings and marshy meadows in the valley added to the wealth of environmental conditions." One specimen is also labeled "base, Mount Pirre." The exact dates of collection are unclear as his field notes indicate that he left Cana on Mar 28 and returned to Cana on Apr 9, yet several USNM specimens are clearly labeled Mar 29-31, which are dates that overlap with those of Marragantí [0808/7744 (FH)], ca. 25 mi (40 km) distant. He either cataloged the Cana specimens late or had someone at Cana collecting for him, 1912 Feb 25-Mar 3, 7-10, 12-20, 22-27, 29 ?, 30 ?, 31 ?, Apr 10-11, May 19-22, 24-31, Jun 1-8, 10-12, 9 mi (14 km) E of Cana at the Río Grande, which according to Goldman (1920:11), is "a local name applied to the upper trunk of the Río Tuyra" where the "Río Cana, Río Setegantí, Río Escucha Ruido and Río Limon unite in the marshy valley," 1912 Mar 3 (USNM, as "Cana"; and as "Santa Cruz de Cana"; Nelson, 1912b; Nelson, 1913a; Chapman, 1917; Wetmore, 1950b; Wetmore, 1951a; Wetmore, 1953; Deignan, 1961:176, as "Santa Cruz de Cana"; Haffer, 1962; Olson, 1981b, as "9 mi E of Cana"; Olson, 1981c; Olson, 1983); Benson, GL = Cerro Pirre (q. v.) indicated on some specimens, below 2,700 feet (800 m), 1928 Mar 9-Apr 26, Jun 23, 26-27, Jul 13-26, 28-31, Aug 1-6, 8 (ANSP; CM; FMNH; MCZ; Chapman, 1929b; Griscom, 1929a; Griscom, 1932a; Howell, 1956; Eisenmann and Howell, 1962; Haffer, 1967b; Wetmore, 1970); Pearson, at 2,000 feet (600 m), 1938 Feb 28-Mar 2, Apr 4, 11-13, 16, 17 (ANSP); Hines and GMLP, GL for Cerro Pirre, 1965 Aug 9 (USNM); Braun, M., O'Neill, and Parker, at 600-650 m, 1982 Feb 4-5, and as a GL for Cerro Pirre, 1982 Feb 6-12 (LSUMZ; Robbins et al., 1985); Allen-Stotz, D. Braun, Lanyon, Robbins, and Sarco, 650-1,300 m, 4-8 km along Boca de Cupe trail, 1982 Jul 12-13, 16, 25, 27, Aug 2-15, and as a GL for Cerro Pirre, 1982 Jul 19-Aug 2 (GML; LSUMZ; Robbins et al., 1985; Brumfield and Braun, 2001); Wolf, 1985 Jan 30 (Braun and Wolf, 1987); Bermingham, Guitton-Mayerma, Klicka, Lelevier, Maley, Miller, and Winker, at 500 m, 2006 Feb 4-10 (STRI; UAM). (Griscom, 1933; Moore and Peters, 1939; Wetmore, 1952:30; Wetmore, 1959b:10-11; Wetmore, 1963a; Parkes, 1976; Olson, 1981d; Hekstra, 1982). See also: Cana, Loma; Nique, Alturas de; and Pirre, Cerro.

CANA, CERRO; see Cana, Loma

CAÑA, ISLA DE; see Cañas, Isla

CANA, LOMA; Darién

ca. 0746/7744

4,900 feet (1,500 m), Fairchild and Handley (1966) do not give coordinates for the locality, but indicate that Loma Cana is near both Cerro Pirre [0749/7744 (GA)] and the Río Setegantí [0746/7738 (USBGN)]. The peak is not shown on our maps, but presumably Cerro Cana and Loma Cana refer to a peak in the Serranía de Pirre [ca. 0749/7744] very near Cana [ca. 0745/7742], and probably to the peak that most collectors called Cerro Pirre. Angehr (in litt., Aug 2005) points out that "the use of this term most likely reflects the realization by investigators, starting with Pearson, that the peak near Cana was not the true Cerro Pirre." Pearson, at 5,100 feet (1,550 m), the distinction between Pearson's "Mt. Pirre" and "Cana Loma" is unclear as the elevations and dates overlap for the two localities, 1938 Feb 22, Mar 28 (ANSP); Hinds and GMLP, cloud forest, trail to camp, 1965 Aug 2-10, 18 (?) (GML, as "Cerro Cana"; USNM, as "Cerro Cana Strip"). See also: Cana; Nique,

Alturas de; and Pirre, Cerro.

CANA, RÍO; Darién
ca. 0745/7742
Short river that originates W of Cana [ca. 0745/7742] and flows E entering the Río Seteganti [0746/7738 (USBGN)] at Cana (Anthony, 1916 (map)). Schneider, 1965 Jul (Ridgely and Gwynne, 1976:265, as "Cana River"; Dungel and Sebela, 1995, as "River Cana"). See also: Cana.

CANA, RÍO; see Cañaveral, Río

CANAL; see Changuinola Canal

CANAL AFUERA, ISLA; see Canal de Afuera, Isla del

CANAL DE AFUERA, ISLA; see Canal de Afuera, Isla del

CANAL DE AFUERA, ISLA DEL; Veraguas (Pacific)
0741/8137 (USBGN)
201 m, island with a land area of 2 km^2 (200 ha), located in the Pacific Ocean 15 km NE of Isla de Coiba [0728/8147 (USBGN)] (JOG, as "Isla de Canal de Afuera"; Olson, 1997a). Batty, specimens at AMNH are dated 1902 Jan 10, but Batty's data were completely fabricated and his specimens labeled "Afuera Island" were actually collected in 1901 near his mainland bases at Boquete [0847/8226 (USBGN)] and Boquerón [0830/8234 (USBGN)] (AMNH, as "Afuera Island"; Zimmer, 1944, as "Afuera Is."; Eisenmann, 1950, as "Afuera I."; Olson, 2008); Wetmore, described the island as heavily forested with central peaks that were steeply sloped and surrounded by a level coastline, landed at a small beach on the W coast near a small stream, up a sloping ridge through tall forest with heavy undergrowth to the top of the island, shrubs on the easternmost beach, 1962 Mar 20-21, 26 (USNM, as "Isla Canal de Afuera"; Olson, 1997a, as "Isla Canal de Afuera"); Garcés and Mena, 1996 Feb 18, 20-21 (MVUP, as "Isla Canal Afuera").

CANAL DE COLORADO; see Colorado, Canal de

CANALES DE TIERRA, ISLA; Veraguas (Pacific)
0745/8134 (USBGN)
161 m, island with a land area of ca. 2 km^2 (200 ha) located 0.5 km from the SW coast of Veraguas province in the waters NE of Isla de Coiba [0728/8147 (USBGN)] and Isla del Canal de Afuera [0741/8137 (USBGN)] (MGRP; MHA, as "Isla Medidor"). Batty, specimens at AMNH are dated 1902 Feb 1, 4, but Batty's data were completely fabricated and his specimens labeled "Medidor Island" were actually collected in 1901 near his mainland bases at Boquete [0847/8226 (USBGN)] and Boquerón [0830/8234 (USBGN)] (AMNH, as "Medidor Island"; Hartert and Goodson, 1917:418, as "Medidor Island"; Olson, 2008).

CANAL ZONE FOREST AND WILDLIFE PRESERVE; see Madden Road Timber Preserve

CANA RIVER; see Cana, Río

CAÑAS, ISLA; Panamá, Archipiélago de las Perlas
0823/7849 (USBGN)

Island with a land area of 500 ha (MacArthur et al., 1973) or 522 ha (Wright et al., 1985) located 0.5 km E of Isla del Rey [0822/7855 (USBGN)] (JOG, as "Isla de Cañas"), and according to a map section in Wetmore's journal, the elevation is 450 feet (140 m) (USNM). Wetmore, sandy beach on the S coast, a mangrove swamp fed by a fresh water stream, remaining forest along a stream, a trail inland through gallery forest and through areas of cultivation, in a low forest with dense undergrowth, inland to clearings on the hill slopes, 1960 Jan 20-23 (USNM, as "Isla de Cañas"; Wetmore, 1970); MacArthur et al., mist nets, observations, "largely covered with the same tall, rather dry forest that completely clothes [Isla de] Puercos," 1971 Jan, Dec, 1972 Jan, Dec (MacArthur et al., 1972—species list; see comments in Bibliography; MacArthur et al., 1973; Wright et al., 1985); Wright and Campbell, mist nets, observations, 1978 Jan-Mar (Wright et al., 1985); Sinclair, 0.5 mi (1 km) off, 1987 Nov 3 (LACM, as "Canas Is").

CAÑAS, ISLA DE; see Cañas, Isla

CAÑAS, RÍO; see Cañazas, Río

CAÑASAS, RÍO; see Cañazas, Río

CAÑAVERAL; Coclé (Pacific)
0831/8026 (USBGN)
ca. 50 m, town 8 km W of Penonomé [0831/8022 (USBGN)] (JOG; MGRP). Trujillo, 1994 Aug 13 (MVUP).

CAÑAVERAL, RÍO; Bocas del Toro
0901/8142 (USBGN)
Originates in Bocas del Toro province in the N foothills of the Serranía de Tabasará [0833/8140 (USBGN)] and flows N entering the Golfo de los Mosquitos [0900/8115 (USBGN)] 3 km S of Cocoplum Point [0902/8143 (USBGN)] (E762, as "Río Cañaveral (Río Caña)"). Benson, GL = Cocoplum Point, at least some of Benson's specimens are labeled "Río Cano," which presumably refers to the Río Cañaveral as USBGN treats "Río Cana" as an older name for Río Cañaveral, 1927 Oct 24-Nov 12 (AMNH, as "Río Cano"). Three small lakes – Laguna Río Diablo [0901/8144 (USBGN)], Laguna de Jugli [0856/8144 (USBGN)], and Laguna de Samani [0856/8143 (USBGN)] – are located near the river's mouth. Laguna de Samaní (MGRP, as "Laguna de Namaní"; JOG, as "Laguna de Damaní") and Laguna de Jugli (MGRP) are both on the Río Cañaveral ca. 8 km S of its mouth, and Laguna Río Diablo (MGRP; JOG) empties directly into the Golfo de los Mosquitos just N of the river's mouth. These lakes are mentioned in Delgado (1983).

CAÑAZA, RÍO; see Cañazas, Río

CAÑAZAS, RÍO; Panamá (E, Pacific)
0904/7828 (USBGN)
Originates in E Panamá province on the continental divide near the Panamá/Darién provincial border. The river flows S then W entering the Río Chepo [0900/7906 (USBGN)] 4 km E of Lago Bayano [0908/7835 (USBGN)] and 83 km NNW of La Palma [0825/7809 (USBGN)] (JOG, as "Río Cañazas o Nargandí"; MGRP, as "Río Cañazas o Nargandí"). Our older map, MRP (as "Río Cañaza") shows the

river originating farther S and flowing due W. Hinds, 1974 Feb 15-16 (GML, as "Río Canas," as "Río Canasas," as "Río Canazas," and as "Río Canozo").

CAÑAZAS, SERRANÍA DE; see Majé, Serranía de

CANCHIGUA; Panamá (E, Pacific)
0911/7853 (USBGN)
ca. 50 m, town on the Río Chepo [0900/7906 (USBGN)] 9 mi (14 km) ESE of El Llano [0913/7859 (USBGN)] and ca. 3 mi (5 km) SE of Boca de Cañita [0912/7855 (USBGN)] (MHA, as "Conchita"; SMN2A, as "Conchita"). Wetmore, GL = Boca de Cañita, Río Chepo, described as a beach on the Río Chepo ca. 3 mi (5 km) upriver from Boca de Cañita, followed a small, 10-30 foot (3-9 m) wide, dry streambed inland to a 1,000 feet (300 m) tall rock cliff, 1962 Feb 7 (USNM).

CANDELA, QUEBRADA; see Candela, Río

CANDELA, RÍO; Chiriquí
0846/8250 (USBGN)
Originates on the S slopes of Cerro Pando [0855/8243 (USBGN)], flows S within 1-6 km of the Costa Rican border and enters the Río Chiriquí Viejo [0820/8241 (USBGN)] 11 km SW of Santa Clara [0850/8246 (USBGN)] (JOG). Wetmore, GL = Río Sereno [0849/8252 (USBGN)], at 3,600-3,700 feet (1,100 m), specimens are labeled "3 miles E of Sereno," although JOG indicates that the direction is NE. Collected 3 mi (5 km) NE of Río Sereno where the river was 20-25 feet (6-8 m) wide with clear water, the land was largely in pasture with stretches of brush and some forest on the hills above the river, a cleared area with cattle, pigs and some *rastrojo*, a level woodland bordered by a newly planted field with bananas and beans, in a recently burned field, 1955 Feb 21-23 (USNM, as "Quebrada Candela").

CANDELARIA; see Candelaria, Estación Hidro el

CANDELARIA, ESTACIÓN HIDRO EL; Panamá (E, Atlantic)
0923/7931 (USBGN)
100 m, water measurement station on the Río Pequení [0921/7933 (USBGN)] 5 km NNE of its mouth, near the mouth of the Quebrada Candelaria [0923/7931 (USBGN)] (JOG). Wetmore, GL for: Quebrada Candelaria, and Río Pequení, 1961 Mar 2-11 (USNM, as "Candelaria Hydrographic Station"; Parkes, 1976, as "Candelaria"; Olson, 1981d, as "Candelaria Hydrographic Station").

CANDELARIA, QUEBRADA; Panamá (E, Atlantic)
0923/7931 (USBGN)
Not shown on our maps, but USBGN indicates a tributary of the Río Pequení [0921/7933 (USBGN)] that discharges near Estación Hidro el Candelaria [0923/7931 (USBGN)]. Wetmore, GL = Estación Hidro el Candelaria, upstream to a 20 foot (6 m) cascade, above in virgin forest, along a small tributary, 1961 Mar 8, 10 (USNM; Wetmore, 1962). See also: Pequení.

CANDELARIA HYDROGRAPHIC STATION; see Candelaria, Estación Hidro el

CANGANDÍ; San Blas
0928/7907 (USBGN)

ca. 50 m, located in W Comarca San Blas ca. 24 km SW of El Porvenir [0934/7857 (USBGN)] on the Río Cangandí [0928/7907 (USBGN)] 2 km S of its junction with the Río Mandinga [0928/7904 (USBGN)] (MGRP). On RDP2 the town is located at the river's mouth on the Colón/San Blas provincial boundary. Blake, at 50 m, mist nets, "premontane wet forest," 1985 May 15-19 (Blake, 1989).

CANGANDÍ, RÍO; San Blas
0928/7907 (USBGN)
Originates on the continental divide in W Comarca de San Blas, flows E then N, and enters the Río Mandinga [0928/7904 (USBGN)] 5 km W of Mandinga [0927/7904 (USBGN)] (JOG). Shown as Río Samgondi on MHA and E501 and as Río Samgandi on SMN2A. Wetmore, GL = Mandinga, on the river, 1957 Feb 13 (USNM); Galindo, GL = Mandinga, at 200 feet (60 m), 1957 May 21 (USNM).

CANGLÓN, RÍO; Darién
0820/7746 (USBGN)
Originates in the Serranía el Filo de Tallo [0827/7757 (USBGN)] and flows E entering the Río Chucunaque [0809/7744 (USBGN)] 12 mi (19 km) NNW of Yaviza [0811/7741 (USBGN)] and 14 mi (23 km) N of El Real de Santa María [0808/7743 (USBGN)] (SMN2B). Baer, GL = Yaviza, Río Chucunaque, 1924 Mar 1, 8 (USNM). For more on Baer see comments under Marsh (1934).

CANGREJO; see Caracoles, Isla

CANGREJO, ISLA; see Caracoles, Isla

CAÑITA; see Boca de Cañita

CAÑITA; see Las Cañitas

CAÑITA, BOCA DE; see Boca de Cañita

CAÑITA, RÍO; Panamá (E, Pacific)
0912/7855 (USBGN)
Originates in central Panamá province and flows WSW entering the Río Chepo [0900/7906 (USBGN)] at Boca de Cañita [0912/7855 (USBGN)] 6 km ESE of El Llano [0913/7859 (USBGN)] (JOG). See also: Boca de Cañita; and Las Cañitas.

CAÑITAS, LAS; see Las Cañitas

CANO, RÍO; see Cañaveral, Río

CAÑO QUEBRADO, RÍO; Panamá (Canal Zone) (W, Atlantic)
0914/8005 (USBGN)
Originates NW of La Chorrera [0853/7947 (USBGN)] and flows N and NE entering the S coast of Lago Gatún (Gatun Lake) [0912/7955 (USBGN)] where the lower course of its original streambed is now submerged. The river formerly continued NE and entered the continuous Río Chagres (see Rio Chagres [0919/8000 (USBGN)]) at Tabernilla [ca. 0908/7949] (IP, as "Cano Quebrado"; RM1). Goldman, specimens are also labeled "near Tabernilla," but his field notes indicate that he traveled by canoe as far upstream as navigable, 1911 Apr 9 (USNM).

CAÑO SADDLE; Colón (Canal Zone) (W)

ca. 0906/8003
100 feet (30 m), a high point on the limits of the Lago Gatún (Gatun Lake) [0912/7955 (USBGN)] watershed, located 23 km SW of Gatún [0916/7955 (USBGN)] on the NW coast of Lago Gatún (Gatun Lake) between Cuipo [0905/8003 (USBGN)] and La Ullama [0907/8002 (USBGN)] (E661, as "Caño Saddles"; E761A). Imhof, observations, 1942 Dec 29-30 (Imhof, 1950; Wetmore et al., 1984:251, 302).

CAÑO SADDLES; see Caño Saddle

CAÑOZO, RÍO; see Cañazas, Río

CANTA GALLO; see Canta Gallo Abajo

CANTA GALLO ABAJO; Chiriquí
0822/8238 (USBGN)
ca. 25 m, town 8 km SW of Alanje [0824/8233 (USBGN)], 1 km E of the Río Escarrea [0820/8241 (USBGN)] and 6 km N of the coast (JOG). Wetmore, GL = Alanje, level, sandy area cut by shallow streams, endless pastures with scattered shrubs and occasional trees, tract of original savanna down to the coast, along the banks of the Río Escarrea (q. v.) through dense growth with creepers, a horse trail inland through low scrub, road bridge over the Río Escarrea, 1960 Mar 4, 12-13 (USNM, as "Canta Gallo").

CANTERA DE MOCAMBO; see Mocambo, Cantera de

CAPANDÍ; see Diablo, Río

CAPANDÍ, RÍO; see Diablo, Río

CAPE MALA; see Azuero, Península de and see Mala, Punta

CAPE MALA PENINSULA; see Azuero, Península de

CAPETANA, LA; see La Capitana

CAPETÍ; Darién
0804/7733 (Fairchild and Handley, 1966)
ca. 50 m, town on the Río Capetí [0805/7735 (USBGN)] 3 mi (5 km) above its mouth and 4 mi (6 km) NE of Boca de Cupe [0803/7735 (USBGN)] (MGRP, as "Río Capeti"; SMN2B). Richardson, GL for Río Capetí, 1915 May 24, 26, 29 (AMNH; Anthony, 1916; Zimmer, 1934:20; Zimmer, 1950:8, 26; Wetmore, 1953).

CAPETÍ, RÍO; Darién
0805/7735 (USBGN)
Originates W of Cerro Tacarcuna [0811/7717 (USBGN)] and flows W joining the Río Tuira [0821/7803 (USBGN)] 1 mi (1.5 km) N of Boca de Cupe [0803/7735 (USBGN)] (SMN2B). Richardson, GL = Capetí, 1915 May 24, 26, 29 (AMNH, as "Capeti River"; Anthony, 1916; Griscom, 1927b, as "Capeti River"; Wetmore, 1952:19).

CAPE VALIENTE; see Valiente, Punta

CAPIRA; Panamá (W, Pacific)
0845/7953 (USBGN)

100 m, town on the Carretera Interamericana [0832/8242 (USBGN)] 17 km SW of La Chorrera [0853/7947 (USBGN)], located 9 km NE of Cerro Campana [0841/7955 (USBGN)] and 9 km ENE of Cerro Trinidad [0844/7958 (USBGN)]. The town is located on the Río Capira [0848/7950 (USBGN)], which originates on the E slopes of Cerro Campana (E501; E761A; MGRP; MHA). Olson et al. (1968) suggest that some of the specimens labeled "Capira" were taken W of town in the "Serranía de la Capira," which is now known as the Cordillera del Llorón [0843/7957 (USBGN)]. Rothschild Collection, in high forest, 1888 Mar 21 (AMNH; Olson et al., 1968); Heyde, 1888 Apr 21 (USNM); Twomey, 1939 Apr 13 (CM; Trimble, 1943; Parkes, 1969); Montenegro, GL for El Aguacate Número Uno, 1970 Aug 21, 1971 Nov 21, 1972 Jun 28, 1973 Jul, 1976 Mar 6, 1978-1981, 1982 Jun 14, 1983 Aug 15-16, 1984 May 16, 25 (GML); Collector ?, 1989 Mar 22 (GML). (Todd, 1942:325).

CAPIRA, LAGO DE; Panamá ? (W, Pacific) ?
ca. 0845/7953 ?
The specimen had no province indicated, but probably came from the area around Capira [0845/7953 (USBGN)] in Panamá province, Collector ?, 1974 Feb 23 (GML).

CAPIRA, SERRANÍA DE LA; see Capira

CAPITANA, LA; see La Capitana

CARACOLES, ISLA; Panamá, Archipiélago de las Perlas
0830/7857 (USBGN)
23 m, small island with a land area of 2 ha (Angehr, 2003:275, as "Isla Cangrejo") located 4 km N of Isla del Rey [0822/7855 (USBGN)] (E501). Described by Angehr (2003:275, as "Isla Cangrejo") as an island within the Caracoles group that is too small for habitation. Described by Rendahl (1920:5-6, pl.1 (map), as "Cangrejo") as a steep sided, desert rock island located 1 mi (1.5 km) N of Isla del Rey, Bovallius, observations, see discussion of Bovallius under Pueblo Nuevo [ca. 0820/8013 ?], 1882 (Rendahl, 1920, as "Cangrejo").

CARACOLES, PUNTA; Darién
0741/7818 (USBGN)
Sea level, point on the SW coast of Darién province 30 mi (48 km) S of Punta Garachiné [0806/7825 (USBGN)] and 15 mi (24 km) NNW of Jaqué [0731/7810 (USBGN)] (SMN2B). Murphy and Correia, Askoy Expedition, 4 mi (6 km) W of, 1941 Feb 26 (AMNH; Murphy, 1944a).

CARBON, QUEBRADA; Bocas del Toro
ca. 0922/8235
An intermittent stream that originates in the hills WSW of Charagre [0923/8234 (USBGN)] and flows SE entering the Río Teribe [0922/8232 (USBGN)] 2.5 km SSW of Charagre (E762). USBGN and E762 indicate another Quebrada Carbon at 0920/8233, which is not the SBE collecting locality, and that originates SW of Charagre and enters the Río Changuinola [0928/8227 (USBGN)] ca. 7 km S of Charagre. SBE, GL = Charagre, Rio Teribe, 3 km SW of Charagre on the N bank of the Río Teribe, 1991 Feb 24, 28 (USNM; Brumfield and Braun, 2001, as "N

bank Río Teribe"; Brumfield et al., 2001).

CARBUNCO, CERRO; Panamá (E, Pacific)
0911/7914 (USBGN)
200 m, not shown on our maps, but USBGN indicates a peak ca. 16 km WNW of Chepo [0910/7906 (USBGN)]. Wetmore, however, applies the name to the entire range (USNM), which is shown on JOG extending ca. 17 km ENE from Cerro Carbunco to a maximum elevation of 620 m. The range is bordered on the W by the Río Pacora [0901/7917 (USBGN)], on the N by the Río Indio [0912/7914 (USBGN)] and on the E by the Río Mamoní [0908/7904 (USBGN)]. The range is not part of the continental divide, which is parallel to it ca. 17 km to the NW. Described by Wetmore as a long, narrow ridge extending E-W, the E was covered with open forest and the W with open grassland, Wetmore and Perrygo, GL = Chepo, at 250-1,000 feet (80-300 m), rolling, grass covered hills, forested area near Río Tranca [ca. 0905/7908], 1949 Apr 14 (USNM; Wetmore, 1951a).

CARDENAS; Panamá (Canal Zone) (E, Pacific)
0900/7934 (USBGN)
40 m, 5 km N of Balboa and ca. 2 km E of the Panama Canal [0920/7955 (USBGN)] (E762, as "Cardenas Townsite"). USBGN treats Cardenas Townsite as an older name for Cardenas. Kilham, observations 1965 Feb, 1970 Nov-1971 Feb (Kilham, 1972, as "Cardenas Village"); Trainer, observations, 1981-1983 (Trainer, 1989).

CÁRDENAS, RÍO; Panamá (Canal Zone) (E, Pacific)
0859/7935 (USBGN)
Originates in Panamá province on the continental divide and flows S into the former Canal Zone, where it crosses Camino Chiva Chiva (Chiva Chiva Road) [0902/7935 (USBGN)] and Las Cruces Trail [ca. 0904/7936] and passes through Fort Clayton Army Reservation [0901/7934 (USBGN)] before it discharges into the Río Guanábano [0901/7934 (USBGN)], which enters the canal near Miraflores Locks [0900/7936 (USBGN)] (E661; E8660; RM1). Wetmore, GL = Chiva Chiva, green, herbaceous vegetation, 1962 Jan 7 (USNM).

CARDENAS VILLAGE; see Cardenas

CAREENING CAY; see Carenero, Cayo

CAREENING KEY; see Carenero, Cayo

CARENERO, CAYO; Bocas del Toro, Archipiélago de Bocas del Toro
0921/8214 (USBGN)
2 m, island with a land area of 0.8 km^2 (80 ha), located 200 m E of Bocas del Toro [0920/8215 (USBGN)] and 2 km W of Isla Bastimentos [0918/8208 (USBGN)] (E762). SBE, forested, but most of the land subject to tidal inundation, 1989 Jan 29 (USNM; Olson, 1993a).

CARIBBEAN SLOPE; see Boquete Trail

CARIBBEAN SLOPE OF VOLCÁN DE CHIRIQUÍ; see Boquete Trail

CARIBBEAN SLOPE OF WESTERN PANAMA; see Almirante

CARIBBEAN SLOPE OF WESTERN PANAMA; see Bocas del Toro

CARRASQUILLA; see La Carrasquilla

CARRASQUILLA, LA; see La Carrasquilla

CARRETERA DE GAMBOA; see Gaillard, Carretera

CARRETERA ESCOBAL; see Gatún-Escobal Road

CARRETERA GUALACA-CHIRIQUÍ GRANDE; see Gualaca-Chiriquí Grande Road

CARRETO, BAHÍA DE; see Permé

CARRIASO, QUEBRADA; see Quebrada Carriaso

CARRIAZO; Panamá (E, Pacific)
0912/7914 (USBGN)
100 m, a town located on the Río Pacora [0901/7917 (USBGN)] 15 km NNE of Pacora [0905/7917 (USBGN)] and 1 km SE of San Miguel [0914/7915 (USBGN)] and 2 km NW of the mouth of the Río Indio [0912/7914 (USBGN)] (MGRP). See also: Quebrada Carriaso.

CASA LARGA; see Calzada Larga

CASA LARGO; see Calzada Larga

CASA TILLEY; Chiriquí
ca. 0852/8235
5,300 feet (1,600 m), settlement located between Bambito [0850/8237 (USBGN)] and Cerro Punta [0852/8235 (USBGN)] just above the mouth of a small tributary of the Río Chiriquí Viejo [0820/8241 (USBGN)] (VCP). Greenwell, GL = Cerro Punta, at 5,300 feet (1,600 m), 1962 Mar 13-14 (USNM).

CASA WINKLOSKI; see Bambito

CASAYA, ISLA; Panamá, Archipiélago de las Perlas
0830/7902 (USBGN)
Sea level, small island with a land area of ca. 3 km^2 (300 ha) located in Bahía de Panamá [0850/7915 (USBGN)] 10 km NW of Isla del Rey [0822/7855 (USBGN)] and 5 km NW of Isla Viveros [0828/7900 (USBGN)] (E501; JOG). Bovallius, the island's W side was covered with "dense primeval forests," and the E side was "well cultivated and richly populated," and "there were great cultivations of cocoas and banana-trees," see discussion of Bovallius under Pueblo Nuevo [ca. 0820/8013 ?], 1882 Apr 13 (NHRM; Rendahl, 1920); Levings et al., observations, 1982-1983 (Levings et al., 1986, as "Cassaya"). (N. Smith, 1971).

CASAYA, RÍO; Panamá (Canal Zone) (E, Atlantic)
0907/7941 (USBGN)
Originates in hills S of Juan Mina [0910/7939 (USBGN)], flows SW then W entering the E bank of the Río Chagres [0907/7942 (USBGN)] 2.5 km ENE of Gamboa [0907/7942 (USBGN)] (E866). See also: Las Cruces.

CASAYETA, ISLA; see Bavarena, Isla

CASCAJAL; Coclé (Atlantic)
0843/8028 (USBGN)
400 m, town on the Río Cascajal [0850/8029 (USBGN)] ca. 29 km NNW of

Penonomé [0831/8022 (USBGN)] (MGRP). Originally there was much confusion about the location of Cascajal. Until 1903 Panamá was a province of Colombia and therefore many of Heyde's specimens, which were collected in the late 1880s, bear the label "Colombia." This may explain why Hellmayr (1925:309), writing after Panamá's independence, mistakenly placed Cascajal in eastern Panamá. Griscom (1932:348) later stated that Cascajal was in western Panamá, but mistakenly describes Coclé as "exclusively an arid tropical savannah country in which the rain forest species [of Heyde and Lux]...do not occur." He also writes that there is no Cascajal in Coclé, and that the birds must have come from either the Río Cascajal just E of the Canal Zone or from the Río Coclé [= Río Coclé del Norte [0905/8035 (USBGN)]] just W of the Canal Zone. The definitive answer, however, comes from an undated, handwritten note in the USNM museum catalogs, presumably written by Heyde regarding his own specimens, indicating that Cascajal was located in the "Department of Coclé, Isthmus of Panamá, Atlantic side," and that specimens were collected in the mountains above the Río Coclé del Norte, which clearly indicates that specimens came from the region near the town that now bears the name Cascajal. According to Wetmore, the region was one of heavy forest, with some savanna areas on the high slopes of the mountain divide, and he believed that specimens labeled "Cascajal" came from the savanna area down through the forest to the Río Coclé del Norte. Heyde and Lux, "virgen mountain," "virgen forest," "deep woods of high trees," and "near river Coclé del Norte, but higher up in the mountains," 1889 Feb 2-Apr 4 (USNM; Hartert, 1898a:498, as "Panama"; Ridgway, 1909; Hellmayr, 1925; Griscom, 1932a; Griscom, 1935b:312, as "Río Coclé"; Olson, 1993b); Collector ? [Heyde and Lux ?], 1889 Feb 5, 9 (AMNH; MCZ; Wetmore, 1963b); Weir, at 350-420 m, 2003 Apr 20-21 (CU; LSUMZ).

CASCAJAL; Colón (E)

ca. 0933/7937

40 feet (10 m), town on the Río Cascajal [0934/7939 (USBGN)] 4 km E of Portobelo [0933/7939 (USBGN)] in an area of grass and marsh (E761A). See also: Cascajal, Río.

CASCAJAL, RÍO; Coclé (Atlantic)

0850/8029 (USBGN)

Originates on the continental divide and flows N joining a tributary of the Río Coclé del Norte [0905/8035 (USBGN)] ca. 29 km NNW of Penonomé [0831/8022 (USBGN)] (JOG). See also: Cascajal.

CASCAJAL, RÍO; Colón (E)

0934/7939 (USBGN)

Originates N of Cerro Bruja [0929/7934 (USBGN)] and flows WNW entering the Caribbean Sea [1500/7500 (USBGN)] 1 km N of Portobelo [0933/7939 (USBGN)] (JOG). Goldman, 7 mi (11 km) E of Portobelo, thick jungle along the river beyond Cascajal [ca. 0933/7937], which according to his journal was the last settlement on the river, 1911 Jun 3 (USNM, as "Cascajal River"; Goldman, 1920); Olson and Mueller, GL = Portobelo, 1966 Jun 26 (UF; USNM); Davis and Graham, observations, 1987 Dec-1988 May (Davis and Graham, 1991, as "Cascajal River"). See also: Bruja, Cerro; and Portobelo.

CASCAJAL RIVER; see Cascajal, Río

CASCAJILLOSO; Veraguas (Pacific)

ca. 0717/8043

Not shown on our maps, but according to Angehr (in litt., Aug 2005), Cascajilloso is the name that was used for the higher camp on the Río Cobachón [0713/8041 (USBGN)] (q. v.) drainage. David Agro provided the coordinates 0717/8043, elevation 650 m. This is not to be confused with the nearby town El Cascajilloso [0723/8053 (USBGN)] that appears on JOG. Agro, Conyne, Pederson, and Ridgely, GL = Península de Azuero, 1996 Feb 18-22 (ANSP; MVUP; Joseph, 2000).

CASCO ANTIGUO; see San Felipe

CASITA; Darién

0801/7722 (Fairchild and Handley, 1966)

According to Fairchild and Handley (1966:17), this was a camp located near Tacarcuna Village [0805/7717 (FH)] and was also called Tacarcuna Casita. Their coordinates indicate that it was located on the Río Topalisa [0801/7728 (USBGN)] ca. 10 km above its mouth. Feinstein, GL = Tacarcuna Village, at 2,700 feet (800 m), 1959 Jan 21-28 (USNM, as "Casita Camp"); Collector ? [Feinstein ?], GL = Tacarcuna Village, at 2,700 feet (800 m), 1959 Jan 24 (USNM, as "Casita Camp"); Hinds and GMLP, 1979 Mar (?) 21-23, Jun 22-30, Jul 1, 3, 6 (GML, as "Mi Casita").

CASITA ALTA; Chiriquí

ca. 0849/8229

Described by Blake (1958:502 (map)) as a section of the Mönniche properties located NE of Cerro Copete [ca. 0848/8231] on the Quebrada El Velo [0849/8228 (USBGN)]. The property was named for a small hut at 7,000 feet (2,150 m) often used by visiting biologists. Mönniche, at 6,800-9,000 feet (2,050-2,750 m), 1932-1933, 1936-1937 (ANSP; CAS; FMNH; MCN; Blake, 1958); Davidson, at 6,800-8,500 feet (2,050-2,600 m), 1933 Jan 24-25, 27 (CAS); Hartman, GL = Boquete, at 7,000 feet (2,150 m), 1949 Feb 9, 14, and GL = Cerro Copete, at 7,000 feet (2,150 m), 1949 Feb 8, 11-12 (USNM).

CASITA CAMP; see Casita

CASTELLO; see Castillo

CASTILLO; Veraguas (Pacific)

0813/8103 (USBGN)

100 m, town 16 km NNW of Santiago [0806/8059 (USBGN)] (JOG, as "El Castillo"). Arcé, 18?? (USNM; Sclater and Salvin, 1868c:388-390, as "Castello"; Sclater and Salvin, 1867:123-124, as "Castillo," and as "Veragua"; Salvin, 1870bb; Berlepsch, 1911). Some of Arcé's specimens from Castillo may have come from La Yeguada (G. Angehr, pers. comm.) See also: Veraguas, Provincia de.

CASTILLO, LAGUNA DEL; see La Yeguada, Laguna de

CASSAYA, ISLA; see Casaya, Isla

CATHEDRAL PARK; Panamá (E, Pacific)

ca. 0857/7932

Presumably refers to the Plaza of the National Cathedral, centrally located in Casco Antiguo in San Felipe [0857/7932 (USBGN)] (PTC, as "Plaza Cathedral Central"). Collector ?, either "Panama Province" or "Panama Railroad" is written on the specimen label, 1964 Apr 4 (GML).

CATIVÁ; see Catival, Campamento de

CATIVÁ, RÍO; see Catival, Río

CATIVAL; see Catival, Campamento de

CATIVAL, CAMPAMENTO DE; Veraguas, Isla de Coiba

0729/8144 (USBGN)

Sea level, prison camp on the central E coast of Isla de Coiba [0728/8147 (USBGN)] on the Río Catival [0729/8143 (USBGN)] (q. v.) 1 km above its mouth (JOG). Wetmore, GL = Isla de Coiba, tall trees with large buttressed bases and gnarled roots on the muddy shore, stands of low growth, inland along the Río Catival to a small lagoon, 100 yards (90 m) by 50 yards (45 m), with 3 foot (1 m) deep water between low, cleared hills, 1956 Jan 11, 13-14, 17, 19, 27, Feb 2 (USNM, as "Catival," and as "Cativá"; Wetmore, 1957).

CATIVAL, RÍO; Veraguas, Isla de Coiba

0729/8143 (USBGN)

Originates in the N central hills of Isla de Coiba [0728/8147 (USBGN)] and flows SE entering Bahía Damas [0727/8141 (USBGN)] 5 km SW of Punta Damas [0730/8141 (USBGN)] (JOG). According to Wetmore, the river near Campamento de Catival [0729/8144 (USBGN)] was 20-30 feet (6-9 m) wide, but downstream on the coast the river opened onto flats where the water was only 4-5 inches deep during low tide, Wetmore, GL = Isla de Coiba, specimens labeled Río Catival were all collected near the river mouth, considerable stand of red mangroves, white mangroves in bloom, sandy beach with *Tilia* and with coconut palms mingled with dense brush, on a trail from Campamento de Catival through an old field into a wetland, then through saw grass along a low, sandy ridge and down to the beach, 1956 Jan 12, 16, 18, 20, 27, Feb 2 (USNM, as "Boca Río Catival," and as "Boca Río Cativá"; Wetmore, 1957). See also: Catival, Campamento de; and La Central.

CAUCHALES; see El Cauchal

CAUCHERO, QUEBRADA; Panamá (E, Pacific)

ca. 0846/7832

Described by Wetmore as a stream that originates on the S slopes of Cerro Chucantí [ca. 0850/7828], flows S and joins the Río Chucantí [0846/7832 (USBGN)] at ca. 450 feet (140 m), Wetmore and Perrygo, GL = Cerro Chucantí, Serranía de Majé, at 450-1,000 feet (140-300 m), the entire region was covered in green, gallery forest, collected near the 50-60 foot (15-20 m) wide mouth and upstream near its headwaters, along tributaries, on the SW base of Cerro Chucantí, along ridges above the river, 1950 Mar 5-9, 11-18 (USNM; Wetmore, 1950b; Wetmore, 1952:19; Wetmore, 1959b:21). Not shown on our maps.

CAUSEWAY; see Fortified Islands

CAVES; see Chilibrillo, Río

CAVIMO, EL; see Finca El Jobito

CAVULLA CAMP; Herrera

ca. 0737/8049

3,000 feet (900 m), described by Aldrich and Bole (1937:4 (map), 8) as a camp on the ridge that separates the E and the W watersheds of Península de Azuero [0740/8035 (USBGN)], located between the headwaters of the Río Negro [0737/8059 (USBGN)] and the Río Mariato [0737/8059 (USBGN)] ca. 1 mi (1.5 km) S of Cerro Viejo [0737/8048 (USBGN)] and ca. 18 mi (29 km) E of Golfo de Montijo [0740/8107 (USBGN)]. According to correspondence in 1971 between Wetmore and the collectors, the camp was well known and dated back to the rubber collection era if not earlier. The collectors indicated that the camp was located on the Veraguas side of the Veraguas/Los Santos provincial border, although more recent maps place the camp in the southernmost section of Herrera province. See Península de Azuero [0740/8035 (USBGN)] for further discussion of the provincial borders. Aldrich and Bole, GL = Golfo de Montijo, Península de Azuero, at 3,000-3,200 feet (900-1,000 m) in subtropical forest, in open savannas on the ridge tops, 1932 Mar 11-14 (CM; Aldrich and Bole, 1937).

CAYMÁN, CALETA; see Naranjo, Ensenada del

CAYMITO RIVER; see Caimito, Río

CAYO AGUA, ISLA; see Agua, Cayo

CAYO DE AGUA, ISLA; see Agua, Cayo

CÉBACO, ISLA; see Cébaco, Isla de

CÉBACO, ISLA DE; Veraguas (Pacific)

0732/8109 (USBGN)

349 m, island with a land area of 77 km^2 (7,700 ha) (Angehr, 2003:170) to 83.9 km^2 (8,390 ha) (Olson, 1997a) in the Pacific Ocean just S of Golfo de Montijo [0740/8107 (USBGN)] and W of Península de Azuero [0740/8035 (USBGN)]. The long, thin island extends 26 km from SW to NE and varies in width from 1 to 6 km, its N shore rises very quickly to 100 m reaching elevations greater than 300 m in the W (JOG, as "Isla Cebaco"). The island is considered an Important Bird Area (Angehr, 2003, as "Cébaco Island"). Batty, specimens at AMNH and at BMNH are dated 1902 Feb 1-7, but Batty's data were completely fabricated and his specimens labeled "Cebaco Island" were actually collected in 1901 near his mainland bases at Boquete [0847/8226 (USBGN)] and Boquerón [0830/8234 (USBGN)] (AMNH, as "Cebaco Island"; BMNH, as "Cebaco Island"; Friedmann, 1935, as "Cebaco Island"; Aldrich and Bole, 1937:6, as "Cebaco Island"; Zimmer, 1941b:14, as "Cebaco Is."; Zimmer, 1944, as "Cebaco Is."; Eisenmann, 1950, as "Cebaco I."; Wetmore, 1957:6, as "Cebaco Island"; Olson, 2008); Wetmore, GL for Ensenada del Naranjo, 1962 Mar 27-28, GL for: Loma de la Bandera, and Platanal, 1965 Jan 11, 13-20 (USNM; Parkes, 1976; Olson, 1981a"; and Olson, 1997a, all as "Isla Cébaco").

CEBACO ISLAND; see Cébaco, Isla de

CEDAR CREEK; see Cider Creek

CEDRA, QUEBRADA; see Cider Creek

CEDRAL; Bocas del Toro

ca. 0854/8228

4,800 feet (1,450 m), described by Blake (1958:502) as one of three collecting sites used by Mönniche along the Atlantic drainage of the Holcomb Trail [ca. 0901/8215] (q. v.). The camp was "adjacent to a small stream in a fine extensive forest" and presumably located ca. 13 km NNW of Boquete [0847/8226 (USBGN)] on the upper reaches of the Río Changuinola [0928/8227 (USBGN)]. Mönniche, at 4,800 feet (1,450 m), 1933 (FMNH; Blake, 1958; Wetmore, 1967). Not shown on our maps.

CEDRO, PUNTA DEL; see Fea, Punta

CELORIO, QUEBRADA; Darién

0805/7818 (USBGN)

A short river that originates on Serranía del Sapo [0750/7817 (USBGN)] S of Garachiné [0804/7822 (USBGN)] and flows N entering Ensenada de Garachiné [0808/7820 (USBGN)] 6 km E of Garachiné and 2 km W of the mouth of the Río Sambú [0805/7818 (USBGN)] (MGRP; MHA, as "Río Celorio"; Barbour and Brooks, 1923, as "Río Celorio"). See also: Jesusito, Río.

CELORIO, RÍO; see Celorio, Quebrada

CEMENTO PANAMÁ; Colón (E)

ca. 0915/7940

Wetmore indicated "Cement Plant" on our maps on the Carretera Boyd-Roosevelt [0917/7943 (USBGN)] ca. 16 km NNE of Gamboa [0907/7942 (USBGN)]. Jaramillo, 1979 Jul 31 (MVUP).

CEMENT PLANT; see Cemento Panamá

CENEGÓN DEL MANGLE; see El Mangle, Ciénaga

CENTINELA, MORRO; Darién

0734/7814 (USBGN)

According to USBGN, this is a rock in the water off the S Darién coast near Bahía Piña [0734/7811 (USBGN)] (MHA, as "Islas Centinelas"). Mentioned in Robins (1958) regarding observations at Bahía Piña.

CENTRAL;

A modifier used by Blanton, and perhaps others, to indicate that collections were made in the interior of the country far from either coast. Presumably this could refer to either slope of the continental divide, although the specimens known to us that are labeled Central were collected on the Atlantic drainage. See also: Pac; and Road C29.

CENTRAL, CORDILLERA;

0838/8150 (USBGN)

Panamá/Coclé/Veraguas/Bocas del Toro/Chiriquí Principal mountain range W of the Panama Canal [0920/7955 (USBGN)] that extends through W and central Panamá creating the continental divide. The range is a continuation of the Cordillera de Talamanca [0930/8340 (USBGN)] (q. v.) in Costa Rica and extends roughly E-W along the Bocas del Toro/Chiriquí provincial border, through N Veraguas province and through central Coclé province ending in W Panamá province (RDPF). In the literature, the Panamanian mountains are sometimes called the Talamanca Range or the Cordillera de Talamanca (Griscom, 1935b:321, as "Talamanca"; Davidson, 1936:232, as "Cordillera de Talamanca"; Eisenmann, 1957:247, as "Talamanca range"; LaBastille, 1974, as "Talamanca Range"). Warszewicz, at 6,000-8,000 feet (1,850-2,450 m), according to Gould's notes that accompany his hummingbird plates (Gould, 1990 (1849-1861):1), "M. Warszewicz, who, while crossing from Bocco del Toro [sic] on the Atlantic side of the Isthmus of Panama to David on that of the Pacific, was induced to deviate in search of novelties to the Rovalo peak [presumably Cerro Róbalo [0853/8220 (USBGN)]]," he probably crossed the divide using the Boquete Trail [ca. 0857/8207] (q. v.) and those labeled "North side of the Cordillera" were presumably collected on the Atlantic slope in Bocas del Toro province, some specimens were also collected at Volcán Barú [0848/8233 (USBGN)] (q. v.), c. 1849 (ANSP, as "Veragua, towards Costa Rica"; BMNH, as "Cordillera of Chiriqui"; Gould, 1850a, as "Cordillera of Chiriqué in Veragua"; Gould, 1850b, as "Cordillera of Chiriquí," and as "North side of the Cordillera, Veragua"; Salvin, 1867b:129, as "Cordillera of Chiriquí"; Salvin and Godman, 1888-1904, vol. 2:145, as "Cordillera of Chiriquí"); Arcé, date ? (FMNH, as "Cordillera Veragua"); Safford, according to Wetmore, these specimens were purchased (not collected) by Safford in May 1887 from Panama City. Many of the specimens in this collection were labeled "Isthmus of Panama," and two were labeled "Mountains of Veragua." Wetmore believed that some, if not all, of the specimens were from Veraguas and Chiriquí provinces and had been collected by E. Arcé, 1887 May (USNM); Cardiff, Dittmann, and Lovette, at ca. 900 m, Bocas del Toro Province, Distrito Chiriquí Grande [0851/8205 (USBGN)] , 4.1 km by road N continental divide, presumably along the Gualaca-Chiriquí Grande Road [ca. 0857/8207] (q. v.), 1996 Mar 3, and as a GL for Fortuna, 1996 Mar 4-8 (LSUMZ; MVUP, as "Cordillera Central," and as "La Cordillera Central"); Thorstrom et al., observations at 0841/8040, which appears on our maps near the continental divide, ca. 10 km W of El Copé [0837/8036 (USBGN)], 2000 Apr 22 (Thorstrom et al., 2002, as "Coclé"). (Griscom, "Mts. of Chiriquí and Veraguas"; Davidson, "Mts. of Chiriquí"). See also: Chucú, Cordillera de; Chiriquí, Provincia de; Los Andes, Cordillera; Panama, Isthmus of; Tabasará, Serranía de; and Veraguas, Provincia de.

CENTRAL, LA; see La Central

CENTRO JARAMILLO; see Jaramilla

CERRO AZUL, LAGO; see Goofy Lake

CERRO CANA; see Cana, Loma

CERRO CANA STRIP; see Cana, Loma

CERRO COPETE; see Copete, Cerro

CERRO DEL SAPO; see Sapo, Cerro

CERRO LARGA; see Cerro Largo

CERRO LARGO; Herrera
0750/8050 (USBGN)
400 m, town in W Herrera province on the ridge separating the E and the W watersheds of Península de Azuero [0740/8035 (USBGN)] located 10 km NW of Las Minas [0748/8045 (USBGN)] (MGRP; SMN2B). Benson, 1925 Jun 29-Jul 14 (AMNH; Aldrich and Bole, 1937; Zimmer, 1941b, as "Cerro Larga"). See also: El Montuoso, Loma.

CERRO MALI CAMP; see Malí, Cerro

CERRO MORENO; see Moreno, Cerro

CERRO PANDO, SILLA DE; see Pando, Silla de Cerro

CERRO PUNTA; Chiriquí
0852/8235 (USBGN)
1,929 m, town in NW Chiriquí province 6 km S of the Bocas del Toro provincial border and 4 km SSE of the peak Cerro Punta [0853/8235 (USBGN)] (q. v.) near the Río Chiriquí Viejo [0820/8241 (USBGN)] (JOG). Hartmann, R., date ? (ANSP); Davidson, at 6,000 feet (1,850 m), 1931 Jan 1-10 (CAS; CM; Davidson, 1938, as "Cerro Punto"); Benson, at 6,100-6,200 feet (1,850-1,900 m), 1931 Dec 9, 1932 Jan 5-8, 10-11, 13-16 (MCZ, as "Cerro Punto"; MVUP; USNM, as "Cerro Punta," and as "Cerro Punto"); Eisenmann, at 5,400-7,300 feet (1,650-2,250 m), observations, high in tree crowns, stream borders, edges of clearings, cut-over open woods, montane pastures, 1958 September 17-21, 1960 Feb, 1961 Apr (Eisenmann, 1962c); Wetmore, worked near Cerro Punta in 1954, 1955, 1960 and 1965, collections were made near town and at secondary localities located within 5 km of town, which was located on a broad, grassy flat with valleys extending to the continental divide where forest remained only on the upper slopes, at 6,100 feet (1,850 m), at 6,800 feet (2,050 m), 1955 Mar 4, 6, 8, 10, at 5,700-6,900 feet (1,700-2,100 m), collected on a small stream, in pastures, along a trail through tall trees with long moss and many epiphytes and much bamboo in the undergrowth, up to the Cerro Punta-Boquete Trail [ca. 0852/8235], on foot along the road with bamboo thickets under tall trees, in a clearing at 6,900 feet (2,100 m) with many tall trees still standing, on the Río Chiriquí Viejo in areas visited in 1955, the higher slopes had been cleared in the previous five years and only occasional thickets of low shrubs with berries remained, along a shaded stream in a partly logged forest, 1960 Feb 25-26, Mar 19, 21, and as a GL for Río Chiriquí Viejo, 1954 Mar 12, 23, GL for: Bajo Chumbago, Bajo Grande, Cerro Punta-Boquete Trail, Cerro Respingo, La Garita, Las Nubes, and Potrero Cuatro Mil, 1955 Feb 28-Mar 5, 7-8, 10-11, GL for Cerro Punta-Boquete Trail, 1965 Mar 19 (LSUMZ; MVUP; USNM; Huff and Wetmore, 1967); Loftin and Belton, banding, 1962-1964 (Loftin, 1975); Means, Koslovsky, and Sawyer, GL = Volcán Barú, at 5,000-6,000 feet (1,500-1,850 m), specimens are also labeled "camp," "Río Chiriquí Viejo," and "tropical biological station," 2.5 mi (4 km) S of Cerro

Punta, 1967 Feb 25-26, 28, Mar 1-2, 4-6, 8-13, 15-20, 23-25, 30, Apr 1, 14 (AMNH; FSU; GML; MCN; USNM); Ambrose (MARU), at 6,000 feet (1,850 m), 1961 Apr 28, and as a GL for: Cerro Punta-Boquete Trail, and Finca Martínez, 1961 Apr 27, May 31 (AMNH); NIID (from Yunker), 1961 Apr 30, May 1, 2, 30 (USNM); Yunker, 1961 May 30 (USNM); Hayward, 1962 Mar 6-14 (BYU; USNM; Wetmore et al., 1984:304); Greenwell, GL for Casa Tilley, 1962 Mar 13-14 (USNM); Quinones, 1963 Mar (GML); Tyson, at 5,800-6,000 feet (1,750-1,850 m), 1965 Feb 17-20, Mar 8, 15 (GML; MCN; USNM); Morton, GL = Volcán Barú, collected within 6 mi (10 km) of Cerro Punta, primarily N and E of, 1965 Dec 30, 1966 Apr 23-24, Jul 20 (AMNH; FMNH; YPM); Johnson, 1 mi (1.5 km) S, 1966 Apr 10-11 (MVZ); Olson, at 5,500-6,200 feet (1,700-1,900 m), near and above, 1966 Jul 3 (UF; USNM); Collector ?, 1967 Sep 10, Oct 2, 20 (YPM); Heatley, 1967 Sep 22, 28 (AMNH); Kalma, 1968 Jan 20 (YPM); Hilty, 1968 Mar 7, 17, 19, 23-24, 26, Apr 2, 6-7, 9, 15-16, 18, 27, May 22, 24 (LSUMZ); Leck, 5,400-5,800 feet (1,650-1,750 m), 1968 Apr 26, Mar 24 (MCN; Leck, 1972b; Leck and Hilty, 1968); Stevenson, GL = Volcán Barú, at 5,400 feet (1,650 m), 1968 Aug 31, Sep 2-5, 7-8 (FSU; LSUMZ; USNM); Buskirk and Powell, at 5,500 feet (1,700 m), based at the FSU Cerro Punta Field Stadium, 1968 Dec 12-13, 15-18, 23-27, 1969 Jan 1-2 (AMNH; LSUMZ; Buskirk et al., 1972); Strauch, 1971-1973 (UMMZ; Strauch, 1977). (Wetmore, 1967; Oniki, 1972; Willson et al., 1982). See also: Barú, Volcán; and Las Nubes.

CERRO PUNTA; see Punta, Cerro

CERRO PUNTA, QUEBRADA; see Caldera, Río

CERRO PUNTA-BOQUETE TRAIL; Chiriquí

ca. 0852/8235

Described by Wetmore as a trail that leads E from Cerro Punta [0852/8235 (USBGN)] through Bajo Grande [0852/8233 (USBGN)] and continues to Boquete [0847/8226 (USBGN)] (USNM, as "Boquete Trail"). E762 and JOG both show an unnamed foot track extending E from Bajo Grande, ascending to ca. 2,500 m through a pass S of the continental divide and N of Volcán Barú [0848/8233 (USBGN)] and continuing E along the upper reaches of the Río Caldera [0837/8221 (USBGN)] to Boquete. The easternmost section of this trail is shown on Blake's map (1958:502) extending W from the Holcomb Trail [ca. 0901/8215] at Agua Caliente [ca. 0851/8229] along the Quebrada Cerro Punta [= Río Caldera]. This route exists today as a combination of foot paths, rough roads and paved roads. The foot path is a well marked and well maintained trail that is commonly known as the Sendero Los Quetzales or as the Respingo Trail. Wetmore, GL = Cerro Punta, at 7,200-7,300 feet (2,200 m), entered the trail near Bajo Grande, forest, huge thistles, an abandoned clearing, pastures, a formerly cultivated slope at 7,300 feet (2,200 m) and up to 7,500 feet (2,300 m) to the E wall of the secondary crater, a steep slope, 1955 Mar 4, 7, GL = Cerro Punta, at 7,000 feet (2,150 m), entered the Boquete Trail [sic] at Cerro Punta and climbed to 7,300 feet (2,200 m), forest cut except on steepest slopes, level land in garden plots, 1965 Mar 19 (USNM, as "Boquete Trail"); Ambrose (MARU), GL = Cerro Punta, at 7,100 feet (2,150 m), 1961 May 31 (AMNH, as "Boquete Trail"); Johnson and Brush, 6,800 feet (2,050 m), 1 mi (1.5 km) E of

Cerro Punta, 1966 Apr 11 (Johnson and Brush, 1972, as "Camino a Boquete"); Cardiff, Dittmann, and Lovette, GL for Paso de Respingo, 1996 Mar 9-11 (LSUMZ). See also: Boquete Trail; and Holcomb Trail.

CERRO PUNTO; see Cerro Punta

CERRO SILVESTRE; Panamá (W, Pacific)
0857/7941 (USBGN)
ca. 50 m, town 3 km W of Arraijan [0857/7939 (USBGN)] (MGRP). Vega, GL = Arraijan, 1978 Jan 17 (GML).

CERRO TIGRE; see Cerro Tigre Army Reservation

CERRO TIGRE ARMY RESERVATION;
0904/7938 (USBGN)
Panamá (Canal Zone) (E, Pacific)
50-200 m, former United States Government military base that covered a land area of ca. 1.6 km^2 at the S terminus of Camino Madden (Madden Road) [0907/7937 (USBGN)] near its junction with Carretera Gaillard [0904/7940 (USBGN)]. At the N end of the base there was a 250 m hill, also named Cerro Tigre (E8660). Meyer, 1941 Mar 30, Apr 10, 13 (USNM, as "Cerro Tigre").

CERRO VIEJO CAMP; Herrera
ca. 0737/8049
2,000 feet (600 m), according to Aldrich and Bole (1937:4 (map), 8), the camp was located on the SW side of Cerro Viejo [0737/8048 (USBGN)] between the Río Mariato [0737/8059 (USBGN)] and the Río Negro [0737/8059 (USBGN)] ca. 16 mi (26 km) E of Golfo de Montijo [0740/8107 (USBGN)] in heavy tropical rain forest. The authors attributed their specimens to Veraguas province, but more recent maps indicate that the region is located in Herrera province. See Península de Azuero [0740/8035 (USBGN)] for further discussion of the provincial borders. Aldrich and Bole, GL = Golfo de Montijo, Península de Azuero, 1932 Mar 6-9 (CM; Aldrich and Bole, 1937).

CERRO VIENTO; Panamá (E, Pacific)
0905/7928 (USBGN)
ca. 50 m, town located 5 km NNW of Juan Díaz [0903/7927 (USBGN)] (JOG). Terrero, 1985 Apr 15 (MVUP); Luque, 1992 Jul 23 (MVUP).

CHADÓ, RÍO; Darién
0731/7810 (USBGN)
Originates in S Darién province on Cerro Piña [0741/7809 (USBGN)] in SE Serranía del Sapo [0750/7817 (USBGN)] and flows S to join the Río Jaqué [0731/7810 (USBGN)] at its mouth (Atlas, 1988:206; SMN2B). Shown as the Río Piña on RDPF, although these are two distinct rivers. See also: Playa Nieve, Laguna.

CHAGRE, RÍO; see Chagres, Río

CHAGRECITO, RÍO; see Chagresito, Río

CHAGRES, CAMPO; see Madden, Represa

CHAGRES, RÍO; Colón (Canal Zone) (W)

0919/8000 (USBGN)

Originates at the NW end of Lago Gatún (Gatun Lake) [0912/7955 (USBGN)] at Represa Gatún (Gatun Dam) [0916/7956 (USBGN)] and empties into the Caribbean Sea [1500/7500 (USBGN)] 11 km WSW of Colón [0922/7954 (USBGN)] (JOG). Before the Panama Canal [0920/7955 (USBGN)] was built, this river was continuous with the river E of the canal that is also called the Río Chagres [0907/7942 (USBGN)]. The original streambed continued roughly N from Gamboa [0907/7942 (USBGN)] to Gatún [0916/7955 (USBGN)] along the path now occupied by the Panama Canal (MIP; PR). Jewel, 4 mi (6 km) below Gatún, near the sea, on the river banks, 1911 Feb 19, Jun 11, 21, 1912 Feb 22 (ANSP, as "Chagres River," and as "Río Chagres"; Jewel, 1913; Stone, 1918, as "Chagres River"); Loftin, GL = Fuerte San Lorenzo, near [Fuerte] San Lorenzo, 1964 Sep (GML); Olson, GL for Piña Beach, 1966 Jul 13, 15-16 (UF, as "Chagres River"; USNM, as "Chagres River"). See also: Fuerte San Lorenzo; and Indio, Río.

CHAGRES, RÍO; Colón/Panamá (Canal Zone) (E, Atlantic)

0907/7942 (USBGN)

Originates in central Panamá province in the mountains NE of Panama City [0858/7932 (USBGN)] and NW of Chepo [0910/7906 (USBGN)], flows SW along the former Canal Zone/Panamá provincial boundary, and enters the SE shore of Lago Alajuela (Madden Lake) [0915/7935 (USBGN)] where a portion of its bed is submerged. The river exits the SW end of the reservoir via Represa Madden (Madden Dam) [0913/7937 (USBGN)] and flows W and SSW along the former Canal Zone/Colón provincial boundary entering the E end of Lago Gatún (Gatun Lake) [0912/7955 (USBGN)] at Gamboa [0907/7942 (USBGN)]. This point is currently considered the river's terminus, but see Río Chagres [0919/8000 (USBGN)] for a discussion of the original stream path. Between Lago Alajuela (Madden Lake) and Gamboa, the river has numerous islands, side channels and estuaries, and below Juan Mina [0910/7939 (USBGN)] the river width varies from 300 m to 1 km (E866; JOG). Collector ?, date ? (ANSP); Busck, upper, 1911 Apr (USNM); According to Goldman, in 1912 between Matachin [ca. 0907/7943] and Alhajuela [ca. 0913/7937] the river valley was quite narrow and bordered by somewhat rocky hills. The river banks were 10-15 feet (3-5 m) deep in most places, formed by alluvial deposits and planted with banana, yuca and sugar cane. Corn was planted on the hill slopes and the uncultivated areas were covered with coarse grass, *caña blanca*, bamboo, *guarumo*, and *Ficus* and *Erythrina* trees. The river near Alhajuela passed through a rocky formation where the uncut banks were smooth and water worn. Chapin, Rogers and Mathews, mouth, 1923 Jul 17, GL for Río Aguardiente, 1923 Jul 17 (FMNH); Weber, some have Gamboa as the GL, 1925 Aug 1, 5-6 (FMNH, as "Chagres River"); Enders, GL for Salamanca, 1935 Jan 7-10, 14 (ANSP, as "River Chagres"); Miller, 1937 Apr 11 (USNM, as "Chagres River"); Fairchild, GL = Juan Mina, near Juan Mina, 1943 Oct 22 (ANSP, as "Chagres River"); Trapido, GL = Juan Mina, near Juan Mina, date ? [1940s ?] (USNM, as "Chagres River"); Noyes, lower, 1947 Sep 5, 21, mouth 1947 Aug 29 (MVZ); Hartman, GL = Juan Mina,

1949 Jan 11-13, 15, 17-19, 21-22, 1951 Jan 9, 12, 16-18, 20, 23, Feb 5, 1953 Jan 3, 4-10, 12-14, 16, 1956 Jan 1, 5, 13, 15, 1957 Dec 15, 1958 Jan 4 (USNM, as "Chagres River," and as "Río Chagres"; Hartman and Brownell, 1961a; Brumfield et al., 2001, as " Juan Mina, Río Chagres"); Clark, 1949 Mar 25 (USNM, as "Chagres Valley"); Hardy, GL = Gamboa, lower Río Chagres, specimens are labeled both Colón and Panamá province, 1949 Aug 22 (MCN, as "Chagres River"); Wetmore, GL = Juan Mina, he described the river section from Juan Mina to Gamboa as an arm of Lago Gatún (Gatun Lake), as the original river bed had been submerged. He generally collected within 3 km of Juan Mina, but he also collected near the mouth at Gamboa (see below) and as far as 6 km NE of Juan Mina ca. 1 km below Represa Madden (Madden Dam). Near Juan Mina in an extensive area of water hyacinths and blooming water lilies, along grassy edges of lagoons, in an old citrus plantation and fair-sized woods, E and NE of Juan Mina along the (former) Canal Zone/Panamá provincial boundary through cultivated areas with bananas, oranges, corn, yuca and *rastrojo* and in small tracts of second growth forest along the creeks, SE of Juan Mina on a hill covered with open gallery forest and growth choked with thin vines near its base, inland to a hill covered with second growth forest, 0.5 mi (1 km) below Juan Mina on a trail through level, forested ground with palms and open forest with some creeping bamboo, 1 mi (1.5 km) S of Juan Mina to a lagoon and from here inland along a rocky creek through forest to an open area with low, forested ridges beyond, upstream to Santa Rosa [0911/7939 (USBGN)] (q. v.) into a swampy area with grass and tall spreading trees standing in water, upstream into Laguna de Guacas (see Las Huacas [0911/7939 (USBGN)]) and other side channels, upstream in an area of strong current near the highway bridge (= Carretera Boyd-Roosevelt [0917/7943 (USBGN)] (q. v.)) with Represa Madden (Madden Dam) in sight, 1955 Jan 6-15, 17-19, 21-22, Dec, 9-10, 12, 14-16, GL = Gamboa, near Gamboa on a small bay fringed with jungle, 1955 Dec 10, GL = Juan Mina, near Juan Mina in channels near the dock, expanse of grass, water hyacinths and other aquatic vegetation, forest at Quebrada Juan Mina [0910/7939 (USBGN)], rough surfaced pasture and swampy shore near Río Honda [0910/7940 (USBGN)], on a trail N from Río Honda to Santa Rosa [0911/7939 (USBGN)], 1959 Jan 27, 30-Feb 1, at Boca de Gatuncillo [0912/7939 (USBGN)] (q. v.) 1959 Jan 28, 30, at Santa Rosa (q. v.) 1959 Jan 29, 31, GL = Gamboa, on Lago Gatún (Gatun Lake) [0912/7955 (USBGN)], 1959 Jan 29, GL = Juan Mina, near Juan Mina in a narrow channel and marsh, in the old citrus orchard and in a cornfield, upriver to Santa Rosa [0911/7939 (USBGN)] and downriver 2 mi (3 km) to the embayment where the Río Aguardiente [0910/7941 (USBGN)] discharges, forest trail inland, 1961 Jan 10-11, 13-15, 18, at Guayabalito (q. v.), 1961 Jan 14 (LSUMZ; USNM); Van Horn, 1963 Apr 2, 1965 Sep 3, Oct 24, 1966 Jan 17, Nov 11, 1968 Sep 19 (GML); De Leon, 1965 Oct 5 (GML); Olson, GL = Gamboa, along, 1987 Mar 6-8 (USNM, as "Chagres River"); Mendoza, GL = Santa Rosa, 1983 Jan 11, 24, Jul 11, Nov 7, 1985 (?), 1988 (?) (GML, as "Chagres"); Cardiff and Lovette, GL = Gamboa, ca. 4 km NE of Gamboa, 1997 Mar 12 (LSUMZ); Birks, Davis, Faucett, and Rohwer, GL for Río Chagresito, 2002 Feb 26-Mar 4 (UWBM). (Rogers, 1939; Eisenmann, 1957; Gorski, 1969; Emlen et al., 1998). See also: Azul, Cerro; Bas Obispo; Juan Mina; and

Sardinilla, Río.

CHAGRES FIELD; see Gamboa

CHAGRESITO, RÍO; Panamá (E, Pacific)
0922/7920 (USBGN)
Originates on the Colón/Panamá provincial border in the mountains NW of Cerro Brewster [0921/7915 (USBGN)] and flows SW entering the Río Chagres [0907/7942 (USBGN)] ca. 50 km NNE of Panama City [0858/7932 (USBGN)] (E761A, as "Río Chagrecito"). Birks, Davis, Faucett, and Rohwer, GL = Río Chagres, at 323 m, at the confluence of the Ríos Chagresito and Chagres, ca. 49 km NNE of Panama City, montane tropical rainforest, 092144/791918, 2002 Feb 26-Mar 4 (UWBM, as "Río Chagrecito").

CHAGRES RIVER; see Chagres, Río

CHAGRES VALLEY; see Chagres, Río

CHALITE, CERRO; see Chalite, Cerros de

CHALITE, CERROS DE; Bocas del Toro
0852/8201 (USBGN)
347 m, peak located 15 km SE of Chiriquí Grande [0857/8207 (USBGN)] and ca. 10 km S of the Laguna de Chiriquí coast (JOG). Lovette and Weir, at 200 m, 08°51'31"/82°03'38", 2003 Mar 23-25 (CU; and LSUMZ, both as "Cerro Chalite").

CHAMÁ, ISLA; Panamá (W, Pacific)
0844/7935 (USBGN)
Sea level, very small island in Bahía de Panamá [0850/7915 (USBGN)] 15 km NE of Punta Chame [0839/7942 (USBGN)] and 6 km SSW of Isla Taboga [0847/7933 (USBGN)] (E501, as "Isla Chame"; JOG; MGRP). Meyer, E and W sides of the island, steep slopes, 1941 Feb 23, 1942 Feb 15, 1943 Feb 21 (USNM, as "Chama Island," and as "Chame Island"). (Angehr, 2003:252).

CHAME; Panamá (W, Pacific)
0835/7953 (USBGN)
20 m, city in SW Panamá province on the Carretera Interamericana [0832/8242 (USBGN)] and on the Río Chame [0834/7950 (USBGN)] 5 km above its mouth (JOG). Wetmore, GL for: Playa Coronado, and Río Lajas, 1953 June 21-22 (USNM).

CHAME; see Hato Chamí

CHAME, BAHÍA DE; Panamá (W, Pacific)
0838/7947 (USBGN)
Shallow bay on the SW coast of Panamá province, the shores are largely covered with mangrove swamps that extend inland for 1-7 km, the bay extends 7 km N-S, and opens on the NE to Bahía de Panamá [0850/7915 (USBGN)], the SE coast is enclosed by a 1 km wide peninsula that extends 10 km from the mainland (E762; JOG). The bay is considered an Important Bird Area (Angehr, 2003). (Butler et al., 1992). See also: Sajalices, Río.

CHAME, CERRO; Panamá (W, Pacific)
0836/7950 (USBGN)

523 m, peak 6 km NE of Chame [0835/7953 (USBGN)] (JOG). Wetmore described the slopes above 750 feet (230 m) as fairly steep with tall grass and exposed rocks while the lower regions were covered with dry forest (USNM). Rogers, at 1,600 feet (500 m), observations, sedges, stunted shrubs, 1942 Apr 22 (Rogers, 1945; Wetmore, 1951a); Wetmore, at 150-500 feet (45-150 m), on a trail from Las Pavas (not located) on the W side of the mountain in dry forest and near recent clearings, 1955 Mar 26-28 (USNM; Wetmore, 1956a; Wetmore, 1970; Huff and Wetmore, 1967); Tortosa, 1964 Aug 30 (GML).

CHAME, CERRO; see Hato Chamí

CHAMÉ, ISLA; see Chamá, Isla

CHAME, PUNTA; Panamá (W, Pacific)
0839/7942 (USBGN)
Sea level, point 22 km ENE of Chame [0835/7953 (USBGN)] on the E tip of the peninsula that forms the SE coast of Bahía de Chame [0838/7947 (USBGN)] (JOG). Abrego, 1984 Jun 26 (MVUP); Castro, Myers, and Ricklefs, 1986 Nov-1987 Jan, 1987 Dec-1988 Jan (ANSP; Castro and Myers, 1988; Castro and Myers, 1990; Castro et al., 1992). (Butler et al., 1992).

CHAME ISLAND; see Chamá, Isla

CHAMÍ; see Hato Chamí

CHANATÍ, RÍO; Darién
0827/7816 (USBGN)
Originates SW of Santa Fé [0839/7809 (USBGN)] in the SE foothills of Serranía de Cañazas [0852/7822 (USBGN)], the river flows S and enters a narrow bay on the N coast of Golfo de San Miguel [0822/7817 (USBGN)] 9 mi (14 km) NW of La Palma [0825/7809 (USBGN)] and immediately W of the Río Cucunatí [0827/7815 (USBGN)] (SMN2B). Festa, "*foreste del*," made several camps on the river, apparently above the long, narrow bay that leads to the gulf, his travel account indicates that he was on the Río Chanatí from Jun 15-23, although he may have collected here again as Salvadori and Festa list a specimen from July, 1895 Jun, Jul ? (Salvadori and Festa, 1899, as "Río Cianati"; Festa, 1909, as "Río Cianati").

CHANGAMÉ, ISLA; Panamá (Canal Zone) (W, Pacific)
0854/7933 (USBGN)
Sea level, very small island with a land area of 1.5 ha (Angehr, 2003:247) located in Bahía de Panamá [0850/7915 (USBGN)] 2 km SE of Punta Batele [0854/7934 (USBGN)] (E761A, as "Changarmi Island"). According to Angehr (2003:247), the flora "consists of coarse sedges and other herbaceous vegetation with a few small scrubby trees." Hallinan, observations, c. 1915 (Hallinan, 1924:307, as "Changarmi Island"); Meyer, 1941 Feb 15, 23, Mar 29, May 10, Jun 7, 1942 Feb 15 (USNM, as "Changame Island").

CHANGARMI ISLAND; see Changamé, Isla

CHANGENA, RÍO; see Changuena, Río

CHANGENA RIVER; see Changuena, Río

CHANGUENA, RÍO; Bocas del Toro

ca. 0908/8230

Described by Wetmore as a main tributary of the Río Changuinola [0928/8227 (USBGN)] that drained both the E and S slopes of Cerro Fabrega [0907/8252 (USBGN)] and the N and W slopes of Cerro Róbalo [0853/8220 (USBGN)] (q. v.) (USNM). The river as described by Wetmore appears on two of our older maps, MHA (as "R. Changena") and MRP (as "R. Chángena"), but on more recent maps the topography and names in the region are different and the major river flowing through this area is labeled the Río Culubre [0908/8230 (USBGN)] (q. v.) and appears to be the same river described above. Most specimens taken on the river appear to have come from one of two camps: Rancho Mojica [ca. 0902/8241] (q. v.) at 4,800 feet (1,450 m) and a lower camp at 2,400-2,600 feet (750-800 m). According to Hinds, the lower camp was 22 mi (35 km) W of Almirante [0918/8224 (USBGN)], and although the area 35 km due W of Almirante is in the Río Teribe [0922/8232 (USBGN)] drainage, the area 35 km SW of Almirante is at 800-1,000 m at the mouth of a tributary of the Río Culubre that drains the continental divide SE of Cerro Fabrega. Wetmore (from Ratibor Hartmann), headwaters of, between Cerro Fabrega on the NW and Cerro Róbalo on the SE, 1960 Jul 9 (USNM); Hinds and Keenan, GL = Río Changuinola, at 2,500 feet (750 m), 1961 Sep (USNM, as "Río Changena"); Hinds, at 2,400 feet (750 m), 22 mi (35 km) W of Almirante, lower camp, 1961 May (?) 13, Sep 7-15, 17-26 (GML, as "Changena River"; USNM, as "Changuena River"); GMLP, lower camp, 1961 Sep 2, 9 (GML, as "Changena River"; USNM, as "Río Changena"; Wetmore, 1962, as "Río Changena"); Tipton, GL for Rancho Mojica, 1961 Sep 10 (USNM, as "Río Changena"); Yunker, GL = Río Changuinola, at 2,600 feet (800 m), 1961 Sep 18 (USNM, as "Río Changena Camp"); Collector ?, at 725-980 m, date ? (Olson et al., 1968; Olson, 1980a, as "Changuena River").

CHANGUENA RIVER; see Changuena, Río

CHANGUINOLA; Bocas del Toro

0926/8231 (USBGN)

Sea level, town in extreme NW Bocas del Toro province, 7 km S of the coast and 21 km NW of Almirante [0918/8224 (USBGN)] in a cultivated area between the lower reaches of the Río San San [0931/8231 (USBGN)] and the Río Changuinola [0928/8227 (USBGN)] (JOG). The region has been cultivated since the late 1800s when fruit companies moved to Changuinola and Almirante. A blight altered the banana cultivation in the 1920's after which bananas were grown in "large diked impoundments, which are artificially flooded periodically to deprive the soil fungus of oxygen. These impoundments, known locally as 'lakes,' are flooded in rotation and provide water birds with excellent habitats" (Eisenmann, 1957:248). A railroad network that was developed during the early 1900s now extends from Changuinola to the surrounding farms and towns and to Almirante (q. v.). Wedel, 1921 (?) Jan 21, 1927 May-1929 Jan (ANSP; FMNH; MCZ; NHRM; Peters, 1929; Peters, 1931a; Peters and Griscom, 1929; Lincoln, 1944; Eisenmann, 1950); Eisenmann, observations at flooded impoundments, 1956 Jun-Jul (Eisenmann, 1957); Wetmore, flooded banana plantations near Farm 8 [= Finca Ocho [0926/8231 (USBGN)]], walked ca. 2 mi (3 km) to a small clump of trees with an undergrowth of *caña blanca,* plantations near the Río Changuinola, along trails in high grass, 1958 Jan 24, 30, Feb 13, 27, Mar 4 (USNM); Handley and Hinds, unmodified, and 7 km SSW, 1960 Feb 29, Mar 6, 9-10 (USNM); Loftin, observations near a citrus grove, 1962 Sep 30 (Loftin, 1963a); Smith, N., observations, 1964 Sep 3 (Wetmore et al., 1984:27); Strauch, 1971-1973 (UMMZ; Strauch, 1977); Martínez, 1979 Feb 17 (AMNH); Booth, 6 km W of the airport, JOG and MGRP both indicate three airports at Changuinola, but according to Angehr (in litt., Aug 2005) the site where Booth's collections were made was almost certainly "c. 6 km WSW of the main city airport located at the north end of the town of Changuinola (this is certainly the one that would have been referred to as "Changuinola Airport"), and thus about 1 km W of the minor "pista de aterrizaje" located between Fincas 32 and 34. Coordinates are UTM 1042000 N 327000 E, or 0925/8235," 1981 Nov 10 (USNM); Braun and Brumfield, at 30 m, at 09°26'54"/82°31'12", at 50 m, at 09°28'18"/82°34'00", ca. 7 km NW of, and at 160 m, at 09°20'24"/82°28'12", 13 km SE of, 2004 Mar 12-13, 17 (LSUMZ). (Olson, 1981a; Olson, 1981d; McDonald et al., 2001). See also: La Amistad, Parque Internacional.

CHANGUINOLA, RÍO; Bocas del Toro

0928/8227 (USBGN)

Originates on the continental divide near Cerro Horqueta [0850/8226 (USBGN)] and flows NNW. The river crosses the railroad from Almirante [0918/8224 (USBGN)] to Changuinola [0926/8231 (USBGN)] 4 km S of Changuinola just SE of El Empalme [0925/8231 (USBGN)], then turns NE and widens entering the Caribbean Sea [1500/7500 (USBGN)] 8 km ENE of Changuinola and 18 km SE of the mouth of the Río Sixaola [0934/8234 (USBGN)] (E762; JOG). According to Delgado (1983), the wetlands on the lower course of the river are extensive and with a diversity of canals and beaches. Some specimens labeled Río Changuinola may have been collected at camps on the Río Changuena [ca. 0908/8230]. Wedel, 1926 Nov-1928 Sep (ANSP, as "Changuinola River"; MCZ, as "Changuinola River"; Peters, 1931a; Olson, 1981c, as "Changuinola River"); Hinds and Keenan, GL for Río Changuena, 1961 Sep (USNM, as "Upper Changuinola"); Hinds, specimens are labeled 2,400-2,600 feet (750-800 m) and lower camp, indicating that these were collected at the camps on the Río Changuena, 1961 Sep (GML); Yunker, GL for Río Changuena, 1961 Sep 18 (USNM, as "Upper Changuinola Drainage"); Loftin, observations on the river ca. 5 mi (8 km) upriver from Empalme [sic], 1962 Sep 30 (Loftin, 1963a, as "Changuinola River"); GMLP, 1960 Oct 5 (?), 1980 Jan 15, 18, Feb 11-12, 29, May 9, 1986 Jan 15 (?) (GML); Hinds, at 50 m, IRHE dam site camp on the lower Changuinola, 1980 Jan 13-22, 31, Feb 1-3, 5-6, 11-12, 24, 26, 28, Mar 10-12, 15, 30, Apr 12, May 9, Nov 12 (?), Dec 3 (?), 1981 Jan 13 (?), 1986 Jan 20 (?) (GML; Wetmore et al., 1984:398); Delgado, is also labeled Corriente Grande [0918/8232 (USBGN)] (q. v.), 1980 Mar 22 (GML); Ridgely, observations on the lower Changuinola at the IRHE dam site camp, in undergrowth at forest edge, 1980 Apr (Wetmore et al., 1984:450, 458, 465); SBE, at 10 m, on the E bank of the river 1/4 (0.25) and 3/5 (0.6) km SE of the railroad bridge, 1991 Feb 25-26, Mar 3

(MVUP; USNM; Parsons et al., 1993; Brumfield et al., 2001, as "Changuinola River"); Brumfield, 5.4 km E, on the road from Changuinola to Almirante, 1994 Jan 20 (USNM); Braun, M., and McDonald, GL = Boquisco, at 100 feet (30 m), collected on the river banks 1 km S of, 2 km SW of, and 2 km W of Boquisco, 1994 Apr 20-23, 25-28 (USNM); Braun and Brumfield, GL = Río Changuinola Arriba, at 09°11'42"/82°28'48", E bank, 2004 Mar 14-15, and GL = Río Riscó [0916/8228 (USBGN)], below Río Riscó, 2004 Mar 15 (LSUMZ). (Olson, 1983, as "Changuinola River"; McDonald et al., 2001). See also: Changuena, Río; Changuinola; and Corriente Grande.

CHANGUINOLA ARRIBA, RÍO; see Río Changuinola Arriba

CHANGUINOLA CANAL; Bocas del Toro

ca. 0919/8221

Described by Kennard and Peters (1928:444, as "Canal") and by Peters (1931) as a canal that was parallel to the coast and only a short distance inland from it. It extended from the mouth of the Río Changuinola [0928/8227 (USBGN)] to Bahía Almirante [0920/8218 (USBGN)] and provided safe passage to barges carrying bananas before the railroad between Changuinola and Almirante [0918/8224 (USBGN)] was built. The canal was 6-15 mi (10-24 km) long and although abandoned by the mid-1920s, strong currents kept it open. According to E762, JOG, and MHA (all of which show an unnamed canal), the canal extends ca. 12 km SE from the mouth of the Río Changuinola parallel to the ocean coast and entered Boca del Drago [0925/8220 (USBGN)] 3 km N of Punta Donato [0923/8221 (USBGN)]. According to Angehr (in litt., Aug 2005), the canal "still exists today and [is] used by small boats traveling between Changuinola and the Almirante area." Kennard et al., GL = Almirante, 1926 (MCZ; Kennard and Peters, 1928:444, as "Canal"); Wedel, 1927 Oct-1928 Feb, 1938 Sep 23 (FMNH, as "Old Canal"; MCZ; Peters, 1931a); Wetmore, GL = Almirante [0918/8224 (USBGN)], Boca del Drago, entered the Boca del Drago side of the canal and followed it for ca. 3 mi (5 km) through black, silt-free swamp water, tall trees and brush, 1958 Feb 20 (USNM). See also: Boca del Drago; Banano, Río; Drago, Boca del; and Tirbi Bight.

CHANGUINOLA-GUABITO ROAD; Bocas del Toro

ca. 0930/8237

The road extends N from Changuinola [0926/8231 (USBGN)] then continues W and NW to Guabito [0930/8237 (USBGN)]. The road crosses the Río San San [0931/8231 (USBGN)] and extends along wet, lowland areas with elevations of 10 m or less (E762). Braun and Brumfield, at 09°28'24"/82°33'54", 2004 Mar 16 (LSUMZ).

CHANGUINOLA RIVER; see Changuinola, Río

CHANIS; see Chanis, Barriada

CHANIS, BARRIADA; Panamá (E, Pacific)

ca. 0901/7929

Sea level, section of NE greater Panama City [0858/7932 (USBGN)] located in NE Corregimiento Parque Lefevre [0901/7929 (USBGN)] (Atlas, 1988:208, as

"Chanis"; E962, as "Reparto Chanis"). Yanguez, 1990 Feb 27 (MVUP).

CHANIS, REPARTO; see Chanis, Barriada

CHANNEL 2; Bocas del Toro

ca. 0922/8222 ?

Not named on our maps, but presumably the name refers to a mainland canal that opens to Boca del Drago [0925/8220 (USBGN)] channel 3.5 km S of the Changuinola Canal [ca. 0919/8221]. This canal is shorter than the Changuinola Canal and extends to the NW for 3 km before terminating in mangrove swamps (E762). See also: Drago, Boca del.

CHAPERA, ISLA; Panamá, Archipiélago de las Perlas

0835/7902 (USBGN)

Island with a land area of 1.78 km^2 (178 ha) (Wright et al., 1985), located 16 km NNW of Isla del Rey [0822/7855 (USBGN)] (JOG, as "Isla Chepra"), and according to a map section in Wetmore's journal, the elevation is 150 feet (45 m) (USNM). Bovallius, "very hilly," "covered with a rich vegetation," "not at all cultivated," see discussion of Bovallius under Pueblo Nuevo [ca. 0820/8013 ?], 1882 Apr 1-2 (NHRM; Rendahl, 1920); Wetmore, gallery forest covered the island and creepers were common, landed on the SW coast and worked in a red mangrove swamp behind a sandy beach, walked up a steep slope to a ridge crest at 150 feet (45 m), and along a freshwater stream, 1960 Jan 18 (USNM; Olson, 1980b); (MacArthur et al., 1972—species list; see comments in Bibliography); Wright, Faaborg, and Campbell, mist nets, observations, 1976 Jun-Aug, 1978 Jan-Mar (Wright et al., 1985); Levings et al., observations, 1982-1983 (Levings et al., 1986); Brawn et al., mist nets, 199? (Brawn et al., 1996). (N. Smith, 1971).

CHARAGRE; Bocas del Toro

0923/8234 (USBGN)

20 m, town on the N bank of the Río Teribe [0922/8232 (USBGN)] 8.5 km SW of Changuinola [0926/8231 (USBGN)] (E762). SBE, GL for: Quebrada Carbon, and Río Teribe, 1991 Feb 24, 28 (USNM; Brumfield et al., 2001).

CHARARE; Panamá (E, Pacific)

ca. 0914/7904

100-200 m, described by Wetmore as a section of the Río Mamoní [0908/7904 (USBGN)] ca. 4 mi (6 km) N of Chepo [0910/7906 (USBGN)], presumably near the mouth of the Río Charare [0914/7904 (USBGN)], Wetmore and Perrygo, GL = Chepo, Río Mamoní, various crops along the river banks, areas of thickets and trees, second growth forest, 1949 Apr 22 (USNM). Not shown on our maps.

CHARARE, RÍO; Panamá (E, Pacific)

0914/7904 (USBGN)

Originates NE of Chepo [0910/7906 (USBGN)] and joins the Río Mamoní [0908/7904 (USBGN)] 7 km NNE of Chepo (JOG). See also: Charare.

CHARCO AZUL, BAHÍA DE; Chiriquí

0812/8245 (USBGN)

Large bay within Golfo de Chiriquí [0800/8220 (USBGN)] on the extreme W coast of Chiriquí province extending ca. 55 km W from the islands at the mouth of the Río Chiriquí [0820/8226 (USBGN)] to Península Burica [0810/8255 (USBGN)] (q. v.) (JOG). See also: Divalá.

CHARCO DEL TORO; see Toro, Charco del

CHARCO NÚMERO DOS; see Número Dos, Charco

CHARCO NÚMERO UNO; see Número Uno, Charco

CHARÓ, RÍO; see Chadó, Río

CHELELÉ, RÍO; see Guaval

CHEPIGANA; Darién
0818/7803 (USBGN)
Sea level, town located 16 km SE of La Palma [0825/7809 (USBGN)] on the S banks of the Río Tuira [0821/7803 (USBGN)] (MGRP; MHA). Richardson, 1914 Nov 28, Dec 12 (AMNH; USNM; Anthony, 1916; Miller and Griscom, 1921; Hellmayr, 1929; Zimmer, 1939; Zimmer, 1941b:14; Zimmer, 1955:5); Festa, passed through Chepigana on Aug 6, en route to Laguna de Matusagratí [0820/7756 (USBGN)], whether or not this locality appears on specimen labels is unclear, 1895 (Salvadori and Festa, 1899; Festa, 1909).

CHEPILLO, ISLA; Panamá (E, Pacific)
0857/7908 (USBGN)
53 m, small island with a land area of ca. 0.8 km^2 (80 ha), located in Bahía de Panamá [0850/7915 (USBGN)] 5 km S of the mainland coast at the mouth of the Río Chepo [0900/7906 (USBGN)] (JOG). Bovallius, observations, see discussion of Bovallius under Pueblo Nuevo [ca. 0820/8013 ?], 1882 (Rendahl, 1920, as "Chepillo Island"). (Reed, 1988). See also: Panamá, Bahía de.

CHEPILLO ISLAND; see Chepillo, Isla

CHEPO; Herrera
0744/8049 (USBGN)
600 m, town in SW Herrera province in Distrito Las Minas [0746/8048 (USBGN)] ca. 12 km SW of Las Minas [0748/8045 (USBGN)], 25 km SSW of Ocú [0757/8047 (USBGN)], and 30 km SW of Pesé [0754/8037 (USBGN)] (JOG; RDP2). Delgado, some labels have GL = Loma El Montuoso, at 900 m, 1977 Aug 1, 29 (GML, as "Chepo de Las Minas").

CHEPO; Panamá (E, Pacific)
0910/7906 (USBGN)
120 feet (37 m) (TM); city in central Panamá province ca. 55 km NE of Panama City [0858/7932 (USBGN)] and 5 km NNW of the mouth of the Río Mamoní [0908/7904 (USBGN)] (MGRP; JOG). According to Wetmore, in 1949 Chepo was located on rolling ground that rose in low knolls, later in 1962 he determined that the savanna region E of Panama City [0858/7932 (USBGN)] terminated S of Chepo, as the area around El Llano [0913/7859 (USBGN)] was originally forested. Goldman (1920:5) also described Chepo as "near the edge of the most easterly of the open savannas which extend at intervals along the Pacific coast to beyond the

Costa Rican frontier." Arcé, from Griscom (1933:302) "it is not generally remembered, however, that Arcé made a brief stop at Chepo before embarking on his highly successful work in Veraguas," 18?? (Salvin and Godman, 1888-1904, vol. 2; Griscom, 1933); Goldman, according to his field notes he collected 6 mi (10 km) N of Chepo on a ridge that was ca. 1,000 feet (300 m) high and extended eastward from the main range along the south side of the Río Mamoní [0908/7904 (USBGN)], collected up to ca. 650 feet (200 m). According to JOG, Goldman probably worked ca. 6 km (not miles) N of Chepo along the ridge called Cerro Carbunco [0911/7914 (USBGN)] (q. v.) by Wetmore, because the area 6 mi (10 km) N of Chepo is more than 2 km N of the Río Mamoní, 1911 Mar 18, 10 mi (16 km) NW of Chepo while traveling to Cerro Azul [0910/7924 (USBGN)], 1911 Mar 20 (USNM; Nelson, 1911a; Goldman, 1920; Aldrich, 1945; Olson, 1980c); Kelly, GL for Gaspar Sabana, 1921 Mar (USNM); Smith, A. P., most specimens in this region were labeled San Antonio [0909/7903 (USBGN)], 1927 Mar 8 (YPM); Wetmore and Perrygo, 1949 Apr 24, 5 mi (8 km) W of Chepo, 1949 Apr 21, the secondary localities attributed to Chepo were located within 8 km of town, GL for: Ana Luz, Boca del Toro, Camarón, Cerro Carbunco, Charare, Chepo Viejo, Gavilán, Paso de Mula, Paso Hondo, Porcada Valdez, Río Chepo, Río Chichebre, Río Mamoní, Río Tranca, San Antonio, and Zanja Limón, 1949 Apr 6-11, 13-18, 20-22, 24, 26 (USNM; Wetmore, 1951a; Eisenmann, 1962a); Loftin and Eisenmann, rice fields near, 1961 Sep, 1962 Nov (Loftin, 1963a; Eisenmann, 1963b); Loftin, 1962 Apr (Loftin and Olson, 1963); Hayward, 1962 Apr 12 (WFVZ); Collector ? [Hayward ?], 1962 Apr 12 (GML); GMLP, 1968 Jul 1, Nov 8, 1969 Sep 9, 1973 Mar 24 (GML, as "Chepo," and as "Chepo, Bayano"); Guerrero, 1975 (GML, as "Chepo, Pacora"); Morton, 5 mi (8 km) W, 1971 Mar 5 (USNM); Tudor, observations, below Chepo along the Río Chepo, 1971 Apr 2 (Wetmore et al., 1984:34); Garcia, 1974 Jun 2 (GML); Braun, M., 1984 Aug 29 (MVUP); Baker, *ciénaga en las cercanías de* Chepo (wetlands near the outskirts of Chepo), 1984 Sep 5 (MVUP); Guevara, 1990 Mar 31 (MVUP); Agro, 1 mi (1.5 km) W of, 1996 Feb 11 (ANSP); Agro, Cardiff, and Lovette, GL for Cordillera de San Blas, 1996 Feb 29 (LSUMZ; MVUP). (Eisenmann, 1955c; Wetmore, 1952:19; Wetmore, 1953; Wetmore, 1970; Strauch, 1975b). See also: Chepo, Río; La Capitana; Majé, Serranía de; Pacora, Río; Platanales; and San Antonio.

CHEPO, RÍO; Panamá (E, Pacific)

0900/7906 (USBGN)

Large river that originates in NE Panamá province on the Cordillera de San Blas [0918/7900 (USBGN)], flows W through a broad valley between the Cordillera de San Blas and the Serranía de Majé [0855/7830 (USBGN)], then turns SW and empties into Bahía de Panamá [0850/7915 (USBGN)] 43 km E of Panama City [0858/7932 (USBGN)] (JOG, as "Río Bayano"). Goldman (1920:5) described the mouth as "broad" and the journey ca. 12 mi (19 km) upstream to the Río Mamoní [0908/7904 (USBGN)] was "between lines of low tidal forest." At the mouth of the Mamoní, the river was "a large, deep stream with low, but usually rather steep, muddy banks left exposed at low tide." Also known as the Rio Bayano (TM, as "Río Chepo (or Bayano)"; MHA, as "Río Chepo ó Bayano"; RDPF, as "Río Chepo ó Bayano"). Griscom (1933) places the Río Chepo in Darién province, but the river originates in Panamá province near the Panamá/Darién province border. Construction in the 1970s of a hydroelectric dam on the river created Lago Bayano [0908/7835 (USBGN)] (q. v.) and Isla Majé [ca. 0909/7851]. Arcé, 18?? (Griscom, 1933:304); Johnson, GL for Río Mamoní, 1914 Jan 19-20, 28, Feb 3, 5 (USNM, as "Bayano River"); Smith, A. P., near San Antonio (q. v.), 1927 Feb 24, Mar 3, 26 (YPM; Griscom, 1933); Imhof, observations, savannas that extend E from the Canal Zone to the Río Chepo, 1942 (Imhof, 1950); Wetmore and Perrygo, GL = Chepo [0910/7906 (USBGN)], at Ana Luz (q. v.), 1949 Apr 15-17, 20, GL = El Llano [0913/7859 (USBGN)], the river near El Llano was ca. 125-150 yards (140 m) wide flowing swiftly below 20-25 foot (6-8 m) earthen banks with occasional rock exposures, collected above El Llano in tracts of original forest bordered by rice fields, 1962 Feb 4-5, GL = Boca de Cañita [0912/7855 (USBGN)], at Juan Pelón (q. v.), 1962 Feb 6, 8, at Canchigua (q. v.), 1962 Feb 7 (USNM, as "Río Bayano"); Guerrero, GL for Serranía de Majé, 1962 Jun 8 (GML); Marciaq, 1962 Dec (GML, as "Río Bayano"); Collector ?, 1962 mid-Dec (USNM, as "Río Bayano"); Galindo, Hinds, Mendez, and GMLP, GL for Serranía de Majé, 1972 Feb 12, Apr 15, Oct 10, Nov 18, Dec 2, 1973 Jan 10, 14, 25-26, Feb 8, 10, 13, Mar 23-25, Apr 4, May 5, Jun 5, 30, Nov 13, 1974 May 16, 1975 Jun 9, 1976 Jan 22, Oct 22, Nov 11, Dec 3, 1977 Jan 15, Feb 24, Mar, Jun 23-24, 1978 Mar 14, 31 (GML, as "Bayano," and as "Bayano River"; LSUMZ, as "Bayano River"); Galindo, GL for Río Majecito, 1973 Jan 25 (LSUMZ, as "Bayano River"); Collector ?, GL for Isla Majé, 1973 Jan 25 (GML, as "Río Bayano"); Galindo and Hinds, GL for Calle Larga, 1973 Jan 26, Feb 4, 9, Apr 2, May 10, Sep 8 (GML, as "Bayano"; LSUMZ, as "Bayano River"); Eisenmann, bird netted and released at Galindo camp (Río Majecito ?), 1973 Jan 26 (Wetmore et al., 1984:553, as "Rio Bayano"); Galindo, GL for Isla Majé, 1973 Nov 9 (GML, as "Río Bayano"); Pineda, GL for Isla Majé, 1974 May 25 (GML, as "Río Bayano"); Mendieta, 1985 Mar 14 (MVUP); Buehler et al., observations near the mouth, 2003 Sep 7-27 (Buehler et al. 2004, as "Río Bayano"). (McKim, 1947, as "Bayano River"; Wetmore, 1952:25, 31, as "Río Bayano"; Bennett, 1962).

CHEPO DE LAS MINAS; see Chepo

CHEPO ROAD; Panamá (E, Pacific)

ca. 0910/7906

Presumably refers to the section of the Carretera Interamericana [0832/8242 (USBGN)] that extends roughly E-W between the turnoff for Pacora [0905/7917 (USBGN)] and Chepo [0910/7906 (USBGN)]. The road appears on SMN2A (1946), and according to SMN2B it was an "all weather, loose surface road." The road continued E of Chepo for only a short distance according to this map. According to Angehr (in litt., Aug 2005), "this older Chepo Road parallels the modern paved Carretera Interamerica about 1 km to its north." Meyer, 1941 Mar 31, May 27, 1944 Apr 9, 19 (USNM).

CHEPO VIEJO; Panamá (E, Pacific)

ca. 0910/7904

100-300 feet (30-90 m), described by Wetmore in 1949 as a region of steep-sided knolls, some with forest and others cleared, located E of Chepo [0910/7906 (USBGN)] across the Río Mamoní [0908/7904 (USBGN)], Wetmore and Perrygo, GL = Chepo, Río Mamoní, 1949 Apr 10 (USNM). Not shown on our maps.

CHEVO, QUEBRADA; see Chevo, Río

CHEVO, RÍO; Chiriquí

0847/8250 (USBGN)

Originates N of Santa Clara [0850/8246 (USBGN)] and flows SW entering the Río Candela [0846/8250 (USBGN)] 3 km SSE of Río Sereno [0849/8252 (USBGN)] (MGRP). Wetmore, GL = Santa Clara, at 3,900 feet (1,200 m), on the road 0.5 mi (1 km) W of Santa Clara, the 12-18 foot (4-5 m) wide stream was clear and rapid with a rocky bed bordered by forest, trees with drupes and underbrush, 1955 Feb 17 (USNM, as "Quebrada Chivo," and as "Quebrada Chevo").

CHICAO; Darién

ca. 0727/7758

ca. 100 m, described by Wetmore as an area of dense forest with tall trees, thick undergrowth and numerous areas of palms bordering the Río Imamadó [ca. 0727/7758] and extending E along the Río Chicao [ca. 0727/7758], Wetmore and Perrygo, GL = Río Jaqué [0731/7810 (USBGN)], 1947 Apr 1, 6, 14-16 (USNM). Not shown on our maps.

CHICAO, RÍO; Darién

ca. 0727/7758

Described by Wetmore as a small, swiftly flowing stream that originates in the Cordillera de Juradó [0728/7755 (USBGN)] and flows NE joining the Río Imamadó [ca. 0727/7758] just above its mouth, Wetmore and Perrygo, GL = Río Jaqué, 1947 Apr 16 (USNM). Not shown on our maps.

CHICAQUE, RÍO; see Camp Creek

CHICHEBRE, RÍO; Panamá (E, Pacific)

ca. 0907/7908

Originates in the hills NW of Chepo [0910/7906 (USBGN)] and flows S crossing the Carretera Interamericana [0832/8242 (USBGN)] 7 km W of Chepo before it enters the Río Tapagrilla [ca. 0907/7909], which is a tributary of the Río Hondo [0901/7906 (USBGN)] (JOG). The USBGN coordinates for this river [0912/7910] are located on JOG at the river's headwaters not its mouth. Wetmore originally wrote "Río Chilibre" on his specimens and in his field catalog, but he later replaced "Chilibre" with "Chichebre." Wetmore and Perrygo, GL = Chepo, 5 mi (8 km) W of Chepo, from the highway walked 1 mi (1.5 km) upstream, the water was very low and largely not running, plantations with fair stands of forest, guava in bloom, 1949 Apr 8 (USNM).

CHICO; see Río Chico

CHICO, RÍO; Coclé (Pacific)

0820/8028 (USBGN)

Originates on the continental divide and flows SSE forming a portion of the

Veraguas/Coclé provincial boundary. The river continues SE and E, crosses the Carretera Interamericana [0832/8242 (USBGN)] 1 km S of Natá [0820/8031 (USBGN)], and enters the Río Grande [0818/8024 (USNM)] 5 km E of Natá (JOG). See also: Natá.

CHICO, RÍO; Darién
0811/7741 (USBGN)
Originates on the Cordillera de San Blas [0918/7900 (USBGN)] and flows SW and W entering the Río Chucunaque [0809/7744 (USBGN)] at Yaviza [0811/7741 (USBGN)] (SMN2B). MGRP mistakenly labels the section of the Río Chucunaque below its junction with the Río Chico, as the Río Chico. Baer, GL = Yaviza, Río Chucunaque, presumably he collected on the lower reaches of the stream between Camp Creek [ca. 0813/7735] (q. v.) and Yaviza, 1924 Feb 22, 24 (USNM); Cavalier, GL = Yaviza, 0.5 mi (1 km) E Yaviza, 1967 Oct 4-5, 7 (MVZ). For more on Baer see comments under Marsh (1934). See also: Three Falls Creek.

CHICO, RÍO; Panamá (E, Atlantic)
0916/7930 (USBGN)
Originates in the mountains NNE of Panama City [0858/7932 (USBGN)] and flows SW entering the Río Chagres [0907/7942 (USBGN)] 27 km NE of Gamboa [0907/7942 (USBGN)] ca. 5 km before it enters Lago Alajuela (Madden Lake) [0915/7935 (USBGN)] (MGRP). Aparicio, observations, 1997 Oct 15 (Adsett and Wege, 1998).

CHICO, RÍO; Panamá (E, Pacific)
0901/7911 (USBGN)
Short river originating in the mangrove swamps and grassy marshes on the W bank of the Río Hondo [0901/7906 (USBGN)], meandering greatly as it flows SW, widening to 2 km at its mouth and entering Bahía de Panamá [0850/7915 (USBGN)] 2 km S of Río Chico [0902/7910 (USBGN)] (TM). Imhof, observations, mouth, 1942 Nov 17-18 (Imhof, 1950); Wetmore and Perrygo, GL = Río Chico, beach, brush, mangrove at the mouth, white mangrove at the mouth of the Río Tereza (not found, but possibly the Río Serezo [ca. 0901/7911] (q. v.)), 1949 Mar 16, 24, 26, 30, GL = Río Chico, mouth of the Río Chico, 1956 Mar 5-6, GL = Río Chico, mouth of the Río Chico, tide was out and a 1 mi (1.5 km) wide expanse of mud-flats was bare, down the beach and into mangroves, *emajagua* blossoms, 1961 Mar 24 (USNM; Olson, 1980b, as "Chico, mouth of Río Chico"). See also: Río Chico.

CHICO, RÍO; Veraguas/Coclé (Pacific)
0820/8028 (USBGN)
Originates on the continental divide and flows SSE along the Veraguas/Coclé provincial boundary, then flows E into Coclé province crossing the Carretera Interamericana [0832/8242 (USBGN)] near Natá [0820/8031 (USBGN)] before entering the Río Grande [0818/8024 (USNM)] (JOG). See also: Chitra.

CHICO, RÍO; see Río Chico

CHICÚ, CERRO; see Chucú, Cordillera del

CHILAR; see El Chilar

CHILAR, EL; see El Chilar

CHILAR, RÍO; see El Chilar, Río

CHILAR, RÍO EL; see El Chilar, Río

CHILIBRE; Panamá (E, Atlantic)
0909/7937 (USBGN)
50 m, town on the Río Chilibre [0910/7939 (USBGN)] 3 km SE of Juan Mina [0910/7939 (USBGN)] and just E of the former Canal Zone border, both Camino Madden (Madden Road) [0907/7937 (USBGN)] and Carretera Boyd-Roosevelt [0917/7943 (USBGN)] pass though the town (E661; E866; E8660). Clark, GL for Camino Madden, 1937 May (USNM); Hespenheide and Morton, SW of, date ? (CM); Java and Rosero, 1984 Jan 21 (MVUP); Morales, 1985 Apr 1, 11 (MVUP).

CHILIBRE, RÍO; Panamá (Canal Zone) (E, Atlantic)
0910/7939 (USBGN)
Originates in Panamá province on the continental divide and flows NW entering the Río Chagres [0907/7942 (USBGN)] less than 1 km N of Juan Mina [0910/7939 (USBGN)] and 1 km S of Santa Rosa [0911/7939 (USBGN)] (E661; E866). Hallinan, on the river and in its valley, floating vegetation, vegetation on the river banks, high trees, forest, 1914 May 31, 1916 Jan 1, 16 (AMNH; Hallinan, 1924); Wetmore, GL = Juan Mina, ca. 2 mi (3 km) above its mouth, the first mile (1.5 km) of the river was low, broad and swampy with extensive lagoons and side channels, above this the tree-lined river narrowed and its hilly banks were cleared for cultivation, 1955 Jan 16 (USNM); Van Horn, 1963 Jun 16 (USNM); De Leon, 1964 Dec 30 (GML); Puga, 1978 Jan 11, 17, Mar 15, Apr 10 (GML, as "Chilibre"); Mendoza, GL = Santa Rosa, 1978 Apr 18, 21 (GML, as "Chilibre"). See also: Las Huacas.

CHILIBRE, RÍO; see Chichebre, Río

CHILIBRILLO, RÍO; Panamá (Canal Zone) (E, Atlantic)
0910/7938 (USBGN)
Originates SE of Calzada Larga [0910/7934 (USBGN)] and flows NW until a point 2 km S of Alhajuela [ca. 0913/7937], then it turns SSW and enters the Río Chilibre [0910/7939 (USBGN)] 3 km E of Juan Mina [0910/7939 (USBGN)] (E761A). Goldman, near the caves S of Alhajuela, followed a recently made trail S, then waded downstream for about 2 mi (3 km), according to his journal, the caves were located on a tributary of the Río Chilibrillo ca. 400 yards (350 m) above its mouth, running water in the caves, which were large rifts in the limestone-like formation with several 30-40 foot (9-10 m) lateral chambers, 1912 Jan 30 (USNM; Goldman, 1920, as "Chilibrillo River").

CHILIBRILLO CAVES; see Chilibrillo, Río

CHIMÁN; Panamá (E, Pacific)
0842/7837 (USBGN)
Sea level, coastal town in SE Panamá province 12 km W of the Darién provincial border, on the NE bank of the Ensenada de Chimán [0843/7839 (USBGN)] just below the mouth of the Río Chimán [0843/7838 (USBGN)] (JOG). Angehr (2003:265-267) describes the Chimán Wetlands [0846/7843 (Angehr, 2003:265)] as

an area of "extensive mudflats (11,400 ha), up to three km wide, and coastal mangroves (18,300 ha) surrounding Chimán. Griscom and Crosby, 1927 Feb-Mar (Griscom, 1927b; Griscom, 1929a); Wetmore and Perrygo, specimens labeled Chimán were collected within 15 km of town in one of two general areas: S of town along the coast or N of town along the Río Chimán and its tributaries. GL for: Juan Avele, La Aguadilla, Playa del Creo, Punta Madroña, Quebrada Escartín, Quebrada Jorón, Río Chimán, Río Corotú, and Tigrillo, 1950 Feb 16-28, Mar 31-Apr 1 (USNM; Wetmore, 1950b; Wetmore, 1952:27; Wetmore, 1962; Wetmore, 1970; Karr, 1971a; Olson, 1981d, as "Chiman, Río Chiman"). (Butler et al., 1992).

CHIMÁN, ENSENADA DE; Panamá (E, Pacific)
0843/7839 (USBGN)
Inlet on the SE coast of Panamá province extending 3 km E-W and 3 km inland, opening on the S to the Gulf of Panama [0800/7930 (USBGN)] (JOG). See also: Chimán; and Chimán, Río.

CHIMÁN, RÍO; Panamá (E, Pacific)
0843/7838 (USBGN)
Originates on the S slopes of Pico Columna [0857/7838 (USBGN)], flows SSE and enters Ensenada de Chimán [0843/7839 (USBGN)] at Chimán [0842/7837 (USBGN)]. Its lower course is a 1 km wide, swampy delta that extends 6 km N of Chimán (JOG). On MHA and SMN2A the river is labeled "Río San Francisco" and the name "Río Chimán" is given to a tributary that enters from the E. On SMN2B the river is labeled "Río Boca Bien." On SMN2B and Atlas (1988:205) the Río Chimán originates E of Pico Columna, not on the S slopes as shown on JOG and RDPF. Griscom and Crosby, from Griscom (1933:302), "I stopped at the mouth of the Río Chimán for two days in 1927, and found the coast forested at this point," 1927 Mar 5-8 (AMNH; Griscom, 1927b; Griscom, 1933; Zimmer, 1941b:14); According to Wetmore, in 1950 tall, white mangroves bordered by red mangroves grew in the delta and the surrounding hills were covered with dry forest, at low tide the delta was exposed mud outside of the main channel, and the river was affected by tides just beyond the mouth of the Río Corotú [0851/7838 (USBGN)], Wetmore and Perrygo, GL = Chimán, ca. 1 mi (1.5 km) N of Chimán on the E bank of the delta at Caño Santa Cruz (not located), muddy mangroves on the shore, inland to a 100 foot (30 m), rocky knoll and a partly cleared flat, 1950 Feb 16, at Quebrada Escartín (q. v.), 1950 Feb 18, 28, at Juan Avele (q. v.), 1950 Feb 18, 20, at Río Corotú (q. v.), 1950 Feb 18, 20, 23-28 (USNM; Wetmore, 1962; Olson, 1981d, as "Chiman, Río Chiman").

CHIMÁN WETLANDS; see Chimán

CHIQUERO; Chiriquí
ca. 0848/8230
Described by Blake as the upper valley of the Río Chiquero [ca. 0849/8229], and according to his map, the elevation rises from ca. 5,000 feet (1,500 m) near the river's mouth to over 9,000 feet (2,750 m) near its headwaters (Blake, 1958:502 (map)). Mönniche, 5,100-5,600 feet (1,550-1,700 m), 1932-1933, 1936, 1938 (FMNH; MCN; Blake, 1958); Davidson, at 5,000 feet (1,500 m), 1933 Dec 11, 14-17, 24

(Davidson, 1938). See also: Alto de Chiquero.

CHIQUERO, ALTO; see Alto de Chiquero

CHIQUERO, ALTO DE; see Alto de Chiquero

CHIQUERO, PLAYA DE; Panamá, Archipiélago de las Perlas
0818/7851 (USBGN)
Not shown on our maps but USBGN indicates a beach on the SE coast of Isla del Rey [0822/7855 (USBGN)] near Punta Chiquero [0818/7851 (USBGN)] just N of Bahía San Telmo [0816/7853 (USBGN)]. Tejera, GL = Isla del Rey, 1981 Mar 15, 17-22 (MVUP, as "Playa Chiquero").

CHIQUERO, RÍO; Chiriquí
ca. 0849/8229
Originates on Cerro Copete [ca. 0848/8231] and flows NE and E joining the Río Caldera [0837/8221 (USBGN)] near Bajo Mono [0848/8228 (USBGN)] and N of the Quebrada El Velo [0849/8228 (USBGN)] (Blake, 1958:502 (map)). According to Blake (1958:503), the S rim of the Chiquero Valley is the N boundary of Peña Blanca and Velo. See also: Chiquero.

CHIQUERO VALLEY; see Chiquero

CHIRIQUÍ; Chiriquí
0824/8219 (USBGN)
40 m, town 13 km ESE of David [0826/8226 (USBGN)] 1.5 km S of the Río Chiriquí [0820/8226 (USBGN)] on the Carretera Interamericana [0832/8242 (USBGN)] ca. 5 km N of the river deltas and mangroves on the NE coast of Bahía de Charco Azul [0812/8245 (USBGN)] (E762; JOG; MHA). Many of the earlier specimens labeled simply "Chiriquí" were certainly collected throughout the province, not necessarily near the town. Collector ?, date ?, 1875, 1898 Jan (FMNH; OC); Berlepsch, date ? (FMNH); Kellett and Wood, c. 1846-49 (BMNH; Seemann, 1853; Salvin, 1882; Salvin and Godman, 1888-1904, vol. 2:451; Davidson, 1938); Arcé, 18??, 186?, 1870, 1875, 1876 (FMNH; USNM as "Chiriquí," as "Chiriquí, Chiriquí," and as "Chiriquí, Veragua"; Davidson, 1932; Davidson, 1938); Nehrkorn, 1885 Nov (FMNH); Collector ? [Watson ?], 1903 Mar 5, Apr 7, May 1, 17, Oct 17, Nov 6, 13, 1905 Oct, Mar 10 (FMNH); Watson, 1903 Mar, 1905 Nov (YPM); Rosenberg collection [Watson ?], 1904 Dec 23 (YPM); Morton, 1965 Jul, Dec, 1966 Apr 24, Jul (YPM); Braun, M., at 10 m, 8 km SE of Chiriquí, 1991 Mar 2 (USNM).

CHIRIQUÍ, CORDILLERA DE; see Central, Cordillera

CHIRIQUÍ, GOLFO DE; Chiriquí/Veraguas (Pacific)
0800/8220 (USBGN)
Gulf extending 125 km E-W along the S coast of Panamá between Punta Burica [0802/8255 (USBGN)] and the westernmost coast of S Veraguas province (JOG). Murphy, observations, according to the map, the ship traveled along the coast passing just S of Isla de Coiba [0728/8147 (USBGN)], therefore "Gulf of Dulce, western Panamá" must refer to Golfo de Chiriquí, 1956 Nov 25 (Murphy, 1958, as "Gulf of Dulce"). See also: Chiriquí, Laguna de.

CHIRIQUÍ, ISLA; Bocas del Toro

0853/8152 (USBGN)

Sea level, an island with a land area of ca. 3 km^2 (300 ha) that is located in the Río Cricamola [0859/8154 (USBGN)] 10 km SSE of Boca de Cricamola [0859/8155 (USBGN)] and just S of the coastal mangroves. The island is formed by Poman Creek [0854/8153 (USBGN)], an anabranch of the Río Cricamola , and the Río Cricamola itself (E762). Wedel, 1937 Feb 13 (YPM, as "Chiriqui Island").

CHIRIQUÍ, ISLA; Chiriquí

ca. 0815/8223 ?

According to correspondence between Wetmore and Davidson in 1953, Watson was a retired ship captain who settled in Boquete [0847/8226 (USBGN)], therefore it seems likely that Chiriquí Island was located somewhere in the Golfo de Chiriquí [0800/8220 (USBGN)], especially because all of his other localities are within Chiriquí province. We could not find Chiriquí Island on our maps, but the name may be a misspelling or a local name for Isla Chirote [0815/8223 (USBGN)], which is located at the mouth of the Río Chiriquí [0820/8226 (USBGN)]. Watson, date ?, 1903 Aug 13, Oct 20, 1904 Nov 6 (YPM, as "Chiriqui Island"). See also: Sevilla, Isla.

CHIRIQUÍ, LAGUNA DE; Bocas del Toro

0903/8200 (USBGN)

Bay on the N coast of Bocas del Toro province extending 45 km from the SW coast of Península Valiente [0905/8151 (USBGN)] W to Isla Popa [0911/8207 (USBGN)]. The bay extends 20 km N from the mainland coast with access to the Caribbean Sea [1500/7500 (USBGN)] by marine channels in the Archipiélago de Bocas del Toro [0920/8210 (USBGN)] (JOG; MGRP). Benson, at Almirante [0918/8224 (USBGN)] (q. v.), and at Bocas del Toro [0920/8215 (USBGN)] (q. v.), 1927 (Griscom, 1928:1, as "Chiriquí Lagoon"); Miller and Richardson, arrived in the bay on February 27 and were based on a house boat, anchored in or near Laguna de Bluefield [0909/8155 (USBGN)], where 60 U. S. Naval personnel were stationed surveying the region. They collected in the bay from the boats, on the islands and on the mainland. Many specimens appear to be labeled only "Chiriquí Lagoon," therefore Miller's field notes served as the primary source for the dates attributed to the secondary localities here. We feel that this is an accurate representation of his work, but must also note that some specimens were collected on the open waters and were therefore accurately labeled "Chiriquí Lagoon." In addition, Miller and Richardson often collected separately, further complicating attempts to create a complete list of dates for each secondary locality. GL for: Cayo Agua, Cayo Toro, Cayos Tiger, Cayos Zapatilla, Isla Escudo de Veraguas, Isla Popa, Laguna de Bluefield, Middle Cay, Nassau Hill, Península Valiente, and Río Cricamola, 1936 Feb 28-Apr 4 (MVZ, as "Chiriquí Lagoon"; UCLA, as "Chiriquí Lagoon"; Miller, 1937, as "Chiriquí Lagoon"); SBE, GL for: Alberrys Creek, Isla Bastimentos, Isla Popa, and Old Point, 1987 Mar 13-15, 17, 23, 28, GL = Punta Róbalo, on the water 7 km E of Punta Róbalo, 1991 Mar 10 (MVUP; USNM, as "Chiriquí Lagoon"; Olson, 1993a). (Angehr, 2003:311, as "Chiriquí Lagoon Wetlands"). See also: Chiriquí, Golfo de.

CHIRIQUÍ, MOUNT; see Barú, Volcán

CHIRIQUÍ, PROVINCIA DE; (Pacific)

0830/8215 (USBGN)

Province formerly included under the old term "Veragua," that extends along the SW coast of Panamá from the Costa Rican border E to the Río Tabasará [0802/8142 (USBGN)], which forms the boundary with Veraguas province. Golfo de Chiriquí [0800/8220 (USBGN)] extends along the entire S coast of the province, and the N border with Bocas del Toro province is formed by the continental divide, which is generally above 2,000 m in Chiriquí province. Some of the earliest ornithological work in the province was conducted in the W near David [0826/8226 (USBGN)], the provincial capital. Other well explored regions in the W include Boquete [0847/8226 (USBGN)], where Tolef B. Mönniche made extensive collections, and Volcán Barú [0848/8233 (USBGN)], the highest point in Panamá. Limited work has been conducted in E Chiriquí province, especially at the higher elevations near the continental divide. The most extensive work conducted in the E was centered around Cerro Flores [0829/8144 (USBGN)], Hato Chamí [0827/8146 (USBGN)], and San Félix [0817/8152 (USBGN)]. Most of the islands in Golfo de Chiriquí remain unexplored by ornithologists especially now that the work conducted by Batty has been proven fraudulent (Olson, 2008). The Carretera Interamericana [0832/8242 (USBGN)] has opened exploration to many coastal areas, whereas the creation of Lago Fortuna [ca. 0842/8216] and the Gualaca-Chiriquí Grande Road [ca. 0857/8207] opened high elevation regions E of Boquete to ornithological exploration.

CHIRIQUÍ, RÍO; Chiriquí

0820/8226 (USBGN)

Originates on the continental divide near Fortuna [0843/8216 (USBGN)] and flows SSW toward the coast. As it enters Bahía de Charco Azul [0812/8245 (USBGN)] ca. 13 km S of David [0826/8226 (USBGN)], the river mouth forms a large delta with numerous islands in the coastal mangroves (E562; JOG). (Delgado, 1983). See also: David, Río; and Fortuna.

CHIRIQUÍ, VOLCÁN DE; see Barú, Volcán

CHIRIQUICITO; Bocas del Toro

ca. 0857/8211

10 m, described by Kennard and Peters (1928:444) and by Peters (1931:295) as a cacao plantation on the Río Guarumo [0900/8211 (USBGN)] located 3-4.5 mi (5-7 km) inland from Chiriquí Grande [0857/8207 (USBGN)] at the terminus of an old railway line. According to E762, the Río Guarumo is 8 km due W of Chiriquí Grande in a lowland area, and a railway line now extends ca. 11 km W and SW of Chiriquí Grande to Punta de Peña [0855/8211 (USBGN)]; presumably Chiriquicito was just S of this current town. Kennard et al., 1926 Mar 12, 22, 24-30 (MCZ; Kennard and Peters, 1928); Smith, 1927 May 6 (Griscom, 1933:301); Wedel, 1928 Apr 16-18, Aug 27 (MCZ; Peters, 1931a); Loftin, 1962 Apr 17 (Loftin and Olson, 1963). (Peters and Griscom, 1929). Not shown on our maps.

CHIRIQUICITO, QUEBRADA; Bocas del Toro
0859/8208 (USBGN)
Originates in the hills SW of Chiriquí Grande [0857/8207 (USBGN)], flows NNE through lowlands and coastal mangroves entering Laguna de Chiriquí [0903/8200 (USBGN)] 5 km NW of Chiriquí Grande (E762, as "Quebrada Chiriquilito"). Kennard et al., "narrow backwaters in the swamps where the vegetation meets close overhead," 1926 Mar 24 (Kennard and Peters, 1928:447, 451-452, as "Chiriquí Creek").

CHIRIQUICITO GRANDE; see Chiriquí Grande

CHIRIQUÍ CREEK; see Chiriquicito, Quebrada

CHIRIQUÍ GRANDE; Bocas del Toro
0857/8207 (USBGN)
Sea level, town on the SW coast of Laguna de Chiriquí [0903/8200 (USBGN)] at the mouths of several small streams. In the region N and W of town the elevation is low with mangroves on the coast, and S of town hills rise above 300 m (E762). Described by Peters (1931:294) as a fishing village located at the terminus of the Boquete Trail [ca. 0857/8207]. The town had formerly been a shipping point for bananas and was connected to Chiriquicito [ca. 0857/8211] by a railroad. Kennard et al., arrived in Chiriquí Grande on 6 Mar 1926 to begin their ascent of the Boquete Trail, whether Chiriquí Grande appears on specimen labels is unclear, although specimens were collected at Chiriquicito (q. v.) and Quebrada Chiriquicito [0859/ 8208 (USBGN)] (q. v.) (MCZ; Kennard and Peters, 1928:444-445, 447, 452-453); Smith, the name on specimen labels is ambiguous, presumably he collected near Chiriquí Grande, although specimens may have come from the nearby Chiriquicito (q. v.), 1927 May 2-6 (YPM, as "Chiriquisita Grande"; Griscom 1933:301, as "Chiriquicito Grande"); SBE, at 10 m, at 100 feet (30 m), 4 km W of, 5-6 km SW of, 1991 Mar 1, 4-9, 13-14 (MVUP; USNM; Olson, 1993a; Parsons et al., 1993; Brumfield et al., 2001); Braun and Brumfield, at sea level, 08°56'48"/82°08'12", and at 28 m, at 08°56'36"/82°09'18", 2004 Mar 11, 18 (LSUMZ); Guitton-Mayerma, Lovette, Miller, and Weir, at ca. 70 m, at 08°59'11"/82°13'50", 13.6 km NW of, 2004 Apr 18-20 (CU; STRI; UAM); DaCosta, Guitton-Mayerma, Klicka, Lelevier, Miller, and Spellman, at ca. 25 m, at 0855/8208, S of Laguna de Chiriquí, 2 km WSW of, 2006 Feb 23-24, and as a GL for Palo Seco Bosque de Protección, 2006 Feb 22-28 (MBM; STRI). See also: Boquete Trail.

CHIRIQUÍ GRANDE ROAD; see Gualaca-Chiriquí Grande Road

CHIRIQUÍ ISLAND; see Chiriquí, Isla

CHIRIQUÍ LAGOON; see Chiriquí, Laguna de

CHIRIQUÍ LAGOON WETLANDS; see Chiriquí, Laguna de

CHIRIQUÍ POINT; see Big Plantain Cay

CHIRIQUISITO GRANDE; see Chiriquí Grande

CHIRIQUÍ VIEJO, RÍO; Chiriquí
0820/8241 (USBGN)

The river originates as two small streams on the continental divide in NW Chiriquí province and flows S entering Bahía de Charco Azul [0812/8245 (USBGN)] 20 km ENE of Puerto Armuelles [0817/8252 (USBGN)]. One branch that Wetmore called the N fork, originates N and NW of Cerro Punta [0852/8235 (USBGN)] and flows SSW past Bambito [0850/8237 (USBGN)]. Three km N of Volcán [0846/8238 (USBGN)] the N fork is joined by another branch, which Wetmore called the S fork. This branch originates W of Volcán Barú [0848/8233 (USBGN)] and flows W through the E end of Llanos del Volcán [ca. 0847/8238] and joins the N fork above Volcán. From Volcán the river continues W past Palo Santo [0849/8241 (USBGN)] and Pando [0849/8242 (USBGN)], then turns S and flows within 8 km of the Costa Rican border until its reaches Progreso [0827/8250 (USBGN)] after which it turns SE. The river mouth flows parallel to the coast for 2 km creating a sandy peninsula and a narrow inlet that merges with the mouth of the Río Escarrea [0820/8241 (USBGN)] (JOG; MGRP). Wetmore's work in NW Chiriquí province in 1954, in 1955, and in 1960 was along the upper drainage of the Río Chiriquí Viejo. The three GLs for this region are Cerro Punta, Volcán, and Santa Clara, and most of the secondary localities in this region are within a few kilometers of the river or one of its tributaries. GL = Volcán, at Palo Santo (q. v.), the river near Palo Santo was 35-40 feet (10 m) wide with a boulder bed, 1954 Feb 24, 26, Mar 5, 8, 10, 12, 21, 26, GL = Volcán, at 4,100-4,700 feet (1,250-1,400 m), collected on and above the river within 5 km of Palo Santo, in slopes draining into the river, in a partly logged forest with dense undergrowth and fallen trees, 1954 Feb 25, Mar 2, GL = Cerro Punta, at 5,800 feet (1,750 m), river valley below Cerro Punta and above Bambito [0850/8237 (USBGN)], along the river in a stretch of *rastrojo* with some timber, on the slopes above the river in forest with dense and tangled undergrowth, 1954 Mar 12, 23, GL = Volcán, SE of Palo Santo in the sandy flood plain with boulders, open areas, low trees and brush and along a side channel forming an island, along small creeks, in an old coffee plantation bordered by *rastrojo*, 1955 Feb 1, 6, 10-11, GL = Volcán, at 4,100-4,200 feet (1,250-1,300 m), pastures, stretch of original forest near Palo Santo, 1965 Mar 2, 7, 20 (USNM); Smith, W., observations along 3 km below Cerro Punta, 1962 Mar 26-28 (W. Smith, 1971); Loftin and Belton, banding, 1962-1964 (Loftin, 1975). See also: Pando, Silla de Cerro; Rico, Estero; and Tisingal.

CHIRÚ, EL FARALLÓN DEL; see El Farallón del Chirú

CHIRÚ, FARALLÓN DEL; see Farallón del Chirú, Isla

CHITABÉ, QUEBRADA; see Chitrabé, Quebrada

CHITRA; Veraguas (Pacific)

ca. 0832/8055

700 m, the only map that we have showing the town Chitra is MHA, which places it in E Veraguas province 23 km NE of Santa Fé [0831/8105 (USBGN)] and 18 km N of La Yeguada [0827/8052 (USBGN)]. On this older map, the town is at ca. 0831/8040 just below the continental divide near the headwaters of the Río Chico [0820/8028 (USBGN)]. According to Angehr (in litt., Aug 2005), map "4041-IV, Gatú, (printed 1981) [E762] shows a "Bajo Chitra" at c. 0831/8055…Bajo Chitra is

about 18 km nearly due E of Santa Fe, not 23 km NE [as shown on MHA]." JOG shows both El Bajo Chitra [0832/8055 (USBGN)] at 800 m, 14 km NW of La Yeguada and ca. 14 km ENE of Santa Fé and an airfield named Chitra [0829/8049 (USBGN)], at 700 m, 4 km N of La Yeguada. Benson's elevations suggest that Chitra itself may have been above what is shown on more modern maps as El Bajo Chitra. Presumably Chitra and Boquete de Chitra were near one another if not the same locality. Arcé, 18??, 1868, 1869 (BMNH; USNM, as "Boquete de Chitra," and as "Chitra"; Sclater and Salvin, 1868c:388-389; Salvin, 1870bb, as "Boqueti de Chitra," and as "Chitra"; Salvadori, 1893, as "Boquete de Chitra," and as "Chitra"; Ridgway, 1906, as "Boquete de Chitra"; Griscom, 1935a); Heyde, 1888 Jul, 1889 Jun 14, 22-23, Jul 2, 10 (USNM, as "Chitra," and as "Chitra, Coclé"); Batty, specimens obtained by Batty from Arcé and the locality and dates are surely fabricated, as some of the specimens could not have come from Veraguas (Olson, 2008), 1901 Mar 5-6, Apr 17, 18, 27 (CM; FMNH, as "Boquete de Chitra"); Benson, at 3,600-5,000 feet (1,100-1,500 m), 1925 Dec 17-31, 1926 Jan 1-Mar 15 (AMNH; USNM; Griscom, 1924b; Griscom, 1927a; Zimmer, 1932:9; Zimmer, 1934:14; Zimmer, 1939; Zimmer, 1941b:10, 14; Zimmer, 1944; Zimmer, 1950:15; Zimmer, 1951:13; Wetmore, 1967; Olson, 1993b); Delgado, 1972 Jul, Aug 30 (GML). (Chapman, 1917; Hellmayr, 1918; Moore and Peters, 1939; Wetmore, 1962; Wetmore, 1967; Wetmore, 1970; Parkes, 1976; Escalante-Pliego and Peterson, 1992). See also: Veraguas, Provincia de.

CHITRABÉ, QUEBRADA; Herrera

ca. 0756/8041

Described by Wetmore as a river that crossed the road between La Trinidad [0755/8043 (USBGN)] and Pesé [0754/8037 (USBGN)], 4 mi (6 km) W of Pesé, presumably a tributary of the Río Parita [0802/8027 (USBGN)]. Wetmore and Perrygo, GL = Pesé, the streambed was bare rock and bordered by tall trees and pastureland, some of which had recently been burned, 1948 Mar 30 (USNM; Wetmore, 1950a; Wetmore, 1952:17, as "Quebrada Chitabé"). Not shown on our maps.

CHITRÉ; Herrera

0758/8026 (USBGN)

ca. 25 m, the capital of Herrera province, located 1 km N of the Herrera/Los Santos provincial boundary and 3 km from the S coast of Bahía de Parita [0808/8024 (USBGN)] (JOG). The Chitré Wetlands are considered an Important Bird Area by Angehr (2003) and the site includes "mangroves, salt pans (*albinas*), brackish and fresh marshes, shrimp ponds, and salt evaporation ponds (*salinas*) at the southern end of Parita Bay near the town of Chitré." Loftin, observations, 1963 November 24 (Loftin, 1963a); Butler et al., banded birds at, 1989 Jan-1995 Apr (Delgado and Butler, 1993; Butler et al., 1996); Guglielmo et al., 1995 Dec-Mar 1996 (Guglielmo et al., 2002). (Loftin, 1966a; Banasch et al., 1992; Butler et al., 1992).

CHITRÉ, ISLA; Panamá, Archipiélago de las Perlas

0836/7904 (USBGN)

Sea level, very small island with a land area of less than 1 ha located 20 km off the

NW coast of Isla del Rey [0822/7855 (USBGN)] (MGRP). (MacArthur et al., 1972—species list; see comments in Bibliography); Levings et al., observations, 1982-1983 (Levings et al., 1986).

CHIVA; see Chiva Chiva

CHIVA CHIVA; Panamá (Canal Zone) (E, Pacific)
0902/7935 (USBGN)
ca. 50 m, town on Camino Chiva Chiva (Chiva Chiva Road) [0902/7935 (USBGN)] 2.1 mi (3.4 km) NE of Miraflores Locks [0900/7936 (USBGN)] (E8660; RMM2). The stone quarry may refer to an abandoned mine shown on E762 at ca. 0903/7934 on Camino Chiva Chiva (Chiva Chiva Road) ca. 1-2 km NE of Chiva Chiva. Benson, at 0-100 feet (30 m), 1931 Jun 30, Jul 9, 17, 24 (USNM); Imhof, observations, 1942 Dec 27 (Imhof, 1950); Keenan and MCU, many specimens are also labeled "quarry," 1955 Jun 13, 20, 22, 27, Jul 20, Sep 6, Nov 21, and as a GL for Río Camarón, 1955 Jun 13 (GML; USNM); Wetmore, specimens labeled Chiva Chiva were collected along the Camino Chiva Chiva (Chiva Chiva Road) at Caimitillo Arm, near the stone quarry, and at Chiva Chiva, 1955 Dec 29, 1956 Jan 1, a tract of natural forest along a little stream valley near the stone quarry, at the lake (see Caimitillo Arm), 1957 Jan 13, and as a GL for Río Caimitillo, 1960 Jan 12, GL for Río Camarón, 1961 Jan 7, Mar 19, GL for Río Cárdenas, 1962 Jan 7, on Camino Chiva Chiva (Chiva Chiva Road) in hills near the former Canal Zone/Panamá provincial border, 1963 Jan 12 (USNM, as "Chiva-Chiva," and as "Chiva Chiva"); Olson, near, 1966 May 23, Jun 3 (UF, as "Chiva," and as "Chiva Experimental Station"); Morton, 1970 Jun 8, Sep 9, 1971 Jan 27 (AMNH, as "Chiva"); Seutin, *5 km al sureste de Paraíso* (5 km SE of Paraiso), 1994 Oct 11 (MVUP, as "Chivo Chivo"). (Olson, 1981d; Olson, 1983). See also: Caimitillo Arm.

CHIVA CHIVA, CAMINO; Panamá (Canal Zone) (E, Pacific)
0902/7935 (USBGN)
An alternate name for Road C21 that extends ca. 8 mi (13 km) NE from the Carretera Gaillard [0904/7940 (USBGN)] at Lago de Miraflores [0901/7936 (USBGN)] to the Carretera Boyd-Roosevelt [0917/7943 (USBGN)] near Santa Rita [0905/7930 (USBGN)]. Within the former Canal Zone the road crosses the Ríos Camarón [0902/7936 (USBGN)] (q. v.) and Cárdenas [0859/7935 (USBGN)] and extends through Chiva Chiva [0902/7935 (USBGN)] (q. v.) and Mocambo Abajo [0904/7933 (USBGN)]. The USBGN coordinates indicate that the road terminates near Chiva Chiva, but RM1 indicates that the road terminates farther N at its junction with the Carretera Boyd-Roosevelt at ca. 0905/7930 (E762; RM1, as "Chiva Chiva Trail," and as "Road C21"). The locality is usually seen in the literature and on specimen labels as Chiva Chiva Road. Scholes and Scholes, observations, the article indicates the date as 1951 Dec 23, although according to the introduction, the date should read 1950 Dec 23 (Scholes and Scholes, 1954); Karr, mist nets, according to Karr (1971a), the study site at 0903/7934 was "composed of second growth woodland mixed with grassy clearings. This site was about 7.5 km from the origin of [Road] C21, about 3.5 km northwest of the main entrance to Fort Clayton," 1968 Dec 14-15 (Karr, 1971a"; Karr, 1971b; and Karr, 1971c, all as "Chiva Chiva

Road"); Morton and Hespenheide, 1970 Jun 8, Sep 9, 14, 1971 Jan 19, 27, Feb 26, Mar 19, 22, Apr 10, 16, "C-21 Police Retreat," 1973 Dec 17, 1974 Mar 6, 8-9 (AMNH; CAS; CM"; and USNM, all as "Chiva Chiva Road"; Morton, 1980); Weathers, GL = Panama City, 12 km NNW of Panama City, off Chiva Chiva Road, 1985 Nov (CAS, as "Chiva Road," and as "Chiva Chiva Road"); Braun, M., 1994 Apr 13 (USNM, as "Chiva Chiva Road"); Agro, 1996 Feb 27 (MVUP, as "El Camino a Chivo Chivo"). (McDonald et al., 2001). See also: Chiva Chiva.

CHIVA CHIVA LAKE; see Caimitillo Arm

CHIVA CHIVA ROAD; see Chiva Chiva, Camino

CHIVA CHIVA TRAIL; see Chiva Chiva, Camino

CHIVA EXPERIMENTAL STATION; see Chiva Chiva

CHIVO, QUEBRADA; see Chevo, Río

CHIVO CHIVO; Panamá (E, Pacific)
0904/7932 (USBGN)
200 m, town on the Camino Chiva Chiva (Chiva Chiva Road) [0902/7935 (USBGN)] just E of the former Canal Zone boundary ca. 3 km SW of Santa Rita [0905/7930 (USBGN)] (MGRP). Benson, at sea level, both specimens are clearly labeled Chivo Chivo, not Chiva Chiva, but they are also labeled Canal Zone with an elevation of "0 feet," therefore may have been collected at Chiva Chiva [0902/7935 (USBGN)] (q. v.), 1931 Jun 30 (USNM); Atencio, 1966 Nov 11 (GML). See also: Chiva Chiva.

CHOLITA, LA; see La Cholita

CHONGO, BOSQUE DE; see Bosque de Chongo

CHORCHA; see Chorcha Abajo

CHORCHA, CERRO; Chiriquí
0839/8207 (USBGN)
2,238 m, peak on the continental divide and on the Bocas del Toro/Chiriquí provincial border located 33 km due S of Chiriquí Grande [0857/8207 (USBGN)], 25 km NE of Gualaca [0832/8218 (USBGN)], and 41 km NE of David [0826/8226 (USBGN)] (JOG; MGRP). Described by Sclater (1856c:138, as "Sierra de Chorcha") as "a flat table-mountain" located N of David on the continental divide. See also: Fortuna.

CHORCHA, GRAN GALERA DE; see Gran Galera de Chorcha

CHORCHA, MESETA DE; see Gran Galera de Chorcha

CHORCHA, MINA DE; see Mina de Chorcha

CHORCHA ABAJO; Chiriquí
0822/8217 (USBGN)
20 m, 18 km SE of David [0826/8226 (USBGN)], 4 km S of the Carretera Interamericana [0832/8242 (USBGN)], and ca. 2 km N of the river deltas and mangroves on the NE coast of Bahía de Charco Azul [0812/8245 (USBGN)] (E762, as "Chorcha"; JOG, as "Chorcha"; MGRP; MHA, as "Chorcha"). See also: Mina de Chorcha.

CHORERA; see La Chorrera

CHORRERA; see La Chorrera

CHORRERA, LA; see La Chorrera

CHORRERA, BAHÍA; Panamá (W, Pacific)

0850/7943 (USBGN)

Bay within Bahía de Panamá [0850/7915 (USBGN)] located N of Bahía de Chame [0838/7947 (USBGN)] where it extends ca. 15 km along the W coast of Panamá province S of La Chorrera [0853/7947 (USBGN)] (E501; E761A, as "Chorrera Bay"; MGRP; MHA). Mentioned by Rendahl (1919:2, as "Chorrera Bay") in his description of Río Caimito [0852/7943 (USBGN)].

CHORRERA, PUERTO DE; see Puerto Caimito

CHORRERA BAY; see Chorrera, Bahía

CHORRO, EL; see El Chorro

CHUCANTÍ, CERRO; Panamá (E, Pacific)

0848/7827 (Angehr and Christian, 2000)

1,439 m, peak on the E end of Serranía de Majé [0855/7830 (USBGN)] (q. v.) 21 km NE of Chimán [0842/7837 (USBGN)] on the Panamá/Darién provincial border. The Río Congo [0825/7822 (USBGN)] originates on the E slopes in Darién province (Atlas, 1988:205; MGRP; RDP2). According to Wetmore, in 1950 the slopes had a few areas of exposed lava rock, but were primarily covered with green, gallery forest. The growth on the exposed ridges was short and the steep declines were covered with creeping bamboo and a few small areas on the highest peaks were in the subtropical zone (USNM). From our maps and the account given by Angehr and Christian (2000), Wetmore apparently worked W of the actual peak itself and probably did not reach the Panamá/Darién border. According to Angehr (in litt., Aug 2005), "map E663, sheet 1188 (1:100,000) shows a low divide between the headwaters of the Rios Chucantí [0846/7832 (USBGN)] and Ipetí [0905/7832 (USBGN)], running NW-SE at elevations between 320-520 m for a distance of about 5 km. Since this is the only ridge anywhere near Wetmore's camp that would have allowed him a view of the Bayano Valley at the elevations he gives, this must be the place he refers to. Coordinates are c. UTM 972000 N 773000 E (0847/7831) at the point on the ridge where Wetmore probably reached it. This is c. 5 km W of the peak of Cerro Chucantí itself, and 6 km W of my camp there. Wetmore clearly was always within Panama province." Wetmore and Perrygo, Serranía de Majé, at 1,700 feet (500 m), from the S climbed to the NW slope and walked several miles along the peak ridge with a view (to the N) of the Río Chepo [0900/7906 (USBGN)] valley, 1950 Mar 13-14, 16, 18, and as a GL for: Cerro Maestro, Quebrada Cauchero, and Río Chucantí, 1950 Mar 5-18 (USNM; Wetmore, 1950b; Wetmore, 1952:19; Wetmore, 1967; Olson, 1983); Angehr, Christian, Castillo, and Jordán, GL = Río Congo, 730-1,489 [sic] m, ascended E slopes via logging roads, cloud forest above 1,150 m, 1996 Aug 27-Sep 3 (MVUP; Angehr and Christian, 2000); DaCosta, Guitton-Mayerma, Klicka, Miller, and Spellman, GL = Serranía de Majé, at 750-850 m, at 0847/7827 in the foothills 18 km SW of Chimán, and at 1,100-

1,400 m, at 0848/7828 on the peak 17 km SW of Chimán, 2006 Feb 15-17 (MBM; STRI).

CHUCANTÍ, QUEBRADA; see Chucantí, Río

CHUCANTÍ, RÍO; Panamá (E, Pacific)

0846/7832 (USBGN)

Originates on the SE end of Serranía de Majé [0855/7830 (USBGN)] and flows SSW entering the Río Majé [0840/7835 (USBGN)] 10 km NNE of Majé [0840/7835 (USBGN)] (JOG). On Atlas (1988:205) and MGRP the Río Majé is named Río Chucantí. Described by Wetmore as a swift stream on the S slopes of Cerro Chucantí [ca. 0851/7831], the river was ca. 100-125 feet (30-40 m) wide with steep banks and many large boulders in its bed, Wetmore and Perrygo, GL = Cerro Chucantí, Serranía de Majé, at 400-1,250 feet (120-400 m), on the Río Chucantí near the mouth of the Quebrada Cauchero [ca. 0846/7832] and farther upstream, in forested flats between the streams, on forested ridges above the river, in small ravines, on the SW base of Cerro Chucantí, 1950 Mar 9-18, GL = Charco del Toro [ca. 0844/7833], on trail from Charco del Toro through gallery forest to the Río Chucantí, near its lower end, 1950 Mar 26 (USNM, as "Quebrada Chucantí"; Wetmore, 1967, as "Quebrada Chucantí").

CHUCHU, CERRO; see Chucú, Cordillera del

CHUCÚ, CORDILLERA DEL; Veraguas (Pacific ?, Atlantic ?)

ca. 0840/8050

Described by Salvin (1870:177) as a locality in the district El Mineral de Veraguas [ca. 0830/8100], and he speculated that the locality was on the Pacific versant, which may or may not be the case. Appears to be a section of the Cordillera Central [0838/8150 (USBGN)] ENE of Santa Fé [0831/8105 (USBGN)] based on the presence of two peaks, Cerro Chuchu and Cerro Chicú, both of which appear on our maps roughly in the region that was considered the District El Mineral de Veraguas. Cerro Chuchu [ca. 0840/8100] appears on an older map (PDV) on the Caribbean slope of the continental divide ca. 28 km SE of Calovébora [0848/8113 (USBGN)] between the headwaters of the Río Guázaro [0848/8105 (USBGN)] and the Río Bejuco [0849/8102 (USBGN)]. PDV does not have coordinates, but from MGRP and JOG we estimated the coordinates at 0840/8100, the elevation at 1,300 m, and the peak to be ca. 18 km NE of Santa Fé. On a newer map (RDP2) in the same region but ca. 10-15 km to the E, Cerro Chicú [ca. 0835/8048] appears at 1,764 m, on the Veraguas/Coclé provincial border on the continental divide ca. 30 km ENE of Santa Fé [0831/8105 (USBGN)]. Wetmore (1968:343) discusses the erroneous use of the name "Santiago de Veraguas" by Salvin (1867:154) and by Salvin and Godman (1879-1904, vol. 2:268) for this locality. Arcé, 18??, 1869 (BMNH; USNM; Sclater and Salvin, 1869c:439-440, as "Cordillera of Chucu"; Salvin, 1870b; Salvin and Godman, 1888-1904, vol. 2:268; Wetmore, 1967; Olson, 1981c; Olson, 1993b). See also: Veraguas, Provincia de.

CHUCUNAQUE, RÍO; Darién

0809/7744 (USBGN)

Originates on the continental divide in the northernmost section of Darién province near the Comarca de San Blas/Panamá provincial border. The river flows SSE through a large valley between the Serranía del Darién [0830/7730 (USBGN)] and the Serranía el Filo de Tallo [0827/7757 (USBGN)] and joins the Río Tuira [0821/7803 (USBGN)] 2 mi (3 km) NW of El Real de Santa María [0808/7743 (USBGN)] (DP; SMN2B). Baer, GL = Yaviza (q. v.), the expedition was based at Camp Townsend [ca. 0810/7741] on the Río Chucunaque from Feb 8-Mar 29, from here trips were made to the Ríos Chico [0811/7741 (USBGN)], Tuquesa [0823/7747 (USBGN)], Canglón [0820/7746 (USBGN)], and Tuira, and to Three Falls Creek [ca. 0810/7741 ?], on Mar 29 the expedition traveled up the Río Chucunaque reaching the mouth of the Río Surcurtí [0844/7759 (USBGN)] in early Apr, 1924 Feb 18, Mar 2, 7, 27, and as a GL for: Camp Creek, Camp Townsend, Rio Canglón, Río Chico, Río Surcurtí, Río Tuquesa, Río Tuira, and Three Falls Creek, 1924 Feb 11, 14, 22-25, 27, Mar 1-5, 8, Apr 13 (USNM), (for more on Baer see comments under Marsh, 1934). ; Wetmore worked from the "Río Tuquesa camp of Panamanian Delhi Petrolera," which was located near the mouth of the Río Tuquesa [0823/7747 (USBGN)]. In this region the Río Chucunaque was 60-125 feet (20-40 m) wide and the shore alternated between rock exposures and steep earthen banks covered with jungle or coarse, green grass. Collected within 10 mi (16 km) of camp along the Río Chucunaque, inland along its tributaries, and in the surrounding forest. Access to the uninhabited and uncultivated forest was facilitated by a series of N-S and E-W bulldozed trails that had been cut by an oil company for exploration. GL = mouth of the Río Ucurgantí [0827/7748 (USBGN)], 0.5 mi (1 km) inland along the Río Ucurgantí, then 5 mi (8 km) S along a trail leading back to camp, 1959 Mar 20-22, 31, GL = mouth of the Río Sansón [0824/7747 (USBGN)], 1959 Mar 21, GL = mouth of Quebrada Félix [0822/7747 (USBGN)], inland along Quebrada Félix through forest that was partly open and partly a thick tangle of undergrowth, 1959 Mar 23, GL = mouth of the Río Tuquesa [0823/7747 (USBGN)], on the river 3 mi (5 km) above the mouth of the Río Tuquesa, along a trail leading N to Río Ucurgantí, inland along the dry ravine of Quebrada Félix, trails near Quebrada Félix, trails through open forest to the Río Tuquesa, good-sized ant swarm, upstream on the Río Chucunaque to an inland trail to a wet season lagoon, walked N parallel to the Río Chucunaque to the Quebrada Cuatro (not located), followed the dry streambed of the Quebrada Cuatro inland through low trees and open undergrowth to pools of water and blossoming guava trees, near various tangles of undergrowth and creepers, 1959 Mar 20-Apr 3 (USNM; Wetmore, 1962, as "Chucunaque Valley, near the mouth of the Río Tuquesa"); Slattery, GL = Yaviza, 10 km N and 5 km W of Yaviza, 1966 Dec 2 (GML). (Wetmore, 1952:16; Delgado, 1983). See also: El Salto; and Yaviza.

CHUCUNAQUE VALLEY; see Chucunaque, Río

CHUMBAGO; Chiriquí

0852/8234 (USBGN)

2,000 m, town located on the Río Chiriquí Viejo [0820/8241 (USBGN)] 2 km N of Cerro Punta [0852/8235 (USBGN)] and 2 km SE of the peak Cerro Punta [0853/8235 (USBGN)] (JOG, as "Guadalupe"; RDPF). See also: Bajo Chumbago.

CHUMBAGO, BAJO; see Bajo Chumbago

CHURUNDU; see Curundu

CIANATI, RÍO; see Chanatí, Río

CIANTÍ, RÍO; see Chanatí, Río

CIDER CREEK; Bocas del Toro

ca. 0918/8224

Originates NW of Almirante [0918/8224 (USBGN)] and flows SSE entering Ambrosia Bight [0916/8222 (USBGN)] at Almirante. The main course of the creek is N of the railroad connecting Almirante and Changuinola [0926/8231 (USBGN)] (E501, as "Q. Cedra"). According to Peters (1931:295, as "Cedar Creek"), the creek was actually a small tidal estuary at Almirante [0918/8224 (USBGN)]. The waterway is not shown on our more recent maps and should not be confused with Cedar Creek [0917/8207 (USBGN)], the name given to both a stream and a town on Isla Bastimentos [0918/8208 (USBGN)] (JOG). Wedel, 1927 Nov (MCZ; Peters, 1931a, as "Cedar Creek," and as "Cider Creek").

CIÉNAGA DE BUHO; see Buho, Ciénaga de

CIÉNAGA EL BRUJO; see El Brujo

CIENAGA LAS MACANAS; see La Macana, Ciénaga de

CILICO CREEK; Bocas del Toro

0904/8217 (USBGN)

Settlement (JOG, E762) on Quebrada Cilico Creek [0904/8215 USBGN], 5 km inland from the W side of Laguna de Chiriquí, ca. 2 km NNW of Palma Real [0903/8217 USBGN]. Billo, GL = "Silico Creek Community," 2007 May 22, 30, Jun 2-3,"2 km N of Silico Creek, W side of road, on ridge above river," (USNM).

CINCO, MANACA CERRO; see Manaca Cerro Cinco

CINIQUE, PLAYA; see Cinique, Playa de

CINIQUE, PLAYA DE; Panamá, Archipiélago de las Perlas

0818/7853 (USBGN)

Not shown on our maps but USBGN indicates a beach on the S coast of Isla del Rey [0822/7855 (USBGN)] and on the N coast of Bahía San Telmo [0816/7853 (USBGN)]. Tejera, GL = Isla del Rey, 1981 Mar 15 (MVUP, as "Playa Cinique").

CIRÍ; Colón (Canal Zone) (W)

ca. 0900/8001

Study area on the SW end of Lago Gatún (Gatun Lake) [0912/7955 (USBGN)] located roughly between Cuipo [0905/8003 (USBGN)] and the mouth of the Río Ciri [0858/8003 (USBGN)], Tejera, observations, pastures, small remnants of secondary forest, on and near the lake, 1992-1993 (Tejera, 1995:11-12).

CIRÍ, RÍO; Colón/Panamá (Canal Zone) (W, Atlantic)

0858/8003 (USBGN)

Originates in Panamá province on the NE slopes of Cerro Gaital [0837/8007 (USBGN)] and flows N entering the S coast of Bahía Trinidad (E561, as "Río Ciri

Grande"; JOG, as "Río Cirí," and as "Río Cirí Grande") where the lower reaches of the original streambed are now submerged. Jewel, 20-25 mi (32-40 km) S of Gatún, "clearing along the bank of the river," "overhanging tree-roots on the river banks," 1911 Nov 19, 1912 Mar 31, Jul 13 (ANSP, as "Río Siri"; Stone, 1918:260, 274, as "Río Siri"); Chapin, Rogers and Mathews, GL for La Ullama, 1923 Aug 15-16, 18, 20-23 (FMNH); Smith, N., observations, date ? (Wetmore et al., 1984:350, as "Río Cirí Grande").

CIRÍ GRANDE, RÍO; see Cirí, Río

CISNE, CAYO; see Swan Cay

CISNE, CAYO DEL; see Swan Cay

CITARO; see Cituro

CITURO; Darién
0800/7736 (Fairchild and Handley, 1966)
1,000 feet (300 m), settlement on the Río Cupé [0804/7735 (USBGN)] 5 mi (8 km) SSW of Boca de Cupe [0803/7735 (USBGN)] and 16 mi (26 km) NNE of Cana [ca. 0745/7742] (SMN2A). According to Anthony's (1916) map, the town was at the mouth of the Río Cituro [ca. 0800/7736], which originated SW of town on Mount Cituro [ca. 0758/7739]. According to Chibas (1898:56), Cituro was "the half way station" between Boca de Cupe and Cana. Richardson, GL for Río Cupe, 1915 May 12-13, 18, 20 (AMNH; Anthony, 1916; Chapman, 1923, as "Citaro"; Chapman, 1929b; Griscom, 1932a; Zimmer, 1932:3; Zimmer, 1939; Zimmer, 1950:15, 50; Wetmore, 1951a; Wetmore, 1959b:21).

CITY OF PANAMA; see Panama City

CIUDAD DE PANAMÁ; see Panama City

CIUDAD RADIAL; Panamá (E, Pacific)
0903/7926 (USBGN)
Section of greater Panama City [0858/7932 (USBGN)] located in E Panamá province in Corregimiento Juan Díaz [0902/7927 (USBGN)] E of Juan Díaz [0903/7927 (USBGN)] (q. v.) and S of Pedregal [0904/7926 (USBGN)] (Atlas, 1988:208). González, 1986 Mar 14 (MVUP).

CLAYTON, FORT; see Fort Clayton Army Reservation

CLUB HOUSE; see La Jagua Hunting Club

COBACHÓN; Veraguas (Pacific)
0713/8041 (GA)
According to Angehr (in litt., Aug 2005), who along with Ridgely led the expedition to Cerro Hoya, "our base camp was at the small town of Cobachón (not shown on maps, UTM c. 798000 N 535500 E, or 0713/8041, elevation near sea level) at the mouth of the Río Cobachón (4037-I, Horcones) and we also camped at about 650 m farther up the drainage. Agro et al.'s specimens were collected primarily in the vicinity of the town of Cobachón and at our high camp, which they called Cascajilloso [ca. 0717/8043] (q. v.)." According to JOG, the Río Cobachón [0713/8041 (USBGN)] is a short river that originates on the S coast of Península de Azuero

[0740/8035 (USBGN)] in S Veraguas province and flows S entering the Pacific Ocean 30 km ESE of Punta Naranjo [0715/8056 (USBGN)]. Agro, Conyne, and Pederson, GL = Península de Azuero, 1996 Feb 15-17, 2 mi (3 km) N of, 1996 Feb 17 (ANSP, as "Cobachon"; MVUP, as "Cobachón").

COBACHÓN, RÍO; see Cobachón

COBRE, RÍO; Veraguas (Pacific)
0801/8118 (USBGN)
Originates in NW Veraguas province on the upper slopes of Cordillera Central [0838/8150 (USBGN)], flows SSE and joins the Río San Pablo [0751/8110 (USBGN)] 1 km NE of Soná [0801/8119 (USBGN)] (JOG). Described by Wetmore as a large effluent of the Río San Pablo, Wetmore, GL = Soná, N of Soná through pastures to the flats on the S shore, just above its mouth, 1953 June 13 (USNM).

COCA, LA; see La Coca

COCAL; see El Cocal

COCAL, EL; see El Cocal

COCALITA; see Cocalito

COCALITO; Darién
0719/7759 (USBGN)
Sea level, town on the S coast of Darién province 19 mi (31 km) SSE of Jaqué [0731/7810 (USBGN)] and 9 mi (14 km) NW of the Colombian border (SMN2B, as "Cocalita"). Garcés, GL = Garachiné, 1987 Oct 6 (MVUP).

COCLÉ; Coclé (Pacific)
0827/8026 (USBGN)
ca. 50 m, town on the Carretera Interamericana [0832/8242 (USBGN)] 10 km SW of Penonomé [0831/8022 (USBGN)] (JOG). Braun, M., 5 km SW, km 166 on the Carretera Interamericana, 1991 Feb 28 (USNM).

COCLÉ, LLANOS DE; Coclé (Pacific)
ca. 0825/8015
Described by Eisenmann as a "badly eroded," "often sandy," and "rather sterile" district in SE Coclé province adjacent to W Panamá province. Presumably refers to the lowland (ca. 25-50 m) region that extends SE of Penonomé [0831/8022 (USBGN)] along the Carretera Interamericana [0832/8242 (USBGN)] S and SW of El Valle [0836/8008 (USBGN)]. Eisenmann, observations, c. 1948-1956 Jun-Jul (Eisenmann, 1962a). Not shown on our maps.

COCLÉ, PROVINCIA DE; (Atlantic, Pacific)
0830/8025 (USBGN)
Province in central Panamá just N of Península de Azuero [0740/8035 (USBGN)] and W of the Canal Zone region. The continental divide is higher than 1,500 m in the W near the Veraguas boundary, above 1,000 m in the E at Cerro Gaital [0837/8007 (USBGN)], but less than 200 m in central Coclé province NNE of Penonomé [0831/8022 (USBGN)], the provincial capital. North of the divide the province is bordered by Colón province and therefore does not have an Atlantic coast. The S

coast is largely mangroves and mudflats towards the W along the Bahía de Parita. The capital is Penonomé [0831/8022 (USBGN)], which is located in the lowland regions, where the Carretera Interamericana [0832/8242 (USBGN)] traverses the province. The W boundary is shared with Veraguas province, the short S boundary is with Herrera province, and the E boundary is with Panamá province. The early terms "Veragua" and "Isthmus of Panama" were also applied to areas of Coclé province in addition to referring to modern Veraguas province and the former Canal Zone, respectively. Early collectors in the province included Arcé, Herrera, and Heyde and Lux. The vast majority of more recent work was conducted by Wetmore and Perrygo, who worked in the province during 1951-1953, and later by Wetmore himself as he investigated new localities over the course of four additional years, 1957-58, 1962-63.

COCLÉ, RÍO; see Coclé del Norte, Río

COCLÉ DEL NORTE, RÍO; Colón/Coclé (W, Atlantic)
0905/8035 (USBGN)
Originates at the merger of several streams and flows N along the Colón/Coclé provincial boundary entering Colón province and discharging into the Caribbean Sea [1500/7500 (USBGN)] at Coclé del Norte [0905/8035 (USBGN)], which is a coastal town in W Colón province ca. 80 km SW of Colón [0922/7954 (USBGN)] (JOG). Griscom (1935b:312) lists mentions "Río Coclé, Caribbean coast," which must be the Río Coclé del Norte and perhaps refers to a Heyde and Lux specimen from Cascajal at MCZ. Wetmore's (1959b:19) mention of the headwaters of the Río Coclé del Norte refers to specimens he collected at the head of the Río Guabal. See also: Cascajal; and Guabal, Río.

COCO, CAYO; see San Cristóbal, Isla

COCO, EL; see El Coco

COCO, ISLA DE; Panamá, Archipiélago de las Perlas
0822/7901 (USBGN)
Small island with a land area of ca. 10 ha located 5 km off the W coast of Isla del Rey [0822/7855 (USBGN)] (MGRP). Wright and Campbell, observations, 1978 Jan-Mar (Wright et al., 1985, as "Cocos Island").

COCO, PUNTA; see Cocos, Punta

COCO CAY; see San Cristóbal, Isla

COCOLÍ; Panamá (Canal Zone) (W, Pacific)
0859/7935 (USBGN)
Sea level, town on Carretera Borinquén [0901/7938 (USBGN)] 4 km NNW of Balboa [0857/7934 (USBGN)] and 4 km N of Farfan [0856/7935 (USBGN)] (E762). Keenan and MCU, 1955 Jul 1, 27, Aug 1, 6, and as a GL for: Río Cocolí, Road K6, and Road K9, 1955 Jul 27, Aug 1, 3, Sep 1, Nov 22 (GML; USNM); Keenan, 1958 Sep 1 (GML); Morton, 7-10 mi (11-16 km) N, 1966 May 15, 19-20, 23, Jun 13 (FMNH). (W. Robinson et al., 2000b). See also: Cocolí, Río; and Road K6.

COCOLÍ, RÍO; Panamá (Canal Zone) (W, Pacific)

0859/7936 (USBGN)
Originates near the continental divide in the former Canal Zone SSW of Paraiso [0902/7938 (USBGN)] and N of Arraijan [0857/7939 (USBGN)], flows S crossing Road K15 [ca. 0901/7939], then turns E, crosses Road K6 [ca. 0900/7939], and flows parallel to Road K9 [ca. 0858/7936] widening before it enters Lago de Miraflores [0901/7936 (USBGN)] ca. 3 km N of Cocolí [0859/7935 (USBGN)] (E762; RM1). MCU, GL = Cocolí, 1955 Aug 3 (USNM, as "Cocolí River"); Olson and Fischer, 1963 Mar 16 (USNM, as "Cocolí River"). See also: Road K6.

COCOLÍ RIVER; see Cocolí, Río

COCONATI, RÍO; see Cucunatí, Río

COCOPLUM; see Cocoplum Point

COCOPLUM POINT; Bocas del Toro
0902/8143 (USBGN)
Sea level, point extending into Golfo de los Mosquitos [0900/8115 (USBGN)] from the SE coast of Península Valiente [0905/8151 (USBGN)] near its base (E762; JOG, as "Punta Cocoplum"; Olson, 1993a:101 (map), as "Cocoplum"). The name has always appeared in the literature and on specimens as simply "Cocoplum." Benson, GL for Río Cañaveral, 1927 Oct 24-Nov 12 (AMNH; Chapman, 1931a; Peters, 1931a:345; Rogers, 1937; Rogers, 1939; Lincoln, 1944; Eisenmann, 1955a; Zimmer, 1941b:14; Zimmer, 1950:26; Zimmer, 1955:5; Wetmore, 1959b:13; Meyer de Schauensee, 1966:87; Parkes, 1976; Olson, 1981d).

COCOS, ISLA; see Coco, Isla de

COCOS, ISLA DE; see Coco, Isla de

COCOS, PUNTA; Panamá, Archipiélago de las Perlas
0813/7854 (USBGN)
The southernmost point on Isla del Rey [0822/7855 (USBGN)], located on the peninsula that extends 8 km off the S coast (JOG, as "Punta Coco"). According to Rendahl (1920:5), a mountain ridge extends N-S along the length of the peninsula. The peninsula's E coast, which encloses Bahía San Telmo [0816/7853 (USBGN)], was a fertile strip of shore with "beautiful vegetation," a fresh water lagoon, and "a very rich bird-life" at low tide, Bovallius, GL = Isla del Rey, 1882 Apr-May (NHRM; Rendahl, 1920); Robins, observations, 1957 Jul 15-26 (Robins, 1958, as "Cocos Point"); Tejera, 1981 Mar 23 (MVUP, as "Punta Coco"). See also: Rey, Isla del.

COCOS ISLAND; see Coco, Isla de

COCO SOLITO; see Majagual-Gold Hill

COCO SOLO; Colón (Canal Zone) (E)
0922/7953 (USBGN)
Sea level, town on the E coast of Bahía Manzanillo [0922/7953 (USBGN)] 2.5 km NE of Colón [0922/7954 (USBGN)], formerly located within the limits of Coco Solo Naval Reservation [0923/7952 (USBGN)] (E8660). In fact, the term Coco Solo may refer more generally to the Naval Reservation. Ambrose, 1956 Jul 23 (Eisenmann, 1957); Brown and Eisenmann, observations at Coco Solo Meadows,

1983 Oct 7, Nov 1 (Brown and Engleman, 1986); Gardner, GL for Galeta Road, 1984 Jul 6 (WFVZ); Engleman, D., GL for Galeta Road, 1984 Nov 6 (WFVZ); Brenes, 1985 Apr 7 (MVUP); SBE, 1987 Mar 10 (USNM). See also: Coco Solo Naval Reservation; and Majagual-Gold Hill.

COCO SOLO MEADOWS; see Coco Solo

COCO SOLO NAVAL RESERVATION; Colón (Canal Zone) (E)
0923/7952 (USBGN)
25 m, former United States Government military base on the E coast of Bahía Manzanillo [0922/7953 (USBGN)] extending N from Fort Gulick [0919/7952 (USBGN)] to the N coast of Isla Galeta [0924/7952 (USBGN)]. France Field [0922/7952 (USBGN)] was located on the base (E866, as "Coco Solo Naval Station"; E8660, as "Rodman Naval Station (Coco Solo Annex)"); Barrett, 1973 May 29 (GML, as "Rodman, CZ, Colón"). See also: Coco Solo; France Field; and Rodman United States Naval Reservation.

COCO SOLO NAVAL STATION; see Coco Solo Naval Reservation

COCOVÍ, ISLA; Panamá (Canal Zone) (W, Pacific)
0852/7936 (USBGN)
Sea level, very small island with a land area of less than 1 ha located in Bahía de Panamá [0850/7915 (USBGN)] 3 km S of Venado Beach [0854/7936 (USBGN)] and 1.5 km S of Isla Venado [0853/7936 (USBGN)] (E761A, as "Cocovi Island"). Hallinan, observations, 1915 Mar 24 (Hallinan, 1924:309, as "Cocovi Island").

COCUYOS DE VERAGUA; see Cucuyos de Veragua

COIBA, ISLA; see Coiba, Isla de

COIBA, ISLA DE; Veraguas
0728/8147 (USBGN)
416 m, the largest island on the Pacific coast of Central America with a land area of 503 km^2 (50,314 ha) (Angehr, 2003:165) located in the Pacific Ocean 23 km off the SW coast of Veraguas province. The island extends 40 km from N to SE and varies in width from 6-20 km. The central regions are hilly and portions of the W coast are steep whereas the E coast is lower in elevation and contains several large bays (JOG; MGRP; MHA, as "Isla Coiba o Quibo"). The island is now within Isla de Coiba Parque Nacional [ca. 0730/8140], a protected area and an Important Bird Area (Angehr, 2003) that extends ca. 80 km N-S and ca. 50 km E-W encompassing Isla de Coiba, the surrounding waters, and the nearby islands (RDP2). The proper name "Isla de Coiba" has perhaps never been used on specimen labels or in the scientific literature (except Olson 2007), where the locality is given simply as "Isla Coiba" or "Coiba Island." Batty, the collection from Isla de Coiba is a mixture of specimens genuinely taken on the island along with others from the mainland, including western and central Panama. Specimens are dated 1901 Apr 4-Jun 29 (although two were found at MCN that appeared to be dated 1902 Jan 24 and 1902 Feb 6), but the dates on many specimens appear to have been fabricated as Batty was probably on Isla de Coiba from only 1901 Apr 12 to 27, if that long (AMNH, as "Coiba"; BMNH, as "Coiba Island"; FMNH, as "Coiba"; MCN, as "Coiba";

Hartert, 1901; Hartert, 1927; Rothschild, 1901; Griscom, 1927a; and Griscom, 1933, all as "Coiba Island"; Peters, 1931a:299; Chapman, 1940, as "Coiba"; Zimmer, 1941a:7; Zimmer, 1955:5; Eisenmann, 1950; Wetmore, 1950a; Wetmore, 1957; and Olson, 1981d, all as "Isla Coiba"; Olson, 2008); Kelsall and Cullingford, GL for: Bajo España, and Colonia Penal, 1924 Aug 31-Sep 4 (BMNH, as "Coiba Island"; Douglas and Johnson, 1926, as "Coiba Island"; Wetmore, 1957:8, as "Isla Coiba"); Méndez, A., observations, 1932 (Wetmore, 1957:8, as "Isla Coiba"); Beebe, GL for Banco Hannibal, 1938 Mar 20 (Beebe, 1942; Wetmore, 1957:8); Wetmore, the secondary localities are all located along the E coast, GL for: Bajo España, Campamento de Catival, Campamento de Juncal, Campamento de María, Campamento de Punta Damas, Campamento de San Juan, La Central, Playa Blanca, Río Catival, and Salinas, 1956 Jan 7- Feb 3 (USNM, as "Isla Coiba"; Wetmore, 1957, as "Isla Coiba"; Eisenmann, 1962c, as "Isla Coiba"; Olson, 1980b, as "Isla Coiba"; Olson, 1981a, as "Isla Coiba"; Olson, 1981d, as "Isla Coiba"; Olson, 1997a, as "Isla Coiba"; Olson 2007, as "Isla de Coiba"); Murphy, observations, W of, 1956 Nov 24-25 (Murphy, 1958, as "Coiba Island"); Northern, GL for Ensenada Hermosa, 1957 Mar 23 (LACM, as "Coiba Is"); Morton, 1965 Oct 8-11 (AMNH, as "Coiba Island"); Ridgely, Hinds, and Morales, primary forest, 1976 Apr 12-13 (GML, as "Coiba," and as "Isla Coiba"); Brumfield and Miller, at 07°36'00"/ 81°43'24", Sendero los Monos (not located), and at sea level, at ANAM headquarters [Autoridad Nacional del Ambiente, 7°37'36"/81°43'42"], 2004 Apr 9-10, 12, and as a GL for: Playa Hermosa, and Bahía de Juncal, 2004 Apr 10-11 (LSUMZ, as "Isla Coiba"). (Shelvocke, 1726; Routledge, 1920; Gould, 1944, as "Quibo"; Wetmore, 1967; Parkes, 1976, as "Isla Coiba"; Graves and Gotelli, 1983; Gotelli and Graves, 1990; Cardiel et al., 1997).

COIBA, PARQUE NACIONAL; see Coiba, Isla de

COIBA ISLAND; see Coiba, Isla de

COIBA PARQUE NACIONAL; see Coiba, Isla de

COIBITA, ISLA; Veraguas (Pacific)
ca. 0739/8143
Rock in the group of rocks called Frijoles [0739/8143 (USBGN)] located in the Pacific Ocean off the coast of Veraguas province (JOG; MHA). Specimens have not been collected here, although the name is frequently used for Isla Ranchería [0738/8142 (USBGN)] (q. v.) where collections have been made (Atlas, 1988:202; E762; Wetmore, 1957:5).

COLOBRE, RÍO; see Utivé, Río

COLOMBIA; see Panamá

COLÓN; Colón (E)
0922/7954 (USBGN)
Sea level, the capital of Colón province, located on the E coast of Bahía Limón [0920/7956 (USBGN)] at the Atlantic entrance of the Panama Canal [0920/7955 (USBGN)] (JOG). Built on the extensive swamps of an island that was originally called "Isla Manzanilla" or "Isla Manzanillo," the city was founded in 1852 during

the construction of the Panama Railroad. The Panama Railroad crew remained on a brig anchored in the bay during the early months of construction as the island was "a virgin swamp, covered with a dense growth of the tortuous, water-loving mangrove, and interlaced with huge vines and thorny shrubs, defying entrance even to the wild beasts common to the country" (Otis, 1859:150). Originally named Aspinwall for William H. Aspinwall of the Panama Railroad Co. (Otis, 1859:153) the name Colón, which is Spanish for Columbus, was adopted c. 1890. According to Sclater and Salvin (1864:343), "the line of railway passes through a forest region that, on the Atlantic side, is only broken by large swampy tracts." The island is now connected to the mainland, by a causeway built during the construction of the Panama Canal [0920/7955 (USBGN)] (E8660; HMCZ; IP). Collector ?, 18??, 1876, 1877 Dec, 1878 (FMNH, as "Aspinwall," and as "Colón"; MCZ, as "Aspinwall"; USNM, as "Aspinwall"; Bangs and Penard, 1921:375, as "Aspinwall," and as "Colon"; Boucard, 1895:124); Rhoads, GL for Lion Hill, 18?? (USNM, as "Aspinwall"); McLeannan, GL for Lion Hill, 18?? (USNM, as "Aspinwall"); Boucard, 1876 Dec, 1877 Jan (USNM; FMNH, as "Aspinwall," and as "Colón"); Harris, 1877 Feb (FMNH); Festa, in the Caribbean Sea, 300 mi (485 km) from Colón, 1895 May (Salvadori and Festa, 1899); Ferry, 1908 Mar 11-14, 16-17 (FMNH; Brodkorb, 1950; Blake, 1958:560); Bailey, 1913 Feb 15, 17-18 (USNM); Jewel, 1913 Feb 25, at sea, 60 mi (97 km) N of, 1910 Dec 4 (ANSP; Stone, 1918:248, 260); Fagan, in the Caribbean 300 mi (485 km) NNE of Colón, 1924 Aug 27 (USNM; Wetmore, 1965:63); Imhof, observations, 5 mi (8 km) off, 1943 Feb 23 (Wetmore, 1965:63); Noyes, 3 mi (5 km) S, 7 mi (11 km) W, 1947 Nov 2, 8 mi (13 km) E, 1948 Jan 31 (MVZ); Loftin, observations in autumn, 1962 (Loftin, 1963a); Pitman, observations, 19 km off Colón, 1985 Sep 20 (Ridgely and Gwynne, 1989:55). (Todd, 1913; Wetmore, 1959b:19, incorrectly listed as western Colon province; Olson, 1981a; Olson, 1981d, as "Aspinwall," and as "Colón"). See also: French Canal; Majagual; and Marajal.

COLÓN, ISLA; Bocas del Toro, Archipiélago de Bocas del Toro
0924/8217 (USBGN)
76 m, island with a land area of 63 km^2 (6,300 ha) (Angehr, 2003:97) in the waters between the Caribbean Sea [1500/7500 (USBGN)] and Bahía Almirante [0920/8218 (USBGN)]. Bocas del Toro [0920/8215 (USBGN)], the capital of Bocas del Toro province, is on its S coast (E762, as "Isla Colón (Isla Bocas)"; JOG; MHA, as "Isla de Colón"). The island is also called Columbus Island, which is the English translation of the name (Kennard and Peters, 1928:443), and is considered an Important Bird Area (Angehr, 2003). Wedel, 1927 Jan 30 (MCZ, as "Columbus"; Peters, 1931a:303, as "Columbus Island"); SBE, GL for: Big Creek, Bocas del Toro, Flat Rock Beach, La Gruta, La Represa, Paunch, and Sandfly Bay, 1987 Mar 12, 1988 Feb 6-23, 1989 Jan 28-30, Feb 27, 1990 Apr 15-16 (USNM; Olson, 1993a); Guitton-Mayerma, Lelevier, Maley, and Miller, Smithsonian Marine Lab, 2005 Dec 28-30 (STRI; UAM); Billo, GL for Boca del Drago, 2007 Apr 6-7 (USNM).

COLÓN, ISLA DE; see Colón, Isla

COLÓN, PROVINCIA DE; (Atlantic)

0910/8000 (USBGN)

Province in N central Panama, extending along the Caribbean coast from Río Belén [0853/8052 (USBGN)] on the Veraguas boundary to the Río Mandinga [0928/7904 (USBGN)] on the Comarca de San Blas boundary. The S provincial boundary is shared with Coclé province in the W and Panama province in the E, where it extends through both Lago Gatún (Gatun Lake) [0912/7955 (USBGN)] and Lago Alajuela (Madden Lake) [0915/7935 (USBGN)]. With the exception of ca. 40 km NE of Lago Alajuela (Madden Lake) [0915/7935 (USBGN)], the provincial boundary is N of the continental divide and like most of the province at low elevations. The province includes the N section of the former Canal Zone as well as the Atlantic entrance to the Panama Canal [0920/7955 (USBGN)], where Colón [0922/7954 (USBGN)], the provincial capital is located. The former Canal Zone, and specifically the continuous Río Chagres (see Río Chagres [0919/8000 (USBGN)]), has been used as a route across the isthmus since the 1500s (HMCZ). Fuerte San Lorenzo [0919/8000 (USBGN)] was the Atlantic terminus for travelers crossing the isthmus until the construction of the Panama Railroad [ca. 0922/7954], which led to the founding of Colón and the shifting of the transisthmian route. Many early specimens collected in the province came from the former Canal Zone and were simply labeled Isthmus of Panama or Panama Railroad. See also: Las Cruces Trail; Panama Canal; Panama Railroad; and Panama, Isthmus of.

COLÓN, PUERTO; Colón (Canal Zone) (E)

0922/7955 (USBGN)

Harbor that extends ca. 1 km N-S along the W coast of Colón [0922/7954 (USBGN)] (E8660). Griscom and Crosby, 1924 Feb 13, 1927 Feb 9, Mar 13 (Griscom, 1927b, as "Colon Harbor"); Imhof, observations, 1942 Dec 17 (Imhof, 1950, as "Colon Harbor").

COLÓN HARBOR; see Colón, Puerto

COLONIA PENAL; Veraguas, Isla de Coiba

ca. 0730/8142

Sea level, a prison settlement on the central E coast of Isla de Coiba [0728/8147 (USBGN)] 1.5 km W of Punta Damas [0730/8141 (USBGN)] (Atlas, 1988:202; E762; JOG; MHA). Not listed in USBGN, which does, however, list a prison named Isla de Coiba at 0730/8142 (USBGN), possibly a recent name change that we could not locate on our maps. Kelsall and Cullingford, GL = Isla de Coiba, vicinity, 1924 Aug 31-Sep 4 (BMNH, as "Coiba Island"; Douglas and Johnson, 1926, as "Coiba Island"; Wetmore, 1957:8, as "Penal Colony"). See also: Bajo España; and La Central.

COLORADA; Chiriquí

ca. 0825/8250

ca. 25 m, town 4 km S of Progreso [0827/8250 (USBGN)], 16 km N of Puerto Armuelles [0817/8252 (USBGN)], and 5 km WNW of La Esperanza [0824/8248 (USBGN)] (MGRP, as "Colorado"). Sánchez, GL = La Esperanza, 1983 Jul 9, 14, 17 (MVUP).

COLORADA, LA; see La Colorada

COLORADO; see Colorada

COLORADO, CANAL DE; Chiriquí

0825/8251 (USBGN)

Canal originating 18 km N of Puerto Armuelles [0817/8252 (USBGN)], flowing SE through a cultivated area, and entering Bahía de Charco Azul [0812/8245 (USBGN)] 11 km NE of Puerto Armuelles [0817/8252 (USBGN)]. A branch of the canal also extends NW and connects to rivers in Costa Rica (JOG). Wetmore, GL = Puerto Armuelles, collected along a sandy beach near the mouth, in dry, fairly open mangroves, near flowering trees, 1966 Feb 28, Mar 9 (USNM). See also: Olivo.

COLORADO, CERRO; Chiriquí

ca. 0832/8148

ca. 1,500 m, not shown on our maps but the peak was the site of an extensive copper mining project from 1970-1981, and according to the maps and descriptions of the copper mines that are presented in Gjording (1991), the peak is located near the continental divide ca. 6 km WSW of Cerro Santiago [0833/8144 (USBGN)]. Myers and Duellman (1982:2-3) indicate that the peak is a short distance W of Cerro Santiago and "is situated in the Serranía de Tabasará just south of the continental divide at approximately 81°48' west longitude." In a footnote (p. 3) Burt states that "Cerro Santiago ... [is in] 'an area that has obviously never heard the ring of a machete but which is only about five kilometers from the Cerro Colorado access road.'" The map in Clark et al. (1977) indicates that at its nearest the divide is ca. 1 km from the peak. JOG and MGRP indicate that the continental divide in this region reaches 1,800 m, but E562 shows the divide at 1,500 m, and Myers and Duellman (1982:3) mention that the elevation of the region has been misreported on earlier maps. This lower elevation is also shown on an unpublished map prepared by the Colorado School of Mines (CO, USA) in 1977. Smith, N., at 1,700 m, observations, in Chiriquí province, and in Bocas del Toro province after crossing the divide near, 1979 Apr (Wetmore et al., 1984:50, 330); Guarnaccia, observations above, 1986 May 16, 23 (Ridgely and Gwynne, 1989:170, 422); Weir, at 1,700 m, at 0830/8146, 2003 Apr 12-13 (CU; LSUMZ). See also: Colorado Camp.

COLORADO, LA; see La Colorada

COLORADO, RÍO; Chiriquí

0849/8243 (USBGN)

Originates on the continental divide 7 km E of the Costa Rican border, flows SSW and joins the Río Chiriquí Viejo [0820/8241 (USBGN)] 6 km W of Palo Santo [0849/8241 (USBGN)] (JOG). Wetmore, GL = Volcán [0846/8238 (USBGN)], at 4,000-4,500 feet (1,200-1,350 m), at the bridge on the main road (according to JOG, the bridge is 5 km WNW of Palo Santo), above the river in a small tract of timber and *rastrojo*, 1954 Mar 21, 26 (USNM). See also: Pando; and Volcán.

COLORADO CAMP; Chiriquí

ca. 0831/8148

GALLITO, EL; see El Gallito

GALLO; Coclé (Pacific)

ca. 0812/8030

Described by Wetmore as vast, open salt ponds between Aguadulce [0815/8033 (USBGN)] and Bahía de Parita [0808/8024 (USBGN)], shown on E761A, as "salt beds," 5 km SE of Aguadulce. Wetmore may have also referred to the region as Salinas Aguadulce. Wetmore, GL = Aguadulce, at the salt ponds, brushy pastures between Gallo and Aguadulce, horse trail to small savannas bordered with low, thorny scrub, 1963 Jan 16-17, 24-25 (USNM); Tejera, *albina de* (salt lake or marsh), *llano de* (savannas of), 1978 Jun 21 (MVUP, as "Albina de Gallo," and as "Llanos del Gallo"); Tejera, 1990 Feb 18 (MVUP, as "Isleta El Gallo"). See also: Aguadulce; and El Salado.

GALLO, ALBINA DE; see Gallo

GALLO, CANTA; see Canta Gallo Abajo

GALLO, ISLETA EL; see Gallo

GALLO, LLANOS DEL; see Gallo

GAMBOA; Colón (Canal Zone) (E)

0907/7942 (USBGN)

20 m, town on the Panama Canal [0920/7955 (USBGN)] on the W bank of the mouth of the Río Chagres [0907/7942 (USBGN)] 25 km NW of Panama City [0858/7932 (USBGN)] and 35 km SE of Colón [0922/7954 (USBGN)] (E866; JOG). Specimens collected at the Police Station or the Penitentiary were collected in Panamá province on the E bank of the river ca. 1 km SE of Gamboa (JOG). Gamboa Woods refers to a "22-ha woodland tract that is approximately 60 years old, having re-grown after being completely cleared in the mid-1930's when the town of Gamboa was constructed." The woods border the town (W. Robinson et al., 2000b). Hallinan, on Lago Gatún (Gatun Lake) [0912/7955 (USBGN)], 1915 Oct 15 (AMNH; Hallinan, 1924; Zimmer, 1941b:3; Zimmer, 1955 5); Ude, 1920 (YPM); Collector ? [Loomis ?], 1921 Feb 3, Mar 11, 15 (CU); Loomis, 1921 Feb 5 (CU); Kelsall and Cullingford, 1924 Jun (BMNH; Douglas and Johnson, 1926); Weber, 1928 Aug 4-5, and as GL for Río Chagres, 1928 Aug 4-5 (FMNH; USNM); Eisenmann, GL for Lago Gatún, 1949 Jul 5 (Eisenmann, 1951); Hardy, GL for Río Chagres, 1949 Aug 22 (MCN); Wetmore, the dock at the Gamboa police station, Balboa [0857/7934 (USBGN)] side of the Río Chagres, 1953 June 24, and as a GL for Río Chagres, 1955 Dec 10, GL for: Las Cruces Trail, and Río Chagres, 1959 Jan 29, GL for Río Frijoles, 1960 Jan 9-10, 13 (USNM; WFVZ; Brumfield et al., 2001); Blanton, 1953 Sep 16, Oct 9, and as a GL for Road C2H, 1953 Sep 16 (USNM); MCU, 1955 Jul 7, 25, Oct 10, and as a GL for Road C29, 1955 Apr 21, 28-29, May 2, 18, 23, Jul 17 (GML; USNM); Alderton, observations, 1955-1961 (Alderton, 1961, as "South Gamboa"; Alderton, 1963); Koford, 1957 (USNM); Keenan, 1959 Jun (GML); Gale, 4.6 mi (7.4 km) NNE, 17 mi (27 km) E, and at the Penitentiary, 1960 Nov 25, 1963 Jan 30, Mar 30, 1964 Jun 1-2 (AMNH; USNM, as "Gamboa," and as "Gamboa Pen."); Wetmore (from Gale), GL for Pipeline

Road, 1960 Dec 27; Wetmore (from Gale) 1961 Mar 5 (USNM); Yunker (from NIID), GL for: Cacao Plantation Road, and Carretera Gaillard, 1961 Oct 26, Nov 20, 1962 Jan 23, 29, Feb 2, 5, 8, Mar 28 (USNM); Hespenheide and Karr, GL for Río Frijolita, 1962 Jul 20 (CM); Loftin and Belton, banding, 1962-1964 (Loftin, 1975; Loftin et al., 1966); Olson et al., 2 mi (3 km) SE, near, edge of woods, 1963 Mar 10, 12, 30, May 2, and as a GL for: Carretera Gaillard, and Pipeline Road, 1963 Mar 30, 1966 May 8-10, 13, 15, 18-21, 25-26, 28-30, Jun 1, 3-10, 13, 15, 23, Jul 11, 20, GL for Río Chagres, 1987 Mar 6-8 (UF; USNM; Loftin and Olson, 1963); Chapman and Tyson, 5-7 mi (8-11 km) N, 1964 Jan 15-16, 19-28, Feb 2-3, 5-6, Dec 22, and as a GL for Río Frijoles, 1964 Jan 15, 27-31, Feb 1-4 (AMNH; ANSP; FSU; LSUMZ; USNM; Wetmore et al., 1984:235 as "Gamboa (Pipeline Road, Limbo Hunt Club)"); Hayward, 1964 Apr 4 (USNM); Todd, GL for Madden Road Timber Preserve, 1964 Jul 15 (LACM, as "Gamboa Vic"); Smith, N., GL for Pipeline Road, 1964 Nov 11 (CU); Morton, collected within 6 mi (10 km) of Gamboa, and along the Pipeline Road (q. v.), 1965 Sep 17, Oct 1, 23, 26, 1966 Jan 5, Feb 21, May 6, 10, 26, 28, Jul 10, 15-17, 20, 25, 1971 Mar 22, observations 1970-1974 (AMNH; FMNH; YPM; Morton, 1976; Morton, 1982, as "Gamboa-Pipeline Road"); GMLP, 1967 Mar 4 (GML); Crossin, 1 mi (1.5 km) W of, 1968 Aug 9, and as GL for Pipeline Road, 1968 Aug 10 (USNM); Stevenson, GL for Pipeline Road, 1968 Sep 10 (LSUMZ); Calderón, GL Juan Mina, 1984 Oct 18 (MVUP); Ossa, GL for Juan Mina, 1985 Jan 5 (MVUP); Gardner, near, 1986 Feb 16 (WFVZ); De Ortega, *área del canal*, 1986 Mar 16 (MVUP); Olson, Angle, and Morton, along the Chagres River, 1987 Mar 6-8 (USNM); Collector ?, 1989 Feb 10, 1995 May 28, 1996 Jun 25, 1997 Jan 24-25, Jun 8, Aug 29, 1998 Apr 26, Jun 7 (LSUMZ); Broce, 1989 Mar 3 (MVUP); Brawn et al., mist nets, 199? (Brawn et al., 1996); SBE, GL = Parque Nacional Soberanía, on Pipeline Road (q. v.), 1991 Feb 14-17 (MVUP; USNM; Brumfield et al., 2001); Braun, M., 1994 May 5-6 (USNM); Cardiff, Dittmann, and Lovette, GL for Río Agua Salud, 1996 Mar 15-18 (LSUMZ; MVUP); Williams, GL for Río Sardinilla, 1997 Feb 6 (LSUMZ); Cardiff and Lovette, GL for Río Chagres, 1997 Mar 12 (LSUMZ). (Barnard, 1956; Gorski, 1969; Greenberg, 1981a; Olson, 1981d; Emlen et al., 1989; Wilczynski et al., 1989, as "Chagres Field"; Karr, 1990; Enstrom, 1992; Poulin and Lefebvre, 1995; Poulin and Lefebvre, 1996; Morton and Derrickson, 1996; Fleischer et al., 1997; Stutchbury et al., 1998; McDonald et al., 2001); Maley, 2006 Jan 14-15, 24-26 (STRI; UAM). See also: Juan Mina; and Limbo Plot.

GAMBOA, CAMINO VIEJO DE; see Old Gamboa Road

GAMBOA, CARRETERA DE; see Gaillard, Carretera

GAMBOA HIGHWAY; see Gaillard, Carretera

GAMBOA PEN; see Gamboa

GAMBOA PIPELINE ROAD; see Pipeline Road

GAMBOA ROAD; see Gaillard, Carretera

GAMBOA ROAD JUNCTION; see Gaillard, Carretera

GARACHÉ, RÍO; see Gariché, Río

GARACHINÉ; Darién

0804/7822 (USBGN)

Sea level, town on the S coast of Golfo de San Miguel [0822/7817 (USBGN)] at the SW end of Ensenada de Garachiné [0808/7820 (USBGN)] 6 km SE of Punta Garachiné [0806/7825 (USBGN)] at the mouth of the Río San Antonio [0804/7821 (USBGN)] (MGRP; MHA). According to Barbour and Brooks (1923), "Garachiné has no harbor but nevertheless is well-protected by the high and arid headland of La Punta de Garachiné [see Punta Garachiné]. This point and the plain on which the town is situated lie in a small, isolated zone of very sparse rainfall, the moisture being precipitated from the trade winds by the high peak of Mt. Sapo." This is quite different from the "luxuriant" vegetation described by Barbour and Brooks for Cerro Sapo [0758/7822 (USBGN)] (q. v.) and the Río Sambú [0805/7818 (USBGN)] valley. Vanderbilt Expedition, Feathers was Vanderbilt's collector and all specimens appear to be labeled simply Garachiné, but Vanderbilt (in Bond and Meyer de Schauensee, 1944:23) indicates that they were based at Punta Garachiné, "our own collecting centered at first around Garachiné Point, which is rather desolate in appearance, being covered with the giant but bare Cuipo trees. At this time of the year the rains had not started, though they threatened daily, and the creek-beds were dry. In a few weeks we were to see a startling change as the gray-brown of the point turned to a luxuriant green with the advent of the rains," 1941 Apr 18-21, 24-26, 28-29, May 2-4, 6-8, 6 mi (10 km) S, 1941 May 7 (ANSP; USNM; Vanderbilt, 1944; Bond and Meyer de Schauensee, 1944; Meyer de Schauensee, 1950a; Wetmore, 1950a; Wetmore, 1950b; Wetmore, 1951a; Wetmore, 1952:19; Bond, 1953); Garcés, 1987 Oct 6, and as a GL for Cocalito, 1987 Oct 6 (MVUP). (Angehr, 2003:299).

GARACHINÉ, CABO; see Garachiné, Punta

GARACHINÉ, ENSENADA DE; Darién

0808/7820 (USBGN)

Large bay within Golfo de San Miguel [0822/7817 (USBGN)] that extends ca. 20 km SW-NE along the mainland between Punta Garachiné and Punta Patiño [0815/7818 (USBGN)], the bay opens on the NW to Golfo de San Miguel (MGRP). The bay is labeled "South Bay," on SMN2A, which also labels the waters above Punta Patiño as "North Bay." See also: Garachiné.

GARACHINÉ, LA PUNTA DE; see Garachiné, Punta

GARACHINÉ, PUNTA; Darién

0806/7825 (USBGN)

Sea level, point located 45 km SW of La Palma [0825/7809 (USBGN)] on the promontory that extends N of Serranía del Sapo [0750/7817 (USBGN)] and separates the S waters of Golfo de San Miguel [0822/7817 (USBGN)] and Ensenada de Garachiné [0808/7820 (USBGN)] from the Gulf of Panama [0800/7930 (USBGN)] (MGRP; SMN2B). According to Griscom (1933:303), the arid tropical zone on the coast near Punta Garachiné is "an open gallery forest, containing a totally different assemblage of birds from the savannah types of Veraguas and Panama City." Griscom Crosby Expedition, "the most fruitful country visited was, however, the coast just back of Cape Garachiné," the coastal region is "arid tropical, not humid tropical, and the forest is largely scrub relieved by the gigantic 'cuipo' tree (*Cavanillesia planifolia*)," open thickets, thorny thickets, low hills, dense red mangrove swamp, ca. 1 mi (1.5 km) offshore, 1927 Feb 25-27, Mar 3-5, 8 (AMNH, as "Cape Garachiné"; Griscom, 1927a, as "Cape Garachine"; Griscom, 1927b, as "Garachiné Cape," and as "Garachiné"; Zimmer, 1934:14, as "Cape Garachine"; Zimmer, 1950:26, 50, as "Cape Garachine"; Rogers, 1939, as "Cape Garachine"); Murphy and Correia, Askoy Expedition, 12 mi (19 km) W of, 1941 May 22 (AMNH; Murphy, 1944a). (Wetmore, 1952:16; Angehr, 2003:297-298). See also: Garachiné.

GARACHINÉ CAPE; see Garachiné, Punta

GARACHINÉ POINT; see Garachiné, Punta

GARAY, QUEBRADA; Bocas del Toro
0912/8221 (USBGN)
Short mainland river that enters Ensenada Shepherd [0913/8221 (USBGN)] 5 km SSE of Punta de Gallinazo [0915/8222 (USBGN)] and 12 km SSE of Almirante [0918/8224 (USBGN)] (E762). Wedel, 1926 Dec 19 (Peters, 1931a:317, as "Gerray Creek"; Wetmore, 1946b:17, as "Garray Creek, Almirante Bay"); Wetmore, GL = Almirante, the mangroves at the mouth were limited and the creek quickly entered high ground, followed the broad valley into the hills, but unable to enter the surrounding private lands, 1958 Feb 26 (USNM, as "Garay Creek").

GARAY CREEK; see Garay, Quebrada

GARCÍA, CAYO; Bocas del Toro, Archipiélago de Bocas del Toro
0914/8220 (USBGN)
Sea level, small island with a land area of ca. 0.06 km^2 (6 ha) located just off the SE coast of Isla Pastores [0914/8220 (USBGN)] in the waters between Ensenada Grande [0914/8219 (USBGN)] and Ensenada Shepherd [0913/8221 (USBGN)] (E762). On less detailed maps the stretch of mangroves and Cayo García appear as a peninsula extending ca. 1 km off the SE coast of Isla Pastores (JOG; MGRP). SBE, GL = Isla Pastores, at SE tip of Isla Pastores, 1989 Mar 1 (USNM).

GARICHÉ, RÍO; Chiriquí
0821/8243 (USBGN)
Originates just NE of Volcán [0846/8238 (USBGN)] and flows SSW crossing the Carretera Interamericana [0832/8242 (USBGN)] 17 km W of La Concepción [0831/8237 (USBGN)] before entering the Río Chiriquí Viejo [0820/8241 (USBGN)] 3 km above its mouth (E762; JOG). Worth and Cresson, observations and collections, Río Garaché [sic] forest, a few miles from El Volcán, 1937 Jun-Jul (ANSP, as "El Volcan"; Worth, 1939a); Wetmore, GL = La Concepción, at 800 feet (240 m), 10 mi (16 km) W of La Concepción, trail above the river leading through a pasture with an open stand of shrubs, trail down to the river where a small stand of forest remained, 1960 Mar 2 (USNM). (Wetmore 1968: 256, as Río Garaché, lapsus)

GARITA, LA; see La Garita

GARRAY CREEK; see Garay, Quebrada

GARZA, QUEBRADA; Bocas del Toro

0907/8217 (USBGN)

Short river that originates in hills NW of Punta Róbalo [0902/8215 (USBGN)] and flows NE entering the Río Uyama [0907/8215 (USBGN)] 4 km above its mouth (E762). Braun, M., Brumfield, and Glen, at 50 m, 2 km SW of the confluence of the Río Uyama and the Quebrada Garza, 1994 Jan 11-13 (USNM; Brumfield et al., 2001).

GASPAR SABANA; Panamá (E, Pacific)

0915/7907 (USBGN)

Not shown on our maps, but USBGN indicates a town at 300 m on the Río Mamoní [0908/7904 (USBGN)] ca. 10 km NNW of Chepo [0910/7906 (USBGN)]. According to Angehr (in litt., Aug 2005), the town of Gaspar Sabanas appears on his "map 4343-I [E762], at UTM 1021900 N 707100 E (0914/7907), about 1 km S of the USGBN coordinates and 7 km (not 10 km) NNW of Chepo." Kelly, GL = Chepo, 1921 Mar (USNM, as "Gaspar Savanna").

GASPAR SABANAS; see Gaspar Sabana

GASPAR SAVANNA; see Gaspar Sabana

GATÚN; Colón (Canal Zone) (E)

0916/7955 (USBGN)

Sea level, town on the original Panama Railroad [ca. 0922/7954] and the continuous Río Chagres (see Río Chagres [0919/8000 (USBGN)]) 11 km SE of Fuerte San Lorenzo [0919/8000 (USBGN)] (IP, as "le Gatun"; PR). The town is now located 9 km SSW of Colón [0922/7954 (USBGN)] on the N coast of Lago Gatún (Gatun Lake) [0912/7955 (USBGN)] ca. 1 km N of its original location just E of Esclusas Gatún (Gatun Locks) [0916/7955 (USBGN)] (E761A; E8660; JOG). According to Jewel (1913:422), Gatún was located at the beginning of "the first foothills of the Isthmus." The town was separated from the coast by "three or four miles of low flat land – some of it very marshy," and salt water reached Gatún via "the old French Canal, the East Diversion and the new American canal." Hallinan, jungle, grove of thorn trees, clearings, swamp, marsh, thorn scrub, high grass, sandbar in a stream, trees and shrubbery on stream banks, along the road, 1908 Aug 26, 30, Sep 3, 6, Oct 12, 18, Nov 18, 22, 24, 1909 Feb 1-3, 5, 10-11, 17, 19, Mar 21, 27-28, 31, Apr 1, 4-5, 7-8, 11, 13, 21-22, 24, 26, 29, May 4-6, 8, 26 (AMNH; Hallinan, 1924); Jewel, within several miles of town, on and near Represa Gatún (Gatun Dam) [0916/7956 (USBGN)], in the hydraulic fill of Represa Gatún (Gatun Dam), along streams, in second growth pastures and thickets, in wild banana thickets, heavy grass in a dry field, and a clearing near forest edge, in low bushes, jungle, wet forest, and heavy forest, 1910 Dec, 1911 Jan-Dec, 1912 Jan-Sep, Nov, 1913 Feb, Mar (AMNH; ANSP; CU; FMNH; LACM; LSUMZ; Jewel, 1913; Stone, 1918; Rogers, 1939; Meyer de Schauensee, 1950a); Goldman, in the hills W of the Río Chagres [0919/8000 (USBGN)], into the woods near New Gatún [ca. 0917/7955] (q. v.), near Represa Gatún (Gatun Dam) (q. v.), 1911 Jan 1-3, 10-12, 14, 16, 20-21, 25, 28-29, 31-Feb 5, 25-28, Mar 2, 6, 8, 10, 15, 31, Apr 5-7, 21-28, 30, May 2-4, 8,

15-16, Jun 18, and as a GL for Río Indio, 1911 Feb 8, 10, 13-23, Mar 3-4 (MVUP; USNM; Todd, 1913; Todd, 1917; Wetmore, 1959b19; Olson, 1981a; Olson, 1981d; Olson, 1983; Brumfield et al., 2001, as "Río Indio, near Gatun," and as "Gatun"); Collector ?, 1911 Mar 2, May 21 (CU); Grant, 1911 Mar 15, 1912 Jan 28 (CU); Shivas Expedition, 1914 Feb 27-28 (AMNH); Collector ?, 1925 Sep (Cooke, 1946); Fischer, 1929 Mar, May 11, 13, 23 (USNM); Arbib and Loetscher, "grassy savannahs, scrub woodland, deep jungle and open plantations," "most of the shore-birds were seen along the Chagres River [Río Chagres [0919/8000 (USBGN)]] below the spillway," according to E8660 the spillway is located at Represa Gatún (Gatun Dam) ca. 2 km SW of Gatun, 1933 Aug 7, 17, 1934 Aug 5-Sep 1 (Arbib and Loetscher, 1935); Allen, 1944 Oct 12 (CU); Noyes, within 5 mi (8 km) of town, 1947 Sep 6, Nov 2, 1948 Jan 18, Feb 19, 23, Mar 13, 18 (MVZ); Johnson, 1948 Mar 8, 10-11 (CU); Scholes and Scholes, observations at the clubhouse, 1951 Jun 3 (Scholes and Scholes, 1954); White, 3 km NNW, 1972 May 6 (ANSP); Gardner, near, is labeled "Gatun Road," which presumably refers to Road R2 [ca. 0917/7955] (q. v.) or possibly Road S2 [ca. 0920/7958] (q. v.) on the W side of the canal, 1973 Nov, 1974, 1985 Nov 2, 1986 Mar 7, 1987 Sep, Nov 15, Dec 25, 1988 Jan 5 (WFVZ, as "Gatun," and as "Gatun Road"); Stabler, 1985 Nov 28 (WFVZ); Boggs, near, 1986 Apr 12 (WFVZ); Engleman, D., 1987 Sep (WFVZ); Brawn et al., mist nets, 199? (Brawn et al., 1996, as "Atlantic"); Seutin, GL for: Gatún Drop Zone, and Tiger Trail, 1994 Sep 20, 24-25 (LSUMZ; MVUP); Guitton-Mayerma and Miller, 20 km W of, at 09°11'43"/79°58'46", 2005 Jan 28-30 (STRI; UAM). (Sclater, 1858d; Zimmer, 1941b:14; Zimmer, 1955:5; Eisenmann, 1955c; Eisenmann, 1952:12; Wetmore, 1952:16, 18, 31; Wetmore, 1962:7, footnote; Wetmore, 1967; Wetmore, 1970; Parkes, 1976; J. Smith, 1977; Morton, 1982; Escalante-Pliego and Peterson, 1992).

GATÚN, ESCLUSAS; Colón (Canal Zone)
0916/7955 (USBGN)
Locks on the Panama Canal [0920/7955 (USBGN)] located on the NW coast of Lago Gatún (Gatun Lake) [0912/7955 (USBGN)] immediately W and NW of Gatún [0916/7955 (USBGN)] (E762, as "Gatun Locks"). The locality has always been written on specimen labels and mentioned in the literature as simply "Gatun Locks." Scholes and Scholes, observations, 1951 Jun 2 (Scholes and Scholes, 1954); Olson and Wiese, near, 1966 Jul 15, on the road (see Road S1 [ca. 0918/8001]) between Piña Beach [ca. 0918/8001] and, 1966 Jul 13-14, 16 (UF; USNM); Engleman, D., Smith, J., Gale, and Ramirez, observations near the French Canal [0920/7955 (USBGN)] (q. v.), 1973 Jan 14 (Wetmore et al., 1984:271); Collector ?, 5 km NW Gatún Locks on road to Fort Sherman [Army Reservation], according to our maps this is Road S2 [ca. 0920/7958] (q. v.), 1997 Mar 4 (LSUMZ, as "Gatun Locks"); Cardiff and Lovette, GL for Fort Sherman Army Reservation, 1997 Mar 19 (LSUMZ); Guitton-Mayerma and Miller, just W of, 2005 Jan 28, Dec 3 (STRI, as "Gatun Locks"; UAM, as "Gatun Locks").

GATÚN, LAGO; Colón/Panamá (Canal Zone) (E, W, Atlantic)
0912/7955 (USBGN)
26 m, large reservoir that was formed ca. 1912 during the construction of the

Panama Canal and that covers an area of 164 mi² (425 km²) surrounding the N end of the Panama Canal [0920/7955 (USBGN)] between Colón [0922/7954 (USBGN)] and Gamboa [0907/7942 (USBGN)]. Numerous rivers feed the reservoir and its highly irregular shoreline surrounds a multitude of islands. Water flow into the reservoir from the Río Chagres [0907/7942 (USBGN)] is controlled by Represa Madden (Madden Dam) [0913/7937 (USBGN)] whereas water flow out to the Caribbean Sea [1500/7500 (USBGN)] is controlled by Represa Gatún (Gatun Dam) [0916/7956 (USBGN)] (E661, as "Gatun Lake"). The locality is often referred to as Gatun Lake. Jewel, small grassy island, dead tree in water, 1911 Jan 29, May 21, Jul 9, 16, Nov 25, 30, Dec 3-13, 17, 1912 Apr 28, Feb 23, Mar 25, 30, Jun 9, 1913 Feb 23, Mar 30 (AMNH; ANSP; Jewel, 1913; Stone, 1918; Eisenmann, 1951; and Eisenmann, 1952:20, all as "Gatun Lake"); Goldman, he was based at Gatún (q. v.), 1911 Feb 1 (MVUP); Chapin, Rogers and Mathews, GL for La Ullama, 1923 Aug 15-16, 18, 20-23 (FMNH, as "Gatun Lake"); Lord, 1924 (LACM, as "Gatun Lake"); Gilliard, 1937 Jan 24, and as a GL for: Bahía Gigante, and Frijoles, 1937 Jan 26, Feb 7, 13, 17 (AMNH, as "Gatun Lake"); Cooper, N of Fairchild Plant (not located), 1941 Jan 20 (FMNH, as "Gatun Lake"); Imhof, observations, near Gatún, near the locks, 1942 Dec 28-29 (Imhof, 1950, as "Gatun Lake"); Eisenmann, GL = Isla Barro Colorado, observations near Isla Barro Colorado, 1949 Jun 28, GL = Gamboa, observations near Gamboa, 1949 Jul 5 (Eisenmann, 1951, as "Gatun Lake"); Murphy, GL for Isla Barro Colorado, observations, 1956 Nov 28 (Murphy, 1958, as "Gatun Lake"); GMLP, 1960 Sep 29 (GML, as "Gatun Lake"); Loftin, autumn 1962 (Loftin, 1963a, as "Gatun Lake"); Acosta, GL = Isla Barro Colorado, *cerca de* (near) Isla Barro Colorado, 1981 Jan 15 (MVUP); Seutin, 1994 Sep 23 (LSUMZ, as "Gatun Lake"). (Shiras, 1915; Zaret and Paine, 1973; Delgado, 1983). See also: Chagres, Río.

GATÚN, REPRESA; Colón (Canal Zone) (W)

0916/7956 (USBGN)

5 m, dam on Lago Gatún (Gatun Lake) [0912/7955 (USBGN)] (q.v.) 11 km SSW of Colón [0922/7954 (USBGN)] and ca. 2 km SW of Gatún [0916/7955 (USBGN)] on the W side of the Panama Canal [0920/7955 (USBGN)]. The dam controls flow out of Lago Gatún (Gatun Lake) and into the Río Chagres [0919/8000 (USBGN)] (E661, as "Gatun Dam"; E866, as "Gatun Dam"). Loftin and Olson, area, 1963 Jun 2 (Loftin and Olson, 1963, as "Gatun Dam"); Morton, 1971 Jan 24, Apr 4, and as a GL for Gatún-Escobal Road, 1974 Mar 13 (CAS, as "Gatun Dam"; USNM, as "Gatun Dam"; YPM, as "Gatun Dam"; Morton, 1971); Strauch, drainage ditch, 1972 Nov 1 (UMMZ; Strauch, 1975a, as "Gatun Dam"); Wright, censuses, 1976-1978 (Wright, 1979, as "Gatun Dam"). (Eisenmann, 1961, as "Gatun Dam"). See also: Gatún.

GATÚN, RÍO; Colón (Canal Zone) (E)

0916/7947 (USBGN)

Originates SW of Cerro Bruja [0929/7934 (USBGN)] and flows SW entering the E coast of Lago Gatún (Gatun Lake) [0912/7955 (USBGN)], where the lower course of its original streambed is now submerged. The river formerly continued SW and entered the continuous Río Chagres (see Río Chagres [0919/8000 (USBGN)]) ca.

0.5 mi (1 km) S of Gatún [0916/7955 (USBGN)] (IP; MGRP; MIP, as "Río Gatuncillo"; PR). Goldman, GL for Monte Lirio, 1911 Jan 24 (USNM); Jewel, "3 mi (5 km) up the river, well into the forest," 1911 Jul 8, 1912 Jun 19 (Jewel, 1913, as "Gatun River"; Stone, 1918, as "Gatun River").

GATUNCILLO; Colón (E)
0914/7939 (USBGN)
100 m, town on the Carretera Boyd-Roosevelt [0917/7943 (USBGN)], 16 km NNE of Gamboa [0907/7942 (USBGN)] on the Río Gatuncillo [0913/7939 (USBGN)] 4 km above its mouth (JOG). See also: Manganese Mine.

GATUNCILLO; see Boca de Gatuncillo

GATUNCILLO, BOCA DE; see Boca de Gatuncillo

GATUNCILLO, RÍO; Colón (Canal Zone) (E)
0913/7939 (USBGN)
Originates in Colón province just W of Lago Alajuela (Madden Lake) [0915/7935 (USBGN)] and flows S joining the Río Chagres [0907/7942 (USBGN)] at Boca de Gatuncillo [0912/7939 (USBGN)] 4.5 km N of Juan Mina [0910/7939 (USBGN)] (E661; E866). Wetmore, GL = Juan Mina, the river narrowed immediately above its mouth and trees from either bank met overhead to shade the river, 1955 Dec 16 (USNM).

GATUN DAM; see Gatún, Represa

GATÚN DROP ZONE; Colón (Canal Zone) (W)
ca. 0917/7956
According to Angehr (personal communication), the name refers to a former military parachute drop training area located ca. 2 km NW of Gatún [0916/7955 (USBGN)], ca. 3 km SW of Fort Sherman Army Reservation [0920/7958 (USBGN)], and adjacent to the SW side of French Canal [0920/7955 (USBGN)]. The area is covered with grass and dissected by channels that drain into French Canal. According to Angehr (in litt., Aug 2005), Howard Drop Zone was located near Howard AFB. Martínez, GL = Fort Sherman Army Reservation, 1992 Sep 13 (MVUP, as "Campo de Paracaidismo"); Seutin, GL = Gatún, 2 km NW of Gatún, 1994 Sep 20, 25 (LSUMZ, as "Drop Zone").

GATÚN-ESCOBAL ROAD; Colón (Canal Zone) (W)
ca. 0916/7955
Road that extends NE-SW along the NW shore of Lago Gatún (Gatun Lake) [0912/7955 (USBGN)] between Gatún [0916/7955 (USBGN)] and Escobal [0909/7958 (USBGN)] (E8660, as "Road S10"). Ridgely and Gwynne (1989) refer to this road as the Escobal Road and often mention it in conjunction with Achiote Road [ca. 0917/8003] (q. v.). Willis, 110 m, observations, "many monocotyledonous plants in the undergrowth and epiphytes on tree trunks despite low elevation and high temperatures," "mixed second growth and high forest on a side road west of the main road along Gatun Lake," at 0914/7958, 1963 Sep 2-3 (Willis, 1967, as "Escobal Road"); Smith, observations, 1964 (Wetmore et al., 1984:336, as "Escobal Road"); Morton, GL = Represa Gatún, 1974 Mar 13 (CAS, as "Escobal Road"); Holness,

1986 Nov 1 (MVUP, as "Carretera Escobal"); Cardiff and Lovette, GL for Road S9, 1997 Mar 16 (LSUMZ); Cardiff, GL = Esclusas Gatún, along Gatún-Escobal (S-10) Road, ca. 4 mi (6 km) SW Gatún Locks [Esclusas Gatún], 1997 Mar 19 (LSUMZ; MVUP, as "Carretera Escobal"). See also: Treinticinco, Quebrada.

GATUN LAKE; see Gatún, Lago

GATUN LOCKS; see Gatún, Esclusas

GATUN ROAD; see Gatún

GAUCHO; Coclé (Pacific)
ca. 0822/8026
20 feet (6 m), described by Wetmore as an area of low scrub along shallow drains near El Gago [0822/8026 (USBGN)] (USNM), presumably the area on E761A of intermittent streams 2 km N of El Gago. Wetmore, GL = El Gago, 1963 Jan 20, 23 (USNM; Olson, 1980c, as "Gago, Gaucho"). Not shown on our maps.

GAVILÁN; Panamá (E, Pacific)
ca. 0911/7903
According to Wetmore, Gavilán is a region E of Chepo [0910/7906 (USBGN)] across the Río Mamoní [0908/7904 (USBGN)] and N of Paso de Mula [ca. 0910/7904] (USNM). The locality is not named on our maps, but the region appears on TM as 100-300 feet (30-90 m) hills amidst intermittent streams. Wetmore and Perrygo, GL = Chepo, 1949 Apr 24 (USNM).

G. D. OMAR TORRIJOS PARQUE NACIONAL; see Omar Torrijos Parque Nacional

GELABERT AIRPORT; see Marcos A. Gelabert, Aeropuerto

GENERAL DE DIVISIÓN OMAR TORRIJOS PARQUE NACIONAL; see Omar Torrijos Parque Nacional

GEORGE GREEN PARK; Panamá (Canal Zone) (E, Pacific)
0905/7937 (USBGN)
150 m, a park located within Madden Road Timber Preserve [0906/7937 (USBGN)] (now part of Parque Nacional Soberanía [0909/7941 (Angehr, 2003:239)]) on the SE section of the preserve at the headwaters of the Río Pedro Miguel [0901/7936 (USBGN)]. Camino Madden (Madden Road) [0907/7937 (USBGN)] extends roughly N-S through the park and E8660 indicates a monument in the N central section of the park on the E side of Camino Madden near one of the small streams that forms the headwaters of Río Pedro Miguel (E8660). Although JOG does not indicate the park, the park monument is indicated as "Monumento." Eaton, GL = Madden Road Timber Preserve, "found along creek at Green Mount," "Green Memorial," 1970 Apr 24, Aug 24 (?) (UF, as "Green Mount," and as "Green Memorial").

GERCHOW KEY; see Gerechow Cay

GERECHOW CAY; Bocas del Toro, Archipiélago de Bocas del Toro
0916/8222 (ADL)
Not named on our maps, but the coordinates given by ADL correspond to a mangrove island with a land area of ca. 0.2 km^2 (20 ha) that is located 2 km SE of

Almirante [0918/8224 (USBGN)] and ca. 200 m off the N coast of Ambrosia Bight [0916/8222 (USBGN)]. Wedel, 1928 Jul 5 (MCZ, as "Gerchow Key"; Peters, 1931a:298, 339, as "Gerchow Key").

GERRAY CREEK; see Garay, Quebrada

GH ROAD; see Gaillard, Carretera

GH-M12; see Gaillard, Carretera

GIGANTE, BAHÍA; Panamá (Canal Zone) (W, Atlantic)
0908/7950 (USBGN)
Bay on the SE coast of Isla Barro Colorado [0909/7951 (USBGN)] (E866, as "Gigante Bay"). Van Tyne, 1927 Aug 18; Gross, 1927 Nov 6; Skutch, 1937 Jan 3 (Eisenmann, 1952, as "Gigante Bay"); Gilliard, GL = Lago Gatún (Gatun Lake) [0912/7955 (USBGN)], 1937 Jan 26 (AMNH, as "Gigante Bay"); Wetmore, 1949 Mar 9; Eisenmann and Ryan, 1950 Jul 8 (Eisenmann, 1952, as "Gigante Bay").

GIGANTE BAY; see Gigante, Bahía

GML CAMP; see Malí, Cerro

GOBERNADOR ISLAND; see Gobernadora, Isla

GOBERNADORA, ISLA; Veraguas (Pacific)
0734/8112 (USBGN)
227 m, island with a land area of 7.6 km^2 (760 ha), located in the Pacific Ocean just S of Golfo de Montijo [0740/8107 (USBGN)], 3 km NW of Isla de Cébaco [0732/8109 (USBGN)] and 4 km S of Punta Brava [0736/8114 (USBGN)] (JOG; Olson, 1997a). Batty, specimens at AMNH and at BMNH are dated 1902 Jan 11-13, but Batty's data were completely fabricated and his specimens labeled "Gobernador Island" were actually collected in 1901 near his mainland bases at Boquete [0847/8226 (USBGN)] and Boquerón [0830/8234 (USBGN)] (AMNH, as "Gobernador Island"; BMNH, as "Gobernador Island"; Aldrich and Bole, 1937:6, as "Gobernador Island"; Eisenmann, 1950, as "Gobernador I."; Wetmore, 1957:6, as "Gobernadora Island"; Olson, 2008); Wetmore, collected near La Punta [0734/8111 (USBGN)], N along the beach to a broad wooded flat facing the sea with no low undergrowth, followed small streams inland, followed a dry stream inland to the base of hills, 1965 Jan 8-11 (USNM; Parkes, 1976; Wetmore, 1967; Olson, 1981d; Olson, 1997a).

GOLF, ALTOS DEL; see Altos del Golf

GONZÁLEZ, LOMA; Darién
ca. 0729/7806
250 m ?, described by Wetmore in 1946 as the first hill S of the airstrip at Jaqué [0731/7810 (USBGN)] (USNM). Presumably a peak in the low range of hills that extends SW of Jaqué parallel to the coast, in which the elevation rises to 100 m 4 km SE of the airstrip and reaches 250 m 6 km SE of the airstrip (JOG). Wetmore and Perrygo, GL = Jaqué, area heavily wooded with dry upper slopes and steep rock faces with small caves in the portion facing SW toward the ocean, 1946 Mar 21, 23, 27, Apr 12 (USNM). See also: El Brazo. Not shown on our maps.

GONZÁLEZ ISLAND; see Pedro González, Isla

GOOFY LAKE; Panamá (E, Pacific)

ca. 0910/7924

600 m, small lake less than 1 km S of Cerro Azul [0910/7924 (USBGN)] (E661; JOG, as "Lago Cerro Azul"; MGRP, as "Lago Cerro Azul"). Morton, GL = Cerro Azul, 0-10.7 km past Goofy Lake, 1971 Feb 7, Mar 5 (AMNH; USNM).

GORDA, PUNTA; Panamá, Archipiélago de las Perlas

0821/7850 (USBGN)

Sea level, point on the SE coast of Isla del Rey [0822/7855 (USBGN)] (q. v.), located 15 km NNE of Punta Cocos [0813/7854 (USBGN)] (JOG). See also: La Puntita.

GORDO, CERRO; Panamá (Canal Zone) (W, Atlantic, Pacific)

?

Meyer indicates a Cerro Gordo in conjunction with Camino Madden (Madden Road) [0907/7937 (USBGN)]. We could only find one peak in the region with this name, but it is located on the W side of the canal (not the E side as the road is) and is located ca. 3.5 km SW of the S terminus of Camino Madden. The coordinates are ca. 0902/7940 (E661; E761A; MIP). Angehr (in litt., Aug 2005) doubts that this Cerro Gordo is the correct locality, but knows of no locality on or near Camino Madden with this name. Meyer, GL for Camino Madden, 1941 May 24 (USNM).

GORGONA; Colón (Canal Zone)

ca. 0907/7944

76 feet (23 m), former station on the original Panama Railroad [ca. 0922/7954] that is now submerged in Lago Gatún (Gatun Lake) [0912/7955 (USBGN)] near what is now Isla Gorgona [0907/7943 (USBGN)]. The station was located on the continuous Río Chagres (see Río Chagres [0919/8000 (USBGN)]) 2.5 mi (4 km) W of Las Cruces [ca. 0907/7941] (MIP). Hallinan, clearing, shrubbery, 1916 Feb 20 (AMNH; Hallinan, 1924). See also: Nueva Gorgona.

GORGONA, ISLA; Panamá (Canal Zone) (E, Atlantic)

0907/7943 (USBGN)

10 m, island with a land area of ca. 0.06 km^2 (6 ha), located 2 km W of Gamboa [0907/7942 (USBGN)] (E866, as "Gorgona Island"). Brown, 1904 Jun 27 (USNM, as "Gorgona Island").

GORGONA ISLAND; see Gorgona, Isla

GRAILLARD HIGHWAY; see Gaillard, Carretera

GRANDE, ENSENADA; Bocas del Toro

0914/8219 (USBGN)

Bay in the Archipiélago de Bocas del Toro [0920/8210 (USBGN)] between Isla Pastores [0914/8220 (USBGN)] and Isla San Cristóbal [0915/8216 (USBGN)] opening on the N to Ambrosia Bight [0916/8222 (USBGN)] and on the S to Cuan Bay [0912/8218 (USBGN)] (JOG).

GRANDE, ENSENADA PLAYA; see Playa Grande, Ensenada

GRANDE, ESTERO BOCA; see Boca Grande, Estero

GRANDE, LAGUNA; see Las Lagunas

GRANDE, PLAYA; see Las Panamas, Playa

GRANDE, QUEBRADA; Chiriquí
0840/8239 (USBGN)
Not shown on our maps, but USBGN indicates a tributary of the Río Escarrea [0820/8241 (USBGN)] that enters ca. 1 km NW of Buena Vista [0839/8238 (USBGN)]. See also: Buena Vista.

GRANDE, RÍO; Coclé (Pacific)
0818/8024 (USBGN)
Originates in NW Coclé province W of El Copé [0837/8036 (USBGN)] and flows SE entering mangrove swamps on the N end of Bahía de Parita [0808/8024 (USBGN)] 20 km NE of Aguadulce [0815/8033 (USBGN)] (JOG). The Manglares del Río Grande [0819/8022 (Angehr, 2003:185)] are considered an Important Bird Area. Wetmore, 1962 Mar 9, GL = El Potrero [0831/8032 (USBGN)], S of El Potrero near Jobo [= El Jobo [0829/8031 (USBGN)]] on a high stream bank through weedy fields and tomato fields, farther S leaving the low hills and entering the savanna region, 1962 Mar 8-9 (USNM). See also: Omar Torrijos Parque Nacional.

GRANDE, RÍO; Panamá (Canal Zone) (W, Pacific)
0902/7938 (USBGN)
River that originates on the continental divide W of Paraiso [0902/7938 (USBGN)] and flows SE. At present, the river enters the Panama Canal [0920/7955 (USBGN)] at Contractors Hill [0902/7939 (USBGN)], but it originally continued SE entering Bahía de Panamá [0850/7915 (USBGN)] ca. 2.5 mi (4 km) WSW of Panama City [0858/7932 (USBGN)] (E8660; PR, as "R. Grande"). Several of the stations located on the original Panama Railroad [ca. 0922/7954] were located on the river until the construction of the Panama Canal, which now occupies most of the original riverbed. See also: Río Grande Station.

GRANDE, RÍO; see Cana

GRANDE, RÍO; see Guabo, Río

GRANDE, RÍO; see Río Grande

GRANDE, RÍO; see San Lorenzo, Río

GRAN GALERA DE CHORCHA; Chiriquí
0826/8213 (USBGN)
494 m, volcanic cone ca. 4 km in diameter located 25 km E of David [0826/8226 (USBGN)] (JOG). Angehr (2003:152) indicates that "a narrow band of semideciduous lowland forest (430 ha) clings to the steep sides," whereas "the top of the mesa… [has] been cleared for cattle pasture." The mesa of the Important Bird Area is privately owned. Ponce, *empanada de la meseta* (slope of the plateau), 1980 Jan 6 (MVUP, as "Meseta de Chorcha"); Angehr, 1996 Aug 30 (MVUP).

GREEN, BAHÍA DE; Bocas del Toro
0912/8214 (USBGN)

Small bay on the mainland coast within the waters of Laguna Bocatorito [0914/8215 (USBGN)], located 5 km SE of Bocatorito [0914/8216 (USBGN)] (E762). SBE, GL for Cerro Brujo, 1989 Feb 27 (USNM).

GREEN MEMORIAL; see George Green Park

GREEN MOUNT; see George Green Park

GUABAL; see Guaval

GUABAL, RÍO; Coclé (Atlantic)

ca. 0846/8036

According to Wetmore, the river originates N of El Copé [0837/8036 (USBGN)] on the Atlantic slope of the continental divide, flows N and enters the Río Limón [0847/8037 (USBGN)] from the E (USNM). Angehr (in litt., Aug 2005) indicates that the river is labeled as such on his map 4042-II (Coclesito) [E762]. According to JOG, the Rio Limón is one of several streams that originates on the divide N of El Copé and eventually discharges into the Río Coclé del Norte [0905/8035 (USBGN)]. Wetmore, GL for: Alto Carvallo, Las Peñitas, and Tigre, 1962 Feb 26-Mar 3 (USNM; Wetmore, 1962; Wetmore, 1970; Olson, 1980a, as "Head of Río Guabal, Tigre"; Olson, 1981c, as "Head of Río Guabal"; Olson, 1981d, as "Tigre, head of Río Guabal").

GUABITO; Bocas del Toro

0930/8237 (USBGN)

Sea level, town 11 km NW of Changuinola [0926/8231 (USBGN)] on the railroad and on the Río Sixaola [0934/8234 (USBGN)] 10 km SSW of its mouth (JOG). SBE, GL for Río Sixaola, 1991 Feb 23 (USNM; Brumfield et al., 2001); Braun, M., at 10 m, 8 km SW, at 0928/8240, 1994 Apr 28-30, May 4 (USNM; Brumfield et al., 2001).

GUABO; see El Guabo

GUABO, RÍO; Bocas del Toro

0854/8211 (USBGN)

Originates on the continental divide near Cerro Hornito [ca. 0839/8211] and flows NW, then N entering the Río Guarumo [0900/8211 (USBGN)] 12 km S of its mouth (E762; JOG; MGRP). On older maps (MHA and MRP) the river originates farther W on the slopes of Cerro Horqueta [0850/8226 (USBGN)], flows ENE, and discharges into the Río Guarumo. These older maps show the Río Guarumo as a much longer river, the upper course of which includes what is now named the Río Guabo on our recent maps. Kennard and Peters (1928:446, 454) refer to the Río Grande in connection with the Boquete Trail, and Peters (1931:294) indicates that the Boquete Trail "crosses the Guaruma River a few times" before winding up the mountains. Neither USBGN nor our maps indicate a Río Grande in the immediate region, and recent maps indicate that the Río Guarumo is a very short river. Therefore the specimens from Kennard et al. were probably collected on the Río Guabo or one of its tributaries in the region S of El Guabo and N of Buena Vista, where the elevation ranges from 60 to 300 m. Kennard et al., GL = Boquete Trail, at 200 feet (60 m), along the river, 1926 Mar 8, 10 (MCZ, as "Río Grande"; Kennard and

Peters, 1928:456, 454, as "Río Grande"). See also: Boquete Trail; Buena Vista; and El Guabo.

GUACAS, LAGUNA DE; see Las Huacas

GUACIMO; see Guásimo

GUADALUPE; see Chumbago

GUAGARAL, CERRO; San Blas
0917/7917 (Adsett and Wege, 1998)
Not shown on our maps, but described by Adsett and Wege (1998) as "contiguous with Cerro Brewster [0921/7915 (USBGN)], western San Blas," presumably located on the Atlantic drainage in Comarca de San Blas in Nusagandí [ca. 0919/7859]. Angehr (in litt., Aug 2005), who accompanied Adsett, Seutin and Ponce in 1994, states that "not having GPS, we were not certain of the exact location, but we believed it to be c. 2 km SW of Cerro Brewster, and close to the same elevation, on the continental divide between the drainages of the Rio Pacora and Rio Piedras, and thus in Panama province." Observers ?, 1992 Feb 28-Mar 2, 1994 Feb, 1995 Apr 29-May 2 (Adsett and Wege, 1998).

GUALACA; Chiriquí
0832/8218 (USBGN)
200 m, town located 18 km NE of David [0826/8226 (USBGN)] (MGRP). Eisenmann and N. Smith, observations, 1965 Sep 27 (Ridgely and Gwynne, 1989:89, as "Guataca, eastern Chiriquí"); Braun and Brumfield, at 1,000 m, at 08°38'06"/82°12'42", ca. 15 km NE of, and at 1,000 m, at 08°36'30"/82°13'06", ca. 12 km E of, 2004 Mar 18 (LSUMZ); DaCosta, Guitton-Mayerma, Klicka, Lelevier, Miller, and Spellman, GL for Fortuna, 2006 Feb 25-28 (MBM; STRI). See also: Gualaca-Chiriquí Grande Road.

GUALACA-CHIRIQUÍ GRANDE, CARRETERA; see Gualaca-Chiriquí Grande Road

GUALACA-CHIRIQUÍ GRANDE ROAD; Bocas del Toro/Chiriquí
ca. 0857/8207
A road that extends roughly N-S between Chiriquí Grande [0857/8207 (USBGN)] on the Bocas del Toro coast and Gualaca [0832/8218 (USBGN)] in central Chiriquí province. The road crosses the continental divide N of Fortuna [0843/8216 (USBGN)] at ca. 1,200 m and is sometimes called the "Fortuna Road." It is also known as the "*oleoducto* road" as the oil pipeline between Puerto Armuelles [0817/8252 (USBGN)] and Chiriquí Grande is roughly parallel to the Bocas del Toro section of the road. The road appears on E562 as a secondary road labeled "Panama 10," and specimens labeled 23-24 road km N of Los Planes [ca. 0838/8213] were collected on or near the continental divide. Graves, M. Braun, and Parsons, GL = Los Planes, at 1,100-1,200 m, collected along the road 6-23.3 road km N of Los Planes, specimens were taken in Chiriquí province and on the continental divide, 1990 May 1-7, 9-11, at 600 m, 6.5 road km N of the continental divide in Bocas del Toro province, 1990 May 10 (MVUP, as "Hornito"; USNM); SBE, GL = Los Planes, at 1,100 m, 24 road km N of Los Planes in Bocas del Toro province, 1991 Mar 3, 11-12 (USNM; Olson, 1993b); Mobley and Robbins, división continental,

1995 Apr 10-11, 15 (MVUP, as "Carretera Gualaca-Chiriquí Grande," and as "Gualaca Continental"); Weir, at 1,150 m, Chiriquí to Chiriquí Grande Road at continental divide, 8°47'29"/82°12'36", Bocas del Toro province, 2003 Mar 2-5 (CU, as "Chiriquí Grande Road"; LSUMZ, as "Chiriquí Grande Road"); Lovette, at 200 m, 0835/8213, N of Chiriquí along, 2003 Mar 22 (Museum ?, as "Fortuna Road"). See also: Fortuna.

GUALACA CONTINENTAL; see Gualaca-Chiriquí Grande Road

GUANÁBANO, RÍO; see Cárdenas, Río

GUANCHE; see Guanche, Río

GUANCHE, RÍO; Colón (E)
0930/7942 (USBGN)
Originates on Cerro Bruja [0929/7934 (USBGN)] and flows WNW entering the Caribbean Sea [1500/7500 (USBGN)] at the town Guanche [0930/7942 (USBGN)] 6 km SW of Portobelo [0933/7939 (USBGN)] (JOG). Morton, GL = Portobelo, near Portobelo, 1971 Apr 4 (USNM, as "Río Guandie"); Graham, observations, along, 1987 Jan 18 (Ridgely and Gwynne, 1989:305, 395).

GUANDIE, RÍO; see Guanche, Río

GUÁNICO, RÍO; Los Santos
0717/8025 (USBGN)
Originates in SW Los Santos province and flows SE entering the Pacific Ocean 15 km S of Tonosí [0724/8027 (USBGN)]. The river appears to be a continuation of the Río Jobero [0721/8031 (USBGN)] and the distinction between the two rivers is unclear on our maps (JOG). Wetmore, GL = Las Palmitas [0720/8030 (USBGN)], collected up- and downstream within 6 km of Las Palmitas, the gravelly streambed was ca. 10-15 m wide near Las Palmitas and the clear water was flowing rapidly. Collected in an area that had not been cleared, in an *espavé* stand along the river bank, in brush with occasional, tall *espavé* trees and many guava trees, mature orange trees grew throughout the valley. Upstream towards Jobero [0722/8033 (USBGN)] (q. v.) the valley had been largely cleared, collected in scattered tracts of second growth and in a small tract of original forest, 1962 Jan 21-28 (USNM; Wetmore, 1965, as "Guanico Arriba"; Wetmore, 1968, as "Guanico Arriba"). See also: Jobero, Río.

GUÁNICO ARRIBA; Los Santos
0720/8030 (USBGN)
Wetmore (1965:93, 405; 1968:27, 33, 86, 95, 113, 116, 136, 152, 171, 283) specifically refers to observations and specimens from Guánico Arriba in Jan 1962, but specimens are labeled Las Palmitas [0720/8030 (USBGN)], Río Guánico [0717/8025 (USBGN)] (q. v.).

GUARARÉ; Los Santos
0749/8017 (USBGN)
ca. 25 m, town in NE Los Santos province 6 km N of Las Tablas [0746/8017 (USBGN)]. La Enea [0750/8016 (USBGN)] is 2 km NNE of Guararé and 2 km from the coast (MGRP; RDP2). Osorio, La Enea-Guararé, 1980 Oct 22 (MVUP);

Garcia, 1985 Aug 20 (MVUP); Alvarado, 1994 Jun 21 (MVUP).

GUARUMAL; Veraguas (Pacific)

ca. 0756/8115

Sea level, described by Wetmore as a small island in the Río San Pablo [0751/8110 (USBGN)] ca. 10 river mi (16 km) below Soná [0801/8119 (USBGN)]), Wetmore and Perrygo, GL = Soná, Río San Pablo, heavy woodland bordered by swampy woods that flood during high tide, the trees were tall and the undergrowth was fairly heavy with much palm, 1953 Jun 3, 11 (USNM). Not shown on our maps.

GUARUMALITO; Veraguas (Pacific)

ca. 0756/8116

Sea level, described by Wetmore as a small island in the Río San Pablo [0751/8110 (USBGN)] 8-9 river mi (13-14 km) below Soná [0801/8119 (USBGN)], Wetmore and Perrygo, GL = Soná, Río San Pablo, landed at a small hill covered with second growth and crossed it to a wet forest beyond, river was muddy and swift, 1953 June 2 (USNM). Not shown on our maps.

GUARUMA RIVER; see Guarumo, Río

GUARUMO, RÍO; Bocas del Toro

0900/8211 (USBGN)

Short river that originates SW of Chiriquí Grande [0857/8207 (USBGN)] at the merger of several streams. The river flows N through a lowland area entering Laguna de Chiriquí [0903/8200 (USBGN)] 10 km NW of Chiriquí Grande (E762). Peters (1931:294) indicates that the Boquete Trail "crosses the Guaruma River a few times" before winding up the mountains, but on older maps the Río Guabo [0854/8211 (USBGN)] (q. v.) and the Río Guarumo were both labeled the Río Guarumo (MHA; MRP).

GUÁSIMO; Chiriquí

0822/8234 (USBGN)

ca. 25 m, town 4 km SSW of Alanje [0824/8233 (USBGN)] (JOG). Wetmore, GL = Alanje, 10 acre (4 ha) stand of 40-50 foot (10-15 m) trees that appeared to be original growth, ant swarms, *rastrojo*, 1960 Mar 8 (USNM, as "Guacimo").

GUATACA; see Gualaca

GUAVAL; Bocas del Toro

ca. 0835/8114

1,500 feet (450 m), described by Griscom (1927a:2-3) as one of two camps on the Río Calovébora [0848/8113 (USBGN)] reached by Benson and Gaffney who crossed the continental divide above Santa Fé [0831/8105 (USBGN)]. According to a map, which accompanied a letter from Benson to Griscom dated Nov. 7, 1926 (AMNH), Guaval was located W of the Bocas del Toro/Veraguas provincial boundary on a tributary of the Río Calovévora and no doubt near the town El Guabal [ca. 0836/8115], shown on MGRP near the junction of the Río Chelelé [0836/8113 (USBGN)] with the Río Calovébora [0848/8113 (USBGN)]. Benson and Gaffney, 1926 Sept 4, Aug 22 (AMNH; Griscom 1927a; Peters, 1931a:297; Howell, 1956; Wetmore, 1959b:10; Eisenmann and Howell, 1962, as "Guaval, Río Calovévora"; Olson,

1981a, as "Guabal, Río Calovévora").

GUAYABALITO; Colón (Canal Zone) (E)
0911/7939 (USBGN)
30 m, town 3 km N of Juan Mina [0910/7939 (USBGN)] on the W bank of the Río Chagres [0907/7942] less than 1 km N of Santa Rosa [0911/7939 (USBGN)] (E866). Wetmore, GL = Juan Mina, Río Chagres, in an area of tall forest that was open underneath due to regular flooding, 1961 Jan 14, 16 (USNM); Quinones, 1968 Jan 11 (GML).

GUAYABO, ENSENADA EL; see Jayabo, Ensenada del

GUAYABO, ENSENADO DE; see Jayabo, Ensenada del

GUAYABO CHIQUITO, ENSENADA; Darién
ca. 0722/7802
Not named on our maps, but must be the small bay just S of Punta Guayabo Chiquito [0724/7802 (USBGN)], a point shown on MGRP 20 km SSE of Jaqué [0731/7810 (USBGN)] on the S coast of Ensenada del Jayabo [0725/7803 (USBGN)]. Murphy and Correia, Askoy Expedition, 1941 May 20 (AMNH; Murphy, 1944a).

GUAYABO CHIQUITO, PUNTA; see Guayabo Chiquito, Ensenada

G U A Y A Q U I L ;
Sundevall (1870:609) first described *Dendroica petechia aequtorialis* from Guayaquil, but through correspondence Peters (1931:334) determined that the specimen was likely collected in Panama City [0858/7932 (USBGN)] (Olson, 1980b:476).

GUINDILLA, ISLA; Panamá (Canal Zone) (W, Atlantic)
0908/7951 (USBGN)
20 m, small island with a land are of ca. 5 ha located 400 m off the S coast of Isla Barro Colorado [0909/7951 (USBGN)] (E762, as "Guindilla Island"). Marcus, observations, 1982 Oct 20 (Marcus, 1985, as "Guindilla (Kikos) Island"). Angehr (1999) mentions observations in Guindilla Cove, which is "on the mainland that extends about 2 SW of Guindilla Island" (in litt., Aug 2005).

GUINDILLA COVE; see Guindilla, Isla

GUINDILLA ISLAND; see Guindilla, Isla

GUISADO; Chiriquí
ca. 0850/8247
1,200 m, town on the main road 4 km NW of Santa Clara [0850/8246 (USBGN)] (JOG). See also: Guisado, Quebrada.

GUISADO, QUEBRADA; Chiriquí
ca. 0847/8249
Originates N of Santa Clara [0850/8246 (USBGN)] and flows SW entering the Río Chevo [0847/8250 (USBGN)] 1 km above its mouth (MGRP, as "Río Guisado"). Described by Wetmore as a stream W of the Río Chevo, 3 road mi (5 km) W of Santa Clara [0850/8246 (USBGN)] with swift, clear water over a 15-35 foot (5-10 m) wide, stony bed surrounded by steep and bare slopes (USNM). Presumably

Wetmore collected near Guisado [ca. 0850/8247] (q. v.), which is on the main road. Wetmore, GL = Santa Clara, at 3,900-4,100 feet (1,200-1,250 m), forested hill above, E of the river on a ridge between the Quebrada Guisado and the Río Chevo in a stand of timber and *rastrojo* and in forest with much undergrowth, 1954 Mar 18, 20, 1955 Feb 18 (USNM); Hartman, GL = Santa Clara, 1958 Jan 26 (USNM).

GULF OF DARIÉN; see Urabá, Golfo de

GULICK, FORT; see Fort Gulick

GUNETA, LA; see La Guneta

HACIENDO SAN JOSÉ; see San José, Sabana de

HANNIBAL, BANCO; Veraguas (Pacific)
ca. 0725/8203
According to Angehr (in litt., Aug 2005), "this is an underwater seamount and well-known sport-fishing locality, minimum depth indicated as 28 fathoms (c. 50 m), about 12.6 nautical mi (23 km) WSW of" Punta Hermosa [0729/8153 (USBGN)], the westernmost land on Isla de Coiba [0728/8147 (USBGN)]. "It is shown on the US Hydrographic Service Map H.O. 1018 (1957) 'Punta Burica to Cape Mala,' and on many other nautical charts." Beebe, GL = Isla de Coiba, 1938 Mar 20 (Wetmore, 1957:15).

HATO; see Hato, Campamento de

HATO, CAMPAMENTO DE; Veraguas, Isla de Coiba
0730/8142 (USBGN)
Sea level, prison camp on the central E coast of Isla de Coiba [0728/8147 (USBGN)] 3 km WSW of Punta Damas [0730/8141 (USBGN)] (JOG). Wetmore, observations, flats near, water beyond, stream, 1956 Jan 20 (Wetmore, 1957:19, 21, as "Hato").

HATO, RÍO; see Río Hato

HATO BAYANO; see Mamoní, Río

HATO CHAMÍ; Chiriquí
0827/8146 (USBGN)
ca. 900 m, town in Reserva Indígena del Tabasará [0825/8142 (USBGN)] that is located 20 km NNE of San Félix [0817/8152 (USBGN)], 20 km SSW of Cerro Santiago [0833/8144 (USBGN)], and near the peak Cerro Chamí [0827/8146 (USBGN)]. The town is 2 km E of the Río San Félix [0810/8151 (USBGN)] and 6 km W of the Río Cuvíbora [0816/8138 (USBGN)] (MGRP). Presumably Davidson's locality "Chame," which she described as the Indian name of a "knife ridge just south of Cerro Flores [0829/8144 (USBGN)] (q. v.)" (Davidson, 1938:255, as "Chame"), is located near this town, Davidson and Hoigaard, at 2,800-3,200 feet (850-1,000 m), 1932 Jan 2-6, 8-15, 17 (CAS; CM, as "Chami"; Davidson, 1938, as "Chame").

HÂVRE DU DARIÉN; see Tuira, Río

HERMOSA, ENSENADA; Veraguas, Isla de Coiba
0731/8152 (USBGN)
Bay on the NW coast of Isla de Coiba [0728/8147 (USBGN)] that extends 3 km

from SW to NE (MGRP, as "Ensenada Playa Hermosa"). Northern, GL = Isla de Coiba, 1957 Mar 23 (LACM, as "Hermosa Bay"); Garcés and Mena, 1996 Feb 10, 12-13, 15 (MVUP, as "Playa Hermosa"); Brumfield and Miller, GL = Isla de Coiba, at sea level, at 07°31'54"/81°51'18", old airstrip, 2004 Apr 10-11 (LSUMZ, as "Playa Hermosa"). See also: Hannibal, Banco.

HERMOSA, PLAYA; see Hermosa, Ensenada

HERMOSA BAY; see Hermosa, Ensenada

HERRERA, PROVINCIA DE; (Pacific)

0750/8045 (USBGN)

Covering the central region of Península de Azuero [0740/8035 (USBGN)], Herrera province has the smallest coastal area of the ten provinces. The 15 km coastline in the NE is primarily mangrove swamps, and Chitré [0758/8026 (USBGN)], the provincial capital, is located 3 km inland from its southernmost shore. Much of the land in the NE region lies below 100 m elevation, whereas in the SW the peaks are just under 1,000 m. The N boundary with Coclé province is formed by the Río Santa María [0806/8029 (USBGN)], the W boundary with Veraguas is formed both by hilly ridges and the Río Conaca [0807/8046 (USBGN)], and the SE boundary with Los Santos province is formed in part by the Río La Villa [0759/8023 (USBGN)]. Early ornithological work in the province was conducted by Benson in 1925 and by Aldrich and Bole in 1932, both of whom worked along the Veraguas/Herrera provincial boundary in a region that was originally considered part Veraguas province. See Península de Azuero [0740/8035 (USBGN)] for further discussion of the provincial boundaries. The only other extensive work in the province was conducted by Wetmore and Perrygo in 1948, and aside from these investigations, only sporadic sightings and collections have been documented.

HICARON, see Jicarón, Isla

HICARON ISLAND; see Jicarón, Isla

HIDO, QUEBRADA; see Quebrada Hido

HIDROELÉCTRICA DEL BAYANO; see Bayano, Lago

HIGOS, LOS; see Los Higos

HIGUERONAL; Panamá (E, Pacific)

ca. 0855/7820

ca. 100 m, town on the Carretera Interamericana [0832/8242 (USBGN)], ca. 23 km S of Piriá [0907/7820 (USBGN)], ca. 45 km NE of Chimán [0842/7837 (USBGN)] and ca. 15 km W of the Darién provincial border (RDP2). DaCosta, Guitton-Mayerma, Klicka, Miller, and Spellman, GL = Piriá, at 125 m, at 08°54'40"/78°20'04", 23 km S of Piriá, 2006 Feb 13-14, 18 (MBM; STRI).

HIGUEROS, ISLA; see Los Higueros, Isla

HIGUEROS ISLAND; see Los Higueros, Isla

HOLCOMB TRAIL; Bocas del Toro/Chiriquí

ca. 0901/8215

Now called Culebra Trail. Described by Blake (1958:502-503) as a trail built c.

1910 by an American engineer named Holcomb that extended N from Boquete into Bocas del Toro province. According to Blake's map, the Holcomb Trail extended NNW from Boquete along the Quebrada Cerro Punta [= Río Caldera [0837/8221 (USBGN)]] and it branched at Agua Caliente [ca. 0851/8229]. Here the Holcomb Trail departed from the Río Caldera and continued N along the W slopes of Cerro Horqueta [0850/8226 (USBGN)] crossing the continental divide at ca. 6,000 feet (1,850 m). According to MRP, this route continued NNE into Bocas del Toro province terminating at the mouth of the Río Róbalo [0901/8215 (USBGN)] on Laguna de Chiriquí [0903/8200 (USBGN)]. E762 indicates that a road still extends ca. 2 km N of Horqueta [0848/8227 (USBGN)] along the Quebrada Horqueta [0848/8227 (USBGN)], which is W of Cerro Horqueta, but terminates before reaching the divide. E762 indicates an entirely roadless section of Bocas del Toro N of the headwaters of the Quebrada Horqueta, but according to the map, a trail that continued N or NNE would have immediately reached the headwaters of the Río Changuinola [0928/8227 (USBGN)] and could have descended NNE to the Río Róbalo by any number of routes. Mönniche collected along the Holcomb Trail in Bocas del Toro province at Camp Cylindro [ca. 0853/8228] (q. v.), at Camp Holcomb [ca. 0853/8228] (q. v.), and at Cedral [ca. 0854/8228] (q. v.) (Blake, 1958:502). Blake indicated that the trail was also called the Boquete Trail, although his map indicates that the trail crosses the continental divide and the Chiriquí/ Bocas del Toro provincial border W of Cerro Horqueta [0850/8226 (USBGN)] and therefore NW of Boquete, quite different from the Boquete Trail [ca. 0857/ 8207] described by Kennard and Peters (1928), which crosses the divide 12-15 mi E of Boquete. Tim Billo (in litt. to Olson 21 Jun 2007) hiked what he considered to be the Holcomb Trail in 2007 and reported as follows (accents added): "At the bottom, it starts at Norteño [not in USBGN] on the new road connecting C[hiriquí] G[rande] [0857/8207 (USBGN)] and Almirante [0918/8224 (USBGN)]. It follows the Río Róbalo initially (the old trail ending at sea level at Palma Real and Punta Róbalo) and then heads up the Caño Sucio [0900/8218-8221 JOG2, not any of the 3 Caño Sucios in USBGN], passing the Ngobe communities of Filo Verde, Loma de Agua [neither in USBGN], Coclecito [0849/8033 (USBGN)] (4 hours walking), and Santos [not in USBGN] (5-6 hours walking), all along the Caño Sucio. Coclecito was established by Chiriqui Ngobe in the 1970s and Santos by the same people in the 1950s and 60s. At Santos, the trail climbs steeply 300m or so up to the ridge at 700m and then descends 300m to the Río Romero [not in USBGN] (4 hours from Santos to Río Romero). It crosses the Río Romero close to where it joins the Río Changuinola, then crosses the Río Changuinola at about 400m (there is a cable crossing the river here—you must bring your own pulley and harness). Here, there is much ranching from ranchers from Chiriquí. The south side of the river has some enormous pastures. The locals call this section of the Río Changuinola, Río Culebra. This is not to be confused with the Río Culubre [0908/8230 (USBGN)], a tributary of the Río Changuinola which is 20km further down river. The trail continues, forested, and along the border of pastures up past some prominent steep peaks, and follows a ridge system up to Camp Cilindro. Cilindro is about 6-7 hours walk above the Río Changuinola. The divide is another 2 hours walking from here.

"

HONDA, BAHÍA; Bocas del Toro, Archipiélago de Bocas del Toro
ca. 0918/8209
A well-protected bay (DB) on the S side of the NE projection of Isla Bastimentos E of Cayo Nancy. Billo, GL = Isla Bastimentos, 2007 Apr 2, "Finca La Loma on N side of Bahia Honda near E end" (USNM).

HONDA, BAHÍA; Veraguas (Pacific)
0744/8132 (USBGN)
A well-protected bay on the mainland coast of SW Veraguas province located 42 km SSW of Las Palmas [0808/8127 (USBGN)] and 30 km NE of Isla de Coiba [0728/8147 (USBGN)]. The bay extends 5 km inland between 2 small peninsulas with Punta Jabalí [0743/8132 (USBGN)] on its S coast (JOG; RDPF; Wetmore, 1957:13). Garth, 1931 Dec 21 (LACM); Ashcraft, 1938 Mar 1, 1939 Mar 27-28 (LACM). (Beebe, 1942).

HONDA, ENSENADA; Panamá, Archipiélago de las Perlas
0824/7906 (USBGN)
According to Angehr (in litt., Aug 2005), this large embayment appears on the NE coast of Isla Pedro González [0824/7906 (USBGN)] on map 4341-II (Isla Pedro Gonzalez) [E762], UTM 930000 N, 710000 E (0825/7906). Feathers, specimens are labeled simply "Pedro Gonzales," but the expedition report indicates that work was conducted at Honda Ensenada, 1941 May 10 (ANSP, as "Pedro Gonzales"; Bond and Meyer de Schauensee, 1944:31, as "Pedro González").

HONDA, LA; see La Honda

HONDA, PEÑON DE LA; see Villa, Isla

HONDA, RÍO; Colón (Canal Zone) (E)
0910/7940 (USBGN)
Originates in Colón province W of Boca de Gatuncillo [0912/7939 (USBGN)], flows SSE joining the Río Chagres [0907/7942 (USBGN)] 2 km S of Santa Rosa [0911/7939 (USBGN)] (E866). See also: Chagres, Río; and Santa Rosa.

HONDITA, LA; see La Hondita

HONDO; see Paso Hondo

HONDO, POZO; see Burunga

HONDO, RÍO; Panamá (E, Pacific)
0901/7906 (USBGN)
Originates in the hills NW of Chepo [0910/7906 (USBGN)], flows S crossing the Carretera Interamericana [0832/8242 (USBGN)] 4 km W of Chepo and joins the Río Chepo [0900/7906 (USBGN)] 3 km above its mouth (JOG). See also: Paso Hondo.

HORCONCITOS; Chiriquí
0819/8210 (USBGN)
40 m, town 33 km SE of David [0826/8226 (USBGN)] and S of the Carretera Interamericana [0832/8242 (USBGN)] in the low, central coastal region of Chiriquí

province (E761A; MGRP). Wilson, 2000 Feb 5 (MVUP); Lelevier, Maley, and Miller, 7 km S of, 2005 Dec 27 (STRI; UAM).

HORNITO, CERRO; Chiriquí

ca. 0839/8211

2,102 m, USBGN treats Cerro Hornito as an older name for Cerro Chorcha [0839/8207 (USBGN)], but according to E562 and E762 Cerro Hornito is 5 km W of Cerro Chorca and S of the continental divide. It is located 19 km NE of Gualaca [0832/8218 (USBGN)], 4 km NE of Los Planes [ca. 0838/8213], and 14 km SSE of Fortuna [0843/8216 (USBGN)]. Angehr (in litt., Oct 2003) confirms that Hornito and Chorcha are in fact two distinct peaks. Hinds, GL = Fortuna, 1976 Aug 14 (GML); Ridgely, observations "on the Hornito ridge south of Fortuna," 1982 Jul 4 (Ridgely and Gwynne, 1989:402); Graves, Braun, M., and Parsons, GL = Los Planes, at 1,600-2,000 m, 4 km NE and 4 km NW of Los Planes, 1990 May 4 (USNM); Mobley and Robbins, Fortuna Reserva, at 0840/8215, 1995 Apr 14, 18, 20 (MVUP); Garcés and Mena, 1995 Apr 18 (MVUP). See also: Fortuna.

HORNITO, RÍO; Chiriquí

ca. 0844/8215

Originates on the N slopes of Cerro Hornito [ca. 0839/8211] and flows NNW within a few km of the Gualaca-Chiriquí Grande Road [ca. 0857/8207]. The river enters the S coast of Lago Fortuna [ca. 0842/8216] ca. 23 km NNE of Gualaca [0832/8218 (USBGN)] (E562). Hinds, GL = Fortuna, 1976 Aug 14 (GML).

HORNITO RIDGE; see Hornito, Cerro

HORQUETA; Chiriquí

0848/8227 (USBGN)

1,300 m, town 4 km NNW of Boquete [0847/8226 (USBGN)] (E762). Mönniche, at 5,100-5,400 feet (1,550-1,650 m), 1932-1933, 1939 (ANSP; CAS; FMNH; MCN; Blake, 1958; Monroe, 1963b); Davidson, GL = Boquete, at 5,400-5,800 feet (1,650-1,750 m), 1934 Feb 1, 4-6, 8-10, 12-13, 15 (CAS; Davidson, 1938).

HORQUETA, CERRO; Bocas del Toro/Chiriquí

0850/8226 (USBGN)

2,231 m, peak on the continental divide 8 km NNW of Boquete [0847/8226 (USBGN)], 12 km E of Cerro Punta [0852/8235 (USBGN)], and 10 km NE of Volcán Barú [0848/8233 (USBGN)] (JOG; MGRP). Mönniche, 1933 May 7 (YPM); Van Hagen, at 5,400 feet (1,650 m), collected on both the Pacific and the Atlantic slopes, 1940 July (UCLA, as "Cerro de Horqueta," and as "Cerro de Orqueta"); Olson et al., at 7,000 feet (2,150 m) and above, 1966 Jul 24 (UF; USNM).

HORQUETA, CERRO DE; see Horqueta, Cerro

HOSPITAL POINT; Bocas del Toro, Archipiélago de Bocas del Toro

ca. 0920/8213

Sea level, point on the NW coast of Cayo Nancy [0919/8211 (USBGN)] where an early hospital was established (E501, as "Hospital"). SBE, GL = Cayo Nancy, 1988 Feb 25-Mar 9 (MVUP; USNM).

HOWARD; see Howard Air Force Base

HOWARD AFB; see Howard Air Force Base

HOWARD AIR FIELD; see Howard Air Force Base

HOWARD AIR FORCE BASE; Panamá (Canal Zone) (W, Pacific)
0855/7936 (USBGN)
ca. 25 m, former United States Government military base that was located 8 km SW of Panama City [0858/7932 (USBGN)] just NW of Fort Kobbe Army Reservation [0854/7935 (USBGN)] and just SE of Cerro Galera [0856/7937 (USBGN)] (E661, as "Howard Field"). Pujals, observations, 1973 Jul 15 (Wetmore et al., 1984:378, as "Howard Air Field"); Trainer, observations at colonies named "Lookout," "Antenna," "Vera Cruz (this is not the nearby town Veracruz)," and "Kobbe Beach (q. v.)," 1981-1982 (Trainer, 1987, as "Lookout Colony"; Trainer, 1988; Trainer, 1989, as "Howard"); Williams, 1994 Dec 20 (LSUMZ, as "Fort Howard Air Force Base"); Collector ?, 1995 Mar 17, 24, 1996 Feb 23 (LSUMZ, as "Fort Howard Air Force Base," and as "Howard Air Force Base"); Arosemena, Gorrichategui, and Rivera, *área del canal*, 1996 Feb 12 (MVUP, as "Howard AFB"). (W. Robinson et al., 2000b).

HOYA, CERRO; Veraguas/Los Santos
0718/8042 (USBGN)
Described by Aldrich and Bole (1937:5) as a peak at 7,000 feet (2,150 m) in S Península de Azuero [0740/8035 (USBGN)]. The peak appears on MHA on the Veraguas/Los Santos provincial border at 200 m 16 km SW of Tonosí [0724/8027 (USBGN)] and 5 km N of the S coast of Península de Azuero. On the more recent map, RDP2, the peak appears at 1,559 m in Veraguas province just W of the Los Santos border and ca. 13 km N of the peninsula's S coast. Also in RDP2, in the table *Principales Elevaciones* the peak is listed as being in both Veraguas and Los Santos provinces with coordinates 07°22'37"/80°37'53". Angehr (2003) also indicates an elevation of 1,559 m. An unnamed peak with elevation 1,559 m appears at the above coordinates on MGRP on the provincial border. Delgado (1985) indicates a slightly higher elevation of 1,660 m for the peak. The peak is now included in Parque Nacional Cerro Hoya [0717/8046 (Angehr, 2003:172)], an Important Bird Area located largely in Veraguas province (Angehr, 2003). The peak is the highest in elevation of three in the region known as Los Tres Cerros (Myers, 1969; Delgado, 1985; Angehr, 2004). Handley and Greenwell, at 2,750 feet (850 m), Los Santos province, cloud forest, Myers (1969:29) indicates that using Handley's sketch map he traveled to what was presumably Handley's camp, he reached a camp at 940 m "on the ridge that lies just east of the summit of Cerro Hoya, across the headwater valley of the Río Pedregal [0714/8037 (USBGN)]," 1962 Feb 7, 11, 13-14, 16, 18, 21-22, 24-26 (USNM; Myers, 1969; Olson, 1983); Delgado and ANSP personnel collected near Cerro Hoya while working on Península de Azuero (q. v.) (Delgado, 1985; Angehr, 2003:172).

HUACAS, LAS; see Las Huacas

HUILE; Panamá (W, Atlantic)
0902/7944 (USBGN)
100 m, town 3 km N of Nuevo Emperador [0900/7944 (USBGN)] and 2 km E of Santa Clara [0902/7945 (USBGN)] (E762; JOG). Araúz, 1984 Apr 19, 21 (MVUP);

Acosta, M. Braun, Brenes, Calderón, Justavino, Lopéz, Madrid, Martín, Mendieta, Pagán, Palacios, Pinto, Ponce, Saldaña, Sandoval, Terreros, Valdespino, and Vargas, 1985 Mar 19, 23-25 (MVUP); Aguilar, Batista, De Leon, Garcia, González, Jaramillo, Mendoza, Quezada, Sánchez, and Vasquez, 1986 Apr 3-4 (MVUP); Araúz, González, Jaramillo, Jeannette, and Tapia, 1986 Jun 27 (MVUP); Araúz, Fernández, Lee, and Serrano, 1986 Jul 4 (MVUP); Araúz, Coronas, Holness, and Lamela 1986 Jul 18 (MVUP); Abrego, Hernández, Izos, and Ledezma (or Ledesma), 1986 Aug 22 (MVUP); Garcés and Holness, 1986 Nov 8 (MVUP); Ah Chu, Araúz, Garcés, and Holness, 1987 Feb 14 (MVUP); Araúz, Candanedo, and Garcés, 1988 May 11 (MVUP); Araúz and Garcés, 1988 Jun 2 (MVUP).

HUMEDALES DE MATUSAGARATÍ; see Matusagratí, Laguna de

ICACO, PUNTA; see Lanza, Punta

ICANDÍ, RÍO; see Aguas Claras, Río

ICANTÍ; see Aguas Claras Arriba

ICANTÍ, RÍO; see Aguas Claras, Río

IGLESIA, CERRO; Chiriquí

0819/8148 (USBGN)

715 m, peak 16 km NW of Tolé [0814/8141 (USBGN)] and 8 km ENE of San Félix [0817/8152 (USBGN)] (JOG). Griscom et al., at 1,500 feet (450 m), near this peak en route to and from Cerro Flores [0829/8144 (USBGN)], 1924 Mar 2, 21 (AMNH; CM; Griscom, 1924a). See also: San Félix, Río.

IGUANA, ISLA; Los Santos

0737/8000 (USBGN)

Sea level, island with a land area of ca. 1 km^2 (100 ha) in the Gulf of Panama [0800/7930 (USBGN)] off the E coast of Península de Azuero [0740/8035 (USBGN)] 12 km NNE of Pedasí [0732/8002 (USBGN)] and 18 km N of Punta Mala [0728/8000 (USBGN)] (JOG; Olson, 1997a). Wetmore (1957:6) misinterpreted a Batty specimen of *Buteo magnirostris* labeled "Iguaros Island" as being from Isla Iguana. See Los Higueros, Isla [0818/8224 (USBGN)]. According to Angehr (2003:209), the island is "mostly covered by low forest and surrounded by coral reefs." In his journal Wetmore described the shore as largely rock exposures that led inland to thin soil covered with grass clumps, and with brush and forest in the center. Wetmore, observations, 1956 Feb 6 (Wetmore, 1957:34), N end in low forest, dense above and open below, S end in thickets of low scrub interspersed with masses of vines, 1957 Feb 26-27 (USNM; Olson, 1980b; Olson, 1997a), observations N of Isla Iguana, at sea between Isla Boná [0834/7936 (USBGN)] and Isla Iguana, 1962 Mar 17, 1963 Feb 8-9 (Wetmore, 1965:433, 435).

IGUAROS, ISLA; see Los Higueros, Isla

IGUAROS ISLAND; see Los Higueros, Isla

IMAMADÓ, RÍO; Darién

ca. 0727/7758

Not shown on our maps, but according to a drawing by Wetmore, the river originates on the Cordillera de Juradó [0728/7755 (USBGN)] in the S tip of Darién province and flows N joining the Río Jaqué [0731/7810 (USBGN)] 17 mi (27 km) ESE of Jaqué [0731/7810 (USBGN)]. According to Angehr (in litt., Aug 2005), the river is labeled as such on the 1:250,000 Special Map of the Darién. Wetmore and Perrygo, GL = Río Jaqué, at ca. 300 feet (90 m), mouth of the Río Imamadó, camp was near the river mouth that was ca. 30 feet (9 m) wide and surrounded by level, forested areas with tall trees, some with thick undergrowth, others with an open understory and areas of palm, "definitely a green forest, but still a gallery forest," 1947 Mar 29-31, Apr 2-12, 14-16 (LSUMZ; USNM; Wetmore, 1951a; Eisenmann and Howell, 1962, as "mouth of the Río Imamadó, Río Jaqué"; Olson, 1980a; and Olson, 1981b, both as "Río Jaqué, mouth of Río Imamadó"); Greenwell, GL = Jaqué, near camp, junction of the Río Jaqué and the Río Imamadó, 1966 Feb 18, 26-27 (USNM). (Myers, 1969). See also: Chicao, Río; Chicao; Mora, Quebrada; and Seca, Quebrada.

INATÍ, RÍO; San Blas

ca. 0926/7904

Described by Wetmore as a 20-30 foot (6-9 m) wide stream, located in a steep-sided valley that was reached via a branch of the trail from Mandinga [0927/7904 (USBGN)] to Río Nergala [0927/7902 (USBGN)] (USNM). Presumably it flows into the Golfo de San Blas [0930/7900 (USBGN)] between the Río Cangandí [0928/7907 (USBGN)] and the Río Nergala, Wetmore, GL = Mandinga, very old second growth, forested hills, downstream in open *rastrojo* and tall bamboo, 1957 Feb 8, 14, 16 (USNM). Not shown on our maps.

INDIA DORMIDA, CERRO LA; see La India Dormida, Cerro

INDIO, RÍO; Colón (Canal Zone) (W)

0917/7958 (USBGN)

Short river that originates W of Gatún [0916/7955 (USBGN)] and enters the Río Chagres [0919/8000 (USBGN)] 5.5 km WNW of Gatún (E8660). Goldman, GL = Gatún, near Gatún, near the mouth of the Río Indio, daily canoe trips were made down the Río Chagres and up the Río Indio, but all specimens were labeled Río Indio, 1911 Feb 8, 10, 13-23, Mar 3-4 (USNM; Goldman, 1911; Todd, 1913, as "Río Indío (near Gatun)"; Friedmann, 1950a, as "Río Indio, near Gatún"; Wetmore, 1970; Payne, 1974; Olson, 1980a, as "Río Indio, near Gatun"; Olson, 1983; Brumfield et al., 2001, as "Río Indio, near Gatun"); Jewel, 4 mi (6 km) from Gatún, 1911 Nov 19 (Stone, 1918:248, as "Río Indis").

INDIO, RÍO; Colón/Panamá/Coclé (W, Atlantic)

0912/8011 (USBGN)

Originates at the merger of several small streams that drain the N and the E slopes of Cerro Gaital [0837/8007 (USBGN)]. The river flows N along the Panamá/Coclé provincial boundary, then along the Colón/Coclé provincial boundary, and enters Colón province discharging into the Caribbean Sea [1500/7500 (USBGN)] at Boca del Río Indio [0912/8011 (USBGN)] (E761B; E762). According to MGRP, the head waters of the river are located in Distrito Capira [0850/8000 (USBGN)]. Wetmore and Perrygo, all of Wetmore's collecting in 1952, except for the Taboga islands, was along the Río Indio and its tributaries in one of three general areas: along the coast at Boca del Río Indio, near the river's mouth; 4 km above its mouth near El Chilar [0909/8012 (USBGN)]; or 25 km above its mouth near Boca de Uracillo [0858/8011 (USBGN)]. According to Wetmore, the mouth was 100 yards (90 m) wide with *tagua* palms growing in the flood plain. The river banks were covered with palms and great trees filled with air plants, and rolling knolls above the river were covered with original forest and thick undergrowth. Above El Chilar the river was 70 yards (65 m) wide in a valley with fertile soil and the river near Boca de Uracillo narrowed to 40-50 yards (35-45 m) with 200-400 foot (60-120 m) hills rising on its banks. Wetmore and Perrygo, 1952 Mar 8, GL = El Chilar, near town among guava trees, open thickets and thick, wet, swampy woodland, dense forest along the river 0.5-1 mi (1-1.5 km) above El Chilar, 1952 Feb 18-19, Mar 7, GL = Boca de Uracillo (q. v.), 1952 Feb 22-Mar 4, and as a GL for: Boca del Río Indio, La Punta de Pilón, and Río Membrillar, 1952 Feb 14-17, 20 (USNM; Wetmore, 1952:31; Wetmore, 1953; Wetmore, 1959b:19; Parkes, 1969, as "Boca del Río Indio"; Payne, 1974; Olson, 1980a, as "El Uracillo, Río Indio," as "Río Indio, Boca del Río Indio," and as "Río Indio, Chilar"; Olson, 1981d, as "Chilar, Río Indio"); Garcés, Jiménez, López, Ramos, and Rodríguez, *cabecera de* (head of), Angehr (in litt., Aug 2005) indicates that this was a Western Watershed study site also known as Río Indio Nacimiento, the camp was located at 650 m, 08°39'06" N, 80°06'45" W, 2001 May 17-18 (MVUP; Louis Berger Group et al., 2003).

INDIO, RÍO; Panamá (E, Pacific)
0912/7914 (USBGN)
Originates NE of Cerro Carbunco [0911/7914 (USBGN)] and flows SW joining the Río Pacora [0901/7917 (USBGN)] 15 km NNE of Pacora [0905/7917 (USBGN)] (JOG). See also: Carbunco, Cerro.

INDIO NACIMIENTO, RÍO; see Indio, Río

INDIS, RÍO; see Indio, Río

INGENIO FELIPILLO; see Felipillo

INSOLETA, ISLA; see La Porcada, Isla

INSOLETA ISLAND; see La Porcada, Isla

INSOLITO, ISLA; see La Porcada, Isla

INSOLITA ISLAND; see La Porcada, Isla

INTERAMERICANA, CARRETERA; (E, W, Pacific)
0832/8242 (USBGN)
Darién/Panamá/Coclé/Herrera/Veraguas/Chiriquí
Road that extends roughly E-W across Panamá. On maps it crosses the Costa Rican border ca. 30 km N of Puerto Armuelles [0817/8252 (USBGN)] and the Colombian border ca. 35 km NNE of Altos de Quía [0735/7727 (USBGN)] (JOG; MGRP), but the section through Darién to Colombia is not complete. Many ornithological localities are located along the road, which is also called the Pan American Highway. See also: Remedios; and Tocumen, Aeropuerto de.

IPETÍ; Panamá (E, Pacific)
ca. 0900/7830
ca. 50 m, town on the Río Ipetí [0905/7832 (USBGN)] 2 km N of the Carretera Interamericana [0832/8242 (USBGN)] and 6 km S of the SE coast of Lago Bayano [0908/7835 (USBGN)] (MGRP). Ridgely, observations, near, above Ipetí in the upper Bayano valley, 1976 Apr 23 (Ridgely and Gwynne, 1989:237, 251). See also: Ipetí, Río.

IPETÍ, RÍO; Panamá (E, Pacific)
0905/7832 (USBGN)
Originates on the N slopes of Serranía de Majé [0855/7830 (USBGN)] and flows N entering the SE coast of Lago Bayano [0908/7835 (USBGN)] 67 km ESE of Chepo [0910/7906 (USBGN)] (MGRP). Hinds, 1974 Feb 18 (GML).

IRHE DAM SITE CAMP; Bocas del Toro
ca. 0917/8233
The site of a proposed hydroelectric dam on the lower Río Changuinola [0928/8227 (USBGN)] (q. v.). The acronym IRHE stands for Instituto de Recursos Hidráulicos y Electrificación. According to Angehr (in litt., Aug 2005), "the site of the proposed dam would have been where the river narrows at UTM 1027000 N 330400 E (0917/8233), just above where the Quebrada El Guabo enters the Rio Changuinola as shown on 3643-I (Corriente Grande) [E762]." Apparently construction never began, and future plans are unclear.

ISLA CANAL AFUERA; see Canal de Afuera, Isla del

ISLA DE BARRO COLORADO; see Barro Colorado, Isla

ISLA DE COIBA; see Coiba, Isla de and see Colonia Penal

ISLA DE COIBA PARQUE NACIONAL; see Coiba, Isla de

ISLA GRANDE; Bocas del Toro
0929/8240 (USBGN)
Described by Peters (1931:295) as an island in the Río Sixaola [0934/8234 (USBGN)] near Victoria [0929/8239 (USBGN)] and ca. 20 mi (32 km) NW of Almirante [0918/8224 (USBGN)]. Our maps and USBGN indicate a town 37 km NW of Almirante and 17 km WNW of Changuinola [0926/8231 (USBGN)]. The town is located on the railroad 2 km WNW of Victoria and 1 km S of the main course of the Río Sixaola [0934/8234 (USBGN)] but surrounded by a small secondary stream channel (E762; JOG). Wedel, 1927 Jul 18, Dec 31, 1928 Jun 23 (MCZ; Peters, 1931a).

ISLA ESCUDO; see Escudo de Veraguas, Isla

ISLETA, LA; see La Isleta

ISLETA EL GALLO; see Gallo

ISTHME DE PANAMA; see Panama, Isthmus of

ISTHMO PANAMA; see Panama, Isthmus of

ISTHMUS; see Panama, Isthmus of

ISTHMUS OF PANAMA; see Panama, Isthmus of

JABALÍ, PUNTA; Veraguas (Pacific)
0743/8132 (USBGN)
Sea level, point located on the SW coast of Veraguas province on the S coast of Bahía Honda [0744/8132 (USBGN)] (MHA). Used by Wetmore (1957:13) as a mainland landmark in his description of Isla de Coiba [0728/8147 (USBGN)].

JACÚ; Chiriquí
0829/8249 (USBGN)
ca. 25 m, town 23 km NNE of Puerto Armuelles [0817/8252 (USBGN)] and 2.5 km E of the Costa Rican border in a cultivated area on the railroad that leads to Puerto Armuelles (JOG). McViel, 18?? (USNM).

JAGUA, RÍO LA; see La Jagua, Río

JAGUA HUNTING CLUB, LA; see La Jagua Hunting Club

JAQUÉ; Darién
0731/7810 (USBGN)
Sea level, coastal town in S Darién province at the mouth of the Río Jaqué [0731/7810 (USBGN)] (SMN2B). Wetmore and Perrygo, specimens without secondary localities were collected in one of two general areas: N and W of Jaqué near the rocky shore and steep hills of Punta Jaqué [0731/7811 (USBGN)], or in the area E and SE of Jaqué in the mangroves at the mouth of the Río Jaqué, in brush bordering the airstrip (2 mi (3 km) SE of town, SMN2B), along the sandy and rocky coastal beaches, and along small streams that flow into the Gulf of Panama [0800/7930 (USBGN)], 1946 Mar 14-21, 23-24, 29, 31, Apr 2-3, 7-10, 11, 13, 15. The secondary localities attributed to Jaqué in 1946 were located in the region extending 6 km E and SE of Jaqué in one of three general areas: along the Río Jaqué; in the low hills ESE of Jaqué; or along the coast SE of Jaqué. GL for: El Brazo, La Tulita, Laguna Playa Nieve, Loma Dolores, Loma González, Loma Playita del Medio, Quebrada Alisa, Quebrada Fundeadero, Quebrada Lisa, and Río Jaqué, 1946 Mar 15-16, 20-23, 25-28, 30-Apr 1, 3-14 (MVUP; USNM; Wetmore, 1951a; Wetmore, 1953; Wetmore, 1959b:10; Wetmore, 1962; Wetmore, 1963a; Olson, 1980a; Olson, 1980b); Vaughan, 1946 Apr 8 (USNM); Greenwell, GL for: Río Imamadó, and Río Jaqué, 1966 Feb 18, 26-27 (USNM); Burton, British Trans-Americas Expedition, 1972 Feb (BMNH; Burton, 1973b; Burton, 1975). (Wetmore, 1970; Olson, 1981d).

JAQUÉ, PUNTA; Darién
0731/7811 (USBGN)
Sea level, point 3 km W of Jaqué [0731/7810 (USBGN)] on a peninsula that separates Bahía Piña [0734/7811 (USBGN)] on the N from the mouth of the Río Jaqué [0731/7810 (USBGN)] on the S (JOG). See also: Jaqué.

JAQUÉ, RÍO; Darién
0731/7810 (USBGN)
Originates in the S tip of Darién province on the SE slopes of the Sierra de Jungurudó [0735/7754 (USBGN)] near the Colombian border. The river flows S, then where

the Río Imamadó [ca. 0727/7758] joins it, the river changes course flowing NW across Darién province parallel to the coast, but separated from it by the Cordillera de Juradó [0728/7755 (USBGN)]. As it empties into the Gulf of Panama [0800/7930 (USBGN)], the Río Chadó [0731/7810 (USBGN)] joins it from the N (JOG; MGRP; SMN2B), and according to Wetmore in 1946, they form a 200-300 yard (180-270 m) wide tidal mouth, bordered with extensive red and white mangrove swamps, the latter 100-150 feet (30-45 m) tall (USNM). The river enters the Pacific Ocean ca. 100 km S of La Palma [0825/7809 (USBGN)] (MGRP). Wetmore and Perrygo, GL = Jaqué, collected near Jaqué and as far as 5 river mi (8 km) upriver in small channels through mangrove swamps, in low hills leading SE to Loma González [ca. 0729/7806], in old clearings overgrown with guava and brush and in tall trees with open undergrowth, 1946 Mar 20, 23, 26, 28, 30, Apr 1, 8, 13-14, at El Brazo (q. v.) 1946 Mar 22. According to Wetmore, in 1947 the Río Jaqué was affected by tides as far as the mouth of the Río Pavarandó [0749/7804 (USBGN)], from here the river valley was largely cultivated with bananas upstream as far as the mouth of the Arroyo Tortadó [ca. 0730/7801], after which the valley narrowed and was forested. Above the mouth of the Río Imamadó [ca. 0727/7758] the river was ca. 30 feet (9 m) wide, Wetmore and Perrygo, collected primarily within 8 km of their camp, which was located at the mouth of the Río Imamadó [ca. 0727/7758 (SMN2B)], but a few specimens were taken along the Río Jaqué on their journeys between Jaqué and camp, GL for: Arroyo Tortadó, Chicao, Las Peñitas, Loma Calle Larga, Loma Corobá, Loma Las Peñitas, Moja Culo, Quebrada Mora, Quebrada Seca, Río Chicao, Río Imamadó, Río Pavarandó, and Tortadó, 1947 Mar 27-Apr 18 (LSUMZ; MVUP; USNM; Wetmore, 1959b:10, 21; Eisenmann and Howell, 1962, as "mouth of the Río Imamadó, Río Jaqué"; Olson, 1980a, as "Río Jaqué, mouth of Río Imamadó"; Olson, 1981b, as "Río Jaqué, mouth of Río Imamadó," and as "Río Jaqué, Las Peñitas"; Olson, 1983); Greenwell, GL = Jaqué, at the Río Imamadó (q. v.), 1966 Feb 18, 26-27 (USNM); Burton, British Trans-Americas Expedition, 1972 Feb (BMNH; Burton, 1973b; Burton, 1975); Phillips, *cabecera de* (head of), 1995 Feb 21 (MVUP). (Myers, 1969; Wetmore, 1970). See also: Delores, Loma; La Tulita; and Playa Nieve, Laguna.

JARAMILLA; Chiriquí

ca. 0847/8226

Refers to a region just E of Boquete [0847/8226 (USBGN)] between 1,000-1,400 m, but the exact location is unclear as our maps show the following towns within 3 km NNE and SSE of Boquete: Alto Jaramillo [ca. 0845/8224]; Centro Jaramillo [0847/8226 (USBGN)]; Jaramillo Abajo [0846/8226 (USBGN)]; and Jaramillo Arriba [0848/8225 (USBGN)] (E562; MGRP). Mönniche, GL = Boquete, below 5,000 feet (1,500 m), 1934 (FMNH; Blake, 1958, as "Jaramilla (Boquete)").

JARAMILLO; see Jaramilla

JARAMILLO ABAJO; see Jaramilla

JARAMILLO ARRIBA; see Jaramilla

JARDIN BOTANICO SUMMIT; see Summit Park

JARDINO, LLANO DEL; Veraguas (Pacific)

ca. 0803/8109

ca. 50 m, described by Wetmore as an area 15 mi (24 km) E of Soná [0801/8119 (USBGN)] on the N side of the highway opposite the dirt road leading to Río de Jesús [0759/8110 (USBGN)]. The 1.5 mi (2.5 km) long expanse was completely level and covered with short grass and clumps of 10-30 foot (9 m) trees and shrubs, Wetmore and Perrygo, GL = Soná, 1 mi (1.5 km) through brush to a small stream, shallow ravines filled with brush, *chumico* and *nance* trees in bloom, 1953 May 18, 20, 26, June 5, 10 (USNM); Poole, GL = Soná, 1953 May 26 (USNM). Not shown on our maps.

JARDÍN OLÍMPICO; Panamá (E, Pacific)

ca. 0902/7928

10 m, section of greater Panama City [0858/7932 (USBGN)] located immediately W of Juan Díaz [0903/7927 (USBGN)] and ca. 1.5 km NE of Parque Lefevre [0901/7929 (USBGN)] (E762; E962). Adams, 1972 Apr 5 (GML).

JAYABO, ENSENADA DEL; Darién

0725/7803 (USBGN)

A cove on the S Darién coast that opens on the SW to Gulf of Panama [0800/7930 (USBGN)] 18 km SSE of Jaqué [0731/7810 (USBGN)] and 28 km NNW of the Colombian border (MGRP, as "Ensenada el Guayabo"; MHA, as "Ensenada el Guayabo"). Murphy and Correia, Askoy Expedition, 1941 Mar 2, 4, 3 mi (5 km) NW of, 1941 Mar 5 (AMNH, as "Ensenado de Guayabo"; Murphy, 1944a).

JEFE, CERRO; Panamá (E, Atlantic, Pacific)

0914/7921 (USBGN)

1,007 m, peak on the continental divide 20 km NNW of Pacora [0905/7917 (USBGN)] (E761A, as "El Jefe"; MGRP; RDPF). Based on a trip to Cerro Jefe in Jan 1965, Myers (1969:27) indicates that the "wind-swept forest is comprised largely of palms," "bromeliads are abundant both on the ground and in the trees," and that "the entire forest is low and dense and the ground is spongy." Handley, GL = Cerro Azul, at 3,000 feet (900 m), cloud forest, 1958 Feb 7 (USNM); Morton, 1970 Nov 12 (AMNH); Strauch, 1971-1973 (UMMZ; Strauch, 1977); Roubik, observations, 1979 Nov 24-Dec 10 (Roubik, 1982); Ah Chu and Araúz, 1986 May 2 (MVUP); Ah Chu, 1986 Aug 28 (MVUP); Araúz and Holness, 1986 Sep 10 (MVUP); Observer ?, c. 1987 (Adsett and Wege, 1998); Cardiff, Dittmann, and Lovette, at 600-800 m, NW slope, 1997 Mar 26-27 (LSUMZ; MVUP); Christian, at 850 m, observations, "in the foothills forest," the coordinates given (c. 80°37' W, 9°15' N) must be a typographical error, 1997 Apr (Christian, 2001); Petit et al., surveys, Dec 1993-Feb 1994 (Petit et al., 1999); Weir, 2003 Feb 17 (CU; LSUMZ); Lovette, 2003 Mar 26 (CU; LSUMZ); Guitton-Mayerma, Miller, and Winker, Sendero Vistamares, at 0913/7925, 2004 Feb 7-8 (STRI; UAM). (Myers, 1969).

JESUCITO; see Jesusito

JESUCITO, RÍO; see Jesusito, Río

JESÚS, RÍO; Darién

0804/7814 (USBGN)

Originates on the NE slopes of the Serranía del Sapo [0750/7817 (USBGN)] just E of the Río Jesusito [0801/7816 (USBGN)] and flows N entering the Río Sambú [0805/7818 (USBGN)] 14 km E of Garachiné [0804/7822 (USBGN)] and 5 km N of Boca de Sábalo [0802/7813 (USBGN)] (MGRP). See also: Sambú, Río.

JESÚS, RÍO DE; Veraguas (Pacific)

0755/8104 (USBGN)

Originates in low hills E of Soná [0801/8119 (USBGN)] just N of Río de Jesús [0759/8110 (USBGN)] and flows SE joining the Río San Pedro [ca. 0750/8105] near its mouth (JOG). According to Wetmore in 1953 the river near Río de Jesús was 50 feet (15 m) wide with exposed rock ledges. See also: Río de Jesús.

JESÚS, RÍO DE; see Río de Jesús

JESUSITO; Darién

ca. 0802/7818

The name of one or more camps in the lowland forest along the Río Jesusito [0801/7816 (USBGN)] used by the Barbour, Brooks, Underwood Expedition in 1922. According to Barbour and Brooks (1923:214), from the Río Jesús [0804/7814 (USBGN)] "we marched to the Río Jesusito and then by stages ascended this stream to where it entered a wide gorge in the Sapo range. We made excursions in various directions using our upper Jesusito camp as a base," 1922 April 6-17 (MCZ, as "Jesusito," and as "Upper Jesusito"; Bangs and Barbour, 1922; Barbour and Brooks, 1923; Griscom, 1929a; Hellmayr, 1929; Wetmore, 1953, as "Jesucito"; Wetmore, 1963a, as "Jesucito"; Haffer, 1962, as "Río Jesusito"; Haffer, 1967b, as "Río Jesusito").

JESUSITO, RÍO; Darién

0801/7816 (USBGN)

Originates on the Serranía del Sapo [0750/7817 (USBGN)] W of the Río Jesús [0804/7814 (USBGN)] and flows N entering the Quebrada Celorio [0805/7818 (USBGN)] 5 km above its mouth and 9 km E of Garachiné [0804/7822 (USBGN)] (MHA, as "Río Jesucito"; MRP, as "Río Jesucito"; Barbour and Brooks, 1923 (map)). See also: Jesusito.

JESÚS MARÍA; Panamá (E, Pacific)

0909/7859 (USBGN)

ca. 50 m, town on the Río Chepo [0900/7906 (USBGN)] 12 km E of Chepo [0910/7906 (USBGN)]. A road extends roughly N-S for ca. 9 km between Platanales [0913/7903 (USBGN)] on the Carretera Interamericana [0832/8242 (USBGN)] and Jesús María (MGRP). See also: Platanales.

JICARITA, ISLA; Veraguas (Pacific)

0713/8148 (USBGN)

122 m, small island in the Pacific Ocean with a land area of ca. 1 km^2 (100 ha) located less than 1 km S of Isla Jicarón [0716/8147 (USBGN)] (JOG, as "Isla de Jicarita"). Stewart, 1937 Feb 10 (SDSNH, as "Jicarita I"; Olson, 1997a).

JICARÓN, ISLA; Veraguas (Pacific)

0716/8147 (USBGN)
342 m, island with a land area of 20 km² (2,000 ha) (RDP2) located in the Pacific Ocean 8 km S of Isla de Coiba [0728/8147 (USBGN)] (JOG). Batty, specimens at AMNH and at BMNH are dated 1902 Jan 14-15, but Batty's data were completely fabricated and his specimens labeled "Hicaron" were actually collected in 1901 near his mainland bases at Boquete [0847/8226 (USBGN)] and Boquerón [0830/8234 (USBGN)] (Olson, 2008) (AMNH, as "Hicaron"; BMNH, as "Hicaron"; Eisenmann, 1950, as "Jicaron I."; Wetmore, 1950a; Wetmore, 1957:7-8, as "Hicaron," and as "Isla Jicarón"); Beaudette and Northern, 1959 Mar 24 (LACM, as "Jicaron Is"; Olson 2007, as "Isla Jicarón").

JINGURUDÍ, SIERRA DE; see Jungurudó, Sierra de

JINGURADO, SERRANÍA DE; see Jungurudó, Sierra de

JOBERO; Los Santos
0722/8033 (USBGN)
100 m, town on the Río Jobero [0721/8031 (USBGN)] 5 km NW of Las Palmitas [0720/8030 (USBGN)] (JOG). See also: Guánico, Río.

JOBERO, RÍO; Los Santos
0721/8031 (USBGN)
Originates in SW Los Santos province flowing SE (JOG) and according to USBGN enters the Río Guánico [0717/8025 (USBGN)] ca. 3 km SE of Jobero [0722/8033 (USBGN)] and ca. 3 km NW of Las Palmitas [0720/8030 (USBGN)], although the distinction between the two rivers in unclear on JOG. In fact, both rivers (the Río Jobero and the Río Guánico) appear as one river with the name "Río Guánico" on our other maps (Atlas, 1988:203; MHA; RDPF; SMN2B), therefore because Wetmore worked near Jobero, some of his specimens labeled Río Guánico may have been collected on the Río Jobero. See also: Guánico, Río.

JOBITO, EL; see Finca El Jobito

JOBITO, FINCA EL; see Finca El Jobito

JOBO; see El Jobo

JOBO, EL; see El Jobo

JOBO, LA PLAYA; see Jobo, Playa del

JOBO, PLAYA; see Jobo, Playa del

JOBO, PLAYA DEL; Chiriquí
0811/8154 (USBGN)
A coastal beach 9 km SSW of Las Lajas [0815/8152 (USBGN)] (q. v.) (MGRP). The beach may be referred to as Las Lajas in the literature. Wetmore described the it as an amazing straight stretch, without a curve in either direction, Wetmore, GL = Las Lajas, a broad, 500-600 yard (450-550 m) stretch of sandy soil covered with tall weeds, scattered thickets and trees, an extensive mangrove swamp with 100-150 foot (30-45 m) trees and the usual arched roots, lagoons, a flooded meadow with grazing cattle, 1956 Feb 15, 22, 24-25 (USNM, as "Playa Jobo," and as "La Playa Jobo"); Lelevier, Maley, and Miller, 5 km N of, 08°10'08"/81°51'21", 2005 Dec 27

(STRI, as "Playa las Lajas"; UAM, as "Playa Las Lajas").

JOHN CROW POINT; see Gallinazo, Punta de

JOHN W. ACKER FIELD STATION; Panamá (W, Pacific)
ca. 0841/7955
750 m, field station maintained by Florida State University on the Pacific slope of Cerro Campana [0841/7955 (USBGN)] (q. v.) (Olson et al., 1968).

JONES, QUEBRADA; Bocas del Toro
0914/8222 (USBGN)
Short mainland river that enters Ensenada Shepherd [0913/8221 (USBGN)] 1.5 km SSW of Punta de Gallinazo [0915/8222 (USBGN)] and 7 km SSE of Almirante [0918/8224 (USBGN)] (E762). Wetmore, GL = Almirante, entered a channel in the mangroves that ended quickly and abruptly at large boulders, worked up the slope and on a trail along the creek, 1958 Feb 12 (USNM, as "Jones Creek").

JONES CREEK; see Jones, Quebrada

JORÓN, QUEBRADA; Panamá (E, Pacific)
ca. 0844/7837
Described by Wetmore as a small stream 1 mi (1.5 km) E of Chimán [0842/7837 (USBGN)] at the edge of a mangrove swamp, reached by a trail from town through second growth, Wetmore and Perrygo, GL = Chimán, 1950 Apr 1 (USNM). Not shown on our maps.

JUAN AVELE; Panamá (E, Pacific)
ca. 0848/7838
ca. 20 m, described by Wetmore as the first area of high hills along the Río Chimán [0843/7838 (USBGN)] located ca. 6 mi (10 km) N of Chimán [0842/7837 (USBGN)], the muddy shores were lined with mangrove swamps in an area influenced by tides, Wetmore and Perrygo, GL = Chimán, Río Chimán, hills on higher slopes, 1950 Feb 18, 20 (USNM). Not shown on our maps.

JUAN DÍAZ; Panamá (E, Pacific)
0903/7927 (USBGN)
10 m, originally a small town, but now a section of greater Panama City [0858/7932 (USBGN)] located ca. 12 km NE of Ancón [0858/7933 (USBGN)] and Calidonia [0858/7933 (USBGN)]. The town is located 4 km NE of Old Panama [0900/7929 (USBGN)] on the Río Juan Díaz [0901/7926 (USBGN)] 3 km N of its mouth (E962). Hallinan, 1915 Oct 17 (AMNH; Hallinan, 1924; Griscom, 1927a, as "Juan Diaz, Canal Zone"; Griscom, 1927b); GMLP, 1962 Jul, Aug 12 (GML; USNM); Muñoz, 1962 Jul 20, Sep 10, 1967 Sep 9 (GML); Herrera, 1963 Sep 4 (GML); Duhrman, 1973 May 13 (GML); Bakes, 1976 Nov 13 (GML); Tejera, *manglar del norte de* (mangroves N of) Juan Díaz, Río Juan Díaz, 1978 May 13, Jun 10, 26, Aug 26, Oct 4, 1979 Jan 11 (MVUP); Tejera and Magallón, *estero del* (estuary of the) Río Juan Díaz, Río Juan Díaz, 1978 Apr 8 (MVUP); Tejera and Pineda, 1978 Nov 7 (MVUP); Tejera and Hernández, 1978 Dec 19 (MVUP); Brooks and Riley, *manglar sur de* (mangroves S of), 1983 Jun 25, Jul 9, Aug 13, 23, 27, Sep 24, Oct 12, 27, Nov 12-13, Dec 11, 28 (MVUP); Guevara, 1985 Apr 13 (MVUP); Joung, *manglar de*

(mangroves of), 1985 Aug 15 (MVUP); Guerra, 1986 Apr 12 (MVUP); Araúz and Holness, 1987 Apr 14 (MVUP); Masters and Samudio, 1994 Aug 10 (MVUP); Brooks, Farrugia, and Ponce, *manglar sur de* (mangroves S of) 1985 Mar 6, 17, 31, Apr 15, 22, 25, May 4 (MVUP). Araúz, Brooks, Farrugia, and Ponce, *manglar sur de* (mangroves S of), 1985 May 4 (MVUP); Lefebvre and Poulin, observations and studies, the coordinates given [09°00' N, 79°04' W] appear to be erroneous, 1990s (Poulin and Lefebvre, 1995; Lefebvre and Poulin, 1996; Lefebvre and Poulin, 1997; Lefebvre and Poulin, 2000). See also: Juan Díaz, Río; and Tapia Número Dos.

JUAN DÍAZ, RÍO; Panamá (E, Pacific)
0901/7926 (USBGN)
Originates on the continental divide W of Cerro Azul [0910/7924 (USBGN)] and flows S entering Bahía de Panamá [0850/7915 (USBGN)] 13 km NE of Panama City [0858/7932 (USBGN)]. Below Juan Díaz [0903/7927 (USBGN)] the river enters an area of tropical grass and mangrove swamps on the coast (E661; E761A), and according to Wiedenfeld (1992) the area near the mouth is "composed almost exclusively of tall black mangroves." Hallinan, vegetation on the river banks, mangroves, tidal swamp, woods, tidal mud flats at the mouth of, 1915 Oct 3, 10, 17, Nov 7 (AMNH; Hallinan, 1924); Ponce, *desembocadura del* (mouth of) Río Juan Díaz, *manglar (sur) de* Juan Díaz, 1984 Jun 21 (MVUP); Justavino and Vásquez, 1985 Apr 13 (MVUP); Anderson, Andrión, Fernández, López, Mendieta, Palacios, Rodríguez, and Valdespino, 1985 Apr 16 (MVUP); González and Madrid, *playa de* Juan Díaz, 1985 Aug 15 (MVUP); Brenes and Vaca, *boca del*, 1985 Sep 1 (MVUP); Wiedenfeld, observations, 1986-1988 (Wiedenfeld, 1992); Collector ?, 0.5 km NW of mouth, 1996 Mar 16 (LSUMZ). See also: Juan Díaz.

JUAN FRANCO; Panamá (E, Pacific)
0859/7931 (USBGN)
Originally a small town located 2 mi (3 km) NE of the Panama City [0858/7932 (USBGN)] (MIP), but now a section of greater Panama City. Eisenmann, observations, open grassland, 1948-1953, 1955-1956 Jun-Jul, 1962 Nov (Eisenmann, 1951; Eisenmann, 1953b; Eisenmann, 1962a, as "suburb of Panama City"; Eisenmann, 1963b, as "eastern suburb located about a half mile north of Panama City"); Wetmore (from I. Robert Eisenmann, Jr.), 1953 June 17 (USNM); Bennett, 1953 Jun 29 (USNM, as "Juan Franco Suburb"; Eisenmann, 1962a, as "suburbs of Panama City").

JUAN GALLEGOS, ISLAS; Colón (Canal Zone) (E)
0913/7952 (USBGN)
128 m, island with a land area of 6.7 km^2 (670 ha) in Lago Gatún (Gatun Lake) [0912/7955 (USBGN)] 7 km SE of Gatún [0916/7955 (USBGN)] (E762, as "Juan Gallegos Island"; Wright, 1988, as "Juan Gallegos Island"). Wright, censuses, 1976-1978, 1980-1981 (Wright, 1979, as "Juan Gallegos Island"; Wright, 1988, as "Juan Gallegos Island"). See also: Monte Lirio.

JUAN MINA; Panamá (Canal Zone) (E, Atlantic)
0910/7939 (USBGN)

20 m, town 8 km NNE of Gamboa [0907/7942 (USBGN)] on a narrow lagoon of the E bank of the Río Chagres [0907/7942 (USBGN)] (E761B). Gorgas Memorial Labs operate a field station here on an overgrown citrus plantation (Sousa and Herman, 1982). Hallinan, forest of large trees, on a trail, orange grove, 1916 Feb 21 (AMNH; Hallinan, 1924); Chapin, Rogers, and Mathews, 1923 Jul 17 (FMNH; Chapman, 1926); Imhof, observations, 1942 Jun 9 (Imhof, 1950); Fairchild, GL for Río Chagres, 1943 Oct 22 (ANSP); Hartman, GL for Río Chagres, 1949 Jan 11-13, 15, 17-19, 21-22, 1951 Jan 9, 12, 16-18, 20, 23, Feb 5, 1953 Jan 3, 4-10, 12-14, 16, 1956 Jan 1, 5, 13, 15, 1957 Dec 15, 1958 Jan 4 (USNM; Hartman, 1955; Hartman, 1956, as "Juan Mina Station"; Hartman and Brownell, 1961a; Hartman and Brownell, 1961b, as "Juan Mina Field Station"; Hartman and Lessler, 1963, as "Juan Mina Field Station"; Brumfield et al., 2001); Koford, 5 mi (8 km) NE Gamboa, 1952 Nov 13-16 (MVZ); Wetmore, on a trail inland from Juan Mina through wet grass, to a small hill covered with second growth, 1955 Dec 10-11. The secondary localities attributed to Juan Mina were located within 7 km of town along the Río Chagres and its tributaries. GL for: Boca de Gatuncillo, Cerro Juan Mina, Quebrada Agua Clara, Quebrada Palenque, Río Chagres, Río Chilibre, and Río Gatuncillo, 1955 Jan 6-22, Dec 9-16, GL for: Boca de Gatuncillo, Río Chagres, and Santa Rosa, 1959 Jan 27-Feb 1, GL for: Boca de Gatuncillo, Guayabalito, Julupa Grande, Quebrada Julupa, Quebrada Santa Clara, Río Aguardiente, Río Chagres, Río Limón, and Santa Rosa, 1961 Jan 10-18 (LSUMZ; USNM; Brumfield et al., 2001); Trapido, GL for Río Chagres, date ? (USNM); Yunker, 1961 Jun 15, 29-30 (FMNH; USNM); GMLP, 1962-1984 (GML); De Leon, 1965 Nov 14, 28 (GML); Galindo, 1969 May 15; 1972 Feb 16 (GML); Quinones, Río Chagres, 1964 Sep 10, 1966 May 9, 1967 Sep 18, Nov 8, 18, 1968 Sep 18, Oct 10 (GML); Van Horn, Río Chagres, 1964 Sep 21, 1970 Apr 21 (GML); Keenan, Río Chagres, 1969 Oct 14 (GML); Mendoza, Río Chagres, 1972 Mar 11, 1977 May-Sep 9 (GML); Chandler, Río Chagres, 1974 Dec 14 (GML); Hinds, 1976 May 13 (GML); Sousa and Herman, blood smears, c. 1980 (Sousa and Herman, 1982); Montenegro, Río Chagres, 1980 Jun 10, Aug 6, 26, Sep 19, Dec 3, 17, 1981 Aug 2, 11, Oct 4, 1983 May 20, Sep 22, 30 Oct 12, 1984 Jan 5 (GML); Calderón, GL = Gamboa, 1984 Oct 18 (MVUP); Ossa, GL = Gamboa, 1985 Jan 5 (MVUP). (Wetmore, 1952:18; Wetmore, 1967; Wetmore, 1970; Parkes, 1976; Olson, 1981a; Olson, 1981d).

JUAN MINA, CERRO; Panamá (Canal Zone) (E, Atlantic)
ca. 0909/7939
Described by Wetmore as a hill on the E bank of the Río Chagres [0907/7942 (USBGN)] S of Juan Mina [0910/7939 (USBGN)] (USNM). The peak is not named on our maps, but according to E866, there are several hills, 150-200 m, in the region 1-2.5 km S of Juan Mina between the Quebrada Juan Mina [0910/7939 (USBGN)] and the Quebrada Julupa [0909/7940 (USBGN)]. Wetmore, GL = Juan Mina, many palms along the Río Chagres, a trail along the edge of a lagoon, a small stream descending from the hills, up through a stand of original forest with a considerable undergrowth of bamboo, 1955 Dec 13 (USNM).

JUAN MINA, QUEBRADA; Panamá (Canal Zone) (E, Atlantic)
0910/7939 (USBGN)

Short river that originates in the hills S of Juan Mina [0910/7939 (USBGN)]. The river flows NNW, then at Juan Mina it turns sharply to the W and enters the Río Chagres [0907/7942 (USBGN)] (E866). See also: Chagres, Río; and Juan Mina, Cerro.

JUAN MINA FIELD STATION; see Juan Mina

JUAN MINA STATION; see Juan Mina

JUAN PELÓN; Panamá (E, Pacific)
0912/7855 (USBGN)
ca. 50 m, not shown on our maps, but USBGN indicates a town on the Río Chepo [0900/7906 (USBGN)] near Boca de Cañita [0912/7855 (USBGN)]. Wetmore, GL = Boca de Cañita, Río Chepo, collected on the Río Chepo 1 mi (1.5 km) upriver from Boca de Cañita, 20 feet (6 m) above the river on stream banks covered with original gallery forest containing mahogany and *espavé* trees that were 100-150+ feet (30-45 m) high, dense undergrowth, stands of bamboo, newly cut trail, 1962 Feb 6, 8 (USNM).

JUGLI, LAGUNA DE; see Cañaveral, Río

JULUPA, QUEBRADA; Panamá (Canal Zone) (E, Atlantic)
0909/7940 (USBGN)
Originates in the hills S of Juan Mina [0910/7939 (USBGN)] and flows NW entering the Río Chagres [0907/7942 (USBGN)] 2.5 km SSW of Juan Mina. The river discharges into a 1 km long lagoon that is separated from the main river channel by a peninsula and several islands (E866). Wetmore, GL = Juan Mina, wooded area, 1961 Jan 11 (USNM).

JULUPA GRANDE; Panamá (Canal Zone) (E, Atlantic)
ca. 0907/7940
Described by Wetmore as a region along the Río Chagres [0907/7942] between Juan Mina [0910/7939 (USBGN)] and the former Las Cruces [ca. 0907/7941] that was reached by side channels (USNM). Presumably he landed on the shore 3 km SSW of Juan Mina in a large embayment on the Río Chagres just below the mouth of the Quebrada Julupa [0909/7940 (USBGN)] (E866). Wetmore, GL = Juan Mina, hill slopes heavily forested and dissected by small streams, 1961 Jan 17 (USNM). Not shown on our maps.

JUNCAL; see Juncal, Campamento de

JUNCAL, BAHÍA; see Juncal, Bahía de

JUNCAL, BAHÍA DE; Veraguas, Isla de Coiba
0734/8143 (USBGN)
Small bay that is located N of Ensenada Arenas [0733/8142 (USBGN)] at the mouth of the Río Juncal [0734/8143 (USBGN)] (MGRP, as "Bahía Juncal"). Brumfield and Miller, GL = Isla de Coiba, 2004 Apr 10 (LSUMZ). See also: Juncal, Campamento de; and Arenas, Ensenada.

JUNCAL, CAMPAMENTO DE; Veraguas, Isla de Coiba
0734/8144 (USBGN)

Sea level, prison camp on the NE coast of Isla de Coiba [0728/8147 (USBGN)] 9 km NW of Punta Damas [0730/8141 (USBGN)]. Located 0.5 km inland on a small, unnamed stream that is just N of the Río Juncal [0734/8143 (USBGN)] (JOG). Bahía de Juncal [0734/8143 (USBGN)] (q. v.) is named on JOG and it appears that the camp was located on its coast. Wetmore, GL = Isla de Coiba, the camp was behind a sandy flat with coconut palms, followed a trail inland past a cornfield to jungle, the forest edge had creepers and bamboo, through pastures back to the coast, along the tide-bare sand, 1956 Jan 30 (USNM, as "Juncal"; Wetmore, 1957).

JUNCAL, RÍO; see Juncal, Bahía de

JUNGLE POINT; see Gallinazo, Punta de

JUNGURUDÓ, SERRANÍA DE; see Jungurudó, Sierra de

JUNGURUDÓ, SIERRA DE; Darién

0735/7754 (USBGN)

1,600 m, mountain range in the S tip of Darién province that extends NW-SE within 20 km of the coast. G. Angehr informs us that the actual local name is Jingurudí. The range is SE of the Serranía del Sapo [0750/7817 (USBGN)] and separates the drainage of the Río Balsas [0815/7759 (USBGN)], the Río Sambú [0805/7818 (USBGN)] and other rivers that flow N into the Golfo de San Miguel [0822/7817 (USBGN)] from the drainage of the Río Jaqué [0731/7810 (USBGN)], which flows directly into the Pacific Ocean (MGRP, as "Serranía de Jungurudó"). According to Angehr et al. (2004), the maximum elevation of the range may be near 1,200 m, several hundred meters lower than that reported on available maps. Christian, at ca. 840-940 m, observations, in the Río Sambú drainage "at a foothills site on the ridge of the Sierra de Jungurudó (07°37' N, 78°00' W)," the coordinates given indicate that he worked ca. 25 km NE of Jaqué [0731/7810 (USBGN)], and according to Angehr (in litt., Aug 2005), these observations were made at the camp described under the entry for Río Antaral [0743/7801 (Angehr et al., 2004)] (q. v.), 1997 Aug (AMNH; LSUMZ; Christian, 2001, as "Serranía de Jungurudó," and as "Sierra de Jungurudó"); Angehr, see Río Antaral, 1997 Aug 17-19 (MVUP, as "Serranía de Jingurado"); Thorstrom et al., observations on cliffs at 0744/7805 and at 0739/7805, which appear on our maps ca. 25 km NNE of Jaqué on the Río Sambú drainage and ca. 20 km NNE of Jaqué on the Río Pavarandó [0749/7804 (USBGN)] drainage, 2000 Apr 13, 26 (Thorstrom et al., 2002, as "Darién").

JURADÓ, CORDILLERA DE; Darién

0728/7755 (USBGN)

Mountain range in the S tip of Darién province that extends NW-SE within 10 km of the coast with peaks above 800 m (MGRP). According to SMN2B, the range extends roughly N-S for 20 mi (32 km) and its peaks are within 8 mi (13 km) of the Colombian border (SMN2B).

JUREL; see Jurel, Playa

JUREL, PLAYA; Chiriquí

0808/8219 (USBGN)

Described by Wetmore as a beach on the NE coast of Isla Parida [0807/8220 (USBGN)], Wetmore, GL = Isla Parida, to a small cornfield, W along the beach to the first headland where a creek came down over rock exposures, the water was in small pools near tall trees, in original forest, *almacigo* trees with ripe berries, a short trail along a creek across the narrow N tip of the island to the NW shore, 1963 Feb 2-3, 5 (USNM, as "Jurel"; Olson, 1997a, as "Jurel"). Not shown on our maps.

K-10 ROAD; see Road K10

K-2 ROAD; see Borinquén, Carretera

K-6 ROAD; see Road K6

K-7 ROAD; see Road K7

K-9 ROAD; see Road K9

KIKOS ISLAND; see Guindilla, Isla

KITCHENER, MOUNT; see Mount Kitchener

KOBBE; see Fort Kobbe Army Reservation

KOBBE, FORT; see Fort Kobbe Army Reservation

KOBBE BEACH; see Fort Kobbe Army Reservation

KUNA YALA, COMARCA; see San Blas, Comarca de

LA AGUADILLA; Panamá (E, Pacific)

ca. 0842/7838

Sea level, described by Wetmore as a cove on the SE shore of Ensenada de Chimán [0843/7839 (USBGN)] ca. 1.5 mi (2.5 km) S of Chimán [0842/7837 (USBGN)], Wetmore and Perrygo, GL = Chimán, red mangroves, between the shore and the woods, followed a small stream inland, 1950 Feb 17 (USNM). Not shown on our maps. See also: Creo, Playa del.

LA ALBINA; see Aguadulce

LA ALBINA LA CALETA; see Aguadulce

LA AMISTAD, PARQUE INTERNACIONAL; Bocas del Toro/Chiriquí

ca. 0910/8240

ca. 800-3,000 m, National and International Park that extends along the continental divide in Bocas del Toro and Chiriquí and into Costa Rica. The Panama section of the park was established in 1988, covers an area of 207,000 ha, and extends E from the Costa Rican border to the region N of Boquete. The upper reaches of the Ríos Changuinola [0928/8227 (USBGN)] and Teribe [0922/8232 (USBGN)] are within the park (RDP2, as "La Amistad Parque Nacional"; Collar and Dunn, 1989). According to Angehr (2003:107-113), the park "contains the highest number of endangered and endemic birds of any IBA [Important Bird Area] in Panama." The acronym PILA is used on some specimen labels. Wilson and Kalar, GL = Santa Clara, presumably worked in the hills N of Santa Clara [0850/8246 (USBGN)], 1999 Jan 17 (MVUP); Jiménez and Ochoa, presumably worked in the hills S of Changuinola [0926/8231 (USBGN)] as specimens are from Corregimiento de

Changuinola [0915/8239 (USBGN)], *montaño*, *submontano* (montane, submontane) 2002 Mar 30, Apr 10, 13 (MVUP, as "PILA").

LA AMISTAD PARQUE NACIONAL; see La Amistad, Parque Internacional

LA ARENA; Veraguas (Pacific)
0800/8126 (USBGN)
200 m, town 14 km W of Soná [0801/8119 (USBGN)] (JOG). Wetmore and Perrygo, GL = Soná, hill near town with a few acres of old second growth forest, some larger trees cut, some creeping bamboo and ferns, but largely open, climbed to ca. 1,200 feet (350 m), 1953 May 16, 21, June 6 (USNM).

LA BANDERA, LOMA DE; Veraguas, Isla de Cébaco (Pacific)
0731/8111 (USBGN)
ca. 300 m, not shown on our maps, but the USBGN coordinates indicate a hill on the E side of the Isla de Cébaco [0732/8109 (USBGN)]. Wetmore, GL = Isla de Cébaco, steep, almost precipitous slopes covered with woods, climbed to ca. 500 feet (150 m), 1965 Jan 11 (USNM; Olson, 1997a).

LA BARQUETA; Chiriquí
0818/8234 (USBGN)
Sea level, coastal town 11 km S of Alanje [0824/8233 (USBGN)] (JOG). Wetmore, GL = Alanje, at ca. 50 feet (15 m) in an area N of the coast in *rastrojo* with some large trees alternating with pastures, areas of sandy soil in a modified savanna region, low country below Alanje in a dry forest and in forest with an undergrowth of palm near a channel with running water, 1960 Mar 3, 10 (USNM).

LA BOCA; Panamá (Canal Zone) (E, Pacific)
0857/7934 (USBGN)
Sea level, town on the E side of the Panama Canal [0920/7955 (USBGN)] near its Pacific entrance between Balboa [0857/7934 (USBGN)] and Fort Amador Army Reservation [0856/7933 (USBGN)] (E661). Hallinan, over the water, a clearing 0.5 mi (1 km) from the shore, trees, 1915 Feb 22, Mar 21, Oct 26, Nov 14 (AMNH; Hallinan, 1924).

LA BOCA, LAGUNA DE; Los Santos
ca. 0733/8001
Sea level, described by Wetmore as a freshwater pond near the coast ca. 2 mi (3 km) NNE of Pedasí [0732/8002 (USBGN)], the pond was ca. 250 yards (230 m) long by 75 yards (70 m) wide with areas of open, deep water and areas of grass-covered water, bordered by extensive thorn shrub, corozo palm, bull horn *Acacia* and spiny creepers mixed with more ordinary growth and occasional good-sized trees, Wetmore, GL = Pedasí, 1957 Mar 12 (USNM). Not shown on our maps.

LA BOCA DE URACILLO; see Boca de Uracillo

LA BODEGA, ENSENADA DE; see Bodega, Ensenada de

LA BONGA; San Blas
0837/7729 (USBGN)
Not shown on our maps but USBGN indicates a settlement ca. 9 km SSW of

Puerto Obaldía [0840/7725 (USBGN)] near the Colombian border. Sheets, observations of *Habia* "in overgrown thickets above Puerto Obaldía, on the trail to La Bonga," 1965 Jun 16-18 Jun (Parkes, 1969:239).

LA CABIMA; Panamá (E, Atlantic)
0908/7932 (USBGN)
ca. 400 feet (120 m), not shown on our maps, but USBGN indicates a town ca. 2 km S of Calzada Larga [0910/7934 (USBGN)], presumably located on or near the Quebrada La Cabima [0910/7934 (USBGN)], which is a tributary of the Río Chilibrillo [0910/7938 (USBGN)]. Busck, specimens are labeled "Pacific Slope," but according to Wetmore's notes, Busck described the locality as "a place on the *Pacific side* [emphasis ours] of the Chagres River region." Correspondence between G. Fairchild and Wetmore indicates that Cabima was located near Calzada Larga [0910/7934 (USBGN)]. Therefore La Cabima, which is within the Río Chagres drainage on the S side (Pacific side) of the river, is probably the locality that Busck called Cabima, even though it is actually located on the Atlantic versant, 24 May 1911 (USNM, as "Cabima").

LA CABUYA; see Cabuya

LA CAJA, BOCA; see Boca La Caja

LA CALDERA; see Caldera

LA CALDERA ROAD; Chiriquí
ca. 0839/8226
Road that extends SSW and NNE of the town Caldera [0839/8223 (USBGN)]. The road terminates at the main Boquete-David road 12 km N of Dolega [0834/8225 (USBGN)] and in the E it terminates 3 km N of Los Planes on the Gualaca-Chiriquí Grande Road [ca. 0857/8207]. The Río Caldera [0837/8221 (USBGN)] crosses the road W of Caldera and the Río Chiriquí [0820/8226 (USBGN)] crosses the road E of the town (E562; E762; RDP2). According to E562, the road branches E of Caldera and a secondary road continues N terminating on the Gualaca-Chiriquí Grande Road N of Lago Fortuna. According to E762, portions of the road overlap with the route "Camino Tres de Noviembre." Ridgely and Richards et al., observations, along La Caldera Road, 1988 Jan 19 (Ridgely and Gwynne, 1989:374).

LA CAMPANA; see Campana

LA CAMPANA; see Copete, Cerro

LA CAMPANA MOUNTAIN; see Campana, Cerro

LA CAPETANA; see La Capitana

LA CAPIRA, SERRANÍA DE; see Capira

LA CAPITANA; Panamá (E, Pacific)
0909/7904 (USBGN)
Sea level, town on the Río Mamoní [0908/7904 (USBGN)] 4 km SE of Chepo [0910/7906 (USBGN)]. The road connecting Chepo and La Capitana extends roughly parallel to the river and within 1 km of its W bank (E761A). Eisenmann

and Pujals, observations on the roadside between Chepo and La Capitana, 1971 Jun 17 (Wetmore et al., 1984:34, as "La Capetana").

LA CARRASQUILLA; Panamá (E, Pacific)
0900/7931 (USBGN)
Not shown on our maps, but USBGN indicates a section of greater Panama City [0858/7932 (USBGN)] ca. 5 km NE of Ancón [0858/7933 (USBGN)]. Muñoz, 1961 Jun 3 (USNM, as "Carrasquilla"); Andrión, 1985 Apr 10 (MVUP, as "Carrasquilla").

LA CENTRAL; Veraguas, Isla de Coiba
ca. 0730/8143
Described by Wetmore as one of 7 sections on Isla de Coiba [0728/8147 (USBGN)]. La Central is the section surrounding Colonia Penal [ca. 0730/8142], Wetmore, GL = Isla de Coiba, trail inland and uphill from Colonia Penal through pastures, cultivated areas, tall grass, and forest being logged, 1 mi (1.5 km) along a ridge in virgin forest with tall trees, open undergrowth and tracts of low bamboo, 1956 Jan 7, 9-12, 14-15, 18, 20, 25, 29, 31, Feb 3, 4 mi (6 km) WNW of La Central, 3 mi (5 km) NW of La Central, followed a trail from Colonia Penal inland and up to 600 feet (180 m), from here walked 1.5 mi (2.5 km) along a ridge where the forested ground sloped down to the Río Catival [0729/8143 (USBGN)], gallery forest with open undergrowth and thickets of 4-5 foot (1-2 m) bamboo, creepers, tall trees filled with berries, 1956 Jan 31, Feb 3 (USNM; Wetmore, 1957).

LA CHOLITA; Panamá (W, Pacific)
ca. 0850/7947
Not shown on our maps but located in W Panamá province in Corregimiento Playa Leona [0850/7947 (USBGN)] near the town Playa Leona [0847/7946 (USBGN)] (q. v.). Salazar, GL = Playa Leona, 1991 Sep 15 (MVUP).

LA CHORRERA; Panamá (W, Pacific)
0853/7947 (USBGN)
ca. 50 m, town on the Carretera Interamericana [0832/8242 (USBGN)] 30 km SW of Panama City [0858/7932 (USBGN)] (JOG). Collector ?, date ? (USNM; Griscom, 1927b); Poisett, date ? (USNM, as "Chorrera"); Chaves, 1914 "early in 1914," "circa May" (USNM, as "Chorera," and as "Chorrera"); Zetek, 1914 (USNM, as "Chorera"); Weber, 1928 Aug 12 (FMNH, as "Chorrera"; USNM, as "Chorrera"); Takos, near, 1945 Aug 29, Nov 19 (MCN, as "Chorrera," and as "La Chorrera"); Scholes and Scholes, observations, 1951 Jan 20 (Scholes and Scholes, 1954, as "Chorrera"); Blanton, 1953 Jul 11, Aug 30 (USNM, as "Chorrera"); Nortiz (or N. Ortiz), 1965 Nov 20 (GML); Dutary, 1968 Jun 30 (GML, as "Chorrera"); Carrasco, GL for Corozales Adentro, 1976 Jun 15-16 (GML, as "Chorrera"); Christie and De Jiménez, 1976 Jul 11 (GML); Gilkes, 1976 Jul 14 (GML); Pinto, 1985 Apr 15 (MVUP); Rivera, 1990 Sep 7, 12 (MVUP). (Zimmer, 1939; Wetmore, 1952:18; Wetmore, 1959b:10, 12, as "Chorrera"; Parkes, 1976; Olson, 1981d). See also: Puerto Caimito.

LA COCA; Panamá (W, Pacific)
0853/7953 (USBGN)

Not shown on our maps, but USBGN indicates a town, at ca. 100 m, ca. 15 km N of Capira [0845/7953 (USBGN)] and ca. 10 km W of La Chorrera [0853/7947 (USBGN)]. Montenegro, most labels indicate GL = Capira, and some also indicate GL = El Aguacate Número Uno, 1978 Aug 24, Oct 15, 19, Nov 20, 23 (GML).

LA COLORADA; Veraguas (Pacific)
0801/8059 (USBGN)
100 m, town 10 km S of Santiago [0806/8059 (USBGN)] near tributaries of the Río Martín Grande [0758/8102 (USBGN)], which empties into the Golfo de Montijo [0740/8107 (USBGN)] (JOG; MGRP). Benson and Seaman, GL = Santiago, 1924 Apr 22-30, May 1, 3-8, 13-16 (AMNH, as "La Colorado, Santiago"; Griscom, 1924b, as "La Colorada, Santiago"; Zimmer, 1941a:7; Zimmer, 1941b:14, as "La Colorado"; Wetmore, 1950a; Olson, 1981d, as "La Colorada, Santiago").

LA COLORADO; see La Colorada

LA CONCEPCIÓN; Chiriquí
0831/8237 (USBGN)
200 m, town in W Chiriquí province on the Carretera Interamericana [0832/8242 (USBGN)] 23 km NW of David [0826/8226 (USBGN)] and 38 km NE of Puerto Armuelles [0817/8252 (USBGN)] (JOG; MHA, as "Concepción"). Davidson, at 1,500 feet (450 m), near, 1929 Dec 2-12 (CAS; CM; Davidson, 1932; Davidson, 1938; and Griscom, 1935a, all as "Concepción"); Wetmore, GL for: Quebrada Santo Domingo, and Río Gariché, 1960 Mar 2 (USNM; and Wetmore, 1970, both as "Concepción"); Collector ?, 9 mi (14 km) N, 1970 Apr 9 (UF, as "Concepción").

LA CONCEPCIÓN; Herrera
0804/8042 (USBGN)
ca. 25 m, town 6 km SW of Santa María [0807/8040 (USBGN)]. The Río Escotá [0806/8033 (USBGN)] passes 1 km W of town, then turns to the E and passes 3 km N of town (JOG). Wetmore and Perrygo, GL = Santa María, W of La Concepción near the Río Escotá, which was bordered by second growth forest, scattered tracts of large trees, stands of bamboo and sawgrass and at Ciénaga de Buho [ca. 0804/8043] (q. v.), specimens labeled 4 mi (6 km) and 5 mi (8 km) SW of Santa María were collected in an area of savanna N of La Concepción, 1948 Mar 9-10, 19 (USNM).

LA CORDILLERA CENTRAL; see Central, Cordillera

LA CORDILLERA DE LOS ANDES; see Los Andes, Cordillera

LA COROCITA, ISLA DE; see Corocita, Isla de la

LA CRESTA; Panamá (E, Pacific)
0859/7932 (USBGN)
50 m, section of greater Panama City [0858/7932 (USBGN)] located NE of Curundu [0859/7933 (USBGN)], and NNE of Calidonia [0858/7933 (USBGN)] (E962). Fairchild, 1969 Nov 23 (GML).

LADRONES, ISLAS; Chiriquí
0752/8226 (USBGN)

70 m, group of more than 10 small islands and islets in Golfo de Chiriquí [0800/8220 (USBGN)] located 63 km S of David [0826/8226 (USBGN)] (JOG; MGRP; MHA). Batty, specimens at AMNH and at BMNH are dated 1902 Jan 8-9, but Batty's data were completely fabricated and his specimens labeled "Ladrones Island" were actually collected in 1901 near his mainland bases at Boquete [0847/8226 (USBGN)] and Boquerón [0830/8234 (USBGN)] (AMNH, as "Ladrones Island"; BMNH, as "Ladrones Island"; Olson, 2008). Observer ?, date ? (Wetmore, 1965:58); Murphy, R., observations, 1956 Nov 25 (Murphy, 1956:107, as "Islas de Ladrones"; Wetmore, 1965:58).

LADRONES, ISLAS DE; see Ladrones, Islas

LADRONES ISLAND; see Ladrones, Islas

LA ENEA; see Guararé

LA ENSENADA; Panamá, Archipiélago de las Perlas
0822/7851 (USBGN)
Sea level, town on the central E coast of Isla del Rey [0822/7855 (USBGN)] ca. 5 km S of Punta Peña Tiburón [0825/7851 (USBGN)] on the S side of a bay that extends 2 km N-S along the coast (JOG). Wetmore, GL = Isla del Rey, landed N of town on the far side of the bay near the N end of Isla Cañas [0823/7849 (USBGN)], at a 350 yard (320 m) wide, sandy beach with 200 foot (60 m) rocky headlands, *almacigo* trees with ripe berries, *rastrojo* and forest near the beach, 1960 Jan 25 (USNM, as "Ensenada"), observations, 1960 Jan 24 (Wetmore et al., 1984:71, as "Ensenada").

LA ESPERANZA; Chiriquí
0824/8248 (USBGN)
ca. 25 m, town 7 km SE of Progreso [0827/8250 (USBGN)] and 17 km NNE of Puerto Armuelles [0817/8252 (USBGN)] (JOG, as "Esperanza"; MGRP). Ridgely, observations, savannas, 1982 Jun 25 (Ridgely and Gwynne, 1989:199); González, 1983 May 31 (MVUP); Sánchez, GL for La Colorada, 1983 Jul 9, 14, 17 (MVUP); Vásquez, 1983 Sep 10, 18 (MVUP); Sánchez, 1983 Oct 29, Nov 9, Dec 10, 12, 15, 19-20, 27, 29 (MVUP).

LA ESPIGADILLA, QUEBRADA; see Espigadilla, Estero

LA ESTRELLA; Chiriquí
0844/8223 (USBGN)
1,060 m, town 7 km SE of Boquete [0847/8226 (USBGN)] (E762). Wetmore, GL = Boquete, at 3,700 feet (1,100 m), drove ca. 6 mi (10 km) E and SE of Boquete, covered 1 mi (1.5 km) in a tract of forest on level ground, pastures and sugar cane fields, 1960 Mar 17 (USNM, as "Estrella").

LA FORTUNA; see Fortuna

LA GARITA; Chiriquí
0852/8235 (USBGN)
1,820 m, town 3 km NW of Cerro Punta [0852/8235 (USBGN)] on the Río Chiriquí Viejo [0820/8241 (USBGN)] (E762). Wetmore, GL = Cerro Punta, at 5,800 feet (1,750 m), a steep valley near Cerro Punta on the W bank of the Río Chiriquí Viejo,

a small tributary bordered by brush, slopes covered in a mixture of pasture and *rastrojo* with areas planted in cabbage, lettuce, peas, and corn, fallow fields with flowering plants, forest at the highest elevations, 1955 Mar 5 (USNM, as "Lagarita"). See also: Las Nubes.

LAGARITA; see La Garita

LAGARTERA; Colón/Panamá (Canal Zone) (W, Atlantic)
ca. 0907/7957
Not clear where this locality is cited, although Wetmore indicated it on our maps, therefore he may have mentioned it in one of his publications. Apparently a village now submerged in Lago Gatún (Gatun Lake) [0912/7955 (USBGN)] ca. 11 km WSW of Isla Barro Colorado [0909/7951 (USBGN)] and 2 km SE of Escobal [0909/7958 (USBGN)].

LA GRUTA; Bocas del Toro, Archipiélago de Bocas del Toro
0924/8216 (USBGN)
20 m, town located roughly in the center of Isla Colón [0924/8217 (USBGN)] 7 km NNW of Bocas del Toro [0920/8215 (USBGN)] (JOG). SBE, GL = Isla Colón, *a orillas del camino* brush (brush at the trail's edge), 1988 Feb 7-23 (MVUP; USNM; Olson, 1993a).

LAGUNA; Chiriquí
ca. 0826/8145
Specimen labels indicate that the locality is near Hato Chamí [0827/8146 (USBGN)], therefore the name likely refers to Cerro Lagunas [0826/8145 (USBGN)] just E of Hato Chamí. Davidson and Hoigaard, 1932 Jan 26 (CAS). Not shown on our maps.

LAGUNA; see Las Lagunas

LAGUNA, LA; see La Laguna

LAGUNA CAMP; see La Laguna

LAGUNA DE AGUA; see Las Lagunas

LAGUNA DE CHIRIQUÍ; see Chiriquí, Laguna de

LAGUNA DE DAMANÍ; see Cañaveral, Río

LAGUNA DE JUGLI; see Cañaveral, Río

LAGUNA DE LA BOCA; see La Boca, Laguna de

LAGUNA DE LA YAGUADA; see La Yeguada, Laguna de

LAGUNA DE NAMANÍ; see Cañaveral, Río

LAGUNA DE PITA; see Matusagratí, Laguna de

LAGUNA DE SAMANÍ; see Cañaveral, Río

LAGUNA DEL CASTILLO; see La Yeguada, Laguna de

LAGUNA DELLA PITA; see Matusagratí, Laguna de

LAGUNA DEL ROSARIO; see El Rosario

LAGUNA GRANDE; see Las Lagunas

LAGUNA PLAYA NIEVE; see Playa Nieve, Laguna

LAGUNA PORRAS; see Bocatorito, Laguna

LAGUNA RÍO DIABLO; see Cañaveral, Río

LAGUNAS; see Las Lagunas

LAGUNAS, CERRO; see Laguna

LAGUNAS, LAS; see Las Lagunas

LAGUNAS DEL VOLCÁN; see Las Lagunas

LAGUNAS DE VOLCAN; see Las Lagunas

LA GUNETA; Chiriquí

Olson recalls seeing this rendering either on a specimen label or in print in reference to a locality in Chiriquí that would presumably be either La Lagunita (q.v.) or Las Lagunas (q.v.).

LAGUNITA, LA; see La Lagunita

LA HONDA; Los Santos

0754/8019 (USBGN)

Sea level, coastal town 11 km ESE of Los Santos [0756/8025 (USBGN)] and 3 km S of La Hondita [0756/8020 (USBGN)] (JOG). Wetmore and Perrygo, GL = Los Santos, 5 mi (8 km) E of Los Santos, long sandy beach bordered by mangroves and thorny brush, near the mouth of the Estero Espigadilla [ca. 0753/8019], on the coast ca. 1 mi (1.5 km) S of La Honda, 1948 Mar 15, 20 (USNM; Olson, 1980b, as "Los Santos, 5 miles E of La Honda," but the specimen was taken at La Honda, 5 mi E of Los Santos); Vasquez, *las salinas de* (salt ponds of) La Hondita, 1991 Aug 31 (MVUP, as "Playa La Honda"). See also: Villa, Isla.

LA HONDA, PLAYA; see La Honda

LA HONDITA; Los Santos

0756/8020 (USBGN)

Sea level, coastal town located 9 km E of Los Santos [0756/8025 (USBGN)] and 3 km N of La Honda [0754/8019 (USBGN)] (JOG). See also: La Honda.

LA INDIA DORMIDA, CERRO; Coclé (Pacific ?)

ca. 0838/8010

ca. 1,000 m, described by Wetmore as a peak W of Cerro Gaital [0837/8007 (USBGN)], Wetmore and Perrygo, GL = El Valle [0836/8008 (USBGN)], at 2,500 feet (760 m), from the W end of the valley followed a streambed into woodlands, along a ridge with a view of the Pacific drainage, extensive grasslands, 1951 Mar 29 (USNM). Not shown on our maps.

LA ISLETA; Coclé (Pacific)

ca. 0814/8029

Described by Wetmore as an inland area on the banks of the Río Pocrí [0813/8028 (USBGN)] ca. 0.5 river mi (1 km) from Puerto Aguadulce [0815/8030 (USBGN)],

Wetmore, GL = Aguadulce [0815/8033 (USBGN)], Río Pocrí, area of tidal influence, muddy banks, old fields, corn and bean fields, small plantations, 1963 Jan 19, 22 (USNM). Not shown on our maps.

LA ISLETA; Veraguas (Pacific)

ca. 0758/8117

Sea level, described by Wetmore as an island in the Río San Pablo [0751/8110 (USBGN)] ca. 3 river mi (5 km) below Soná [0801/8119 (USBGN)], Wetmore and Perrygo, GL = Soná, Río San Pablo, landed on the mainland opposite the island in open, second growth forest with 50 foot (15 m) trees, near an estuary and pastures, cattle trails through thorn scrub and creepers, 1953 June 1-4 (USNM). Not shown on our maps.

LAJA, QUEBRADA; Chiriquí

ca. 0819/8153

Described by Wetmore as a swift creek flowing over a bare exposure of rock W of El Macano [0819/8152 (USBGN)], presumably a tributary of the Río Dupí [0816/8156 (USBGN)], Wetmore, GL = San Félix [0817/8152 (USBGN)], steep slope above the creek covered in uncut forest, thick brush, 1956 Feb 17 (USNM). Not shown on our maps.

LA JAGUA; see La Jagua Hunting Club

LA JAGUA, RÍO; Panamá (E, Pacific)

0902/7909 (USBGN)

Short river that appears to be a continuation of the Río Santo [0903/7911 (USBGN)], although the course widens at their junction where several small tributaries enter. The river flows SE meandering greatly before joining the Río Chico [0901/7911 (USBGN)] 3 km NE of Río Chico [0902/7910 (USBGN)] (TM). The La Jagua Hunting Club [ca. 0903/7911] was located on the river near its origin, Wetmore and Perrygo, GL = Río Chico, 1949 Mar 28-29, Apr 1-2, at La Jagua Hunting Club (q. v.), 1949 Mar 13-15, 20, 22, 24-25, 27 (USNM); Hartman, 1951 Feb 8, 1956 Feb 1 (USNM, as "La Jagua," and as "La Jagua River"; Hartman, 1955, as "La Jagua River"); Moreno, GL = Río Chico, 1959 Jan 22 (USNM); Wetmore, GL = La Jagua Hunting Club, along the edge of the woods and into the dry marsh, a pool in the river section called Calabaso [ca. 0903/7911], side channel called Corre Falso (not located), channels varied from open pools to those choked by heavy growth, 1961 Mar 18, 20-21, 23-24, GL = La Jagua Hunting Club, 1962 Jan 14, 1964 Mar 24 (USNM); Ridgely, observations, at Finca San José (probably near San José [0904/7916 (USBGN)]), 1967 Jan 11 (Wetmore et al., 1984:626, as "La Jagua Marshes"). See also: Santo, Río.

LA JAGUA GUN CLUB; see La Jagua Hunting Club

LA JAGUA HUNT CLUB; see La Jagua Hunting Club

LA JAGUA HUNTING CLUB; Panamá (E, Pacific)

ca. 0903/7911

Described by Wetmore as a hunting club that was located 10 km SE of Pacora [0905/7917 (USBGN)] and in existence from 1928 until 1964. The clubhouse stood on a long peninsula of high ground above the Río La Jagua [0902/7909 (USBGN)] and the members leased 5,000 acres (2,020 ha) of marshy wetlands with additional access to the surrounding 20,000 acres (8,090 ha). Narrow bands of scrub and low forest bordered the extensive marshes and swamps of the river (USNM). The locality is not shown on our maps, but the area described by Wetmore appears on JOG and TM near the junction of the Río Santo [0903/7911 (USBGN)] (q. v.) with the Río La Jagua ca. 12 km ESE of Pacora and 3 km NNW of Río Chico [0902/7910 (USBGN)]. The region is wetlands and savannas and the elevation is ca. 60 feet (20 m). In his notes for the gazetteer Wetmore indicates that beginning in 1948 he was allowed use of the club facilities whenever desired and in 1960 he was granted honorary membership. His extended studies and collections of the rich birdlife here were of great advantage to his work, in addition to the pleasant association with Mr. Curtis and the other members. Dr. Frank Chapman was an early visitor and La Jagua was known to numbers of other naturalists. The Register of the Club with its record of waterfowl has been placed in the Canal Zone library in Balboa. According to Angehr (in litt., Aug 2005), "the area was largely drained and converted to a sugar cane plantation in the 1970s, and now consists mainly of rice farms, shrimp ponds, and pasture." Curtis, 1936 May 17, Jun 14, observations, 1949 May 20-21 (MCZ, as "La Jagua Swamp"; Griswold, 1936b, as "La Jagua Swamp"; Wetmore, 1951b), reports of banded birds, 1954 Dec 15 (Wetmore, 1965:140); Miller, GL = Río Chico, 1937 Mar 6 (USNM, as "Duck Club," as "La Jagua Gun Club," as "La Jagua Hunt Club," and as "La Jagua Hunting Club"); Wetmore and Perrygo, GL = Río Chico, Río La Jagua, along and near the Río La Jagua in forest, marsh, and savannas, 1949 Mar 13-15, 20, 22, 24-25, 27 (USNM; Eisenmann, 1962a, as "La Jagua Gun Club"); Moreno, 1949 May 29 (USNM; Wetmore, 1951b); Trapido, 1947 Feb (USNM, as "La Jagua Swamp"); Wetmore and Perrygo (from Trapido), GL = Río Chico, 1951 Mar 1 (USNM); Wetmore, GL = Río Chico, on the marsh clogged with grass and water hyacinths, in open water, in forest along the marsh edge, in small groves of trees isolated in the savannas, 1953 Jun 26-28, GL = Río Chico, timber along the marsh, which had considerable water, 1955 Mar 30, observations at a *pozo* above, 1955 Jan 2, GL = Río Chico, in cattle grazing areas, along the 2 mi (3 km) trip SE to Gallinero [ca. 0902/7911] through river channels that were mid-thigh deep, 1958 Mar 19-21, at Gallinero (q. v.), 1958 Mar 21, GL for: Ciénaga de Santo Domingo, and Río La Jagua, 1961 Mar 18, 20-24, at the Club House, 1962 Jan 12, GL for: Calabaso, Ciénaga de Santo Domingo, and Río La Jagua, 1962 Jan 12-14, GL for Río La Jagua, 1964 Mar 24 (USNM, as "La Jagua," as "La Jagua Gun Club," and as "La Jagua Hunting Club"; Wetmore, 1965:140; Payne, 1974, as "La Jagua"); Olson and Loftin, 1963 Jan 26 (USNM, as "La Jagua Hunt Club"); Moreno, 1965 Jun 28 (GML, as "La Jagua, Pacora"); Means, Koslovsky, and Sawyer, GL = Pacora on some specimens, 1967 Feb 17-22 (FSU, as "La Jagua"; MCN, as "La Jagua"; USNM, as "La Jagua"); Karr 1968-1969 (Karr, 1971b, as "La Jagua"); Dickerman, 1968 Feb 7 (CU, as "La Jagua"); Lanyon, 12 mi (19 km) E of airport, 1969 Apr 19 (AMNH, as "La Jagua"). (Wetmore, 1967, as "La Jagua"; Gorski, 1969, as "La Jagua"; Angehr, 2003:258-260, as "La Jagua"). See also: San Antonio Swamp.

LA JAGUA MARSHES; see La Jagua, Río

LA JAGUA RIVER; see La Jagua, Río

LA JAGUA SWAMP; see La Jagua Hunting Club

LA JARCIA; see Sapo, Cerro

LAJAS, LAS; see Las Lajas

LAJAS, PLAYA LAS; see Jobo, Playa del

LAJAS, RÍO; Panamá (W, Pacific)
0831/7954 (USBGN)
Originates in the SE foothills of Cerro Gaital [0837/8007 (USBGN)], flows SE and enters the Gulf of Panama [0800/7930 (USBGN)] 6 km SSW of Chame [0835/7953 (USBGN)] (JOG). Wetmore, GL = Chame, mouth of, at Playa Coronado (q. v.), 1953 Jun 21-22 (USNM, as "Río Las Lajas").

LAJAS, RÍO LAS; see Lajas, Río

LA JOLLA 1; Panamá (E, Pacific)
ca. 0906/7915
174 feet (53 m), a small airport and radio tower located 8 km NE of Pacora [0905/7917 (USBGN)] (AC, as "La Joya No. 1"; CARP, as "La Joya No. 1"). Blanton, GL = Pacora, 1953 Oct 8 (USNM); Karr, 0907/7916, NE of old La Joya airport (whether this means La Jolla 1 or La Jolla 2 is unclear), 1968-1969 (Karr, 1971b, as "La Joya"). See also: La Jolla 2.

LA JOLLA 2; Panamá (E, Pacific)
ca. 0905/7917
ca. 10 m, a small airport located just W of Pacora [0905/7917 (USBGN)] (AC, as "La Joya No. 2"). Blanton, GL = Pacora, 1953 Sep 23, 30 (USNM); GMLP, GL = Pacora, unclear if collections were made at La Jolla 1 or at La Jolla 2, 1958 Sep 4, Oct 12 (GML, as "La Joya"). See also: La Jolla 1.

LA JOYA; see La Jolla 2

LA JOYA NO. 1; see La Jolla 1

LA JOYA NO. 2; see La Jolla 2

LAKES; see Las Lagunas

LA LAGUNA; Darién
0804/7719 (Fairchild and Handley, 1966)
ca. 3,000 feet (900 m), described by Wetmore as a camp SW of Tacarcuna Village [0805/7717 (FH)], on a round, fresh water lake surrounded by low, dense scrub (USNM). He later describes the lake as a "crater lake at 575 meters on a ridge leading to Cerro Tacarcuna" (Wetmore, 1972:149). According to Fairchild and Handley (1966:17), this locality was near Tacarcuna Village and was also known as Tacarcuna Laguna. Handley, GL = Tacarcuna Village, at 4,000 feet (1,200 m), 1959 Mar 10 (USNM, as "Laguna Camp"); Galindo, Mendez, and GMLP, 1963 May 23-31, Jun 2-17, 23, 25-27, 29, Jul 3, 7 (GML; USNM; Adsett and Wege, 1998); Wetmore, at 2,800 feet (850 m), on the lake, 1964 Mar 19 (USNM). (Olson, 1980a;

Olson, 1983; Hekstra, 1982).

LA LAGUNA; Panamá (E, Atlantic)
0908/7931 (USBGN)
120 m, town located 4.5 km ENE of Alcalde Díaz [0907/7933 (USBGN)] (JOG). Described by Freed as a rural village 25 km NE of Panama City [0858/7932 (USBGN)], Freed, observations, 1982-1986 (Freed, 1986a; Freed, 1986b; Freed, 1987a; Freed, 1987b; Freed, 1988).

LA LAGUNA; Panamá (W, Pacific)
0836/8003 (USBGN)
600 m, town located ca. 9 km ESE of Cerro Gaital [0837/8007 (USBGN)] and ca. 43 km SW of La Chorrera [0853/7947 (USBGN)], located at the end of a 17 km road heading NW from the Carretera Interamericana [0832/8242 (USBGN)] towards Cerro Gaital peak (JOG; MGRP). Bacon and Miller, at ca. 150 m, at 8°37'23"/ 80°02'56", 2002 Sep 14 (STRI; UAM); Lelevier, Maley and Miller, at ca. 150 m, at 8°37'23"/80°02'56", 2005 Dec 26 (STRI; UAM).

LA LAGUNA DE POCRÍ; Coclé (Pacific)
ca. 0813/8028 ?
Not shown on our maps but the name probably refers to a region on the lower reaches of the Río Pocrí [0813/8028 (USBGN)] (q. v.), and possibly to the salt beds ca. 5 km E of Pocrí [0816/8033 (USBGN)] and Aguadulce [0815/8033 (USBGN)] (q. v.). Casas, GL = Aguadulce, 1989 Nov 22 (MVUP).

LA LAGUNITA; Chiriquí
ca. 0846/8240
ca. 1,300 m, described by Wetmore as a 10-30 foot (3-9 m) wide stream or channel that winds through pastures creating a wetland during the rainy season, located 0.5 mi (1 km) NE of Las Lagunas [0846/8241 (USBGN)] (q. v.), presumably a branch of the stream that drains Las Lagunas if not the stream itself, Wetmore, GL = Volcán, at 4,200 feet (1,300 m), marshy borders, slightly elevated banks, neglected pasture grown in bracken, in low scrub and in a flooded woods, 1955 Mar 15 (USNM).

LA LOCERÍA; Panamá (E, Pacific)
ca. 0900/7932
40 m, section of N greater Panama City [0858/7932 (USBGN)] W of San Miguelito and N of Betania [0901/7932 (USBGN)] (E762). Altamar, Bonadies, and Visuetti, 1983 Jul 31 (MVUP); Velotti, 1983 Aug 5 (MVUP); Telesca, 1983 Aug 22 (MVUP).

LA LOMA; Panamá (E, Pacific)
ca. 0905/7917
Alt. ?, apparently a hill on the savannas between Pacora [0905/7917 (USBGN)] and San José [0904/7916 (USBGN)], Wetmore, GL = Pacora, area dry and partly burned, 1961 Feb 15 (USNM). Not shown on our maps.

LA LOMA. FINCA; see Bahía Honda, Bocas del Toro

LA MACANA, CIÉNAGA DE; Herrera
0807/8036 (USBGN)

ca. 25 m, a small, freshwater lake 2 km ESE of El Rincón [0807/8037 (USBGN)], S of the Río Santa María [0806/8029 (USBGN)] and 4 km W of the mouth of the Río Escotá [0806/8033 (USBGN)] (JOG; Delgado, 1983, as "Ciénaga Las Macanas"). Described by Angehr (2003, as "Ciénaga de las Macanas") as a "small slightly brackish lake, a maximum of 6 m deep during the rainy season." Described by Wetmore as a marsh with low bushes, woodlands, and brush that bordered the water, Wetmore and Perrygo, GL = El Rincón, areas of water covered by water hyacinth and other vegetation, and patches of shallow, open water, some that extended over 1 mi (1.5 km), 1948 Mar 1, 3, 6, 17, 25 (USNM, as "Ciénaga Macana"; Wetmore, 1950a, as "Ciénaga Macana"; Wetmore, 1950c, as "Ciénaga Macana"); Delgado, observations, 1981 Jun 17, 1982 Aug 10 (Ridgely and Gwynne, 1989:69, 127, as "Cienaga Las Macanas").

LA MAREA; Veraguas (Pacific)

ca. 0754/8100

ca. 50 m, the locality La Marea, which appears on specimens from Benson and Seaman, does not appear on our maps, but presumably was located near the present towns Marea Arriba [0754/8100 (USBGN)] and Marea Abajo [0754/8101 (USBGN)], and the stream Quebrada de la Marea [0753/8101 (USBGN)]. The region is ca. 23 km S of Santiago [0806/8059 (USBGN)] on the N coast of Golfo de Montijo [0740/8107 (USBGN)] between the lower reaches of the Río San Pedro [ca. 0750/8105] and the Río Piña [0753/8104 (USBGN)]. Benson and Seaman, 1924 May 24-Jun 6, Sep 27 (AMNH; Zimmer, 1941b:14; Zimmer, 1950:21, 26; Wetmore, 1972:348, as "La Maria").

LA MAREA, QUEBRADA DE; see La Marea

LA MARÍA; see La Marea

LA MARTINA; Chiriquí

0821/8233 (USBGN)

Not shown on our maps, but USBGN indicates a town at ca. 25 m, ca. 5 km S of Alanje [0824/8233 (USBGN)]. Wetmore, GL = Alanje, tract of low deciduous trees interspersed with palms, 1960 Mar 8 (USNM; Wetmore, 1962, as "Alanje").

LA MESA; Panamá (E, Pacific)

0911/7916 (USBGN)

100 m, town 20 km WNW of Chepo [0910/7906 (USBGN)] and 11 km NNE of Pacora [0905/7917 (USBGN)] between the Río Pacora [0901/7917 (USBGN)] and the Río Cabobré [0906/7917 (USBGN)] (JOG) in an area shown on TM as open grasslands. Presumably the area S of La Mesa is the Llano de la Mesa to which Goldman's journal referred in 1911 under Cerro Azul [0910/7924 (USBGN)] (USNM). Although La Mesa is not named on TM, a small unnamed settlement is shown at the corresponding coordinates. According to TM, the elevation rises to 800 feet (240 m) 2 mi (3 km) NW of town in an area of hills dissected by small tributaries of the Río Cabobré, and Cerro Prominente [0914/7917 (USBGN)] is located 4 mi (6 km) NW of the town. See also: Azul, Cerro.

LA MESA, LLANO DE; see La Mesa

LA MINA, ISLA; Panamá, Archipiélago de las Perlas
0829/7901 (USBGN)
Sea level, small island with a land area of ca. 0.8 km² (80 ha) located in Bahía de Panamá [0850/7915 (USBGN)] 5 km NW of Isla del Rey [0822/7855 (USBGN)] and just NW of Isla Viveros [0828/7900 (USBGN)] (E501, as "Isla Mina Grande"; JOG). According to Rendahl (1920:4, as "Mina"), the island was "low-lying and devoid of fresh water but very richly wooded by cocoas." Wright and Faaborg, observations, 1976 Jun-Aug (Wright et al., 1985, as "Mina Island").

LA MINA DEL RÍO INDIO; see Las Minas

LANDING FIELD; see Albrook Air Force Base

LANZA, PUNTA; Chiriquí
0807/8219 (USBGN)
USBGN treats Punta Icaco as an older name for Punta Lanza. The map in Wetmore's journal indicates that the point is on the E coast of Isla Parida [0807/8220 (USBGN)], Wetmore, GL = Isla Parida, collected on the coast both N and S of Punta Lanza, and inland behind the point, on a sandy beach bordered by 2 headlands, the edge of a partly dry mangrove, along a trail inland through extensive *rastrojo*, old clearings, uncut gallery forest with 100-125 foot (30-40 m) trees, and up to higher elevations behind the point, along the bay S of the point on a beach near a mangrove swamp, in a rice field, on low hills on the point in virgin jungle with 40-60 foot (10-20 m) trees, creeks and dense, low undergrowth, 1963 Feb 3-4, 7 (USNM, as "Punta Icaco"; Olson, 1997a, as "Punta Icaco").

LA PALMA; Darién
0825/7809 (USBGN)
Sea level, the capital of Darién province, located on a point projecting into the SE end of Golfo de San Miguel [0822/7817 (USBGN)] at the mouth of the Río Tuira [0821/7803 (USBGN)] (SMN2B). Burton, British Trans-Americas Expedition, 1972 Feb (BMNH; Burton, 1973b; Burton, 1975); Altamar, 1983 Aug 4 (MVUP); Cesar, 1990 Nov 17 (MVUP).

LA PALMITA; see Las Palmitas

LA PINTADA; Coclé (Pacific)
0836/8027 (USBGN)
ca. 50 m, town 14 km NW of Penonomé [0831/8022 (USBGN)] (JOG). Heyde, GL = Isthmus of Panama, 1889 Mar 7, May 14 (USNM, as "Pintada"; Wetmore, 1951a).

LA PITA; Chiriquí
0826/8235 (USBGN)
50 m, 11 km N of Palo Grande [0820/8236 (USBGN)] (JOG). Miranda, 1983 Sep 25 (MVUP).

LA PITA, LAGUNA DE; see Matusagratí, Laguna de

LA PLAYA EL AGALLITO; see El Agallito, Playa

LA PLAYA JOBO; see Jobo, Playa del

LA PORCADA, ISLA; Chiriquí
0807/8144 (USBGN)
Sea level, island with a land area of ca. 12 km^2 (1,200 ha) located in Golfo de Chiriquí [0800/8220 (USBGN)] 15 km SW of Tolé [0814/8141 (USBGN)] within 1 km of the mainland coast (MGRP; MRP, as "I. Insolito"; on some maps as Isla Silva de Adentro). Batty, specimens at AMNH are dated 1902 Jan 1-2, but Batty's data were completely fabricated and his specimens labeled "Insoleta Island" were actually collected in 1901 near his mainland bases at Boquete [0847/8226 (USBGN)] and Boquerón [0830/8234 (USBGN)] (AMNH, as "Insoleta Island"; Eisenmann, 1950, as "Insoleta I."; Wetmore, 1957:6, as "Insolita Island"; Olson, 2008).

LA PUENTE, RÍO; Panamá (Canal Zone) (E, Atlantic)
0912/7933 (USBGN)
Originates E of Calzada Larga [0910/7934 (USBGN)] and flows NW. Originally the river entered the Río Chagres [0907/7942 (USBGN)] at El Vigía [ca. 0914/7935], but now it enters the SE coast of Lago Alajuela (Madden Lake) [0915/7935 (USBGN)]. According to E761A, the main channel and one of its tributaries flow underground 2.5 km and 3.5 km NE of Calzada Larga, respectively. This region is marked "Natural Bridge" on the map and is located 8 km E of Alhajuela [ca. 0913/7937] (E761A, as "Río Puente"). Goldman, specimens are labeled "Natural Bridge N of Alhajuela," but, as stated above, E761A shows the Natural Bridge 8 km E of Alhajuela. From Alhajuela Goldman followed a trail through forest and several small savannas near El Vigía, then traveled to Calzada Larga then E to the Río La Puente, waded downstream for ca. 1 mi (1.5 km), the natural bridge was ca. 30 feet (9 m) high and covered a 100 foot (30 m) section of the river, 1912 Jan 31 (USNM, as "Río del Puente").

LA PULIDA; Panamá (E, Pacific)
0903/7929 (USBGN)
40 m, section of greater Panama City [0858/7932 (USBGN)] 1.5 km WNW of Parque Lefevre [0901/7929 (USBGN)] (E762). Tejera, 1977 Sep 26, 1978 Jan 30, 1979 Oct 14 (MVUP); Tejera, 1990 Feb 19 (MVUP).

LA PUNTA; Veraguas, Isla Gobernadora (Pacific)
0734/8111 (USBGN)
Sea level, town on the central E coast of Isla Gobernadora [0734/8112 (USBGN)] (JOG). See also: Gobernadora, Isla.

LA PUNTA DE GARACHINÉ; see Garachiné, Punta

LA PUNTA DE PILÓN; see Pilón, La Punta de

LA PUNTITA; Panamá, Archipiélago de las Perlas
0821/7851 (USBGN)
Sea level, town on Isla del Rey [0822/7855 (USBGN)] located 2 km S of La Ensenada [0822/7851 (USBGN)] and 1.5 km W of Punta Gorda [0821/7850 (USBGN)] on a river mouth (JOG). Wetmore, GL = Isla del Rey, beach, inland along a stream leading into a small mangrove swamp and into forest with trees of moderate height, spiny palm, *Agave* and creepers, 1960 Jan 24 (USNM, as "Manzanillo"), observations

1960 Jan 25 (Wetmore et al., 1984:71, as "Manzanillo," and incorrectly as "June 25").

LARA, RÍO; Darién

0837/7809 (USBGN)

Originates E of Santa Fé [0839/7809 (USBGN)] and flows SW crossing the Carretera Interamericana [0832/8242 (USBGN)] and entering the Río Sabanas [0832/7808 (USBGN)] estuary 10 mi (16 km) N of La Palma [0825/7809 (USBGN)](MGRP; MHA; SMN2B). Festa, "*foreste presso*" (forest near), according to his travel account (Festa, 1909), he set up camp on the Río Lara a short distance above the river's mouth, 1895 Aug 28-Sep 4 (Salvadori and Festa, 1899; Hellmayr, 1936:314; Parkes, 1969).

LA REPRESA; Bocas del Toro, Archipiélago de Bocas del Toro

ca. 0922/8215

Sea level, a small reservoir on Isla Colón [0924/8217 (USBGN)] ca. 0.5 km N of Big Creek [0922/8215 (USBGN)]. Not named on our maps, but E761A shows a short road extending N from town to a "Bomba de Agua" at the location of the reservoir. SBE, GL = Isla Colón, Big Creek, 1989 Jan 28 (USNM, as "La Repressa," and as "La Represa").

LA REPRESA; Chiriquí

ca. 0815/8252

ca. 50 m, town 3 km SW of Puerto Armuelles [0817/8252 (USBGN)] on the Río Corutú [0816/8254 (USBGN)] 1 km above its mouth (JOG). Described by Wetmore as a narrow reservoir on a small stream ca. 2 mi (3 km) from Puerto Armuelles [0817/8252 (USBGN)], presumably near the town La Represa, Wetmore, GL = Puerto Armuelles, on the water, pastures and old fields along the trail from Puerto Armuelles to Represa, 1966 Feb 6, 8, Mar 4 (USNM, as "Represa").

LA REPRESA, QUEBRADA DE; San Blas

ca. 0840/7725

Described by Wetmore as a stream near Puerto Obaldía [0840/7725 (USBGN)], so named because there was an old concrete dam on the stream that had been built in the 1930s, Wetmore, GL = Puerto Obaldía, collected along the stream and near the reservoir, 1963 Mar 16-18 (USNM); Handley, GL = Puerto Obaldía, 1963 Apr 3 (USNM, as "Quebrada Repressa").

LA REPRESSA; see La Represa

LARGA, CERRO; see Cerro Largo

LARGA, PLAYA; see Long Beach

LARGO, CERRO; see Cerro Largo

LAS BÓVEDAS; Panamá (E, Pacific)

ca. 0857/7932

Sea level, located on the southeasternmost tip of Casco Antiguo in San Felipe [0857/7932 (USBGN)] (PTC). According to Angehr (it litt., Oct 2003) the name refers to the "old arched vaults under the sea side of the old city wall." Quezada,

GL = San Felipe, 1986 Mar 19 (MVUP).

LAS BRUJAS, ISLAS; see Trinidad, Río

LAS CAÑITAS; Panamá (E, Pacific)

ca. 0917/7851

Not shown on our maps but located in E Panamá province in Corregimiento Cañita [0917/7851 (USBGN)] near the town Cañita [0914/7953 (USBGN)]. According to MGRP (as "Cañita o Tamogantí"), Cañita is on the Río Cañita [0912/7855 (USBGN)] 4 km above its mouth and 9 km E of El Llano [0913/7859 (USBGN)]. Cañita and Boca de Cañita [0912/7855 (USBGN)] (q. v.) are two distinct localities. Grimaldo, 1983 Jul 31 (MVUP).

LAS CASCADAS; Panamá (Canal Zone) (W, Atlantic)

ca. 0905/7941

200 feet (60 m), former station on the original Panama Railroad [ca. 0922/7954] (PR, as "Las Cascada") that appears on map E761A as a town 4 km SSE of Gamboa [0907/7942 (USBGN)] on the Panama Canal [0920/7955 (USBGN)] section named Las Cascadas Reach [0905/7941 (USBGN)]. Hallinan, in the jungle, at the edge of a small clearing, 1908 May 10, 17, 22, 24 (AMNH; Hallinan, 1924); Maxon, 2-3 mi (3-5 km) E, 1911 Feb 19 (USNM).

LAS CRUCES; Panamá (Canal Zone) (E, Atlantic)

ca. 0907/7941

20 m, originally a town 2 km NE of Gamboa [0907/7942 (USBGN)] on the E bank of the Río Chagres [0907/7942 (USBGN)] (IP, as "Crucés"; MHA, as "Cruces"; MIP, as "Cruces"). Described by Wetmore in 1959 as ruins partly submerged by Lago Gatún (Gatun Lake) [0912/7955 (USBGN)] at the W terminus of Las Cruces Trail [ca. 0904/7936] (USNM). The ruins appear on E866 just N of the Río Casaya [0907/7941 (USBGN)]. See also: Las Cruces Trail.

LAS CRUCES, SENDERO; see Las Cruces Trail

LAS CRUCES TRAIL; Panamá (Canal Zone) (E, Atlantic, Pacific)

ca. 0904/7936

Old road through the former Canal Zone, extending roughly E-W from Panama City [0858/7932 (USBGN)] to the former Las Cruces [ca. 0907/7941] on the Río Chagres [0907/7942 (USBGN)]. The trail passes through Curundu [0859/7933 (USBGN)], Chiva Chiva [0902/7935 (USBGN)], and Madden Road Timber Preserve [0906/7937 (USBGN)] (E661, as "Panama Cruces Trail"; MHA, as an unnamed pack road). The route was in use from 1530-1855 for both commercial and passenger traffic that arrived at Las Cruces from Fuerte San Lorenzo [0919/8000 (USBGN)] via boat on the continuous Río Chagres (see Río Chagres [0919/8000 (USBGN)]) (HMCZ, as "Cruces Trail"). According to Wetmore, in 1959 the trail section near the Las Cruces ruins was 6-12 feet (2-4 m) wide with stretches that were paved with stones. The section of the trail that extends through Parque Nacional Soberanía [0909/7941 (Angehr, 2003:239)] is known as Sendero Las Cruces. Collector ? [Hallinan ?], near Corozal, 1915 Dec 24 (AMNH); Hallinan, GL = Corozal, near Corozal, trees, thickets, low shrubbery, thick shrubbery, swamp, 1915 Sep 5, 19,

Dec 24 (AMNH; CU; Hallinan, 1924, as "Las Cruces"); Benson, at 100 feet (30 m), 1931 Aug 8, and as a GL for Camino Madden (Madden Road), 1931 July 26, Aug 23 (USNM, as "Cruces Trail"); Miller, 1937 Feb 10, 13, 16, 18-20, 26, Mar 2 (USNM, as "Los Cruces Trail"); Meyer, GL =Madden Road Timber Preserve, 1941 Sep 1 (USNM, as "Cruces Trail"); Wetmore, GL = Gamboa, 1.5 mi (2.5 km) inland along a ridge above flooded valleys, through low, fairly open gallery forest, 1959 Jan 29 (USNM; Brumfield et al., 2001); Agro, Conyne, and Lovette, 1996 Feb 12-13 (MVUP, as "Camino de Cruces").

LAS CUMBRES; Panamá (E, Pacific)
0905/7932 (USBGN)
100 m, town 15 km N of Panama City [0858/7932 (USBGN)] on the Carretera Boyd-Roosevelt [0917/7943 (USBGN) just S of the continental divide (E661; JOG; RM1). Wetmore, observations, 1959 Apr 7 (Wetmore, 1965:177); Fairchild, 1959 Nov 7, 1962 Apr 1-2, 1964 Jun 17, Jul 3, 1967 Oct 16, 1968 Oct 31 (GML); GMLP, 1960 Jun 3 (GML); Fairchild, 1962 Mar 26 (GML); Correa, 1965 Aug 18 (GML); Harper, 1983 Aug 18 (MVUP); Barria and Sánchez, 1985 Apr 10 (MVUP); Sánchez, 1986 Mar 20 (MVUP); Rodríguez, 1986 Apr 9 (MVUP).

LA SEDA; Panamá (W, Pacific)
ca. 0857/7951
Not shown on our maps but located in W Panamá province in Distrito de La Chorrera [0857/7951 (USBGN)] near La Chorrera [0853/7947 (USBGN)] (q. v.). Castillo, 1994 Aug 13 (MVUP).

LAS GUACAS; see Las Huacas

LAS HUACAS; Panamá (E, Atlantic)
0911/7939 (USBGN)
35 m, town 2.5 km NNE of Juan Mina [0910/7939 (USBGN)] on the E bank of the Río Chagres [0907/7942 (USBGN)], but cut off from the main river by islands and side channels that form at the mouth of Quebrada Moja Polla [0911/7939 (USBGN)] (E762; E866, as "Las Guacas"). Hallinan, cultivated areas, forests, in the valley of the Río Chilibre [0910/7939 (USBGN)], 1915 Nov 28 (AMNH; Hallinan, 1924, as "Las Guacas"). The section of the Río Chagres near Las Huacas is presumably the "Laguna de Guacas" mentioned in Wetmore's journal. See also: Chagres, Río.

LAS LAGUNAS; Chiriquí
0846/8241 (USBGN)
1,300 m, shown on maps as two lakes located 4 km SW of Volcán [0846/8238 (USBGN)], each covering an area of ca. 1 km^2 (100 ha). Delgado (1983) indicates that three lakes comprise the Lagunas del Volcán, and in 1960 Wetmore (USNM) mentions a third lake that is seasonal. The lakes drain via a small stream that flows less than 1 km NE into Quebrada Barriles [0848/8241 (USBGN)] (E562, as "Laguna del Volcán"; E762; JOG, as "Lagunas del Volcán," and as "Son Dos Lagunas"). According to Wetmore, the lakes are located on the S edge of Llanos del Volcán [ca. 0847/8238], the region was originally forested, but by 1954 the forest had been

replaced with *rastrojo*. Laguna de Agua is the smaller lake and 300 yards (270 m) to the S of it lies the larger lake, Laguna Grande (USNM). Wetmore also refers to the lakes as "Volcan Lakes" and as "the lakes." Ridgely and Gwynne (1989) also use the term "Volcán Lakes." Wetmore, GL = Volcán, at 4,200-4,500 feet (1,300-1,350 m), collected on both lakes, each only one-third full of water in the dry season, the smaller lake was slightly marshy on one end and boggy on the other, 1954 Mar 2, 6, 9, 27, extensive logging since 1954, canoe on the larger lake and on the road along the smaller lake, dead trees along the water's edge, 1955 Feb 9, Mar 14, GL = Volcán, at 4,200 feet (1,300 m), timber remained around the two lakes, but elsewhere the land had been cleared, berry-filled trees, logging road to the N end, rough trail to a circular basin, dry with tall grass that was evidently a third lake in the rainy season, 1960 Feb 12, 15, 19, 23, 27, Mar 20, GL = Volcán, forest cleared to the lake shore, 1965 Mar 22 (MVUP, as "Laguna Grande"; USNM, as "Laguna," as "Lagunas," as "Lagunas del Volcán," as "Laguna Grande," and as "Las Lagunas"; Olson, 1981c, as "El Volcán, Laguna Grande"); Holmes, GL = Volcán, 1964 May (GML, as "Lagunas"); DaCosta, Guitton-Mayerma, Klicka, Lelevier, Miller, and Spellman, at 1,340 m, ca. 2 km SW of Volcan, 2006 Feb 27-28 (MBM, as "Lagunas de Volcan"; STRI, as "Lagunas de Volcan"). (Eisenmann, 1957, as "Volcan Lakes"). See also: La Lagunita.

LAS LAJAS; Chiriquí
0815/8152 (USBGN)
ca. 50 m, city in E Chiriquí province, 22 km W of Tolé [0814/8141 (USBGN)] and 3 km S of the Carretera Interamericana [0832/8242 (USBGN)] (JOG). The name Las Lajas may also refer to the beach 9 km to the S, which is properly known as Playa del Jobo [0811/8154 (USBGN)] (q. v.). Described by Wetmore as a straggling settlement divided into 2 sections: Las Lajas Adentro and Las Lajas Afuero, the latter being the more formal town. Wetmore, the secondary localities attributed to Las Lajas were located in the region extending toward the coast 5 km SW and 9 km SSW of town, GL for: Santa Cruz, and Playa del Jobo, 1956 Feb 14-15, 20, 22, 24-26 (USNM; Eisenmann, 1957; Wetmore, 1970; Olson, 1983); Ridgely, forest adjacent to mangroves S of, 1976 (Ridgely and Gwynne, 1898:77); Engleman and Guarnaccia, 1984 Aug 19 (Ridgely and Gwynne, 1989:423). (Angehr, 2003:311-312, as "Manglares de San Lorenzo y Las Lajas").

LAS LAJAS; Panamá (W, Atlantic ?)
0855/7956 (USBGN)
Not shown on our maps but USBGN indicates a town in NW Panamá province near the continental divide ca. 10 km NE of Santa Rita [0852/7953 (USBGN)]. Brenes and Gómez, 1985 Apr 16 (MVUP).

LAS LAJAS, MANGLARES DE; see Las Lajas

LAS LAJAS, PLAYA; Jobo, Playa del

LAS LAJAS, RÍO; see Lajas, Río

LAS LOMAS; Chiriquí
0826/8224 (USBGN)
ca. 50 m, town 4 km E of David [0826/8226 (USBGN)] (E562). Rodrigo, 1987

(MVUP).

LAS MACANAS, CIÉNAGA; see La Macana, Ciénaga de

LAS MACANAS, CIÉNAGA DE; see La Macana, Ciénaga de

LAS MARGARITAS; Panamá (E, Pacific)
0911/7905 (USBGN)
ca. 50 m, town on the Carretera Interamericana [0832/8242 (USBGN)] 2 km NNE of Chepo [0910/7906 (USBGN)]. The town is 1 km W of the Río Mamoní [0908/7904 (USBGN)] and 7 km SSE of the mouth of the Río Corpus Christi [0915/7910 (USBGN)] (MGRP). Guerrero, Hernández, and Rovi, 1985 Mar 16 (MVUP); Guerrero, Mendieta, and Rodríguez, GL for: Río Corpus Christi, and Río Mamoní, 1985 Mar 16 (MVUP); Guevara, 1985 Apr 16 (MVUP).

LAS MENDIETA; Herrera
ca. 0750/8045 ?
Not located on our maps but the specimen label indicates that the locality is within Herrera province. Rodríguez, 1977 Jun 15 (MCN).

LAS MERCEDES; see El Dorado

LAS MINAS; Coclé (Atlantic)
0856/8009 (GA)
According to Angehr (in litt., Aug 2005), this locality is one of the Western Watershed sites. "Garcés'specimens are...from our locality La Mina del Río Indio, GPS, 08°56'17" N 80°08'57" W, elevation 20 m. This is a small town not shown on maps, but is in the area marked as La Sardina on 4142-I (La Boca del Uracillo) [E762], a few 100 m W of the Rio Indio. It is about 3 km SE of Wetmore's El Uracillo [Boca de Uracillo [0858/8011 (USBGN)]] locality, in Coclé close to the point where the provinces of Coclé, Colón, and Panamá meet. Garcés specimens were taken from our net lines just W of these coordinates." Garcés, Jiménez, López, Ramos, and Rodríguez, 2001 May 3-4 (MVUP; Louis Berger Group et al., 2003).

LAS MINAS, BAHÍA; Colón (E)
0924/7950 (USBGN)
Inlet located on the N coast of Colón province 10 km E of Colón [0922/7954 (USBGN)]. The inlet is less than 1 km wide in areas, but extends inland for 6 km (E761A). Goldman, landed at a small ranch and walked back (W) to Majagual [ca. 0921/7953]. Because specimens are labeled "near Santa Rita" [0919/7947 (USBGN)], Goldman probably landed on the S coast of the bay. His journal entry "country heavily forested to eastward" probably refers to the S coast of the bay E of Majagual, 1911 Mar 13 (USNM, as "Bay of Las Minas"). See also: Marajal; and Majagual.

LAS MINAS BAY; see Las Minas, Bahía

LAS MINERAS; Coclé (Pacific)
0816/8035 (USBGN)
50 m, town 5 km WNW of Aguadulce [0815/8033 (USBGN)] (JOG). Tejera, 1976 (MVUP, as "Mineras").

LAS NUBES; Chiriquí

0853/8236 (USBGN)

2,000 m, town 4.5 km NW of Cerro Punta [0852/8235 (USBGN)] near the headwaters of the Río Chiriquí Viejo [0820/8241 (USBGN)] (E762). Wetmore, GL = Cerro Punta, at 5,800-6,000 feet (1,750-1,850 m), from La Garita [0852/8235 (USBGN)] walked ca. 3 mi (5 km) through an open valley on the N fork the Río Chiriquí Viejo to the W side of Cerro Punta peak [0853/8235 (USBGN)] where the valley was mostly in pasture with some *rastrojo*, scattered timber, and cultivation, near the foot of Cerro Punta, 1955 Mar 11 (USNM); DaCosta, Guitton-Mayerma, Klicka, Lelevier, Miller, and Spellman, at 2,305 m, at 08°53'43"/82°36'59", ca. 5 km NNW of Cerro Punta, Las Nubes ANAM [Autoridad Nacional del Ambiente] Site, 2006 Feb 25-27 (MBM; STRI).

LAS NUBES ANAM SITE; see Las Nubes

LAS PALMAS; Veraguas (Pacific)

0808/8127 (USBGN)

200 m, city in W Veraguas province 24 km NW of Soná [0801/8119 (USBGN)] and 27 km SE of Tolé [0814/8141 (USBGN)] (JOG). Used to locate: Honda, Bahía; Querqué; and Vidal, Río.

LAS PALMITAS; Los Santos

0720/8030 (USBGN)

ca. 50 m, town on the Río Guánico [0717/8025 (USBGN)] 12 km NW of its mouth, 6 km NW of Ave María [0719/8027 (USBGN)], and 5 km SE of Jobero [0722/8033 (USBGN)] (MGRP, as "La Palmita o El Rompío"). According to Myers (1969:29), this town is known locally as Guánico Arriba, and although the town is listed in USBGN as a distinct locality, its coordinates are identical with those for Las Palmitas. Wetmore, GL for Río Guánico, 1962 Jan 21-28 (USNM; Wetmore, 1965, as "Guanico Arriba"; Wetmore, 1968, as "Guánico Arriba"; Wetmore, 1970).

LAS PANAMAS, PLAYA; Panamá (W, Pacific)

ca. 0827/7959

Not shown on our maps, but the coordinates correspond to a region on JOG where the town Playa Grande [0827/7959 (USBGN)] is located. Playa Grande is located on the coast 3 km SW of Río Mar [0827/7958 (USBGN)] and 21 km NE of Río Hato [0823/8010 (USBGN)]. Several maps also indicate a beach named Playa Grande at this same location. Bacon and Miller, at 08°26'41"/79°59'05", 2002 Sep 15 (STRI; UAM).

LAS PEÑITAS; Coclé (Atlantic)

ca. 0841/8036

Described by Wetmore as a small settlement with a few houses 1 km SE of Tigre [ca. 0842/8036] near the head of the Río Guabal [ca. 0846/8036]. JOG, which does not show the settlement, indicates an elevation of ca. 1,100 m for this region; but according to Angehr (in litt., Aug 2005), map 4042-II (Coclesito) [E762] indicates that Wetmore's elevation of ca. 650 m is correct. Wetmore, GL = Río Guabal, head of the Río Guabal, at 650 m, high forest, 1962 Mar 3 (USNM).

LAS PEÑITAS; Darién

ca. 0728/7759

ca. 100-300 m, described by Wetmore as an area on the Río Jaqué [0731/7810 (USBGN)] ca. 1 mi (1.5 km) downstream from its junction with the Río Imamadó [ca. 0727/7758] that extends to the extensive base of Loma Sierpe (not located), Wetmore and Perrygo, GL = Río Jaqué, collected in gallery forest with gravelly soil, with open undergrowth and occasional thorny creepers on the steep slopes, on gravel beaches, in trees adjacent to a plantain plantation, 1947 Mar 28, Apr 7, 9, 11-14, 17 (USNM; WFVZ; Wetmore and Galindo, 1972, as "Los Peñitas"; Olson, 1981b, as "Río Jaqué, Las Peñitas"). Not shown on our maps.

LAS PEÑITAS, LOMA; Darién

ca. 0729/7758

ca. 1,000 feet (300 m), described by Wetmore as a steeply sloped hill in Las Peñitas [ca. 0728/7759], Wetmore and Perrygo, GL = Río Jaqué, 1947 Apr 12 (USNM). Not shown on our maps.

LAS PERLAS, ARCHIPIÉLAGO DE; see Perlas, Archipiélago de las

LAS PIRAGUALES; see El Cortezo

LAS SALINAS; see Aguadulce

LAS TABLAS; Los Santos

0746/8017 (USBGN)

ca. 25 m, the capital of Los Santos province, located in the NE region of the province 9 km from the W shore of the Gulf of Panama [0800/7930 (USBGN)] and 27 km SSE of Chitré [0758/8026 (USBGN)] (JOG). See also: Tonosí.

LAS TRANCAS; see La Tranca

LA TRANCA; Chiriquí

0830/8225 (USBGN)

120 m, town 6 km S of Dolega [0834/8225 (USBGN)] and 7.5 km N of David [0826/8226 (USBGN)] on a tributary of the Río David [0822/8224 (USBGN)] (E762). Watson, 1905 Nov (FMNH, as "Las Trancas").

LA TREINTICINCO; see Treinticinco, Quebrada

LA TRINIDAD; Herrera

0755/8043 (USBGN)

ca. 75 m, town 10 km W of Pesé [0754/8037 (USBGN)] (JOG). The town is S of Pesé on MHA. See also: Chitrabé, Quebrada; and Tejel, Quebrada.

LA TULITA; Darién

ca. 0732/7806

Described by Wetmore as an old channel, 1 river mi (1.5 km) above El Brazo [ca. 0732/7808] and ca. 4.5-5.0 river mi (7-8 km) upstream from the mouth of the Río Jaqué [0731/7810 (USBGN)], Wetmore and Perrygo, GL = Jaqué [0731/7810 (USBGN)], inland on slightly higher ground, knolls still covered with timber, 1946 Apr 6 (USNM). Not shown on our maps.

LA ULLAMA; Colón (Canal Zone) (W)
0908/8206 (USBGN)
USBGN indicates a town on the W coast of Bahía Trinidad [0905/8000 (USBGN)]. The town is not shown on our maps but E761A does show Cerro Ullama [0907/8002 (USBGN)], which has the same coordinates, on the coast 4 km NNE of Cuipo [0905/8003 (USBGN)] and 8 km SW of Escobal [0909/7958 (USBGN)]. Chapin, Rogers and Mathews, GL = Escobal, Cuipo, Lago Gatún, Río Ciri, between Cuipo and Escobal, 1923 Aug 15-16, 18, 20-23 (FMNH, as "La Uyama").

LAUREL, PUNTA; Bocas del Toro, Archipiélago de Bocas del Toro
ca. 0909/8200
Sea level, community (E762) on the SE coast of Isla Popa, not to be confused with the point of the same name at the SE tip of the island [0908/8206 USBGN]. Billo, GL = Isla Popa, 2007 May 7, at distances of 2, 3, 4, 5, 6, and 7 km N (USNM).

LA UYAMA; see La Ullama

LA VENTA BEACH; Los Santos
ca. 0740/8010
Described by Lincoln (1936:146) as a beach on the Pacific coast 87 mi (140 km) SW of Panama City [0858/7932 (USBGN)], which places the beach in NE Los Santos province on the central E coast of Península de Azuero [0740/8035 (USBGN)]. Collector ?, 1934 Oct 1 (Lincoln, 1936; Eisenmann, 1951). Not shown on our maps.

LA VERBENA; Colón ? (Canal Zone) (E ? W ?, Atlantic)
ca. 0917/7953 ?
Not located on our maps. Imhof indicated that it was located in the Canal Zone on the Atlantic drainage, therefore we assumed that it was in Colón province although the Atlantic drainage extends into N Panamá province. He made observations at Fort William D. Davis Army Reservation [0917/7953 (USBGN)] on Oct 20-28, therefore it may be near the former base. Imhof, observations, 1942 Oct 20 (Imhof, 1950).

LA VILLA, RÍO; Herrera/Los Santos
0759/8023 (USBGN)
Originates in S central Herrera province, flows NE along the Herrera/Los Santos provincial boundary and empties into the Bahía de Parita [0808/8024 (USBGN)] 5 km NE of Chitré [0758/8026 (USBGN)] (JOG; MGRP). USBGN treats Río De La Villa as an older name for Río La Villa, although that source lacks an entry for the latter name. Tejera, Viané, and Adames, collections were made ca. 6-7 km S of the river's mouth near the town Santa Ana Abajo [0757/8022 (USBGN)], "*entre las albinas de la Villa de Los Santos y el mar*" (between the salt flats of the town of Los Santos and the sea), 1979 Jan 5, 1980 Aug 2 (AMNH; MVUP, as "Salinas de Ernesta," and as "Salinas Villalaz"; Tejera and Rincón, 1982).

LA VILLA, RÍO DE; see La Villa, Río

LA VIVIENDA, ISLA; see Vivienda, Isla La

LA VIVIENDA ISLAND; see Vivienda, Isla La

LA YAGUADA, LAGUNA DE; see La Yeguada, Laguna de

LA YEGUADA; see La Yeguada, Laguna de

LA YEGUADA, LAGUNA DE; Veraguas (Pacific)
0828/8052 (USBGN)
700 m, a lake with a surface area of 1.12 km² (112 ha) (RDP2, as "Laguna La Yeguada") located in NE Veraguas province 42 km NNE of Santiago [0806/8059 (USBGN)] and 1 km N of the town La Yeguada [0827/8052 (USBGN)] at the headwaters of the Río San Juan [0812/8049 (USBGN)] (JOG, as "Laguna La Yeguada"; MGRP, as "Laguna La Yeguada"). According to Delgado (1983:114), the region near the lake has been extensively reforested and is the site of a hydroelectric project. According to correspondence between Mr. and Mrs. Eustorgio Méndez and Wetmore, this lake is at the base of a hill known as Castillo and was formerly called Laguna del Castillo (USNM). Arcé, 18?? (Salvin, 1870b, as "Laguna del Castillo"); Dutary, 1969 Jun 2 (GML, as "La Yeguada"); Delgado, observations, 1981 Aug (Ridgely and Gwynne, 1989:423, as "Laguna de la Yaguada"). See also: Veraguas, Provincia de.

LA ZUMBADORA; Panamá (E, Pacific)
ca. 0915/7920
Wetmore placed this on our maps 8 mi (13 km) NNE of Cerro Azul [0910/7924 (USBGN)] near the headwaters of the Río Pacora [0901/7917 (USBGN)], although the designation (town, camp, etc.) is unclear. Wetmore (from GML), GL = Cerro Azul, at 2,000 feet (600 m), 1961 Jan 31 (USNM); Schneider, 1964 Jan 25 (GML).

LEALES; Los Santos
0745/8031 (USBGN)
ca. 50 m, town in W Los Santos province 24 km SSW of Los Santos [0756/8025 (USBGN)] (JOG). Casas, GL = Macaracas, 1990 Jan 23 (MVUP, as "Los Leales").

LEALES, LOS; see Leales

LEANDRO TRAIL; see Bajo Mono

LE GATUN; see Gatún

LEONA, PLAYA; see Playa Leona

LEONES, ISLA; Veraguas (Pacific)
0744/8107 (USBGN)
194 m, island with a land area of ca. 8 km² (800 ha) centrally located in Golfo de Montijo [0740/8107 (USBGN)] (JOG). Batty, specimens at AMNH are dated 1902 Jan 7, but Batty's data were completely fabricated and his specimens labeled "Leones Island" were actually collected in 1901 near his mainland bases at Boquete [0847/8226 (USBGN)] and Boquerón [0830/8234 (USBGN)] (AMNH, as "Leones Island"; Olson, 2008).

LEONES ISLAND; see Leones, Isla

LÉRIDA; see Finca Lérida

LÉRIDA, FILA; Chiriquí
ca. 0847/8230

Described by Blake (1958:502 (map)) as a narrow ridge that extends from the S base of Cerro Copete [ca. 0848/8231] eastward along the S banks of the Quebrada El Velo [0849/8228 (USBGN)]. Mönniche, 1932-1933, 1936-1937 (FMNH: MCN). See also: Finca Lérida.

LÉRIDA, FINCA; see Finca Lérida

LÉRIDA RIVER; see Finca Lérida

LIMBO; see Limbo Plot

LIMBO, RÍO; see Limbo Plot

LIMBO HUNT CAMP; see Limbo Plot

LIMBO HUNT CLUB; see Limbo Plot

LIMBO PLOT; Colón (Canal Zone) (E)
0909/7944 (Karr, 1971a)
Described by Karr (1971a) as a study area in a "structurally mature, humid forest area about 8 km north of Gamboa in the middle Chagres River basin adjacent to Gatun Lake on the Caribbean Slope of the Canal Zone...adjacent to the cabin of the Limbo Hunt Club on the Navy Pipeline Reservation (commonly called 'The Pipeline Road')." W. Robinson et al. (2000b) described the site as a 100-ha plot on the Pipeline Road [ca. 0915/7948] within Parque Nacional Soberanía [0909/7941 (Angehr, 2003:239)] that is covered by 60-120 year old forest with an average canopy height of 30-40 m and located "at least 3.5 km from the nearest non-forest edge." The coordinates correspond to the area on JOG ca. 6 km NW of Gamboa [0907/7942 (USBGN)] along the Río Frijoles [0909/7946 (USBGN)]. Kilham, observations 1965 Feb, 1970 Nov-1971 Feb (Kilham, 1972, as "Limbo Hunt Club"); Karr, mist nets, 1968 Jul-1982 Mar (Karr, 1971a, as "Limbo Hunt Club"; Karr, 1971b, as "Limbo Hunt Club"; Karr, 1971c, as "Navy Pipeline Reservation"; Karr, 1982a, as "Limbo Hunt Club"; Karr, 1982b, as "Limbo Hunt Club"; Karr, 1990; Karr and Freemark, 1983); Schemske and Brokaw, mist nets, (Schemske and Brokaw, 1981, as "Limbo Hunt Club"); Bertin and Wilzbach, observations, 1977 Jun 27-29, Jul 2 (Bertin and Wilzbach, 1979, as "Limbo Hunt Club"); Collector ?, GL = Pipeline Road, 1996 Nov 22 (LSUMZ); Bard et al., observations and recordings, 1997 Jun 18-Aug 20 (Bard et al., 2002). Wetmore et al. (1984:235, as "Limbo Hunt Club") indicate that in 1964 Chapman and Tyson collected in Limbo Plot. (Oniki, 1972, as "Limbo Hunt Camp"; Loiselle and Hoppes, 1983, as "Limbo Hunt Club"; W. Robinson et al., 1998; W. Robinson et al., 2000a; T. Robinson et al., 2000, as "Río Limbo"). See also: Gamboa.

LIMÓN; Los Santos
0730/8002 (USBGN)
50 m, town 3 km SSW of Pedasí [0732/8002 (USBGN)] (JOG). See also: Escotá, Quebrada; and Los Higos.

LIMÓN, BAHÍA; Colón (Canal Zone) (E, W, Atlantic)
0920/7956 (USBGN)
Bay surrounding the Atlantic entrance of the Panama Canal [0920/7955 (USBGN)].

The bay extends ca. 5 km E-W between Fort Sherman Army Reservation [0920/7958 (USBGN)] and Colón [0922/7954 (USBGN)] and breakwaters separate the opening of the bay on the N from the Caribbean Sea [1500/7500 (USBGN)] (E761A, as "Limon Bay"). The bay was also known as Navy Bay (Otis, 1859).

LIMÓN, EL; see El Limón

LIMÓN, PUERTO; see Puerto Limón

LIMÓN, PUNTA; Bocas del Toro, Archipiélago de Bocas del Toro
0909/8204 (USBGN)
Sea level, point located on the westernmost coast of Cayo Agua [0909/8202] (JOG). SBE, GL = Cayo Agua, near Punta Limón, 1990 Apr 5-13 (MVUP; USNM); Billo, GL = Cayo Agua, 2007 May 4, "1/2 km NE" (USNM).

LIMÓN, QUEBRADA; Herrera
ca. 0755/8030
Described by Wetmore as a creek very close to El Barrero [0756/8031 (USBGN)] (USNM), presumably a tributary of the Río La Villa [0759/8023 (USBGN)]. Wetmore and Perrygo, GL = El Barrero, the river was dry with occasional small pools and bordered by cattle trails, trees, and dense, thorny scrub, 1948 Feb 28 (USNM). Not shown on our maps.

LIMÓN, QUEBRADA; see Limón, Río

LIMÓN, RÍO; Colón (E)
0913/7939 (USBGN)
Originates NE of Boca de Gatuncillo [0912/7939 (USBGN)] near the W shore of Lago Alajuela (Madden Lake) [0915/7935 (USBGN)] and flows SSW entering the Río Chagres [0907/7942] 200 m N of Boca de Gatuncillo (E661; E866). Wetmore, GL = Juan Mina, 1961 Jan 12 (USNM, as "Quebrada Limón").

LIMÓN, RÍO; Darién
ca. 0747/7741
Described by Goldman (1920:11-12) as a river on the E slopes of Cerro Pirre [0749/7744 (GA)]. According to the maps presented in Woakes (1900) and in Anthony (1916), the stream originates a short distance N of Cana [ca. 0745/7742] on the SE slopes of Cerro Pirre and is one of several small streams just N of Cana that flows E forming the headwaters of the Río Tuira [0821/7803 (USBGN)]. Goldman, GL for Cerro Pirre, 1912 Mar 5-6, Apr 12-14, 16-20, 22-27, 29-May 3, 5-6, Jun 7 (USNM). See also: Pirre, Cerro.

LIMÓN, RÍO; Veraguas (Pacific)
0746/8134 (USBGN)
Not shown on our maps but USBGN indicates a river that discharges into Canal de Bahía Honda [0742/8136 (USBGN)] ca. 2 km W of Bahía Honda [0744/8132 (USBGN)]. Cambra, *cerca de* (near the) Río Limón, 2001 Aug 8-10 (MVUP).

LIMÓN, ZANJA; Panamá (E, Pacific)
ca. 0910/7905
Described by Wetmore as a small tributary of the Río Mamoní [0908/7904

(USBGN)] with a rocky bed in tall, gallery forest (USNM). The stream presumably originates NW of Chepo [0910/7906 (USBGN)] and flows SE. Wetmore and Perrygo, GL = Chepo, collected upstream, NW of Chepo, high gallery forest, dry streambed, trail to low hills in forest, 1949 Apr 9, 26 (USNM). Not shown on our maps.

LIMON BAY; see Limón, Bahía

LIMONES; Chiriquí

0806/8252 (USBGN)

Sea level, coastal town located on Península Burica [0810/8255 (USBGN)] 19 km S of Puerto Armuelles [0817/8252 (USBGN)] (JOG). According to USBGN, Puerto Limones is an older name for Limones. Weir, GL = Península Burica, at 100-160 m, 2003 Feb 26-28 (CU, as "Puerto Limones"; LSUMZ, as "Puerto Limones").

LINE OF PANAMA RAILROAD; see Panama Railroad

LINE OF R. R. see Panama Railroad

LINE P. R. R.; see Panama Railroad

LION HILL; Colón (Canal Zone)

ca. 0913/7954

Former station on the original Panama Railroad [ca. 0922/7954] that was located 4 km SE of Gatún [0916/7955 (USBGN)] on the continuous Río Chagres (see Río Chagres [0919/8000 (USBGN)]) (PR). Isla Lion Hill [0914/7954 (USBGN)] remains above water in Lago Gatún (Gatun Lake) [0912/7955 (USBGN)] near the former station (E8660, as "Lion Hill Island"). According to Jewel (1913:422), Lion Hill was located in "the heart of Black Swamp country," and at the time of his writing, as Lago Gatún was being flooded, it was only accessible via canoe. According to Sclater and Salvin (1864:343), before the formation of Lago Gatún, "Lion Hill Station [was] situated in a district of the densest tropical forest; the ground [was] undulating, the hollows being swampy in the wet season, and even in the dry the tracks through the forest [were] confined to the low hills. The difficulties of collecting [were] greatly enhanced by the dense undergrowth of a species of *Agave*, the barbed leaves of which [caught] one at every step…I should say that birds were very numerous both in species and individuals. [We followed the tracks which led] into the forest, [until] stopped by an impenetrable swamp. All along these paths birds abounded, and we had no difficulty in securing in a couple of hours enough work to occupy the rest of the day." Rhoads, GL = Colón, 18?? (USNM, as "Leon Hill"); McLeannan, various spellings of his name appear in the literature, many of his specimens which are labeled "Lion Hill" are also labeled "near Aspinwall" [see Colón], according to Sclater and Salvin (1864:343), McLeannan was the stationmaster here and made extensive collections in this area, presumably many of his specimens that are simply labeled "Panama," "Isthmus of Panama," or "Line of Panama Railroad" were collected near Lion Hill, although the exact localities of these specimens are unclear as McLeannan worked on both sides of the divide [see Panama Railroad], mid 1850s-1860s (BMNH, as "Lion Hill Station"; MCZ, as "Line of Panama R. R."; USNM, as "Panama," and as "Lion Hill"; Lawrence, 1865a, as

"Lion Hill Station"; Lawrence, 1867f, as "Lion Hill Station"; Sclater and Salvin, 1876; Salvadori, 1891, as "Lion Hill Station"; Salvin and Godman, 1879-1904, vol. 1-3, as "Lion Hill," and as "Lion Hill Station"; Ogilvie-Grant, 1893; Ridgway, 1906, as "Lion Hill Station ?"; Ridgway, 1908, as "Lion Hill Station ?"; Ridgway, 1911a, as "Lion Hill Station"; Chubb, 1919; Griscom, 1932b; Howell, 1956; Wetmore, 1959b:6, as "Lion Hill Station"); Brown, 1900 Mar 2-30 (ANSP; MCZ; UCLA; and USNM, all as "Loma del Leon"; Salvin and Godman, 1897-1904, vol. 3:319, as "Lion Hill Station"; Bangs, 1900; Bangs, 1901a; Bangs, 1901c; Bangs, 1902a; Bangs, 1903b; Thayer and Bangs, 1905:146; Thayer and Bangs, 1906; Carriker, 1910:379; Todd, 1912; Todd,; Bangs and Penard, 1918:78; Bangs and Penard, 1919; Bangs and Penard, 1921:374-5, 380; and Darlington, 1931, all as "Loma del Leon"; Griscom, 1932a; Wetmore, 1963a, as "Loma del Leon (Lion Hill)"; Parkes, 1969 as "Loma del Leon"; Olson, 1981d, as "Lion Hill (Loma del León"); Goldman, 1911 Jan 13, 17-19, 22, Feb 6-7, Mar 9, Apr 4, 11, 14, 19, May 1, 1912 Feb 6 (USNM; Nelson, 1911b; Todd, 1913; Todd, 1929; Brumfield et al., 2001). (Berlepsch, 1911, as "Loma del Leon"; Chapman, 1917; Chapman, 1929b; Griscom, 1929a; Hellmayr, 1918; Zimmer, 1934:13, 20; Zimmer, 1939, as "Lion Hill ?"; Zimmer, 1941b:7, 10, 14; Zimmer, 1950:8, 50; Zimmer, 1955:5, 19; Griscom, 1933; Wetmore, 1953; Wetmore, 1959b:19; Wetmore, 1967; Wetmore, 1970; Olson, 1983).

LION HILL, ISLA; Colón (Canal Zone) (E)
0914/7954 (USBGN)
82 m, island with a land area of 50.6 ha (Wright, 1985, as "Lion Hill Island") located 6 km SE of Gatún [0916/7955 (USBGN)] in Lago Gatún (Gatun Lake) [0912/7955 (USBGN)] NW of Isla Barro Colorado [0909/7951 (USBGN)] (E762, as "Lion Hill Island"). Wright, censuses, 1976-1980 (Wright, 1979, as "Lion Hill Island"; Wright, 1985, as "Lion Hill Island"; Wright, 1988, as "Lion Hill Island"). See also: Lion Hill.

LION HILL STATION; see Lion Hill

LISA, QUEBRADA; Darién
ca. 0730/7808 ?
Described by Wetmore as a stream ca. 2 mi (3 km) SE of Jaqué [0731/7810 (USBGN)], unclear where the stream begins or empties and unclear whether Quebrada Lisa and Quebrada Alisa are 2 distinct localities, Wetmore and Perrygo, GL = Jaqué, hills covered with a dense growth of white cane and 8-10 foot (2-3 m) *Tilia*, along the stream in gallery forest with 75-100 feet (25-30 m) trees, 1946 Mar 15-16 (USNM). Not shown on our maps.

LITTLE ZAPATILLA KEY; see Zapatilla, Cayos

LLANO, EL; see El Llano

LLANO, EL; see Volcán, Llanos del

LLANO DE LA MESA; see La Mesa

LLANO DEL JARDINO; see Jardino, Llano del

LLANO DE PIEDRA; Los Santos
0740/8034 (USBGN)
100 m, town 30 km NNW of Tonosí [0724/8027 (USBGN)] on the main road that connects Tonosí and Chitré [0758/8026 (USBGN)] (JOG; MGRP). Eisenmann and Smith, N, observations, gallery woodland S of, wooded stream border between Tonosí and, 1964 July 24 (Wetmore et al., 1984:201; Ridgely and Gwynne, 1989:244).

LLANOS; see Volcán, Llanos del

LLANOS DE CACICÓN; see Cacicón, Llanos de

LLANOS DEL GALLO; see Gallo

LLANOS DEL VOLCÁN; see Volcán, Llanos del

LLORÓN, CORDILLERA DEL; Panamá (W, Pacific)
0843/7957 (USBGN)
Mountain range in W Panamá province on the E end of Cordillera Central [0838/8150 (USBGN)] ca. 15 km NE of Cerro Gaital [0837/8007 (USBGN)] extending ca. 7 km N-S from Cerro Trinidad [0844/7958 (USBGN)] to Cerro Campana [0841/7955 (USBGN)] (E762). See also: Capira.

LLORONA, SIERRA; Bocas del Toro
0910/8155 (USBGN)
213 m, a hill on the northernmost land of Península Valiente [0905/8151 (USBGN)] located near the coast ca. 1 km SSE of Punta Valiente [0911/8155 (USBGN)] (JOG). SBE, GL = Península Valiente, at the summit, forested, 1990 Apr 19 (USNM). See also: Nassau Hill.

LOCERÍA, LA; see La Locería

LOMA, LA; see La Loma

LOMA DE LA BANDERA; see La Bandera, Loma de

LOMA DE LÉON, see Lion Hill

LOMA DEL LEON; see Lion Hill

LOMA DELORES; see Delores, Loma

LOMA GONZÁLEZ; see González, Loma

LOMA PARTIDA, ISLA; Bocas del Toro, Archipiélago de Bocas del Toro
0909/8210 (USBGN)
100 m, island with a land area of ca. 3 km^2 (300 ha) located just SW of Isla Popa [0911/8207 (USBGN)] and separated from it by Canal Sumwood [0909/8209 (USBGN)] (E762, as "Isla Loma Partida (Split Hill Island)"; Olson, 1993a:101 (map)). SBE, observations, mangrove mudflats on the NE shore, along Canal Sumwood, 1991 Mar 5 (Olson, 1993a). See also: Sumwood, Canal.

LOMA PLAYITA DEL MEDIO; see Playita del Medio, Loma

LOMAS, LAS; see Las Lomas

LOMA SIERPE; see Las Peñitas

LONG BAY POINT; see Paunch

LONG BEACH; Bocas del Toro, Archipiélago de Bocas del Toro
0920/8208 (USBGN)
Sea level, sandy beach extending 4 km along the NE coast of Isla Bastimentos [0918/8208 (USBGN)]. Its N end is 7 km E of Bastimentos [0921/8212 (USBGN)], the island's largest town (E762). SBE, GL = Isla Bastimentos, high forest between Long Beach and Sal Creek [0918/8206 (USBGN)], 1987 Mar 24 (USNM). Billo, GL = Bastimentos, 2008 Apr 18, as "Playa Larga".

LOOKOUT; see Howard Air Force Base

LORENZO, FORT; see Fuerte San Lorenzo

LOS ANDES, CORDILLERA; Bocas del Toro/Chiriquí
0848/8220 (USBGN)
Mountains within the Cordillera Central [0838/8150 (USBGN)] that extend along the Bocas del Toro/Chiriquí provincial border N and NE of Boquete [0847/8226 (USBGN)] (E762, as "La Cordillera de Los Andes"). Collector ? [Mönniche ?], at 5,000 feet (1,500 m), 1933 Jun 26 (ANSP, as "Cordillera, Bocas del Toro"); Mönniche, according to Blake (1958:502), specimens were taken at or near the point where the Holcomb Trail crossed the continental divide, GL = Boquete, 1930, 1933, 1938-39, in Bocas del Toro province, at 6,500 feet, 1932, 1933 (FMNH, as "Cordilleras"; MCN, as "Cordillera"; Blake, 1958, as "Cordilleras").

LOS ANDES, LA CORDILLERA DE; see Los Andes, Cordillera

LOS ASIENTOS; Los Santos
0731/8008 (USBGN)
100 m, city in E Península de Azuero [0740/8035 (USBGN)] 17 km NW of Punta Mala [0728/8000 (USBGN)] (JOG). The town is incorrectly placed on the coast N of Pedasí [0732/8002 (USBGN)] on MHA and SMN2A. Wetmore, GL for Río Oria, 1957 Mar 18-19 (USNM; Wetmore, 1962; Wetmore, 1970).

LOS BAJOS DEL COBRE; see Congal

LOS CRUCES TRAIL; see Las Cruces Trail

LOS FRAILES; see Frailes del Sur

LOS FRAILES DEL SUR; see Frailes del Sur

LOS HATILLOS; Herrera
0756/8033 (USBGN)
ca. 50 m, town 2 km W of El Barrero [0756/8031 (USBGN)] (JOG). See also: Agua Caliente, Quebrada.

LOS HIGOS; Los Santos
0731/8006 (USBGN)
100 m, town 3 km E of Los Asientos [0731/8008 (USBGN)] and 9 km W of Pedasí [0732/8002 (USBGN)] (JOG). Described by Wetmore as a settlement with houses strung for 1 mi (1.5 km) along the road between Limón [0730/8002 (USBGN)] and Los Asientos, Wetmore, GL = Pedasí, on the road through town, 1957 Mar 18 (USNM).

LOS HIGUEROS, ISLA; Chiriquí

0818/8224 (USBGN)

Sea level, island with a land area of ca. 4.5 km^2 (450 ha) located 15 km S of David [0826/8226 (USBGN)] in Bahía de Charco Azul [0812/8245 (USBGN)] at the mouth of the Río Chiriquí [0820/8226 (USBGN)] (JOG; MGRP; MHA, as "I. Iguaros"). Batty, specimens at AMNH, at BMNH, and at FMNH are dated 1902 Feb 4-5, but Batty's data were completely fabricated and his specimens labeled "Iguaros Island" were actually collected in 1901 near his mainland bases at Boquete [0847/8226 (USBGN)] and Boquerón [0830/8234 (USBGN)] (AMNH, as "Iguaros Island"; BMNH, as "Iguaros Island"; FMNH, as "Iguaros Island"; Hartert and Goodson, 1917:418, as "Iguaros Island"; Wetmore, 1957:6, as "Iguana Island"; Olson, 2008).

LOS LEALES; see Leales

LOS MOGOTES, ISLA; Colón (E)

0932/7941 (USBGN)

80 feet (25 m), small island with a land area of ca. 0.1 km^2 (10 ha) located in Bahía de Buenaventura [0932/7941 (USBGN)] 3 km SW of Portobelo [0933/7939 (USBGN)] and 0.5 km off the N coast of Colón province (E761A, as "Buenaventura I."). Goldman, GL = Portobelo, near Portobelo, 1911 May 29-30 (USNM, as "Buenaventura Island"; Olson, 1981d, as "Buenaventura Island, near Portobelo").

LOS MOLEJONES, QUEBRADA; see Molejon

LOS MONOS, SENDERO; see Coiba, Isla de

LOS PÁJAROS, QUEBRADA DE; Herrera

ca. 0758/8035

Described by Wetmore as a tributary of the Río Parita [0802/8027 (USBGN)] with headwaters near Bayano [0755/8035 (USBGN)], farther W of El Barrero [0756/8031 (USBGN)] than Quebrada Agua Caliente [ca. 0759/8034], Wetmore and Perrygo, GL = El Barrero, N of Bayano where the water was standing in pools, bordered by stands of large trees, pastures, and thickets, 1948 Mar 22, SW of Bayano along its headwaters, 1948 Mar 23 (USNM). Not shown on our maps.

LOS PEÑITAS; see Las Peñitas

LOS PIRAGUALES; see El Cortezo

LOS PLANES; Chiriquí

ca. 0838/8213

1,000 m, town ca. 12 km NE of Gualaca [0832/8218 (USBGN)] on the Gualaca-Chiriquí Grande Road [ca. 0857/8207]. The town is 4 km SW of Cerro Hornito [ca. 0839/8211] (E562; E762, as "Los Planes de Hornito"). According to Angehr (in litt., Oct 2003), the town was built to house the workers who constructed the Fortuna Dam. Graves, Braun, M., and Parsons, GL for: Cerro Hornito, and Gualaca-Chiriquí Grande Road, 1990 May 4-7, 9-11 (USNM); SBE, GL for Gualaca-Chiriquí Grande Road, 1991 Mar 3, 11-12 (USNM; Olson, 1993b); MacDougall-Shackleton and Harbison, observations along Quebrada Aleman Trail, 1997 Feb (MacDougall-Shackleton and Harbison, 1998). See also: Fortuna.

LOS PLANES DE HORNITO; see Los Planes

LOS PLAYONES; see Sardinilla

LOS QUETZALES, SENDERO; see Cerro Punta-Boquete Trail

LOS SANTOS; Los Santos

0756/8025 (USBGN)

ca. 25 m, city in N Los Santos province on the Río La Villa [0759/8023 (USBGN)] 7 km SSW of its mouth and 3 km S of Chitré [0758/8026 (USBGN)] (JOG). Wetmore and Perrygo, GL for: La Honda, and Monagre, 1948 Mar 15-16, 20 (USNM; Wetmore, 1953; Wetmore, 1970; Parkes, 1976; Olson, 1980b, as "Los Santos, 5 miles E of La Honda," but the specimen was taken at La Honda, 5 mi (8 km) E of Los Santos); Cigarruista, 1994 Aug 11 (MVUP).

LOS SANTOS, PROVINCIA DE; (Pacific)

0735/8020 (USBGN)

Province covering the SE region of Península de Azuero [0740/8035 (USBGN)]. Its NE coast is on the Gulf of Panama [0800/7930 (USBGN)] and its S coast lies on the Pacific Ocean. The N boundary with Herrera province is formed by the Río La Villa [0759/8023 (USBGN)] and its SW boundary with Veraguas province is formed both by hilly ridges and rivers. The provincial boundaries vary between older and more recent maps. See Península de Azuero for a full description of the changes. The NE coast, the location of the provincial capital Las Tablas [0746/8017 (USBGN)], quickly rises to low, 50-100 m hills with few large river estuaries or areas of mangrove swamp. The peaks in the central hills are just below 1,000 m and reach 1,500 m in the SW. The vast majority of work, as well as the earliest work, in the province was conducted by Wetmore, who methodically explored new regions in 1948 (with Perrygo), in 1956, and in 1957.

LOS TRES CERROS; see Hoya, Cerro

LOS VOLADORES; Herrera

ca. 0802/8036

Described by Wetmore as a section of hills extending from the road connecting Paris [0803/8033 (USBGN)] and Portobelillo [0801/8036 (USBGN)] NW to Potuga [0804/8037 (USBGN)], the hills were covered with secondary forest with some very tall trees (USNM). Not shown on our maps, but the region described by Wetmore appears on JOG as 50-200 m hills surrounding Cerro Picacho [0802/8036 (USBGN)]. Wetmore and Perrygo, GL = Portobelillo, on hills ca. 2-3 mi (3-5 km) NE of Portobelillo and in a valley called El Fical (not located), 1948 Feb 26 (USNM).

LUCHA FRANCO; Panamá (E, Pacific)

0905/7932 (USBGN)

100 m, town on the Carretera Boyd-Roosevelt [0917/7943 (USBGN)] 1.5 km S of Las Cumbres [0905/7932 (USBGN)] (E762). Burke, 1977 Dec 6 (MVUP); Castillo and Burke, 1992 Jul 23 (MVUP).

MACANA, CIÉNAGA; see La Macana, Ciénaga de

MACANA, CIÉNAGA DE LA; see La Macana, Ciénaga de

MACANAS, CIÉNAGA DE LAS; see La Macana, Ciénaga de

MACANO; see El Macano

MACANO, EL; see El Macano

MACARACAS; Los Santos
0744/8033 (USBGN)
ca. 50 m, town in W Los Santos province 27 km SSW of Los Santos [0756/8025 (USBGN)] (JOG). Casas, *Bosque Común* (communal woodland), and as a GL for Leales, 1990 Jan 23 (MVUP).

MADDEN, AERÓDROMO; Panamá (E, Atlantic)
0910/7933 (USBGN)
100 m, an abandoned landing field located 2 km SE of Calzada Larga [0910/7934 (USBGN)], near several tributaries of the Río Chilibrillo [0910/7938 (USBGN)] (E8660, as "Madden Field"). RM1, RMM2, RMM1 and RMM2 (all as "Madden Field") show the field ca. 1 mi (1.5 km) NNE of Calzada Larga. Yunker et al., 1961 Nov 28 (USNM, as "Madden Field").

MADDEN, CAMINO; Colón/Panamá (Canal Zone) (E, Atlantic, Pacific)
0907/7937 (USBGN)
A road that extends roughly N-S for 12.6 mi (20 km) between Summit Park [0904/7939 (USBGN)] and Represa Madden (Madden Dam) [0913/7937 (USBGN)]. One km beyond the dam, where it crosses into Colón province, the road terminates at its junction with the Carretera Boyd-Roosevelt [0917/7943 (USBGN)]. Also known as Madden Road and as Road C25 (E661, as "Madden Road"; E8660, as "Madden Road," and as "Road C25"; RM1, as "Madden Road," and as "Road C25"), and on our more recent maps it appears as a section of Panama 81 (JOG; MGRP). Approximately the S 7 km of the road are now within Parque Nacional Soberanía [0909/7941 (Angehr, 2003:239)]. Benson, at 100 feet (30 m), GL = Las Cruces Trail, at the intersection [see Madden Road Timber Preserve [0906/7937 (USBGN)]], 1931 Jul 26, Aug 23 (USNM, as Madden Road"); Miller, near, 1937 Feb 9 (USNM, as "Madden Road"); Clark, GL = Chilibre, 1937 May (USNM, as "Madden Dam Road"); Meyer, 1941 Mar 8, 18, Jun 23, GL = Cerro Gordo [ca. 0902/7940] (q. v.), near Cerro Gordo, 1941 May 24, GL = Madden Road Timber Preserve, 1941 Jul 16, 19-20 (USNM, as "Madden Road"); MCU, 1955 Aug 15, Oct 2, 20, 25, Nov 11, 17, Dec 2 (GML, as "Madden Dam Road"; USNM, as "Madden Dam Road"); Collector ? [MCU ?], 1955 Aug 15 (UF, as "Madden Dam Road"); Belton, woods near mile post 1, 1964 Aug 30 (GML, as "Madden Dam Road").

MADDEN, LAGO; see Alajuela, Lago

MADDEN, REPRESA; Colón/Panama (Canal Zone) (E, Atlantic)
0913/7937 (USBGN)
Dam on the Río Chagres [0907/7942 (USBGN)] that was completed in 1935 forming Lago Alajuela (Madden Lake) [0915/7935 (USBGN)]. Located 14 km NE of Gamboa [0907/7942 (USBGN)] and 6 km NE of Juan Mina [0910/7939 (USBGN)]. Camino Madden (Madden Road) [0907/7937 (USBGN)] extends along the dam (E661, as "Madden Dam"; E8660, as "Madden Dam"). The Boy Scout Camp

appears to be on the S side of the dam (Morton, 1980:439, as "Madden Lake Boy Scout Camp"). According to Angehr (in litt., Aug 2005), the boy scout camp was 2 km due E of the dam "on the large peninsula that juts into the lake from the SE." The area is now seen as a collection of buildings with the label "tanque de agua" on 4243-I (Buena Vista) [E762] at UTM 1C14000 N 654000 E (0910/7936). "The installation was turned over to Panama and is now known as 'Campo Chagres' and serves as the headquarters of Chagres National Park." Arbib and Loetscher, observations, "primaeval rain forest off the [Canal] Zone up the Chagres river," 1934 Aug 21 (Arbib and Loetscher, 1935, as "Madden Dam"); MCU, 1955 Oct 25, Dec 2 (USNM, as "Madden Dam"); Collector ?, 1959 Jul 10 (GML, as "Madden Dam"); Morton, Boy Scout Camp, 1971 Feb 26, Mar 25, 1974 Jan 6-7, Mar 9, observations, studies 1970-1975 (CAS, as "Madden Boy Scout Camp"; USNM, as "Madden Dam"; Morton, 1976; Morton, 1980, as "Madden Lake Boy Scout Camp"). See also: Chagres, Río.

MADDEN BOY SCOUT CAMP; see Madden, Represa

MADDEN DAM; see Madden, Represa

MADDEN DAM ROAD; see Madden, Camino

MADDEN FIELD; see Madden, Aeródromo

MADDEN FOREST; see Madden Road Timber Preserve

MADDEN FOREST PRESERVE; see Madden Road Timber Preserve

MADDEN FOREST RESERVE; see Madden Road Timber Preserve

MADDEN LAKE; see Alajuela, Lago

MADDEN LAKE BOY SCOUT CAMP; see Madden, Represa

MADDEN ROAD; see Madden, Camino

MADDEN ROAD TIMBER PRESERVE;
0906/7937 (USBGN)
Panamá (Canal Zone) (E, Atlantic, Pacific)
224 m, former United States Government establishment that was located 9 km SE of Gamboa [0907/7942 (USBGN)] and 6 km N of Paraiso [0902/7938 (USBGN)]; the area is now part of Parque Nacional Soberanía [0909/7941 (Angehr, 2003:239)]. According to Wetmore's notes, the preserve was originally established by Canal Zone Governor Harry Burgess as "Forest Preserve" by an order dated 19 Apr 1931, and later named "Canal Zone Forest and Wildlife Preserve" by executive regulation number 75, dated 13 Feb 1964. The official name given by USBGN is Madden Road Timber Preserve. The preserve covered a land area of ca. 15 km^2 (1,500 ha) and the continental divide extended through the N portion of the park. Camino Madden (Madden Road) [0907/7937 (USBGN)] extended through the central region of the preserve in a roughly N-S direction, and Las Cruces Trail [ca. 0904/7936] extended through the N region of the park in a roughly NW-SE direction. The two roads intersected in the N central region of the preserve (E661, as "Forest Reserve"; E8660). Meyer, 1941 May 30, Pacific side of, 1941 May 3, and as a GL for: Camino Madden, and Las Cruces Trail, 1941 Jul 16, 19-20, Sep 1 (USNM, as

"Forest Reserve"); Allen, 1944 Dec 4 (AMNH, as "Madden Forest Preserve"); Keenan and MCU, 1955 Jul 5 (GML, as "Madden Forest"; USNM, as "Madden Forest"); Willis, observations, "tall secondary forest on the hills of the continental divide," 1960-1981 (Willis, 1966a, as "Forest Reserve"; Willis, 1967, as "Madden Forest Reserve"; Willis, 1972:36, as "Forest Reserve"; Willis, 1983a, as "Madden Forest Reserve"; Karr, 1971a, as "Madden Forest Reserve"); Todd, GL = Gamboa, 1964 Jul 15 (LACM, as "U. S. Forest Preserve"); Morton, 1965 Oct 28, 1966 Apr 29, May 4, 6, 10, 15, Jun 12, 19, Jul 17, 19, studies 1973-1975 (AMNH, as "Madden Forest"; FMNH, as "Madden Forest"; YPM, as "Madden Forest"; Morton, 1980, as "Madden Forest Preserve"; Morton, 1982, as "Summit Gardens-Madden Forest"); Means, Koslovsky, and Sawyer, "1.7 mi (2.7 km) N on Panama-Colon Rd from Gamboa turnoff," according to our maps, they collected on Camino Madden 2.7 km N of Summit Park [0904/7939 (USBGN)], 1967 Feb 14 (FSU, as "Madden Forest"; USNM, as "Madden Forest"); Eaton, 1970 Apr 24, and as a GL for George Green Park, 1970 Apr 24, Aug 24 (?) (UF, as "Madden Forest"); Gardner, 1986 Jan 28 (WFVZ, as "Madden Forest Reserve"). (Kilham, 1972, as "Madden Forest Reserve").

MADROÑA, PUNTA; Panamá (E, Pacific)
0842/7840 (USBGN)
Sea level, point on the SW bank of Ensenada de Chimán [0843/7839 (USBGN)] 5 km SW of Chimán [0842/7837 (USBGN)] (JOG). According to Wetmore, cuipo trees covered the steep-sided hills on the point, Wetmore and Perrygo, GL = Chimán, a sandy strip covered with palms and other trees, brush, and a mangrove swamp, 1950 Mar 31 (USNM, as "Madroño"; Wetmore, 1950b, as "Punta Madroño").

MADROÑO; Panamá (E, Pacific)
0917/7908 (USBGN)
200 m, town on the upper reaches of the Río Mamoní [0908/7904 (USBGN)] 16 km NW of Chepo [0910/7906 (USBGN)] (MGRP). Wetmore indicated that the locality was "in the Río Bayano [sic] Valley" (Wetmore et al., 1984:473, as "Saino Madronio"). In his notes for the gazetteer he does not mention Saino Madronio, but "Salto de Madroño," which he placed on the Río Mamoní near Chepo. From these different notes, it seems likely that the specimens were collected near the present town of Madroño. Hinds and GMLP, 1975 Mar 20-21 (GML, as "Sainio Madronio," as "Saino Madroño," and as "Zaino Madronio"). In addition, GML has specimens labeled San José Madroño, which presumably refers to the region NW of Madroño where our most recent map indicates a town named San José [ca. 0917/7906]. According to RDP2, San José is due N of Chepo on the continental divide at 300 m and ca. 8 km NE of Madroño. Hinds, 1975 Mar 15-16, 18, 24 (GML, as "Madroño," and as "San Jose Madroño").

MADROÑO; see Madroña, Punta

MADROÑO, PUNTA; see Madroña, Punta

MADROÑO, SALTO DE; see Madroño

MAESTRO, CERRO; Panamá (E, Pacific)
ca. 0849/7830

ca. 1,200 m, described by Wetmore as a peak covered with green, gallery forest on the E end of Serranía de Majé [0855/7830 (USBGN)] just SE of Cerro Chucantí [ca. 0851/7831], Wetmore and Perrygo, GL = Cerro Chucantí, Serranía de Majé, at 1,000 feet (300 m), base of Cerro Maestro, forest, dry creeks strewn with large boulders, 1950 Mar 16 (USNM). Not shown on our maps.

MAFAFA; see Mafofa

MAFOFA; Panamá, Archipiélago de las Perlas

ca. 0815/7854

Sea level, town on the E coast of the peninsula on the S coast of Isla del Rey [0822/7855 (USBGN)]. The town is located on the W coast of Bahía San Telmo [0816/7853 (USBGN)] 3 km N of Punta Cocos [0813/7854 (USBGN)] (E501, as "Mafafa"; MHA, as "Mafafa"). Wetmore, observations, 1960 Jan (Wetmore, 1972:183). (Murphy, 1944b, as "Mafofa").

MAJAGUAL; Colón (Canal Zone) (E)

ca. 0921/7953

Presumably a town along the Río Majagual [0921/7953 (USBGN)] and possibly the present town of Majagual-Gold Hill [0921/7953 (USBGN)] (q. v.), which is located ca. 2 km SE of Colón [0922/7954 (USBGN)]. Goldman, near Colón, 1911 Mar 13 (USNM). See also: Las Minas, Bahía; and Marajal. Not shown on our maps.

MAJAGUAL, RÍO; Colón (Canal Zone) (E)

0921/7953 (USBGN)

Short river that originates S of France Field [0922/7952 (USBGN)] and flows N and W between Coco Solo Naval Reservation [0923/7952 (USBGN)] and Fort Gulick [0919/7952 (USBGN)] entering mangrove swamps on the E coast of Río Folk [0921/7954 (USBGN)] (E8660). See also: Majagual; and Majagual-Gold Hill.

MAJAGUAL-GOLD HILL; Colón (Canal Zone) (E)

0921/7953 (USBGN)

USBGN treats Coco Solito as an older name for both Majagual-Gold Hill and Coco Solo [0922/7953 (USBGN)]. The towns are very near one another, therefore when either the literature or specimen labels refer to Coco Solito, it is unclear which town is correct. We placed information regarding Coco Solito under Majagual-Gold Hill for two reasons: Coco Solo appears on our maps as Coco Solo, not as Coco Solito; and secondly because when Coco Solito appears on our maps, it is closer to the Río Majagual [0921/7953 (USBGN)] than Coco Solo is. Town 1 mi (1.5 km) E of Rainbow City [0921/7953 (USBGN)] and 1 mi (1.5 km) S of Coco Solo Naval Reservation [0923/7952 (USBGN)] (RM2, as "Coco Solito"). Morton, year ? Aug 5 (Wetmore et al., 1984:22, as "Coco Solito"); Ambrose, observations, 1955 (Wetmore, 1972:417, as "Coco Solito"); Ambrose, 1955 Aug 5, 1956 Aug 3 (AMNH, as "Coco Solito"; Eisenmann, 1959, as "near Colón"; Wetmore et al., 1984:24, as "Coco Solito"). See also: Coco Solo.

MAJÉ; Panamá (E, Pacific)

0840/7835 (USBGN)

Sea level, not shown on our maps, but USBGN indicates a town near the mouth of the Río Majé [0840/7835 (USBGN)] ca. 7 km SE of Chimán [0842/7837 (USBGN)]. Wetmore and Perrygo, GL for Río Majé, 1950 Mar 2-3, 30 (USNM; Wetmore, 1952:27; Olson, 1980b, as "Majé, mouth of the Río Majé").

MAJÉ; see Majé, Isla

MAJÉ, ISLA; Panamá (E, Pacific)
ca. 0909/7851
ca. 50 m, island in the W end of Lago Bayano [0908/7835 (USBGN)] near the dam with a land area of 1,433 ha (Angehr, 2003:280) and the site of a former GML research station (Candanedo and D'Croz, 1983:4 (map)). The island is not named on our maps although RDP2 indicates an island and MGRP indicates a group of islands in the W end of the reservoir near the dam. These two maps indicate that Isla Majé is ca. 2 km SE of the dam, less than 1 km off the W coast of the reservoir, and ca. 30 km E of Chepo [0910/7906 (USBGN)]. The few specimens at GML that were labeled simply Majé were presumed to have been collected here, although Majé could refer to Isla Majé, to Serranía de Majé [0855/7830 (USBGN)] (q. v.), or to Río Majé [0906/7845 (USBGN)] (q. v.). GMLP, GL = Lago Bayano, 1972 Oct 4, 1977 Mar 17-18, Jul 22, 1978 Jul 30 (GML); Collector ?, GL = Río Chepo, near Majé, 1973 Jan 25 (GML, as "Majé"); Galindo, GL = Río Chepo, near Majé, 1973 Nov 9 (GML, as "Majé"); Pineda, GL = Río Bayano, 1974 May 25 (GML, as "Majé"); Greensmith, observations, 1976 Dec 5 (Wetmore et al., 1984:271, as "Bayano River Research Station"); Lowe, observations, 1977 May 28 (Wetmore et al., 1984:269, as "Maje Island"); Sousa and Herman, blood smears, c. 1980 (Sousa and Herman, 1982, as "Bayano"); Polshek and Hinds, GL = Lago Bayano, 1989 Jan 14, Feb 13, 20, 22, Mar 5, 17, 20, 28, Apr 15, 19 (UF); Bacon, Guitton-Mayerma, and Miller, 2002 Oct 20 (STRI; UAM). (Ridgely and Gwynne, 1989:266, 490-491, as "Majé Island," and as "Isla Majé").

MAJÉ, RÍO; Panamá (E, Pacific)
0840/7835 (USBGN)
Originates on the S slopes of Cerro Chucantí [ca. 0851/7831] and flows SSW forming a narrow, 2 km long coastal inlet of mangrove swamps and mudflats as it enters the Gulf of Panama [0800/7930 (USBGN)] (Atlas, 1988:205, as "Río Chucantí"; JOG). Wetmore and Perrygo, GL = Majé [0840/7835 (USBGN)], mouth of the Río Majé, mangroves, rocky knoll at the edge of the swamp, upstream beyond the mangroves where the river narrowed and its banks were covered with grass and other vegetation, 1950 Mar 2-3, 30, GL = Charco del Toro [ca. 0844/7833], near Charco del Toro the river is affected by tides, on gravel bars, on the river banks in woods, overgrown banana plantations and a recent clearing, fig trees, along dry creeks into wooded areas, on trails through gallery forest leading N to the mountains, 1950 Mar 3, 20-25, 27-28 (USNM; Wetmore, 1951a; Wetmore, 1963a; Olson, 1980b, as "Majé, mouth of the Río Majé"; Olson, 1981d, as "Charco del Toro, Río Majé").

MAJÉ, RÍO; Panamá (E, Pacific)
0906/7845 (USBGN)
Originates on the N slopes of the Serranía de Majé [0855/7830 (USBGN)] and

flows N entering the SW coast of Lago Bayano [0908/7835 (USBGN)]. Originally the river continued N and entered the Río Chepo [0900/7906 (USBGN)] near the town of Majé [0910/7849 (USBGN)] (JOG; MGRP). (Angehr, 2003:282). See also: Majé, Isla.

MAJÉ, SERRANÍA DE; Panamá (E, Pacific)
0855/7830 (USBGN)
Mountain range in SE Panamá province extending ca. 80 km NW from the Darién provincial border and parallel to the Pacific coast with peaks ca. 30 km inland. Rivers originating on the N and NW slopes drain into the Río Chepo [0900/7906 (USBGN)], while those originating on the S slopes flow directly into the Gulf of Panama [0800/7930 (USBGN)] (Atlas, 1988:205; MHA). MHA and E501 show the peaks on this range at 300-360 m, but more recent maps, including JOG, CARP, and Atlas (1988), indicate that the peaks are 925-1,439 m. Shown on SMN2B as "Serranía de Cañazas," but our other maps and USBGN indicate that Serranía de Cañazas [0852/7822 (USBGN)] is a separate range. Considered an Important Bird Area by Angehr (2003:285-287). Wetmore, used as an additional locator for: Cerro Chucantí (q. v.), Cerro Maestro (q. v.), Quebrada Cauchero (q. v.), and Río Chucantí (q. v.), 1950 Mar 5-18 (USNM; Wetmore, 1950b); Guerrero, GL = Río Chepo, 1962 Jun 8 (GML); Galindo, Hinds, Mendez, and GMLP, GL = Río Chepo, presumably they collected on the N and NW slopes near the Río Majecito [0909/7849 (USBGN)] (q. v.), some specimens are also labeled Chepo, a few say Darién, and one is labeled Altos de Bayano, in addition a few specimens are labeled only Bayano and may have come from lower elevations on Lago Bayano, 1972 Apr 15, Oct 10, Nov 18, Dec 2, 1973 Jan 14, 26, Feb 8, 10, 12-13, Mar 23-25, Apr 4, May 5, Jun 5, 14, Nov 13, 1974 May 16, 1975 Jun 9, 1976 Jan 22, Oct 22, Nov 11, Dec 3, 1977 Jan 15, Feb 24, Mar, Jun 23-24 (GML, as "Altos de Majé"; LSUMZ, as "Altos de Majé"); Eisenmann, observations, date ? (Wetmore et al., 1984:62, as "Altos de Majé"); DaCosta, Guitton-Mayerma, Klicka, Miller, and Spellman, GL for Cerro Chucantí, 2006 Feb 15-17 (MBM; STRI). (Myers, 1969, as "Serranía de Cañazas"). See also: Calle Larga; Majé, Isla; and Toro, Charco del.

MAJECITO, RÍO; Panamá (E, Pacific)
0909/7849 (USBGN)
Originates on the W end of Serranía de Majé [0855/7830 (USBGN)]. The river flows first NW then NE and originally entered the Río Majé [0906/7845 (USBGN)] just above its mouth (RDPF). The river now enters the SW coast of Lago Bayano [0908/7835 (USBGN)] 24 km ESE of Chepo [0910/7906 (USBGN)] (MGRP). Galindo and GMLP, GL = Río Chepo, at 150 feet (45 m), presumably he collected near the river's mouth, 1973 Jan 25 (GML, as "Boca de Majecito"; LSUMZ).

MALA; see Mala, Punta

MALA, CAPE; see Azuero, Península de

MALA, CAPE; see Mala, Punta

MALA, PUNTA; Los Santos
0728/8000 (USBGN)
Sea level, cape located on the easternmost land of Península de Azuero [0740/

8035 (USBGN)] in eastern Los Santos province where the Río Caldera [0728/8000 (USBGN)] enters the Gulf of Panama [0800/7930 (USBGN)] (JOG). Griscom (1927a) used the term "Cape Mala Peninsula" in reference to Península de Azuero (q. v.). Kelsall and Cullingford, Gulf of Panama, 30 mi (48 km) E of Mala, 1924 Sep (BMNH; Douglas and Johnson, 1926); Wetmore and Perrygo, at a 38 acre (15 ha) coast guard station, 50 foot (15 m) cliffs dropped into the ocean, small sandy beaches bordered by *Tilia*, a small stream mouth bordered by mangroves, gallery forest with thick undergrowth bordering the station, an extensive open meadow, 1948 Mar 27 (USNM); Fagan, "abeam Cape Mala, aboard SS Santa Elisa at midnight," 1922 Sep 24 (USNM, as "Cape Mala"; Wetmore, 1923; Wetmore, 1957:34); Wetmore, GL for Frailes del Sur, 1956 Feb 6, 1962 Mar 18, GL for Ensenada Venado, 1957 Feb 25-26 (USNM; Wetmore, 1957), observations N of Punta Mala with Islas Boná [0834/7936 (USBGN)] and Otoque [0836/7936 (USBGN)] in sight, 1956 Feb 6 (Wetmore, 1965:433). (Wetmore, 1970). See also: Caldera, Río.

MALA, PUNTA; Panamá (Canal Zone) (E, Pacific)
0857/7933 (USBGN)
Sea level, coastal point that is now part of greater Panama City [0858/7932 (USBGN)] (E761A, as "Mala Pt."). Described by Schneider and Mallory as a "protected mudflat lying between Fort Amador [Army Reservation [0856/7933 (USBGN)]] and Panama City," observations, 1978 Jan-May (Schneider and Mallory, 1982).

MALA PENINSULA; see Azuero, Península de

MÁLAGA, ISLA; Panamá, Archipiélago de las Perlas
0830/7903 (USBGN)
Sea level, according to Angehr (in litt., Aug 2005), the island is labeled on 4341-II [E762]. It is the smaller and more northern of the 2 islands just E of Isla Bayoneta [0829/7904 (USBGN)], the southern one being Isla La Vivienda [0829/7903 (USBGN)]. Wetmore, a small island separated from the E coast of Isla Bayoneta by a narrow channel that becomes a mud flat during low tide, sandy beach, mangroves, low original forest, 1960 Jan 29 (USNM, as "Isleta Malaga"; Olson, 1980b, as "Isleta Málaga"); (MacArthur et al., 1972—species list; see comments in Bibliography).

MALAGA, ISLETA; see Málaga, Isla

MALATUPO; see Mulatupo Sasardí

MALÍ, CERRO; Darién
0807/7714 (Fairchild and Handley, 1966)
ca. 5,000 feet (1,500 m), described by Wetmore as a rocky peak on the continental divide in the Serranía del Darién [0830/7730 (USBGN)] SSE of Cerro Tacarcuna [0811/7717 (USBGN)], camp was located on a narrow ridge 4 mi (6 km) W of the peak (USNM). Handley, at 4,100-5,900 feet (1,250-1,800 m), a few specimens have Tacarcuna Village listed as the GL, 1959 Mar 15, 18, 1964 Jan 30-31, Feb 5-12, 14, 17-18, Mar 3 (USNM, as "Cerro Mali," and as "Mali Camp"; Hekstra, 1982; Browning, 1989); Galindo, at 4,700-5,100 feet (1,400-1,550), specimens have various labels, but most were collected at the camp located 4 mi (6 km) W of the peak,

several are also labeled "auxiliary ridge of Cerro Mali," 1962 May 2-3, 21, 1963 May 22-25, 27-28, 30-31, Jun 1-5, 7-9, and as a GL for Tacarcuna Village, 1963 Jul 4 (LSUMZ; USNM, as "Cerro Mali," as "Cerro Mali Camp," and as "GML Camp"; Wetmore, 1963b; Olson, 1981c); Hinds, Quintero, and GMLP, at 4,800 feet (1,450 m), camp 4 mi (6 km) W of top, 1963 May 22-31, Jun 1, 3-7, 10 (GML, as "Cerro Mali," and as "Cerro Mali Camp"; USNM, as "Cerro Mali," and as "Cerro Mali Camp"); Hinds and GMLP at 1,400 m, 1975 Jan 17-24 (GML, as "Cerro Mali," and as "Pico Mali"; LSUMZ, as "Cerro Mali," and as "Pico Mali"); Wetmore, at 4,200-4,700 feet (1,300-1,400 m), collections made near the camp in green forest with little undergrowth, in cloud forest alternating with more open, gallery type forest, to the E along the ridge trail that joins the continental divide, and on trails to the headwaters of the Río Púcuro [0801/7733 (USBGN)], 1964 Feb 19-Mar 3, at 4,700 feet (1,400 m), E end of Cerro Malí at the continental divide, palms among the trees, comparatively open undergrowth, cloud forest predominates along the E end of Cerro Malí, 1964 Feb 26, and as a GL for Río Púcuro, 1964 Feb 23, 25, Mar 1-2 (USNM; Wetmore, 1972:351). (Wetmore, 1967; Wetmore, 1970; Myers, 1969). See also: Tacarcuna Village; and Tacarcuna, Cerro.

MALÍ, PICO; see Malí, Cerro

MAMONÍ, RÍO; Panamá (E, Pacific)

0908/7904 (USBGN)

Originates in N central Panamá province near the Comarca de San Blas/Panamá provincial border and flows E, then S joining the Río Chepo [0900/7906 (USBGN)] 5 km SSE of Chepo [0910/7906 (USBGN)] and 6 km SSE of Las Margaritas [0911/7905 (USBGN)] (JOG; MGRP). Goldman (1920:5) refers to a landing called Hato Bayano at the mouth of the Río Mamoní. Johnson, GL = Río Chepo, 10 mi (16 km) above mouth, 1914 Jan 20 (USNM, as "Manoni River"); Wetmore and Perrygo, GL = Chepo, ca. 1 mi (1.5 km) outside of Chepo in thickets and tall trees along the river, 1949 Apr 6, 8, 11, 13-14, 21, near the river in the area extending 8 km N and 8 km S of Chepo at the following secondary localities: at Camarón (q. v.), 1949 Apr 7, at Chepo Viejo (q. v.), 1949 Apr 10, at Charare (q. v.), and at Boca del Toro (q. v.), 1949 Apr 22 (USNM; Wetmore, 1950a; Wetmore, 1950b); Guerrero, Mendieta, and Rodríguez, GL = Las Margaritas, 1985 Mar 16 (MVUP). See also: Limón, Zanja.

MANACA; Chiriquí

ca. 0820/8249

ca. 25 m, two of our maps indicate a town named "Manaca Civil" 14 km S of Progreso [0827/8250 (USBGN)] and 8 km NE of Puerto Armuelles [0817/8252 (USBGN)] (E562, as "Manaca Civil"; MGRP, as "Manaca Civil"). E562 also indicates Manaca Norte 1 km NNE of Manaca Civil, and Wetmore collected in this region at Manaca Cerro Cinco (q. v.). Sánchez, specimens are labeled both Corregimiento Progreso [0823/8248 (USBGN)] and Corregimiento Puerto Armuelles [0820/8255 (USBGN)], *llano* (plains), *zona agrícola de* (agricultural zone of) Manaca Norte , *cerca al colegio* (near the school), *tecales* (teak trees) de Manaca, 1983 Jun 18, 24-26, Jul 4 (MVUP); Morales, 1983 Jun 25 (MVUP); Vásquez, 1983 Sep 10 (MVUP).

MANACA CERRO CINCO; Chiriquí

ca. 0823/8251

Described by Wetmore as a tract of forest on level land (without a mountain in sight) owned by the Chiriquí Land Company, located on the highway 9 mi (14 km) N of Puerto Armuelles [0817/8252 (USBGN)]. Not shown on our maps, but presumably this is near, if not the same as Manaca [ca. 0820/8249] (q. v.). Wetmore, GL = Puerto Armuelles, forest with heavy undergrowth and areas of more open undergrowth, 1966 Jan 30 (USNM).

MANACA CIVIL; see Manaca

MANACA NORTE; see Manaca

MANANTÍ, RÍO; Bocas del Toro

0856/8158 (USBGN)

Originates on the continental divide W of the Río Cricamola [0859/8154 (USBGN)], flows N, then turns W and NW flowing through an expanse of coastal mangroves before entering Laguna de Chiriquí [0903/8200 (USBGN)] 7 km SW of Boca de Cricamola [0859/8155 (USBGN)] (JOG; MHA, as "Río Manatí"; Peters, 1931a:295, as "Manati River"). Wedel, 1927 Nov 11 (MCZ, as "Manati R."; Peters, 1931a, as "Manati River").

MANATÍ, RÍO; see Manantí, Río

MANATI RIVER; see Manantí, Río

MANDANGO RIVER; see Mandinga, Río

MANDINGA; Colón (Canal Zone)

?

Not located on our maps. Specimen labels indicate Colón or Canal Zone as the province or they indicate no province. Montenegro did extensive collecting at Pantano Mojinga (Mojinga Swamp) [0918/7958 (USBGN)] (q. v.), and these 4 labels may be a misspelling or a misreading of Mojinga. Another possibility is the Río Mandinga [0906/7942 (USBGN)] (q. v.), although this river is within Panamá province. Montenegro, 1979 May 11 (GML).

MANDINGA; San Blas

0928/7905 (USBGN)

Sea level, airfield just W of Mandinga [0927/7904 (USBGN)] (SMN2B). According to Wetmore in 1957, the airstrip appeared abandoned, as the buildings were not in use and were largely hidden by tall brush (USNM). See also: Mandinga [0927/7904 (USBGN)].

MANDINGA; San Blas

0927/7904 (USBGN)

Sea level, town in extreme NW Comarca de San Blas, 30 km N of Chepo [0910/7906 (USBGN)]. Located on the W coast of Golfo de San Blas [0930/7900 (USBGN)] near the mouth of the Río Mandinga [0928/7904 (USBGN)] (MHA; SMN2B). Wetmore, collected within 8 km of town primarily near the airfield, also named Mandinga [0928/7905 USBGN)] (q. v.), on the coast and along trails leading

E to the Río Cangandí [0928/7907 (USBGN)], [S ?] to the Río Inatí [ca. 0926/7904 ?], and SW to the Río Nergala [0927/7902 (USBGN)]. Near town and near the airfield Wetmore collected in recently cleared areas, in dense, tall grass and thick *rastrojo*, in savanna with shoulder-high saw grass, in 3-4 year old *rastrojo*, in old second growth forest, in swampy woods and wet forests, and near creeks and cornfields. Wetmore traveled along the coast by boat on the Río Mandinga just above its mouth, through 50-70 foot (15-20 m) mangroves, to the old wharf and inside channels, to the SE end of the bay past the mangroves and into a wet, lowland forest. He also walked along trails to the nearby rivers and although Mandinga is also a GL for these secondary localities, some of the specimens collected along the rivers were labeled only Mandinga. Trail along the Río Mandinga through tall cane with clumps of trees, on the road to Río Cangandí (the road from Mandinga to Río Cangandí is the old Fruit Company Rail Road) in low *rastrojo*, in tracts of original lowland forest with 60 foot (20 m) trees and dense undergrowth and many creepers, 2 mi (3 km) to the Río Cangandí in brush and scattered forest along the river, off the road to Río Cangandí to a lake that was probably formed when the road was built, trail to Río Nergala in new *rastrojo* with tall *Heliconia*, up a long narrow hill to a stand of original forest, near the Río Inatí in very old second growth, 1957 Jan 22-Feb 17, and as a GL for: Río Cangandí, Río Inatí, Río Mandinga, Río Nergala, and Río Piedra, Jan 23, Feb 3, 6, 8, 13-14, 16 (LACM; USNM; Wetmore, 1959b:10, 20; Wetmore, 1962; Wetmore, 1967; Parkes, 1976; Olson, 1980a; Olson, 1981a; Olson, 1981d; Olson, 1983); Galindo, GL for Río Cangandí, at 200 feet (60 m), 1957 May 21 (USNM).

MANDINGA, RÍO; Colón/San Blas (E)

0928/7904 (USBGN)

Originates on the continental divide and flows NE along the Colón/San Blas boundary. It enters Comarca de San Blas for its last 2 km where it flows past Mandinga [0927/7904 (USBGN)] and forms a large delta among mangroves on the W end of Golfo de San Blas [0930/7900 (USBGN)] (E501; JOG). Wetmore, GL = Mandinga, landed in the bay on a stretch of sand near one of the river's mouths, which was 20-30 yards (20-30 m) wide, 1957 Feb 6 (USNM). Mentioned in Parkes (1969).

MANDINGA, RÍO; Panamá (Canal Zone) (W, Atlantic)

0906/7942 (USBGN)

Originates in the former Canal Zone on the continental divide near Nuevo Emperador [0900/7944 (USBGN)] and flows N entering Lago Gatún (Gatun Lake) [0912/7955 (USBGN)] 1 km S of Gamboa [0907/7942 (USBGN)] (E661; E8660). Olson and Funchess, 1963 May 12 (USNM, as "Mandango River").

MANENÉ; Darién

ca. 0741/7749

ca. 50 m, town at the W foot of Alturas de Nique [0740/7745 (USBGN)] on the Río Balsas [0815/7759 (USBGN)] 43 km ENE of Jaqué [0731/7810 (USBGN)] and ca. 85 km SSE of La Palma [0825/7809 (USBGN)] (MGRP). Seutin, observations, 1996 Jul 6-13 (Seutin, 1998).

MANGANESE MINE; Colón (E)

ca. 0927/7933

ca. 100 m, described by Wetmore as an abandoned manganese mine that was originally connected to Nombre de Dios [0935/7928 (USBGN)] via railroad. In 1957 the mine was reopening and roads were under construction. The mine does not appear on our maps, but the index sheet of E661 indicates a mine on the Río Boquerón [0921/7933 (USBGN)] ca. 11 mi (18 km) SSW of Nombre de Dios and 5 mi (8 km) N of the river mouth. Wetmore, GL = Río Boquerón, at 800 feet (240 m), on a ridge W of the Río Boquerón and E of the Río Salamanca [0920/7935 (USBGN)], 15 road mi (24 km) N of Gatuncillo [0914/7939 (USBGN)] and 3 mi (5 km) S of the mine, virgin forest with many tall trees along the newly bulldozed road, 1957 Mar 3 (USNM).

MANGLE, CENEGÓN DEL; see El Mangle, Ciénaga

MANGLE, CIÉNAGA EL; see El Mangle, Ciénaga

MANGROVE; Panamá (Canal Zone) (W, Pacific)

ca. 0856/7935

Described by Schneider and Mallory (1982) as a "protected mudflat with mangrove trees" on the W bank of the Panama Canal [0920/7955 (USBGN)] and N of the Thatcher Ferry Bridge (Bridge of the Americas) [0856/7933 (USBGN)]. According to our maps, the locality is either just S of or within Rodman United States Naval Reservation [0857/7935 (USBGN)], observations, 1978 Jan-May (Schneider and Mallory, 1982).

MANONI RIVER; see Mamoní, Río

MANZANILLA, ISLA; see Colón

MANZANILLA ISLAND; see Colón

MANZANILLO; see La Puntita

MANZANILLO, BAHÍA; Colón (Canal Zone) (E)

0922/7953 (USBGN)

Bay on the N coast of Colón province just E of Colón [0922/7954 (USBGN)] that extends ca. 2 km E-W and opens on the S to the Río Folk [0921/7954 (USBGN)] and on the N to the Caribbean Sea [1500/7500 (USBGN)] (E761A, as "Manzanillo Bay").

MANZANILLO, ISLA; see Colón

MANZANILLO BAY; see Manzanillo, Bahía

MANZANILLO ISLAND; see Colón

MAR, RÍO; see Río Mar

MARAJAL; Colón (Canal Zone) (E)

ca. 0921/7953

Sea level, a town on the E coast of Río Folk [0921/7954 (USBGN)] just S of the mouth of the Río Majagual [0921/7953 (USBGN)] 1.5 mi (2.5 km) SW of Colón [0922/7954 (USBGN)] (PR). Not shown on recent maps, but estimated from E866 that it was located just N of what is now the town of Rainbow City [0921/7953 (USBGN)]. Goldman, 1911 Mar 13 (USNM; Wetmore, 1959b, incorrectly listed as western Colón province). See also: Las Minas, Bahía; and Majagual.

MARCOS A. GELABERT, AEROPUERTO; Panamá (E, Pacific)
0859/7931 (USBGN)
Sea level, an airport located just N of Punta Paitilla [0858/7931 (USBGN)] (E661, as "Gelabert Airport"; E962, as "Aeropuerto Marcus Gelabert"). After the U. S. military left Panamá, the airport relocated to the former site of Albrook Air Force Base [0859/7933 (USBGN)]. Eisenmann, observations, 1961 Apr 9-May 16 (Eisenmann, 1962a, as "Paitilla Airfield," and as "Punta Paitilla Airfield"); Quinn, 1981 Dec 12 (GML, as "Paitilla Airport").

MARÉ, RÍO; see Moré, Río

MAREA, LA; see La Marea

MAREA ABAJO; see La Marea

MAREA ARRIBA; see La Marea

MARGARITA; Colón (Canal Zone) (E)
0920/7954 (USBGN)
Sea level, town 1 km S of Rainbow City [0921/7953 (USBGN)]. The dump is located on the W end of town just S of Mount Hope [0920/7954 (USBGN)] (E8660). Olson and Wiese, Margarita dump, 1966 Jul 15 (UF); Auzins, near, 1983 Dec 5 (WFVZ); Gómez, near, 1983 Oct 9 (WFVZ).

MARGARITA, ISLA; see Margarita Island

MARGARITA ISLAND; Colón (Canal Zone) (E)
0923/7953 (USBGN)
A peninsula located 3.5 km NE of Colón [0922/7954 (USBGN)] on the NE coast of Bahía Manzanillo [0922/7953 (USBGN)]. The N end of the peninsula is rounded like an island and is connected to the mainland by a narrow, 300-m strip of land. Fort Randolph Army Reservation [0923/7953 (USBGN)] was formerly located on the peninsula (E661, as "Margarita I"; E8660). Engleman, D. and Scott, observations, 1983 Nov 19-1984 Jan 2 (Ridgely and Gwynne, 1989:373, as "Isla Margarita"); Engleman D. and Engleman, L., observations, 1983 Nov 26 (Ridgely and Gwynne, 1989:158, as "Isla Margarita"); Engleman, D. and Wong, 1985 Oct 5 (WFVZ, as "Margarita I."); Gardner, 1985 Dec 1, 1986 Feb 2 (WFVZ, as "Margarita I."); Araúz, Garcés, and Mate, 1988 Sep 10, 23 (MVUP, as "Isla Margarita"); Candanedo, Garcés, Ponce, Mate, and Vergara, 1988 Oct 8-9, 28-29, Nov 11, 1989 Jan 9, Feb 3, Mar 4 (MVUP, as "Isla Margarita"); Batista, 1990 Jul 7 (MVUP, as "Isla Margarita").

MARGARITAS, LAS; see Las Margaritas

MARÍA; see María, Campamento de

MARÍA, CAMPAMENTO DE; Veraguas, Isla de Coiba
0725/8142 (USBGN)
Sea level, prison camp on the SE coast of Isla de Coiba [0728/8147 (USBGN)] 1.5 km SSW of Punta Felipa [0724/8142 (USBGN)] (JOG). Wetmore, GL = Isla de

Coiba, near camp on a small sandy beach bordering a rocky stretch of coast, walked S near the beach in a narrow stretch of mangroves, through *rastrojo* inland to the forest, 1956 Jan 24 (USNM, as "María"; Wetmore, 1957).

MARÍA, ENSENADA; Veraguas, Isla de Coiba
0724/8141 (USBGN)
Bay on the SE coast of Isla de Coiba [0728/8147 (USBGN)] just S of Bahía Damas [0727/8141 (USBGN)] (q. v.) extending ca. 8 km S and E along the shore from Punta Felipa [0724/8142 (USBGN)] to Estero Boca Grande [0724/8139 (USBGN)] (E762; JOG). See also: Blanca, Playa.

MARÍA CHIQUITA; Colón (E)
0927/7945 (USBGN)
Sea level, coastal town 20 km NE of Colón [0922/7954 (USBGN)] (JOG). Chapelle, observations, near a dairy farm, 1954 Aug 21 (Eisenmann, 1955c). See also: Piedras, Río.

MARÍA LETICIA, REPARTO; Panamá (W, Pacific)
ca. 0857/7951
Not shown on our maps but located in W Panamá province in Distrito de La Chorrera [0857/7951 (USBGN)] near La Chorrera [0853/7947 (USBGN)] (q. v.). Villarreal, 1983 Aug 30 (MVUP).

MARIATO, RÍO; Veraguas/Herrera (Pacific)
0737/8059 (USBGN)
Originates on the S and SE slopes of Cerro Alto Higo [0743/8051 (USBGN)] and flows SW entering Veraguas province, the river discharges into the E shore of Golfo de Montijo [0740/8107 (USBGN)] NE of Isla de Cébaco [0732/8109 (USBGN)] (JOG). Aldrich and Bole, drainage of, used to describe Altos Cacao [ca. 0739/8049], Cerro Viejo Camp [ca. 0737/8049], and Cavulla Camp [ca. 0737/8049], 1932 Feb 28-Mar 9, 11-14 (CM; Aldrich and Bole, 1937:7-8).

MARIATO RIVER; see Mariato, Río

MARIATO RIVER CAMP; Veraguas (Pacific)
ca. 0740/8054
250 feet (75 m), described by Aldrich and Bole as a camp that was also known as the Mariato Rubber Camp, located on the Río Mariato [0737/8059 (USBGN)] ca. 10 mi (16 km) E of its mouth "at the point where it emerges from its steep-sided mountain valley on to the coastal plain." It should be noted that the current town, Campamento de Mariato [0740/8101 (USBGN)], which is located on the coast at the mouth of the river, is not the site of the former camp. Aldrich and Bole, GL = Golfo de Montijo, Península de Azuero, in heavy semi-deciduous forest, in second growth jungle, and in heavy semi-deciduous forest on the lower mountain slopes, 1932 Feb 21-26 (CM; Aldrich and Bole, 1937:4 (map) 8, as "Mariato River (or Rubber) Camp").

MARIATO RUBBER CAMP; see Mariato River Camp

MARINA, RÍO; Panamá, Archipiélago de las Perlas
ca. 0815/7906

Described by Wetmore as a stream on Isla San José [0815/7907 (USBGN)] that originates in the N central hills, flows S and empties into a mangrove swamp on the N end of Ensenada Playa Grande [0815/7906 (USBGN)] (USNM). Murphy and Correia, Askoy Expedition, observations, 1945 (Murphy, 1944a; Wetmore, 1946b:14); Wetmore and Morrison, 1944 Feb-Oct 3 (Wetmore, 1946b). See also: San José, Isla. Not shown on our maps.

MARRAGANTÍ; Darién
0808/7744 (Fairchild and Handley, 1966)
Described by Goldman as the Darién Gold Mining Company's station on the Río Tuira [0821/7803 (USBGN)] located ca. 1 mi (1.5 km) above El Real de Santa María [0808/7743 (USBGN)], Goldman, 150 mi (240 km) E of Panama City [0858/7932 (USBGN)], near town, jungle, 1912 Mar 30-Apr 1, 4-6, May 11, Jun 21 (USNM; Nelson, 1912b; Nelson, 1913b). Not shown on our maps.

MARROQUIN, FILO DE; Los Santos
0729/8042 (USBGN)
Not named on our maps, but USBGN indicates the ridge 11 km NW of El Cortezo [0726/8038 (USBGN)] that gives rise to the Río Marroquincito [0725/8039 (USBGN)] and several other tributaries of the Río Tonosí [0721/8022 (USBGN)]. The ridge extends ca. 5 km E from the Veraguas/Los Santos provincial border into SW Los Santo province, and its elevation ranges from 700-1,100 m (JOG). Delgado collected on the ridge near El Cortezo [0726/8038 (USBGN)] (q. v.). Mentioned in Wetmore et al. (1984:635, as "Marroquin Mountains").

MARROQUIN MOUNTAINS; see Marroquin, Filo de

MARTINA, LA; see La Martina

MARTÍNEZ, FINCA; see Finca Martínez

MARTÍN GRANDE, RÍO; Veraguas (Pacific)
0758/8102 (USBGN)
Originates near Santiago [0806/8059 (USBGN)] and flows SSW joining the Río San Pedro [ca. 0750/8105] near Puerto Mutis [0755/8103 (USBGN)] (RDPF). See also: San Pedro, Río; and La Colorada.

MATA AHOGADA, RÍO; see Mata Ahogado, Río

MATA AHOGADO, RÍO; Panamá (W, Pacific)
0828/7957 (USBGN)
Originates in the hills NE of El Valle [0836/8008 (USBGN)] on the continental divide and flows SE entering the Gulf of Panama [0800/7930 (USBGN)] at San Carlos [0829/7957 (USBGN)] (E761B; MGRP). According to Wetmore, the headwaters of the Río Mata Ahogado are separated from the nearby headwaters of the Río Indio [0912/8011 (USBGN)] on the Atlantic drainage by the low ridge that connects Cerro Gaital [0837/8007 (USBGN)] and Cerro Agua [ca. 0838/8005], Wetmore and Perrygo, GL = El Valle, at 1,600 feet (480 m), head of the Río Mata Ahogado, SE of El Valle, on a trail along the Río Mamey (not located) to the Río Mata Ahogado, which was a clear, rapid stream in a narrow valley bordered by shadowing cliffs, abundant tree growth, small orange and cabbage plots, 1951 Mar

23, at 2,000 feet (600 m), head of the Río Mata Ahogado, below the ridge connecting Cerro Gaital and Cerro Agua, the stream was 6-10 feet (2-3 m) wide in a broad open valley with clearings for vegetables, sugar cane, water cress, patches of secondary growth brush, and creepers, 1951 Mar 23, 30-31 (USNM, as "Río Mata Ahogada"; Wetmore, 1959b, as "Río Mata Ahogada").

MATACHIN; Panamá (Canal Zone) (W, Atlantic)
ca. 0907/7943
Former station on the original Panama Railroad [ca. 0922/7954] that was located 1.5 mi (2.5 km) W of Gamboa [0907/7942 (USBGN)] on the W side of the continuous Río Chagres (see Río Chagres [0919/8000 (USBGN)]) (PR). Matachin Island [0907/7942 (USBGN)] remains above water in Lago Gatún (Gatun Lake) [0912/7955 (USBGN)] near the former station (E866). Goldman, 1911 May 6 (USNM). See also: Alhajuela; Bas Obispo; and Chagres, Río.

MATACHIN ISLAND; Panamá (Canal Zone) (W, Atlantic)
0907/7942 (USBGN)
30 m, small island with a land area of ca. 0.02 km^2 (2 ha) located in Lago Gatún (Gatun Lake) [0912/7955 (USBGN)] 1 km S of Gamboa [0907/7942 (USBGN)] just W of the Panama Canal [0920/7955 (USBGN)] (E866). See also: Matachin.

MATADERO, PUNTA; Panamá, Archipiélago de las Perlas
0827/7852 (USBGN)
The USBGN coordinates and Rendahl (1920:pl 1) indicate a point on the NE coast of Isla del Rey [0822/7855 (USBGN)]. Bovallius, 1882 May 2 (NHRM; Rendahl, 1920, as "Pta. Matadoro").

MATADORO, PUNTA; see Matadero, Punta

MATA PUERCO, RÍO; Panamá, Archipiélago de las Perlas
0815/7907 (USBGN)
Described by Wetmore as a stream that originates in S central Isla San José [0815/7907 (USBGN)] and flows NW discharging into Ensenada de Bodega [0816/7908 (USBGN)] (USNM). See also: San José, Isla. Not shown on our maps.

MATÍAS HERNÁNDEZ, RÍO; Panamá (E, Pacific)
0901/7929 (USBGN)
Originates in greater Panama City [0858/7932 (USBGN)] and flows SSE entering Bahía de Panamá [0850/7915 (USBGN)] just E of Parque Lefevre [0901/7929 (USBGN)] (E962). González, 1983 Mar 10 (MVUP). See also: Costa del Este.

MATISNILLA, RÍO; see Matisnillo, Río

MATISNILLO, RÍO; Panamá (E, Pacific)
0859/7931 (USBGN)
Short river that originates in Panama City [0858/7932 (USBGN)] and flows S entering Bahía de Panamá [0850/7915 (USBGN)] W of Punta Paitilla [0858/7931 (USBGN)] (E762). Goldman, 1912 Jan 20 (USNM, as "Río Matisnilla").

MATUGANTÍ; see Matugantí, Quebrada

MATUGANTÍ, QUEBRADA; Darién

ca. 0745/7730

Not shown on our maps, but according to Wetmore's notes, the stream is a tributary of the Río Tuira [0821/7803 (USBGN)] in S Darién province. Wetmore et al. (1984:344) indicate that Matugantí is above the mouth of the Río Púcuro [0801/7733 (USBGN)] and Wetmore's notes indicate that the stream originates W of the Río Tuira, probably near Cerro Setetule [0747/7736 (USBGN)]. Angehr (in litt., Aug 2005) indicates that "a town labeled 'Maturgandí' is shown on the 1:250,000 Special Map of Darién close to the coordinates" we have specified for Quebrada Matuganti. The town is on the Río Tuira ca. 4 km NW of the town Punusa. Through correspondance with Ridgely, he confirmed that "Matugantí" was near the location of "Maturgandí" on his map. Ridgely, 50 m, edge of second growth and forest, 1975 Jul 20 (GML; Wetmore et al., 1984:344, as "Matugantí," and 635, as "Quebrada Matujanti"); Ridgely and N. Smith, observations (Ridgely and Gwynne, 1989:316, as "Matugantí").

MATUJANTI, QUEBRADA; see Matugantí, Quebrada

MATURGANTÍ; see Matugantí, Quebrada

MATUSAGARATÍ, HUMEDALES DE; see Matusagratí, Laguna de

MATUSAGRATÍ, LAGUNA DE; Darién

0820/7756 (USBGN)

A wetland area that extends ca. 20 mi (32 km) NW from El Real de Santa María [0808/7743 (USBGN)] between the E bank of the Río Tuira [0821/7803 (USBGN)] estuary and the Serranía el Filo de Tallo [0827/7757 (USBGN)] (MRP, as "Laguna de la Pita"; SMN2A, as "Pita Lagoon"; SMN2B). Festa, according to his travel account, he entered the swamp from the E bank of the Río Tuira ca. 7 km N of Chepigana [0818/7803 (USBGN)] from a spot called Punta Pita [ca. 0821/7802], and from here traveled a few miles E to the N end of the swamp, 1895 Aug 7-20 (Salvin and Godman, 1897-1904, vol. 3:319, 328, as "Laguna de Pita"; Salvadori and Festa, 1899, as "Laguna della Pita"; Salvadori and Festa, 1900:41-42, as "Laguna della Pita"; Festa, 1909, as "Laguna della Pita"). (Angehr, 2003:312, as "Humedales de Matusagaratí").

MAURO, RÍO; Panamá (E, Atlantic)

0925/7933 (USBGN)

Not shown on our maps, but USBGN indicates a tributary of the Río Boquerón [0921/7933 (USBGN)] that discharges ca. 4 km NNE of Estación Hidro el Peluca [0923/7934 (USBGN)]. See also: Número Uno, Campamento; and Pequení, Río.

MAXON RANCH; see Cuchirvo

MEDIA, RÍO; see Medio, Río

MEDIDOR, ISLA; see Canales de Tierra, Isla

MEDIDOR ISLAND; see Canales de Tierra, Isla

MEDIO, RÍO; Colón (Canal Zone) (W)

0916/8003 (USBGN)

Originates near the NW coast of Lago Gatún (Gatun Lake) [0912/7955 (USBGN)] and flows NW entering the Río Piña [0917/8003 (USBGN)] just above its mouth 18 km SW of Colón [0922/7954 (USBGN)] (E661, as "Rio Media or South Branch Río Piña"; E761A, as "Río Media"). Miller, near, unclear if this is the correct river as Miller's specimen labels do not indicate a province. He also was working at Cerro Corozal [0859/7934 (USBGN)] and on Las Cruces Trail [ca. 0904/7936] at this time, 1937 Feb 3, 11 (USNM); Willis, observations, 1963 Sep 2-3 (Karr, 1971a; Willis, 1972). See also: Road S9.

MEMBRILLAR, QUEBRADA; see Membrillar, Río

MEMBRILLAR, RÍO; Colón (W)
0911/8011 (USBGN)
Originates SSE of El Chilar [0909/8012 (USBGN)], flows N and joins the Río Indio [0912/8011 (USBGN)] less than 1 km above its mouth (E761B, as "Quebrada Membrial"). Wetmore and Perrygo, GL = Río Indio, 0.25-1 mi (0.5-1.5 km) upstream in an area of green and lush second growth with some tall trees remaining, clearings for rice, 1952 Feb 16, 20 (USNM, as "Quebrada Membrillar").

MENDIETA, LAS; see Las Mendieta

MENDOSA, RÍO; see Frijolita, Río

MENDOZA, RÍO; see Frijolita, Río

MENSABÉ, PUERTO; see Puerto Mensabé

MENSABÉ, RÍO; see Puerto Mensabé

MERCADO PÚBLICO; see San Felipe

MERCEDES, LAS; see El Dorado

MESETA DE CHORCHA; see Gran Galera de Chorcha

METROPOLITAN NATURE PARK; see Metropolitano, Parque Natural

METROPOLITANO, PARQUE NACIONAL; see Metropolitano, Parque Natural

METROPOLITANO, PARQUE NATURAL;
0859/7933 (Angehr, 2003:245)
Panamá (Canal Zone) (E, Pacific)
Park adjacent to the W edge of Betania [0901/7932 (USBGN)] and S of Parque Nacional Soberanía [0909/7941 (Angehr, 2003:239)] (E762, as "Parque Nacional Metropolitano"). Some of the land was formerly included in Fort Clayton Army Reservation [0901/7934 (USBGN)]. According Engleman et al. (1989:479-480, as "Metropolitan Nature Park"), the park contains tropical semideciduous forest and covers over 500 acres (200 ha) of land. Arosemena and Herrera, 1990 Apr 7 (MVUP, as "Parque Metropolitano," and as "Parque Natural Metropolitano").

MI CASITA; see Casita

MIDDLE CAY; Bocas del Toro, Archipiélago de Bocas del Toro
ca. 0913/8156
Described by Miller as an island in Cayos Tiger [0913/8156 (USBGN)], that was 100 yards (90 m) long with an elevation of ca. 40 feet (10 m) (UCLA). According

to Miller's description and the name "middle," Miller collected on a small island located between the two largest islands in the group. According to E762, the island is 300 m WNW of Tiger Rock [0912/8155 (USBGN)] and just E of the 2nd largest island in the group. Miller and Richardson, GL = Cayos Tiger, landed on the lee side, coconut palms and *Cecropia* trees grew along the margins and the center was covered with head-tall grass that was as "tough as a barbed wire entanglement." The specimen labels have "Tiger Rock, Tiger Cays" penciled in, whereas the field notes refer to Middle Cay, and the original label appears to be simply "Chiriqui Lagoon," 1936 Mar 12 (UCLA); SBE, GL = Península Valiente, Cayos Tiger, 1990 Apr 24 (USNM, as "Middle Tiger Rock"). See also: Tiger, Cayos; and Tiger Rock.

MIDDLE TIGER ROCK; see Middle Cay

MILE 2; see Milla 2

MILE 5; see Milla 5

MILITARY INSTRUCTION CENTRE; see C. I. M.

MILLA 1; Bocas del Toro

ca. 0918/8225

Sea level, railroad stop on the W outskirts of Almirante [0918/8224 (USBGN)] (E762). According to Galindo et al. (1963), GMLP banded birds near the ball ground "on the edge of the 'One-Mile' sector of Almirante. This plot consist[ed] of an open field of short grass crossed by two drainage ditches and bordered by second-growth thickets of thorny bushes." See also: Almirante.

MILLA 2; Bocas del Toro

ca. 0918/8225

10 m, railroad stop located 3 km NW of Almirante [0918/8224 (USBGN)] (E762). According to Galindo et al. (1963), GMLP banded birds near the railroad in a plot that was "located in a small abandoned cattle farm, mostly covered by grass reverting to second growth in various stages of development. To the North the plot enclos[ed] a hill covered by thinned-out lowland tropical rainforest." Most of the specimens collected by GMLP are simply labeled Almirante. Hinds, GL = Almirante, 1961 Jan (USNM, as "2 Millas"); Mendez (GMLP), GL = Almirante, 1963 Aug 23 (USNM, as "Milla 2 (Mile 2)"); Olson, GL = Almirante, vicinity of Milla 2, 1963 Apr 6-12 (USNM, as "Almirante"; Loftin and Olson, 1963). See also: Almirante.

MILLA 5; Bocas del Toro

ca. 0921/8226

10 m, railroad stop located 6.5 km NW of Almirante [0918/8224 (USBGN)]. Several small, intermittent tributaries of the Río Banano [920/8223 (USBGN)] cross the railroad near this stop (E762). According to Galindo et al. (1963), GMLP banded birds in a plot that was located "in an open field bordered by banana and cacao plantations." Hinds, GL = Almirante, 1962 Mar 28 (USNM, as "Mile 5"). See also: Almirante.

MILLA 7; Bocas del Toro

ca. 0921/8226

Sea level, railroad stop located 8.5 km NW of Almirante [0918/8224 (USBGN)] (E762). Wedel, GL = Almirante, 1926 Aug 14 (ANSP, as "7 mi"); Benson, GL = Almirante, at 300 feet (90 m), 1927 Jul 2 (USNM, as "7 mi"). See also: Almirante.

MINA, ISLA; see La Mina, Isla

MINA DE CHORCHA; Chiriquí

ca. 0825/8216

Alt. ?, a locality of Enrique Arcé that was placed very generally by Salvin (1870:177) on the S slopes of Volcán de Chiriquí [= Volcán Barú]. Wetmore (1968:124, 461) believed that the locality was a short distance E of David [0826/8226 (USBGN)]. Based on the species composition of Arce's specimens from this locality (Salvin, 1870b), Angehr (in litt., Oct 2003, Aug 2005) believes the locality was likely the town Cerro de La Mina. USBGN indicates a town Cerro de la Mina at 0825/8216 that appears on E762 just W of Gran Galera de Chorcha [0826/8213 (USBGN)] 6 km NNE of Chorcha Abajo [0822/8217 (USBGN)]. Arcé, 18?? (USNM, as "Mina Chorcha"; Salvin, 1870b; Berlepsch, 1911; Hellmayr, 1918; Davidson, 1938).

MINA GRANDE, ISLA; see La Mina, Isla

MINA ISLAND; see La Mina, Isla

MINAS, LAS; see Las Minas

MINDI; Colón (Canal Zone) (E)

ca. 0918/7954

ca. 10 feet (3 m), station on the original Panama Railroad [ca. 0922/7954] that was located 2 mi (3 km) NE of Gatún [0916/7955 (USBGN)] just off the S coast of Bahía Limón [0920/7956 (USBGN)] near both a swamp and a river also named Mindi (PR). Mindi Dairy Farm [0918/7954 (USBGN)] appears on E761A 4 km NNE of Gatún and ca. 1 km E of Mindi Dock [ca. 0918/7955], which is located on the Panama Railroad. USBGN indicates that Mindi Dairy Farm is a populated place, but according to Angehr (in litt., Aug 2005), USBGN commonly lists small farms as populated places. Hallinan, on trails and roads, above a swamp, grass, thick shrubbery, 1909 Jan 28, Feb 19, 1915 Jul 11 (AMNH; Hallinan, 1924); Jewel, grassy marshes, tree in a pasture, 1911 Sep 17, Oct 7-22, 29, Nov 3, 1912 Sep 1 (AMNH; ANSP; Jewel, 1913; Stone, 1918, as "Mindi," and as "Minti"); Eisenmann and Chapelle, observations, at the dairy farm, 2 mi (3 km) S of the dairy near wet pastures near Carretera Bolivar [0918/7953 (USBGN)] between Fort William D. Davis Army Reservation [0917/7953 (USBGN)] and Gatún (Eisenmann, 1955c, as "Mindi Dairy").

MINDI DAIRY; see Mindi

MINDI DAIRY FARM; see Mindi

MINDI DOCK; see Mindi

MINERAL; Veraguas (Atlantic)

ca. 0845/8058

200 m, town in NE Veraguas province that was connected to Santa Fé [0831/8105 (USBGN)] by a foot path and located on the Río Santiago [0848/8058 (USBGN)]

ca. 36 km N of Santa Fé and ca. 14 km S of the coast (MHA; Salvin, 1870b (map)). The town is not shown on recent maps but the corresponding region appears ca. 30 km NNE of Santa Fé on JOG, which was used to estimate the lat/long coordinates for the town. See also: El Mineral de Veraguas.

MINERAS; see Las Mineras

MINERAS, LAS; see Las Mineras

MIÑON, CERRO; see Miñón, Sierra

MIÑÓN, SIERRA; Panamá (Canal Zone) (W, Pacific)
0856/7937 (USBGN)
660 feet (200 m), hill located 7.5 km WSW of Panama City [0858/7932 (USBGN)] (E761A). Christian, at ca. 30 m, 1 km N, 1997 Feb 17 (LSUMZ).

MINTI; see Mindi

MIRADOR; Chiriquí
ca. 0850/8229 ?
A lookout point on Finca Lérida Estate, Mönniche's former plantation that is still in operation. Blake indicates that Mönniche's specimen of *Empidonax atriceps* was collected at Mirador on the Caribbean Slope, but he does not provide a description of the locality. Ridgely and Gwynne (1989:229) refer to "the 'Mirador' on Finca Lérida," and Engleman et al. (1989:498) indicate that the birding on Finca Lérida is "especially good [at] the El Mirador area." Mirador translates into lookout point, and if the specimen came from Bocas del Toro province, Mönniche probably traveled over the divide from this area of his plantation. Mönniche, below 9,500 feet (2,900 m), "Caribbean slope," 1933 July 17 (FMNH; Blake, 1958:540).

MIRAFLORES; Colón (Canal Zone)
ca. 0915/7955
Former town that was located 1 km S of Gatún [0916/7955 (USBGN)] on the continuous Río Chagres (see Río Chagres [0919/8000 (USBGN)]) before it was submerged in Lago Gatún (Gatun Lake) [0912/7955 (USBGN)] (IP). Goldman, 1911 Apr 30, May 13 (USNM); Custer, unclear if the label Miraflores refers to the former town, which was probably submerged by 1915, or to Miraflores Locks [0900/7936 (USBGN)], which are located on the Pacific drainage of the Panama Canal [0920/7955 (USBGN)], 1915 Feb 15 (USNM); Specimens examined, Collector ?, specimens may be from Miraflores Locks, date ? (Wetmore, 1970; Olson, 1981d). See also: Miraflores, Lago de; and Miraflores Locks.

MIRAFLORES, LAGO DE; Panamá (Canal Zone) (Pacific)
0901/7936 (USBGN)
One km wide lake that surrounds the Panama Canal [0920/7955 (USBGN)] for ca. 2 km SSE of Pedro Miguel [0901/7937 (USBGN)] (E8660). Murdoch, 1954 Nov 5 (GML, as "Miraflores Lake"); Loftin, 1961, autumn 1962 (Loftin, 1963a, as "Miraflores Lake"). See also: Miraflores Locks.

MIRAFLORES LAKE; see Miraflores, Lago de

MIRAFLORES LOCKS; Panamá (Canal Zone) (Pacific)

0900/7936 (USBGN)

Locks on the Panama Canal [0920/7955 (USBGN)] just S of Lago de Miraflores [0901/7936 (USBGN)] and 4 mi (6 km) NNW of Balboa [0857/7934 (USBGN)] (E661; RM1). The locks are accessed from Panama City [0858/7932 (USBGN)] by Carretera Gaillard [0904/7940 (USBGN)], therefore most specimens were probably collected on the E side of the Panama Canal. Jewel, worked primarily near Gatun, although Stone (1918:241) indicates that Miraflores is "nearer to Panama," 1911 Mar 5 (ANSP, as "Miraflores"; Stone, 1918, as "Miraflores"), "small lake near," 1911 Nov 26 (Jewel, 1913, as "Miraflores (Pacific Side)"); Deignan, 1932 Jul 16 (FMNH); Griswold, 1936 Apr 28-29 (MCZ); Collector ?, captive bird from National Zoological Park, 196? (USNM); Olson, mostly open with a power line cut, guinea grass, mangrove swamps nearby, 1966 May 1-3, 7, 12, 14, 17, 30, Jun 1-2, 21-24, 27-29, Jul 8, 28, Aug 2 (FMNH; UF; USNM; Olson, 1968; Wetmore, 1972:417, as "Miraflores Lake"); Mueller, 1966 Jun 21 (USNM); Wolf, NW of Balboa, 1966 Jul 27 (MVZ); De Leon, 1996 Nov 15 (GML). See also: Miraflores.

MIRAMAR; Bocas del Toro

0900/8215 (USBGN)

Sea level, town on the SW coast of Laguna de Chiriquí [0903/8200 (USBGN)] 3 km S of the mouth of the Río Róbalo [0901/8215 (USBGN)] and 15 km NW of Chiriquí Grande [0857/8207 (USBGN)] (E762; JOG). The town is N of Cerro Miramar [0859/8215 (USBGN)] and SE of Miramar Arriba [0900/8215 (USBGN)] (E762). Martínez, the only specimen from here is a coastal species (*Rallus longirostris*), so the distance indicated on the label "22 mi (35 km) W of Chiriquí Grande," cannot be correct even if intended as measured along the coast rather than due W, which is far inland (ANSP, as "Miramar Village").

MIRAMAR; Chiriquí

ca. 0848/8226 ?

Wetmore believed the locality was on the southern slopes of Volcán Barú [0848/8233 (USBGN)] (Wetmore, 1972:24-25). Our maps do not indicate a Miramar in Chiriquí province, but the name is used in Bocas del Toro province for Miramar [0900/8215 (USBGN)] (q. v.), Cerro Miramar [0859/8215 (USBGN)], and Miramar Arriba [0900/8215 (USBGN)], all of which are near the coast and at low elevations. Six km SSW of Cerro Miramar, which is at 400 m, the elevation rises to 1,100 m; and although our maps indicate that this area is roadless, several streams originate here and flow NNE into Laguna de Chiriquí [0903/8200 (USBGN)] (E762). Lara, A. at 1,100 m, 1897 Sep (AMNH; Wetmore, 1972:24).

MIRAMAR, CERRO; see Miramar

MIRAMAR ARRIBA; see Miramar

MIRAMAR VILLAGE; see Miramar

MOCAMBO, CANTERA DE; Panamá (E, Pacific)

ca. 0903/7934

Presumably a quarry located near Mocambo Abajo [0904/7933 (USBGN)] and Mocambo Arriba [0904/7934 (USBGN)], which are 11 km and 13 km, respectively,

N of Panama City [0858/7932 (USBGN)]. The name may refer to an abandoned mine shown on E762 at ca. 0903/7934 on Camino Chiva Chiva (Chiva Chiva Road) [0902/7935 (USBGN)] 3 km SW of Mocambo Abajo and ca. 2 km NE of Chiva Chiva [0902/7935 (USBGN)]. This quarry is also mentioned under the description for Chiva Chiva. Tejera, Araúz, Ponce, Riley, *área del canal*, 1984 Oct 6, 14, 21, Nov 18 (MVUP); Araúz and Durán, 1984 Nov 25 (MVUP). See also: Chiva Chiva.

MOCAMBO, RÍO; see Mocambo, Cantera de

MOCAMBO ABAJO; see Mocambo, Cantera de

MOCAMBO ARRIBA; see Mocambo, Cantera de

MOGO MOGO, ISLA; Panamá, Archipiélago de las Perlas
0834/7902 (USBGN)
Sea level, small island with a land area of ca. 0.8 km^2 (80 ha) located 15 km NW of Isla del Rey [0822/7855 (USBGN)] (E501, as "Isla Pajaros"; JOG). Bovallius, very hilly, covered with tropical forest, see discussion of Bovallius under Pueblo Nuevo [ca. 0820/8013 ?], 1882 April (NHRM; Rendahl, 1920, as "Pajaros (or Mogo Mogo)"); Levings et al., observations, 1982-1983 (Levings et al., 1986).

MOJA CULO; Darién
ca. 0729/7759
ca. 100 m, described by Wetmore as a section of the Río Jaqué [0731/7810 (USBGN)] between the mouth of the Arroyo Tortadó [ca. 0730/7801] and the mouth of the Río Imamadó [ca. 0727/7758], where the river was ca. 40 feet (10 m) wide and shallow with gravelly rapids, the valley had little cultivation, and the steep-sided hills were covered with heavy forest, Wetmore and Perrygo, GL = Río Jaqué, 1947 Apr 6 (USNM). Not shown on our maps.

MOJICA, RANCHO; Bocas del Toro
ca. 0902/8241
4,800 feet (1,450 m), described by Wetmore as a camp used by MCU that was located on the N slope of Cerro Pando [0855/8243 (USBGN)] near the headwaters of the Río Changuena [ca. 0908/8230], Tipton, GL = Río Changuena, at 4,800 feet (1,450 m), 1961 Sep 10 (USNM). Not shown on our maps.

MOJINGA, PANTANO; Colón (Canal Zone) (W)
0918/7958 (USBGN)
ca. 10 m, swamp located on the S border of Fort Sherman Army Reservation [0920/7958 (USBGN)] 6.5 km NW of Gatún [0916/7955 (USBGN)]. The Río Mojinga [0918/7959 (USBGN)] originates in the swamp and flows SW for 1 km before entering the Río Chagres [0919/8000 (USBGN)] 7 km NW of Gatún (E762, as "Mojinga Swamp"). According to Angehr (2003:217, as "Mojinga Swamp"), the swamp "includes cativo (*Prioria copaifera*) forest and an open marsh dominated by herbaceous vegetation and palms" and is included in the Important Bird Area Fuerte San Lorenzo [0916/7958, Angehr, 2003:217)]. The locality appears always to have been written on specimen labels as "Mojinga Swamp." Rojas, 1967 Mar 2, 1968 May 14 (GML); Montenegro, 1979 May 24, Jun 5, Jul 5, 18, Sep 28, Oct 17, Nov 1, 8, Dec 14, 1980 Jan 11, Feb 1, Mar 13, Apr 10, May 2, 8, 12, 15, Jul 2, Nov 11, Dec

19, 1981 Feb 18, Apr 13, May 28, Jun 18, Sep 24, Nov 4, 1982 Jan 7, 1983 May 12, Dec 29 (GML); Collector ? [Montenegro ?], 1979 Sep 28, 1982 Jul 1, 1983 Apr 7, 14, 19, Jun 23, Jul 7 (GML); GMLP, 1981 Jun 28 (GML); Hinds, 1983 Sep 1 (GML).

MOJINGA, RÍO; see Mojinga, Pantano

MOJINGA SWAMP; see Mojinga, Pantano

MOLEJON; Coclé (Atlantic)

ca. 0847/8032

Not shown on our maps but the coordinates indicate a region on our maps (JOG; MGRP) on the upper reaches of the Río Coclé del Norte [0905/8035 (USBGN)] at 200-400 m, ca. 18 km NE of El Copé [0837/8036 (USBGN)], and ca. 5 km S of Coclecito [0849/8033 (USBGN)]. According to Miller (personal communication), it is located outside of Omar Torrijos Parque Nacional [0841/8040 (Angehr, 2003:177)] along the road to Coclecito. Bacon, Guitton-Mayerma, and Miller, Finca Moreno (a private farm), 08°47'19"/80°32'28", 2002, Sep 24-30, Oct 07-15, 25-30 (STRI; UAM).

MOLINO COVE; see Piña, Bahía

MONAGRE; Los Santos

0758/8021 (USBGN)

Sea level, coastal town 8 km NE of Los Santos [0756/8025 (USBGN)] and 5 km SE of the mouth of the Río La Villa [0759/8023 (USBGN)] (JOG). Collections made at the "Estero" probably came from the Río La Villa, and those from the "Playa" either came from the river or the coast. Wetmore and Perrygo, GL = Los Santos, sandy clay coastline bordered by thorny scrub and cactus, ca. 0.5 mi (1 km) N of Monagre at Estero Monagre, which was narrow and winding and bordered by mangroves "with extensive barrens that were flooded at high tide," 1948 Mar 16 (USNM); Bernal and Vasquez, 1978 Apr 8, 1991 Aug 24 (MVUP, as "Playa Monagre").

MONAGRE, ESTERO; see Monagre

MONAGRE, PLAYA; see Monagre

MONAGRILLO; Herrera

0759/8026 (USBGN)

ca. 25 m, town 1 km N of Chitré [0758/8026 (USBGN)] (JOG). Wetmore and Perrygo, GL for Río Parita, 1948 Feb 25 (USNM; Olson, 1980b, as "Monagrillo, Río Parita"); Quintero and Saez, 1984 Aug 30 (MVUP); Saucedo, 1985 Aug 22 (MVUP).

MONO, RÍO; Darién

0743/7733 (Fairchild and Handley, 1966)

Originates in S Darién province on the Colombian border and flows N entering the Río Tuira [0821/7803 (USBGN)] N of the Río Mangle [0741/7734 (USBGN)] (SMN2A). Galindo, 1969 Feb 18 (Wetmore et al., 1984:635); Hinds and GMLP, at 2,400-2,600 feet (750-800 m), 1969 Jan 28-31, Feb 2-3, 8, 22, Mar 3-4, 1970 Mar 8-9, 1971 Jan 14, Feb 17-28, Mar 1-7, 9-18, 21, 1972 Mar 11, 18-23, 25, 27, 30-31,

Apr 1, 3-7, 9-12, 1974 ?, 1976 ?, 1979 ?, 1985 Aug 21-24, 27, Sep 4-5 (GML); Galindo and GMLP, GL for Altos de Quía, 1970 Feb 21, 1971 Feb 21, 23, Mar 15 (USNM; Wetmore and Galindo, 1972); Collector ? [GMLP ?], 1972 Apr 8 (LSUMZ); Ridgely, observations, confluence with the Río Tuira, 1975 Jul (Wetmore et al., 1984:482).

MONO, RÍO; see Azul, Cerro

MONOS, SENDERO LOS; see Coiba, Isla de

MONTE LIRIO; Colón (Canal Zone) (E)
0914/7951 (USBGN)
20 m, a town on the present Panama Railroad [ca. 0922/7954] 9.5 km NW of Frijoles [0910/7948 (USBGN)] on the NE coast of Islas Juan Gallegos [0913/7952 (USBGN)] in Lago Gatún (Gatun Lake) [0912/7955 (USBGN)] (E866). The town was not located on the original Panama Railroad line, but on the Río Gatún [0916/7947 (USBGN)] ca. 5 mi (8 km) E of its mouth and 4 mi E of Tiger Hill [ca. 0914/7954] (MIP). When Lago Gatún (Gatun Lake) was filled, the course of the railroad moved E to include the town as a station stop. Goldman, GL = Río Gatún, 1911 Jan 24 (USNM).

MONTE MABOBO; Panamá (E, Pacific)
ca. 0902/7914
ca. 20 feet (6 m), described by Wetmore as a tract of forest a short distance SE of Ciénaga de Santo Domingo [ca. 0903/7915], Wetmore and Perrygo, GL = Río Chico [0902/7910 (USBGN)], tall, green forest with an undergrowth of palms, vines, and shrubs, 1949 Mar 22 (USNM). Not shown on our maps.

MONTERIA; Panamá (E, Pacific)
ca. 0905/7926
Not shown on our maps but located in central Panamá province in Corregimiento Pedregal [0905/7926 (USBGN)] near the town Pedregal [0904/7926 (USBGN)] (q. v.). Moore and Lima, 1990 Feb 28 (MVUP); Moore and Morris, 1990 Apr 3 (MVUP); Moore, 1990 Sep 4-5 (MVUP).

MONTE TEODORO; Darién
ca. 0754/7727
Described by Wetmore in 1959 as several acres of land cleared ca. 3 year previously without having been cultivated, located on the Río Paya [0757/7730 (USBGN)] 7 river mi (11 km) above its mouth. See also: Tuira, Río. Not shown on our maps.

MONTE VERDE; Chiriquí
ca. 0817/8252
100 m, town 3 km NW of Puerto Armuelles [0817/8252 (USBGN)] (JOG). Wetmore, GL = Puerto Armuelles, along the road between Puerto Armuelles and Monte Verde through pastures with occasional stands of trees, 2 year old *rastrojo* thick with *Heliconia* and occasional dead trees, 1966 Feb 24, Mar 2 (USNM).

MONTIJO; Veraguas (Pacific)
0759/8103 (USBGN)
ca. 50 m, town 16 km SSW of Santiago [0806/8059 (USBGN)] just N of the N

coast of Golfo de Montijo [0740/8107 (USBGN)] and 15 km N of the mouth of the Río San Pedro [ca. 0750/8105] (JOG; RDP2). Guevara and Ureña, 1990 Feb 23 (MVUP); DaCosta, Guitton-Mayerma, Klicka, Lelevier, Miller, and Spellman, N of Golfo de Montijo, at 66 m, 08°00'15"/81°04'36", 4 km NW of, 2006 Mar 4 (MBM; STRI).

MONTIJO, GOLFO DE; Veraguas (Pacific)
0740/8107 (USBGN)
Bay that extends 27 km N into the S coast of Veraguas province just W of Península de Azuero [0740/8035 (USBGN)], its shores are largely islands amid mangrove swamps receiving the discharge of numerous rivers including the Río San Pablo [0751/8110 (USBGN)] and the Río San Pedro [ca. 0750/8105]. The bay opens on the S to the Pacific Ocean, where it extends 27 km from E to W just N of Isla Gobernadora [0734/8112 (USBGN)] and Isla de Cébaco [0732/8109 (USBGN)] (JOG; Wetmore, 1957). Aldrich and Bole, GL for: Altos Cacao, Cavulla Camp, Cerro Viejo Camp, Mariato River Camp, and Paracoté, 1932 Feb 5-19, 21-26, 28-Mar 5-9, 20-Apr 2 (CM; Aldrich and Bole, 1937, as "Montijo Bay"; Olson, 1981d, as "Montijo Bay"). See also: Montijo; Palo Seco, Río; and Puerto Mutis.

MONTIJO BAY; see Montijo, Golfo de

MONTIJO GULF; see Montijo, Golfo de

MONTOSA, CERRO; see El Montuoso, Loma

MONTUOSA, CERRO; see El Montuoso, Loma

MORA, QUEBRADA; Darién
ca. 0725/7759 ?
Described by Wetmore as a tributary of the Río Imamadó [ca. 0727/7758] upstream beyond the mouth of the Río Chicao [ca. 0727/7758], Wetmore and Perrygo, GL = Río Jaqué, 1947 Apr 10 (USNM). Not shown on our maps.

MORAVIA; Chiriquí
ca. 0850/8246 ?
Not located on our maps but must be near Santa Clara [0850/8246 (USBGN)] on the Quebrada Santa Clara [ca. 0848/8251]. Wetmore (from Alois Hartmann), GL = Santa Clara, Quebrada Santa Clara, 1954 Mar 2 (USNM).

MORÉ, RÍO; Colón (E)
0934/7939 (USBGN)
Described by Goldman (1920:7) as one of several streams that discharges into the bay near Portobelo [0933/7939 (USBGN)] (q. v.). Not shown on our maps but the coordinates indicate that the stream originates N of Portobelo entering the Caribbean Sea [1500/7500 (USBGN)] very near the Río Cascajal [0934/7939 (USBGN)]. Wetmore (1972:408, as "Río Maré") indicates that Goldman collected a nest here, although specimens appear to be labeled simply "Porto Bello." Goldman, GL = Portobelo, near Portobelo, banks of, 1911 Jun 1 (USNM, as "Porto Bello"; Goldman, 1920; Wetmore, 1972:408, as "Río Maré").

MORENA, ISLA; Panamá, Archipiélago de las Perlas

ca. 0820/7900

According to Angehr (in litt., Aug 2005), Isla Morena appears on the DMA nautical chart 21605 "Panama Canal to Golfo de San Miguel" (1978) as "a small island about 5 km W of Punta Trapiche [0820/7858 (ADL)] on Isla del Rey [0822/7855 (USBGN)]. There is some confusion about the name, however, because on 4341-III [E762], this same island is called Isla Aposentro (UTM 922000 N 719500 E; 0820/7900). DMA 21605, on the other hand, uses the name 'Isleta Aposentro' for a smaller island about 2 km E of 'Isla Morena;' this island is unnamed on 4341-III...Morena/Aposentro is about 500 m long and well forested, and certainly the site where Morrison would have collected. Given Morrison's use of the name Morena for the island in 1944, I would assume that DMA 21605 has the names correct and 4341-III has it wrong." In addition, Isla Morena is shown incorrectly 1 km N of Isla San José [0815/7907 (USBGN)] on two of our older maps (E501; MHA). Morrison, 1944 May 19 (USNM, as "Moreno Island"; Wetmore, 1946b:1, as "Moreno Island"; Olson, 1980b, as "Isla Moreno"); (MacArthur et al., 1972—species list; see comments in Bibliography).

MORENITA; see Moreno, Cerro

MORENO, CERRO; Coclé (Atlantic)

ca. 0845/8032

Not shown on our maps but the coordinates indicate a region (JOG; MGRP) on the upper reaches of the Río Coclé del Norte [0905/8035 (USBGN)] at ca. 400 m, ca. 16 km NE of El Copé [0837/8036 (USBGN)], and ca. 9 km S of Coclecito [0849/8033 (USBGN)]. According to Miller (personal communication), the locality is outside of Omar Torrijos Parque Nacional [0841/8040 (Angehr, 2003:177)] along the road to Coclecito. USBGN indicates a town named Moreno [0837/8028 (USBGN)] just S of this region. Bacon, Guitton-Mayerma, and Miller, at the small community Morenita, 2002 Oct 31-Nov 5 (STRI; UAM).

MORENO, ISLA; see Morena, Isla

MORGAN'S GARDENS; Panamá (Canal Zone) (E, Pacific)

ca. 0859/7934

Dyrcz (1983) describes the locality as a "private plant nursery, with many high trees, open grassy patches and clumps of bushes, surrounded by deforested areas with some settlements. A branch of the Panama Canal (influenced by tides) borders the study area." He states that the gardens are 5 km N of Balboa [0857/7934 (USBGN)], which corresponds roughly with Cardenas [0900/7934 (USBGN)] and Fort Clayton Army Reservation [0901/7934 (USBGN)]. According to Angehr (in litt., Aug 2005), Morgan's Gardens "was just W of the Gaillard Highway and the Railroad tracks opposite Corozal Cemetery. It would be closer to 4 km north of Balboa than 5 km, as Dyrcz has it." Boyden, observations at flowering bushes, 1974 Jun 23-25 (Boyden, 1978); Dyrcz, observations of nests, 1979 Mar-Jun (Dyrcz, 1982; Dyrcz, 1983); Morton, observations, 1970s (Morton, 1982).

MORRO ISLAND; see Taboga, Morro de

MORTÍ; see Mortí Hydro

MORTÍ, RÍO; Darién

0851/7759 (USBGN)

Originates on the continental divide in N Darién province and flows S then SW entering the upper reaches of the Río Chucunaque [0809/7744 (USBGN)] 52 km NNE of La Palma [0825/7809 (USBGN)] (MGRP). See also: Mortí.

MORTÍ HYDRO; Darién

0852/7754 (Myers, 2003)

According to Myers (2003), this site was a temporary helipad on the Río Mortí, for which he gives the coordinates 08°52'28" N/77°54'19" W, and an elevation of ca. 100-200 m. The site was 2.7 km NNE of the Cuna village Mortí Arriba, which USBGN names Mortí [0851/7755 (USBGN)]. According to Angehr (in litt., Aug 2005), no hydroelectric projects exist on the Río Mortí, therefore the helipad was likely for surveying a proposed dam site. Quintero, forest, 1967 Oct 27, 28, Nov 22, 24, 29, Dec 2, 6, 8 (GML).

MOSQUITOS, GOLFO DE LOS; Bocas del Toro/Veraguas/Colón

0900/8115 (USBGN)

Gulf on the NW coast of Panamá extending along the NE coast of Península Valiente [0905/8151 (USBGN)] to the N coast of Veraguas province and the SW coast of Colón province. The gulf opens on the N to the Caribbean Sea [1500/7500 (USBGN)] (JOG).

MOUNT CHIRIQUÍ; see Barú, Volcán

MOUNT HOPE; Colón (Canal Zone) (E)

0920/7954 (USBGN)

Sea level, town located 2 km S of Colón [0922/7954 (USBGN)] just W of Rainbow City [0921/7953 (USBGN)] on the NE side of French Canal [0920/7955 (USBGN)] (E761A, as "Mt. Hope"; E8660). Jewel, 1912 Feb 9 (?), 1913 Feb 9, 20, 22, Apr 13, Jun 1, Aug 31, Sep 13 (ANSP, as "Mt Hope"; Stone, 1918, as "Mt. Hope"); Ambrose, observations, 1955 (Wetmore, 1972:417).

MOUNT KITCHENER; Darién

ca. 0759/7735 ?

Described by Goldman in 1912 as a station along the route from El Real de Santa María [0808/7743 (USBGN)] to Cana [ca. 0745/7742] (q. v.) located ca. 12 mi (19 km) by train from Boca de Cupe [0803/7735 (USBGN)] and ca. 12 mi (19 km) by train from Paca [0759/7735 (USBGN)], presumably located along the lower reaches of the Río Paca [0759/7732 (USBGN)] in the NE foothills of Serranía de Pirre [ca. 0749/7744] as the train ran by gravity from the Paca train station down to Mount Kitchener, Goldman, 1912 Jun 6 (USNM). Not shown on our maps.

MOUNT PIRRE; see Pirre, Cerro

MOUNT SAPO; see Sapo, Cerro

MOUNT TACARCUNA; see Tacarcuna, Cerro

MOUNT TATARCUNA; see Tacarcuna, Cerro

MT. CHIRIQUÍ; see Barú, Volcán

MT. PIRRE; see Pirre, Cerro

MT. SAPO; see Sapo, Cerro

MT. TACARCUNA; see Tacarcuna, Cerro

MUERTA, PLAYA; see Muerto, Playa

MUERTO, PLAYA; Darién

ca. 0753/7821

Sea level, beach on the S coast of Darién province, ca. 10 mi (16 km) S of Cerro Sapo [0758/7822 (USBGN)] and ca. 22 km NNW of Punta Caracoles [0741/7818 (USBGN)] (MHA, as "Playa Muerta"; SMN2A, as "Playa Muerta"). Vanderbilt Expedition, apparently specimens at ANSP are labeled only 3 mi (5 km) SW of Cerro Sapo, but according to Vanderbilt (1944:3) and to Bond and Meyer de Schauensee (1944:22-23, as "Death Beach"), the expedition worked for 2 days along the coast between Punta Garachiné and Playa Muerto as they searched for access to Cerro Sapo. According to Vanderbilt's account (Bond and Meyer de Schauensee, 1944:22), they "'discovered a beach with a small valley behind it that led straight up to the Cerro Sapo, only three miles away,'" which appears to be the locality where they collected, 1941 Apr 15-17 (ANSP).

MULATUPO; see Mulatupo Sasardí

MULATUPO SASARDÍ; San Blas

0857/7745 (USBGN)

Sea level, settlement on a small island within Islas Sasardí [0855/7742 (USBGN)] that is located less than 1 km E of the SE coast of Comarca de San Blas, 2 km S of Punta Sasardí [0858/7745 (USBGN)], and ca. 50 km NW of Cabo Tiburón [0842/7721 (USBGN)] (MGRP; RDP2). Eisenmann, observations on the San Blas coast and the nearby island of Mulatupo Sasardí, 1962 Dec 1-2 (Wetmore et al., 1984:19, as "Malatupo"). See also: Sasardí.

MUÑOZ; see Bajos del Muñoz

MUÑOZ, RÍO; see Bajos del Muñoz

MURQUI; Darién

ca. 0758/7722

550 feet (170 m), described by Pedro Galindo in 1958 as the first rest camp along the trail cut from the Paya Camp [ca. 0757/7730] to Tacarcuna Village [0805/7717 (FH)]. The camp was located on the Río Murqui (not located) in the Paya river drainage, between the Río Paya [0757/7730 (USBGN)] and the Río Topalisa [0801/7728 (USBGN)]. According to Wetmore, mammals were being collected at Murqui while he was below at the Paya Camp, Wetmore, GL = Río Paya, specimen brought to Wetmore, 1959 Feb 22 (USNM). Not shown on our maps.

MUTIS, PUERTO; see Puerto Mutis

MUTURI; Bocas del Toro

0907/8155 (USBGN)

Sea level, town located on the NW coast of Península Valiente [0905/8151 (USBGN)] 3.5 km SE of Punta Bluefield [0908/8157 (USBGN)] (JOG). SBE, collected NW of town at Playa Verde [0908/8155 (USBGN)] (q. v.), 1990 Apr 25, 27 (USNM; Brumfield et al., 2001).

NAMANÍ, LAGUNA DE; see Cañaveral, Río

NANCY, CAYO; Bocas del Toro, Archipiélago de Bocas del Toro
0919/8211 (USBGN)
47 m, a long and narrow island with a land area of ca. 7 km^2 (700 ha), located parallel to the NW shore of Isla Bastimentos [0918/8208 (USBGN)] with the NW tip of the island located 2.5 km SE of Bocas del Toro [0920/8215 (USBGN)] (E762, as "Cayo Nancy (Isla Solarte)"; JOG). SBE, the island was forested in 1988, but reported to have been planted entirely in bananas in the early 1900s, on the SE end of the island, 1988 Mar 2, and as a GL for Hospital Point, 1988 Feb 25-Mar 9 (MVUP; USNM; Olson, 1993a); Billo, GL " Isla Solarte (Cayo Nancy)", 2007 May 17, 29, "1 km SSE of northernmost point" (USNM).

NAOS, ISLA; Panamá (Canal Zone) (E, Pacific)
0855/7932 (USBGN)
20 m, island with a land area of ca. 0.2 km^2 (20 ha) located 3.5 km S of Panama City [0858/7932 (USBGN)]. Materials excavated during the construction of the Panama Canal [0920/7955 (USBGN)] were used to build a 2 km long causeway that connects Fort Amador Army Reservation [0856/7933 (USBGN)] and the island (E762, as "Naos Island"; PC, as "Naos Island"). Hallinan, on the water, 1915 Jun 8 (AMNH; Hallinan, 1924, as "Naos Island"; Griscom, 1935b:292, as "off Naos Island"; Eisenmann and Serventy, 1962, as "Naos Island"); Muñoz, 1984 Nov 24 (GML); Seutin, *en el causeway*, 1994 Oct 21 (MVUP); Lovette, GL = Panama City, 1995 Jun 4 (LSUMZ); Collector ?, GL = Panama City, 1995 Aug 13 (LSUMZ, as "Naos Island"). (Caldwell and Rubinoff, 1983). See also: Fortified Islands; and Panamá, Bahía de.

NAOS ISLAND; see Naos, Isla

NARANJAS, PUNTA; see Naranjo, Punta

NARANJO, ENSENADA DEL; Veraguas, Isla de Cébaco (Pacific)
0729/8114 (USBGN)
Bay on the extreme W coast of Isla de Cébaco [0732/8109 (USBGN)], extending 2 km E-W and opening on the S to the Pacific Ocean (JOG). USBGN treats Caleta Cayman as an older name for the bay. Described by Wetmore as a well-protected bay on the W coast of the island, Wetmore, GL = Isla de Cébaco, along a small stream with only scattered pools, in a swampy area with bits of high woodland, followed a creek inland to low hills, in a considerable area of level ground cleared for farming, 1962 Mar 27-28 (USNM, as "Caleta Cayman"; Wetmore, 1972:393, as "Caleta Caimán"; Olson, 1997a, as "Caleta Cayman").

NARANJO, PUNTA; Veraguas (Pacific)
0715/8056 (USBGN)
Sea level, point in S Veraguas province on the SW coast of Península de Azuero [0740/8035 (USBGN)] extending W into the Pacific Ocean (JOG, as "Punta Narango"; MGRP, as "Punta Naranjas"; MHA, as "Punta Naranjas"). Beaudette,

1959 Mar 23 (LACM, as "Punta Naranjas").

NARGALA, RÍO; see Nergala, Río

NARGANÁ, ÁREA SILVESTRE DE; see Nusagandí

NARGANDÍ, RÍO; see Cañazas, Río

NASAGANDÍ; see Nusagandí

NASSAU HILL; Bocas del Toro

ca. 0910/8155 ?

Apparently on Península Valiente [0905/8151 (USBGN)], and possibly equivalent to Sierra Llorona [0910/8155 (USBGN)] (q. v.) as this is the only area on the peninsula where the elevation exceeds 200 m, Miller and Richardson, GL = Laguna de Chiriquí [0903/8200 (USBGN)], Península Valiente, 1936 Mar 16 (UCLA). Not shown on our maps.

NATÁ; Coclé (Pacific)

0820/8031 (USBGN)

ca. 25 m, town on the Carretera Interamericana [0832/8242 (USBGN)] 10 km NNE of Aguadulce [0815/8033 (USBGN)] (JOG). Heyde and Lux, at least was collected on the banks of the Rio Chico [0820/8028 (USBGN)], "open woods," 1888 Oct 28, Dec 11-13, 21, 28, 1889 Jan 3, 5, 7-9, 13, 20, 24-25, Feb 3, Mar 13, 31, May 16, 20, 28, Jun 8, 1890 Jan 14, 1893 Jan 5 (SDSNH; USNM; Ridgway, 1909; Ridgway, 1912; Chapman, 1915a; Eisenmann, 1958; Monroe and Howell, 1966); Collector ? [Heyde and Lux ?], 1889 Jan 19 (UF); Davis, observations, savanna, 1960 Apr 20-21 (Eisenmann, 1962a).

NATURAL BRIDGE; see La Puente, Río

NAVY BAY; see Limón, Bahía

NAVY COVE; Panamá, Archipiélago de las Perlas

ca. 0813/7908

Apparently a cove on the S coast of Isla San José [0815/7907 (USBGN)]. The cove is not named on our maps, but a map in Wetmore's journal indicates a "Navy Station" on the S coast of the island that is accessed by "Navy Road," and presumably this coastal area was also called Navy Cove (USNM). Murphy and Correia, Askoy Expedition, 1945 Nov 23 (Murphy, 1944a; Wetmore, 1946b).

NAVY PIPELINE RESERVATION; see Pipeline Road

NAVY PIPELINE ROAD; see Pipeline Road

NECCATEGUA, RÍO; see Nergala, Río

NEGRO, RÍO; Coclé (Atlantic)

ca. 0858/8011

Described by Wetmore as a swift, shallow stream 40-60 yards (35-55 m) wide located very near Boca de Uracillo (USNM). Unclear where the stream begins or empties as our maps show no sizable rivers in the region aside from the Río Uracillo and the Río Indio. Wetmore and Perrygo, GL = Boca de Uracillo (q. v.), 1952 Feb 22-28 and possibly through Mar 2 (USNM; Olson, 1981d, as "El Uracillo, Río Negro").

Not shown on our maps.

NEGRO, RÍO; Veraguas/Herrera (Pacific)
0737/8059 (USBGN)
Originates in Herrera province near Cerro Viejo [0737/8048 (USBGN)], flows NW and W entering Veraguas province and the Río Mariato [0737/8059 (USBGN)] ca. 6 km above its mouth (MGRP). Aldrich and Bole, drainage of, used to describe Altos Cacao [ca. 0739/8049] (q. v.), Cerro Viejo Camp [ca. 0737/8049] (q. v.), and Cavulla Camp [ca. 0737/8049] (q. v.), 1932 Feb 28-Mar 9, 11-14 (CM; Aldrich and Bole).

NERGALA, RÍO; San Blas
0927/7902 (USBGN)
Originates in W Comarca de San Blas, flows NNE entering the Golfo de San Blas [0930/7900 (USBGN)] 6 km ESE of Mandinga [0927/7904 (USBGN)] (JOG). Shown as Río Neccategua on E501, MHA, and SMN2A. Wetmore, GL = Mandinga, 5 mi (8 km) SE of Mandinga, near the Río Nergala, 1957 Jan 23 (USNM, as "Río Nargala").

NEW CRISTOBAL; see Nuevo Cristóbal

NEW CULEBRA; see Culebra

NEW EMPIRE ROAD; see Nuevo Emperador

NEW GATÚN; Colón (Canal Zone) (E)
ca. 0917/7955
ca. 100 feet (30 m), town 1 km NNE of Gatún [0916/7955 (USBGN)] and 1 km SW of Fort Davis Army Reservation [0917/7953 (USBGN)] (E761A). See also: Gatún.

NEW GRENADA; see Panamá

NEW GRENADA; see Panama, Isthmus of

NEW SAN JUAN; see San Juan

NIEVE, LAGUNA PLAYA; see Playa Nieve, Laguna

NIGUA, QUEBRADA; Bocas del Toro
0917/8224 (USBGN)
Originates W of Almirante [0918/8224 (USBGN)] and flows ESE entering Ambrosia Bight [0916/8222 (USBGN)] just S of Almirante (E762). Wedel, GL = Almirante on a few specimens, 1927 Jul, Oct, Nov, 1929 Jan, 1938 Oct-Dec, 1939 Mar (ANSP; CMNH; FMNH; MCZ; Peters, 1931a); Wetmore, GL = Almirante, entered the shallow water at the mouth and followed a narrow, winding channel lined with red mangroves to an area of slightly higher banks and to wet open pastures studded with bushes and forest beyond, followed upriver as far as navigable by canoe, from Fruitdale [0918/8225 (USBGN)] continued 3 mi (5 km) on foot along the swift and shallow stream, through small plantations of cacao, bananas and corn in a great area of *rastrojo* that was green, dense and impenetrable except along the trails, 1958 Feb 1, 10, 21, Mar 5 (USNM).

NIQUE, ALTURAS DE; Darién
0740/7745 (USBGN)
1,568 m, mountains in S Darién province at the S end of Serranía de Pirre [ca. 0749/7744] 29 mi (47 km) S of El Real de Santa María [0808/7743 (USBGN)] on the Colombian border and on the continental divide (SMN2B). According to Angehr (in litt., Aug 2005), "it would appear that the use of this name began (1972) just after the publication of the 1:250,000 'Special Map' of the Darién, dated 1968, on which the name is applied to the highest peak near Cana...It reflects the realization that the peak near Cana was not the true Cerro Pirre." Hinds, Galindo, and GMLP, at 2,000-4,200 feet (600-1,300 m), 1972 Mar 18, 23, 25, Apr 4, 6-8, 11, 24, May (?), Jun (?) (GML, as "Alturas de Nique," as "Cerro Nique," as "Nique Camp," and as "Nique Range"; LSUMZ, as "Cerro Nique"; USNM as "Cerro de Nique," and as "Cerro Nique"). See also: Cana; Cana, Loma; Pirre, Cerro; Pirre, Río; and Seteganti, Río.

NIQUE, CERRO; see Nique, Alturas de

NIQUE, CERRO DE; see Nique, Alturas de

NIQUE CAMP; see Nique, Alturas de

NIQUE RANGE; see Nique, Alturas de

NOMBRE DE DIOS; Colón (E)
0935/7928 (USBGN)
Sea level, coastal town 43 km NE of Colón [0922/7954 (USBGN)] (JOG). Graham, observations, eastern Colón province east to, 1987 Jan-Jun (Ridgely and Gwynne, 1989:108). See also: Manganese Mine.

NORTE, PUNTA; Bocas del Toro, Archipiélago de Bocas del Toro
0911/8203 (USBGN)
Sea level, point located on the northernmost coast of Cayo Agua [0909/8202] (JOG). Handley, GL = Cayo Agua, 1963 Feb 12-14 (USNM); Billo, GL = Cayo Agua, 2007 May 4, "1 km S" (USNM).

NORTH BAY; see Garachiné, Ensenada de

NUBES, LAS; see Las Nubes

NUEVA CALIFORNIA; Chiriquí
0847/8239 (USBGN)
1,300 m, town located 2 km NW of Volcán [0846/8238 (USBGN)] (JOG). See also: Volcán.

NUEVA GORGONA; Panamá (W, Pacific)
0833/7952 (USBGN)
Sea level, coastal town 3 km SSE of Chame [0835/7953 (USBGN)] and 4 km WSW of the mouth of the Río Chame [0834/7950 (USBGN)] (JOG). Wetmore and Perrygo, 0.5 mi (1 km) E of town along steeply sloped beach with black sand, trees along a crest above the beach, inland for 2 mi (3 km) along cattle trails in 10-30 foot (3-9 m) thorny scrub with mangroves in the distance, 1951 Mar 17 (USNM); Wetmore, 1953 June 23 (USNM; Wetmore, 1953; Wetmore, 1962; Wetmore,

1970); Galindo, 1964 Apr 15, 1972 Nov 12 (GML, as "Gorgona," and as "Nueva Gorgona"); GMLP, 1967 May 13 (GML); Smith, N., observations, 1971 (Wetmore et al., 1984:198, as "Nueva Gorgona Beach"); De Gil del Real, *cerca de la playa* (near the beach), 1986 Nov 23 (MVUP).

NUEVA GORGONA BEACH; see Nueva Gorgona

NUEVA RIVERA; Panamá (E, Pacific)

ca. 0905/7926

Not shown on our maps but located in central Panamá province in Corregimiento Pedregal [0905/7926 (USBGN)] near the town Pedregal [0904/7926 (USBGN)] (q. v.). Gómez, 1990 Feb 23 (MVUP).

NUEVA SUIZA; Chiriquí

0851/8236 (USBGN)

1,800 m, town on the Río Chiriquí Viejo [0820/8241 (USBGN)] 2 km W of Cerro Punta [0852/8235 (USBGN)] and 8 km NW of Volcán Barú [0848/8233 (USBGN)] (E762; MGRP). Means, Koslovsky, and Sawyer, GL = Volcán Barú, at 5,500 feet (1,700 m), 1967 Mar 15-17 (AMNH; FSU; USNM); Rabenold, observations, 1977 Jan (Rabenold, 1980). (N. Smith, 1975).

NUEVO, PUERTO; see Puerto Nuevo

NUEVO CHORRILLO; Panamá (W, Pacific)

ca. 0856/7943

Not shown on our maps but located in Corregimiento Juan Demóstenes Arosemena [0856/7943 (USBGN)], which is ca. 19 km W of Panama City [0858/7932 (USBGN)] roughly midway between La Chorrera [0853/7947 (USBGN)] and Arraijan [0857/7939 (USBGN)]. Bradvica, Chandler, González, and Isaza, 1983 Feb 3 (MVUP).

NUEVO CRISTÓBAL; Colón (E)

0922/7954 (USBGN)

The NE section of the city Colón [0922/7954 (USBGN)], located on the W coast of Bahía Manzanillo [0922/7953 (USBGN)] (E761A, as "New Cristobal"). Eisenmann and Ambrose, observations, 1956 Jun 19 (Wetmore, 1972:417, as "New Cristobal").

NUEVO EMPERADOR; Panamá (W, Atlantic)

0900/7944 (USBGN)

100 m, town 15 km NNE of La Chorrera [0853/7947 (USBGN)] and 10 km NW of Arraijan [0857/7939 (USBGN)] on the N slope of the continental divide (JOG; RM2, as "Nuevo Emperador (Paja)"; RMM2, as "Paja"). From town Road K19 [ca. 0853/7947] (q. v.) extends 8 mi (13 km) SSW to La Chorrera, and in the other direction extends 7 mi (11 km) E into the Canal Zone terminating at Empire Known Distance Range [0904/7940 (USBGN)], which is located on the W side of the Panama Canal [0920/7955 (USBGN)] (RM2). MCU, 1955 Sep 26, Oct 5, Nov 28-29 (GML, as "New Empire"; MCN, as "New Empire Road"; USNM, as "New Empire Road"); Mendoza, possibly the specimen referred to by Wetmore (1968:45) from Santa Clara [0902/7945 (USBGN)] (q. v.), 1955 Oct 5 (GML, as "New Empire (Nuevo Emperador)"; Wetmore, 1968:45, as "Nuevo Emperador Road"); GMLP,

1958 Nov 10 (GML); MARU, 1961 Aug 10 (UF; USNM); Blum, near, 1966 Jun 25 (UF).

NUEVO EMPERADOR ROAD; see Santa Clara

NUEVO VIGÍA; Colón (E)
0915/7936 (USBGN)
200 m, town on the central W coast of Lago Alajuela (Madden Lake) [0915/7935 (USBGN)] 17 km NNE of Gamboa [0907/7942 (USBGN)] and 35 km SE of Colón [0922/7954 (USBGN)] (JOG). Young, 1983 Aug 15 (MVUP); Samudio, 1983 Aug 19 (MVUP).

NÚMERO DOS, CHARCO; Colón/Panamá (E, Atlantic)
ca. 0924/7933
Described by Wetmore in 1961 as an area on the Río Boquerón [0921/7933 (USBGN)] above Charco Número Uno [ca. 0923/7933] and below Campamento Número Uno [ca. 0924/7933] (USNM). Not shown on our maps, but the name possibly refers to the second stretch of river marked on E661 as a canyon, located below a falls 2 km NNE of Estación Hidro el Peluca [0923/7934 (USBGN)]. See also: Boquerón, Río.

NÚMERO UNO, CAMPAMENTO; Colón ? Panamá ? (E, Atlantic)
ca. 0924/7933
Described by Wetmore in 1961 as a camp on the Río Boquerón [0921/7933 (USBGN)] in an area that was mostly forested, located above Charco Número Dos [ca. 0924/7933], but below the mouth of the Río Mauro [0925/7933 (USBGN)] (USNM). Not shown on our maps, but estimated to be ca. 3 km N of Estación Hidro el Peluca [0923/7934 (USBGN)]. See also: Boquerón, Río.

NÚMERO UNO, CHARCO; Colón/Panamá (E, Atlantic)
ca. 0923/7933
Described by Wetmore in 1961 as a deep area on the Río Boquerón [0921/7933 (USBGN)] above Estación Hidro el Peluca [0923/7934 (USBGN)], its head marked by a 20-25 foot (6-8 m) cascade (USNM). Not shown on our maps, but the name possibly refers to the first stretch of river marked on E661 as a canyon, 1 km N of Estación Hidro el Peluca. See also: Boquerón, Río.

NUSAGANDÍ; San Blas
ca. 0919/7859
450 m, according to Engleman et al. (1989:490), Nusagandí is a forest reserve within western Comarca de San Blas that protects ca. 40,000 ha of primary forest and is accessed via the El Llano-Cartí Road [ca. 0928/7858]. The reserve is located just N of the continental divide along the S border of Comarca de San Blas and Panamá province and the elevation in the reserve ranges from 650 m on Cerro Brewster [0921/7915 (USBGN)] to near sea level. According to Whitney and Rosenberg (1993), the forest is "largely undisturbed, with an irregular canopy ranging in height from 15 to 25 m. The canopy is rather open...allowing sunlight to penetrate to the forest floor, although it is denser locally." According to Angehr (in litt., Aug 2005), "the name properly applies to the facilities established by the Kuna Indians

on the El Llano-Cartí Road near the boundary of the Comarca de San Blas to prevent colonization of the Comarca after the road was built." The name is sometimes applied to the reserve, which is also known as the PEMASKY Reserve (*Proyecto de Estudio para el Manejo de Areas Silvestres de Kuna Yala*); but its official name is Área Silvestre de Narganá. Blake, at 350 m, mist nets, "tropical wet forest," 1985 Apr 8-13, May (?) 13 (Blake, 1989; Ridgely and Gwynne, 1989:89; Adsett and Wege, 1998); Ridgely and Guarnaccia, observations, 1986 Mar 7 (Ridgely and Gwynne, 1989:431); Whitney and Rosenberg, at 180-390 m, observations and recordings near the tourist lodge (ca. 0913/7858), 1987 Feb, 1992 Jan (CU; Whitney and Rosenberg, 1993; Adsett and Wege, 1998, they mistakenly site the year as 1989); Braun, M., at 300 m, 1 km N, 1994 Jan 25-26 (USNM, as "Nasagandi"); Observer ?, 1995 Mar 18, Apr 15-16, 1996 Feb 10, 1997 Jan 31-Feb 1, 1998 Jan 25 (Adsett and Wege, 1998); Christian, at 390 m, observations, humid forest ca. 0920/7903, 1996 May (Adsett and Wege, 1998; Christian, 2001, as "Nusagundi"). See also: Brewster, Cerro; and El Llano-Cartí Road.

NUSAGUNDÍ; see Nusagandí

NUSUGANDÍ ROAD; see El Llano-Cartí Road

OBALDÍA; see Puerto Obaldía

OBALDÍA, PUERTO; see Puerto Obaldía

OBARRIO; see Obarrio, Barriado

OBARRIO, BARRIADO; Panamá (E, Pacific)
ca. 0859/7931
20 m, section of greater Panama City [0858/7932 (USBGN)] located in E Corregimiento Bella Vista [0859/7931 (USBGN)] (E962, as "Obarrio"). Quintero, 1990 Apr 24, Aug 31, Nov 26 (MVUP).

OBISPO; see Bas Obispo

OBISPO STATION; see Bas Obispo

OBSERVATORIO, PUNTA; see Felipa, Punta

OCCIDENTE, RÍO; see Oeste, Río

OCHO, FINCA; see Finca Ocho

OCÚ; Herrera
0757/8047 (USBGN)
100 m, town in NW Herrera province 38 km W of Chitré [0758/8026 (USBGN)] (JOG). White, observations, 1950 Jan 20 (Wetmore, 1972:374); Scholes, observations, 1951 May 27 (Eisenmann, 1962a); Vega, 1992 Sep 28 (MVUP).

OESTE, RÍO; Bocas del Toro
0916/8224 (USBGN)
Originates in the hills WSW of Almirante [0918/8224 (USBGN)] and flows SE entering the W shore of Ambrosia Bight [0916/8222 (USBGN)] 4 km S of Almirante (JOG; MHA, as "Río Occidente"). Collectors often used the English equivalent, "Western River," on their specimen labels. Wedel, some specimens are labeled with

GL = Almirante or with GL = Bahía Almirante [0920/8218 (USBGN)], 1923 Jul, 1927 Jul-Nov, 1928 May-Jul, Sep, Oct, 1936 Aug, 1938 Oct-Dec, 1939 Mar-Apr (ANSP; CMNH; FMNH; MCZ; MVZ; UCLA; USNM; YPM; Peters, 1931a (all as "Western River"); Kennard et al., GL = Almirante, 1926 Feb 11, 20, 28, Mar 2, 11 (MCZ; Kennard and Peters, 1928; Peters, 1929; and Peters and Griscom, 1929, all as "Western River"); Wetmore, GL = Almirante, the river valley had long been cleared of its original vegetation, collected in Ambrosia Bight between the mouth of the Río Oeste and Almirante, near the river mouth in mangroves and cultivated fields, along the stream banks to the head of navigation in swampy areas, cacao and banana plantations, and pastures with some tall trees and dense growth, and inland on trails along the base of steep hills and up ridges through a tract of original forest, 15-20 year old *rastrojo*, and a dry upland forest, 1958 Jan 15, 21, 23, 25, 28, Feb 5, 8, 15-16, 18, 24, 28, Mar 3 (USNM, as "Western River"); Munch, GL = Almirante, 1958 Feb 25 (USNM, as "Western River"); SBE, at the mouth of the river near Bahía Almirante, 1-1.5 mi (1.5-2.5 km) above its mouth, 1989 Feb 28, at 5 m, 0.5 km N of the bridge, according to our maps, the road crosses the river ca. 8 km WNW of its mouth, 1991 Mar 1, 3 (MVUP; USNM; Olson, 1993a; Parsons et al., 1993; Brumfield et al., 2001). (Olson, 1981c, as "Western River").

OJO DE AGUA; Chiriquí

ca. 0848/8239

1,400 m, town 1 km NNW of Nueva California [0847/8239 (USBGN)] (E762, as "Finca Ojo de Agua"). Wilson, 2000 Mar 7 (MVUP).

OJO DE AGUA, FINCA; see Ojo de Agua

OLD CANAL; see Changuinola Canal

OLD FRENCH CANAL; see French Canal

OLD GAMBOA ROAD; Panamá (Canal Zone) (E, Atlantic, Pacific)

ca. 0905/7940

According to Angehr (personal communication) the name refers to the unnamed road that appears on E762 branching to the W from Carretera Gaillard [0904/7940 (USBGN)] less than 1 km NW of Paraiso [0902/7938 (USBGN)] and extending parallel to it. The road crosses various small streams and terminates 3 km SSW of Gamboa [0907/7942 (USBGN)]. Engleman et al. (1989:480) indicate that the road extends through second-growth woodland and returns to Carretera Gaillard. Collector ? [Seutin ?], GL = Paraiso, some specimens are labeled 5 km NW of Paraiso Summit and others as simply Old Gamboa Road, 1989 Mar 6, 10, 12 (LSUMZ); Braun, M., GL for Summit Park, 1994 Apr 13, May 6 (USNM); Seutin, 4-5 km NW of Paraiso, 1994 Sep 19-20, 22 (LSUMZ); Cardiff, Christian, and Dittmann, GL = Paraiso, 5 km NW of Paraiso, 1996 Mar 14, 1997 Mar 14 (LSUMZ; MVUP, as "Camino Viejo de Gamboa"); Babin, GL = Summit Park, on the road 4.5 km NW of Summit Park, 1999 Mar 20, 22-25 (LSUMZ); Lelevier and Maley, at 0907/7943, 2004 Dec 24-25, 2005 Jan 28-31 (STRI; UAM). (Fraga, 1989, as Old Gamboa Road (Road C2F)").

OLD PANAMA; Panamá (E, Pacific)

0900/7929 (USBGN)

Sea level, ruins of the Spanish town that served as the headquarters for South American conquests from 1519-1671, located on the N coast of Bahía de Panamá [0850/7915 (USBGN)] 7 mi (11 km) NE of Panama City [0858/7932 (USBGN)] (HMCZ). According to USBGN, the correct Spanish name for this locality is Panamá La Vieja, which sometimes is written as Panamá Vieja or as Panamá Viejo. According to Buehler et al. (2004, as "Panama Viejo"), the coast is a "broad mudflat with soft fine sediments, flanked to the east by a tourist center and to the west by a rocky outcrop." Bovallius, 1883 Feb 26 (NHRM; Rendahl, 1919:4, 10, 11, 25, as "Panama Viejo"); Hallinan, rocks and sand along the shore, in the ruins, offshore, 1907 Aug 11, 1915 Feb 21 (AMNH; Hallinan, 1924, as "Old Panama," and as "Old Panama City"); Goldman, at the ruins, 1912 Feb 7 (USNM); Osgood and Anderson, 1912 Feb 7 (FMNH); Tevis, captive bird, 1913 Mar 10 (MVZ, as "Old Town of Panama"); Kelsall and Cullingford, 1924 Jun (BMNH; Douglas and Johnson, 1926); Weber, 1928 Aug 1, 11 (FMNH, as "Old Panama City," and as "Panama City, Ruins"; USNM, as "Panama Ruins"); Arbib and Loetscher, observations, "beach with a broad expanse of mud flat," 1933 Aug 13, 1934 Aug 7, 28 (Arbib and Loetscher, 1935, as "Panama Vieja"); Griswold, 1936 Apr 25 (MCZ, as "Old Panama Ruins"); Eisenmann, observations, mud flats near the ruins, and between San Francisco de la Caleta [0859/7930 (USBGN)] and, 1948-1950 late Jun-early Jul, 1962 Nov (Eisenmann, 1948; Eisenmann, 1951; Eisenmann, 1963b); Eisenmann and Chapelle, observations, mud flats, rocky islets, 1954 May, Aug-Oct, 1955 Apr-Jun (Eisenmann, 1955c); Grocott, 1954 (Caballero y C. and Hidalgo, 1955, as "Panama Viejo"); Loftin, mangroves near, 1962 Aug-Nov (Loftin, 1963a, as "Panama Vieja"); Loftin and Olson, beach, rocks, 1963 Mar 3, 5, 10, Apr 12 (Loftin and Olson, 1963, as "Panama Viejo"); Rodríguez and Zambrano, *cerca del bohío turístico* (near the tourist kiosk), 1991 Aug 27 (MVUP, as "Panamá La Vieja"); Sánchez, 1994 Aug 3 (MVUP). See also: San Francisco de la Caleta.

OLD PANAMA CITY; see Old Panama

OLD PANAMA RUINS; see Old Panama

OLD PIPELINE ROAD; see Pipeline Road

OLD POINT; Bocas del Toro, Archipiélago de Bocas del Toro

0917/8205 (USBGN)

Sea level, the easternmost land on Isla Bastimentos [0918/8208 (USBGN)] located on the central E coast of the island. A settlement with the same name and coordinates is located on the point (JOG; MGRP, as "Punta Vieja"; MHA, as "Punta Patiño"). Handley, GL = Isla Bastimentos, 1963 Feb 3-4, 7-8, 10 (USNM, as "Punta Vieja"); SBE, GL = Isla Bastimentos, Laguna de Chiriquí, mostly swamp forest that in 1987 was completely dry from drought, small areas of pasture and cultivation around the settlement, 1987 Mar 13-29 (MVUP, as "Punta Vieja"; USNM).

OLD TACARCUNA SITE; see Tacarcuna Village

OLD TACARCUNA VILLAGE SITE; see Tacarcuna Village

OLD TOWN OF PANAMA; see Old Panama

OLD VILLAGE SITE; see Tacarcuna Village

OLEODUCTO ROAD; see Gualaca-Chiriquí Grande Road

OLIVO; Chiriquí

ca. 0819/8247

Sea level, town 10 km NE of Puerto Armuelles [0817/8252 (USBGN)] (MGRP, as "Olivo Civil"). Described by Wetmore as an area NE of Puerto Armuelles between the Canal de Colorado [0825/8251 (USBGN)] and Aguacatón [ca. 0820/8248]. The forested area had an extensive series of drainage ditches, some affected by tides, that had been dug ca. 20 years previously, Wetmore, GL = Puerto Armuelles, tracts of original forest, a broad expanse of second growth, areas being cleared, cutover forest scattered with extensive plantings of rice and corn, through mangroves to the coast and to an open area that was a lagoon in the rainy season, mangroves edged with low, flowering shrubs, on slightly higher ground with fig trees, along the sandy beach near the mouth of Canal de Colorado, 1966 Jan 31, Feb 3, 7, 11, 14, 16-17, 19, 21, 23, 26, 28, Mar 3, 7, 12-13 (USNM; Olson, 1980b, as "Puerto Armuelles, Olivo").

OLIVO CIVIL; see Olivo

OMAR TORRIJOS, PARQUE NACIONAL; Coclé (Atlantic, Pacific)

0841/8040 (Angehr, 2003:177)

Also known as El Copé National Park, the reserve includes an area along the continental divide N of El Copé [0837/8036 (USBGN)] (RDP2, as "G. D. Omar Torrijos Parque Nacional"). The park covers 25,257 ha and ranges in elevation from 80 m on the Río Belén [0853/8052 (USBGN)] to 1,710 m (an unnamed peak on the W end of the park). The park contains the upper watersheds of the Río Grande [0818/8024 (USBGN)], the Río Belén and the Río Coclé del Norte [0905/8035 (USBGN)] and can be accessed from El Copé in the W or from Coclecito [0849/8033 (USBGN)] in the E (Angehr, 2003:177, as "Parque Nacional General de División Omar Torrijos Herrera"). Bacon and Miller, entrance, at 08°47'10"/80°35'25", 2002 Sep 20 (STRI; UAM); Weir, at 600-900 m, at N 08°40'11", W 80°35'35", 2003 February 12-14, Apr 21-22, May 2 (CU, as "El Cope National Park"; LSUMZ, as "El Cope National Park"). See also: Molejon; and Moreno, Cerro.

ORCHID, ISLA; Panamá (Canal Zone) (W, Atlantic)

0911/7951 (USBGN)

70 m, island in Lago Gatún (Gatun Lake) [0912/7955 (USBGN)] with a land area of 16 ha (Gradwohl and Greenberg, 1982) to 17.1 ha (Wright, 1985) located 200 m off the N coast of Isla Barro Colorado [0909/7951 (USBGN)] and 1.5 km W of Punta Buena Vista [0911/7950 (USBGN)] (E866, as "Orchid Island"). Willis, observations, old forests, 1960-1970 (Willis, 1972:8, as "Orchid Island"); Wright, censuses, 1976-1978 (Wright, 1979, as "Orchid Island"; Wright, 1985, as "Orchid Island"); Worthington, 1978-1979 (Worthington, 1983); Gradwohl and Greenberg, observations, 1980 Aug 1-Sep 11 (Gradwohl and Greenberg, 1982, as "Orchid Island").

ORCHID ISLAND; see Orchid, Isla

ORIA, RÍO; Los Santos
0726/8008 (USBGN)
Originates NW of Los Asientos [0731/8008 (USBGN)] and flows SE entering the Pacific Ocean 15 km WSW of Punta Mala [0728/8000 (USBGN)] (JOG). Wetmore, GL = Los Asientos, 6 mi (10 km) W of Los Asientos where the river was 35-50 feet (10-15 m) wide and deep with good current, downstream to virgin forest with fairly open undergrowth, creepers and palms, 1957 Mar 18-19 (USNM; Wetmore, 1962).

ORQUETA, CERRO; see Horqueta, Cerro

ORQUETA, CERRO DE; see Horqueta, Cerro

OTOQUE, ISLA; Panamá (Pacific)
0836/7936 (USBGN)
190 m, island in Bahía de Panamá [0850/7915 (USBGN)] with a land area of ca. 3 km^2 (300 ha) located 37 km SSW of Panama City [0858/7932 (USBGN)] and 11 km SE of Punta Chame [0839/7942 (USBGN)] (JOG). Wetmore, observations at sea near the island, 1956 Feb 6 (Wetmore, 1957:32, 34). See also: Mala, Punta.

PAC; Panamá (Canal Zone)
A modifier used by Blanton, and perhaps others, to indicate that collections were made on the Pacific drainage. See also: Farfan Beach; Palo Seco Leprosarium; Pacora Road; Road K6; and West Ferry Ramp.

PACA, RÍO; Darién
0759/7732 (USBGN)
Originates on the E slopes of Serranía de Pirre [ca. 0749/7744] and flows NNE between Serranía de Pirre and Cerro Setetule [0747/7736 (USBGN)] entering the Río Tuira [0821/7803 (USBGN)] 7.5 km SSE of Boca de Cupe [0803/7735 (USBGN)] (MGRP). According to Goldman, the train from Cana [ca. 0745/7742] to Boca de Cupe passed near the head of the Río Paca in the Cana Valley (see Cana). See also: Kitchener, Mount.

PACA; Darién
0759/7735 (USBGN)
ca. 50 m, town 12 km S of Boca de Cupe [0803/7735 (USBGN)] on the Río Paca [0759/7732 (USBGN)] ca. 4 km above its mouth (DP). Described by Goldman as a train station located 6 mi (10 km) by train from Cana [ca. 0745/7742] and 24 mi (39 km) by train from Boca de Cupe [0803/7735 (USBGN)] (USNM). From this description the train station was probably located on the upper reaches of the Río Paca SW of the town of Paca. See also: Cana; and Kitchener, Mount.

PACHECA; see Pacheca, Isla

PACHECA, ISLA; Panamá, Archipiélago de las Perlas
0840/7903 (USBGN)
ca. 200 feet (60 m), island with a land area of 0.62 km^2 (62 ha) (Wright et al., 1985), located 24 km NNW of Isla del Rey [0822/7855 (USBGN)] (JOG). The island, which is mostly forested, is an Important Bird Area (Angehr, 2003:267). See Isla

Saboga [0837/7904 (USBGN)] for comments about Brown's collections on these two islands. Bovallius, a rock island with a thinly wooded plateau, see discussion of Bovallius under Pueblo Nuevo [ca. 0820/8013 ?], 1882 Mar (NHRM; Rendahl, 1920); Brown, 1904 Apr 14-16 (MCZ, as "Pacheca I."; Thayer and Bangs, 1905, as "Pacheca Island"); Griscom and Crosby, 1927 Mar 9 (AMNH, as "Pacheco Island"); Cresson and Worth, 1937 Jun 11-12 (ANSP, as "Pacheca"); Murphy and Correia, Askoy Expedition, N end of the island, 1941 Feb 10 (AMNH, as "Pacheca Island"; Murphy, 1944a; Murphy, 1944b, as "Pacheca Island"; Murphy, 1945, as "Pacheca Island"); Wetmore, 1960 Jan 20 (USNM); (MacArthur et al., 1972—species list; see comments in Bibliography); Wright and Faaborg, mist nets, observations, 1976 Jun-Aug (Wright et al., 1985); Reed, photographs, 1983 Jun 17 (Reed, 1988, as "Pacheca Island"); Brawn et al., mist nets, 199? (Brawn et al., 1996); Agro, Cardiff, Dittmann, and Lovette, off Isla Pacheca, 1996 Feb 28 (LSUMZ).

PACHECA ISLAND; see Pacheca, Isla

PACHECILLA ISLAND; see Pachequilla, Isla

PACHECO ISLAND; see Pacheca, Isla

PACHEQUILLA, ISLA; Panamá, Archipiélago de las Perlas
0840/7904 (USBGN)
Very small island with a land area of 5.4 ha (Angehr, 2003:267) located just NW of Isla Pacheca [0840/7903 (USBGN)] (JOG), and according to a map section in Wetmore's journal, the elevation is 80 feet (20 m). The island, which is "mostly covered with herbaceous vegetation with a few scattered shrubs," is an Important Bird Area (Angehr, 2003). Meyer, 1944 Apr 23 (USNM, as "Pachecilla Island"); Wetmore, rocky island, hunted from the boat, 1960 Jan 16 (USNM); Reed, photographs, 1983 Jun 17 (Reed, 1988, as "Pachequilla Island").

PACIFIC SADDLE CLUB; see Pedro Miguel

PACIFIC TERMINUS; Panamá (Canal Zone) (E, Pacific)
ca. 0859/7933
Term used by Loftin that refers to greater Panama City [0858/7932 (USBGN)], specifically the area including Panama City [0858/7932 (USBGN)], Balboa [0857/7934 (USBGN)], Curundu [0859/7933 (USBGN)] and Fort Clayton Army Reservation [0901/7934 (USBGN)], Loftin, observations, 1961 Nov, 1962 Nov; Loftin and Olson, 1963 Apr 1, 12 (Loftin, 1963a; Loftin and Olson, 1963).

PACORA; Panamá (E, Pacific)
0905/7917 (USBGN)
40 feet (10 m) (TM); town 24 km SW of Chepo [0910/7906 (USBGN)] on the Río Pacora [0901/7917 (USBGN)] 7 km N of its mouth (JOG). Brown, Cpt., hunting on an open plain near, 1922 Apr (MCZ; Bangs and Barbour, 1922:193); Chapin, 1923 Jul 15 (FMNH); Weber, 1928 Jul 28-31 Aug 1, 1930 Jul 15, 31 (FMNH; USNM); Benson, at 100 feet (30 m), 1931 Jul 6, 20 (USNM); Chapman, observations and photographs, 1920s (Chapman, 1929a, as "Pacora Marshes"); Curtis, 1938 Jan 2, 15 (ANSP; Bond and Meyer de Schauensee, 1944:27, as "Pecora"); Meyer, 1944 Mar 26 (USNM); Wetmore and Perrygo, GL for: Río Cabobré, Río Pacora, and

Sabana de San José, 1949 Mar 15, 19, 21, 26-27, Apr 2-3, 21 (USNM); Moreno, 1950 Sep 18 (USNM); Blanton, 1953 Oct 16, 21, and as a GL for: La Jolla 1, and La Jolla 2, 1953 Sep 23, 30, Oct 8 (USNM); GMLP, 1958 Apr 16, Oct 10, 14, 21, 28-30, Nov 8, 14, 16, 28, 1960 Oct 21-22, and as a GL for: La Jolla No. 2, and San Antonio Swamp, 1958 Sep 4, 15, Oct 12 (USNM); Wetmore, GL for: San José, and Sabana de San José, 1958 Mar 19, 22, GL for: La Loma, Río Santo, and Río Pacora, 1961 Feb 15, Mar 23, 25 (USNM); Rojas, 1960 May 2, 1968 Feb 20 (GML); Yunker, 1961 May 29 (USNM); Powers, 1964 Jul 7 (GML); Moreno, 1964 Sep 10, 1965 Mar 10 (GML); Means et al., GL for La Jagua Hunting Club, 1967 Feb 18-22 (FSU; USNM); Fretwell, 1968 Jan 11 (AMNH); Mora, 1968 Jun 30 (GML); Strauch, GL for Sabana de San José, 1972 Aug 30 (LSUMZ); Kourney, 1973 Sep 30 (GML); Johnson, 1976 Apr 18 (GML); Collector ?, 1976 May 6 (GML); Justiniani, 1990 Sep 23 (MVUP); Braun and Brumfield, at 09°02°24"/79°15'06", ca. 7 km SE of, 2004 Mar 26-28 (LSUMZ). (Wetmore, 1952:18; Wetmore, 1953; Wetmore, 1967; Wetmore, 1970; Parkes, 1976; Olson, 1981d). See also: La Jagua Hunting Club; and Pacora, Río.

PACORA; see Trinidad Arriba

PACORA, RÍO; Panamá (E, Pacific)
0901/7917 (USBGN)
Originates on the NW slopes of Cerro Prominente [0914/7917 (USBGN)], flows E around its base then turns S crossing the Carretera Interamericana [0832/8242 (USBGN)] 2.5 km N of Pacora [0905/7917 (USBGN)], continues S entering Bahía de Panamá [0850/7915 (USBGN)] 19 km W of the mouth of the Río Chepo [0900/7906 (USBGN)] and 20 km E of Panama City [0858/7932 (USBGN)] (JOG). According to Buehler et al. (2004), the region just W of the river's mouth is "broad mudflats of soft fine sediments, flanked inland successively by savannah-like grassland with scattered trees and extensive rice fields." Bovallius, presumably he collected near the mouth, 1883 Mar 2-4 (NHRM; Rendahl, 1919, as "Pacore River," as "Pacora River," and as "Pacora"); Goldman, at 100 feet (30 m), 10 mi (16 km) W of Chepo [0910/7906 (USBGN)] while returning from Cerro Azul [0910/7924 (USBGN)], 1911 Mar 28 (USNM; Aldrich, 1945; Olson, 1980c); Weber, 1928 Jul 30-Aug 1 (FMNH, as "Pacora Marshes," and as "Pacora River"; USNM, as "Pacora Marshes," and as "Pacora River"); Wetmore and Perrygo, GL = Pacora, 2 mi (3 km) N of Pacora, where the river was 100 feet (30 m) wide with a sandy, gravelly bed near its junction with the Río Cabobré [0906/7917 (USBGN)], in forest with heavy undergrowth, 1949 Mar 15, 1-2 mi (1.6-3 km) S of Pacora in the region W of San José [0904/7916 (USBGN)], forest with large trees, from the pastures in San Joaquin (not located) down to the river, 1949 Apr 2-3, 5 mi (8 km) NNE of Pacora in old river channels, trees and savannas, 1949 Apr 21, GL = Cerro Azul [0910/7924 (USBGN)], at Bajo Grande (q. v.), 1949 Mar 23, at San Miguel (q. v.), 1949 Apr 19 (USNM); Blanton, 1953 Sep 18, Oct 7 (USNM); Handley, GL = Cerro Azul (q. v.), head of the river, 1958 Jan 25, 27 (USNM); Wetmore, GL = Pacora, hunted above the bridge just E of Pacora where the woodland had been cut and the vegetation was dry, 1961 Mar 25 (USNM; Parkes, 1969); Buehler et al., observations,

2003 Sep 7-27 (Buehler et al. 2004). (Gorski, 1969). See also: C. I. M.

PACORA MARSHES; see Pacora, Río

PACORA ROAD; Panamá (E, Pacific)
ca. 0905/7917 ?
Presumably refers to the section of the Carretera Interamericana [0832/8242 (USBGN)] that extends ca. 15 km between Pedregal [0904/7926 (USBGN)] and Pacora [0905/7917 (USBGN)], as well as the road that extends 2 km S of the Carretera Interamericana to Pacora (JOG). According to SMN2B this section of the highway was an "all weather, loose surface road." Blanton, 1953 Oct 2, 7-8, 14, 29, GL = Pedregal, 1953 Sep 23 (USNM).

PACORE RIVER; see Pacora, Río

PAITILLA, PLAYA DE; see Paitilla, Punta

PAITILLA, PUNTA; Panamá (Canal Zone) (E, Pacific)
0858/7931 (USBGN)
80 feet (25 m), point 3 km NE of Panama City [0858/7932 (USBGN)] extending ca. 1 km S from the mainland on a small peninsula that was included in the former Canal Zone (E661, as "Paitilla Pt."; E761A, as "Paitilla Pt."). Benson, at 100 feet (30 m), 1931 Jul 20 (USNM, as "Paitilla Point"); Meyer, GL = Panama City, 1942 Feb 14 (USNM, as "Patilla Point"); Canto, 1984 Sep 5 (MVUP, as "Playa de Paitilla"). See also: Marcos A. Gelabert, Aeropuerto.

PAITILLA AIRFIELD; see Marcos A. Gelabert, Aeropuerto

PAITILLA POINT; see Paitilla, Punta

PAITILLA PT; see Paitilla, Punta

PAJA; see Nuevo Emperador

PAJA, RÍO; Panamá (Canal Zone) (W, Atlantic)
0905/7948 (USBGN)
Originates near Nuevo Emperador [0900/7944 (USBGN)] and flows NNW entering the S coast of Lago Gatún (Gatun Lake) [0912/7955 (USBGN)] 11 km WSW of Gamboa [0907/7942 (USBGN)] (E761A; JOG). Originally continued NNW and entered the Río Caño Quebrada [0914/8005 (USBGN)] 2 mi (3 km) S of Tabernilla [ca. 0908/7949] (PR). Garella, observations of *Pelecanus occidentalis* crossing the divide between the headwaters of the Río Caimito [0852/7943 (USBGN)] and the Río Paja, 1845 (Garella, 1845:73; Wagner, 1870:86, 88; Wetmore, 1965:51).

PAJA BLANCA; Chiriquí
0819/8235 (USBGN)
Sea level, town 9 km S of Alanje [0824/8233 (USBGN)] and 2 km NNW of La Barqueta [0818/8234 (USBGN)] (JOG). Wetmore, GL = Alanje, walked for 2 mi (3 km) to a swamp that bordered the ocean, 1960 Mar 5 (USNM).

PAJARONCITO, BOCA DE; Veraguas (Pacific)
0756/8139 (USBGN)
A short, one kilometer wide stretch of river formed as various rivers including the

Río Vidal [ca. 0801/8138] and the Río Bubí [0759/8138 (USBGN)] merge opposite the E coast of Isla de la Corocita [0758/8139 (USBGN)] (JOG). See also: Corocita, Isla de la.

PAJAROS, ISLA; see Mogo Mogo, Isla

PALENKA, ISLA; see Bóquita, Isla

PALENQUE; Colón (E)
0913/7941 (USBGN)
50 m, town 6.5 km NNW of Juan Mina [0910/7939 (USBGN)] (E866). See also: Palenque, Río.

PALENQUE, ISLA; see Bóquita, Isla

PALENQUE, QUEBRADA; see Palenque, Río

PALENQUE, RÍO; Colón (Canal Zone) (E)
0910/7940 (USBGN)
Originates in Colón province NW of Palenque [0913/7941 (USBGN)], flows S and enters a 1 km wide bay on the W bank of the Río Chagres [0907/7942 (USBGN)]. The river's mouth is 2 km W of Juan Mina [0910/7939 (USBGN)] and 200 m SE of the mouth of the Río Aguardiente [0910/7941 (USBGN)]. A foot trail extends E-W between Aguas Claras [0911/7941 (USBGN)] and Santa Rosa [0911/7939 (USBGN)] crossing the river 2 km N of its mouth (E866). Wetmore, GL = Juan Mina, collected near the river ca. 2 mi (3 km) above its mouth in Bajo Palenque (ca. 0911/7940, not shown on our maps, but must refer to a region ca. 3 km SSE of Palenque) in a small tract of forest en route from Aguas Claras to Santa Rosa, collections made near the former Canal Zone/Colón provincial boundary, 1955 Jan 20 (USNM, as "Quebrada Palenque"); Reyna, GL = Santa Rosa, 1971 Apr 28 (GML).

PALENQUE ISLAND; see Bóquita, Isla

PALMA, LA; see La Palma

PALMARAZO; Coclé (Atlantic)
0844/8040 (USBGN)
USBGN indicates a town in NW Coclé province ca. 40 km NW of Penonomé [0831/8022 (USBGN)], N of the continental divide. According to Angehr (in litt., Aug 2005), this was one of the Western Watershed study sites and it appears on map 4042-II (E762). He provides the GPS coordinates 08°44'02" N 80°38'48" W, elevation 80 m and states that "Garcés'specimens were collected from our net lines just to the south of these coordinates." Garcés, Jiménez, López, Ramos, and Rodríguez, 2001 Jul 15-17 (MVUP; Louis Berger Group et al., 2003).

PALMA REAL; Bocas del Toro
0903/8217 (USBGN)
10 m, town on the N banks of the Río Róbalo [0901/8215 (USBGN)] 4 km W of Punta Róbalo [0902/8215 (USBGN)] (E762). Braun, M., Brumfield, and Glenn, at 30-50 m, above Río Róbalo, 1 km W of Palma Real, and 3 km NE of Palma Real, 1994 Jan 15-16, 18 (USNM, as "Palma Real del Róbalo," and as "Palma Real";

Brumfield et al., 2001).

PALMAS, LAS; see Las Palmas

PALMIRA; Chiriquí
0843/8227 (USBGN)
1,000 m, town 4 km SSW of Boquete [0847/8226 (USBGN)] (E562, as "Palmira Centro"). The town Rovira Arriba [0839/8231 (USBGN)] is at 400 m and is 16 km SSW of Boquete (E562). As both Palmira and Rovira appear on the label, the specimen may have been collected SW of Palmira. Moreno, 1983 Dec 25 (MVUP, as "Palmira (Rovina)").

PALMITAS, LAS; see Las Palmitas

PALO ALTO; Chiriquí
0848/8225 (USBGN)
1,280 m, town 3 km NNE of Boquete [0847/8226 (USBGN)] (E762). Wetmore, GL = Boquete, at 5,000 feet (1,500 m), considerable uncut forest on steep slopes that were cleared near their bases, several small, rapid streams, 1960 Mar 16 (LSUMZ; USNM; Olson, 1981c, as "Boquete, Palo Alto").

PALO BLANCO, RÍO; Chiriquí
0818/8249 (USBGN)
Originates in the hills W of Puerto Armuelles [0817/8252 (USBGN)] near the Costa Rican border and follows a semicircular path N of Puerto Armuelles entering Bahía de Charco Azul [0812/8245 (USBGN)] 6 km NE of Puerto Armuelles (JOG). Wetmore, GL = Puerto Armuelles, reached the river from the highway 6 mi (10 km) N of Puerto Armuelles, sandy flood plain, small area of second growth with a few tall trees remaining, 1966 Jan 29 (USNM).

PALO GRANDE; Chiriquí
0820/8236 (USBGN)
ca. 25 m, capital of Distrito de Alanje [0824/8237 (USBGN)] located 21 km SW of David [0826/8226 (USBGN)] and 3 km N of the coast (MGRP). See also: Corozo.

PALOMETA; Panamá (Canal Zone) ? (E ? W ?, Pacific)
?
Not located on our maps. According to the specimen label, the name refers to a section of coast in the Bahía de Panamá [0850/7915 (USBGN)], Pacific Project, 1968 Dec 16 (GML; Loftin, 1991).

PALO SANTO; Chiriquí
0849/8241 (USBGN)
1,300 m, town 6 km NW of Volcán [0846/8238 (USBGN)] and 1 km N of the Río Chiriquí Viejo [0820/8241 (USBGN)]. The town is in the area between the river's wide, open valley and the higher elevations of Cerro Pando [0849/8242 (USBGN)] and other peaks to the N (JOG). Wetmore (1965:20) uses the term Finca Palo Santo, which must refer to a farm near the town. Starting in 1955 in his journals Wetmore referred to the entire region surrounding Palo Santo as Tisingal [0848/8240 (USBGN)] (q. v.), Wetmore, GL = Volcán, at 4,200-4,400 feet (1,300-1,350 m), in 1954 most specimens were also labeled Río Chiriquí Viejo (q. v.), collected

within 5 km of Palo Santo in the cultivated valley bordered by woodland, cane, extensive potato fields, brush and with bananas growing in the gravelly flood plain, in pastures, near a pond and a swampy meadow, on the steeply cut slopes down to the Río Chiriquí Viejo, along small tributaries of the Río Chiriquí Viejo, up a logging road into the forest above, 1954 Feb 24, 26, Mar 4-5, 8, 10, 12, 21, 26, 1955 Jan 31, Feb 2-3, 9-10, 12, Mar 14, 16, GL = Volcán, at 4,200-4,800 feet (1,300-1,450 m), in a boggy field, in low brush and grass along channels draining the area, in tall weeds, to the river through a narrow belt of woodland, 1960 Feb 6, 8-11, 13, 17, 19, 22-23, 26, Mar 22, GL = Volcán, at 4,200-4,500 feet (1,300-1,350 m), along an old logging road at 4,500 feet (1,350 m) just above Palo Santo, in a wooded area near the Río Chiriquí Viejo, on a flat area above a steep, wooded bluff dropping 200 feet (60 m) to the Río Chiriquí Viejo, 1965 Feb 26, Mar 15-18, 23-24 (LSUMZ; MVUP; USNM; Olson, 1983, as "El Volcán, Palo Santo"); Hartman, GL = Volcán, at 4,500 feet (1,350 m), 1958 Feb 13, 15-17, 19-21 (USNM; Hartman, 1955).

PALO SECO; see Palo Seco Bosque de Protección

PALO SECO; see Palo Seco Leprosarium

PALO SECO, RÍO; Veraguas (Pacific)
0737/8058 (USBGN)
Originates on the W slope of Península de Azuero [0740/8035 (USBGN)] flowing NW then SW into Golfo de Montijo [0740/8107 (USBGN)] E of Isla de Cébaco [0732/8109 (USBGN)] (JOG). Aldrich and Bole (1937:7) indicate that collecting occurred on the uppermost areas of the drainage, although the name does not appear on specimen labels. See also: Montijo, Golfo de; and Azuero, Península de.

PALO SECO, BOSQUE DE PROTECCIÓN; Bocas del Toro
ca. 0855/8230
According to Angehr (2003:114, as "Palo Seco Protection Forest"), the forest covers an area of 167,000 ha, is considered an Important Bird Area, and ranges in elevation from 40 to 2,300 m forming a buffer zone for Parque Internacional La Amistad [ca. 0910/8240]. The protected forest extends ca. 100 km along the N borders of Parque Internacional La Amistad and Fortuna Reserva Forestal [ca. 0840/8210] (RDP2). Jiménez and Ochoa, at Velario (not located), presumably collected in the hills S of Changuinola [0926/8231 (USBGN)] as specimens are from Corregimiento de Changuinola [0915/8239 (USBGN)], 2002 Apr 20-21 (MVUP, as "Palo Seco"); DaCosta, Guitton-Mayerma, Klicka, Lelevier, Miller, and Spellman, GL = Chiriquí Grande, at 450-500 m, 08°47'36"/82°11'20", 17 km SSW of Chiriquí Grande, and at 1,325-1,375 m, 08°47'03"/82°12'50", continental divide 20 km SSW of Chiriquí Grande, 2006 Feb 22-28 (MBM, as "Bosque Protector de Palo Seco," and as "Palo Seco"; STRI, as "Bosque Protector de Palo Seco," and as "Palo Seco").

PALO SECO LEPROSARIUM; Panamá (Canal Zone) (W, Pacific)
0855/7934 (USBGN)
80 feet (25 m), on the W side of the Panama Canal [0920/7955 (USBGN)] near its Pacific entrance 1 km N of Punta Batele [0854/7934 (USBGN)] and 3 km W of Isla Naos [0855/7932 (USBGN)] (E761A, as "Palo Seco (Leper colony)"). Hallinan,

0.5 mi (1 km) offshore, 1915 Mar 24, 27 (AMNH; Hallinan, 1924, as "Palo Seco"); Imhof, observations, across the bay, 1942 Nov 8 (Imhof, 1950, as "Palo Seco"); Blanton, GL = Farfan (on some), 1953 Sep 27 (USNM, as "Pac, Palo Seco Road"); Loftin, 1962 Oct 2 (Loftin, 1963a); Hespenheide, GL for Road K1B, 1969 Oct-1970 Mar (Hespenheide, 1980, as "Palo Seco"); Morton, 1971 Jan 5 (USNM, as "Palo Seco"; Morton, 1971).

PALO SECO PROTECTION FOREST; see Palo Seco Bosque de Protección

PALO SECO ROAD; see Palo Seco Leprosarium

PANAMÁ; Panamá

ca. 0910/7940

In the earliest days of Spanish exploration and colonization the isthmus now known as Panamá went under the names Veraguas and Darién. The first colony on the isthmus was in the area of eastern Darién and the Golfo de Urabá as part of a royal concession called Tierra Firme (q.v.). Panamá was the name of an Amerindian fishing village on the Pacific coast until the founding of Panama City in 1519 as a terminus for the overland transportation of riches from Peru via the Las Cruces Trail (q.v.) first to the Caribbean port of Nombre de Dios (q.v.) and later Portobelo (q.v.). The name Panamá then took on a more general sense connoting the middle part of the isthmus, sometimes as far west as the province of Cocle (see Panamá, Isthmus of). This sense took on additional meaning in the mid-19th century with the emergence of three geopolitical factions: Azuero (based in Los Santos—q.v.), Veraguas, to the west, and the remaining central portion that included the capital and was known as Panamá. Prior to independence in 1903 as the Republica de Panamá, the isthmus was part of Colombia in its various configurations as Gran Colombia, Republic of New Grenada (1830-1886), and Republica de Colombia (1886-1903). The term Panamá was used in the early literature and by early collectors in reference to the central region of Panama or the area that is now located in both Panama and Colón provinces. For example in Sclater and Salvin (1873:162) the habitat of *Tinamus ruficeps* is given as both Panama and Chiriquí, indicating that Panama signifies a region, not just Panama City. Today the term Panamá refers to Panama City, which is listed as a separate entry for clarity. Dow, 18?? (USNM); Roome, J. H., 18?? (USNM); Suckley, G. B., 1855 Dec 28 (USNM); McLeannan, McLeannan was the stationmaster at both Frijoles [0910/7948 (USBGN)] and at Lion Hill [ca. 0913/7954] (q. v.) on the Panama Railroad [ca. 0922/7954] (q. v.) and made collections both by himself and with Galbraith from 1850 through the 1860s. He collected on both sides of the continental divide, although most of his work was done on the Atlantic drainage, 18??, 1862 (USNM; Sclater and Salvin, 1873, as "Panama"); Hopkins, specimens were prepared by Hicks, and one includes the label "Williamstown" (not located), 18?? (USNM); Hicks, 18??, 1864 Dec 20-23, 1865 Jan 3-21 (USNM); Boucard, 18??, 1875-1877, 1877 Jan (USNM); Ridgway, R., 1887 Oct 4 (USNM); Townsend, specimen labels indicate that he was on board the U. S. S. Albatross, which was on expedition exploring the Pacific waters off the coast of Mexico and central America, and which specifically stopped in Panama to embark Alexander Agassiz. The CM specimen has the additional label "Panama

Canal Colombia." According to Avery (1915), by 1891 the French Canal Company had dredged a channel from deep waters in Bahía de Panamá [0850/7915 (USBGN)] to wharves at Balboa, which is likely the locality referenced by this label, 1891 Mar 17 (CM, as "Panama"; USNM, as "Panama"; Townsend, 1890). See also: Cascajal; Panama Canal; Panama City; Panama Railroad; Panamá, Bahía de; Panama, Isthmus of; and Panama, Savanna of.

PANAMÁ; see Panama City

PANAMÁ, BAHÍA DE; Panamá (E, W, Pacific)
0850/7915 (USBGN)
Bay that extends ca. 80 mi (130 km) E-W along the S coast of Panamá province and opens on the S to the Gulf of Panama [0800/7930 (USBGN)] (SMN2B). Townsend, see also Panamá, 1888 (Townsend, 1890, as "Panama Bay"; Griscom, 1935b:292, as "Bay of Panama"); Goldman, worked in the vicinity of Isla Naos [0855/7932 (USBGN)] and landed on Roca San José [0854/7931 (USBGN)], 1912 Feb 4 (USNM, as "Bay of Panama"); Osgood and Anderson, 1912 Feb 4 (FMNH, as "Panama Bay"); Miller and Richardson, observations near Balboa, 1936 (Miller, 1937); Lea, anchored off Canal Zone, 1946 Mar 11 (CU; Wetmore, 1972:391, as "Panama Bay"); Eisenmann, GL = Balboa, observations on a launch trip not far from Balboa, 1948 Jun 28 (Eisenmann, 1951, as "Panama Bay"); Wetmore, observations offshore, c. 1950 (Wetmore, 1952:7, 9); Olson, between Isla Flamenco [0854/7931 (USBGN)] and Isla Taboga [0847/7933 (USBGN)], 1963 Apr 13 (Loftin and Olson, 1963, as "Bay of Panama"); Watson, 1965 Apr 27 (USNM, as "Bay of Panama"); Pacific Project, see Gulf of Panama for details of the expedition, 1968-1969; Reed, observations, SE of Panama City at 0850/7920, 1983 May 31, between Chepillo and Taboguilla Islands, 1983 Jun 14 (Reed, 1988, as "Bay of Panama"); Rodríguez, GL = San Felipe, *terraplén* (embankment), 1991 Aug 1 (MVUP); Rowlett, Panama City, offshore, 1992 Nov 2 (LACM, as "Panama City"); Agro, GL = Panamá City, ca. 50 km SE of Panama City, 1996 Feb 28-29 (ANSP, as "Gulf of Panama"; MVUP, as "Golfo de Panamá"); Cardiff, ca. 50 km SE of Panama City, and as a GL for Isla Taboguilla, 1996 Feb 28 (LSUMZ, as "Panama Bay"); Guitton-Mayerma, Lelevier, and Maley, at 09°01'36"/79°21'21", 2006 Jan 22, 25 (STRI, as "Bay of Panama"; UAM, as "Bay of Panama"). (Murphy, 1944a; Murphy, 1944b; Murphy, 1944c; Montgomery and Murcia, 1982; Buehler et al., 2004, as "Upper Bay of Panama"). See also: Boná, Isla; Chamá, Isla; Changamé, Isla; Chepillo, Isla; Cocoví, Isla; Farallón, Isla; Flamenco, Isla; Fortified Islands; Otoque, Isla; Perlas, Archipiélago de las Perlas; San José, Roca; Taboga, Isla; Taboga, Morro de; Taboguilla, Isla; Tórtola, Isla; Uravá, Isla; and Venado, Isla.

PANAMA, BAY OF; see Panamá, Bahía de

PANAMÁ, CIUDAD DE; see Panama City

PANAMA, GULF OF;
0800/7930 (USBGN)
Los Santos/Herrera/Coclé/Panamá/Darién (Pacific)
Large gulf on the S coast of Panamá, enclosed on the W by Península de Azuero

[0740/8035 (USBGN)], the N by the isthmus and the E by Darién province. Bahía de Panamá [0850/7915 (USBGN)] is in its N waters (JOG; RDPF). Lien, at sea 140 mi (225 km) SW of Panama, 1919 Oct (UCLA); Kelsall and Cullingford, at sea, 20-300 mi (32-480 km) S of Panama [City], off Península Azuero (q. v.), and S of Archipiélago de las Perlas [0825/7900 (USBGN)] (q. v.), 1924 Sep, Oct (BMNH; Douglas and Johnson, 1926); Miller and Richardson, observations rounding Cape Mala [= Peninsula de Azuero], 1936 (Miller, 1937); Murphy, observations, 1956 Nov-Dec (Murphy, 1958); Pacific Project, cruises began and ended in the Bahía de Panamá 11 km SE of Balboa [0857/7934 (USBGN)] and continued through the gulf for 3-5 days. Boats traveled between Archipiélago de las Perlas and Bahía Piña [0734/7811 (USBGN)], and the following localities were specifically mentioned in Loftin (1991) or on specimen labels: Fortified Islands [ca. 0855/7931], Isla Camote [0817/7848 (USBGN)], Isla del Rey [0822/7855 (USBGN)], Isla Galera [0812/7847 (USBGN)], Isla San José [0815/7907 (USBGN)], Isla San Pablo [0828/7851 (USBGN)], Islote Valladolid [0843/7936 (USBGN)], Palometa (not located), and Punta de Cruz [0813/7908 (USBGN)], 1968 Aug 15-18, Sep 17-20, Oct 11-15, Nov 2-4, Dec 17-20, 1969 Jan 24-27, Feb 28-Mar 2, Apr 25-28, Jun 19-22, Aug 29-31 (GML, as "Panama Bay"; MCN, as "Bay of Panama," and as "Gulf of Panama"; USNM, as "Bay of Panama," as "Panama Bay," and as "Gulf of Panama"; Loftin, 1991); Holthius, 1967 May 8 (FMNH); Pitman, on the water, 1987 Oct 1, 1988 Oct 4-5 (LACM; SDSNH). See also: Panamá, Bahía de.

PANAMA, ISTHME DE; see Panama, Isthmus of

PANAMA, ISTHMO; see Panama, Isthmus of

PANAMA, ISTHMUS OF;

0900/8000 (USBGN)

Colón/Panamá (Canal Zone) (Atlantic, Pacific)

The central regions of Colón and Panamá provinces, primarily, but not limited to, the former Canal Zone area between Colón [0922/7954 (USBGN)] and Panama City [0858/7932 (USBGN)]. Two USNM specimens collected by Heyde are clearly labeled "Pintada, Coclé, Isthmus of Panama" indicating that by 1889 the term "Isthmus of Panama" was applied to an extensive area. Selfridge (1874:7-8) defines the eastern limit of the Isthmus of Panama as the Cordillera de San Blas [0918/7900 (USBGN)] and the headwaters of the Río Chagres [0907/7942 (USBGN)] and uses the term Isthmus of Darién [ca. 0810/7745] (q. v.) for the region east of this. Early collections made along both the Panama Railroad [ca. 0922/7954] and the Panama Canal [0920/7955 (USBGN)] were sometimes labeled Isthmus of Panama or simply Panama [ca. 0910/7940] (q. v.). Dow, 18?? (USNM); Maack, collected from Colón to Panama City, and as a GL for Bas Obispo, San Pablo, and Río Grande Station, 1871 (MCZ, as "Atlantic Side of Isthmus of Panama," and as "Isthmus of Panama"; Maack, 1872; Bangs and Penard, 1921:391, as "Isthmus, Atlantic side"); Hassler Expedition, date ? [1872 ?] (USNM); Collector ?, 1890 (FMNH, as "Isthmus"); Heyde, GL for Pintada, 1889 Mar 7, May 14 (USNM); McLeannan and Galbraith, collected primarily along the Panama Railroad (q. v.), 1850s-1860s (Lawrence, 1862b; Lawrence, 1862d; Lawrence, 1864; Lawrence,

1874; Sclater, 1862b, as "Isthmo Panama"; Sclater, 1863a); Safford, Wetmore added this note to Safford's collection: "These skins were obtained by Safford in Panama City, the date, May, 1887 being the one when he secured them, not the date of collection. While Safford was told that they were taken on the Isthmus, several are unquestionably from Western Panama. It is probable that all come from Veraguas or Chiriquí and that they may have been collected by Enrique Arcé," 1887 May, Jun 4 (USNM, as "Isthmus of Panama," and as "Panama Isthmus"; see also Central, Cordillera). (Sclater, 1864b; Sclater, 1865; De Lattre and Bourcier, 1846, as "Isthme de Panama"; Dubois, 1875, as "Isthme de Panama"; Ridgway, 1902b; Darlington, 1931, as "Isthmus of Panama, Atlantic Side"). See also: Darién, Isthmus of; and Panama Railroad.

PANAMÁ, PROVINCIA DE; (Atlantic, Pacific)

0905/7905 (USBGN)

Province in central Panama that extends E from the Río Las Guias [0825/8003 (USBGN)] on the Coclé provincial boundary to the Darién provincial boundary on the N coast of Golfo de San Miguel [0822/7817 (USBGN)]. The NW provincial boundary with Colón province extends through Lago Gatún (Gatun Lake) [0912/7955 (USBGN)] and Lago Alajuela (Madden Lake) [0915/7935 (USBGN)] N of the continental divide then continues along the divide in E Colón province and along the border with Comarca de San Blas. The continental divide is generally ca. 500 m in E Panamá, but in the W, aside from Cerro Trinidad [0844/7958 (USBGN)] and the E slopes of Cerro Gaital [0837/8007 (USBGN)], the divide is quite low, with elevations dropping below 150 m near the Panama Canal [0920/7955 (USBGN)]. The province includes the S section of the former Canal Zone including the Pacific entrance of the Panama Canal, where the nation's capital, Panama City [0858/7932 (USBGN)], is located. Many of the islands located in Gulf of Panama [0800/7930 (USBGN)], including the Archipiélago de las Perlas [0825/7900 (USBGN)], are included within the provincial boundaries. The former Canal Zone, and specifically Las Cruces Trail [ca. 0904/7936], has been used as a route across the isthmus since the 1500s (HMCZ). Many early specimens collected in the province were simply labeled Panamá, Isthmus of Panama, or Panama Railroad. See also: Las Cruces Trail; Old Panama; Panama Canal; Panama Railroad; Panama, Isthmus of; and Panamá.

PANAMA, SAVANNA OF; Panamá (E, Pacific)

ca. 0859/7931

A savanna that extended several miles N and NE from Panama City [0858/7932 (USBGN)] to Pacora [0905/7917 (USBGN)] (q. v.) (Chapman, 1929a), and according to Wetmore (USNM), as far as Chepo [0910/7906 (USBGN)] (q. v.). In a letter from Brown, who collected on the savanna from his base at Calidonia [0858/7933 (USBGN)], the region of low flat country extended "for some four or five miles, gradually rising, to the hills. This [was] a grassy plain, very dry and burnt in appearance, especially in the dry season, with little patches of wood – island like – scattered about here and there. Near the city of Panama there [were] several orange groves, where I collected Blue-creepers and some Tanagers that I did not see elsewhere" (Bangs, 1906a). McLeannan and Galbraith, on the Pacific drainage in the "Savannahs near Panama," 1860-1861 winter (Lawrence, 1862e:322, as "Savannah of Panama"; Lawrence, 1865a, as "Savanna near Panama City"); Collector ? [André ?], 1899 Feb 23 (FMNH, as "Savanna"); André, 1899 Feb 22-24 (FMNH, as "Savanna," and as "Savannah"; MVZ, as "Savanna near Panama"); Brown, worked from his base at Calidonia, mangroves, woodlands, grasslands, some specimens are labeled simply "near Panama" [see Panama City], which probably refers to the Savanna of Panama, 1904 May 2-26 (MCZ; Bangs, 1906a; Thayer and Bangs, 1906; Bangs and Penard, 1921:380, as "Savanna, near Panama"); Griswold, 1936 Apr 22 (MCZ, as "Panama:savannahs"). (Zimmer, 1941b:3, 7, 14, as "Savanana near Panama"; Parkes, 1976, as "Savanna near Panama"; Olson, 1981d, as "Savanna").

PANAMA, SAVANNAH OF; see Panama, Savanna of

PANAMA 21; see Gualaca-Chiriquí Grande Road

PANAMA 71; see El Espino

PANAMA 71; see El Valle

PANAMA 75; see Road K19

PANAMA 81; see Madden, Camino

PANAMA 84; see Road K10

PANAMA 85; see Portobelo Road

PANAMA CANAL; Colón/Panamá (Canal Zone) (Atlantic, Pacific)
0920/7955 (USBGN)
The Panama Canal that exists today is the result of several international efforts including the construction of the original Panama Railroad [ca. 0922/7954] completed in 1855, the French effort led by Ferdinand de Lesseps from 1882-1889 to build a sea level canal, and finally the U. S. construction from 1904-1914 of the lock style canal that exists today. However, before the U. S. began digging the canal itself, efforts focused on improving sanitation and ridding the Canal Zone of mosquitoes, known at the time to carry both yellow fever and malaria. Previous work in the region, building the Panama Railroad [ca. 0922/7954] and attempting to build the sea level canal, had resulted in tremendous death tolls due to the diseases that plagued the Canal Zone. Once health standards were improved, the original Panama Railroad was reinforced to move workers and supplies as well as the earth excavated along the canal's route. At the Atlantic entrance the Panama Canal extends S through Bahía Limón [0920/7956 (USBGN)] and enters Lago Gatún (Gatun Lake) [0912/7955 (USBGN)] through Esclusas Gatún (Gatun Locks) [0916/7955 (USBGN)], then turns SSE exiting the reservoir and Colón province near Gamboa [0907/7942 (USBGN)]. In Panamá province the canal cuts through the continental divide, passes through Lago de Miraflores [0901/7936 (USBGN)] and arrives at Balboa [0857/7934 (USBGN)] on the Pacific coast. The canal extends along portions of the former streambeds of the continuous Río Chagres (see Río Chagres [0919/8000 (USBGN)]), the Río Camacho [0906/7941 (USBGN)] and the Río Grande

[ca. 0857/7934] and the larger part of the original Panama Railroad was shifted to the E during the canal's construction. Collector ?, Panamá province, Panama Canal area, 1995 Dec 2-3 (LSUMZ); Collector ?, GL for Gaillard Cut, 1998 Jun 24 (LSUMZ). (Tejera, 1995; Condit et al., 2001).

PANAMA CITY; Panamá (E, Pacific)

0858/7932 (USBGN)

Sea level, Panama's capital city is located at the Pacific entrance of the Panama Canal [0920/7955 (USBGN)] and just E of the former Canal Zone border adjacent to Balboa [0857/7934 (USBGN)] and Ancón [0858/7933 (USBGN)]. Greater Panama City now extends E beyond Old Panama and includes many regions that were small towns and settlements as recently as the 1980s. In Spanish the city is called Panamá, but we chose to use the English name Panama City in the text to distinguish it from both the country and the province (E661, as "Panama"; E962, as "Ciudad de Panama"). Hodgetts, near the city of Panama, 18?? (Sclater and Salvin, 1868a); Markham, 18?? (Salvin and Godman, 1879-1904, vol. 1:326); Hinds, 1839 (BMNH, as "Panama"; Gould, 1843; Gould, 1844, as "Panama"; Wetmore, 1959b:6, 9, as "Panama = Panama City"; Wetmore, 1972:337); Seeman, 185? (BMNH; Gould, 1856, as "at Panama"); Collector ? [Seeman ?], date ? (BMNH, as "Panama"; Gould, 1861a, as "Panama"); Bell, observations, 18?? (Gould, 1861b:47); Hicks, 1864 Dec 12, 24 (USNM); Rhodes, near, date ?, 1865 Dec (UF; USNM; Olson, 1980b, as "near Panama City"); Safford, 20 mi (32 km) from, most of the specimens in this collection were labeled "Isthmus of Panama," (q.v.) 1887 May (USNM); Festa, 1895 (Stone, 1918:240); Brown, he was based at Calidonia [0858/7933 (USBGN)] (q. v.) while collecting for E. A. and O. Bangs, therefore the term "near Panama" probably refers to the Savanna of Panama [ca. 0859/7931] (q. v.), 1904 May 2-26 (ANSP, as "Panama near Panama"; MCZ, as "Panama near Panama"; USNM, as "near Panama"; Thayer and Bangs, 1905:152, 156; Thayer and Bangs, 1906, as "City of Panama"; Bangs and Penard, 1918:90, as "near Panama City"; Bangs and Penard, 1919; Chapman, 1929b, as "Canal Zone, near Panama"; Griscom, 1929a:156, as "near Panama City"; Griscom and Greenway, 1937:424); Curl, 1906 (USNM); Hallinan, on the shore near, 1907 Aug 11 (AMNH; Hallinan, 1924; Zimmer, 1955:11); Deignan, observations, 1932 Jul (Deignan, 1933); Chapman and Clark, observations at Gorgas Memorial Institute, 1938 Jan (Chapman, 1941); Meyer, GL for Punta Paitilla, 1942 Feb 14 (USNM); Eisenmann, year ? Jun-Jul, 1950 Jul 15 (Eisenmann, 1952:48, 55, 59); Mendez, 1956 Apr 1, and as a GL for El Carmen, 1972 Oct 23, 1973 Dec 27 (GML, as "Ciudad de Panamá"); Collector ?, 1959 Aug 24 (GML); Stevenson, 1959 autumn (UF); Loftin, observations in the area, on the waterfront, 1962 Aug-Nov (Loftin, 1963a); Rojas, 1965 Jan 15 (GML); Olson et al., waterfront, 1966 (Olson et al., 1967); Telford, 1970 Jun 12 (UF); Morton, 1971 Jun 11 (USNM; Morton, 1971); Fonseca and Ordonez, 1973 Dec 8 (GML, as "Ciudad de Panama"); Rodríguez, *caja del Seguro Social* (Social Security Office), 1985 Apr 16 (MVUP); Weathers, GL for Camino Chiva Chiva (Chiva Chiva Road), 1985 Nov (CAS); Pitman, GL for Rodman United States Naval Reservation, 1987 Oct 8, Nov 9 (LACM); Branca, 1990 Jun 18 (GML, as "Ciudad de Panama"); Rowlett, offshore, 1992 Nov 2 (LACM); Overton, GL for Río Farfan, 1994 Sep 28 (LSUMZ); Lovette, GL for Isla Naos, 1995 Jun 4 (LSUMZ); Collector ?, GL for Isla Naos, 1995 Aug 13 (LSUMZ); Agro, GL for Bahía de Panamá, 1996 Feb 28-29 (ANSP; MVUP, as "Ciudad de Panamá"); Christian, GL for Cerro Ancon, 1997 Mar 26 (LSUMZ); Weir, on road to Gamboa, 2003 Mar 21 (CU, as "Panama"). (Griscom, 1933; Eisenmann, 1952:58-59; Eisenmann, 1955c; Eisenmann, 1963a, as "Panama"; Eisenmann and Howell, 1962, as "Panama"; Olson, 1981d as "near Panama," and as "Panama"; N. Smith, 1980; N. Smith, 1982; Tejera and Tejera, 2001; Tejera et al., 2001). See also: Ancón; Balboa; Juan Franco; La Venta Beach; and Panama, Savanna of.

PANAMÁ LA VIEJA; see Old Panama

PANAMA R.R. LINE; see Panama Railroad

PANAMA RAILROAD; Colón/Panamá (Canal Zone) (Atlantic, Pacific)
ca. 0922/7954

The original route of the Panamá Railroad, along with its numerous stations, was completed in 1855, although the world's first transcontinental railroad began operating before the tracks actually connected Colón [0922/7954 (USBGN)] and Panama City [0858/7932 (USBGN)]. During the early stages of planning and construction, prospectors seeking the shortest possible route to the gold in California hired boatmen to navigate the continuous Río Chagres (see Río Chagres [0919/8000 (USBGN)]), after which they continued by various means across the divide to Panama City. But in Oct 1851 weather inhibited the safe travel of two steamships to the Río Chagres landing forcing the ships to retreat to the harbor at Bahía Limón [0920/7956 (USBGN)]. When the prospectors learned that a railway had been completed from Colón to Gatún [0916/7955 (USBGN)], they "proposed that…[they] should be transported over the railroad to Gatun…[and despite the absence of passenger cars] the anxious emigrants, over one thousand in number,…were then disembarked and safely transported on a train of working cars to the Río Chagres at Gatun" (Otis, 1859:152). By 1854 the railroad extended as far as Culebra [ca. 0903/7940] and from here passengers traveled by mule "through heavy sloughs and rapid streams, along the border of deep ravines and over precipitous mountains…[to] arrive in Panama [City], only 12 miles distant" (Otis, 1859:166). Once completed, the route extended 47 mi (71 km) from Colón to Panama City. From the N terminus the route extended SSW from Colón to Gatún, then continued roughly SE along the continuous Río Chagres to Gamboa [0907/7942 (USBGN)]. This N section (Atlantic drainage) of the original route was built on extremely deep swamps that were filled, or, when they were seemingly bottomless at 180 feet (55 m), the tracks were floated along causeways that required continuous attention. Leaving the Río Chagres at Gamboa [0907/7942 (USBGN)], the route continued along the Río Camacho [0906/7941 (USBGN)] up to Culebra [ca. 0903/7940] at 262 feet (80 m) on the continental divide. From here the remaining 10 mi (16 km) of track entered the Pacific drainage and descended along the Río Grande [ca. 0857/7934] down to sea level at Panama City. Approximately 30 towns and railroad station stops are indicated on the original line (MIP; PR), and the names of these stops appear on the labels of early ornithological specimens. Beginning in 1904, the original route was reinforced and regularly shifted during the construction of the Panama Canal [0920/7955 (USBGN)] and from ca. 1908-1912 the tracks were moved to higher ground E of the original route to accommodate both Lago Gatún (Gatun Lake) [0912/7955 (USBGN)] and the Panama Canal. At the same time many of the train stations were also relocated, although some of those submerged in Lago Gatún (Gatun Lake) were lost or replaced with entirely new stations. Train service was suspended in the 1980s until the Panama Canal Railway Company began renovations of the line and reopened regular service again in 2001. Arcé, 18?? (Salvin and Godman, 1879-1904, vol. 1:302, 305, vol. 2:35, etc., as "Line of Railway"; Akhurst, 18?? (USNM); McLeannan and Galbraith, according to Sclater and Salvin (1864), McLeannan was the stationmaster at Lion Hill, although according to an earlier article that appeared in Harper's New Monthly Magazine (Otis, 1859:161), the Frijoles [0910/7948 (USBGN)] station "was superintended by a Mr. M'Clellan [sic], famed along the line as a taxidermist and general collector of curios." As far as we know, these two stations are the only ones that specifically appear on McLeannan's labels, and therefore it seems likely that he was stationmaster at each location for a period of time. In addition to collections made at these two stations, Lawrence (1862c) indicates that McLeannan collected along the railroad throughout the Atlantic drainage whereas Lawrence (1862e) indicates that McLeannan and Galbraith also collected a few specimens on the Pacific drainage, therefore specimens that bear a general label such as "Line of Panama Railroad" could have been collected anywhere between Colón and Panama City [0858/7932 (USBGN)]. Lawrence (1862b:286) indicates that specimens were "obtained at a locality about equidistant from both oceans," which definitely indicates the Atlantic drainage and perhaps Frijoles Station, mid 1850s-1860s, and as a GL for Frijoles, 185? (ANSP, as "Panama RR," and as "Line of Panama RR"; MCZ, as "Line of Panama Rail Road"; USNM; Lawrence, 1862b, as "Isthmus of Panama," and as "Line of Panama Railroad"; Lawrence, 1862c; Lawrence, 1862e; Lawrence, 1862f; Lawrence, 1862g; Lawrence, 1867b; Lawrence, 1865a; Lawrence, 1867c; Lawrence, 1867f; Lawrence, 1867g; Salvin and Godman, 1888-1904, vol. 2:57-8, as "Line of Railway," and as "Panama Railway"; Baird, 1872, as "Panama R. R."; Bangs, 1907:307, as "Line of R. R."; Ridgway, 1908, as "Line of Panama Railway"; Chapman, 1915a, as "Panama R. R. Line"; Chapman, 1929b, as "Line P. R. R."; Bangs and Penard, 1921:375, 380, 391, as "Line of Panama R. R.," as "Line of Panama Railroad," and as "Near Panama R. R."; Bangs and Penard, 1922:200, as "Line of R. R."; Zimmer, 1940; Parkes, 1969). (Oberholser, 1904; Todd, 1913; Chapman, 1917, as "Panama R. R."; Olson, 1981d, as "Line of Panama Railway"). See also: Bas Obispo; Bohío; Colón; Corozal; Culebra; Darién Railroad Stop; Empire; Frijoles; Gatún; Gorgona; Las Cascadas; Lion Hill; Matachin; Mindi; Monte Lirio; Paraiso; Río Grande Station; San Pablo; Tabernilla; and Tiger Hill.

PANAMA RAIL ROAD; see Panama Railroad

PANAMA RAILROAD, LINE OF; see Panama Railroad

PANAMA RAILWAY; see Panama Railroad

PANAMA RUINS; see Old Panama

PANAMÁ VIEJA; see Old Panama

PANAMÁ VIEJO; see Old Panama

PAN AMERICAN HIGHWAY; see Interamericana, Carretera

PANCHILLO; Coclé (Pacific)

ca. 0813/8033

Not shown on our maps but located in S Coclé province in Corregimiento Aguadulce [0813/8033 (USBGN)] near Aguadulce [0815/8033 (USBGN)] (q. v.). Tejera, 1977 Dec 26 (MVUP).

PAN DE AZUCAR; see Pan de Azucar, Cerro

PAN DE AZUCAR, CERRO; Coclé (Pacific)

0823/8026 (USBGN)

Described by Wetmore as a small, low, rocky hill covered with scrub located near El Gago [0822/8026 (USBGN)] (USNM). The hill is not named on our maps, but the USBGN coordinates correspond to a 160 foot (50 m) hill on E761A located 3 km N of El Gago. Wetmore, GL = El Gago, 1963 Jan 20 (USNM, as "Pan de Azucar").

PANDO; Chiriquí

0849/8242 (USBGN)

1,200 m, town 8 km NW of Volcán [0846/8238 (USBGN)] and 2 km E of the mouth of the Río Colorado [0849/8243 (USBGN)] (JOG). The term is used by Wetmore for the valley that extends 1 km S of town to the steep banks of the Río Chiriquí Viejo [0820/8241 (USBGN)], 2 km NE of town to Silla de Cerro Pando [ca. 0849/8241] and 2 km NW of town to the Río Colorado (USNM). Elevation ranges from 1,100 m near the Ríos Colorado and Chiriquí Viejo up to 1,500 m at Silla de Cerro Pando (JOG). Wetmore, GL = Volcán, at 4,000-4,500 feet (1,200-1,350 m), in an area of forest and *rastrojo*, in an old coffee plantation, on a trail along the upper slopes of Pando where tall trees were in bloom, at 4,000 feet (1,200 m) near the Río Chiriquí Viejo and a small stream with deep banks and forested slopes, up steep slopes to a plateau-like top covered in dense, creeping bamboo, through a series of abandoned pastures with fruiting blackberry bushes, 1954 Feb 28, Mar 4, 7, 11, 24, 29, GL = Volcán, at 4,200-4,300 feet (1,300 m), noted extensive clearing since 1954, trail through an old coffee plantation, recent clearings, and pastureland partly overgrown with brush, 1955 Feb 4 (LSUMZ; USNM). See also: Pando, Cerro; and Pando, Silla de Cerro.

PANDO, CERRO; Chiriquí

0849/8242 (USBGN)

1,700 m, peak 8 km NW of Volcán [0846/8238 (USBGN)] and 2 km S of Silla de Pando [0850/8241 (USBGN)] (JOG). Hartmann, L., Cerro Pando district, date ? (CAS); Hartmann, R., unclear which of the two peaks named Cerro Pando was visited, 1937 (ANSP); Hartman, Wetmore penciled El Volcán [= Volcán] on most of these specimen labels, at 4,000-4,100 feet (1,200-1,250 m), 1951 Mar 5-8, 10-13 (USNM); Wetmore called this peak Cerro Tisingal (see Tisingal [0848/8240 (USBGN)]) after 1954, GL = Volcán, at 5,000 feet (1,500 m), 1954 Mar 1, at 5,500-5,800 feet (1,700-1,750 m), from Silla de Cerro Pando [ca. 0849/8241] up a steep logging road, smaller growth remained but the timber trees had been removed, trails on the summit radiated in all directions providing good access, 1960 Feb 8-9 (USNM, as "Tisingal," and as "Cerro Pando"). (Myers, 1969). See also: Pando, Silla de Cerro.

PANDO, CERRO; Chiriquí
0855/8243 (USBGN)
2,482 m, peak 17 km NW of Volcán [0846/8238 (USBGN)] in the most northwestern region of Chiriquí province on the Bocas del Toro province/Costa Rican border (JOG). Enders, 10 mi (16 km) from El Volcán, 1935 Jul (ANSP).

PANDO, SILLA DE CERRO; Chiriquí
ca. 0849/8241
1,500-1,700 m, described by Wetmore as a ridge or a saddle that extends along the N and NE sides of Cerro Pando [0849/8242 (USBGN)] separating Cerro Pando from Cerro Picacho [0854/8238 (USBGN)] and accessible by road (USNM). The ridge that Wetmore called Silla de Cerro Pando is not named on our maps, but the town Silla de Pando [0850/8241 (USBGN)] appears on JOG at 1,600 m 13 km W of Cerro Punta [0852/8235 (USBGN)] and 9 km NW of Volcán [0846/8238 (USBGN)], and it appears that Wetmore collected near the dirt track that branches N from the main road to the town Silla de Pando. Olson (1981c:366) indicates that Silla de Cerro Pando is 17 km WNW of the peak Volcán Barú [0848/8233 (USBGN)], which also corresponds with the town Silla de Pando. Wetmore, GL = Volcán, at 5,000-5,500 feet (1,500-1,700 m), the area was being logged in 1954, yet the steep slopes and higher elevations were still forested and the borders were a dense tangle of tree ferns, weeds and thorny shrubs. In 1955, however, he noted extensive clearing, collected along a logging road with one side in weeds and berry-filled brush and the other side in forest with dense tangle, up a steep slope into untouched woodland where huge trees towered over smaller growth and the forest floor was fairly open, in a cleared area grown with tall weeds and shrubs, in new clearings, 1954 Feb 27, Mar 1, 3, 7, 11, 13, 22, 28, 1955 Feb 2, 4, at 5,500-6,000 feet (1,700-1,850 m) on a steep slope E of Silla de Cerro Pando on a narrow ridge, forest generally intact, at 6,000 feet (1,850 m) undergrowth cleared and coffee planted, 1954 Mar 5, GL = Volcán, at 4,800-6,000 feet (1,450-1,850 m), mountain slopes denuded to make extensive pastures, 1960 Feb 6, 8, 10-13, 15-16, 18, 25, GL = Volcán, pastures dotted with stumps and trees, at 4,600-5,000 feet (1,400-1,500 m), below Silla de Cerro Pando in open pastures on the flat above the Río Chiriquí Viejo [0820/8241 (USBGN)], 1965 Feb 26, 28, Mar 9 (LSUMZ; USNM; Olson, 1981c, as "El Volcán, Silla de Cerro Pando"). See also: Pando; Pando, Cerro; Picacho, Cerro; and Volcán.

PANTERA, ISLA; Colón (Canal Zone) (E)
0914/7955 (USBGN)
65 m, small island with a land area of 15.5 ha (Wright, 1985, as "Pantera Island") located 3.5 km S of Gatún [0916/7955 (USBGN)] in Lago Gatún (Gatun Lake) [0912/7955 (USBGN)] NW of Isla Barro Colorado [0909/7951 (USBGN)] (E762, as "Pantera Island"). Wright, censuses, 1976-1977 (Wright, 1979, as "Pantera Island";

Wright, 1985, as "Pantera Island"; Wright, 1988, as "Pantera Island").

PANTERA ISLAND; see Pantera, Isla

PARACAIDISMO, CAMPO DE; see Gatún Drop Zone

PARACOTÉ; Veraguas (Pacific)
ca. 0740/8101
Sea level, described by Aldrich and Bole (1937:7-8) as the local name for a small settlement located at the headquarters of the Boston-Panama Company, located on the E shore of Golfo de Montijo [0740/8107 (USBGN)] about 0.5 mi (1 km) inland and about 1 mi (1.5 km) S of the mouth of the Río Ángulo [0742/8102 (USBGN)]. Aldrich and Bole, GL = Golfo de Montijo, Península de Azuero, within 1 mi (1.5 km) of the settlement, at sea level along brushy roadsides, in tidal mud flats, on sandy beaches, in mangrove swamps, bottomland forest, and coconut groves, at 50 feet (15 m) in second growth and virgin semi-deciduous forest, brushy and grassy savannas, 1932 Feb 5-19, Mar 20-Apr 2 (CM; Aldrich and Bole, 1937; Wetmore, 1952:22; Wetmore, 1963a; Olson, 1981a; Olson, 1981d, as "E Shore of Montijo Bay, 1 mile S of Angulo River mouth")

PARAISO; Panamá (Canal Zone) (E, Pacific)
0902/7938 (USBGN)
Sea level, originally a station on the Panama Railroad [ca. 0922/7954] located on the Río Grande [0902/7938 (USBGN)] 1 km NW of Pedro Miguel [0901/7937 (USBGN)] and 2 km SE of Culebra [ca. 0903/7940] (MIP, as "Paraison Station"; PR). Currently, the name refers to a town at the same location, although the Panama Canal [0920/7955 (USBGN)] now occupies the original channel of the Río Grande (E8660). Hughes, 18?? (Salvin and Godman, 1879-1904, vol. 1-3, as "Paraiso," and as "Paraiso Station"; Sharpe, 1885, as "Paraiso Station"); Meyer, 1941 May 17 (USNM, as "Camp Paraiso"); MCU, 1955 May 11, Jun 10, and as a GL for: Cucaracha Hill, and Snake Hill, 1955 Jul 11, Aug 25 (GML); Collector ? [MCU ?], 1955 Jun 10 (MCN); Loftin and Olson, 1963 Mar 22 (Loftin and Olson, 1963); Collector ? [Seutin ?], GL for Old Gamboa Road, 1989 Mar 6 (LSUMZ, as "Paraiso Summit"); Seutin, GL for Old Gamboa Road, 1994 Sep 19-20, 22 (LSUMZ); Cardiff, Christian, and Dittmann, GL for Old Gamboa Road, 1996 Mar 14, 1997 Mar 14 (LSUMZ); Cardiff, Christian, Dittmann, Lovette, and Naoki, GL for Empire Known Distance Range, 1997 Mar 23 (LSUMZ; MVUP). (Berlepsch, 1911, as "Stationen Paraiso").

PARAISO, CAMP; see Paraiso

PARAISO STATION; see Paraiso

PARAISO SUMMIT; see Paraiso

PARIDA, ISLA; Chiriquí
0807/8220 (USBGN)
95 m, island with a land area of 14.5 km^2 (1,450 ha), centrally located in the Golfo de Chiriquí [0800/8220 (USBGN)] 18 km SSE of the mouth of the Río Chiriquí [0820/8226 (USBGN)] and 35 km SSE of David [0826/8226 (USBGN)] (JOG; MHA; Olson, 1997a). Batty, specimens at AMNH and at BMNH are dated 1902 Jan 18-19, 23, but Batty's data were completely fabricated and his specimens labeled

"Parida Island" were actually collected in 1901 near his mainland bases at Boquete [0847/8226 (USBGN)] and Boquerón [0830/8234 (USBGN)] (AMNH, as "Parida Island"; BMNH, as "Parida Island"; Zimmer, 1941b:14, as "Parida Is."; Eisenmann, 1950, as "Parida I."; Olson, 2008); Wetmore, GL for: Playa Jurel, and Punta Lanza, 1963 Feb 2-5, 7 (USNM; Olson, 1980b; Olson, 1997a).

PARIDA ISLAND; see Parida, Isla

PARIS; Herrera
0803/8033 (USBGN)
ca. 25 m, town 14 km SE of Santa María [0807/8040 (USBGN)] and 9 km SW of the mouth of the Río Santa María [0806/8029 (USBGN)] (JOG). Perrygo, 1948 Feb 14, 20, and as a GL for Río Santa María, 1948 Feb 17 (LSUMZ; USNM); Wetmore and Perrygo, near Paris, 1948 Feb 24, Mar 1. The secondary localities attributed to Paris were located in the region extending 6 km N and 6 km NE of town, GL for: Alvina, Río Escotá, and Río Santa María, 1948 Feb 24, Mar 4-5, 11 (USNM; Wetmore, 1953; Olson, 1980b, as "Paris, 4 miles NE of Alvina," but the specimen was taken at Alvina, 4 mi (6 km) NE of Paris); Delgado, 1977 Jul 5 (GML).

PARITA; Herrera
0800/8031 (USBGN)
ca. 25 m, town in eastern Herrera province 10 km NW of Chitré [0758/8026 (USBGN)] (JOG). Twomey, 1939 Apr 13 (CM; Trimble, 1943); Perrygo, 1948 Feb 2-7, 9-10, 12-14, 16-21, 23, and as a GL for Río Parita, 1948 Feb 13, 19 (LSUMZ; USNM; Monroe and Howell, 1966); Wetmore and Perrygo, 1948 Feb 26-29, Mar 2-9, 12-13, 15, 22-23, 25, 30, 3 mi (5 km) S of Parita, 1948 Mar 23, 29, and as a GL for: Alvina, and Río Parita, 1948 Feb 27-28, Mar 24 (USNM; Wetmore, 1950a; Wetmore, 1953; Wetmore, 1970); Delgado, 1977 Jul 11 (GML).

PARITA, BAHÍA DE; Coclé/Herrera/Los Santos
0808/8024 (USBGN)
Shallow bay extending ca. 35 km N-S along the S coast of Coclé province, the E coast of Herrera province and the NE tip of Los Santos province. The coastline is largely marsh, mangrove swamp and tropical grass, and mudflats extend 1-5 km offshore (JOG; RDPF, as "Golfo de Parita"; SMN2B, as "Golfo de Parita"). The bay is considered an Important Bird Area (Angehr, 2003). (Butler et al., 1992).

PARITA, GOLFO DE; see Parita, Bahía de

PARITA, RÍO; Herrera
0802/8027 (USBGN)
Originates in W central Herrera province and flows NE entering the Bahía de Parita [0808/8024 (USBGN)] 6 km N of Monagrillo [0759/8026 (USBGN)] and 9 km NE of Parita [0800/8031 (USBGN)] (JOG). Perrygo, GL = Parita, bank of, 1948 Feb 13, 19 (USNM); Wetmore and Perrygo, GL = Monagrillo, edge of a mud flat N of Monagrillo, 1948 Feb 25, GL = Parita, river section 1 mi (1.5 km) S of Parita, 1948 Feb 28, Mar 24 (USNM; Olson, 1980b, as "Monagrillo, Río Parita").

PARQUE LEFEVRE; Panamá (E, Pacific)

0901/7929 (USBGN)

Sea level, section of greater Panama City [0858/7932 (USBGN)] located just N of Old Panama [0900/7929 (USBGN)] on the E bank of the Río Abajo [0901/7929 (USBGN)] (E962). Yunker (from NIID), 1962 Apr 18 (USNM); Kourney, 1973 Jan 1 (GML); Pineda, 1978 Oct 31 (MVUP).

PARQUE METROPOLITANO; see Metropolitano, Parque Natural

PARQUE NACIONAL ALTOS DE CAMPANA; see Campana, Cerro

PARQUE NACIONAL GENERAL DE DIVISIÓN OMAR TORRIJOS HERRERA; see Omar Torrijos, Parque Nacional

PARQUE NACIONAL METROPOLITANO; see Metropolitano, Parque Natural

PARQUE NACIONAL PORTOBELO; see Portobelo

PARQUE NACIONAL SOBERANÍA; see Soberanía, Parque Nacional

PARQUE NACIONAL VOLCÁN BARÚ; see Barú, Volcán

PARQUE NATURAL METROPOLITANO; see Metropolitano, Parque Natural

PARQUE URRACÁ; see Bella Vista

PASIGA, RÍO; Panamá (E, Pacific)

0855/7855 (USBGN)

Originates in the western hills of Serranía de Majé [0855/7830 (USBGN)], flows SW then S emptying into the Bahía de Panamá [0850/7915 (USBGN)] 40 km NW of Chimán [0842/7837 (USBGN)] (JOG; MGRP; RDP2). According to Buehler et al. (2004), the coast at the river's mouth is a "an undisturbed intertidal flat composed of gravel, sand, and silt." Buehler et al., observations, 2003 Sep 7-27 (Buehler et al. 2004).

PASO ANCHO; Chiriquí

0849/8238 (USBGN)

1,500 m, town on the Río Chiriquí Viejo [0820/8241 (USBGN)] 5 km N of Volcán [0846/8238 (USBGN)] and 3 km SSW of Bambito [0850/8237 (USBGN)] (JOG). Tribaldos, 1986 Mar 29 (MVUP).

PASO BLANCO, FINCA; see Finca Paso Blanco

PASO DE MULA; Panamá (E, Pacific)

ca. 0910/7904

ca. 100-300 feet (30-90 m), town 3 km E of Chepo [0910/7906 (USBGN)] across the Río Mamoní [0908/7904 (USBGN)] (RDP; SMN2B). Wetmore and Perrygo, GL = Chepo, traversed ca. 3 mi (5 km) in an area of rolling hills with old second growth and new clearings, 1949 Apr 18 (USNM).

PASO DE RESPINGO; see Respingo, Paso de

PASO HONDO; Panamá (E, Pacific)

ca. 0911/7907 ?

ca. 20 m, not shown on our maps, but presumably located near the Río Hondo

[0901/7906 (USBGN)] and possibly the town Hondo [ca. 0911/7907], which according to E761A is located 3.5 km WNW of Chepo [0910/7906 (USBGN)]. Wetmore and Perrygo, GL = Chepo [0910/7906 (USBGN)], specimen purchased from a hunter, 1949 Apr 8 (USNM).

PASTORES, ENSENADA; see Shepherd, Ensenada

PASTORES, ISLA; Bocas del Toro, Archipiélago de Bocas del Toro

0914/8220 (USBGN)

95 m, island with a land area of ca. 3 km^2 (300 ha) located just over 1 km E of Punta de Gallinazo [0915/8222 (USBGN)] between Ensenada Shepherd [0913/8221 (USBGN)] on the W and Ensenada Grande [0914/8219 (USBGN)] on the E (E762, as "Isla Pastores (Shepherd Island)"; JOG). Wedel, 1928 May 17 (MCZ, as "Shepherd Isl."; YPM, as "Shepherd Island"; Peters, 1931a, as "Sheperd Island"; Peters and Griscom, 1929, as "Sheperd Id."); Wetmore, GL = Almirante [0918/8224 (USBGN)], in 1958 the rocky island rose abruptly from the coconut palm-lined coast to the higher elevations that were covered with brush and creepers, collected on the E coast S of Vergayo Point [ca. 0914/8219] in the brush, on the S coast through coconut palms with extensive tangled vegetation and through cacao plantations, 1958 Feb 22 (USNM, as "Shepherd Island"); SBE, at sea level, extensive pasture with large trees, *rastrojo*, 1989 Mar 3, and as a GL for: Cayo García, and Cayo Roldán, 1989 Mar 1 (USNM); Billo, GL = Isla Pastores, 2007 May 18, "1 km S of northernmost point," (USNM). See also: Almirante, Bahía.

PASTORES, QUEBRADA; Bocas del Toro

0913/8221 (USBGN)

Short mainland river that enters Ensenada Shepherd [0913/8221 (USBGN)] 4 km S of Punta de Gallinazo [0915/8222 (USBGN)] and 10 km SSE of Almirante [0918/8224 (USBGN)] (E762, as "Quebrada Pastores (Shepherd Creek)"). Wetmore, GL = Almirante, upriver through the mangroves at the mouth, the creek immediately became shallow with a rocky bottom, through cacao and *rastrojo* to an area of high trees, inland to the hills W of the stream with scattered trees and houses, through small patches of cacao, bananas and *rastrojo*, 1958 Feb 14, Mar 7 (USNM, as "Shepherd Creek"); SBE, more pasture than in Wetmore's time, 1989 Feb 26 (USNM; Parsons et al., 1993; Brumfield et al., 2001).

PATILLA POINT; see Paitilla, Punta

PATIÑO, PUNTA; see Old Point

PATTERSON, CAYO; Bocas del Toro, Archipiélago de Bocas del Toro

0906/8154 (USBGN)

14 m, small island with a land area of ca. 0.5 km^2 (50 ha) located 1 km W of the NW coast of Península Valiente [0905/8151 (USBGN)] in Laguna de Chiriquí [0903/8200 (USBGN)], 3 km SSE of Muturi [0907/8155 (USBGN)] (E762; JOG). SBE, GL = Península Valiente, some forest but any place not in pasture was uprooted and muddy due to the activities of humans and pigs, 1990 Apr 22 (USNM).

PAUNCH; Bocas del Toro, Archipiélago de Bocas del Toro

0923/8214 (USBGN)

Not shown on our maps, but USBGN indicates a point on the central W coast of Isla Colón [0924/8217 (USBGN)] located ca. 5 km N of Bocas del Toro [0920/8215 (USBGN)] and just N of the town Paunch [0923/8215 (USBGN)]. SBE, GL = Isla Colón, Bocas del Toro, ca. 2 mi (3 km) N of Bocas del Toro, 4 km N of Bocas del Toro, 1987 Mar 12, 1988 Feb 17, 1990 Apr 16 (USNM, as "Long Bay Point," and as "Paunch").

PAVARANDÓ, RÍO; Darién

0749/7804 (USBGN)

Originates in SE Serranía del Sapo [0750/7817 (USBGN)] E of Cerro Piña [0741/7809 (USBGN)] and flows S entering the Río Jaqué [0731/7810 (USBGN)] 6 mi (10 km) E of Jaqué [0731/7810 (USBGN)] (SMN2B). Wetmore and Perrygo, GL = Río Jaqué, a gravel bar at the mouth of the Río Pavarandó, 1947 Mar 27 (USNM, as "Río Pavarondó").

PAVARANDÓ, BOCA DE; see Sambú, Río

PAVARONDÓ, RÍO; see Pavarandó, Río

PAVO, RÍO; Veraguas

0722/8054 (USBGN)

Originates on SW Península de Azuero [0740/8035 (USBGN)] near the Veraguas/Los Santos provincial border and near Cerro Hoya [0718/8042 (USBGN)]. The river flows N, then W entering the Pacific Ocean ca. 30 km SE of Isla de Cébaco [0732/8109 (USBGN)] (MGRP). The mangroves surrounding the river's mouth are considered an Important Bird Area (Angehr, 2003). Delgado, observations, date ? (Ridgely and Gwynne, 1989:77: Angehr, 2003:171).

PAYA; Darién

ca. 0753/7724

100 m, town on the Río Paya [0757/7730 (USBGN)] 8 mi (13 km) above its mouth and 28 mi (45 km) SE of El Real de Santa María [0808/7743 (USBGN)] (RDPF; SMN2B). USBGN lists the settlement at latitude 0758, but Atlas (1988:206), DP, MGRP, RDPF, and SMN2B and all show it between 0753 and 0755, and Fairchild and Handley (1966:17, as "Paya Village") list it at 0753/7724. Wetmore, GL for Río Paya, 1959 Mar 10-11 (USNM); Feinstein, 1959 Mar 22 (USNM, as "Paya Village").

PAYA, RÍO; Darién

0757/7730 (USBGN)

Originates on the continental divide on the Serranía del Darién [0830/7730 (USBGN)] in the easternmost section of Darién province and flows SW and W joining the Río Tuira [0821/7803 (USBGN)] 8 mi (13 km) W of Paya [0758/7724 (USBGN)] (SMN2B). Fairchild and Handley (1966) indicate that several of their localities are "near Boca de Río Paya," which refers to the river's mouth or to a settlement with this name at the river's mouth. GMLP, mouth of, 1958 (USNM); According to Wetmore, in 1959 the lower section of the river was clear with numerous rapids and steep banks dropping 10-30 feet (3-9 m) to the water's edge, and the banks were largely forested with only a few cultivated areas, Wetmore, GL = Paya, worked from the Paya camp located near the mouth of the Río Paya, although these specimens were brought to Wetmore and were presumably collected upstream near the town of Paya, 1959 Mar 10-11, and as a GL for: Río Tuira, and Murqui, 1959 Feb 13-26, 28-Mar 17 (USNM; Haffer, 1962, as "Boca de Paya"; Olson, 1981d, as "mouth of Río Paya, Río Tuira"); Burton, British Trans-Americas Expedition, specimens labeled "Boca de Paya (Río Tuira)" were given the coordinates 0756/7732, and those labeled "Río Paya" were given coordinates ca. 0756/7727, 1972 Feb (BMNH; Burton, 1973b; Burton, 1975).

PAYA CAMP; Darién

ca. 0757/7730

ca. 400 feet (120 m), a letter from Pedro Galindo dated 2 Sep 1958 seems to indicate that Paya Camp was located near the mouth of the Río Paya [0757/7730 (USBGN)] (USNM). Hinds and GMLP, 1959 Mar 29, Apr 11-15 (GML; USNM). Not shown on our maps.

PAYA VILLAGE; see Paya

PEAR LAKE; see Cuatro Calles

PEARL ISLANDS; see Perlas, Archipiélago de las

PECORA; see Pacora

PEDASÍ; Los Santos

0732/8002 (USBGN)

ca. 25 m, city on the SE coast of Península de Azuero [0740/8035 (USBGN)] 7 km NNW of Punta Mala [0728/8000 (USBGN)] S of the Río Pedasí [0734/8002 (USBGN)] on one of its tributaries 3 km W of the coast (JOG). Wetmore, 1957 Mar 17. The secondary localities attributed to Pedasí were located within 9 km of town, GL for: Bajadero, Laguna de la Boca, Los Higos, Quebrada del Toro, Quebrada Escotá, Río Caldera, and Río Pedasí, 1957 Mar 8-18, 20 (USNM; Wetmore, 1962; Wetmore, 1970; Olson, 1980b, as "Pedasí, Bajadero").

PEDASÍ, RÍO; Los Santos

0734/8002 (USBGN)

Originates near Los Asientos [0731/8008 (USBGN)], flows NE and enters the Gulf of Panama [0800/7930 (USBGN)] 4 km N of Pedasí [0732/8002 (USBGN)] (JOG). Wetmore, GL = Pedasí, the road crossing 2 mi (3 km) N of Pedasí and 1 mi (1.5 km) above the river mouth, cultivated areas and scanty forest along the stream, upriver to a 2-3 ha area of original forest, 1957 Mar 14 (USNM).

PEDREGAL; Chiriquí

0822/8226 (USBGN)

10 m, town 7 km S of David [0826/8226 (USBGN)] in a lowland area of mangroves and river deltas on the coast of Bahía de Charco Azul [0812/8245 (USBGN)] (E762; JOG; MHA). Brown, 1901 Aug 21 (MCZ, "Bogaba, Pedregal"; Bangs, 1902a; Thayer and Bangs, 1906).

PEDREGAL; Panamá (E, Pacific)

0904/7926 (USBGN)
ca. 25 m, town 10 km NE of Panama City [0858/7932 (USBGN)] on the Carretera Interamericana [0832/8242 (USBGN)] (JOG). Blanton, GL for Pacora Road, 1953 Sep 23 (USNM); Moore and Olivardia, 1990 Apr 6 (MVUP).

PEDREGAL, RÍO; Los Santos
0714/8037 (USBGN)
Originates on the Veraguas/Los Santos provincial border on the E slopes of Cerro Hoya [0718/8042 (USBGN)] and flows SE, then S entering the Pacific Ocean 22 km S of El Cortezo [0726/8038 (USBGN)] (MGRP). See also: Hoya, Cerro.

PEDRO DE COCAL; see El Cocal

PEDRO GONSALES, ISLA; see Pedro González, Isla

PEDRO GONSALES ISLAND; see Pedro González, Isla

PEDRO GONZÁLES, ISLA; see Pedro González, Isla

PEDRO GONZÁLEZ; see El Cocal

PEDRO GONZÁLEZ, ISLA; Panamá, Archipiélago de las Perlas
0824/7906 (USBGN)
109 m, third largest island in the Archipiélago de las Perlas [0825/7900 (USBGN)], 3 km by 6 km, located 14 km W of Isla del Rey [0822/7855 (USBGN)] (JOG). Bovallius, inhabited by ca. 100 people, interior covered with a rich vegetation, see discussion of Bovallius under Pueblo Nuevo [ca. 0820/8013 ?], 1882 Apr 4 (?), 22, 24 (NHRM; Rendahl, 1920); Brown, 1900 May 8 (MCZ, as "Pedro Gonzales Island"; Bangs, 1901a, as "Pedro Gonsales Island"; Bangs, 1901c, as "Pedro Gonsales Isd."; Thayer and Bangs, 1905:145, as "Pedro Gonzales Island"); Griscom and Crosby, 1927 Feb 15-19 (AMNH, as "Pedro González Island"; Griscom, 1927b, as "Pedro Gonzales," and as "Pedro Gonzales Island"; Wetmore, 1946b); Feathers, GL for Ensenada Honda, 1941 May 10 (ANSP, as "Pedro Gonzales"; Bond and Meyer de Schauensee, 1944:31, as "Pedro González"); Wetmore, GL for El Cocal, 1944 Mar 9, 11 (USNM, as "Pedro González Island"; Wetmore, 1946b, as "Pedro González Island"; Eisenmann, 1962c, as "Pearl Islands"; Olson, 1980b); Morrison, 1944 May 16, Jun 8, Jul 21 (USNM, as "Pedro Gonzales Island"; Wetmore, 1946b; Wright et al., 1985, refers to Wetmore 1947, which is actually 1946); (MacArthur et al., 1972—species list; see comments in Bibliography); Wright and Campbell, mist nets, observations, 1978 Jan-Mar (Wright et al., 1985); Levings et al., observations, 1982-1983 (Levings et al., 1986); Guitton-Mayerma and Miller, 2006 Apr 27-28 (STRI; UAM). (Zimmer, 1941a:7, as "Pedro Gonzales"; Wetmore, 1952:19, 22, 25, 27, 31; Wetmore, 1970; Monroe and Howell, 1966; N. Smith, 1971; Angehr, 2003:271-273). See also: Honda, Ensenada; and San Sebastian.

PEDRO GONZÁLEZ ISLAND; see Pedro González, Isla

PEDRO MIGUEL; Panamá (Canal Zone) (E, Pacific)
0901/7937 (USBGN)
Sea level, town on the Panama Canal [0920/7955 (USBGN)] 10 km NNW of Balboa [0857/7934 (USBGN)] (JOG). According to Barnard (1954), the town was

surrounded by "many low hills, some grass-covered and some jungle-clad, [that brought] various forms of wildlife to the very edge of the landscaped areas about the residences." The town was bordered by "a narrow river and small lake, on the south and east" and was only 2.5 miles S of the continental divide. Jewel, second growth, "tree in a wild banana jungle," 1911 Apr 23, 1912 May 5 (ANSP; Stone, 1918); Benson, at 200 feet (60 m), 1931 Jun 19, 24 (USNM, as "P. Miguel"); Harrower, observations, trail through second growth, 193? Jul 17 (Willis, 1972:14); Arbib and Loetscher, observations, "open country, second growth woods," 1934 Aug 1 (Arbib and Loetscher, 1935); Barnard, observations "on a hillside beside a second growth jungle," 1938-1939, 1952-1953, 1955 (Chapman, 1941; Barnard, 1954; Barnard, 1956); MCU, 1955 May 4 (GML; USNM); Willis, 8 km N, 1961 Jan 2, Sep 29-28, Nov 2, 9 (MVZ); Olson, 1966 May 4-5, 16, 22 (UF; USNM; and Olson et al., 1967, all as "Pedro Miguel Dump"); Morton, 1970 Jul 20, Nov 7, 9 (AMNH); Strauch, Pacific Saddle Club (not located), 1971 Sep 9 (LSUMZ). (Eisenmann, 1957; Wetmore, 1970).

PEDRO MIGUEL, ESCLUSAS; Panamá (Canal Zone) (Pacific)
0901/7937 (USBGN)
Locks on the Panama Canal [0920/7955 (USBGN)] 5 mi (8 km) NNW of Balboa [0857/7934 (USBGN)] on the N end of Lago de Miraflores [0901/7936 (USBGN)] (E762, as "Esclusa de Pedro Miguel"; RM1, as "Pedro Miguel Locks"). The locks can be accessed from Panama City [0858/7932 (USBGN)] by Carretera Gaillard [0904/7940 (USBGN)], therefore most collections were probably made on the E side of the Panama Canal. Greenway, 1932 Jul (Greenway, 1934, as "San Miguel Locks"; Eisenmann, 1952:25); Griswold, 1936 Apr 29-30 (MCZ, as "Pedro Miguel Locks"); Meyer, 1941 May 18 (USNM, as "Pedro Miguel Locks"); Morton, observations, 1974 Jan-Mar, 1975 Mar (Morton, 1979, as "Pedro Miguel Locks").

PEDRO MIGUEL, RÍO; Panamá (Canal Zone) (E, Pacific)
0901/7936 (USBGN)
Originates in Madden Road Timber Preserve [0906/7937 (USBGN)] and flows S entering the N end of Lago de Miraflores [0901/7936 (USBGN)] at Pedro Miguel [0901/7937 (USBGN)] (E761A; E8660). Benson, at 200 feet (60 m), 1931 Jun 19 (USNM, as "Río P. Miguel"); Keenan and MCU, 1955 May 4, 9, Jul 18 (GML; USNM).

PEDRO MIGUEL DUMP; see Pedro Miguel

PEDRO MIGUEL LOCKS; see Pedro Miguel, Esclusas

PELADO, CERRO; see Bald Hill

PELADO, ISLA; Panamá (E, Pacific)
ca. 0838/7839
Sea level, small, rocky and barren island in the Gulf of Panama [0800/7930 (USBGN)] located ca. 10 km SW of Chimán [0842/7837 (USBGN)] (MHA, as "Islote El Pelado"; Wetmore, 1952:5). Wetmore, observations, 1950 Feb-Mar (Wetmore, 1965:61, 73).

PELÓN, RÍO; Colón (Canal Zone) (E)

0911/7947 (USBGN)

Originates NW of Gamboa [0907/7942 (USBGN)] and flows SW entering Lago Gatún (Gatun Lake) [0912/7955 (USBGN)] ca. 1.5 km NE of Frijoles [0910/7948 (USBGN)]. The river crosses the Pipeline Road [ca. 0915/7948] 3 km NE of Frijoles and 11 km NW of Gamboa (E866). Ridgely and Forster, observations on the upper Río Pelón, 1976 Jan 31 (Ridgely and Gwynne, 1989:67-68).

PELUCA; see Peluca, Estación Hidro el

PELUCA, ESTACIÓN HIDRO EL; Panamá (E, Atlantic)

0923/7934 (USBGN)

100 m, a water measurement station near the former Canal Zone border on the Panamá side of the Río Boquerón [0921/7933 (USBGN)], 3 km above its mouth and just below the mouth of the Quebrada Peluca [0923/7934 (USBGN)] (E661, as "Peluca Hydrographic Station"; JOG). Wetmore, at 400 feet (120 m), 1961 Feb 17-19, 21, and as a GL for: Quebrada Peluca, and Río Boquerón, 1961 Feb 17-27 (USNM, as "Peluca Hydrographic Station"; Wetmore, 1962, as "Peluca Hydrographic Station"; Olson, 1980a, as "Río Boqueron, Peluca Hydrographic Station"; Olson, 1981d, as "Peluca Hydrographic Station"; Olson, 1983, as "Peluca Hydrographic Station").

PELUCA, QUEBRADA; Panamá (E, Atlantic)

0923/7934 (USBGN)

Originates just N of Estación Hidro el Candelaria [0923/7931 (USBGN)] and flows W entering the Río Boquerón [0921/7933 (USBGN)] just N of Estación Hidro el Peluca [0923/7934 (USBGN)]. A trail extends between the two water measurement stations within 0.5 km of the river's S bank (E661). Wetmore, GL = Estación Hidro el Peluca, at 500 feet (150 m), followed the trail above the Quebrada Peluca to Estación Hidro el Candelaria for ca. 1 mi (1.5 km) through old second growth and uncut forest 1961 Feb 17 (USNM, as "Río Peluca"; Wetmore, 1962).

PELUCA, RÍO; see Peluca, Quebrada

PELUCA HYDROGRAPHIC STATION; see Peluca, Estación Hidro el

PEÑA, PUNTA; Chiriquí ?

?

A skull of *Tinamus major* in USNM, formerly in the personal skeletal collection of A. Wetmore, has the collection information Punta Peña, Panama, Nov 1906, R. E. Black. In Wetmore's original catalog it is listed as *Tinamus castaneiceps*, a subspecies that in Panama is restricted to the Pacific slope W of the Canal. We have been unable to locate any Punta Peña or Punta de Peña in this area, nor have we found any information on R. E. Black.

PEÑA, PUNTA DE; see Punta de Peña

PEÑA BLANCA; Chiriquí

ca. 0848/8229

Described by Blake (1958:502 (map), 503) as a section of the Mönniche properties that extends NE from the NE base of Cerro Copete [ca. 0848/8231] along the land between the upper reaches of the Río Chiquero [ca. 0849/8229] and the Quebrada

El Velo [0849/8228 (USBGN)]. Mönniche, at 5,900-9,300 feet (1,800-2,850 m), 1932-1934, 1936, 1938-1940 (ANSP; FMNH; MCN); Davidson, GL = Boquete, Quiel, 1933 Jan 19 (CAS).

PEÑA BLANCA, CERRO; see Peñas Blancas

PENAL COLONY; see Colonia Penal

PEÑAS BLANCAS; Coclé (Atlantic, Pacific)
0839/8037 (GA)
According to Angehr (in litt., Aug 2005), this was a Western Watershed study site. The proper name is "Cerro Peña Blanca, 1314 m, shown on 4041-I [E762] on the continental divide about 4 km NW of the town of El Copé [0837/8036 (USBGN)]...[They] accessed this site by helicopter. GPS coordinates of [their] camp were 08°39'33" N 80°37'21" W, elevation 1280 m." This site is not the USBGN locality, Cerro Peña Blanca [0838/8014 (USBGN)]. Garcés specimens were collected from net lines below camp, all on the Pacific side of the peak. Garcés, Jiménez, López, Ramos, and Rodríguez, 2001 Aug 24-25 (MVUP; Louis Berger Group et al., 2003).

PEÑA TIBURÓN, PUNTA; Panamá, Archipiélago de las Perlas
0825/7851 (USBGN)
The map in Wetmore's field notes indicates a point on the NE coast of Isla del Rey [0822/7855 (USBGN)] ca. 2 km NW of the NW coast of Isla Cañas [0823/7849 (USBGN)] (USNM, as "Punta de Peña Tiburón"). Used by Wetmore (Wetmore et al., 1984:71, as "Punta de Peña Tiburón") to locate La Ensenada [0822/7851 (USBGN)] (q. v.).

PENENOME; see Penonomé

PENÍNSULA BOHÍO; see Bohío, Península

PENÍNSULA VALIENTE; see Valiente, Península

PEÑITAS, LAS; see Las Peñitas

PEÑITAS, LOMA LAS; see Las Peñitas, Loma

PENOME; see Penonomé

PEÑONCITO; see Alcalde Díaz

PEÑON DE LA HONDA ISLAND; see Villa, Isla

PEÑON DE LA HONDA WILDLIFE REFUGE; see Villa, Isla

PENONOMÉ; Coclé (Pacific)
0831/8022 (USBGN)
ca. 75 m, the provincial capital centrally located in Coclé on the Carretera Interamericana [0832/8242 (USBGN)] (JOG). Scholes and Scholes, observations on the roadside between Río Hato [0823/8010 (USBGN)] and, 1951 Jan 28 (Wetmore, 1953; Scholes and Scholes, 1954); Eisenmann and Bull, observations, 1952 July (Wetmore, 1953); Wetmore, 4 mi (6 km) W of Penonomé, 1953 June 20, GL for Puerto Posada, 1963 Jan 23 (USNM; Wetmore, 1953; Wetmore, 1956a; Olson, 1980c, as "4 miles W of Penonomé," and as "Penonomé, Puerto Posada");

Loftin, observations, 1961 Oct 20, 1962 mid-autumn, Nov 24 (Loftin, 1963a, as "Penome," and as "Penonome"); Loftin and Olson, observations, 1963 Mar 8 (Loftin and Olson, 1963); González, 1977 Mar 20, Apr 2, May 26, Jul 12, Aug 12-13, 20, 1978 (?) May 5, 1979 (?) Aug 12, 20, and as a GL for El Rosario, 1977 Nov 26 (GML); Vilchez, 1977 Jun (GML); Tejera, 1978 (MVUP); De Leon, GL for El Rosario, 1984 Sep 11 (GML); Weir, at less than 60 m, at 0837/8021, 2003 Apr 1 (CU, as "Penenome").

PEQUENÍ; Panamá (Canal Zone) (E, Atlantic)
ca. 0919/7933
200 m, according to Tejera (1995:10-11), Pequení refers to a study region near San Juan de Pequení [ca. 0922/7931] with transects located between ca. 0914/7930 and 0923/7936. Study areas included: the N waters of Lago Alajuela (Madden Lake) [0915/7935 (USBGN)] and the Río Pequení [0921/7933 (USBGN)], the coasts of which are primarily used for grazing and agriculture; secondary forests mixed with *potreros* and cultivated areas; and open areas consisting of *potreros* between the Quebradas Tiarillo (not located) and Candelaria [0923/7931 (USBGN)]. Tejera, *potrero*, 1993 May 23 (MVUP; Tejera, 1995).

PEQUENÍ, RÍO; Panamá (Canal Zone) (E, Atlantic)
0921/7933 (USBGN)
Originates in the hills on the Colón/Panamá provincial border, flows SW through Panamá province entering the former Canal Zone just below Estación Hidro el Candelaria [0923/7931 (USBGN)]. The lower course of the original streambed is now submerged in Lago Alajuela (Madden Lake) [0915/7935 (USBGN)], and the river currently enters the N end of the reservoir 2 km E of the Río Boquerón [0921/7933 (USBGN)] (JOG). Enders, 1935 Jan 17, and as a GL for Salamanca, 1935 Jan 14 (ANSP); Griswold, GL for Salamanca Hydrographic Station, 1936 Feb 3, 5, Mar 15, 24-25, 27, Apr (ANSP; MCZ; Griswold, 1936a, as "Río Pequeni, Salamanca Hydrographic Station"); Wetmore, GL = Estación Hidro el Candelaria, in an abandoned field grown in brush, followed an old trail along a forested ridge then descended into a deep stream, up a stream past old and new clearings to a forested area, forest trail leading W to the Río Mauro [0925/7933 (USBGN)], a short distance downstream to deep water where the river banks bordered a newly cleared farm with a few shade trees, downriver to a large clearing, forest and a small stream, banks covered with trees and brush, 1961 Mar 2-11 (USNM; Wetmore, 1962).

PERESENICO, RÍO; see Pirre Station

PERICO, ISLA; see Fortified Islands

PERISENICO, QUEBRADA; see Pirre Station

PERISENICO, RÍO; see Pirre Station

PERLAS, ARCHIPIÉLAGO DE LAS; Panamá (Pacific)
0825/7900 (USBGN)
Group of more than 30 islands, the largest of which is Isla del Rey [0822/7855 (USBGN)]. The islands are located in the Gulf of Panama [0800/7930 (USBGN)]

ca. 90 km SE of Panama City [0858/7932 (USBGN)] (JOG). According to Wetmore, in 1960 these islands were generally low in elevation with occasional hills, they were covered with open gallery forest with some creepers, and several of them lacked fresh water (USNM). Kelsall and Cullingford, they anchored in "St. Elmo's Bay" [= Bahía San Telmo] and collected on Isla del Rey, 1924 Jun (BMNH, as "Pearl Islands"; Douglas and Johnson, 1926, as "Pearl Islands"); Kellett and Wood, according to Sharpe (1906:406) the data on the specimen labels is questionable, "when in command of the Herald, in 1850, Captain Kellett and Lieutenant Wood made considerable collections of birds...Apparently all were properly labeled, but the smaller birds were put into paper cones which came off in transit, and very few of the original labels were preserved," Thayer and Bangs (1905) also discuss the reliability of the labels, 1848 May, 1850 (BMNH, as "Pearl Island, Bay of Panama"; Seemann, 1853; Salvin, 1882; Thayer and Bangs, 1905:140, 147-148; Rendahl, 1920:1); Eugenie Expedition, 1852 Apr (NHRM, as "Las Perlas Isl.," and as "Perlöarna"; Rendahl, 1920:1, 27); Hole, 1932 Feb 16 (LACM, as "Las Perlas Is"); Brown, collected on Islas Saboga (q. v.) and Rey (q. v.) 1900, 1904 (Todd, 1913, as "Pearl Island"; Bangs and Penard, 1921:380, as "Pearl Islands"); Griscom and Crosby, 1927 (AMNH; Griscom, 1927b); Murphy, vicinity of, 1937 Sep, 1941 Feb 9-15, May 23-26, 1945 Nov (Murphy, 1945, as "Pearl Islands"; Wetmore, 1946b); Smith, R. W., some specimens without further data, and as a GL for: Isla Bartolomé, and Isla Contadora, 1937 Feb 2-3 (ANSP); Northern, 1957 Mar 18 (LACM, as "Perlas Is"); Loftin, date ? (GML); Batista and Montgomery, observations, 1981 (Batista and Montgomery, 1982). (Wicker, 1917; Griscom, 1929a; Griscom, 1933; Rogers, 1939; Zimmer, 1939, as "Pearl Islands"; Wetmore, 1952; Graves and Gotelli, 1983; Wright et al., 1985; Gotelli and Graves, 1990; Seutin et al., 1993; Angehr, 2003:267-268). See also: Afuera, Isla de; Bajo Boyarena, Isla; Bartolomé, Isla; Bavarena, Isla; Bayoneta, Isla; Bolaños, Isla; Camote, Isla; Cañas, Isla; Caracoles, Isla; Casaya, Isla; Chapera, Isla; Chitré, Isla; Coco, Isla de; Contadora, Isla; Espiritu Santo, Isla de; Galera, Isla; La Mina, Isla; Málaga, Isla; Mogo Mogo, Isla; Morena, Isla; Pacheca, Isla; Pachequilla, Isla; Pedro González, Isla; Platania, Isleta; Puerco, Isla de; Rey, Isla del; Saboga, Isla; San José, Isla; San Pablo, Isla; San Telmo, Isla; Señora, Isla; Señorita, Isla; Trapiche, Isleta; Viveros, Isla; and Vivienda, Isla La.

PERMÉ; San Blas

ca. 0845/7734

Not shown as such on our maps but Peters and Griscom (1929:48) indicate that Permé is "shown on maps as Anachucuna [sic]." Yet according to a more detailed description provided by Griscom (1932:303-304), Permé was "a small plantation of the United Fruit Co.," located 5 mi (8 km) W of Anachukuna [0843/7733 (USBGN)] and "not more than 25 mi (40 km) from the Colombian border." The land "in the immediate vicinity of the station [was] largely cleared for agriculture," but "virgin forest" began ca. 1 mi (1.5 km) behind the village. Wedel's access to the interior and higher elevations was limited by "the 'bush' [that was] much denser and more impenetrable than at Almirante." Angehr (in litt., Aug 2005) suggests that if Permé was not at Anachukuna but W of it, then it was probably located on Bahía de Carreto [0847/7734 (USBGN)], where a cultivated area is shown on his copy of

E663 sheet 1388. According to MGRP, the central coast of Bahía de Carreto is 23 km NW of Cabo Tiburón [0842/7721 (USBGN)] (on the Colombian border) and 5 km NW of Anachukuna. A small peninsula separates Bahía de Carreto from Bahía Anachucuna [ca. 0842/7729]. Wedel, at 0-1,500 feet (450 m), according to Griscom (1932), he arrived 14 Mar 1929 and departed c. January 1930 for Puerto Obaldía, therefore the records for 1930 beyond Jan and February are probably typos or misreadings of specimen labels, 1929 Mar-Dec, 1930 Jan-Jun, Sep-Nov (ANSP; FMNH; MCZ; OC; YPM; Peters and Griscom, 1929; Griscom, 1931; Griscom, 1932a; Griscom, 1933; Griscom and Greenway, 1937:418; Rogers, 1939; Conover, 1944; Brodkorb, 1950; Wetmore, 1951a; Wetmore, 1959b:10, 20; Eisenmann and Howell, 1962, as "Permé, Colon"; Olson, 1980a; Olson, 1981d; Escalante-Pliego and Peterson, 1992; Adsett and Wege, 1998).

PERRECENEGA; see Pirre Station

PESÉ; Herrera
0754/8037 (USBGN)
ca. 75 m, city located in central Herrera province (JOG). Wetmore and Perrygo, GL for: Quebrada Chitrabé, Quebrada Tejel, and Río Esquiguita, 1948 Mar 23-24, 29-30 (FMNH; USNM; Wetmore, 1950a; Wetmore, 1952:17; Payne, 1974). See also: Borrola de Pesé.

PESÉ, BORROLA DE; see Borrola

PICACHO, CERRO; Chiriquí
0854/8238 (USBGN)
2,986 m, a peak just S of the continental divide ca. 8 km NW of Cerro Punta [0852/8235 (USBGN)] and 13.5 km N of Volcán [0846/8238 (USBGN)] (E762; MGRP; RDPF). Wetmore, GL = Volcán, at 5,300-8,300 feet (1,600-2,550 m), from Silla de Cerro Pando [ca. 0849/8241] walked 4-5 mi (6-8 km) NNE, at the lower elevations through forest with dense undergrowth, at 5,600 feet (1,700 m) the forest was planted in coffee, above 6,000 feet (1,850 m) the area was steep and the forest had large trees with open undergrowth, bamboo thickets, and an extensive area recently planted in coffee, at 6,300 feet (1,900 m) at the base of the final high ridge that was covered with virgin forest, to a peak at 8,325 feet (2,540 m) with heavy moss, ground plants, and few trees, 1955 Feb 3, 7 (USNM; Olson, 1981c, as "El Volcán, Cerro Picacho").

PICACHO, CERRO; see Los Voladores

PIEDRA, CUESTRA DE; see Piedra, Cuestra de

PIEDRA, LLANO DE; see Llano de Piedra

PIEDRA, QUEBRADA; see Quebrada de Piedra

PIEDRA, QUEBRADA DE; see Quebrada de Piedra

PIEDRA, RÍO; San Blas
ca. 0926/7901 ?
Described by Wetmore as a river E of the Río Nergala [0927/7902 (USBGN)] (USNM), not named on our maps, but possibly the small, unnamed river shown on

MHA and E501 that enters the Golfo de San Blas [0930/7900 (USBGN)] 2 km E of the Río Nergala, Wetmore, GL = Mandinga (q. v.), 1957 Feb 3 (USNM).

PIEDRAS, RÍO; Colón (E)
0927/7944 (USBGN)
Originates on Cerro Bruja [0929/7934 (USBGN)] and flows SW, then turns NNW as it enters the Caribbean Sea [1500/7500 (USBGN)] 21 km NE of Colón [0922/7954 (USBGN)] (JOG). According to our recent maps (JOG and MGRP), the Portobelo Road [ca. 0933/7939] crosses the river near its mouth 2 km NE of María Chiquita [0927/7945 (USBGN)], but in the 1960s the road was a trail at best and presumably terminated before reaching the river. Smith, N., GL = Portobelo Road, 2.2 mi (3.5 km) N of Portobelo Road, W Bank, closed forest lacking undergrowth, 1965 Mar 19 (USNM); Smith, N. et al., observations in the María Chiquita/Río Piedras area, c. 1980 (Wetmore et al., 1984:358).

PIJI; see Piji Baisal

PIJI BAISAL; Darién
ca. 0803/7744
Described by Engleman et al. (1989:491-2, as "Piji," and as "Piji Baisal") as a settlement located on the Río Pirre [0808/7744 (USBGN)] en route to Pirre Station [0801/7744 (Croat, 1997)], but outside of the boundary of the Darién Parque Nacional [0749/7744 (GA)]. According to Angehr (in litt., Aug 2005), this is a small Embera village, and the name should perhaps be spelled as one word, Pijibaisal.

PILA; see La Amistad, Parque Internacional

PILÓN, CERRO; Panamá (E, Pacific ?)
ca. 0912/7924
2,650 feet (808 m), described by Wetmore as a bare, grass-covered point on Cerro Azul [0910/7924 (USBGN)], 3 mi (5 km) on the road beyond his more usual stopping points (USNM), presumably 1-2 km N or NE of Cerro Ultima [ca. 0911/7923] and possibly the same area that he described in 1957 while collecting on Cerro Azul (q. v.). Wetmore, GL = Cerro Azul, at 2,700 feet (800 m), and at 2,800 feet (850 m), 2 mi (3 km) NE of Cerro Pilón in dry woodland, 1957 Mar 2 (USNM). Not shown on our maps.

PILÓN, LA PUNTA DE; Colón (W)
0911/8013 (USBGN)
Sea level, not shown on our maps, but USBGN indicates a point on the Atlantic coast ca. 2 km W of Boca del Río Indio [0912/8011 (USBGN)]. Wetmore and Perrygo, GL = Río Indio, 1952 Feb 20 (USNM, as "Punta Pilón").

PILÓN, PUERTO; see Puerto Pilón

PILÓN, PUNTA; see Pilón, La Punta de

PIÑA; Colón (W)
0917/8003 (USBGN)
Sea level, coastal town at the mouth of the Río Piña [0917/8003 (USBGN)] on the former Colón/Canal Zone boundary 17 km SW of Colón [0922/7954 (USBGN)] (JOG). Eisenmann, observations, date ? (Wetmore, 1972:517); Quiroz, El Cacal

(not located), 1984 Sep 13 (MVUP). See also: Achiote Road; Camp Piña; Piña Beach; and Río Indio-Piñas Road.

PIÑA; Veraguas (Pacific)
ca. 0749/8105
Sea level, according to a map that accompanied a letter from Benson to Griscom dated Nov. 7, 1926 (AMNH), Piña was located on the NE coast of Golfo de Montijo [0740/8107 (USBGN)] just opposite Isla Verde [0750/8105 (USBGN)] and ca. 9 km SSW of the mouth of the Río Piña [0753/8104 (USBGN)] (AMNH). Benson and Seaman, 1924 Oct 2-5, 7-8, 10 (AMNH, as "Pina, Santiago").

PIÑA, BAHÍA; Darién
0734/7811 (USBGN)
A bay that extends 4 km NW-SE along the S Darién coast just N of Jaqué [0731/7810 (USBGN)], the bay opens on the SW to the Gulf of Panama [0800/7930 (USBGN)] and its N and S coasts are formed by 2 peninsulas that extend a few kilometers from the mainland (JOG; MGRP; MHA, as "Puerto Piñas"). Murphy and Correia, Askoy Expedition, Molino Cove (not located), 1941 Feb 25 (AMNH, as "Piñas Bay"; Murphy, 1944a; Murphy, 1944c, as "Piñas Bay"); Vanderbilt Expedition, according to Fowler (1944:254, as "Piñas Bay") fishes were collected in the bay and "some specimens were taken from the mouth of the tidal river behind the village," the only river and town shown on our maps are Río Piña [0735/7811 (USBGN)] and Puerto Piña [0735/7811 (USBGN)], therefore the birds that Feathers collected were probably collected over the open water of the bay or near Puerto Piña, 1941 May 1 (ANSP, as "Piñas Bay"; Bond and Meyer de Schauensee, 1944:41, as "Piñas Bay"); Northern, 1957 Mar 8-13, 16, 20-21 (LACM, as "Piñas Bay"); Robins, observations, at sea between the bay, a deep reef 10 mi (16 km) NE of the bay, and a drift line ca. 25 mi (40 km) SW of the bay, 1957 Jul 15-26 (Robins, 1958, as "Piñas Bay"); Paulson, 1961 Sep 9-12 (FMNH); Loftin, 1964 Dec 20 (USNM, as "Piñas Bay"); Pacific Project, GL = Gulf of Panama, near, 1968 Oct 12 (AMNH, as "Piñas Bay"; Loftin, 1991). (Escalante-Pliego and Peterson, 1992, as "Piñas Bay") See also: Jaqué, Punta; and Panama, Gulf of.

PIÑA, CAMP; see Camp Piña

PIÑA, EL PUERTO DE; see El Puerto de Piña

PIÑA, PUERTO; see El Puerto de Piña and Puerto Piña

PIÑA, RÍO; Colón (Canal Zone) (W)
0917/8003 (USBGN)
Originates near the NW coast of Lago Gatún (Gatun Lake) [0912/7955 (USBGN)] and flows NNW entering the Caribbean Sea [1500/7500 (USBGN)] at Piña [0917/8003 (USBGN)]. The lower reaches of the river formed the boundary between Colón province and the former Canal Zone (E661; JOG). See also: Camp Piña; Medio, Río; and Piña Beach.

PIÑA, RÍO; Darién
0735/7811 (USBGN)
Originates on the S end of Serranía del Sapo [0750/7817 (USBGN)] W of Cerro

Piña [0741/7809 (USBGN)] and enters Bahía Piña [0734/7811 (USBGN)] at Puerto Piña [0735/7811 (USBGN)] (SMN2B). Murphy and Correia, Askoy Expedition, 1941 Feb 26-Mar 2 (Murphy, 1944a; AMNH, as "Río Piñas"); Bermingham, Tropic Star Lodge, Piñas-Sambu Trail, at 07°41'27"/78°11'53", 2004 May 13 (STRI, as "Río Piñas"; UAM, as "Río Piñas"); Bermingham, Guitton-Mayerma, Miller, and Weir, Tropic Star Lodge, 0736/7812, 2006 May 5-14 (LSUMZ, as "Río Piñas"; STRI, as "Río Piñas"; UAM, as "Río Piñas").

PIÑA, RÍO; Veraguas (Pacific)
0753/8104 (USBGN)
Originates in NW Península de Azuero [0740/8035 (USBGN)] and flows SW entering the NE coast of Golfo de Montijo [0740/8107 (USBGN)] (JOG). See also: La Marea; and Piña.

PIÑA BEACH; Colón (Canal Zone) (W)
ca. 0918/8001
Sea level, beach on the Caribbean coast 3.5 mi (5.6 km) E of Piña [0917/8003 (USBGN)] and W of the mouth of the Río Chagres [0919/8000 (USBGN)] (RM2). Olson and Wiese, GL = Río Chagres, mouth of, 1966 Jul 13, 15-16, and as a GL for Esclusas Gatún (Gatún Locks), 1966 Jul 13-14, 16 (UF; USNM).

PIÑAS, BAHÍA; see Piña, Bahía

PIÑAS, MORRO DE; Darién
0733/7813 (USBGN)
Not shown on our maps, but USBGN indicates a rock in the water off the S Darién coast near Bahía Piña [0734/7811 (USBGN)]. Mentioned in Robins (1958) regarding observations at Bahía Piña.

PIÑAS, RÍO; see Piña, Río

PIÑAS BAY; see Piña, Bahía

PIÑAS-SAMBU TRAIL; see Piña, Río

PINOGANA; Darién
0808/7741 (USBGN)
Sea level, town 4 mi (6 km) E of El Real de Santa María [0808/7743 (USBGN)] on the Río Tuira [0821/7803 (USBGN)] (SMN2B). Burton, British Trans-Americas Expedition, 1972 Feb (BMNH; Burton, 1973b; Burton, 1975). See also: Tuira, Río; and Urutí.

PINTADA; see La Pintada

PINTADA, LA; see La Pintada

PINUGUILLA; Darién
ca. 0802/7820
Not shown on our maps but located in W Darién province in Corregimiento Garachiné [0802/7820 (USBGN) near Garachiné [0804/7822 (USBGN)] (q. v.). Garcés, 1987 Sep 24 (MVUP).

PIPELINE ROAD; Colón (Canal Zone) (E)
ca. 0915/7948

A secondary road in the former Canal Zone, also labeled Road 842, that extends along a section of pipeline for roughly 22 km along the SE coast of Lago Gatún (Gatun Lake) [0912/7955 (USBGN)]. From Gamboa [0907/7942 (USBGN)] the road extends NNW, passes through Limbo Plot [0909/7944 (Karr, 1971a)], crosses many small rivers and follows a sinuous route terminating on the central E coast of the reservoir 2 km E of Isla Lirio [0915/7950 (USBGN)], a small island in the reservoir. In Gamboa the road connects to Carretera Gaillard [0904/7940 (USBGN)] (E866). According to Karr (1971a), "elevations along this road range from about 30 m to less than 170 m," and canopy height is 40 m or more with generally open undergrowth. Willis, observations, 1960-1970 (Willis, 1972:36); Wetmore (from Gale and Keenan), GL = Gamboa, 1960 Nov 25 (USNM); Wetmore (from Gale), 1960 Dec 27 (USNM, as "Navy Pipeline Road"); Hinds, 1962 May 14 (USNM, as "Gamboa Pipeline Road"); Loftin and Belton, banding, 1962-1964 (Loftin, 1975); Olson et al., GL = Gamboa, 1963 Mar 30, 1966 May 18-21, 28-30, Jun 1, 3-10, 23, Jul 11, 20 (UF, as "Navy Pipeline Road"; USNM, as "Navy Pipeline Road," and as "Gamboa Pipeline Road"; FMNH, as "Navy Pipeline Road"; Olson, 1966, as "Navy Pipeline Road"); Gale, 1963 Mar 31 (GML, as "Gamboa Pipeline Road"); Smith, N., near Gamboa, 6 mi (10 km) N of Gamboa, 1964 Nov 11 (CU); Crossin, GL = Gamboa, 4 mi (6 km) NW, 1968 Aug 10 (USNM); Karr, at Limbo Plot (q. v.), 1968-1969 (Karr, 1971a; Karr, 1971c, as "Navy Pipeline Reservation"); Stevenson, GL = Gamboa, near Gamboa, 1968 Sep 10 (LSUMZ, as "Old Pipeline Road"); Hespenheide, censuses, 1969 Oct-1970 Mar (Hespenheide, 1980); Morton, 5 mi (8 km) NW of Gamboa, near hunters camp [= Limbo Plot ?], 1971 Mar 3, 22, 26, 29, near beginning, hill path, 3 mi (5 km) past Limbo, near Gamboa, near marker 23, 1974 Mar 7-8, 14, Dec 18, studies, 1973-1975 (AMNH; CAS; FMNH; USNM; Morton, 1980; Morton, 1982, as "Gamboa-Pipeline Road"); Strauch, 1971-1973 (UMMZ; Strauch, 1977); Brawn, Karr and Nichols, mist nets, 1977-1986 (Brawn et al., 1995); Beehler, observations at a flowering tree, 1978 Jan 12-15 (Beehler, 1980); Karr and Freemark, mist nets at 4 study plots (Ridge, Limbo, Road, and Valley) along the road, 1979-1982 (Karr and Freemark, 1983); Roubik, observations, 1979 Nov 24-Dec 10 (Roubik, 1982); Braun, M., 1982 Feb 16 (LSUMZ); Olson et al. (SBE), GL = Parque Nacional Soberanía, Gamboa, at 20 m, 0-2 km from Gamboa on the road, 1991 Feb 14-17 (USNM; Brumfield et al., 2001); Agro et al., ca. 10-12 km past the gate, 1996 Feb 11-13 (ANSP); Collector ?, GL for Limbo Plot, 1996 Nov 22 (LSUMZ). (West, 1975, as "Gamboa Pipeline Road"; Karr, 1977; Karr, 1984; Karr, 1990; Karr et al., 1990b; Wright, 1979; Greenberg and Gradwohl, 1980a; Olson, 1981d, as "Gamboa Pipeline Road"; Willson et al., 1982; McDade, 1983; Roubik et al., 1985; Houston, 1987; Sieving, 1992; Poulin and Lefebvre, 1996; Stutchbury et al., 1996; Hau et al., 1998; Mabey and Morton, 1992; Neudorf and Blanchfield, 1995; Wikelski et al., 1999; Wikelski et al., 2000; Roper, 2000). See also: Limbo Plot; and Soberanía, Parque Nacional.

PIRAGUALES, LAS; see El Cortezo

PIRIÁ; Panamá (E, Pacific)

0907/7820 (USBGN)

ca. 50 m, town in E Panamá province ca. 25 km W of the Darién provincial border, ca. 85 km E of Chepo [0910/7906 (USBGN)] on the Río Chepo [0900/7906 (USBGN)] ca. 15 km E of the eastern coast of Lago Bayano [0908/7835 (USBGN)]. The town is N of the Carretera Interamericana [0832/8242 (USBGN)] ca. 17 km SE of Cerro Diablo [0915/7825 (USBGN)], which is located on the continental divide (JOG; MGRP). DaCosta, Guitton-Mayerma, Klicka, Miller, and Spellman, GL for Higueronal, 2006 Feb 13-14, 18 (MBM; STRI).

PIRRE, CERRO; Darién

0746/7744 (GA)

The official Cerro Pirre [0756/7743 (USBGN)] is a peak in S central Darién province at the N end of Serranía de Pirre [ca. 0749/7744] ca. 20 km S of El Real de Santa María [0808/7743 (USBGN)] (RDPF; RDP2). According to Angehr (in litt., Oct 2003), the official Cerro Pirre is not the same as the ornithological Cerro Pirre. Beginning with Goldman and continued by all subsequent collectors, the name Cerro Pirre has been used for a peak ca. 5-6 mi from Cana [ca. 0745/7742] (Goldman, 1920:12), which corresponds to the S end of the Serranía de Pirre very near the Colombian border and the Alturas de Nique [0740/7745 (USBGN)] ca. 20 km S of the actual peak with this name. The confusion arises from the paucity of maps covering Darién province. Among those available — especially the older ones — Cerro Pirre, the upper course of the Río Tuira, and the Colombian border are lacking or displayed incorrectly, therefore separating two peaks, each with the same name, has been difficult. Woakes (1900), Anthony (1916) and Goldman (1920) all contain maps depicting the ornithological Cerro Pirre. Angehr confirmed that Pearson's impressions in 1938 (see below) about Goldman's Cerro Pirre being near the Colombian border and far from the official Cerro Pirre were correct. This same situation is reported by Myers (1969:19) following his work in the region. According to Angehr, the current trail from Cana leads to an area on Serranía de Pirre ca. 5 km N of Alturas de Nique, where he obtained an elevational reading of 1,575 m and the coordinates 07°46'24"/77°43'52", which have been used throughout this text to describe the ornithological Cerro Pirre. When the official Cerro Pirre has been used in a locality description, the USBGN coordinates have been used. Goldman, GL = Cana, at 1,800 feet (550 m), near the base at Cana, 1912 Apr 11, GL = Río Limón [ca. 0747/7741], at 3,000-5,200 feet (900-1,600 m), according to his journal and to Goldman (1920), he made camp at 5,100 feet (1,550 m) on the E slope near the head of the Río Limón, collected near camp, along the top of the range just above camp, and several miles NE of camp along a ridge where they had cut a trail that followed the crest, the small streams below camp were narrow with rocky beds in deep, narrow canyons, 1912 Mar 5-6, Apr 12-14, 16-20, 22-27, 29-May 3, 5-6, Jun 7 (USNM, as "Mt. Pirre," and as "Mount Pirre"; Nelson, 1912b, as "Mount Pirri"; Nelson, 1913a, as "Mount Pirri"; Chapman, 1917, as "Mt. Pirre"; Aldrich, 1945, as "Mt. Pirri"; Wetmore, 1950b; Wetmore, 1951a; Wetmore, 1967); Benson, GL for Cana (q. v.), at 2,600 feet (800 m), 1928 (MCZ; Griscom, 1927a, as "Mt. Pirre"; Griscom, 1929a, as "Mt. Pirre"; Wetmore, 1970); Pearson, 4,700-5,400 feet (1,400-1,650 m), according to a letter from Pearson, dated 22 Dec 1965 (USNM), they camped at the headwaters of the Río Limón, "we believe, right at

Goldman's old clearing. Traces of the trail were still visible. This was Goldman's Mt. Pirre location 5,100 ft. If I remember correctly, the half-dozen residents of Cana at the time of my visit did not consider this Mt. Pirre at all; they applied the name Mt. Pirre to an isolated peak farther north (?)," 1938 Mar 8, 11, 13, 15-16, 18-20, 24, 29, Apr 9, at 5,000 feet (1,500 m), on ridge 15 mi (24 km) S of, according to Pearson's letter, this region had "very rich wet forest with the tallest palms [he had] ever seen" and he believed the site was roughly on the Colombian border, 1938 Apr 6-7 (ANSP, as "Mt Pirri," and as "Mt. Pirre"; Bond and Meyer de Schauensee, 1944:29, as "Mt. Pirre"); Wetmore, GL for Río Seteganti, 1961 Jan 27-Feb 6 (FMNH; USNM); Koepcke, 1961 Feb 6 (FMNH); Hinds, GL = Cana, 1965 Aug 9 (USNM); Burton, British Trans-Americas Expedition, 1972 Mar (BMNH; Burton, 1973b; Burton, 1975); Braun, M., O'Neill, and Parker, GL = Cana, at 600-1,575 m, ca. 9 km NW of Cana on slopes of Cerro Pirre, according to Robbins et al. (1985:200), in 1982 the forest above Cana was "taller (averaging 15-20 m) than at the settlement clearing [Cana]. Palms were common and conspicuous at all levels…From about 700 to 1400 m the understory was open. Between 1000 and 1200 m, the forest was more humid, and arboreal epiphytes were abundant. An extensive stand of bamboo was present at 1200 m, and tree ferns were evident from 1450 m to the summit. At 1480 m, along the main ridge, the forest was stunted ('elfin forest'…), with an average canopy height of 8 to 10 m." 1982 Feb 6-12 (LSUMZ; Robbins et al., 1985); Allen-Stotz, D. Braun, Lanyon, Robbins, and Sarco, GL = Cana, at 650-1,500 m, 1-9 km NW of Cana, E slopes of Cerro Pirre, 1982 Jul 19-Aug 2 (LSUMZ; Robbins et al., 1985); Baker and Hafner, at 1,200 m, E slope, photographs, 1983 Jan 31 (Capparella, 1986). (Griscom, 1924b; Parkes, 1954a, as "Mt. Pirre"; Wetmore, 1962; Wetmore, 1967; Myers, 1969, as "Serranía de Pirre"; Olson, 1981b; Olson, 1983). See also: Cana; Cana, Loma; Nique, Alturas de; and Urutí.

PIRRE, ESTACIÓN; see Pirre Station

PIRRE, MOUNT; see Pirre, Cerro

PIRRE, RÍO; Darién

0808/7744 (USBGN)

Originates NW of Cerro Pirre [0756/7743 (USBGN)] and flows N entering the Río Tuira [0821/7803 (USBGN)] just W of El Real de Santa María [0808/7743 (USBGN)] (SMN2B). Heyde and Lux, GL = El Real de Santa María, "on the banks of the Río Pirre," 1889 Sep 3 (USNM, as "Río Pirri"; Deignan, 1961:221, as "Río Pirri"); Wetmore, GL = El Real de Santa María, near El Real de Santa María where the river was 25-35 feet (8-11 m) wide with clear, quickly moving water in the dry season, ca. 2 mi (3 km) along a trail parallel to the river through old pastures, occasional stands of trees, and cultivated fields with coarse grass and weeds, along a trail to Uruseca [ca. 0807/7744], in old rice fields, along a newly bulldozed path through *rastrojo* to the lower end of the Río Pirre, where the tide brought in murky and silty water, through an extensive pasture and low scrub, 1964 Jan 8, 11, 22-25, Feb 16-17 (USNM); Cavalier, GL = El Real de Santa María, 0.4 km SE and 4 km SW of El Real de Santa María, 1967 Sept 14 (MVZ, as "El Real"). See also: Pirre Station.

PIRRE, SERRANÍA DE; Darién

ca. 0749/7744

Mountain range in S central Darién province that extends N-S from the official Cerro Pirre [0756/7743 (USBGN)] to Alturas de Nique [0740/7745 (USBGN)] (RDPF; Atlas, 1988:206). The ornithological Cerro Pirre [0746/7744 (GA)] (q. v.) is located in the S end of the range near Cana [ca. 0745/7742]. Robbins et al. (1985) worked on the E slope of Serranía de Pirre along the trail to Cana (q. v.); Bermingham, Guitton-Mayerma, Lelevier, Maley, Miller, and Winker, at 1,000 m, at 07°45'48"/77°43'19", Ridge Trail, 2006 Feb 8-9 (STRI; UAM). (Wege, 1996, as "Pirre Ridge"). See also: Balsas, Río; and Tuira, Río.

PIRRE RANGER STATION; see Pirre Station

PIRRE RIDGE; see Pirre, Serranía de

PIRRE STATION; Darién

0801/7744 (Croat, 1997)

According to Engleman et al. (1989:491), Rancho Frio Station, also known as Pirre Station, is a ranger station and campsite just within the SW limits of Darién Parque Nacional [0749/7744 (GA)]. It is the most accessible area of the park located in the lowlands on the W slopes of the official Cerro Pirre [0756/7743 (USBGN)] (q. v.) and is reached from El Real de Santa María [0808/7743 (USBGN)] by road or by ascending the Río Pirre [0808/7744 (USBGN)]. Angehr (in litt., May 2004) informs us that Pirre Ranger Station and Rancho Frio Campsite are two distinct localities, and that the Pirre Ranger Station "is commonly referred to [as] 'Rancho Frio.' However, that name properly belongs to a campsite on the trail up Cerro Pirre." Both are located roughly NNW of Cerro Pirre, but on foot they are located several hours from one another. Pirre Ranger Station was constructed more recently, is at a lower elevation, and is located on the Río Peresenico [ca. 0801/7744], a tributary of the Río Pirre that originates in the N foothills of Cerro Pirre. Presumably the Pirre Station is very near the town Perrecenega [ca. 0801/7744] shown on RDP2 on the Río Pirre ca. 10 km S of El Real de Santa María and 10 km NNW of Cerro Pirre. The Rancho Frio Campsite is located farther S of El Real de Santa María, is "the last site with permanent water on the route" to Cerro Pirre, and is located at an elevation of ca. 500 m. Croat (1997:482-483) gives the coordinates 8°01' N, 77°44' W for botanical specimens taken at Angehr's Pirre Ranger Station. The data from two *Philodendron* specimens was given as: 1) "Parque Nacional Cerro Pirre [sic], vicinity of station along Río Perisenico, 110 m," and 2) "Estación Rancho Frío, N base of Cerro Pirre, ca. 9 km S of El Real, Quebrada Perisenico, 70-270 m." We propose that the two sites be officially named Pirre Station (Estación Pirre) [ca. 0801/7744] and Rancho Frio Campsite (Campamento de Rancho Frio) [ca. 0758/7743] to avoid further confusion. Based solely on elevation and locality names, it is likely that Jiménez and Seutin worked at Pirre Station, whereas Babin worked near the Rancho Frio Campsite. That said, as "Rancho Frio" and "Pirre" have been treated as synonyms, two separate entries were not created for the gazetteer. Jiménez, 1989 Oct 9 (MVUP, as "Estación Pirre"); Seutin, observations at the station, the coordinates 0852/7751 are apparently erroneous as they appear on our maps N of the park's boundaries, 1996 Jul 6 (Seutin, 1998, as "Darién National Park Station"); Babin, at ca. 400 m, near, 1999 Mar 29-30, Apr 4-5, 8 (LSUMZ, as "Rancho Frio (Pirre) Station"); Braun and Brumfield, at 08°01'06"/77°43'54", ca. 10 km S El Real [de Santa María], 2004 Mar 30-Apr 4 (LSUMZ, as "Rancho Frio").

PIRRE UNO; Darién
ca. 0803/7745
According to Brumfield (personal communication), Pirre Uno refers to a police post located ca. 7 km SSW of El Real de Santa María [0808/7743 (USBGN)]. Not shown on our maps, but the corresponding region on MGRP is below 100 m and very near a town named Pirre (ca. 0803/7747). Braun and Brumfield, at 08°03'06"/ 77°45'18", 2004 Apr 4 (LSUMZ).

PIRRI, CERRO; see Pirre, Cerro

PIRRI, MOUNT; see Pirre, Cerro

PIRRI, RÍO; see Pirre, Río

PIRRUGANDÍ, RÍO; see Punorgandí, Río

PITA, LA; see La Pita

PITA, LAGUNA DE; see Matusagratí, Laguna de

PITA, LAGUNA DE LA; see Matusagratí, Laguna de

PITA, LAGUNA DELLA; see Matusagratí, Laguna de

PITA, PUNTA; see Matusagratí, Laguna de

PITA LAGOON; see Matusagratí, Laguna de

PITO, RÍO; San Blas
0842/7732 (USBGN)
Originates on the continental divide on the N face of Cerro Anachucuna [0836/ 7735 (USBGN)] and flows N into the Caribbean Sea [1500/7500 (USBGN)] at Pito [0841/7732 (USBGN)] (SMN2B), which is a coastal town 8 mi (13 km) W of Puerto Obaldía [0840/7725 (USBGN)] (SMN2A). Wetmore, GL = Armila, from the camp at Quebrada Venado [ca. 0840/7729] (q. v.) followed a trail for ca. 5 km W through tall, green forest to a tributary (of the Río Pito) that was ca. 25 feet (8 m) wide with clear water and a rocky streambed, and to several good-sized streams presumed to be the headwaters of the Río Pito, 1963 Mar 4-6, 8 (USNM); Handley, GL = Armila, mouth of, 1963 Mar 29 (USNM).

PLANTAIN CAY; see Big Plantain Cay

PLANTATION, CAMINO; see Cacao Plantation Road

PLANTATION LOOP; see Cacao Plantation Road

PLANTATION ROAD; see Cacao Plantation Road

PLATANAL; Veraguas, Isla de Cébaco (Pacific)
0735/8104 (USBGN)
Sea level, coastal town on the NE coast of Isla de Cébaco [0732/8109 (USBGN)] (JOG). Wetmore, GL = Isla de Cébaco, worked within 5 mi (8 km) of town on the

main trail that extends the length of the island, up to ca. 300 feet (90 m), in fallow fields covered with dense *rastrojo*, through grass, *rastrojo*, yuca, rice and corn fields, along a stream in a deeply cut, steep-sided valley with original forest, along small streams to low *rastrojo* and old second growth, berry-filled trees, a section of original forest with 90-125 foot (25-40 m) tree crown and scanty undergrowth, swamp behind the house, 1965 Jan 13-20 (USNM; Olson, 1997a).

PLATANAL; see Platanales

PLATANAL, ISLA DE; see Platania, Isleta

PLATANALES; Panamá (E, Pacific)

0913/7903 (USBGN)

ca. 50 m, town on the Carretera Interamericana [0832/8242 (USBGN)] 6 km W of El Llano [0913/7859 (USBGN)] and 7 mi (11 km) NE of Chepo [0910/7906 (USBGN)]. The town is on a tributary of the Río Chepo [0900/7906 (USBGN)] (MRP, as "Platanal"; RDP, as "Platanal"; SMN2B, as "Platanal"). According to MGRP, a road extends roughly S for ca. 10 km from Platanales to Jesús María [0909/7859 (USBGN)], and this road is sometimes referred to in the literature as the Platanares-Jesús María Road. Wetmore, GL = El Llano, extensive stand of gallery forest reaching across level land toward the mountains, the forest had been cleared near the Río Chepo, 1962 Feb 9 (USNM, as "Platanar"); Eisenmann, Ridgely, Pujals, and N. Smith, observations, hills E of Chepo and *cuipo* forest on the road to Jesús María, 1973-1975 (Ridgely and Gwynne, 1989:225, as "Platanares"; Wetmore et al., 1984:121, 259, 368, as "Platanares").

PLATANALES; see Platania, Isleta

PLATANALES, ISLA; see Platania, Isleta

PLATANAR; see Platanales

PLATANARES; see Platanales

PLATANARES-JESÚS MARÍA ROAD; see Platanales

PLATANIA, ISLETA; Panamá, Archipiélago de las Perlas

0827/7900 (USBGN)

Sea level, small island with a land area of 7 ha (Wright et al., 1985, as "Platanales") located 4 km W of Isla del Rey [0822/7855 (USBGN)] and 1.5 km S of Isla Viveros [0828/7900 (USBGN)] (JOG). In the same region E501 indicates a group of small islands named "Islas Platonas." According to USBGN Isleta Platania is both an island and an alternate name for yet another island, Isla de Platanal [0827/7901 (USBGN)], which is not shown on our maps, but from the USBGN coordinates is located just W of Isleta Platania. Wright, Faaborg, and Campbell, mist nets, observations, 1976 Jun-Aug, 1978 Jan-Mar (Wright et al., 1985, as "Platanales").

PLATITO; Herrera

ca. 0750/8044

Not shown on our maps, but located in SW Herrera province in Distrito Las Minas [0748/8045 (USBGN)] in Reserva Forestal El Montuoso [0743/8050 (Angehr, 2003:203)], which is a protected area surrounding Loma El Montuoso [0743/8050

(USBGN)] (q. v.). Jiménez and Araúz, 2002 Nov 10 (MVUP).

PLATONAS, ISLAS; see Platania, Isleta

PLAYA BARCO QUEBRADA; see Barco Quebrada, Isla

PLAYA BIQUE; see Bique, Bahía

PLAYA BLANCA; see Blanca, Playa

PLAYA BLANCA; see Playa Blanca, Campamento de

PLAYA BLANCA, CAMPAMENTO DE; Veraguas, Isla de Coiba
0723/8140 (USBGN)
Sea level, prison camp on the SE coast of Isla de Coiba [0728/8147 (USBGN)] 6 km SE of Punta Felipa [0724/8142 (USBGN)] on the W end of Playa Blanca [0723/8139 (USBGN)] (JOG, as "Campamento de Playa Blanco"; Wetmore, 1957:9 as "Playa Blanca"). USBGN lists the official name as Campamento de Playa Blanco, which is probably a typographical error as the camp is located on Playa Blanca. See also: Blanca, Playa.

PLAYA BLANCO, CAMPAMENTO DE; see Playa Blanca, Campamento de

PLAYA CALETA; see Caleta Playa

PLAYA CHIQUERO; see Chiquero, Playa de

PLAYA CINIQUE; see Cinique, Playa de

PLAYA CORONADO; Panamá (W, Pacific)
0831/7953 (USBGN)
Sea level, coastal town 6 km S of Chame [0835/7953 (USBGN)] and 1 km NE of the mouth of the Río Lajas [0831/7954 (USBGN)] (JOG). Eisenmann, observations in tall, planted trees, 1947 Aug 19-26, 1971 May 29-31, 1973 Jan 31, 1974 Mar 27 (Eisenmann, 1948; Wetmore, 1972:417; Wetmore et al., 1984:198, 306); Wetmore, GL = Chame, mouth of the Río Lajas, on the beach at the surf's edge, 1953 Jun 21-22 (USNM). See also: San Carlos.

PLAYA CORTATITA; see Cortatita, Playa

PLAYA DE CHIQUERO; see Chiquero, Playa de

PLAYA DE CINIQUE; see Cinique, Playa de

PLAYA DEL CREO; see Creo, Playa del

PLAYA DEL JOBO; see Jobo, Playa del

PLAYA DE PAITILLA; see Paitilla, Punta

PLAYA EL AGALLITO; see El Agallito, Playa

PLAYA GRANDE; Panamá (E, Atlantic)
ca. 0923/7921
Presumably refers to the region near the Río Playa Grande [ca. 0923/7922], a river that originates on the Colón/Panamá provincial border in the mountains NW of Cerro Brewster [0921/7915 (USBGN)] and flows roughly S entering the Río Esperanza [0920/7923 (USBGN)] ca. 50 km NNE of Panama City [0858/7932

(USBGN)] (E761A). Birks, Davis, Faucett, and Rohwer, GL = Río Esperanza, at 356 m, ca. 51 km NNE of Panama City [0858/7932 (USBGN)], montane tropical rainforest, at 09°23'25"/79°20'59", 2002 Mar 5-11 (UWBM, as "Playa Grande").

PLAYA GRANDE; see Las Panamas, Playa

PLAYA GRANDE; see Puerto Caimito

PLAYA GRANDE, ENSENADA; Panamá, Archipiélago de las Perlas
0815/7906 (USBGN)
Bay centrally located on the E coast of Isla San José [0815/7907 (USBGN)] (JOG). Northern, GL = Isla San Jose, 1957 Mar 18 (LACM).

PLAYA GRANDE, RÍO; see Playa Grande

PLAYA HERMOSA; see Hermosa, Ensenada

PLAYA JOBO; see Jobo, Playa del

PLAYA JUREL; see Jurel, Playa

PLAYA LARGA; see Long Beach

PLAYA LA HONDA; see La Honda

PLAYA LEONA; Panamá (W, Pacific)
0847/7946 (USBGN)
Sea level, town on the SW coast of Bahía Chorrera [0850/7943 (USBGN)] 10 km S of La Chorrera [0853/7947 (USBGN)] (MGRP). Paz, 1984 Jun 7 (MVUP); Pérez, 1984 Aug 1 (MVUP); Salazar, 1991 Sep 15, and as a GL for La Cholita, 1991 Sep 15 (MVUP).

PLAYA MONAGRE; see Monagre

PLAYA MUERTA; see Muerto, Playa

PLAYA MUERTO; see Muerto, Playa

PLAYA NIEVE, LAGUNA; Darién
ca. 0732/7809
Sea level, described by Wetmore as a former channel of the Río Jaqué [0731/7810 (USBGN)], which is now an inlet 1 mi (1.5 km) upriver from Jaqué [0731/7810 (USBGN)], Wetmore and Perrygo, GL = Río Jaqué, low growth with guavas, open forest that floods during the rainy season, and channels through mangroves to the mouth of the Río Charó (= Río Chadó ? [0731/7810 (USBGN)]), 1946 Apr 10-11 (USNM). Not shown on our maps.

PLAYA VENADO; see Venado, Ensenada

PLAYA VERACRUZ; see Veracruz

PLAYA VERDE; Bocas del Toro
0908/8155 (USBGN)
Sea level, town located on the NW coast of Península Valiente [0905/8151 (USBGN)] 1.5 km SE of Punta Bluefield [0908/8157 (USBGN)] and 2 km NW of Muturi [0907/8155 (USBGN)] (JOG). SBE, GL = Península Valiente, Muturi, 1990 Apr 25, 27 (USNM; Brumfield et al., 2001).

PLAYITA DEL MEDIO, LOMA; Darién

ca. 0729/7808

Alt. ?, described by Wetmore as the highest hill ca. 4 mi (6 km) SE of Jaqué [0731/7810 (USBGN)] on Punta Fundeadero [ca. 0729/7808], Wetmore and Perrygo, GL = Jaqué, 1946 Apr 3 (USNM). Not shown on our maps.

PLAYONES, LOS; see Sardinilla

POCRÍ; Coclé (Pacific)

0816/8033 (USBGN)

ca. 25 m, town 2 km N of Aguadulce [0815/8033 (USBGN)] and 1 km S of the Río Pocrí [0813/8028 (USBGN)] (JOG). De Leon, GL for Río Pocrí, 1985 Aug 24 (MVUP).

POCRÍ, LA LAGUNA DE; see La Laguna de Pocrí

POCRÍ, RÍO; Coclé (Pacific)

0813/8028 (USBGN)

Originates NW of Aguadulce [0815/8033 (USBGN)] and flows roughly SSE crossing the Carretera Interamericana [0832/8242 (USBGN)] 3 km NE of Pocrí [0816/8033 (USBGN)]. The river enters mangrove swamps near Puerto Aguadulce [0815/8030 (USBGN)] where it widens as it discharges into Bahía de Parita [0808/8024 (USBGN)] 8 km E of Aguadulce (JOG). Wetmore, GL = Puerto Aguadulce, broad channels bordered with mangroves, 1962 Mar 11-12, GL = Aguadulce, by canoe ca. 0.5 mi (1 km) from Puerto Aguadulce, mangroves, 1963 Jan 18-19, at La Isleta (q. v.), 1963 Jan 19 (USNM; Olson, 1980b, as "Puerto Aguadulce, Río Pocrí"); De Leon, GL = Pocrí, 1985 Aug 24 (MVUP).

POINT DIRGADO; see Dirgado, Punta

POINT VALIENTE; see Valiente, Punta

PONDSOCK, PUNTA; Bocas del Toro

0918/8221 (USBGN)

Sea level, point 6 km E of Almirante [0918/8224 (USBGN)] on a mainland peninsula that separates Bahía Almirante [0920/8218 (USBGN)] and Ambrosia Bight [0916/8222 (USBGN)] (JOG). Wetmore, GL = Almirante, landed a few hundred yards N of Punta Pondsock where the shore was elevated only 2 feet (0.6 m) above the high tide mark and was covered with swampy woods, the 40-70 foot (10-20 m) trees had an undergrowth of small palms, creepers and vines, 1958 Jan 29, Feb 19 (USNM, as "Pondsock Point").

PONDSOCK POINT; see Pondsock, Punta

PONUGA; Veraguas (Pacific)

0753/8058 (USBGN)

ca. 50 m, town 23 km S of Santiago [0806/8059 (USBGN)] (JOG). Rivera, 1990 Aug 30 (MVUP); Faucett, at sea level, 2003 Feb 22 (UWBM); Miller and Weir, 2003 Apr 18 (CU; STRI; UAM).

POPA, ISLA; Bocas del Toro, Archipiélago de Bocas del Toro

0911/8207 (USBGN)

115 m, large island in Archipiélago de Bocas del Toro [0920/8210 (USBGN)] with a land area of 54 km^2 (5,375 ha) (Angehr, 2003:100), located in NW Laguna de Chiriquí [0903/8200 (USBGN)] ca. 2 km E of the W headlands of the bay (E762). One of the boatmen employed by SBE indicated that Isla Popa used to be, or maybe sometimes still is, called Cayo Agua [0909/8202]. The island is considered an Important Bird Area (Angehr, 2003). Miller and Richardson, GL = Laguna de Chiriquí, mangrove thicket, 1936 Mar 9 (UCLA); SBE, GL = Laguna de Chiriquí, collected on the N side of the island inland from Isla Deer [0914/8208 (USBGN)] and 1 km SE of the channel between Isla Popa and Isla Deer, 1987 Mar 9, 30-Apr 13, on the E side, 1991 Feb 28, on the S coast 1 km E of Canal Sumwood [0909/8209 (USBGN)], 1991 Feb 24-27, Mar 1-4, 6-13 (MVUP; USNM; Olson, 1993a); Cooper, surveys in a 200-ha private forest reserve, 1997 Feb-Mar (Cooper, 1999); Billo, GL for "Popa 2 community, ½ km E", 2007 Apr 26, 28, and Punta Laurel, 2007 May 7 (USNM).

POPA 2; Bocas del Toro, Archipiélago de Bocas del Toro

ca. 0913/8208

Community on the NW side of Isla Popa [0911/8207 (USBGN)]; Billo, GL= Isla Popa, 2007 Apr 26, 28, "½ km E" (USNM).

PORCADA, ISLA; see La Porcada, Isla

PORCADA VALDEZ; Panamá (E, Pacific)

0908/7906 (USBGN)

80 feet (20 m), town 5 km S of Chepo [0910/7906 (USBGN)] and 3 km WNW of Ana Luz [ca. 0908/7905] (E761A; MHA, as "La Porcada"). Wetmore and Perrygo, GL = Chepo, 1949 Apr 11 (USNM).

PORRAS, LAGUNA; see Bocatorito, Laguna

PORT ANTONIO; see San Antonio

PORT OF LA CHORRERA; see Puerto Caimito

PORT POSADA; see Puerto Posada

PORT SAN ANTONIO; see San Antonio

PORTOBELILLO; Herrera

0801/8036 (USBGN)

ca. 25 m, town 6 km SW of Paris [0803/8033 (USBGN)] and 6 km SSE of Potuga [0804/8037 (USBGN)] (JOG). Wetmore and Perrygo, GL for Los Voladores, 1948 Feb 26 (USNM, as "Portobellilo"; Wetmore, 1970, as "Portobellilo").

PORTOBELLILO; see Portobelillo

PORTO BELLO; see Portobelo

PORTO BELLO ROAD; see Portobelo Road

PORTOBELO; Colón (E)

0933/7939 (USBGN)

Sea level, coastal town 35 km NE of Colón [0922/7954 (USBGN)] and 1 km S of the mouth of the Río Cascajal [0934/7939 (USBGN)] (JOG). The name Porto Bello appears on older maps (HMCZ). The town is now included in Parque Nacional

Portobelo [0931/7954 (Angehr, 2003:221)], an Important Bird Area that extends from the coast inland to the mountains (Angehr, 2003). Goldman, vicinity of town, according to Goldman (1920:7), "at the government quarry...ample facilities afforded for our investigations in the immediate vicinity. Mountains with peaks, including the Cerro Brujo, rise rather steeply to over 3,000 feet, and closely parallel the coast line southward and westward; a lower spur to the north terminating in the rock quarry, encircles the watersheds of the Río Cascajal, Río Moré and other short streams converging to the head to the bay. The general shore line is rugged, but mangrove lagoons and swamps occur near the mouth and along the lower course of the Río Cascajal," 1911 May 23-27, 29-Jun 1, and as a GL for: Isla Los Mogotes, and Río Moré, 1911 May 29-30, Jun 1 (USNM, as "Porto Bello"; Wetmore, 1952:18; Wetmore, 1959b:10, 20, as "Portobello"; Olson, 1980a, as "Porto Bello"; Olson, 1981a, as "Portobello"; Olson, 1983, as "Porto Bello"; Olson, 1981d, as "Buenaventura Island, near Portobelo"); Chapman, observations at sea 2.5 hours out of Cristóbal [0921/7955 (USBGN)] and 10 mi (16 km) N of the mainland near, 1937 Apr 18 (Chapman, 1937, as "Porto Bello"; Rogers, 1939, as "Porto Bello"; Lincoln, 1944, as "Porto Bello"); Koford, 1957 Feb 9 (Eisenmann, 1963b); Olson and Mueller, GL for Río Cascajal, 1966 Jun 26 (UF; USNM); Hespenheide and Ritte, near, 1970 Jun (CM, as "Portobello"); Morton, GL for Río Guanche, 1971 Apr 4 (USNM); Gardner, 1985 Dec (WFVZ, as "Porto Bello"); Wong, 1990 Mar 23 (MVUP). See also: Cascajal, Río.

PORTOBELO, PARQUE NACIONAL; see Portobelo

PORTOBELO, RÍO; Veraguas (Pacific)
0713/8040 ?
Mentioned in Ridgely and Gwynne (1989:106) as being a locality in the southernmost Azuero Peninsula. According to Angehr (in litt., Aug 2005), this seems "to be an error for the Río Puerto de Vela [0713/8040 (ADL)], just W of the Veraguas border in the S Azuero."

PORTOBELO ROAD; Colón (E)
ca. 0933/7939
Presumably refers to the section of Panama 85 that extends 27 km NNE along the Caribbean coast from Sabanita [0921/7948 (USBGN)] on the Carretara Boyd-Roosevelt [0917/7943 (USBGN)] to Portobelo [0933/7939 (USBGN)] (MGRP; RDP2). According to our recent maps (JOG and MGRP), the Portobelo Road crosses the Río Piedras near its mouth, but E661 (as "Porto Belo Trail") shows the road becoming a trail ca. 3 km E of the Carretera Boyd-Roosevelt less than 1 km N of Puerto Pilón [0922/7948 (USBGN)]. Smith, N., GL for Río Piedras, 1965 Mar 19 (USNM, as "Porto Bello Road").

POSADA, PUERTO; see Puerto Posada

POTRERILLOS ABAJO; Chiriquí
0839/8229 (USBGN)
900 m, town 13 km SSW of Boquete [0847/8226 (USBGN)] (JOG, as "Potrerillos Arriba"). See also: El Banco.

POTRERO, EL; see El Potrero

POTRERO, RÍO; see El Potrero, Río

POTRERO, RÍO EL; see El Potrero, Río

POTRERO CORDILLERA; see Cordillera, Potrero

POTRERO CUATRO MIL; see Cuatro Mil, Potrero

POTRERO SECO; Chiriquí
0848/8236 (USBGN)
Not shown on our maps, but USBGN indicates a settlement at ca. 1,700 m, at the W base of Volcán Barú [0848/8233 (USBGN)] ca. 5 km NE of Volcán [0846/8238 (USBGN)]. See also: Seco, Potrero.

POTRERO SECO; see Seco, Potrero

POTRERO VERDE; see Verde, Potrero

POTUGA; Herrera
0804/8037 (USBGN)
ca. 50 m, town 7 km SE of Santa María [0807/8040 (USBGN)] and 6 km NNW of Portobelillo [0801/8036 (USBGN)] at the NW end of Los Voladores [ca. 0802/8036] (JOG). Wetmore and Perrygo, 1 mi (1.5 km) NW of Potuga near a small, shallow stream, and on rolling hills cleared of forest, 1948 Mar 2 (USNM; Wetmore, 1953).

POZO HONDO; see Burunga

POZOS DE SALUD DE GALÍQUE; Chiriquí
ca. 0818/8155
ca. 50 m, described by Wetmore in 1956 as a small resort on warm water springs, reached by traveling W on the Carretera Interamericana [0832/8242 (USBGN)] 0.5 mi (1 km) past the Río Dupí [0816/8156 (USBGN)], then following a country road 2 mi (3 km) N (USNM). See also: Dupí, Río. Not shown on our maps.

PRODUCCIÓN, CAMPO; see Campo Producción

PROGRESO; Chiriquí
0827/8250 (USBGN)
ca. 25 km, town 25 km N of Puerto Armuelles [0817/8252 (USBGN)] (q. v.) on the Río Chiriquí Viejo [0820/8241 (USBGN)] and on the N end of a large cultivated area (JOG: MGRP). Greenwell, 1966 Mar 26 (USNM).

PROMINENTE, CERRO; Panamá (E, Pacific)
0914/7917 (USBGN)
2,950 feet (900 m), peak in the SW extension of the Cordillera de San Blas [0918/7900 (USBGN)] 16 km N of Pacora [0905/7917 (USBGN)] and 23 km WNW of Chepo [0910/7906 (USBGN)] (TM). See also: Azul, Cerro.

PROVIDENCIA, RÍO; Colón (Canal Zone) (W)
0913/8002 (USBGN)
Originates on the NW coast of Lago Gatún (Gatun Lake) [0912/7955 (USBGN)] just N of La Treinticinco [0911/7958 (USBGN)] and flows W entering a stream that ultimately discharges into the Caribbean Sea [1500/7500 (USBGN)]. The stream crosses Achiote Road [ca. 0917/8003] 5 km NW of La Treinticinco (E8660).

Cardiff, Christian, Dittmann, Naoki, and Lovette, GL = Achiote Road, at 25 m, at Río Providencia, 2 km E of Río Providencia, 1997 Mar 16-22 (LSUMZ; MVUP).

PROVISIÓN, ISLA; see Bastimentos, Isla

PROYECTO DE FORTUNA; see Fortuna

PUCORA RIVER; see Púcuro, Río

PUCRO; see Púcuro

PUCRO, RÍO; see Púcuro, Río

PÚCURO; Darién
0801/7730 (USBGN)
ca. 50 m, town 18 mi (29 km) SE of El Real de Santa María [0808/7743 (USBGN)] on the Río Púcuro [0801/7733 (USBGN)] 3 mi (5 km) above its mouth (SMN2B, as "Pucro"). Described by Wetmore in 1964 as a small settlement with 12-15 houses. Wetmore, through a forest trail to a series of hills in the area known as Cabecera de Camote (not located), through a short stretch of *rastrojo* then into an extensive area of virgin forest with very high trees and considerable undergrowth, 1964 Feb 4, and as a GL for: Río Púcuro, and Quebrada Camarón, 1964 Jan 29-Feb 3, 5-11 (USNM, as "Pucro"; Olson, 1981d, as "Pucro, Río Pucro"); GMLP, 1964 Mar 21 (GML, as "Pucro").

PÚCURO, RÍO; Darién
0801/7733 (USBGN)
Originates on the continental divide S of Cerro Tacarcuna [0811/7717 (USBGN)] and flows WSW joining the Río Tuira [0821/7803 (USBGN)] 6 km SSE of Boca de Cupe [0803/7735 (USBGN)] (CARP; RDPF; SMN2B, as "Río Pucro," and as "Río Tapalisa"). Fairchild and Handley (1966, as "Río Pucro") assigned the coordinates 0759/7734 to the river. Weber, 1928 Aug 1 (USNM, as "Pucora River"); Galindo, GL for Tacarcuna Village, 1963 Jul 4 (USNM, as "Río Pucro"; Wetmore, 1963b, as "Río Pucro"); Wetmore, GL = Púcuro [0801/7730 (USBGN)], specimens were collected up- and downstream from Púcuro where the river was 15-20 feet (5-6 m) wide with clear water, the level land along the river had been largely cleared and was covered with 5-20 year old *rastrojo*, along the river banks on beaches, in a thorny thicket, in a heavy stand of huge *Heliconia*, in forest with deep creeks, ant swarm, in a cleared area, on short trails through forest, scrub, and areas cultivated with cacao, plantain, banana, rice, sugar cane, corn, and squash, 1964 Jan 29-Feb 3, 5-11, according to Wetmore, the river originates as 2 small streams, specimens labeled "head of" and "S fork of" both refer to the small stream that originates close to Cerro Malí [0807/7714 (FH)], whereas those labeled "N fork of" refer to a stream N of Cerro Malí on the S face of Cerro Tacarcuna, these streams originate ca. 2 mi (3 km) from one another, GL = Cerro Malí, at 4,200 feet (1,300 m), head of the Río Púcuro, S fork of the Río Púcuro, along the river in a clearing next to tall, forest trees, 1964 Feb 23, 25, Mar 1-2, GL = Cerro Tacarcuna, at 4,100-4,800 feet (1,200-1,450 m), N fork of the Río Púcuro, 1964 Feb 23, 25, 28, Mar 1-2 (USNM, as "Río Pucro"; Wetmore, 1967, as "Río Pucro"; Olson, 1981d, as "Pucro, Río Pucro"); Hinds, GL = Cerro Tacarcuna, at 650 m, base of, N base of, 1975 Jan

26-30 (GML, as "Río Pucro," and as "Río Pucuro"; LSUMZ, as "Río Pucro"). See also: Topalisa, Río.

PUEBLO, QUEBRADA DE; see Pueblo, Quebrada del

PUEBLO, QUEBRADA DEL; San Blas

ca. 0810/7727

Described by Wetmore as one of two streams bearing this name. This stream is a tributary of the Río Armila [0840/7728 (USBGN)] and is separated from the other Quebrada del Pueblo [ca. 0840/7725] (q. v.) by a 350 foot (110 m) ridge covered in *rastrojo*, Wetmore, GL = Puerto Obaldía, 1963 Mar 18 (USNM).

PUEBLO, QUEBRADA DEL; San Blas

ca. 0840/7725

Described by Wetmore as one of two streams bearing this name. This stream is small, has a sand and gravel bed and discharges into the Caribbean Sea [1500/7500 (USBGN)] 0.25 mi (0.5 km) from Puerto Obaldía [0840/7725 (USBGN)], Wetmore, GL = Puerto Obaldía, inland along the river valley through coconut palms and a tract of timber, a fig tree bearing fruit, along a trail through forest that leads to Colombia, 1963 Feb 16-18, Mar 15-18 (USNM, as "Quebrada de Pueblo," as "Quebrada del Pueblo," and as "Quebrada el Pueblo").

PUEBLO, QUEBRADA EL; see Pueblo, Quebrada del

PUEBLO NUEVO; Bocas del Toro

0910/8215 (USBGN)

Sea level, town located on the mainland on the coast of Ensenada de Boquete [0910/8215 (USBGN)], a bay along the NW shores of Laguna de Chiriquí [0903/8200 (USBGN)]. The town is located 23 km SE of Almirante [0918/8224 (USBGN)] and 14 km N of Punta Róbalo [0902/8215 (USBGN)] (JOG). DaCosta, Guitton-Mayerma, Klicka, Lelevier, Miller, and Spellman, GL = Punta Róbalo, at 37 m, at 09°07'00"/82°17'12", 8 km NNW of Punta Róbalo, 2006 Feb 23 (MBM; STRI).

PUEBLO NUEVO; Coclé (Pacific)

ca. 0820/8013 ?

According to Bovallius (1877), he sailed S from Bahía Chorrera [0850/7943 (USBGN)] past Punta Cocinero (not located) to the mouth of the Río San Carlos (not located), a deep but narrow river. The port Pueblo Nuevo was above the river mouth on a rich plain with palms, where ca. 1,500 inhabitants lived. Despite his descriptions, the location of this port remains unclear. MHA indicates a Pueblo Nuevo on the S coast of Coclé province, ca. 30 km SW of the town San Carlos [0829/7957 (USBGN)] not on any sizable river, but between the mouths of the Río Antón [0818/8016 (USBGN)] and the Río Chico [0819/8011 (USBGN)]. No rivers bearing the name San Carlos were found, although San Carlos town, 30 km NE of Pueblo Nuevo, is located in Distrito San Carlos [0831/8001 (USBGN)]. Wetmore (1946b:4) noted that Bovallius probably wrote his locality descriptions after his return to Europe, as evidenced by the fact that Bovallius' descriptions of the rivers on Isla San José [0815/7907 (USBGN)] did not coincide with Wetmore's findings on the island. In the present case, the description of Pueblo Nuevo is not supported by our maps, and its position remains uncertain. Bovallius, 1883 Jun 13 (NHRM; Bovallius, 1877:94-95; Rendahl, 1919:4, 28).

PUENTE, RÍO; see La Puente, Río

PUENTE, RÍO DEL; see La Puente, Río

PUENTE, RÍO LA; see La Puente, Río

PUERCO, ISLA DE; Panamá, Archipiélago de las Perlas
0822/7849 (USBGN)
106 m, island with a land area of 70 ha (MacArthur et al., 1972, as "Puercos Island") located less than 1 km off the SE coast of Isla Cañas [0823/7849 (USBGN)] (JOG). MacArthur et al., mist nets, observations, "covered by partly deciduous tropical forest whose tallest trees are about 30 m in height," 1971 Jan (MacArthur et al., 1972—species list; see comments in Bibliography; MacArthur et al., 1973; and Wright et al., 1985, as "Puercos Island"). Ridgely and Gwynne (1989:343, as "Isla Puercos").

PUERCOS, ISLA; see Puerco, Isla de

PUERCOS ISLAND; see Puerco, Isla de

PUERTO; see Puerto Aguadulce

PUERTO AGUADULCE; Coclé (Pacific)
0815/8030 (USBGN)
Sea level, town on the Río Pocrí [0813/8028 (USBGN)] 6 km E of Aguadulce [0815/8033 (USBGN)] (JOG). Wetmore, GL for Río Pocrí, 1962 Mar 11-12 (USNM; Olson, 1980b, as "Puerto Aguadulce, Río Pocrí"); Tejera, GL = Aguadulce, 1978 Dec 30, 1979 Nov 2, 1980 Apr 25 (MVUP, as "El Puerto," and as "Puerto").

PUERTO ARMUELLES; Chiriquí
0817/8252 (USBGN)
Sea level, town on the W coast of Bahía de Charco Azul [0812/8245 (USBGN)] 27 km N of Punta Burica [0802/8255 (USBGN)] (JOG). After the Panama disease hit the large banana plantations near Changuinola and Almirante, the United Fruit Company began operating on the Pacific slope. In 1927-1928, using land purchased from the Panama Sugar Company, operations began at Puerto Armuelles and Progreso [0827/8250 (USBGN)], and throughout the region from the Río Chiriquí Viejo [0820/8241 (USBGN)] to the Costa Rican border a network of railroads was built (Yudin and Stephens, 1989:28). Davidson, 1929 Nov 7-19, 21-28, and as a GL for Río Rabo de Puerco, 1929 Nov (CAS; CM; Davidson, 1938); Smith, 1937 Jun 30 (ANSP); Wetmore, specimens without secondary localities were collected by local hunters, 1966 Feb 3-8, 12, 21, 22, 26, 28. The secondary localities attributed to Puerto Armuelles in 1966 were located within 14 km of town, GL for: Aguacatón, Canal de Colorado, Corotú, Manaca Cerro Cinco, Monte Verde, Olivo, Represa, Río Corutú, Río Palo Blanco, and Río San Bartolo, 1966 Jan 29-Feb 12, 14-19, 21-24, 26, Feb 28-Mar 4, 7-13 (USNM; Wetmore, 1959b:18; Wetmore, 1970; Olson, 1980b, as "Puerto Armuelles, Olivo"; Olson, 1981d; Olson, 1983); Braun, M., and Parsons, 12.5 road km N, 1991 Mar 15-17 (USNM; Brumfield and Braun, 2001);

DaCosta, Guitton-Mayerma, Klicka, Lelevier, Miller, and Spellman, GL for Península Burica, 2006 Mar 1-2 (MBM; STRI). (Eisenmann, 1956; Parkes, 1976).

PUERTO CAIMITO; Panamá (W, Pacific)
0852/7943 (USBGN)
Sea level, coastal town located 19 km SW of Panama City [0858/7932 (USBGN)] and 7 km SE of La Chorrera [0853/7947 (USBGN)] at the mouth of the Río Caimito [0852/7943 (USBGN)] (E761A, as "Port of La Chorrera"; JOG). Takos and Middlekauf, the specimen at MCN is also labeled Playa Grande (not located), which presumably refers to a coastal region near Puerto Caimito, 1945 Oct 4, 9 (CM, as "Puerto de Chorrera"; MCN, as "Puerto La Chorrera"); Serrano, *cercanías del* (outskirts of), 1984 Aug 7 (MVUP); Vargas, *cercanías del* (outskirts of), 1984 Aug 23 (MVUP); Gil, *cercanías del* (outskirts of), 1984 Aug 25 (MVUP); Justavino, 1985 Mar 25 (MVUP).

PUERTO DE CHORRERA; see Puerto Caimito

PUERTO DE PIÑA, EL; see El Puerto de Piña

PUERTO DE VELO, RÍO; see Portobelo, Río

PUERTO EL TIGRE; Panamá (W, Pacific)
ca. 0838/7953
Sea level, described by Wetmore as an area N of Bejuco [0836/7953 (USBGN)] accessible to canoes and immediately adjacent to the highway as it passes between low hills and mangrove swamps (USNM). Not shown on our maps, but Wetmore's description corresponds with the area on E762 that is located 4-5 km N of Bejuco with the Carretera Interamericana [0832/8242 (USBGN)] narrowly passing by 150 m hills on the W and mangrove swamps on the E. Wetmore and Perrygo, GL = Bejuco, hills covered with scrub and many *cuipo* trees, mangroves, 1951 Mar 15 (USNM; Olson, 1980b, as "Bejuco, Puerto el Tigre").

PUERTO GAGO; Coclé (Pacific)
ca. 0820/8026 ?
Not found, but Angehr (in litt., Aug 2005) suggests that this may be on the Río Grande [0818/8024 (USBGN)] 3 km S of El Gago [0822/8026 (USBGN)] at the end of the road. Herrera, 1980 May 6 (GML).

PUERTO LA CHORRERA; see Puerto Caimito

PUERTO LIMÓN; Herrera
0800/8029 (USBGN)
ca. 25 m, town 8 km NW of Chitré [0758/8026 (USBGN)] near the Río Parita [0802/8027 (USBGN)] and 5 km SW of its mouth (JOG). Vega, 1991 Jul 29 (MVUP).

PUERTO LIMONES; see Limones

PUERTO MENSABÉ; Los Santos
0746/8010 (USBGN)
Sea level, town on the E coast of Los Santos province at the mouth of the Río Mensabé [0746/8010 (USBGN)] 37 km SE of Chitré [0758/8026 (USBGN)] and

12 km E of Las Tablas [0746/8017 (USBGN)] (JOG). MGRP uses the name for a port at the mouth of the Río Mensabé. Wetmore, observations, 1948 Mar 26 (Wetmore, 1965:386).

PUERTO MUTIS; Veraguas (Pacific)

0755/8103 (USBGN)

Sea level, town 30 km SE of Soná [0801/8119 (USBGN)] on the Río San Pedro [ca. 0750/8105] just above the mouth of the Río de Jesús [0755/8104 (USBGN)] (JOG). Written by Wetmore with the "i" accented, but USBGN and our maps do not indicate an accent. Wetmore, GL for: Isla Verde, and Río San Pedro, 1953 Jun 11 (USNM; Huff and Wetmore, 1967; Olson, 1980b, as "Puerto Mutís, Isla Verde, mouth of Rio San Pedro"); Valdés, 1984 Aug 26 (MVUP); DaCosta, Guitton-Mayerma, Klicka, Lelevier, Miller, and Spellman, N of Golfo de Montijo, at 28 m, 07°56'15"/81°08'00", 8 km WNW of, 2006 Mar 4 (MBM; STRI).

PUERTO NUEVO; Veraguas (Pacific)

ca. 0806/8141

Loftin (personal communication) indicated that he took his locality data from the Esso Tourist Map (ME), therefore Puerto Nuevo is likely a typographical error for either Puerto Vidal [0803/8136 (USBGN)] (q. v.), a coastal town in Veraguas province, or for Pueblo Nuevo [ca. 0806/8141], a coastal town in Chiriquí province just 5 km W of the Veraguas border, located on ET and MHA ca. 10 km NW of Puerto Vidal and ca. 25 km SE of Remedios [0814/8151 (USBGN)]. Tyson and Loftin, observations, 1961 Nov 11 (Loftin, 1963a).

PUERTO OBALDÍA; San Blas

0840/7725 (USBGN)

Sea level, coastal town in extreme E Comarca de San Blas 2 mi (3 km) W of Cabo Tiburón [0842/7721 (USBGN)] (SMN2A). Wedel moved c. January 1930 from Permé [ca. 0845/7734] to Puerto Obaldía, which was the site of a United States naval radio station. From here "Wedel collected to within a half mile of the Colombian boundary," but he was unable to penetrate the hills as the "higher altitudes in the interior [were] cut off from the coast by one or two intervening ridges, which… [were] covered with a jungle so dense that a trail [had to be] cut, if none exist[ed]" (Griscom, 1932a:304). According to Wetmore, in 1963 coconut palms covered most of the lowland, coastal areas. Access to the interior – containing areas of steep pastureland, *rastrojo* and forest – was restricted by steep, rocky hills, a description that reaffirms the reports by Wedel. Wedel, at 0-1,500 feet (0-450 m), according to Griscom (1932), Wedel collected in Puerto Obaldía from early 1930 until 1931, which suggests that records for 1929 may be typos or misreadings of labels or that Griscom was in error. Apparently Wedel returned to the region during 1932-1935 as Rogers (1937 and 1939) and several museums report specimens from this period, 1929 May-Jul, 1930 Jan-Mar, May-Nov, 1931 Jan-Dec, 1932 Feb-Oct, Dec, 1933 Jan-Oct, 1934 Feb-Apr, Sep-Dec, 1935 Jan-Nov (ANSP, as "Obaldia"; CM; CU, as "Port Obaldia"; FMNH, as "Obaldia," as "Port Obaldia," and as "Puerto Obaldia"; LSUMZ, as "Obaldia"; MCZ, as "Obaldia"; NHRM, as "Obaldia"; OC, as "Obaldia"; UCLA; USNM, as "Port Obaldia," and as "Puerto

Obaldia"; YPM, as "Obaldia"; Griscom, 1931, as "Obaldia"; Griscom, 1932a, as "Obaldia"; Rogers, 1937, as "Port Obaldia"; Rogers, 1939, as "Port Obaldia"; Brandt, 1938, as "Port Obaldia"; Lincoln, 1944, as "Port Obaldia"; Zimmer, 1945, as "Port Obaldia"; Brodkorb, 1950, as "Obaldia"; Eisenmann, 1959, as "Obaldia"; Wetmore, 1959b:10, as "Obaldía"; Wetmore, 1963b; Eisenmann and Howell, 1962, as "Puerto Obaldía, Colón"; Olson, 1980a; Adsett and Wege, 1998, as "Obaldía"); Handley, 1962 Apr 3-4, and as a GL for Quebrada de la Represa, 1963 Apr 3 (USNM, as "Pto. Obaldia," and as "Puerto Obaldia"); Wetmore, 1963 Mar 14, and as a GL for: Quebrada del Pueblo [ca. 0810/7727], Quebrada del Pueblo [ca. 0840/7725], and Quebrada de la Represa, 1963 Feb 16-18, Mar 14-18 (USNM); Sheets, observations E of, 1965 Jun (Ridgely and Gwynne, 1976:309; Wetmore et al., 1984:361). (Wetmore, 1962; Olson, 1981d; Escalante-Pliego and Peterson, 1992). See also: Armila; La Bonga; and Ranchón.

PUERTO PILÓN; Colón (E)
0922/7948 (USBGN)
Sea level, town located 12 km E of Colón [0922/7954 (USBGN)] on the SE coast of Bahía Las Minas [0924/7950 (USBGN)] (E761a, as "Port Pilone"; JOG). Noyes, 1947 Aug 29-30 (MVZ).

PUERTO PIÑA; Darién
0735/7811 (USBGN)
Sea level, town on the N coast of Bahía Piña [0734/7811 (USBGN)] located at the mouth of the Río Piña [0735/7811 (USBGN)] 8 km N of Jaqué [0731/7810 (USBGN)] (JOG; MHA, as "Santa Dorotea"). Weir, at 40-90 m, at 0738/7811, 2003 May 5-9 (CU; LSUMZ). See also: Piña, Bahía; and Piña, Río.

PUERTO PIÑA; see El Puerto de Piña

PUERTO PIÑAS; see Piña, Bahía

PUERTO POSADA; Coclé (Pacific)
0821/8023 (USBGN)
Sea level, town on the Río Grande [0818/8024 (USBGN)] in an area of mangrove swamp 4 km SE of El Gago [0822/8026 (USBGN)] and 20 km SSW of Penonomé [0831/8022 (USBGN)] (E761A, as "Port Posada"). Wetmore, GL = Penonomé, along an old road raised above the mangrove swamps, near Puerto Posada extensive areas of mangroves were cut and planted in rice, 1963 Jan 23 (USNM; Olson, 1980c, as "Penonomé, Puerto Posada"); Johnson, Marqués, Quirós, Sánchez, and Serrano, 1986 Sep 9 (MVUP).

PUERTO SALADO; see El Salado

PUERTO SAN ANTONIO; see San Antonio

PUERTO VIDAL; Veraguas (Pacific)
0803/8136 (USBGN)
ca. 50 m, town in SW Veraguas province on the Río Vidal [ca. 0801/8138] 23 km SSE of Tolé [0814/8141 (USBGN)] and 4 km E of the Chiriquí provincial boundary, located between hills on the N and lowland areas with tropical grass and mangrove swamps on the S (JOG; MGRP). Wetmore and Perrygo, the secondary localities

attributed to Puerto Vidal were located in the area extending 10 km W and S of town, GL for: Isla de la Corocita, Río Tabasará, and Río Vidal, 1953 May 29-30, June 6, 8 (USNM; Huff and Wetmore, 1967; Olson, 1980b, as "Puerto Vidal, Río Vidal, Isla Corocita").

PULIDA, LA; see La Pulida

PUMA, ISLA; Colón (Canal Zone) (E)
0914/7954 (USBGN)
40 m, small island with a land area of 76.5 ha (Wright, 1985) located 4 km SSE of Gatún [0916/7955 (USBGN)] in Lago Gatún (Gatun Lake) [0912/7955 (USBGN)] NW of Isla Barro Colorado [0909/7951 (USBGN)] (E762, as "Puma Island"). Wright, censuses, 1979-1981 (Wright, 1985, as "Puma Island"; Wright, 1988, as "Puma Island").

PUMA ISLAND; see Puma, Isla

PUNORGANDÍ, RÍO; Panamá (E, Pacific)
0914/7846 (USBGN)
Originates on the lower slopes of the Cordillera de San Blas [0918/7900 (USBGN)] and flows S. Originally the river entered the Río Chepo [0900/7906 (USBGN)] 37 km ENE of Chepo [0910/7906 (USBGN)], but it now enters the N end of Lago Bayano [0908/7835 (USBGN)] (JOG, as "Río Espavé; MHA, as "Río Pirrugandí"). Burton, 1972 Jan (BMNH, as "Río Espave"; Burton, 1973, as "Río Espave (Bayano Valley)"; Burton, 1975, as "Río Espave (Bayano Valley)").

PUNTA, CERRO; Chiriquí
0853/8235 (USBGN)
2,375 m, peak in NW Chiriquí province 4 km NNW of Cerro Punta [0852/8235 (USBGN)] and 3 km S of the Bocas del Toro provincial border (JOG). See also: Cerro Punta; and Las Nubes.

PUNTA, CERRO; see Cerro Punta

PUNTA, LA; see La Punta

PUNTA BALSA; see Balsa, Punta

PUNTA BLANCA; Bocas del Toro
0910/8230 (USBGN)
100 m, town on the Río Changuinola [0928/8227 (USBGN)] located 19 km SW of Almirante [0918/8224 (USBGN)] (JOG). (Wetmore, 1972:365).

PUNTA BRAVA; see Brava, Punta

PUNTA BURICA; see Burica, Punta

PUNTA CALAFATE; see San Felipe

PUNTA COCO; see Cocos, Punta

PUNTA COCOS; see Cocos, Punta

PUNTA DAMAS; see Damas, Punta and Punta Damas, Campamento de

PUNTA DAMAS, CAMPAMENTO DE; Veraguas, Isla de Coiba
0731/8141 (USBGN)

Sea level, prison camp on the central E coast of Isla de Coiba [0728/8147 (USBGN)] on the peninsula 1 km N of Punta Damas [0730/8141 (USBGN)] (q. v.) (JOG). Wetmore, GL = Isla de Coiba, along a coastal beach, along a trail inland through the peninsula W of the camp, the trail climbed steeply through a pasture with cattle, across cleared land to the edge of a forest with the large trees removed and with a heavy growth of 6-8 foot (2 m) creeping bamboo and tall brush, along logging trails into a broad forested flat and to a stand of gallery forest, open grassland, to the steep slopes and broken land N toward Campamento de Juncal [0734/8144 (USBGN)], 1956 Jan 8-9, 13, 17, 19, 22, 26, Feb 1 (USNM, as "Punta Damas"; Wetmore, 1957).

PUNTA DE GALLINAZO; see Gallinazo, Punta de

PUNTA DE PEÑA; Bocas del Toro
0855/8211 (USBGN)
20 m, town on the Río Guarumo [0900/8211 (USBGN)] 8.5 km SW of Chiriquí Grande [0857/8207 (USBGN)] at the terminus of the railroad (E762, as "Punta de Peña (Chiriquicito Número 3)"). See also: Boquete Trail.

PUNTA DE PEÑA TIBURÓN; see Peña Tiburón, Punta

PUNTA FELIPA; see Felipa, Punta

PUNTA FUNDEADERO; see Fundeadero, Punta

PUNTA GORDA; see Gorda, Punta

PUNTA ICACO; see Lanza, Punta

PUNTA JAQUÉ; see Jaqué, Punta

PUNTA LANZA; see Lanza, Punta

PUNTA LAUREL; see Laurel, Punta

PUNTA MADROÑA; see Madroña, Punta

PUNTA MADROÑO; see Madroña, Punta

PUNTA MALA; see Mala, Punta

PUNTA PAITILLA; see Paitilla, Punta

PUNTA PATIÑO; see Old Point

PUNTA PEÑA; see Peña, Punta

PUNTA PEÑA TIBURÓN; see Peña Tiburón, Punta

PUNTA PILÓN; see Pilón, La Punta de

PUNTA RÓBALO; Bocas del Toro
0902/8215 (USBGN)
Sea level, town located on the central W coast of Laguna de Chiriquí [0903/8200 (USBGN)] 1 km N of the mouth of the Río Róbalo [0901/8215 (USBGN)] (E762). SBE, GL for: Laguna de Chiriquí, and Río Róbalo, 1991 Mar 10 (USNM, as "Róbalo"); DaCosta, Guitton-Mayerma, Klicka, Lelevier, Miller, and Spellman, GL for Pueblo Nuevo, 2006 Feb 23 (MBM; STRI).

PUNTA VIEJA; see Old Point

PUNTITA, LA; see La Puntita

PUNTO, CERRO; see Cerro Punta

PUNUSA; see Punusa, Río

PUNUSA, BOCA DE RÍO; see Punusa, Río

PUNUSA, RÍO; Darién
0748/7732 (USBGN)
Originates in S Darién province near the Colombian border N of Altos de Quía [0735/7727 (USBGN)]. The stream flows SW, then NW entering the Río Tuira [0821/7803 (USBGN)] ca. 95 km SE of La Palma [0825/7809 (USBGN)] and ca. 25 km ENE of Alturas de Nique [0740/7745 (USBGN)] (MGRP). GMLP, 1959 Apr 16-17 (GML, as "Punusa"). Mentioned in Fairchild and Handley (1966, as "Boca de Río Punusa").

PUPRUGANDI, RÍO; San Blas
ca. 0908/7756
Apparently a river that enters the Caribbean Sea [1500/7500 (USBGN)] opposite Ustupo [0908/7756 (USBGN)], Loftin, GL = Ustupo, observations, "far up" the river, 1962 Oct 7 (Loftin, 1963a). Not shown on our maps.

QUARRY; see Chiva Chiva

QUARRY HEIGHTS Panamá (Canal Zone) (E, Pacific)
0857/7933 (USBGN)
Sea level, town that is part of greater Panama City [0858/7932 (USBGN)] located on the SW edge of the city, immediately E of Balboa [0857/7934 (USBGN)] and N of Fort Amador Army Reservation [0856/7933 (USBGN)] near the former Panamá/Canal Zone border (E762; RMM2). Meyer, 1941 Mar 20, 29, May 19 (USNM).

QUEBRADA ALEMAN TRAIL; see Los Planes

QUEBRADA CARRIASO; Panamá (E, Pacific)
ca. 0914/7914
ca. 200-400 m, described by Wetmore as a region near a small tributary of the Río Pacora [0901/7917 (USBGN)], presumably the region near Carriazo [0912/7914 (USBGN)] (q. v.), Wetmore and Perrygo, GL = Cerro Azul [0910/7924 (USBGN)], level forested flat, tall, gallery forest, clearings, untouched forest, 4 mi (6 km) along a ridge trail and up a mountain stream with a rocky bed to ca. 1,200 feet (370 m), 1949 Apr 23, 25 (USNM). Not shown on our maps.

QUEBRADA DE PIEDRA; Chiriquí (Pacific)
0806/8141 (USBGN)
100 m, town 10 km NW of Puerto Vidal [0803/8136 (USBGN)] (JOG, as "Quebrada Piedra"). Wetmore and Perrygo, GL for Río Tabasará, 1953 Jun 6 (USNM, as "Quebrada Piedra"; Wetmore, 1970, as "Quebrada Piedra"); Hinds, 1981 Sep 5 (GML).

QUEBRADA HIDO; Bocas del Toro

ca. 0909/8153

According to Handley's journal, this is an Amerindian village on Península Valiente [0905/8151 (USBGN)] located on a stream also named Quebrada Hido. The stream emptied into Laguna de Bluefields [0909/8155 (USBGN)] and the village was located above the head of tidewater. The stream mouth was surrounded by mangroves that quickly gave way to deep *potrero*, in which walking was very difficult, SBE, GL = Península Valiente, on the stream, boggy areas, in the *potrero* where cattle grazed, brush, fenced areas, hillsides nearby, 1990 Apr 24, 26-28 (USNM). Not shown on our maps.

QUEBRADA PARTICHEQUE; see Tuquesa, Río

QUEBRADA PIEDRA; see Quebrada de Piedra

QUEBRADA SÁBALO; Chiriquí

ca. 0820/8300

400 m, town on the Río Palo Blanco [0818/8249 (USBGN)] 16 km WNW of Puerto Armuelles [0817/8252 (USBGN)] (E562, as "Quebrada Sábalos"). See also: Bosque de Chongo.

QUEBRANCHA, RÍO; see Cabiancho, Río

QUERQUÉ; Veraguas (Pacific)

0802/8124 (USBGN)

ca. 50 m, town 10 km WNW of Soná [0801/8119 (USBGN)] and 12 km SSE of Las Palmas [0808/8127 (USBGN)] (MGRP). Casas, 1990 Feb 3 (MVUP).

QUETZALES, SENDERO LOS; see Cerro Punta-Boquete Trail

QUÍA, ALTOS DE; Darién

0735/7727 (USBGN)

1,361 m, mountains in S Darién province 44 mi (71 km) E of Jaqué [0731/7810 (USBGN)] on the Colombian border and on the continental divide (SMN2B). According to USBGN, Cerros de Quía is the Colombian name for the range. Galindo, Hinds, and GMLP, at 2,400-2,600 feet (750-800 m), some specimens are also labeled Ridge Station, 1970 Mar 8, 10, 13, 1971 Feb 23, 27-28, Mar 2, 13, 15-16, 1972 Apr 4 (GML; LSUMZ; and USNM, all as "Cerro Quía"); Galindo and GMLP, GL = Río Mono, on the trail from Río Mono to, above Río Mono, 1970 Feb 21, 1971 Feb 21, 23, Mar 15 (USNM, as "Cerro Quía"; Wetmore and Galindo, 1972); Hinds and Ridgely, at 530-600 m, and at 2,700 feet (810 m), specimens and observations, heavy forest, 1975 Jul 16-18 (GML; Wetmore et al., 1984:208; and Ridgely and Gwynne, 1989:289, 388, all as "Cerro Quía"). (Wege, 1996, as "Cerros de Quía").

QUÍA, CERRO; see Quía, Altos de

QUIBO; see Coiba, Isla de

QUIEL; Chiriquí

ca. 0848/8228

Described by Griscom (1935a:202, as "El Quiel") as a coffee plantation on Volcán de Chiriquí [= Volcán Barú [0848/8233 (USBGN)]]. Described by Blake (1958:502-

503) as an outlying section of Boquete [0847/8226 (USBGN)], whose S border is the N rim of Callejón Seco [0848/8228 (USBGN)]. Quiel itself is not shown on our maps, but apparently the region is located at 1,500 m, ca. 4-5 km NW of Boquete, S of Quebrada El Velo [0849/8228 (USBGN)] near the current towns of Alto Quiel [0848/8228 (USBGN)] and Bajo Quiel [0848/8227 (USBGN)]. Benson, GL = Boquete, at 4,800-9,000 feet (1,450-2,750 m), 1931 Oct 24-Dec 20 (ANSP; LSUMZ; MCZ; MVUP; USNM; WFVZ; Griscom, 1935a, as "El Quiel"; Olson, 1981c, as "Boquete, Quiel"); Davidson, GL = Boquete, at 5,300-7,800 feet (1,600-2,400 m), 1931 Dec 19-20, 24, 1933 Jan 9, 16, 19-22, 24-27, Dec 6 (?), and as a GL for: Alta Peña, Casita Alta and Peña Blanca, 1933 Jan 19, 24-27 (CM; Davidson, 1938); Mönniche, at 5,100-5,300 feet (1,550-1,600 m), 1932-1939 (FMNH, as "Alto de Quiel," and as "Quiel"; MCN, as "Alto de Quiel," and as "Quiel"; Blake, 1958; Monroe, 1963b, as "Alto de Quiel"); Smith, N. and Eisenmann, at 1,450 m, observations, low shrubby growth at the edge of a potato field, 1964 Jul 17 (Olson et al., 1968).

QUIPO; see Cuipo

QUIVETTA, ISLA; see Ranchería, Isla

RABO DE PUERCO, RÍO; Chiriquí

ca. 0817/8252

Originates W of Puerto Armuelles [0817/8252 (USBGN)] near the Costa Rican border and flows E entering Bahía de Charco Azul [0812/8245 (USBGN)] at Puerto Armuelles (E562). Davidson, GL = Puerto Armuelles, observations, near town on a sandbar, 1929 Nov (Davidson, 1938:256, as "Rabo de Puerco River").

RABO DE PUERCO RIVER; see Rabo de Puerco, Río

RADIO HILL; Panamá (Canal Zone) (W ?, Atlantic)

ca. 0904/7940 ?

Probably refers to the 100 m hill labeled "U. S. Naval Radio Station Summit" on E8660, located on the W side of the Panama Canal [0920/7955 (USBGN)] at the E edge of Empire Known Distance Range [0904/7940 (USBGN)]. E8660 indicates "numerous towers" and "numerous radio towers" in this region on both sides of the canal between Summit Park [0904/7939 (USBGN)] and Empire Known Distance Range. MCU, 1955 Nov 14 (USNM).

RAILWAY, LINE OF; see Panama Railroad

RAINBOW CITY; Colón (Canal Zone) (E)

0921/7953 (USBGN)

Sea level, town on the SE coast of Río Folk [0921/7954 (USBGN)] 2 km SE of Colón [0922/7954 (USBGN)] (E8660). Ambrose, observations, 1955 (Wetmore, 1972:417). See also: Fort Gulick; Marajal; Margarita; and Mount Hope.

RANCHERÍA, ISLA; Veraguas (Pacific)

0738/8142 (USBGN)

130 m, island with a land area of 2.2 km^2 (220 ha) in the Pacific Ocean 3 km NE of Isla de Coiba [0728/8147 (USBGN)], the island extends 3 km E-W and varies in width from 500 m to 2 km (JOG; Olson, 1997a). Shown as Isla Coibita on many

maps (Atlas, 1988:202; E762). For more on the name of the island see Wetmore (1957:5). Wetmore, steep slopes behind the shoreline, sandy beach with coconut palms, extensive swamp bordering a small, wooded valley, 1956 Feb 4 (USNM; Wetmore, 1957; Wetmore, 1967; Olson, 1997a); Garcés and Mena, 1996 Feb 11 (MVUP). (Shelvocke, 1726, as "Quivetta")

RANCHO FRIO CAMPSITE; see Pirre Station

RANCHO FRIO (PIRRE) STATION; see Pirre Station

RANCHO FRIO STATION; see Pirre Station

RANCHÓN; San Blas

ca. 0840/7725

We have not been able to find a suitable description of Ranchón in the literature. According to Griscom (1932), Wedel began his work in San Blas at Permé [ca. 0845/7734] (a "small plantation of the United Fruit Co."), but the limited facilities and difficult terrain in this region prompted him to relocate c. January 1930 to Puerto Obaldía [0840/7725 (USBGN)] (the site of a U. S. naval radio station). Griscom also indicates that Wedel sent a sketch map of the region with his August 12 report, but no such map was found in the archives at MCZ or at AMNH. In our notecard file, Bea Wetmore had assigned the coordinates 0835/7728 to Ranchón with no further indication. The name initially suggests a ranch or other agricultural facility associated with the United Fruit Co., but the dates on specimen labels suggest that Ranchón is nearer to Puerto Obaldía than to Permé. Further examination in the Larrouse English-Spanish Dictionary (García-Pelayo y Gross, 1993) reveals that a maritime usage of *rancho* is "crew's quarters." As the dates overlap with those of Puerto Obaldía and the name suggests a settlement or quarters for naval personnel, it appears likely that Ranchón was located near Puerto Obaldía in the coastal, lowland areas of San Blas. Wedel, 1930 Dec 25, 1931 Jan-Feb (ANSP; FMNH; MCZ; OC; YPM; Griscom, 1932a; Meyer de Schauensee, 1950a; Eisenmann and Howell, 1962, as "Ranchón, Colón").

RANCHO RON; Darién

?

Not located. (Willson et al., 1982).

RANDOLPH, FORT; see Fort Randolph Army Reservation

RATÓN, CERRO; see Flores, Cerro

RAY ISLAND; see Rey, Isla del

REAL DE SANTA MARÍA; see El Real de Santa María

REAL DE SANTA MARÍA, EL; see El Real de Santa María

RED TANK; Panamá (Canal Zone) (E, Pacific)

ca. 0901/7936

Sea level, town located on the E side of the Panama Canal [0920/7955 (USBGN)] on the NE end of Lago de Miraflores [0901/7936 (USBGN)] between Pedro Miguel [0901/7937 (USBGN)] and Fort Clayton Army Reservation [0901/7934 (USBGN)] (RMM2). Benson, at 200 feet (60 m), 1931 Jun 22-23 (MVUP; USNM); Olson and

Fischer, near, 1963 May 23 (USNM). See also: Caimito Arm.

REFUGIO DE VIDA SILVESTRE PEÑON DE LA HONDA; see Villa, Isla

REFUGIO DE VIDA SILVESTRE TABOGA; see Taboga, Isla

REMEDIOS; Chiriquí
0814/8151 (USBGN)
ca. 50 m, town 17 km WSW of Tolé [0814/8141 (USBGN)] on the Río San Félix [0810/8151 (USBGN)] S of the Carretera Interamericana [0832/8242 (USBGN)] and 8 km N of the coast (JOG; MGRP). Brown, 1901 Jul 10 (MCZ; Bangs, 1902a); Griscom et al., 1924 Feb 28, Mar 23-26, 28, and as a GL for Río San Félix, 1924 Mar 26 (AMNH; CM; Griscom, 1933); Loftin and Tyson, observations, 1961 Nov 11 (Loftin, 1963a); Pitman, nearby on the Pan American Hwy, 1987 Nov 8 (LACM); Lelevier, Maley, and Miller, 3 km W of, 2005 Dec 27 (STRI; UAM).

RENACER JAIL; see Gaillard, Carretera

REPARTO CHANIS; see Chanis, Barriada

REPARTO MARÍA LETICIA; see María Leticia, Reparto

REPRESA; see La Represa

REPRESA, LA; see La Represa

REPRESA, QUEBRADA DE LA; see La Represa, Quebrada de

REPRESSA, QUEBRADA; see La Represa, Quebrada de

RESERVA FORESTAL EL MONTUOSO; see El Montuoso, Loma

RESERVA FORESTAL FORTUNA; see Fortuna

RESPINGO, CERRO; Chiriquí
0851/8232 (USBGN)
2,807 m, peak located 4 km E of Cerro Punta [0852/8235 (USBGN)] and 2 km E of Bajo Grande [0852/8233 (USBGN)]. The peak is 1.5 km N of the Cerro Punta-Boquete Trail [ca. 0852/8235] (E762). Wetmore, GL = Cerro Punta, Bajo Grande (q. v.), 1955 Feb 28 (USNM).

RESPINGO, PASO DE; Chiriquí
0850/8232 (USBGN)
Located on the Cerro Punta-Boquete Trail [ca. 0852/8235] 5 km ESE of Cerro Punta [0852/8235 (USBGN)] and 1.5 km S of Cerro Respingo [0851/8232 (USBGN)]. The locality is designated by USBGN as a mountain pass and is located at 2,300-2,480 m between the headwaters of the Río Caldera [0837/8221 (USBGN)] and the Quebrada Bajo Grande [0851/8234 (USBGN)] (E762). Angehr (in litt., Oct 2003) indicates that the Cerro Punta-Boquete Trail is also known as the Respingo Trail and that "'Paso de Respingo' refers to the 'jumping off spot' for the trail above Cerro Punta." Cardiff, Dittmann, and Lovette, GL = Cerro Punta-Boquete Trail, W base at ca. 2,000 m, and at 2,475 m on the pass, 1996 Mar 9-11 (LSUMZ; MVUP).

RESPINGO TRAIL; see Respingo, Paso de

RESTINGA; Panamá (W, Pacific)

ca. 0848/7934

Described by Wetmore (1952:1) as a tourist development on a sandy beach at the N end of Isla Taboga [0847/7933 (USBGN)]. Eisenmann, observations, 1948 Jun 24, 1951 Jul 5 (Wetmore, 1952:32); Wetmore, observations, 1952 Mar 15-23 (Wetmore, 1952).

REY, CONDADO DEL; see Betania

REY, ISLA; see Rey, Isla del

REY, ISLA DEL; Panamá, Archipiélago de las Perlas

0822/7855 (USBGN)

223 m, largest island in the Archipiélago de las Perlas [0825/7900 (USBGN)] with a land area of 249 km^2 (24,900 ha) (Wright et al., 1985) extending roughly 20 km N-S and 14 km E-W. The island is located in the Gulf of Panama [0800/7930 (USBGN)] 65 km W of Golfo de San Miguel [0822/7817 (USBGN)] and 95 km SE of Panama City [0858/7932 (USBGN)]. Aside from the central region, the elevation is generally below 100 m. Apparently the island had a number of different names, one of which was Isla San Miguel, a name that is retained by the island's largest town, San Miguel [0827/7856 (USBGN)], located on the N coast (E501, as "Isla del Rey o de San Miguel"; JOG). According to Angehr (in litt., Aug 2005), the island was called Isla Rica by Balboa. Bovallius, sailed around it and collected on the coast, at Punta Matadero [0827/7852 (USBGN)], and found "a rich bird life…at its southern point, Punta Cocos [0813/7854 (USBGN)]," see discussion of Bovallius under Pueblo Nuevo [ca. 0820/8013 ?], 1882 Apr 30, May 2 (NHRM; Rendahl, 1920, as "San Miguel (or El Rey)"); Brown, 1900 Apr 17-May 8, 1904 Feb 21-Mar 20, 28, Apr 6 (ANSP, as "San Miguel Is.," and as "San Miguel"; CM, as "Isla San Miguel"; FMNH, as "San Miguel"; MCZ, as "San Miguel I."; USNM; Bangs, 1901a; Bangs, 1901b; Bangs, 1901c; Bangs, 1902b; Bangs, 1903b; Thayer and Bangs, 1905; Oberholser, 1912; Todd, 1913; Bangs and Penard, 1918:35, 41; Bangs and Penard, 1919; and Swann, 1922, all as "San Miguel Island"; Peters and Griscom, 1929, as "San Miguel, El Rey"; Griscom, 1932a, as "San Miguel, El Rey" but in error for a San Miguel in the Santa Marta mountains of Colombia (Wetmore 1968:310); Moore and Peters, 1939, as "San Miguel Island"; Olson, 1980b, as "Isla San Miguel"); Griscom and Crosby, 1927 Feb 20-22 (AMNH, as "Isla El Rey"; Griscom, 1927b, as "El Rey"); Fahnenstock Expedition, 1935 Mar 20 (AMNH); Murphy and Correia, Askoy Expedition, GL for Bahía San Telmo, 1941 Feb 14 (AMNH; Murphy, 1944a, as "San Miguel Island"; Murphy, 1945, as "San Miguel Island"); Robins, observations, 1957 Jul 15-26 (Robins, 1958, as "Rey Island"); Beaudette, 1959 Mar 21, and as a GL for: Isla San Telmo, and Río Cacique, 1959 Mar 22 (LACM, as "Del Rey Is"); Wetmore, GL for: La Ensenada, La Puntita, and Río Cacique, 1960 Jan 24-25, 27 (USNM; Wetmore, 1970); Pacific Project, GL = Gulf of Panama, at Isla San Pablo, 1969 Aug 30 (USNM, as "Rey Island"; Loftin, 1991); (MacArthur et al., 1972—species list; see comments in Bibliography); Wright and Campbell, mist nets, observations, 1978 Jan-Mar (Wright et al., 1985); Tejera, GL for: Playa de Chiquero, and Playa de Cinique, 1981 Mar 15, 17-22 (MVUP);

Levings et al., observations, 1982-1983 (Levings et al., 1986); Brawn et al., mist nets, 199? (Brawn et al., 1996). (Douglas and Johnson, 1926; Wetmore, 1945, as "Isla El Rey"; Wetmore, 1952:19, 31, as "Isla El Rey"; Wetmore, 1963a; Zimmer, 1955:5, as "San Miguel Island"; N. Smith, 1971). See also: Cacique, Río; Cocos, Punta; La Ensenada; La Puntita; Mafofa; Matadero, Punta; San Miguel; and San Telmo, Bahía.

REY, ISLA EL; see Rey, Isla del

REY ISLAND; see Rey, Isla del

RICA, ISLA; see Rey, Isla del

RICARDO J. ALFARO, VIA; see Tumba Muerto

RICO, ESTERO; Chiriquí

ca. 0820/8242

Described by Wetmore in 1960 as a coastal area popular with sunbathers and reached by a narrow road that led away from Divalá [0825/8243 (USBGN)]. The area was located a short distance from the bridge over the Río Escarrea [0820/8241 (USBGN)] near Canta Gallo Abajo [0822/8238 (USBGN)] (USNM). From this description Wetmore likely worked ca. 20 km S of La Concepción [0831/8237 (USBGN)] just S of El Cabimo [ca. 0821/8242] (q. v.) near the lower sections of the Río Duablo [0820/8241 (USBGN)] and the Río Chiriquí Viejo [0820/8241 (USBGN)] (q. v.). USBGN lists another Estero Rico, a tidal creek at 0816/8220, but this is more than 35 km from Canta Gallo Abajo, and according to our maps, not accessible by road (JOG). Wetmore, GL = Alanje, a few thatched huts, sandy soil, a trail to the estuary bordered by mangroves, 1960 Mar 13 (USNM; Olson, 1980b); Ridgely, observations, mangroves and seashore, 1976 Mar 6 (Wetmore et al., 1984:362; Ridgely and Gwynne, 1989:284).

RINCÓN, EL; see El Rincón

RINCÓN, PUNTA; Colón (E)

0935/7921 (USBGN)

Sea level, point on the Atlantic coast of NE Colón province 15 km E of Nombre de Dios [0935/7928 (USBGN)] (JOG). Wedel, 1931 Mar 7 (FMNH).

RÍO ABAJO; Panamá (E, Pacific)

0901/7930 (USBGN)

50 m, section of greater Panama City [0858/7932 (USBGN)] located 2 km N of Old Panama [0900/7929 (USBGN)] on the E banks of the Río Abajo [0901/7929 (USBGN)] (E962). Pinzón, 1994 Aug 6 (MVUP). See also: Abajo, Río.

RÍO CHANGENA CAMP; see Changuena, Río

RÍO CHANGUINOLA ARRIBA; Bocas del Toro

ca. 0910/8230

Not shown on our maps but according to Brumfield (personal communication), this was the name of a community upstream from their collecting site. Braun and Brumfield, GL for Río Changuinola, 2004 Mar 14-15 (LSUMZ).

RÍO CHICO; Panamá (E, Pacific)
0902/7910 (USBGN)
Sea level, coastal town 13 km SE of Pacora [0905/7917 (USBGN)] on the Río Chico [0901/7911 (USBGN)] 2 km above its mouth (TM, as "Chico"). Miller, GL for La Jagua Hunting Club, 1937 Mar 6 (USNM, as "Chico"); Wetmore and Perrygo, the secondary localities attributed to Río Chico were located in the area extending 11 km W and NNW of town, GL for: Ciénaga Campana, Ciénaga de Santo Domingo, La Jagua Hunting Club, Monte Mabobo, Río Chico, Río La Jagua, and Sabana de San José, 1949 Mar 13-16, 18, 20, 22, 24-Apr 2, GL for La Jagua Hunting Club, 1951 Mar 1 (LSUMZ, as "Chico"; USNM, as "Chico"); Wetmore, GL for La Jagua Hunting Club, 1953 June 26-28, 1955 Mar 30, GL for Río Chico, 1956 Mar 5-6, 1961 Mar 24, GL for: La Jagua Hunting Club, and Gallinero, 1958 Mar 19-21 (USNM, as "Chico"; Wetmore, 1970, as "Chico"; Olson, 1980b, as "Chico, mouth of Río Chico"; Olson, 1981a, as "Chico"); Moreno, GL for Río La Jagua, 1959 Jan 22 (USNM, as "Chico"). (Wetmore, 1952:27, as "Chico"; Parkes, 1976, as "Chico").

RÍO CHICO; see Chico, Río

RÍO COCLÉ; see Coclé del Norte, Río

RÍO DE JESÚS; Veraguas (Pacific)
0759/8110 (USBGN)
ca. 50 m, town 17 km ESE of Soná [0801/8119 (USBGN)] in low hills between the Río San Pedro [ca. 0750/8105] and the Río San Pablo [0751/8110 (USBGN)] (JOG). Wetmore and Perrygo, collected near town along the Río de Jesús [0755/8104 (USBGN)] (q. v.) in pastures with brush-grown fences, along a small stream, on a hill covered with low brush, downstream in a thorn thicket, brush and second growth, 1953 May 22, 27 (USNM; Wetmore, 1953; Wetmore, 1970).

RÍO DE JESÚS; see Jesús, Río de

RÍO DIABLO; Panamá (E, Pacific)
0911/7832 (USBGN)
ca. 50 m, town on the Río Diablo [0907/7834] ca. 2 km N of its mouth and the NE coast of Lago Bayano [0908/7835 (USBGN)] and 65 km E of Chepo [0910/7906 (USBGN)] (MGRP). Hinds, is labeled Colon province, which is presumably a misprint or a misreading, 1974 Feb 20, 26 (GML).

RÍO DIABLO, LAGUNA; see Cañaveral, Río

RÍO FOLK; Colón (Canal Zone) (E)
0921/7954 (USBGN)
Inlet on the N coast of Colón province that extends ca. 2 km SE from Colón [0922/7954 (USBGN)] to Rainbow City [0921/7953 (USBGN)] and opens on the N to Bahía Manzanillo [0922/7953 (USBGN)] (E8660, as "Folk River").

RÍO GRANDE; Coclé (Pacific)
0825/8029 (USBGN)
ca. 50 m, city on the Carretera Interamericana [0832/8242 (USBGN)] 18 km SW of Penonomé [0831/8022 (USBGN)] (JOG). See also: Grande, Río.

RÍO GRANDE STATION; Panamá (Pacific)

ca. 0900/7935

Sea level, former station of the original Panama Railroad [ca. 0922/7954] located on the Río Grande [0902/7938 (USBGN)] 6 mi (10 km) NW of Panama City [0858/7932 (USBGN)] and 1.5 mi (2.5 km) SE of Pedro Miguel [0901/7937 (USBGN)] (MIP). Maack, 1871 (MCZ, as "Río Grande Station, Isthmus of Panama"; Maack, 1872).

RÍO HATO; Coclé (Pacific)

0823/8010 (USBGN)

ca. 25 m, city on the Carretera Interamericana [0832/8242 (USBGN)] 27 km SE of Penonomé [0831/8022 (USBGN)] and 4 km N of the coast. JOG indicates two airfields near town, one is located 4 km E of town on the Carretera Interamericana and the other is located 5 km NE of town (JOG). According to Angehr (in litt., Aug 2005), the "airfield referred to in the literature is certainly the one on the highway 3 km E of town, which was the site of the US military's Rio Hato Air Base in the 1930s-1940s." Imhof, observations, shore, 1942 Jul 31, Aug 1 (Imhof, 1950, as "Río Hato," and as "Río Hato Air Base"); Eisenmann, observations, sandy areas, c. 1948-1956 Jun-Jul (Eisenmann, 1962a, as "Río Hato Airfield"); Scholes and Scholes, observations on the roadside between Penonomé and, 1951 Jan 28 (Scholes and Scholes, 1954); Wetmore and Perrygo, old airstrip, 1953 May 13 (USNM; Wetmore, 1953); Loftin, 1962 Sept 16 (Loftin, 1963a). See also: Arraijan.

RÍO HATO AIR BASE; see Río Hato

RÍO HATO AIRFIELD; see Río Hato

RÍO INDIO, BOCA DEL; see Boca del Río Indio

RÍO INDIO-PIÑAS ROAD; Colón

ca. 0912/8011

Road that extends roughly NE to SW along the N coast of Colón province from Piña [0917/8003 (USBGN)] to Salud [0912/8008 (USBGN)] to Boca del Río Indio [0912/8011 (USBGN)]. Guitton-Mayerma, Lelevier, and Miller, at 9°16'26"/ 80°03'00", 2005 Dec 3 (STRI; UAM).

RÍO INDIO VILLAGE; see Boca del Río Indio

RÍO MAR; Panamá (W, Pacific)

0827/7958 (USBGN)

Sea level, coastal town 2 km SW of San Carlos [0829/7957 (USBGN)] just S of the Carretera Interamericana [0832/8242 (USBGN)] (JOG). Wetmore, GL = San Carlos, at night on the road into Río Mar, 1953 Jun 21, GL = San Carlos, on the beach, 1958 Mar 16-17 (USNM, as "Riomar"). See also: San Carlos.

RIOMAR; see Río Mar

RÍO MUÑOZ; see Bajos del Muñoz

RÍO SERENO; Chiriquí

0849/8252 (USBGN)

900 m, district capital located 20 km SW of Cerro Pando [0855/8243 (USBGN)]

on the Costa Rican border near the mouth of the Río Sereno [ca. 0849/8251] (MGRP). Wetmore, GL for Río Candela, 1955 Feb 21-23 (USNM, as "Sereno"; Wetmore, 1970, as "Sereno"; Wetmore, 1972:463, as "El Sereno"; Parkes, 1976, as "Sereno"). See also: Sereno, Río.

RÍO TUYRA DELTA; see Tuira, Río

RÍO TUYRA VALLEY; see Tuira, Río

RISACUA, RÍO; Chiriquí

ca. 0827/8226

Not shown on our maps but located in Corregimiento David [0827/8226 (USBGN)] near David [0826/8226 (USBGN)] (q. v.). Presumably a small tributary of the Río David [0822/8224 (USBGN)] (q. v.). Martín and Saldaña, GL = David, 1985 Apr 11 (MVUP).

RISCÓ, BOCA DEL RÍO; see Risco, Río

RISCÓ, RÍO; Bocas del Toro

0916/8228 (USBGN)

Originates NW of Punta Róbalo [0902/8215 (USBGN)] and flows N entering the Río Changuinola [0928/8227 (USBGN)] 9 km WSW of Almirante [0918/8224 (USBGN)] (JOG). Braun and Brumfield, GL for Río Changuinola, 2004 Mar 15 (LSUMZ). Wetmore (1972:365, as "Boca del Río Risco"). See also: Valle de Riscó.

RIVER CANA; see Cana, Río

ROAD 842; see Pipeline Road

ROAD C, K, R, or S;

Roads with these labels are military roads located in and around the former Canal Zone. Angehr (in litt., Aug 2005) suggested that the initials represent the main fort in each sector — Clayton, Kobbe, Randolph, and Sherman. Roads labeled C are in Panamá province E of the Panama Canal [0920/7955 (USBGN)], while those labeled K are in Panamá province W of the canal. These roads may run through the Atlantic drainage, the Pacific drainage or both, as the divide crosses through Panamá province below Lago Gatún (Gatun Lake) [0912/7955 (USBGN)]. R and S Roads are in Colón province and all of them occur on the Atlantic drainage. Those labeled R are located E of the canal, and those labeled S are located W of the canal. Even numbered roads extend roughly N-S and odd numbered roads extend roughly E-W. Some of the larger roads have side roads and short extensions that are labeled alphabetically, for example, Road C2A, Road C2B, etc. (RM1; RM2; RMM1; RMM2).

ROAD C2; see Gaillard, Carretera

ROAD C2H; Panamá (Canal Zone) (E, Atlantic)

ca. 0906/7942

The road is not shown on our maps but by its name it must connect to Carretera Gaillard [0904/7940 (USBGN)], which is also called Road C2, and is presumably within 2 km of Gamboa [0907/7942 (USBGN)], as Road C2F appears on our maps 2 km SE of Gamboa. Blanton, GL = Gamboa, 1953 Sep 16 (USNM, as "C-2H Road").

ROAD C12E; Panamá (Canal Zone) (E, Pacific)

ca. 0900/7934

The road branches from Road C12 [ca. 0901/7934] and extends ca. 0.5 mi (1 km) NE. Its SW terminus is ca. 1.5 mi (2.5 km) ESE of Fort Clayton Army Reservation [0901/7934 (USBGN)] and ca. 1.5 mi (2.5 km) N of Curundu [0859/7933 (USBGN)] (RMM1). Blanton, GL = Fort Clayton Army Reservation, 1953 Sep 17, 22, Oct 13 (USNM, as "C-12-E Road," as "C-12-E Rd," and as "Rd C-12-E").

ROAD C15; Panamá (Canal Zone) (E, Pacific)

ca. 0902/7932

Road that extends 4 mi (6 km) NNE from the NE end of Albrook Air Force Base [0859/7933 (USBGN)] to Road C16 [ca. 0905/7937] (RM1; RMM1). Blanton, GL = Albrook Air Force Base, 1953 Sep 20, Oct 22 (USNM, as "C-15 Road").

ROAD C20; see Boyd-Roosevelt, Carretera

ROAD C21; see Chiva Chiva, Camino

ROAD C25; see Madden, Camino

ROAD C29; Panamá (Canal Zone) (E, Atlantic)

ca. 0907/7938

An unimproved, dirt road that extends ca. 6 km N-S between Las Cruces Trail [ca. 0904/7936] and Carretera Gaillard [0904/7940 (USBGN)]. The S terminus of the road meets Carretera Gaillard [0904/7940 (USBGN)] 6 km SE of Gamboa [0907/7942 (USBGN)] and 2 km NW of Summit Park [0904/7939 (USBGN)]. From here it extends ca. 6 km N, and after it crosses Las Cruces Trail, the road becomes a foot path that continues ca. 6 km N to Juan Mina [0910/7939 (USBGN)]. The approximate coordinates noted above indicate the N terminus of the road, not the foot path (E866). Blanton, Central, 1953 Nov 12 (USNM, as "C-29 Road"); MCU, 1953 [1955 ?] Aug 24 (MCN, as "C-29 Road"); MCU, GL = Gamboa, 1955 Apr 21, 28, May 2, 18, 23 (USNM, as "C-29 Road").

ROAD K1; see Farfan Road and Road K1B

ROAD K1B; Panamá (Canal Zone) (W, Pacific)

ca. 0855/7934

Road that extends ca. 1 km NW-SE between Road K1 [ca. 0857/7934] and Punta Batele [0854/7934 (USBGN)]. The road is located 1 km S of Palo Seco Leprosarium [0855/7934 (USBGN)] (E761B). Hespenheide, GL = Palo Seco Leprosarium, censuses along the unpaved road, disturbed forest in advanced stage of regeneration, 1969 Oct-1970 Mar (Hespenheide, 1980).

ROAD K2; see Borinquén, Carretera

ROAD K5; see Thatcher, Carretera

ROAD K6; Panamá (Canal Zone) (W, Pacific)

ca. 0900/7939

Road that extends ca. 10 km NNW-SSE between Road K15 [ca. 0901/7939] and Cerro Galera [0856/7937 (USBGN)]. In the area 3 km NE of Arraijan [0857/7939 (USBGN)] the Río Cocolí [0859/7936 (USBGN)] crosses the road, whereas the town Cocolí [0859/7935 (USBGN)] is more than 3 mi (5 km) from any point on Road K6, therefore specimens labeled Cocolí and Road K6, were probably collected along the river, not near the town (E762; E8660; RMM2). Blanton, 1953 Sep 14, 27, Oct 8, 12, 19, 26, Nov 2, 9, Wetmore added the label "Cocolí" to , 1953 Sep 27 (USNM, as "K-6 Road," and as "Pac, K-6 Road"; Olson, 1981a, as "Pac, K-6 Road"); MCU, GL = Cocolí, 1955 Jul 27, Aug 1, Nov 22 (GML, as "K-6 Road"; USNM, as "K-6 Road"), GL = Arraijan, 1955 Aug 22 (GML, as "K-6 Road"); Wetmore, GL = Cerro Galera, secondary growth near the summit of Cerro Galera, 1960 Jan 14 (USNM); Morton, studies, 1973-1975 (Morton, 1980, as "K-6 Road"). See also: Galera, Cerro.

ROAD K7; Panamá (Canal Zone) (W, Pacific)

ca. 0856/7935

Presumably a short road that extends E-W through Farfan [0856/7935 (USBGN)]. Blanton, GL = Farfan, 1955 Oct 17 (USNM, as "K-7 Road"). Not shown on our maps.

ROAD K9; Panamá (Canal Zone) (W, Pacific)

ca. 0858/7936

Road that extends 2.4 mi (3.9 km) roughly E-W along the lower section of the Río Cocolí [0859/7936 (USBGN)] between Road K6 [ca. 0900/7939] and Carretera Borinquén [0901/7938 (USBGN)]. The E terminus of the road is just N of the town Cocolí [0859/7935 (USBGN)] (RM1; RMM1). MCU, GL = Cocolí, 1955 Sep 1 (GML, as "K-9 Road"; USNM, as "K-9 Road").

ROAD K10; Panamá (Canal Zone) (W, Atlantic, Pacific)

ca. 0901/7942

Road that extends 6.6 mi (10.6 km) NNW-SSE between Road K19 [ca. 0853/7947] near the origin of the Río Mandinga [0906/7942 (USBGN)] and Arraijan [0857/7939 (USBGN)] (RM1; RMM1; E8660). The road appears as "Panama 84" on MGRP. MCU, GL = Empire Known Distance Range, 1955 Oct 3 (MCN, as "K-10 Road"; USNM, as "K-10 Road"). See also: Arraijan.

ROAD K15; Panamá (Canal Zone) (W, Pacific)

ca. 0900/7939

Road that extends 3 mi (5 km) NW-SE between Carretera Borinquén [0901/7938 (USBGN)] and Road K10 [ca. 0901/7942], the junction with Carretera Borinquén is ca. 2 mi (3 km) NW of Esclusas Pedro Miguel (Pedro Miguel Locks) [0901/7937 (USNM)], and the junction with Road K10 is near the origin of the Río Cocolí [0859/7936 (USBGN)] (RM1; RMM1). See also: Cocolí, Río; and Road K6.

ROAD K16; see Santa Clara

ROAD K19; Panamá (Canal Zone) (W, Atlantic)

ca. 0853/7947

Road, now overgrown and no longer maintained, that extends 8 mi (13 km) NNE from La Chorrera [0853/7947 (USBGN)] to Nuevo Emperador [0900/7944 (USBGN)] and then continues roughly E for 7 mi (11 km) to Empire Known Distance Range [0904/7940 (USBGN)]. E8660 indicates that the road is "fair weather, loose material" between Nuevo Emperador and the headwaters of the Río Mandinga [0906/7942 (USBGN)], where it meets the N terminus of Road K10 [ca. 0901/7942]. From here the road is marked as an "abandoned" cartrack that continues E to Empire Known Distance Range where it eventually terminates at a junction with Carretera Borinquén [0901/7938 (USBGN)] (E8660; RMM2). The section of road between La Chorrera and Nuevo Emperador is also marked as "Route 7" on RM2, as "Chorrera Empire Road" on E761A, and as "Panama 75" on MGRP. See also: Empire Known Distance Range.

ROAD R2; Colón (Canal Zone) (E)

ca. 0917/7955

Short road that extends 2 km N from Gatún [0916/7955 (USBGN)] along the E side of the Panama Canal [0920/7955 (USBGN)] and terminates at its junction with Road R6 [ca. 0922/7954], which continues N to Colón [0922/7954 (USBGN)] (E8660). See also: Gatún.

ROAD R12; see Galeta Road

ROAD R20; see Boyd-Roosevelt, Carretera

ROAD S1; Colón (Canal Zone) (W)

ca. 0918/8001

Road that extends 12 mi (19 km) SE from Piña Beach [ca. 0918/8001] terminating at its junction with Road S10 [ca. 0909/7958] 3 km SW of Esclusas Gatún (Gatun Locks) [0916/7955 (USBGN)]. Road S1 is generally within 1 mi (1.5 km) of the W bank of the Río Chagres [0919/8000 (USBGN)] (E8660; RM2). See also: Gatún, Esclusas.

ROAD S2; Colón (Canal Zone) (W)

ca. 0920/7958

Road that extends ca. 8 mi (13 km) N-S along the W coast of Bahía Limón [0920/7956 (USBGN)] between Gatún [0916/7955 (USBGN)] and Fort Sherman Army Reservation [0920/7958 (USBGN)]. At its S terminus the road crosses the Panama Canal [0920/7955 (USBGN)] at Esclusas Gatún (Gatun Locks) [0916/7955 (USBGN)] and terminates at its junction with Road R2 [ca. 0917/7955] just N of Gatún (RM1; RMM1; E8660). See also: Fort Sherman Army Reservation; and Gatún, Esclusas.

ROAD S8; Colón (Canal Zone) (W)

ca. 0912/8008

Road that extends ca. 17 mi (27 km) SW along the Caribbean from Fort Sherman Army Reservation [0920/7958 (USBGN)] (RM1). See also: Fort Sherman Army Reservation.

ROAD S9; Colón (Canal Zone) (W)

ca. 0914/7958

Road that extends W for ca. 1.5 km from the Gatún-Escobal Road [ca. 0916/7955] to the upper reaches of the Río Medio [0916/8003 (USBGN)]. The E terminus of the road is located on the NW coast of Lago Gatún (Gatun Lake) [0912/7955 (USBGN)] ca. 5.5 km SW of Gatún [0916/7955 (USBGN)] near La Punta de Los

Chivos [0914/7956 (USBGN)] (E8660). Cardiff and Lovette, GL = Gatún-Escobal Road, at 100 m, W off the Gatún-Escobal Road, ca. 6 km SW of Gatún, 1997 Mar 16 (LSUMZ).

ROAD S10; see Gatún-Escobal Road

ROAD S11; see Achiote Road

RÓBALO; see Punta Róbalo

RÓBALO, CERRO; Bocas del Toro
0853/8220 (USBGN)
1,740 m, peak N of the continental divide. On RDPF this peak is located NE of Boquete [0847/8226 (USBGN)] and Cerro Horqueta [0850/8226 (USBGN)] near the headwaters of the Río Róbalo [0901/8215 (USBGN)] consistent with the coordinates given by USBGN. However, on older maps the peak is at 2,175 m and located ca. 0910/8235 directly N of Volcán Barú [0848/8233 (USBGN)] and much closer to Cerro Fabrega [0907/8252 (USBGN)] and the continental divide (MHA; MRP). See also: Changuena, Río, and Central, Cordillera.

RÓBALO, PUNTA; see Punta Róbalo

RÓBALO, RÍO; Bocas del Toro
0901/8215 (USBGN)
Short river that originates at the merger of several streams ca. 5 km W of Punta Róbalo [0902/8215 (USBGN)]. The river flows E entering the W coast of Laguna de Chiriquí [0903/8200 (USBGN)] 1 km S of Punta Róbalo (E762). Presumably named for the fish of the same name, which is properly accented on the first syllable, although local residents accent the second syllable in speaking of the river. SBE, GL = Punta Róbalo, 0-2 m, on the river 1-3 km W of Punta Róbalo, 1991 Mar 10 (USNM); Braun, M., Brumfield, and Glenn, GL = Palma Real (q. v.), 1994 Jan 15-16, 18 (USNM; Brumfield et al., 2001).

ROBLE, EL; see El Roble

ROBLES; see El Roble

ROCA TIGRE; see Tiger, Cayos

RODMAN; see Rodman United States Naval Reservation

RODMAN, FORT; see Rodman United States Naval Reservation

RODMAN MARINE BASE; see Rodman United States Naval Reservation

RODMAN N. S.; see Rodman United States Naval Reservation

RODMAN NAVAL BASE; see Rodman United States Naval Reservation

RODMAN NAVAL STATION; see Coco Solo Naval Reservation and see Rodman United States Naval Reservation

RODMAN NAVY BASE; see Rodman United States Naval Reservation

RODMAN UNITED STATES NAVAL RESERVATION;
0857/7935 (USBGN)
Panamá (Canal Zone) (W, Pacific)

Sea level, former United States Government military base that was located on the W side of the Panama Canal [0920/7955 (USBGN)] 1.5 mi (2.5 km) W of Balboa [0857/7934 (USBGN)] and 4 km W of Panama City [0858/7932 (USBGN)] (E661, as "Rodman"; RM1, as "Rodman Naval Station"; RM2, as "Rodman N. S."). Numerous variations of the proper name occur in the literature and on specimen labels, and the name also applied to a base near the Atlantic entrance of the Canal [see: Coco Solo Naval Reservation]. Wetmore (from Tipton), 1961 Dec 23 (USNM, as "Rodman Navy Base"); Morton, 5 mi (8 km) NW, 1965 Nov 2, entrance, 1971 Mar 19 (AMNH, as "Rodman Navy Base"; USNM, as "Fort Rodman"); Renner, 1973 Feb 23 (GML, as "Rodman"); Barrett, 1973 Jul 10, Aug 9, 14, Oct 8, Nov 3, 1974 Feb 14, May 2 (GML, as "Rodman"); Dawson, 7 mi (11 km) W of Balboa, 1973 Aug 7, Sep 30, Nov 23 (GML, as "Rodman"); De Leon, 1974 Aug 22 (GML, as "Rodman"); Trainer, observations, 1981-1983 (Trainer, 1989, as "Rodman"); Fleischer and Smith, 1986 (Fleischer and Smith, 1992, as "Rodman Marine Base"); Pitman, GL = Panama City, just E of entrance, 1987 Oct 8, Nov 9 (LACM, as "Rodman Naval Base"); Christian, GL for Cerro San Juan, 1997 Jan 31 (LSUMZ, as "Rodman Naval Base"); Collector ?, W side of Panama Canal, N of Pan-American Hwy, 1994 Apr 26, 1995 Feb 8 (LSUMZ, as Rodman"). See also: Mangrove.

RODRÍGUEZ, PUNTA; Bocas del Toro, Archipiélago de Bocas del Toro
0914/8218 (USBGN)
Sea level, point on the central W coast of Isla San Cristóbal [0915/8216 (USBGN)] 13 km SE of Almirante [0918/8224 (USBGN)] (E762, as "Punta Rodríguez (Jack Point)"; JOG, as "Jack Point"). Wetmore, GL = Almirante [0918/8224 (USBGN)], Isla San Cristóbal, landed on a swampy coast S of Punta Rodríguez and entered a pasture with short grass that extended over a steep-sided hill and into a little valley, scattered guava trees ranged over the slopes with a few taller trees on the crest, 1958 Feb 23 (USNM, as "Rodríguez Point").

RODRÍGUEZ POINT; see Rodríguez, Punta

ROLDÁN, CAYO; Bocas del Toro, Archipiélago de Bocas del Toro
0913/8219 (USBGN)
43 m, small island with a land area of ca. 0.2 km^2 (20 ha) located 300 m off the SE tip of Cayo García [0914/8220 (USBGN)] in the waters between Ensenada Grande [0914/8219 (USBGN)] and Ensenada Shepherd [0913/8221 (USBGN)] (E762, as "Cayo Roldán (Pelican Point)"). According to Wetmore, in 1958 the island rose steeply on the E and more gradually on the W to 75-100 feet (25-30 m) and the center had limited clearing with only one house noted on the shore, Wetmore, GL = Almirante, collected in the S in mangroves, some forest, a scattered stand of cacao and some dense old *rastrojo*, 1958 Feb 11 (USNM, as "Roldán Key"); SBE, GL = Isla Pastores [0914/8220 (USBGN)], forest with dense understory of brush and vines, SE of Isla Pastores on the N end of the island, 1989 Mar 1 (USNM; Olson, 1993a).

ROLDÁN KEY; see Roldán, Cayo

ROMPIO, EL; see El Rompio

ROSARIO; see El Rosario

ROSARIO, EL; see El Rosario

ROSARIO, LAGUNA DEL; see El Rosario

ROVINA; see Palmira

ROVIRA ARRIBA; see Palmira

RUBBER CAMP, see Mariato River Camp

RUBBER TREE; Bocas del Toro

ca. 0923/8229

Sea level, a stop on the railroad connecting Almirante [0918/8224 (USBGN)] and Changuinola [0926/8231 (USBGN)] midway between the stops labeled Milla 9 and Milla 11 (see Almirante). The stop is located 8 km SE of Changuinola and 14 km NW of Almirante between swamps that extend E to the coast and low hills that rise above 200 m just W of the railway (E762). Kennard et al., plantation 9 mi (14 km) from Almirante, tall trees with ripe fruit, 1926 Feb 19 (MCZ; Kennard and Peters, 1928:443-444, 446, 458).

S. G.; see Simón Gómez, Río

SABAGO ISLAND, see Saboga, Isla

SÁBALO; see Boca de Sábalo

SÁBALO, RÍO; Darién

0802/7812 (USBGN)

Originates on the NE slopes of Serranía del Sapo [0750/7817 (USBGN)] and flows NE entering the Río Sambú [0805/7818 (USBGN)] 3 km SE of Boca de Sábalo [0802/7813 (USBGN)] (MGRP; SMN2B). See also: Boca de Sábalo.

SÁBALO RIVER; see Sábalo, Río

SABANA, PUNTA; Darién

0825/7806 (USBGN)

Sea level, point in Golfo de San Miguel [0822/7817 (USBGN)] 4 km E of La Palma [0825/7809 (USBGN)] on the S end of the peninsula that forms the W bank of the mouth of the Río Sabanas [0832/7808 (USBGN)] (MGRP, as "Punta Sabanas"). Festa, used as the base for his collecting in Golfo de San Miguel [0822/7817 (USBGN)] (q. v.), from here he made excursions throughout the N waters of the bay, 1895 Jun-Sep (Salvadori and Festa, 1899, as "Punta de Sabana"; Stone, 1918, as "Punta de Sabana"; Griscom, 1933:304).

SABANA, PUNTA DE; see Sabana, Punta

SABANA, RÍO; see Sabanas, Río

SABANAS, PUNTA; see Sabana, Punta

SABANAS, RÍO; Darién

0832/7808 (USBGN)

Originates in NW Darién province near the Panamá/Darién provincial border and flows S just W of the Carretera Interamericana [0832/8242 (USBGN)], which extends roughly N-S through Darién province. The lower 15-20 km of the river form a 1-3 km wide estuary meeting the mouth of the Río Tuira [0821/7803 (USBGN)] at Punta Sabana [0825/7806 (USBGN)] (MGRP, as "Río Sabana," and as "Río Sabanas"; SMN2B, as "Río Sabania"). Festa, based at Punta Sabana throughout much of his trip, on Aug 27 he stayed at Islas Bellas [0832/7808 (USBGN)], islands in the estuary, and from here he continued to the Río Lara [0837/7809 (USBGN)] on Aug 28, 1895 Jul-Aug (Salvadori and Festa, 1899, as "Río Sabana"; Festa, 1909, as "Río Sabana"). Delgado (1983, as "Río Sabana").

SABANETA; Coclé (Pacific)

ca. 0838/8037

ca. 800 m, described by Wetmore as an area NW of El Copé [0837/8036 (USBGN)], small creeks, coffee interspersed with forest, old fields, orange trees, Wetmore, GL = El Copé, 1962 Feb 22-24 (USNM). Not shown on our maps.

SABANETA, ISLA; Chiriquí

0817/8223 (USBGN)

Sea level, according to USBGN, Isla Sabaneta is the central portion of an island subdivided into Isla Bóquita [0817/8222 (USBGN)], Isla Sabaneta and Isla Sabino [0816/8224 (USBGN)], although the narrow connections between these islets are omitted on many maps. USBGN treats both "Isla Almijas" and "Isla Almejas" as older names for Isla Sabaneta, but according to MHA and MRP, these older names were applied to the portion that now includes both Isla Sabaneta and Isla Sabino. Isla Sabaneta is located in Bahía de Charco Azul [0812/8245 (USBGN)] at the mouth of the Río Chiriquí [0820/8226 (USBGN)] 18 km S of David [0826/8226 (USBGN)] just N of Isla Sevilla [0814/8224 (USBGN)]. According to JOG, all three island sections are largely mangrove swamp, (JOG, as "Isla Sabino"; MGRP, as "Isla Sabino"; MHA, as "I. Almijas"; MRP, as "I. Almejas"). Batty, specimens at AMNH and at BMNH are dated 1902 Feb 2, 6, but Batty's data were completely fabricated and his specimens labeled "Almijas Island" were actually collected in 1901 near his mainland bases at Boquete [0847/8226 (USBGN)] and Boquerón [0830/8234 (USBGN)] (AMNH, as "Almijas Island"; BMNH, as "Almejas Island"; Hartert and Goodson, 1917:418, as "Almijas Island"; Zimmer, 1950:3, as "Almijas Is."; Olson, 2008).

SABANIA, RÍO; see Sabanas, Río

SABOGA, ISLA; Panamá, Archipiélago de las Perlas

0837/7904 (USBGN)

68 m (Angehr, 2003:269), island with a land area of 2.88 km^2 (288 ha) (Wright et al., 1985) located 20 km NNW of Isla del Rey [0822/7855 (USBGN)] (JOG). The island, which is mostly forested, is an Important Bird Area (Angehr, 2003). Angehr (in litt., Aug 2005) indicates that Brown's collections from Isla Saboga my have actually come from Isla Pacheca [0840/7903 (USBGN)]. "Thayer and Bangs (1905) reported that 'Saboga is the bird rock of the group, and here cormorants, boobies, man-o'-war birds, and terns breed in great numbers,' and that Brown collected specimens of Brown Booby, Neotropic Cormorant, and Magnificent Frigatebird there. Although Brown visited Pacheca, he apparently neither reported nor collected

any seabirds there. Bovallius, on the other hand, who collected on both Saboga and Pacheca in 1882, found Brown Booby, Neotropic Cormorant, and Brown Pelican to be abundant on Pacheca, which he described as a 'rock-island,' but made no mention of their presence on Saboga, which he described as forested (Rendahl 1920). All subsequent observers who visited either island at the right time of year have reported seabirds breeding abundantly on Pacheca, but none have ever reported any nesting on Saboga. On the other hand, Brown collected at least 21 resident terrestrial species on his 'Saboga,' and only 6 on 'Pacheca,' which would tend to support Saboga being the larger island (which it is). In any case, a word of caution is in order with regard to Brown's records from these two islands." Bovallius, tropical forests, cultivated areas, see discussion of Bovallius under Pueblo Nuevo [ca. 0820/8013 ?], 1882 Mar 29-31 (NHRM; Rendahl, 1920); Brown, 1904 Apr 1-13 (CM; MCZ, as "Saboga I."; USNM, as "Sabago Island," and as "Saboga Island"; Thayer and Bangs, 1905; Todd, 1913, as "Sabogo Island"; Bangs and Penard, 1919, as "Saboga Island"); Griscom and Crosby, 1927 Feb-Mar (Griscom, 1927b, as "Saboga"); Murphy and Correia, Askoy Expedition, 1941 Feb 12, May 25 (AMNH, as "Saboga Island"; Murphy, 1944a; Murphy, 1944b, as "Saboga Island"; Murphy, 1945, as "Saboga Island"); Wetmore, beach on the SE coast, inland along a steep, forested, 60-70 foot (20 m) bank to slopes covered in gallery forest, farther inland to *rastrojo* near the island's center, clearings and cultivated areas, 1960 Jan 19 (USNM; Olson, 1980b); (MacArthur et al., 1972—species list; see comments in Bibliography); Wright and Faaborg, mist nets, observations, 1976 Jun-Aug (Wright et al., 1985); Levings et al., observations, 1982-1983 (Levings et al., 1986); Brawn et al., mist nets, 199? (Brawn et al., 1996); Agro, 2 mi (3 km) off island, 1996 Feb 28 (ANSP); Guitton-Mayerma and Miller, 2006 Apr 26 (STRI; UAM). (Wicker, 1917; Ridgway, 1911b:347 is a typographical error for Isla Taboga; Wetmore, 1952:19, 31; Angehr, 2003:269-271). See also: Taboga, Isla.

SAGUI, CERRO; see Flores, Cerro

SAINO MADROÑIO; see Madroño

SAINO MADROÑO; see Madroño

SAINT MIGUEL, BAY OF; see San Miguel, Golfo de

SAJALICES, RÍO; Panamá (W, Pacific)

0838/7950 (USBGN)

Originates S of Cerro Campana [0841/7955 (USBGN)], flows ENE and crosses the Carretera Interamericana [0832/8242 (USBGN)] 4.5 km S of Campana [0843/7954 (USBGN)], then it turns SE widening greatly as it enters mangroves and mudflats at the SE end of Bahía de Chame (E762). In 1951 Wetmore collected SE of Campana in a river that crossed the Carretera Interamericana. He labeled his specimens Río Camarón [0842/7950 (USBGN)] (q. v.) citing it as the local name, but based on available maps he believed that the river was actually the Río Sajalices (USNM). The Chame quadrangle of E761B, printed in 1945, does not show the Río Camarón crossing the Carretera Interamericana (E761B, as "Río Camarón Abajo"), but the newer series, E762, printed in 1964, shows the Río Camarón as an intermittent stream that crosses the highway 3 km S of Campana. Therefore from his base at Campana, Wetmore would have come to the Río Camarón before the Río Sajalices, and as this name was also given to him by local residents, it seems likely that he actually collected on the Río Camarón and not on the Río Sajalices. Furthermore, he did not change his specimen labels or his field catalog entries from Río Camarón to Río Sajalices.

SAL, QUEBRADA; see Sal Creek

SALADO; see El Salado

SALADO, RÍO ESTERO; see Estero Salado, Río

SALAMANCA; Colón (Canal Zone) (E)
0920/7935 (USBGN)
100 m, town 26 km NNE of Gamboa [0907/7942 (USBGN)] on the NW coast of Lago Alajuela (Madden Lake) [0915/7935 (USBGN)] at the mouth of the Río Salamanca [0920/7935 (USBGN)] (JOG; MGRP). Enders, GL = Río Pequení, Río Chagres, 1935 Jan 7-10, 14 (ANSP). See also: Salamanca, Río; and Salamanca Hydrographic Station.

SALAMANCA, RÍO; Colón (Canal Zone) (E)
0920/7935 (USBGN)
Originates W of the Río Boquerón [0921/7933 (USBGN)] and flows S entering the NW coast of Lago Alajuela (Madden Lake) [0915/7935 (USBGN)] at Salamanca [0920/7935 (USBGN)] (JOG; MGRP). See also: Manganese Mine.

SALAMANCA HYDROGRAPHIC STATION; Colón (Canal Zone) (E)
ca. 0919/7936
100 m, water measurement station located on the NW coast of Lago Alajuela (Madden Lake) [0915/7935 (USBGN)] 24 km NNE of Gamboa [0907/7942 (USBGN)] and 10 km SW of the mouth of the Río Pequení [0921/7933 (USBGN)], and ca. 2 km S of the mouth of the Río Salamanca [0920/7935 (USBGN)] (E661, as "Salamanca Hydrographic Sta."). Griswold, GL = Río Pequení, 1936 Feb 3, 5, Mar 24-25, 27, Apr (MCZ, as "Salamanca H. S."; Griswold, 1936a, as "Río Pequeni, Salamanca Hydrographic Station").

SAL CREEK; Bocas del Toro, Archipiélago de Bocas del Toro
0918/8206 (USBGN)
Creek on Isla Bastimentos [0918/8208 (USBGN)] that originates NW of Old Point [0917/8205 (USBGN)] in hills SE of Long Beach [0920/8208 (USBGN)] and flows S entering the Caribbean Sea [1500/7500 (USBGN)] 2.5 km W of Old Point (E762). According to Handley's journal, the creek was one of several salt water creeks on the S coast of the island, SBE, GL = Isla Bastimentos, 1987 Mar 17, 22, 24 (USNM, as "Salt Creek"). See also: Long Beach.

SALINAS; Veraguas, Isla de Coiba
ca. 0727/8143
Described by Wetmore as a beach adjacent to the mouth of the Río San Juan [0727/8143 (USBGN)] on the E coast of Isla de Coiba [0728/8147 (USBGN)], Wetmore, GL = Isla de Coiba, alternating sand and rock on the beach, trail leading inland to steep hills behind the beach, to the mangrove swamp bordering

Campamento de San Juan [0728/8144 (USBGN)], walked 1 mi (1.5 km) to a hot springs with a hint of sulfur, old *rastrojo*, 1956 Jan 23, 28 (USNM; Wetmore, 1957). Not shown on our maps.

SALINAS; see Aguadulce

SALINAS AGUADULCE; see Aguadulce

SALINAS DE AGUADULCE; see Aguadulce

SALINAS DE ERNESTA; see La Villa, Río

SALINAS VILLALAZ; see La Villa, Río

SALINAS VILLAS; Los Santos
ca. 0757/8022
Not named on our maps, but the name may refer to the salt ponds shown on E761A and JOG in N Los Santos province 7 km ESE of Chitré [0758/8026 (USBGN)], 5 km ENE of Los Santos [0756/8025 (USBGN)], 3 km inland from the coast and a few km E and S of the lower reaches of the Río La Villa [0759/8023 (USBGN)]. According to E761A, the salt ponds are surrounded by mangrove swamps and a drainage ditch extends through the region N of the ponds. Tejera, 1977 Dec 30 (MVUP; Wetmore et al., 1984:634). For a discussion of *salinas* see Aguadulce.

SALTA; see El Salto

SALT CREEK; see Sal Creek

SALTO, EL; see El Salto

SALTO DE MADROÑO; see Madroño

SALUD; Colón (W)
0912/8008 (USBGN)
Sea level, town on the W coast of Colón province at the mouth of the Río Salud [0912/8008 (USBGN)] 30 km SW of Colón [0922/7954 (USBGN)] (JOG). Wiese and Pugh, W of, 1966 Jul 14 (UF; USNM).

SALUD, PUNTA; Panamá (Canal Zone) (W, Atlantic)
0910/7950 (USBGN)
Sea level, point on the NE coast of Isla Barro Colorado [0909/7951 (USBGN)] opposite Punta Buena Vista [0911/7950 (USBGN)] (E661). Skutch, an estuary W of the point, 1935 Apr-May (Eisenmann, 1952:12, 18, 53, as "Salud Point"); Eisenmann, "north shore at coves between Fuertes Estero and," 1950 (Eisenmann, 1952:23, as "Salud Point").

SALUD, RÍO; Colón (W)
0912/8008 (USBGN)
Originates in the low hills 6 km W of Lago Gatún (Gatun Lake) [0912/7955 (USBGN)] and flows N entering the Caribbean Sea [1500/7500 (USBGN)] at Salud [0912/8008 (USBGN)] (JOG). Miller, 1965 Feb 22 (MVZ).

SALUD POINT; see Salud, Punta

SAMANÍ, LAGUNA DE; see Cañaveral, Río

SAMBÚ, RÍO; Darién

0805/7818 (USBGN)

Originates NE of Jaqué [0731/7810 (USBGN)] in the S section of Sierra de Jungurudó [0735/7754 (USBGN)] and flows NW draining the E side of the Serranía del Sapo [0750/7817 (USBGN)]. The river enters Ensenada de Garachiné [0808/7820 (USBGN)] 7 km E of Garachiné [0804/7822 (USBGN)] (MGRP). According to Barbour and Brooks (1923, as "Sambú River"), "the lower reaches of the Sambú are flanked by high mangrove forests which gradually give way to a varied growth of hardwoods as the banks rise slightly higher. As on the other rivers draining the Pacific slope of eastern Panama, the great rise and fall of the tide is felt many miles inland…we steamed up the river on the swiftly rising tide to the head of navigation at the junction of the [Río] Sabalo with the [Río] Sambú." The expedition entered the Serranía del Sapo from Boca de Sábalo [0802/7813 (USBGN)], collected near Cerro Sapo [0758/7822 (USBGN)], and returned to the Río Sambú when they descended the Río Jesús [0804/7814 (USBGN)], but apparently neither Río Sambú nor Río Jesús appears on their specimen labels. Vanderbilt Expedition, Feathers, according to Fowler (1944:184), the river was "tidal at least as far up as [Boca de] Sabalo and the water extremely muddy," and according to Vanderbilt's account (Bond and Meyer de Schauensee, 1944:23), "'we made one excursion up the Sambú River, reaching a point just below the town of [Boca de] Sabalo, where we had excellent collecting in a deserted banana plantation,'" 1941 Apr 20-21, 26, 28 (ANSP, as "Sambu," and as "Sambu R."; USNM, as "Sambu River"; Bond and Meyer de Schauensee, 1944; Wetmore, 1951a; Olson, 1981d); Griscom Crosby Expedition, ascended the river to the "upper limit of the tide," collected in "great swamps at the mouth of," dark, heavy forest, and near a settlement in flowering trees near a clearing, and holes in the river banks, 1927 Feb 24-25, 27 (AMNH; Griscom, 1927b, as "Sambú River"); Christian, observations in the Río Sambú drainage "at a lowland site near the river (c. 07°48' N, 78°03' W)," the coordinates indicate that he worked ca. 45 km SE of Garachiné and ca. 35 km NNE of Jaqué [0731/7810 (USBGN)]. Angehr (in litt., Aug 2005) indicates that these observations were made on the trip described in Angehr et al. (2004), at "an Embera village called Boca de Pavarando, shown on the Special Map of the Darien, for which I have GPS coordinates of 7°47'31" N, 78°02'13", slightly different from those given in his article," 1997 Aug (Christian, 2001). See also: Antaral, Río; and Jungurudó, Sierra de.

SAMBÚ RIVER; see Sambú, Río

SAMGANDI, RÍO; see Cangandí, Río

SAMGONDI, RÍO; see Cangandí, Río

SAN ANTONIO; Panamá (E, Pacific)

0901/7929 (USBGN)

80 feet (25 m), originally a settlement 4 km SW of Pedregal [0904/7926 (USBGN)] and 2 km S of Cerro Viento [0905/7928 (USBGN)] (E761A) that is now considered by USBGN as a neighborhood on the E edge of greater Panama City [0858/7932 (USBGN)]. Acosta, 1985 Mar 15 (MVUP, as "Barriada San Antonio").

SAN ANTONIO; Panamá (E, Pacific)
0909/7903 (USBGN)
20 feet (6 m), town 6 km SE of Chepo [0910/7906 (USBGN)] on the N bank of the Río Chepo [0900/7906 (USBGN)] 2 km above the mouth of the Río Mamoní [0908/7904 (USBGN)] in an area where several small rivers enter the Río Chepo. San Antonio is connected to Chepo by a trail that crosses the Río Mamoní and extends through an area of low hills ranging from 50-200 feet (15-60 m) (E761A; MHA). Smith, A. P., used several variations of the name San Antonio on his specimen labels and it is unclear whether or not these represent distinct localities within the region. According to Griscom (1933), Smith collected "in the vicinity of Port San Antonio, Río Chepo," 1927 late Feb-early Apr (MCZ, as "Port Antonio," and as "San Antonio de Chepo"; YPM, as "Port Antonio," as "San Antonio," and as "San Antonio de Chepo"; Griscom, 1933, as "Port San Antonio"; Zimmer, 1940, as "Port Antonio, Río Chepo"; Howell, 1956, as "Port Antonio"); Wetmore and Perrygo, GL = Chepo, heavy forest, 1949 Apr 13, 24 (USNM; Wetmore, 1963a).

SAN ANTONIO; see San Antonio Swamp

SAN ANTONIO, BARRIADA; see San Antonio

SAN ANTONIO, RÍO; Darién
0804/7821 (USBGN)
Originates on the N slopes of Cerro Sapo [0758/7822 (USBGN)] and flows N entering the sea at Garachiné [0804/7822 (USBGN)] (MHA, as "Río Antonio"; Barbour and Brooks, 1923 (map)). SMN2A shows the river originating on the SE slopes of the peak. Both the Barbour, Brooks, Underwood Expedition and the Vanderbilt Expedition ascended the Río San Antonio to reach Cerro Sapo (q. v.) (Bangs and Barbour, 1922:193; Bond and Meyer de Schauensee, 1944).

SAN ANTONIO DE CHEPO; see San Antonio

SAN ANTONIO SWAMP; Panamá (E, Pacific)
ca. 0905/7912
Probably refers to the region near the town San Antonio [0905/7912 (USBGN)] and perhaps even the area in or near La Jagua Hunting Club [ca. 0903/7911]. The town San Antonio is located 8 km E of Pacora [0905/7917 (USBGN)] on high ground (80 feet (25 m)) ca. 3 km NW of La Jagua Hunting Club. The town is 5 km S of the Carretera Interamericana [0832/8242 (USBGN)], surrounded by small tributaries of both the Río Chico [0901/7911 (USBGN)] and the Río Chepo [0900/7906 (USBGN)], and the region S, SE and NE of town is covered with mangrove swamps (E761A, as "San Antonio"). GMLP, GL = Pacora, 1958 Sep 15 (USNM).

SAN BARTOLO, RÍO; Chiriquí
0817/8251 (USBGN)
Originates in the hills W of Puerto Armuelles [0817/8252 (USBGN)] near the Costa Rican border, flows E then follows a semicircular path N of Puerto Armuelles entering Bahía de Charco Azul [0812/8245 (USBGN)] 2 km NE of Puerto Armuelles (JOG). Wetmore, GL = Puerto Armuelles, worked ca. 12 km W of Puerto Armuelles near the Costa Rican border in an area that was separated from the headwaters of

the Río Corutú [0816/8254 (USBGN)] by a low divide, from here traveled ca. 1 mi (1.5 km) upriver and along its tributaries in a valley covered with tall bunch grass and clumps of shrubs and scattered trees, tall trees along the stream, gravel bars, forested areas, weed-grown pasture, 1966 Feb 18, Mar 1, 10 (USNM).

SAN BLAS, COMARCA DE; San Blas
0915/7815 (USBGN)
Legally formed in 1938, Comarca de San Blas is a partially self-governing state within Panama regulated by the Kuna Indians. According to RDP2, in Law 99 of 23 Dec 1998 the name was changed to Comarca Kuna Yala. The Comarca is entirely within the Atlantic drainage and extends over a roughly 10-20 km wide strip of land along the easternmost Caribbean coast. Numerous islands are located near the mainland and the Comarca is bordered on the W by Colón province and on the E by Colombia. Its S border with Panamá and Darién provinces is formed by the continental divide, which is generally between 200 and 500 m with a few regions above 700 m. For the most part, ornithological work in San Blas has been limited to either the extreme W or E coasts. The earliest collecting was conducted in the late 1920s-mid 1930s by Wedel, who worked in the E near Puerto Obaldía [0840/7725 (USBGN)] and farther W at Permé [ca. 0845/7734] and Ranchón [ca. 0840/7725]. This region was later visited by Wetmore and Handley in the 1960s. Wetmore, Galindo of GML, and a few others have worked in the W along the coast of Golfo de San Blas [0930/7900 (USBGN)]. Beginning in the 1980s, the El Llano-Cartí Road [ca. 0928/7858] allowed access to the interior of W San Blas at Nusagandí [ca. 0919/7859], although the middle regions of the comarca remain unexplored ornithologically.

SAN BLAS, CORDILLERA DE; Panamá/San Blas
0918/7900 (USBGN)
Mountain range that creates the continental divide along the Comarca de San Blas/ Panamá provincial border (RDPF, as "Serranía de San Blas"). Agro, Cardiff, and Lovette, GL = Chepo, W end, 21 mi (34 km) by road NE of Chepo, 1996 Feb 29 (LSUMZ, as "Serrania de San Blas"). Mentioned by Selfridge (1874:7-8, as "San Blas Mountains") in his description of the Isthmus of Darién [ca. 0810/7745].

SAN BLAS, GOLFO DE; San Blas
0930/7900 (USBGN)
Bay on the extreme W coast of Comarca de San Blas formed by Península de San Blas [0933/7903 (USBGN)] on the N and by the mainland on the W and S. The N and the W shores are largely mangrove swamps from which the bay extends 25 km to the E opening to the Caribbean Sea [1500/7500 (USBGN)] (JOG). See also: Mandinga; and Mandinga, Río.

SAN BLAS, PENÍNSULA DE; San Blas
0933/7903 (USBGN)
The northernmost land in Comarca de San Blas, 4 km wide and extending 15 km E from the mainland (E501; MHA). See also: El Porvenir.

SAN BLAS, SERRANÍA DE; see San Blas, Cordillera de

SAN BLAS MOUNTAINS; see San Blas, Cordillera de

SAN CARLOS; Panamá (W, Pacific)
0829/7957 (USBGN)
Sea level, coastal town in W Panamá province at the mouth of the Río Mata Ahogado [0828/7957 (USBGN)], on the Carretera Interamericana [0832/8242 (USBGN)] 14 km SSW of Chame [0835/7953 (USBGN)] (JOG). Wetmore, 1953 June 22, and as a GL for Río Mar, 1953 June 21, along the Carretera Interamericana [0832/8242 (USBGN)] W of San Carlos, 1958 Mar 15, and as a GL for Riomar, 1958 Mar 16-17 (USNM); Eisenmann and Wetmore, observations over the highway near San Carlos between Playa Coronado [0831/7953 (USBGN)] and Río Mar [0827/7958 (USBGN)], 1953 Jun 20-23 (Eisenmann, 1962a); Collector ?, 1999 Mar 16 (LSUMZ). (Salvin and Godman 1897-1904, vol. 3:295).

SAN CARLOS, RÍO; see Pueblo Nuevo

SAN CRISTÓBAL, ISLA;
0915/8216 (USBGN)
Bocas del Toro, Archipiélago de Bocas del Toro
117 m, island with a land area of 37 km² (3,700 ha) (RDP2, as "Isla Cristóbal") located in the S waters of Bahía Almirante [0920/8218 (USBGN)] 12 km SE of Almirante [0918/8224 (USBGN)] just N of Laguna de Tierra Oscura [0912/8216 (USBGN)] where sections of its S coast are less than 1 km from the mainland (E762, as "Isla San Cristóbal (Coco Key)"; JOG, as "Isla Cristóbal"; MHA, as "Isla S. Cristóbal"). Many names have been used for the island including Coco Key, Coco Cay, Cayo Coco, Isla Cristóbal, and Isla San Cristóbal. The island is an Important Bird Area (Angehr 2003:97-98, as "Isla Cristóbal"). Described by Wetmore in 1958 as an island of considerable extent with limited cultivation, Wetmore, GL = Almirante, at Punta Rodríguez (q. v.), 1958 Feb 23 (USNM, as "Coco Cay"); SBE, at sea level on the NW corner, 1989 Feb 16, and as a GL for: Bocatorito, and Cerro de Bocatorito, 1989 Jan 30-31, Feb 3-16, 24, Mar 3, 15 (USNM, as "Isla Cristóbal"; Olson, 1993a, as "Isla Cristóbal"). See also: Almirante, Bahía.

SANDFLY BAY; Bocas del Toro, Archipiélago de Bocas del Toro
ca. 0921/8215
Bay on the S coast of Isla Colón [0924/8217 (USBGN)] on the E coast of the peninsula between Bocas del Toro [0920/8215 (USBGN)] and Big Creek [0922/8215 (USBGN)], SBE, GL = Isla Colón, Bocas del Toro, ca. 2 mi (3 km) N of Bocas del Toro, 1990 Apr 15 (USNM). Not shown on our maps.

SAN FELIPE; Panamá (E, Pacific)
0857/7932 (USBGN)
Sea level, section of greater Panama City [0858/7932 (USBGN)] located on the coast 1.5 km SE of Ancón [0858/7933 (USBGN)] (E762). The old district of San Felipe sits on a small peninsula that is commonly known as Casco Viejo. The *mercado público* (public market—no longer existing) is located on the N coast of the peninsula (E962). Rodríguez, *mercado público*, 1984 Jun 8 (MVUP); Ríos, *mercado público*, 1984 Jun 18 (MVUP); Vargas, *mercado público*, 1984 Jul 2 (MVUP); Quezada,

GL for Las Bóvedas, 1986 Mar 19 (MVUP); Rodríguez, GL for Bahía de Panamá, 1991 Aug 1 (MVUP); Arosemena, at Punta Calafate, shown as the S tip of the peninsula in Atlas (1988:208), 1992 Aug 16 (MVUP); Guerra, 1993 Jul 17 (MVUP).

SAN FÉLIX; Chiriquí

0817/8152 (USBGN)

100 m, city 23 km WNW of Tolé [0814/8141 (USBGN)], 6 km N of Las Lajas [0815/8152 (USBGN)] and 1 km W of the Río San Félix [0810/8151 (USBGN)] (JOG). Davidson, at 100 feet (30 m), near, 1931 Nov 28-30, Dec 1-10 (CAS; CM; Davidson, 1932; Davidson, 1938); Wetmore, the secondary localities attributed to San Félix were located within 6 km of town, GL for: El Macano, Quebrada Laja, Río Dupí, and Río San Félix, 1956 Feb 13, 16-19, 21, 23, 27 (USNM; Parkes, 1976; Olson, 1983); Pine and Izor, collected a number of bird and mammal specimens all of which were labeled 23-25 km NNE of San Félix and several of which contained the additional labels of "Escopeta Camp," "Colorado Camp," and "Cerro Bollo." Pine and Izor indicate that they worked in both Chiriquí and Bocas del Toro provinces, but the distances and directions are difficult to reconcile with our available maps. Pine's journal entry on May 24 states that they worked along the continental divide, but he then writes that "the mining company's map shows the area where we've been working as largely, if not completely, on the Pacific slope. This is in error." Whether or not the mining company's map was in error is unclear. If specimens were collected in Bocas del Toro province (Atlantic slope), they were taken from the headwaters of the Río Cricamola [0859/8154 (USBGN)]. Some specimen labels lack a secondary locality, therefore the elevation and date of collection were used to determine the most appropriate secondary locality, GL for: Cerro Bollo, Colorado Camp, and Escopeta Camp, 1980 May 15-17, 21, 29, Jun 1, 8, 13, 16, 19, 22-27, 29, Jul 1, 3, 5-6 (USNM). (Wetmore, 1970).

SAN FÉLIX, RÍO; Chiriquí

0810/8151 (USBGN)

Originates on the continental divide in E Chiriquí province and flows S entering the Golfo de Chiriquí [0800/8220 (USBGN)] 10 km SSE of Las Lajas [0815/8152 (USBGN)] (JOG). Griscom et al., GL = Remedios [0814/8151 (USBGN)] (q. v.), below Remedios, 1924 Mar 26 (AMNH); Wetmore, GL = San Félix, collected N of El Macano [0819/8152 (USBGN)] (q. v.) where the clear and shallow water flowed over a bed of gravel and small boulders, down 200 feet (60 m) over broken slopes to the river, followed a trail for 1 mi (1.5 km) upstream to the W base of Cerro Iglesia [0819/8148 (USBGN)], forest cleared for pastures leaving only scattered trees and occasional tracts of brush along the river, on gravel beaches, along a bank grown with heavy brush and morning glories, continued farther upstream to steep slopes planted in grass, 1956 Feb 21, 23 (USNM).

SAN FRANCISCO; Veraguas (Pacific)

0815/8058 (USBGN)

ca. 50 m, town in central Veraguas province 16 km N of Santiago [0806/8059 (USBGN)] and ca. 2 km N of the Río Santa María [0806/8029 (USBGN)] (JOG). Described by Wetmore (1956:124) as being 25 km W of Calobre [0819/8051

(USBGN)], although according to JOG the town is ca. 17 km SW of Calobre. Benson, at 300 feet (90 m), 1931 Mar 1-7, 9-12, 16 (MCN; USNM; Wetmore, 1956a; Aldrich and Duvall, 1958); Child, 2 mi (3 km) S, 1965 Apr 18-23 (AMNH; Child, 1969).

SAN FRANCISCO; see San Francisco de la Caleta

SAN FRANCISCO DE LA CALETA; Panamá (E, Pacific)
0859/7930 (USBGN)
10 m, section of greater Panama City [0858/7932 (USBGN)] on the coast N of Punta Paitilla [0858/7931 (USBGN)] and 2 km SW of Old Panama [0900/7929 (USBGN)] (E962, as "San Francisco"). Imhof, observations, near Old Panama, 1942 Sep 2 (Imhof, 1950); Eisenmann, observations in the mud flats between Old Panama and, 1948-1950 late Jun-early Jul (Eisenmann, 1951); Bradvica and González, *bodega* (store), 1983 May 30 (MVUP); Cabal, and González, *cercano al Centro de Convenciones* (near the convention center) *ATLAPA*, 1999 Apr 3 (MVUP, as "San Francisco"). See also: Old Panama.

SAN JOAQUIN; see Pacora, Río

SAN JOSÉ; Panamá (E, Pacific)
0904/7916 (USBGN)
60 feet (20 m), town 2.5 km SE of Pacora [0905/7917 (USBGN)] on the Río Pacora [0901/7917 (USBGN)] and on the SW edge of Sabana de San José [ca. 0905/7915] (TM). Wetmore, GL = Pacora, savannas, 1958 Mar 19 (USNM). See also: La Jagua, Río; and San José, Sabana de.

SAN JOSÉ; Veraguas (Pacific)
0800/8120 (USBGN)
ca. 25 m, town 1 km W of Soná [0801/8119 (USBGN)] (JOG). Wetmore and Perrygo, GL = Soná, at Río Tribique (q. v.), 1953 June 12, 15 (USNM).

SAN JOSÉ; see Madroño

SAN JOSÉ; see San José, Isla

SAN JOSÉ, FINCA; see La Jagua, Río

SAN JOSÉ, HACIENDO; see San José, Sabana de

SAN JOSÉ, ISLA; Panamá, Archipiélago de las Perlas
0815/7907 (USBGN)
135 m, second largest island in the Archipiélago de las Perlas [0825/7900 (USBGN)] with a land area of 47.2 km^2 (4,720 ha) (Wright et al., 1985) located 14 km SW of Isla del Rey [0822/7855 (USBGN)]. The elevation is generally less than 50 m with the highest point at 135 m in the N central section of the island (JOG, as "Isla de San José"). Bovallius, "luxuriant vegetation," numerous watercourses, uninhabited, see discussion of Bovallius under Pueblo Nuevo [ca. 0820/8013 ?], 1882 Apr 26-27 (NHRM; Rendahl, 1920); Griscom and Crosby, 1927 Feb 16-17 (AMNH, as "San Jose Island"; Wetmore, 1946b, as "San Jose Island"); Fahnenstock Expedition, 1935 Mar 17-18 (AMNH, as "San Jose Island"); Miller, 1937 Mar 16 (USNM); Wetmore and Morrison, described the island as "pristine," he worked throughout

the island on Río Mata Puerco [0815/7907 (USBGN)] (q. v.), on Río Marina [ca. 0815/7906] (q. v.), in freshwater streams, in areas of red and white mangroves reaching ca. 80 feet (20 m), in well shaded forests with trees reaching ca. 100 feet (30 m), in areas of dense, tangled growth with creepers and vines and along steep and dry streams in low forest in the N, Wetmore departed in Mar, Morrison remained until Oct and visited other islands during this period, 1944 Feb 9-Oct 3 (USNM, as "San José Island"; Wetmore, 1946a; Wetmore, 1946b, as "San José Island"; Eisenmann, 1962c, as "Pearl Islands"); Murphy, 1945 Nov 8-10, 12-13, 15-19, 21, 23-26 (AMNH, as "San Jose Island"; Wetmore, 1946b, as "San Jose Island"); Northern, GL for Ensenada Playa Grande, 1957 Mar 18 (LACM, as "San Jose Is"); Robins, observations, 1957 Jul 15-26 (Robins, 1958); Pacific Project, GL = Gulf of Panama, S coast, and in the water near Punta de Cruz [0813/7908 (USBGN)] (q. v.), 1968 Nov 9, 1969 Jun 19, Aug 29 (GML, as "San Jose Island"; USNM, as "Punta Cruz"; Loftin, 1991); (MacArthur et al., 1972—species list; see comments in Bibliography).; Wright and Campbell, mist nets, observations, 1978 Jan-Mar (Wright et al., 1985); Merritt and Gale, E side of airstrip, N side of clearing along airstrip, 1982 Nov 1, 5-6, 12-13 (GML); Greenwell, 1986 Jun 8, 10, 29, Jul 7 (USNM); Braun and Brumfield, at 08°15'12"/79°06'18", 2004 Mar 23-25 (LSUMZ). (Zimmer, 1941a:7, as "San José"; Wetmore, 1945; Wetmore, 1952:19, 22, 25, 27, 31; Eisenmann, 1963a; Monroe and Howell, 1966; N. Smith, 1971; Olson, 1980b; Olson, 1981a; Angehr, 2003:273-274). See also: Bald Hill; Cruz, Punta de; East Harbor; and Navy Cove.

SAN JOSÉ, ROCA; Panamá (Canal Zone) (E, Pacific)
0854/7931 (USBGN)
20 m, very small island with a land area of less than 1 ha located in Bahía de Panamá [0850/7915 (USBGN)] 6 km SSE of Panama City [0858/7932 (USBGN)] and 1 km SE of Isla Flamenco [0854/7931 (USBGN)] (E761B, as "San Jose Rock"). Hallinan, near Isla Naos [0855/7932 (USBGN)], thorn, trees, cacti, vines, 1915 Mar 21 (AMNH; Hallinan, 1924, as "San Jose Rock"); Reed, observations, 1983 Jun 27 (Reed, 1988, as "San Jose Rock"). See also: Panamá, Bahía de.

SAN JOSÉ, SABANA DE; Panamá (E, Pacific)
ca. 0905/7915
40-60 feet (10-20 m), the grassland labeled Sabana de San José is 1 km N of San José [0904/7916 (USBGN)], 3 km ESE of Pacora [0905/7917 (USBGN)] and 11 km NW of Río Chico [0902/7910 (USBGN)], although the map indicates extensive grassland areas throughout the region E of Pacora and S of Chepo [0910/7906 (USBGN)] (E761A, as "Sabana San José"). Wetmore and Perrygo, GL = Río Chico, 1949 Mar 18, 20, GL = Pacora, 1949 Mar 19, 26, 27 (USNM); Wetmore, GL = Pacora, 1958 Mar 22 (USNM); Curtis, "the savanna region a few miles from Pacora," 1950 Sep 18 (USNM, as "San José"; Wetmore, 1951b, as "San José"); Strauch, GL = Pacora, 6 km E of Pacora, 1972 Aug 30 (LSUMZ, as "Haciendo San José").

SAN JOSÉ ISLAND; see San José, Isla

SAN JOSÉ MADROÑO; see Madroño

SAN JOSÉ ROCK; see San José, Roca

SAN JUAN; Colón (Canal Zone) (E)
ca. 0915/7935
According to Wetmore, San Juan is a former village that is now submerged in Lago Alajuela (Madden Lake) [0915/7935 (USBGN)]. The name is now used for Corregimiento San Juan [0915/7938 (USBGN)], which is located along the central section of Lago Alajuela (Madden Lake), and for the town New San Juan [0913/7940 (USBGN)] located 5 km WNW of Represa Madden (Madden Dam) [0913/7937 (USBGN)]. Tejera, GL for Lago Alajuela, 1978 May 6, Jun 3, Aug 19, Sep 21, Oct 31, Nov 30, Dec 30, 1979 Jan 11, Feb 11, 1980 Nov 30 (MVUP); Hernández, GL for Lago Alajuela, 1978 Aug 19, Dec 30 (MVUP); Pineda, GL for Lago Alajuela, 1978 Nov 7 (MVUP).

SAN JUAN; see San Juan, Campamento de

SAN JUAN, CAMPAMENTO DE; Veraguas, Isla de Coiba
0728/8144 (USBGN)
Sea level, prison camp on the central E coast of Isla de Coiba [0728/8147 (USBGN)] on the Río San Juan [0727/8143 (USBGN)] (q. v.) 2 km above its mouth (JOG). According to Wetmore, the camp marks the head of tide water on the Río San Juan, Wetmore, GL = Isla de Coiba, low expanse of pasture and low *rastrojo* leading back to steep-faced and forested hills, followed the Río San Juan inland, 1956 Jan 13, 21, 23, 26-28, 31 (USNM, as "San Juan"; Wetmore, 1957). See also: Salinas.

SAN JUAN, CERRO; Panamá (Canal Zone) (W, Pacific)
0857/7934 (USBGN)
120 m, hill 5 km WSW of Panama City [0858/7932 (USBGN)] just N of the Carretera Interamericana [0832/8242 (USBGN)] and 2 km NW of Cerro Farfan [0856/7934 (USBGN)] (E762, as "San Juan Hill"). Christian, GL = Rodman United States Naval Reservation, ca. 25 m, 1997 Jan 31 (LSUMZ).

SAN JUAN, RÍO; Veraguas, Isla de Coiba
0727/8143 (USBGN)
Originates in the N central hills of Isla de Coiba [0728/8147 (USBGN)], flows E then SE and enters the central shore of Bahía Damas [0727/8141 (USBGN)] where it widens in mangrove swamps that extend 2 km inland (E762; JOG). See also: San Juan, Campamento de; and Salinas.

SAN JUAN DE PEQUENÍ; Panamá (E, Atlantic)
ca. 0922/7931
Not shown on our maps, but according to Tejera (1995:10-11) San Juan de Pequení is a settlement on the Río Pequení [0921/7933 (USBGN)]. The coordinates indicate that the settlement is located NNE of Lago Alajuela (Madden Lake) [0915/7935 (USBGN)] on the Río Peqeuní ca. 3 km above its mouth in a region where the elevation is ca. 100 m. Martínez, 1992 May 10 (MVUP). See also: Pequení.

SAN JUAN HILL; see San Juan, Cerro

SAN LORENZO; Chiriquí
ca. 0818/8206
ca. 50 m, town 39 km ESE of David [0826/8226 (USBGN)] on the Carretera

Interamericana [0832/8242 (USBGN)] 10 km N of the coast (JOG). Although USBGN does not list the town San Lorenzo, it does include a bluff named Barranca de San Lorenzo [0813/8204 (USBGN)], a bay named Bahía de San Lorenzo [0812/8205 (USBGN)], and a beach named Playa de San Lorenzo [0813/8202 (USBGN)] in the surrounding area. Pitman, 1987 Nov 8 (LACM). (Angehr, 2003:311-312, as "Manglares de San Lorenzo y Las Lajas").

SAN LORENZO; see Fuerte San Lorenzo

SAN LORENZO; see San Lorenzo, Río

SAN LORENZO, FORT; see Fuerte San Lorenzo

SAN LORENZO, RÍO; Veraguas (Pacific)
0740/8118 (USBGN)
Originates in SW Veraguas province on the mainland W of Golfo de Montijo [0740/8107 (USBGN)] and flows SE discharging into the Pacific Ocean ca. 10 km NW of Punta Brava [0736/8114 (USBGN)] (JOG, as "Río Grande"; MGRP, as "Río Grande"; MHA; MRP; SMN2B). Griscom et al., at sea 5 mi (8 km) off, 1924 Apr 10 (AMNH, as "San Lorenzo R."), at Wilcox Camp [ca. 0741/8118] (q. v.), 1924 Apr 2-12 (AMNH, as "Wilcox Camp, San Lorenzo River"; CM, as "San Lorenzo"; Griscom, 1924a; Griscom, 1924b, as "Wilcox Camp, San Lorenzo River," and as "San Lorenzo River, near camp of Wilcox"; Griscom, 1927a, as "San Lorenzo River"; Griscom, 1933, as "Wilcox Camp"; Olson, 1981a, as "Wilcox Camp, San Lorenzo River [Caribbean Slope]" [sic]).

SAN LORENZO RIVER; see San Lorenzo, Río

SAN MIGUEL; Chiriquí
?
Unclear where Galindo collected. He may have collected in W Chiriquí province at San Miguel del Yuco [0835/8238 (USBGN)], which seems to be a locality of Morales (see San Miguel [ca. 0835/8238 ?]). He may also have collected at San Miguel [0816/8135 (USBGN)], which is located at 400 m near Cerro Viejo [0816/8136 (USBGN)] (E562), which appears to be one of Hind's localities. Galindo, 1971 (GML).

SAN MIGUEL; Chiriquí
ca. 0835/8238 ?
Probably refers to San Miguel del Yuco [0835/8238 (USBGN)], a town at 400 m, located 7 km N of La Concepción [0831/8237 (USBGN)] on the main road to Volcán [0846/8238 (USBGN)] (JOG). This town is 3 km NW of El Calvario [0834/8237 (USBGN)], another locality of Morales in 1968. Morales, 1968 Jan 30-31, Feb 9 (GML).

SAN MIGUEL; Panamá (E, Pacific)
0914/7915 (USBGN)
100 m, town 17 km NNE of Pacora [0905/7917 (USBGN)] on the Río Pacora [0901/7917 (USBGN)] (JOG). Wetmore and Perrygo, GL = Cerro Azul [0910/7924 (USBGN)], Río Pacora, at 600 feet (180 m), along the Río Pacora and its tributaries in tall, green, gallery forest and in recently cleared areas, 1949 Mar 29,

Apr 19 (USNM).

SAN MIGUEL; Panamá, Archipiélago de las Perlas
0827/7856 (USBGN)
Sea level, town on the N coast of Isla del Rey [0822/7855 (USBGN)] (q. v.) (JOG). San Miguel often appears in the literature in reference to the entire island, not just the settlement.

SAN MIGUEL, GOLFO DE; Darién
0822/7817 (USBGN)
Bay that cuts 28 mi (25 km) E into the NW coast of Darién province just below the Panamá provincial border. The Río Tuira [0821/7803 (USBGN)] enters the bay on the E and the bay opens to the Gulf of Panama [0800/7930 (USBGN)] on the W where it extends 17 mi (27 km) from N to S (SMN2A, as "San Miguel Bay"; SMN2B). According to Griscom (1927b:2), the "whole coastal region of the Gulf of San Miguel is arid tropical, not humid tropical, and the forest is largely scrub, relieved by the gigantic 'cuipo' tree (*Cavanillesia planifolia*)," and from Garachiné [0804/7822 (USBGN)] they were able to look inland and see "the sharp cleavage on the lower slopes where the 'cuipo' forest left off, and the heavy rain-forest began." Festa collected throughout the N region of the bay in 1895 (Salvadori and Festa, 1899, as "Golfo di San Miguel"), the Griscom Crosby Expedition worked near Punta Garachiné [0806/7825 (USBGN)] in 1927 (Griscom, 1927b), and the Vanderbilt Expedition worked near Punta Garachiné in 1941 (Bond and Meyer de Schauensee, 1944). See also: Chanatí, Río; Chepigana; Cucunatí, Río; El Real de Santa María; Garachiné, Punta; La Palma; Matusagratí, Laguna de; Lara, Río; Pinogana; Sabana, Punta; Sabanas, Río; Sambú, Río; Sapo, Cerro; and Tuira, Río.

SAN MIGUEL, ISLA; see Rey, Isla del

SAN MIGUEL ARRIBA; Coclé (Atlantic)
0843/8009 (USBGN)
400 m, town in E Coclé province N of the continental divide 30 km NE of Penonomé [0831/8022 (USBGN)] and 8 km NNW of Cerro Gaital [0837/8007 (USBGN)] (MGRP; JOG). Jiménez, López, Ramos, and Rodríguez, according to Angehr (in litt., Aug 2005), this was a Western Watershed study site. Collections were made at the camp, 420 m, coordinates 08°41'28" N, 80°08'50" W, 2001 Jun 30, Jul 1 (MVUP; Louis Berger Group et al., 2003).

SAN MIGUEL DE YUCO; see San Miguel

SAN MIGUEL DEL YUCO; see San Miguel

SAN MIGUEL ISLAND; see Rey, Isla del

SAN MIGUEL LOCKS; see Pedro Miguel, Esclusas

SAN PABLO; Panamá ? (Canal Zone) (Atlantic)
ca. 0907/7948
62 feet (19 m), station on the original Panama Railroad [ca. 0922/7954] that is now submerged in Lago Gatún (Gatun Lake) [0912/7955 (USBGN)], the station was located 7 mi (11 km) W of Gamboa [0907/7942 (USBGN)] and 1 mi (1.5 km) SE

of Tabernilla [ca. 0908/7949] near what is now Isla San Pablo [0907/7947 (USBGN)] (MIP; PR). This locality is near what is now the Colón/Panamá provincial boundary, and it is estimated to lie within Panamá province. Salvin, 18?? (Salvin and Godman, 1879-1904, vol. 1:25, vol. 3:266, as "San Pablo Station"); Maack, 1871 (MCZ, as "San Pablo, Isthmus of Panama"; Maack, 1872).

SAN PABLO; see San Pablo Nuevo Abajo

SAN PABLO, ISLA; Panamá, Archipiélago de las Perlas
0828/7851 (USBGN)
Sea level, small island just 1 km off the NE coast of Isla del Rey [0822/7855 (USBGN)] (E501; JOG). Pacific Project, GL = Gulf of Panama, Isla del Rey, near Isla San Pablo, 1969 Aug 30 (USNM, as "San Pablo Island"; Loftin, 1991).

SAN PABLO, ISLA; see San Pablo

SAN PABLO, RÍO; Veraguas (Pacific)
0751/8110 (USBGN)
Originates in NW Veraguas province on the upper slopes of Cordillera Central [0838/8150 (USBGN)], flows SE and S crossing the Carretera Interamericana [0832/8242 (USBGN)] 33 km NW of Santiago [0806/8059 (USBGN)]. After passing Soná [0801/8119 (USBGN)], it turns SE and widens greatly as it enters the Golfo de Montijo [0740/8107 (USBGN)] (JOG). Brown, collected at Soná (q. v.) on Jul 21, therefore the "Sona River" presumably refers to the Río San Pablo or a tributary that enters near Soná, 1901 Jul 21 (Bangs, 1902a, as "Sona River"); Griscom et al., below Soná, 1924 Mar 28-29 (AMNH); Wetmore and Perrygo, GL = Soná, near Soná in open pastureland bordered by trees and brush that extend into the hills, small tract of timber on a hilltop, 1953 May 14-15, 23, 28, at La Isleta (q. v.), 1953 Jun 1-4, at Guarumalito (q. v.), 1953 Jun 2, at Guarumal (q. v.), 1953 Jun 3, 11, at Congal (q. v.), 1953 Jun 13 (USNM; Wetmore, 1953).

SAN PABLO ISLAND; see San Pablo, Isla

SAN PABLO NUEVO ABAJO; Chiriquí
0826/8230 (USBGN)
ca. 50 m, town 7 km W of David [0826/8226 (USBGN)] (E562; JOG). Hurtado, 1990 Aug 27 (MVUP, as "San Pablo").

SAN PABLO STATION; see San Pablo

SAN PEDRO, RÍO; Veraguas (Pacific)
ca. 0750/8105
Originates in the foothills of Cordillera Central [0838/8150 (USBGN)] and flows SSE crossing the Carretera Interamericana [0832/8242 (USBGN)] 9 mi NW of Santiago [0806/8059 (USBGN)]. After it passes Puerto Mutis [0755/8103 (USBGN)], the river widens forming a 2-3 km wide inlet on the NE shore of Golfo de Montijo [0740/8107 (USBGN)] (JOG). USBGN coordinates [0758/8102] locate the river mouth ca. 5 km above Puerto Mutis, near its junction with the Río Martín Grande [0758/8102 (USBGN)], but Wetmore refers to the river section below Puerto Mutis as the Río San Pedro. Because we could not find another name for this 14

km stretch of river, we considered it the Río San Pedro and estimated coordinates accordingly. Wetmore, GL = Puerto Mutis, mouth of the Río San Pedro, at Isla Verde (q. v.), 1953 June 11 (USNM; Olson, 1980b, as "Puerto Mutís, Isla Verde, mouth of Río San Pedro").

SAN SAN, RÍO; Bocas del Toro
0931/8231 (USBGN)
Originates W of Changuinola [0926/8231 (USBGN)] and flows ENE entering the Caribbean Sea [1500/7500 (USBGN)] 9 km SE of the mouth of the Río Sixaola [0934/8234 (USBGN)] (JOG). Eisenmann, observations, at the mouth of, 1956 Jul 3 (Eisenmann, 1957, as "San San River"); McDonald et al., 1995-1996, experiments at 0927/8240, (McDonald et al., 2001). (Delgado, 1983).

SAN SEBASTIAN; Panamá, Archipiélago de las Perlas
ca. 0824/7906
According to Wetmore (1946b:39), San Sebastian was a settlement on Isla Pedro González [0824/7906 (USBGN)], Morrison, high trees behind, 1944 Jun 8 (USNM, as "Pedro González Island"; Wetmore, 1946b). See also: Pedro González, Isla. Not shown on our maps.

SANSÓN, RÍO; Darién
0824/7747 (USBGN)
Originates in central Darién province on the S end of Serranía el Filo de Tallo [0827/7757 (USBGN)], flows NE and enters the Río Chucunaque [0809/7744 (USBGN)] 2 mi (3 km) NNW of the Río Tuquesa [0823/7747 (USBGN)] (DP; SMN2B). Wetmore, GL for Río Chucunaque, 1959 Mar 21 (USNM).

SANTA ANA ABAJO; see La Villa, Río

SANTA CLARA; Chiriquí
0850/8246 (USBGN)
1,300 m, town 14 km NW of Volcán [0846/8238 (USBGN)] and 14 km E of the Costa Rican border (JOG). According to MGRP, the town is located 18 km NW of Volcán between the Río Chevo [0847/8250 (USBGN)] and the Quebrada Guisado [ca. 0847/8249], but this does not match the description provided by Wetmore, who indicated that the town was E of both rivers. The region is an Important Bird Area (Angehr, 2003:123). Hartman, at 4,000-4,700 feet (1,200-1,400 m), 15 mi (24 km) from the Costa Rican border, 1953 Feb 22-26, 28, Mar 2-4, 1958 Feb 28, Mar 1, 3-5, 1959 May 17, 1960 Mar 2, and as a GL for Quebrada Guisado, 1958 Jan 26 (USNM, as "Costa Rican Border," and as "Santa Clara"; Hartman, 1955); Wetmore, trail leading inland near the Quebrada Santa Clara [ca. 0848/8251], 1955 Feb 15, 19. The secondary localities attributed to Santa Clara (except for Moravia) were tributaries of the Río Candela [0846/8250 (USBGN)] that flowed across the road leading WNW from Santa Clara. GL for: Quebrada Guisado, Quebrada Santa Clara, and Río Chevo, 1954 Mar 15-20, 1955 Feb 16-18, 21-22, 24 (USNM; Wetmore, 1970); Wetmore (from Alois Hartmann), GL for Quebrada Santa Clara at Moravia, 1954 Mar 2; Wetmore (from Armaguedón Hartmann), 1959 May 17, 1960 Mar 2 (USNM); Mendez and GMLP, 1973 Mar 23, 29-30, Apr 1, 3-5, May (?) 4 (GML);

Collector ? [Mendez ?], 1973 Mar 26, 30, Apr 1, 5 (GML); Christian, 4 km N of, 1996 Jan 16 (MVUP); Roberts, at ca. 1,200 m, 1997 Mar 8 (LSUMZ); Christian and Roberts, at 1,600 m, observations 4 km N of, 1997 May 13 (Christian and Roberts, 2000); Wilson and Kalar, GL for Parque Internacional La Amistad, 1999 Jan 17 (MVUP). (Wetmore, 1967).

SANTA CLARA; Coclé (Pacific)
0823/8007 (USBGN)
Sea level, coastal town on the Carretera Interamericana [0832/8242 (USBGN)] 6 km E of Río Hato [0823/8010 (USBGN)] (JOG; MGRP). Used by Wetmore (1965:62) to locate Isla Farallón del Chirú [0821/8007 (USBGN)]. Eisenmann, observations, 1947 ? (Wetmore et al., 1984:306, as "Santa Clara Beach"). See also: Cotón, Río.

SANTA CLARA; Panamá (W, Atlantic)
0902/7945 (USBGN)
100 m, town near the Río Paja [0905/7948 (USBGN)] 5 km NNW of Nuevo Emperador [0900/7944 (USBGN)]. The town is at the terminus of a road (JOG), and according to RM2, this is Road K16A, which branches from Road K16. Road K16 extends roughly N-S between rivers on the SE coast of Lago Gatún (Gatun Lake) [0912/7955 (USBGN)] and a point on Road K19 [ca. 0853/7947] just E of Nuevo Emperador. According to our maps, this is the only road connection to the town. Wetmore (1968:45) mentions specimens of *Leptotila cassini* collected by MCU near Santa Clara on the Nuevo Emperador Road presumably referring to a specimen at GML collected in 1955 by Mendoza at Nuevo Emperador (q. v.).

SANTA CLARA, QUEBRADA; Chiriquí
ca. 0848/8251
Originates as an intermittent stream just N of Santa Clara [0850/8246 (USBGN)] and flows SSW crossing the main road at Santa Clara less than 1 km before discharging into a tributary of the Río Chiriquí Viejo [0820/8241 (USBGN)] (E762). MGRP indicates a Río Santa Clara E of Santa Clara, but this river is much larger than the creek just described and it does not pass near town as indicated by Wetmore. The Río Santa Clara on MGRP appears to correspond with the river named the Río Cotito [0848/8244 (USBGN)] on JOG. Wetmore, GL = Santa Clara, at 3,600-4,300 feet (1,100-1,300 m), at 5,000 feet (1,500 m), the clear stream near Santa Clara was rapid with a stony bed that was 12-18 feet (4-5 m) wide, collected along the stream within 7 km of Santa Clara primarily N of Santa Clara, on a horse trail to a forest with a coffee plantation in the understory, brush around the forest borders, trails farther N to good forest, up to the transition between the tropical zone and the subtropical zone, low tangled brush and occasional trees on the stream banks, pastures and a steeply sloping forested hill bordering the stream, stretch of high brush, 1954 Mar 15-17, 19-20, 1955 Feb 16-17, 21-22, 24 (USNM).

SANTA CLARA BEACH; see Santa Clara.

SANTA CRUZ; Chiriquí
0814/8155 (USBGN)

Sea level, town 5 km SW of Las Lajas [0815/8152 (USBGN)] and 2 km E of Estero El Cabuyal [0813/8157 (USBGN)] (JOG). Wetmore, GL = Las Lajas, an extensive level savanna with many cattle, the area was bordered by lowland coastal forest, trees of moderate height with some that were quite tall, open, semi-xerophytic undergrowth similar to the forests bordering the savannas in Panamá province, stream flowing through a rocky knoll out between a brushy pasture and an open savanna, creeks through a tract of lowland forest, along a deep creek in woodland, 1956 Feb 14, 20, 26 (USNM; Wetmore et al., 1984:18, as "Savannah Santa Cruz").

SANTA CRUZ, CERRO; Panamá (W, Pacific)
0900/7939 (USBGN)
200 m, hill S of the continental divide, 7 km SSE of Empire Known Distance Range [0904/7940 (USBGN)] and 5 km NNE of Arraiján [0857/7939 (USBGN)] (E8660). Collector ?, GL = Empire Known Distance Range, at 80-100 m, 1.5 km SE of Cerro Santa Cruz, 1995 Dec 3 (LSUMZ); Christian, GL = Empire Known Distance Range, at 80-100 m, 1.5 km SE of Cerro Santa Cruz, 1997 Feb 28 (LSUMZ).

SANTA CRUZ, SAVANNAH; see Santa Cruz

SANTA CRUZ DE CANA; see Cana

SANTA DOROTEA; see Puerto Piña

SANTA ESPIRITU MOUNTAINS; see Darién, Serranía del

SANTA FÉ; Darién
0839/7809 (USBGN)
ca. 50 m, town 17 mi (27 km) N of La Palma [0825/7809 (USBGN)] on the Río Sabanas [0832/7808 (USBGN)] just W of the Carretera Interamericana [0832/8242 (USBGN)] (SMN2B). Collector ?, 1905 Jun 9 (MCN); Child, 1963 Mar 30 (Wetmore, 1972:232); Morton, 1966 Jun 26 (YPM); Schneider, 1966 Aug 12 (GML); Collector ?, 1967 Feb (GML); Quintero, forest, 1967 Apr 8, Sep 20, 30 (GML; LSUMZ); Martínez, 1992 Nov 25 (MCN); Simone, 1992 Nov 25 (MCN).

SANTA FÉ; Veraguas (Pacific)
0831/8105 (USBGN)
700 m, town in N central Veraguas province near the headwaters of the Río Santa María [0806/8029 (USBGN)], ca. 7 km E of the Bocas del Toro provincial border and ca. 10 km S of the continental divide. The town is located at the end of highway 33, which extends NNW from Santiago [0806/8059 (USBGN)] (JOG). Arcé, Griscom (1928:3) indicates that Arcé's specimens may have also been collected on the Caribbean slope, 18?? (BMNH; USNM; Salvin, 1866; Salvin, 1867b; Berlepsch, 1911, as "Sta. Fé"; Eisenmann and Howell, 1962; Olson, 1983); Benson, 1,600-2,200 feet (500-650 m), reached the continental divide, cloud forest, strip of heavy tropical rainforest at the mountain base, according to Griscom, collecting was done in the late fall and winter of 1924, although specimens are labeled as follows: 1925 Feb 17-18, 20-28, Mar 1-6, 8, 10-20, 22-28, 30-31, Apr 1-9, 11-16 (AMNH; Hellmayr, 1918; Griscom, 1927a; Peters, 1931a:330; Zimmer, 1939; Zimmer, 1941b:5, 7, 10, 14; Zimmer, 1944; Zimmer, 1950:16, 50; Zimmer, 1951:13; Chapman, 1940; Eisenmann and Howell, 1962; Parkes, 1969; Olson, 1993b);

Quintero, 1967 Sep 30 (GML); Collector ?, 1973 Jan 3 (GML); Hinds and GMLP, 1974 Jan 3 (GML); Petit et al., surveys, plantations near, Dec 1993-Feb 1994 (Petit et al., 1999); Faucett and Weir, at 800 m, 2003 Feb 18-22 (CU; MVUP; UWBM). (Wetmore, 1959b:12; Olson et al., 1968).

SANTA MARÍA; Herrera
0807/8040 (USBGN)
ca. 25 m, city in N Herrera province 20 km W of the coast and less than 2 km S of the Río Santa María [0806/8029 (USBGN)] on the main road to Chitré [0758/8026 (USBGN)] (JOG). Wetmore and Perrygo, 1948 Mar 10, 2 mi (3 km) SW of Santa María, 1948 Mar 9, and as a GL for: La Concepción, and Río Escotá, 1948 Mar 8-10, 19 (USNM; Wetmore, 1950a).

SANTA MARÍA; Panamá (E, Pacific)
ca. 0901/7933
60 m, section of greater Panama City [0858/7932 (USBGN)] in Corregimiento Betania [0900/7932 (USBGN)] (E762). Cedeño, Fabrega, Janguez, Magaña, Miranda, Serrano, and Yanguez, GL for Tumba Muerto, 1989 Dec 10 (MVUP); Fabrega and Miranda, GL for Tumba Muerto, 1990 Mar 21 (MVUP).

SANTA MARÍA, RÍO; Veraguas/Herrera/Coclé (Pacific)
0806/8029 (USBGN)
Originates on the continental divide in Veraguas province N of Santa Fé [0831/8105 (USBGN)] and flows S, then E along the Coclé/Veraguas provincial boundary then along the Coclé/Herrera provincial boundary discharging into the S central shore of Bahía de Parita [0808/8024 (USBGN)] (JOG). Perrygo, GL = Paris [0803/8033 (USBGN)], 6 mi (10 km) N of Paris, 1948 Feb 17; Wetmore and Perrygo, GL = Paris, 4 mi (6 km) N of Paris, above the mouth of the Río Escotá [0806/8033 (USBGN)] in an area affected by tides, stands of large trees on the river banks, open savanna and secondary growth, 1948 Feb 24, 5 mi (8 km) NE of Paris, ca. 5-6 mi (8-10 km) above the mouth in tidal channels, brush-grown pastures, mangroves, 1948 Mar 4-5 (USNM; Wetmore, 1953); Wetmore, GL = El Roble [0810/8039 (USBGN)], near the Carretera Interamericana [0832/8242 (USBGN)] 4 km S of El Roble, 1963 Jan 24 (USNM). See also: Alvina; and El Mangle, Ciénaga.

SANTA MARTA, QUEBRADA; see Santa Marta, Río

SANTA MARTA, RÍO; Chiriquí
ca. 0842/8235 ?
Not shown on our maps but it may refer to the Quebrada Santa Marta [0842/8235 (USBGN)], a small intermittent stream located 10 km SSE of Volcán [0846/8238 (USBGN)] (E762, as "Quebrada Santa Marta"). Watson, at 500 feet (150 m), 1903 Oct 3 (FMNH).

SANTA MÓNICA, FINCA; see Finca Santa Mónica

SANTA RITA; Colón (E)
0919/7947 (USBGN)
840 feet (255 m), town located 13.5 km SSE of Colón [0922/7954 (USBGN)] and 4 km SSE of the S coast of Bahía Las Minas [0924/7950 (USBGN)] (E761A). See

also: Las Minas, Bahía.

SANTA RITA; Panamá (E, Pacific)
0905/7930 (USBGN)
320 feet (100 m), town 14 km NNE of Panama City [0858/7932 (USBGN)] (E661; E761A). Morton, 9 mi (14 km) by road, 1971 Mar 12 (USNM). See also: Chiva Chiva, Camino.

SANTA RITA; Panamá (W, Pacific)
0852/7953 (USBGN)
ca. 100 m, town in Distrito de La Chorrera [0857/7951 (USBGN)] 11 km WSW of La Chorrera [0853/7947 (USBGN)] (MGRP). Morales, Chorrera District, 1968 Jul 20 (GML).

SANTA RITA, CERRO; Colón (E)
ca. 0920/7947
880 feet (270 m), hill located 13.5 km ESE of Colón [0922/7954 (USBGN)] and 2 km E of the Carretera Boyd-Roosevelt [0917/7943 (USBGN)] (E661) near the town Santa Rita [0920/7948 (USBGN)] (JOG). Smith, N., observations, date ? (Ridgely and Gwynne, 1989:236).

SANTA RITA, CERRO; Panamá (E, Pacific)
ca. 0907/7930
Presumably a peak 1-3 km N of Santa Rita [0905/7930 (USBGN)] (q. v.), near the continental divide, Morton, 1971 Mar 12 (AMNH; USNM; Morton, 1971). Not shown on our maps.

SANTA ROSA; Colón (Canal Zone) (E)
0911/7939 (USBGN)
30 m, town 2 km N of Juan Mina [0910/7939 (USBGN)] on the W bank of the Río Chagres [0907/7942 (USBGN)] (E866). Wetmore, GL = Juan Mina, Río Chagres, along a trail from Río Honda [0910/7940 (USBGN)] leading N to Santa Rosa, 1959 Jan 29, 31, GL = Juan Mina, 1961 Jan 16 (USNM); De Leon, 1966 Apr 7 (GML); Reyna, GL for Río Palenque, 1971 Apr 28 (GML); Rojas, 1974 Jan 14 (GML); Mendoza, GL for Río Chilibre, 1978 Apr 18, 21, GL for Río Chagres, 1983 Jan 11, 24, Jul 11, Nov 7, 1985, 1988 (GML). See also: Chagres, Río.

SAN TELMO, BAHÍA; Panamá, Archipiélago de las Perlas
0816/7853 (USBGN)
Bay on the S coast of Isla del Rey [0822/7855 (USBGN)], the N shore is formed by the mainland, the W shore is formed by the peninsula that extends down to Punta Cocos [0813/7854 (USBGN)], and Isla San Telmo [0817/7851 (USBGN)] lies on the E end of the bay (JOG). Murphy and Correia, Askoy Expedition, GL = Isla del Rey, "our San Miguel [Isla del Rey] specimens were all taken in the Santelmo Bay district, at the S end of the island," 1941 Feb 14-15 (AMNH, as "Santelmo Bay"; Murphy, 1944a; Murphy, 1944b, as "Santelmo Bay"; Murphy, 1945, as "Santelmo Bay District"); Robins, observations, 1957 Jul 15-26 (Robins, 1958, as "Bahía Santelmo").

SAN TELMO, ISLA; Panamá, Archipiélago de las Perlas
0817/7851 (USBGN)
70 m, island with a land area of 1.62 km² (162 ha) (Angehr, 2003:277) located 2 km off the S coast of Isla del Rey [0822/7855 (USBGN)] in the E end of Bahía San Telmo [0816/7853 (USBGN)] 7 km NE of Punta Cocos [0813/7854 (USBGN)] (JOG). Angehr also indicates that the island is now a private reserve that is mostly forested. Bovallius, the S side is rocky and the N side slopes gently with rich vegetation, see discussion of Bovallius under Pueblo Nuevo [ca. 0320/8013 ?], 1882 (NHRM; Rendahl, 1920, as "San Elmo Island"); Beaudette, GL = Isla del Rey, 1959 Mar 22 (LACM, as "San Telmo Is"); Wetmore, the island was covered in original forest without trails, worked on a beach on the NW shore, walked inland along a shallow, dry creek, up to 200 feet (60 m) on steep slopes, and across the island to its E coast, 1960 Jan 26 (USNM, as "Isla Santelmo"; Olson, 1980b, as "Isla Santelmo"); (MacArthur et al., 1972—species list; see comments in Bibliography); Tejera, 1981 Mar 16, 20 (MVUP). See also: Rey, Isla del; and San Telmo, Bahía.

SANTELMO, ISLA; see San Telmo, Isla

SANTELMO BAY; see San Telmo, Bahía

SANTELMO BAY DISTRICT; see San Telmo, Bahía

SANTELMO ISLAND; see San Telmo, Isla

SANTIAGO; Veraguas (Pacific)
0806/8059 (USBGN)
ca. 50 m, the capital of Veraguas province, located on the Carretera Interamericana [0832/8242 (USBGN)] in the central eastern region of the province ca. 20 km NNE of the coast of Golfo de Montijo [0740/8107 (USBGN)] (JOG). The town was on the old overland trail to Chiriquí [0824/8219 (USBGN)] and therefore visited by several early naturalists. Collector ?, 18?? (FMNH); Arcé, for many years used Santiago "as his headquarters with various local collectors scouring the province for him. Many of the birds are labeled very generally 'Santiago', and none of the heavy-forest species labeled 'Santiago' ever occurred there, but were taken in the forest near the coast or to the south or in the forest farther in the interior at the base of the mountains" (Griscom, 1927a:17), 18??, (BMNH, as "Santiago de Veraguas"; Salvin, 1866, as "Santiago de Veraguas"; Salvin, 1867b; Salvin, 1870b; Griscom, 1927a; Peters, 1931a:341, as "Santiago de Veragua"; Eisenmann and Howell, 1962, as "Santiago de Veraguas"); Benson and Seaman, vicinity of, and southward to the coastal forest at the head of Golfo de Montijo, specimens labeled 15 mi (24 km) SE of Santiago were collected at El Villano, GL for: El Villano, and La Colorada, 1924 Feb 21, Mar 5, Apr 16, Oct 20-22, 24-25, 27, 31, 1925 Jun 8, 11-12 (AMNH; Griscom, 1924b, as "El Villano, Santiago," and as "La Colorado, Santiago"; Griscom, 1927a; Griscom, 1927b; Wetmore, 1956a; Parkes, 1976, as "15 miles SE Santiago"; Olson, 1981d, as "La Colorada, Santiago"); Bradvica, 1983 Feb 20 (MVUP); Bustamante, *despues de la antena de* (after the antennae for) Ondas Centrales, 1983 Mar 10 (MVUP); Moreno, 1985 Apr 3 (MVUP); Godinez, along the Carretera

Interamericana W of Santiago between kilometer markers 249 and 263 (according to MGP, Santiago is at km marker 248), 1994-1996 (Godinez, 1999, as "Santiago City"). See also: Chucú, Cordillera del; and Veraguas, Provincia de.

SANTIAGO, CERRO; Chiriquí
0833/8144 (USBGN)
2,121 m, peak 27 km NNE of San Félix [0817/8152 (USBGN)] and 30 km NNW of Tolé [0814/8141 (USBGN)] in the Serranía de Tabasará [0833/8140 (USBGN)]) on the continental divide (E562; RDP2). Many older maps indicate a much higher elevation – above 2,800 m – for the peak, but Myers and Duellman (1982:3) indicate that the lower elevation is correct. The region around the peak is considered an Important Bird Area (Angehr, 2003:135-138), and it appears that at least some of the collecting by Griscom et al. at the Cerro Flores [0829/8144 (USBGN)] (q. v.) was actually conducted on or near Cerro Santiago. (Myers, 1969). See also: Bollo, Cerro; and Colorado, Cerro.

SANTIAGO, RÍO; Veraguas (Atlantic)
0848/8058 (USBGN)
Originates in NE Veraguas province and joins the Río Concepción [0850/8059 (USBGN)] 5 km above its mouth (JOG). See also: Mineral.

SANTIAGO CITY; see Santiago

SANTIAGO DE VERAGUAS; see Santiago

SANTO, RÍO; Panamá (E, Pacific)
0903/7911 (USBGN)
Originates in the Sabana de San José [ca. 0905/7915] and flows SE joining the Río La Jagua [0902/7909 (USBGN)] (q. v.) 4 km N of Río Chico [0902/7910 (USBGN)] and 10 km SE of Pacora [0905/7917 (USBGN)] (TM). TM and JOG indicate that the river is very near the location of La Jagua Hunting Club [ca. 0903/7911] (q. v.) and considering that the club had access to 20,000 acres (8090 ha), collections might have been made near this river without collectors distinguishing it from the Río La Jagua. Wetmore, GL = Pacora, dry except for a few small pools, timber along the stream near 2 foot (0.5 m) tall grass, 1961 Mar 23 (USNM).

SANTO DOMINGO; Herrera
ca. 0801/8037
ca. 50 m, town 3 km S of Cabuya [0802/8038 (USBGN)] (MHA). Wetmore and Perrygo, GL = Cabuya, along a small, winding creek with taller trees along the banks, bordered by dense thickets interspersed with open areas, 1948 Mar 13 (USNM).

SANTO DOMINGO, CIÉNAGA DE; Panamá (E, Pacific)
ca. 0903/7915
20 feet (6 m), a marshy wetland 8 km NW of Río Chico [0902/7910 (USBGN)] and 5 km SE of Pacora [0905/7917 (USBGN)], just S of Sabana de San José [ca. 0905/7915] (TM, as "Ciénaga de San Domingo"). Wetmore and Perrygo, GL = Río Chico, shallow mud and water, largely dry, 1949 Mar 16, 30 (LSUMZ, as "Ciénaga Santo Domingo"; USNM, as "Ciénaga Santo Domingo," and as "Ciénaga de Santo Domingo"; Wetmore, 1951b, as "Ciénaga Santo Domingo"); Wetmore, GL = La

Jagua Hunting Club, vast, level, open marsh bordered by considerable tracts of gallery forest interspersed with savanna, 1961 Mar 22, GL = La Jagua Hunting Club, full of water and grown with a tall *Heliconia* type plant, 1962 Jan 13 (USNM, as "Ciénaga Santo Domingo").

SANTO DOMINGO, QUEBRADA; Chiriquí

ca. 0820/8241 ?

Described by Wetmore as a stream 4 mi (6 km) W of La Concepción [0831/8237 (USBGN)]. Not shown on our maps, but presumably its course runs near the town Santo Domingo [0830/8243 (USBGN)], a town located at 200 m and 11 km W of La Concepción [0831/8237 (USBGN)] (JOG). Possibly a tributary of the Río Duablo [0820/8241 (USBGN)] (q. v.) or of the Río Soy [0825/8244 (USBGN)]. Wetmore, GL = La Concepción, at 800 feet (240 m), 4 mi (6 km) W of La Concepción, a rough-surfaced ravine, 1960 Mar 2 (USNM).

SAPO, CERRO; Darién

0758/7822 (USBGN)

1,300 m, peak at the N end of Serranía del Sapo [0750/7817 (USBGN)] located 6 km S of Garachiné [0804/7822 (USBGN)] (MHA). Barbour, Brooks, Underwood Panama Expedition, according to Barbour and Brooks (1923:213), "Mt. Sapo is covered with high rain forest which extends in all its luxuriant splendor over the whole Sambú valley and reaches the coast at the mouths of the Taimatí and Sambú Rivers." The expedition traveled from Garachiné "up the valley of the San Antonio to the lower slopes of Mt. Sapo. From the camp [they] made excursions in various directions," and "made no attempt to climb Mt. Sapo itself; but from a spur, called La Jarcia, above one of [their] camps (1,700 feet) [they] got several splendid views of the range." According to Myers (1969:12), the crest of La Jarcia is 560 m (1,850 feet) and can be reached from Garachiné on foot in 5-6 hours, but "from here the summit of Cerro Sapo is a two- to three-hour walk." Bangs and Barbour (1922:193) indicate that "Mt. Sapo refers to stations on the headwaters of the Río San Antonio, [which]...drains the western slopes of Mt. Sapo itself. The collections are all representative of the fauna of the lower Tropical Zone, the stations being all in the rain forest below 1,500 feet (450 m)," 1922 Apr 19-26 (ANSP, as "M. Sapo," and as "Mt. Sapo"; FMNH, as "Mt. Sapo"; MCZ, as "Mt. Sapo"; Bangs and Barbour, 1922, as "Mt. Sapo"; Barbour and Brooks, 1923, as "Cerro de Sapo," and as "Mt. Sapo"; Griscom, 1929a:153, 171, as "Mt. Sapo"; Griscom, 1932a, as "Mt. Sapo"; Griscom, 1933, as "Mt. Sapo"; Hellmayr, 1929:84, as "Mount Sapo"; Wetmore, 1950b; Haffer, 1967b, as "Mt. Sapo"); Vanderbilt Expedition, Feathers was the collector for the expedition, at 3,000 feet (900 m), at the summit at 4,262 feet (1,300 m), according to Vanderbilt (in Bond and Meyer de Schauensee, 1944:24-26), from Garachiné they ascended the Río San Antonio [0804/7821 (USBGN)], but left the river to climb several hours to the base camp, which was "situated on the ridge at an altitude of about three thousand feet...was generally encased in fog...[and] the forest trees were gigantic and could not be felled with machetes," from the camp, which had no water, "two men made two trips daily far down the ridge to a river known as the Jesusito," they cut a trail to the summit of the peak and "not ten minutes walk from camp, after a climb of perhaps two hundred feet, we were in the sub-tropical forest. There were stunted trees, covered with moss, completely different vegetation, and we saw many new birds...the summit itself proved to be completely overgrown with heavy brush, small trees, and roots," 1941 Apr 21-30, May 1-4, 6-7, 3 mi (5 km) SW of, 1941 Apr 15-17 (ANSP, as "Cerro Sapo," and as "Mt. Sapo"; USNM, as "Mount Sapo"; Vanderbilt, 1944, as "Cerro del Sapo," and as "Mt. Sapo"; Bond and Meyer de Schauensee, 1944, as "Cerro Sapo," as "Mt. Sapo," and as "Mount Sapo"; Olson, 1981b; Olson, 1981c). (Meyer de Schauensee, 1950a, as "Mt. Sapo"; Myers, 1969; Escalante-Pliego and Peterson, 1992, as "Mt. Sapo"; Angehr, 2003:297-298). See also: Muerto, Playa.

SAPO, CERRO DE; see Sapo, Cerro

SAPO, CERRO DEL; see Sapo, Cerro

SAPO, SERRANÍA DEL; Darién

0750/7817 (USBGN)

Mountain range in S Darién province extending NW to SE along the coast between Garachiné [0804/7822 (USBGN)] and Puerto Piña [0735/7811 (USBGN)] (JOG). See also: Piña, Cerro; and Sapo, Cerro.

SAPODILLY CAY; see Zapatilla, Cayos

SAPO HIGHLANDS; see Sapo, Cerro

SARDINILLA; Colón (E)

0918/7938 (USBGN)

70 m, town W of Lago Alajuela (Madden Lake) [0915/7935 (USBGN)] 23 km NNE of Gamboa [0907/7942 (USBGN)] and 4.5 km SW of Salamanca [0920/7935 (USBGN)]. Los Playones [0919/7939 (USBGN)] is located 1.5 km NNW of Sardinilla (E762). Garcés, specimens are labeled Los Playones, Sardinilla, 1991 Nov 17 (MVUP).

SARDINILLA, RÍO; Panamá (Canal Zone) (E, Atlantic)

0905/7941 (USBGN)

Originates in Parque Nacional Soberanía [0909/7941 (Angehr, 2003:239)] SE of Gamboa [0907/7942 (USBGN)] and NW of Paraiso [0902/7938 (USBGN)] and flows SW crossing Carretera Gaillard [0904/7940 (USBGN)] 5.5 km SE of Gamboa before entering the Panama Canal [0920/7955 (USBGN)] (E762). The river does not enter the Río Chagres directly, but indirectly via the Panama Canal. Williams, GL = Gamboa, Río Chagres, 4 km SE of Gamboa, along the Río Sardinilla, 1997 Feb 6 (LSUMZ).

SASARDÍ; San Blas

ca. 0855/7742

Sasardí refers to a point, a bay, a river, a mountain pass, and a group of islands all located near one another on or near the SE coast of Comarca de San Blas ca. 50 km NW of Cabo Tiburón [0842/7721 (USBGN)] on the San Blas/Colombian border. Punta Sasardí [0858/7745 (USBGN)] is located on the coast 52 km NW of Cabo Tiburón (MGRP); Río Sasardí [0900/7748 (USBGN)] enters the Caribbean Sea [1500/7500 (USBGN)] near Punta Sasardí (SMN2B; MRP); Paso Sasardí [0856/7751] is located on the continental divide 11 km S of Punta Sasardí presumably near the peak Cerro Sasardí [ca. 0856/7751] (MGRP, as "Cerro Sasardí); and Islas Sasardí [0855/7742 (USBGN)] are a group of ca. 8 islands located within Bahía de Sasardí [0857/7745 (USBGN)] less than 1 mi (1.5 km) off the coast and 1-5 mi (1.5-8 km) SE of Punta Sasardí (SMN2B). Van Horn, 1969 Jun 19 (GML). See also: Mulatupo Sasardí.

SASARDÍ, BAHÍA DE; see Sasardí

SASARDÍ, CERRO; see Sasardí

SASARDÍ, ISLAS; see Sasardí

SASARDÍ, PASO; see Sasardí

SASARDÍ, PUNTA; see Sasardí

SASARDÍ, RÍO; see Sasardí

SAVANNA; see Panama, Savanna of

SAVANNAH OF PANAMA; see Panama, Savanna of

SAVANNAH SANTA CRUZ; see Santa Cruz

SAVANNA OF PANAMA; see Panama, Savanna of

SCOUT CAMP; see Alajuela, Lago; and Madden, Represa

SCHEELEA ISLAND; Colón (Canal Zone) (E, Atlantic)

ca. 0912/7952

Not shown on our maps, but Roubik et al. (1985) indicate that it is a 7 ha island located ca. 5 km N of Barro Colorado Island [0909/7951 (USBGN)], which places it near Islas Juan Gallegos [0913/7952 (USBGN)] or Islas Las Brujas [0913/7954 (USBGN)], Roubik et al., observations (Roubik et al., 1985).

SECA, QUEBRADA; Darién

ca. 0725/7759 ?

Described by Wetmore as a tributary of the Río Imamadó [ca. 0727/7758] upstream beyond the mouth of the Río Chicao [ca. 0727/7758], Wetmore and Perrygo, GL = Río Jaqué, 1947 Apr 10 (USNM). Not shown on our maps.

SECAS, ISLAS; Chiriquí

0758/8202 (USBGN)

109 m, group of more than 10 islands and islets in the Golfo de Chiriquí [0800/8220 (USBGN)] ca. 45 km NW of Isla de Coiba [0728/8147 (USBGN)] and ca. 65 km SE of David [0826/8226 (USBGN)] (JOG; MGRP; MHA). Batty, specimens at AMNH are dated 1902 Jan 4-5, but Batty's data were completely fabricated and his specimens labeled "Secas Island" were actually collected in 1901 near his mainland bases at Boquete [0847/8226 (USBGN)] and Boquerón [0830/8234 (USBGN)] (AMNH, as "Secas Island"; Olson, 2008); Ashcraft, 1939 Mar 27 (LACM, as "Secas Is," Olson 2007, as "Islas Secas").

SECAS ISLAND; see Secas, Islas

SECO, POTRERO; Chiriquí

ca. 0848/8236

ca. 1,700 m, described by Wetmore as an old crater, covering an area of ca. 0.5 km^2

(50 ha), located on the upper end of Llanos del Volcán [ca. 0847/8238] and below broken hills at the W base of Volcán Barú [0848/8233 (USBGN)]. The elliptical space was covered with short grass with occasional low bushes and the entire area was surrounded by forest (USNM). Presumably this area is located near the town Potrero Seco [0848/8236 (USBGN)] (q. v.). Wetmore, GL = Volcán, Volcán Barú, at 5,600 feet (1,700 m), forest of relatively few species, trees were 40-50 feet (10-15 m) high and their trunks were covered with a white moss, a few trees with berries, 1955 Feb 26 (USNM). Not shown on our maps.

SEDA, LA; see La Seda

SENDERO LAS CRUCES; see Las Cruces Trail

SENDERO LOS MONOS; see Coiba, Isla de

SENDERO LOS QUETZALES; see Cerro Punta-Boquete Trail

SENNORA ISLAND; see Señora, Isla

SENNORITA ISLAND; see Señorita, Isla

SEÑORA, ISLA; Panamá, Archipiélago de las Perlas
0826/7907 (USBGN)
Sea level, small island located in Bahía de Panamá [0850/7915 (USBGN)] 3 km N of Isla Pedro González [0824/7906 (USBGN)] (JOG; MGRP, as "Isleta Señora"). Described by Rendahl (1920:4, as "Sennora") as a rocky holm, "descending abruptly into the sea." (Murphy, 1944b).

SEÑORITA, ISLA; Panamá, Archipiélago de las Perlas
ca. 0826/7907
According to Rendahl (1920:4, as "Sennorita," pl. 1 (map), as "Señorita"), the "rocky holm" is located N of Isla Pedro González [0824/7906 (USBGN)] just E of Isla Señora [0826/7907 (USBGN)].

SEREÑA, QUEBRADA; Colón (W)
ca. 0908/8012
Described by Wetmore as a small tributary of the Río Indio [0912/8011 (USBGN)] that enters from the W ca. 3 km S of El Chilar [0910/8012 (USBGN)], Wetmore and Perrygo, GL = El Chilar, near its mouth in pasture, thickets of brush and tall grass, upstream to an old channel along the base of a hill, tall forest, 1952 Mar 10-11 (USNM). Not shown on our maps.

SERENO; see Río Sereno

SERENO, RÍO; Chiriquí
ca. 0849/8251
Originates W of Santa Clara [0850/8246 (USBGN)], flows S within 1 km of the Costa Rican border and turns E entering the Río Candela [0846/8250 (USBGN)] 10 km W of Santa Clara (JOG). See also: Río Sereno.

SEREZO, RÍO; Panamá (E, Pacific)
ca. 0901/7911
Originates 1 km E of the mouth of the Río Pacora [0901/7917 (USBGN)], flows E within 1 km of the coast entering the Río Chico [0901/7911 (USBGN)] 1 km S

of Río Chico [0902/7910 (USBGN)] (TM). Possibly the stream, Río Tereza, described in Wetmore's journals in 1949. See also: Chico, Río.

SETEGANTÍ; Darién
0818/7805 (USBGN)
Sea level, town 14 km SSE of La Palma [0825/7809 (USBGN)] near the mouth of the Río Tuira [0821/7803 (USBGN)] and 4 km W of Chepigana [0818/7803 (USBGN)] (MGRP; RDP2). Sugasti, 1990 Aug 12-13 (MVUP).

SETEGANTÍ, RÍO; Darién
0746/7738 (USBGN)
Originates on the SE slopes of Serranía de Pirre [ca. 0749/7744] near the Alturas de Nique [0740/7745 (USBGN)] and the ornithological Cerro Pirre [0749/7744 (GA)], and flows N joining the Río Tuira [0821/7803 (USBGN)] near its origin at Cana [ca. 0745/7742] (DP; SMN2B; Anthony, 1916 (map)). Wetmore, GL = Cerro Pirre, at 1,500-3,000 feet (450-900 m), head of the Río Seteganti, camp was located on the upper reaches of the river along a small tributary. Wetmore collected primarily along two trails: The first led N of camp for 2 mi (3 km) to the main river, and the other led W of camp up to 3,200 feet (1,000 m) and the N peaks of the Alturas de Nique near the Colombian border. The forest at 1,500-1,600 feet (450-500 m) was old brush, old second growth, and gallery forest, at the higher elevations *Heliconia* was not present and the forest was virgin, largely open underneath with some dense areas of 6-10 foot (2-3 m) wild pineapple, collected on the main river on a few open acres with tall cane grass and morning glories, 1961 Jan 27-Feb 6 (FMNH; USNM). See also: Cana.

SETEGANTÍ RIVER; see Setegantí, Río

SETETULE, CERRO; Darién
0747/7736 (USBGN)
1,220 m, peak located 12 mi (19 km) SSE of the official Cerro Pirre [0756/7743 (USBGN)] (SMN2B). See also: Kitchener, Mount.

SETIGANTÍ, RÍO; see Setegantí, Río

SEVILLA, ISLA; Chiriquí
0814/8224 (USBGN)
Sea level, shown on MHA and MRP as an island continuous with Isla Chalapas [0816/8222 (USBGN)] and Isla Chirote [0815/8223 (USBGN)], but these three islands are distinct and separated from one another by narrow marine channels. Isla Sevilla has a land area of ca. 15 km^2 (1,500 ha) and is located in Bahía de Charco Azul [0812/8245 (USBGN)] at the mouth of the Río Chiriquí [0820/8226 (USBGN)] 20 km S of David [0826/8226 (USBGN)] (JOG; MGRP). Batty, specimens at AMNH and at BMNH are dated 1902 Jan 12, 20-24, 28, but Batty's data were completely fabricated and his specimens labeled "Sevilla Island" were actually collected in 1901 near his mainland bases at Boquete [0847/8226 (USBGN)] and Boquerón [0830/8234 (USBGN)] (AMNH,; BMNH; and Hartert and Goodson, 1917:418, all as "Sevilla Island"; Eisenmann, 1950, as "Sevilla I."; Wetmore, 1957:6, as "Sevilla Island"; Olson, 2008). (Delgado, 1983).

SEVILLA ISLAND; see Sevilla, Isla

SHEPERD ID.; see Pastores, Isla

SHEPHERD, ENSENADA; Bocas del Toro
0913/8221 (USBGN)
A bay between Isla Pastores [0914/8220 (USBGN)] and the mainland coast S of Valle de Agua [0915/8223 (USBGN)] (MHA, as "E. Shepherd"). Wetmore, GL = Almirante [0918/8224 (USBGN)], 1958 Feb 4 (USNM, as "Shepherd Harbor").

SHEPHERD CREEK; see Pastores, Quebrada

SHEPHERD HARBOR; see Shepherd, Ensenada

SHEPHERD ISLAND; see Pastores, Isla

SHERMAN, FORT; see Fort Sherman Army Reservation

SIBUBE; Bocas del Toro
0936/8249 (USBGN)
ca. 50 m, town in the most N tip of Bocas del Toro province 37 km NW of Changuinola [0926/8231 (USBGN)] on the Río Sixaola [0934/8234 (USBGN)] 28 km W of its mouth (JOG). Smith, 1927 Feb 9-10, 12 (YPM); Handley, 1962 Jan 16, 1963 Jan 22, 25-26 (USNM).

SICSOLA, RÍO; see Sixaola, Río

SIERPE, LOMA; see Las Peñitas

SILENCIO, EL; see El Silencio

SILENCIO, QUEBRADA EL; see El Silencio, Quebrada

SILICO CREEK; see Cilico Creek

SILLA DE CERRO PANDO; see Pando, Silla de Cerro

SILLA DE PANDO; see Pando, Silla de Cerro

SILVA DE ADENTRO, ISLA, see La Porcada, Isla

SILVESTRE, CERRO; see Cerro Silvestre

SIMÓN GÓMEZ; see Simón Gómez, Río

SIMÓN GÓMEZ, RÍO; Coclé (Pacific)
0811/8035 (USBGN)
Originates W of Aguadulce [0815/8033 (USBGN)] and flows S entering the Río Estero Salado [0811/8029 (USBGN)] 8 km WSW of Aguadulce (MGRP). Tejera, GL = Aguadulce, 1976, 1977 Dec 26, 1978 Apr 10 (MVUP, as "S. G.," and as "Simón Gómez").

SIRI, RÍO; see Cirí, Río

SITIO, EL; see El Sitio

SITIO SIERRA; Coclé (Pacific)
ca. 0810/7940
According to their map, the locality is in SW Coclé province near the Herrera boundary, perhaps 10-15 km SW of Aguadulce [0815/8033 (USBGN)], Cooke,

archeological digs, c. 1980 (Cooke and Olson, 1984). Not shown on our maps.

SIXAOLA; see Sixaola, Río

SIXAOLA, RÍO; Bocas del Toro
0934/8234 (USBGN)
Originates at the merger of several rivers on the Bocas del Toro/Costa Rican boundary and flows SE then NE entering the Caribbean Sea [1500/7500 (USBGN)] 15 km NNW of Changuinola [0926/8231 (USBGN)]. The stream meanders greatly and has numerous small islands in its main course (E501; JOG). Carriker, S of, SE of Cuabre [0936/8247 (GG)] (in Costa Rica), many of these specimens were taken on the Costa Rican side (N banks) of the river, although both Carriker and Eisenmann (1957:247-248) indicate that at least some of the birds came from the Panamanian side (S banks) of the river, 1904 Aug 6, 8-9, 13, 17, 22-23, Sep 1-3, 5, 7-8, 10, 12, 21-27, Oct 6-8, 10, 12 (CM, as "Río Sicsola"; Carriker, 1910:353, as "Sicsola or Sixola"; Todd, 1942; Eisenmann, 1957, as "Río Sicsola"; Eisenmann, 1962a, as "Río Sicsola"; Eisenmann and Howell, 1962, as "Río Sicsola [= Río Sixaola, probably Panama]"); Wedel, 1927 Nov 1, 16, and as a GL for Elena, 1927 Jan 15 (MCZ, as "Sixaola," and as "Sixaola River"; Peters, 1931a:302, as "Sixaloa, Elena"); SBE, GL = Guabito [0930/8237 (USBGN)], at 20 m, 25 km W of Guabito, hills on the W bank of the Río Sixaola, 1991 Feb 23 (USNM, as "Sixaola River"; Parsons et al., 1993; Brumfield et al., 2001). (Delgado, 1983).

SIXOLA; see Sixaola, Río

SLOTHIA ISLAND; Panamá (Canal Zone) (W, Atlantic)
ca. 0910/7950
According to Willis and Eisenmann (1979:3 (map)), the small island is located less than 100 m off the NE coast of Isla Barro Colorado [0909/7951 (USBGN)] in the cove near the laboratory. Sexton, observations, 1973 Feb-Mar (Sexton, 1975). Van Tyne, 1927 Apr 21 (Eisenmann, 1952: 23, 44).

SMITHSONIAN TRACT; see Frijoles, Río

SMITHSONIAN TROPICAL RESEARCH INSTITUTE LABORATORY; see Barro Colorado

SMITHSONIAN TROPICAL RESEARCH INSTITUTE OFFICE; see Ancón

SNAKE HILL; Panamá (Canal Zone) (E, Pacific)
ca. 0903/7939
Not named on our maps, but as culebra is the Spanish word for snake, Snake Hill probably refers to a 75 m hill immediately adjacent to Culebra Reach [0903/7939 (USBGN)]. The hill is 3 km NW of Paraiso [0902/7938 (USBGN)] and 2 km S of Summit Park [0904/7939 (USBGN)]. MCU, GL = Paraiso, specimens are labeled "Paraiso, Summit Snake Hill." Whether summit refers to the hill's summit or to Summit Park is unclear, 1955 Aug 25 (GML).

SOBERANÍA, PARQUE NACIONAL;
0909/7941 (Angehr, 2003:239)
Colón/Panamá (Canal Zone) (E, Atlantic, Pacific)

ca. 200 m, a 22,000-ha national park that was established in May 1980 by Decreto Ley No. 13. The park is located in the former Canal Zone on the E side of the Panama Canal [0920/7955 (USBGN)] and extends roughly NW-SE for ca. 30 km. The park's N boundary is in the NE waters of Lago Gatún (Gatun Lake) [0912/7955 (USBGN)] and its S boundary is roughly parallel to and N of Camino Chiva Chiva (Chiva Chiva Road) [0902/7935 (USBGN)] (E762; RDP2; Angehr, 2003). Many ornithological localities — Pipeline Road [ca. 0915/7948], Camino Madden (Madden Road) [0907/7937 (USBGN)], Madden Road Timber Preserve [0906/7937 (USBGN)], a section of Carretera Gaillard [0904/7940 (USBGN)], the lower section of the Río Chagres [0907/7942 (USBGN)], Río Frijoles [0909/7946 (USBGN)], and Summit Park [0904/7939 (USBGN)], to name just a few — are now located within the park's boundaries. Martin and Karr, netting and banding, 1980-1981 (Martin and Karr, 1986); SBE, GL for Pipeline Road, 1991 Feb 14-17 (MVUP; USNM, as "Soberanía National Park"; Brumfield and Braun, 2001, as "Soberanía National Park"; Brumfield et al., 2001, as "start of Pipeline Road, near Gamboa"); Christian and Roberts, at 90 m, observations 3 km SE of the junction of the Río Chagres and the Panama Canal, 1996 Jul 8 (Christian and Roberts, 2000). (Karr et al., 1990a; Sieving, 1990; Poulin and Lefebvre, 1995; Poulin and Lefebvre, 1996; Poulin et al., 1999; Poulin et al., 2001; Roper and Goldstein, 1997; Hau et al., 2000; Morton and Stutchbury, 2000; W. Robinson et al., 2000a; Robinson and Robinson, 2001).

SOBERANÍA NATIONAL PARK; see Soberanía, Parque Nacional

SOLARTE, CAYO; see Nancy, Cayo

SONÁ; Veraguas (Pacific)

0801/8119 (USBGN)

ca. 25 m, city in W central Veraguas province on the Río San Pablo [0751/8110 (USBGN)] 19 km above its mouth, the Río Cobre [0801/8118 (USBGN)] passes N of town and the Río Tribique [0800/8119 (USBGN)] passes S of town (JOG). Heyde and Lux, "open woods," 1888 Feb 19, Mar 17, Apr 19 (USNM, as "Sona, Veraguas, Colombia"); Brown, 1901 Jul 10, 21, 25, 28, Aug 10 (MCZ; Bangs, 1902a; Thayer and Bangs, 1905:143); Griscom et al., 1924 Mar 29, Río San Pablo below Soná, 1924 Mar 28-29 (AMNH); Wetmore and Perrygo, Wetmore collected with Perrygo until Jun 10 after which Wetmore collected alone, specimens without secondary localities were presumably taken near town, 1953 May 16-18, 21, 27, Jun 7, 8 mi (13 km) E of Soná, 1953 May 27, 8 mi (13 km) W of Soná, 1953 May 30. The secondary localities attributed to Soná were located within 25 km of town. GL for: Congal, Guarumal, Guarumalito, Isla Verde, La Arena, La Isleta, Llano del Jardino, Río San Pablo, Río Tribique, Río Tuancle, San José, and Zapotillo, 1953 May 14-16, 18-21, 23, 25-28, 30, June 1-7, 9-13, 15 (USNM; Wetmore, 1953; Wetmore, 1956a; Wetmore, 1962; Wetmore, 1963a; Wetmore, 1967; Wetmore, 1970; Aldrich and Duvall, 1958; Huff and Wetmore, 1967; Olson, 1981a; Olson, 1983); Poole, GL for Llano del Jardino, 1953 May 26 (USNM). (Parkes, 1976; Olson, 1981d). See also: Río de Jesús.

SONA RIVER; see San Pablo, Río

SOSA HILL; Panamá (Canal Zone) (E, Pacific)

0857/7934 (USBGN)

320 feet (98 m), hill on the E side of the Panama Canal [0920/7955 (USBGN)] in Balboa [0857/7934 (USBGN)] 1 km W of Cerro Ancón (Ancón Hill) [0857/7933 (USBGN)] (E761A, as "Sosa High"). Hallinan, thick jungle, dense vines and shrubs, mango tree in a clearing, 1907 Sep 11, 13, 1914 May 7, 14, 20, 25-26, 1915 Feb 26, Jul 18, Sep 26, Oct 31, 1916 May 23 (AMNH; Hallinan, 1924).

SOUTH BAY; see Garachiné, Ensenada de

SPLIT HILL ISLAND; see Loma Partida, Isla

STA. FÉ; see Santa Fé

STRI; see Ancón

SUAI RIVER; see Suay, Río

SUAY, RÍO; Veraguas/Herrera (Pacific)

0744/8102 (USBGN)

Originates in Herrera province in N central Península de Azuero [0740/8035 (USBGN)] and flows WSW entering Veraguas province and the E coast of Golfo de Montijo [0740/8107 (USBGN)] 8 km E of Isla Leones [0744/8107 (USBGN)] (JOG). Mentioned by Aldrich and Boles (1937:7, as "River Suai") in describing the land held by the Boston-Panamá Company. See also: El Montuoso, Loma.

SUBESTACIÓN HIDROELÉCTRICA DEL BAYANO; see Bayano, Lago

SUBURB OF PANAMA CITY; see Juan Franco

SUCUBTÍ; see Surcurtí

SUCUBTÍ, RÍO; see Surcurtí, Río

SUMMIT; see Summit Park

SUMMIT, JARDIN BOTANICO; see Summit Park

SUMMIT GARDENS; see Summit Park

SUMMIT GOLF CLUB; see Summit Park

SUMMIT HILLS GOLF AND COUNTRY CLUB; see Summit Park

SUMMIT PARK; Panamá (Canal Zone) (E, Atlantic)

0904/7939 (USBGN)

75 m, town just N of the continental divide on Carretera Gaillard [0904/7940 (USBGN)] 4.5 km NW of Paraiso [0902/7938 (USBGN)] and 8.5 km SE of Gamboa [0907/7942 (USBGN)]. The S terminus of Camino Madden (Madden Road) [0907/7937 (USBGN)] is at Summit Park, and Summit Hills Golf and Country Club is on Carretera Gaillard 2 km SE of Summit Park (E8660). This is also the location of Summit Gardens "a botanical garden with a small zoo" (Engleman et al., 1989:480). Benson, at 100 feet (30 m), 1931 Jul 26 (USNM, as "Summit"); Meyer, 1940 Sep 14, 1941 Apr 6, 11, Jul 9, May 17, Jun 22-23, Jul 6, Aug 10, Sep 14, 20, Nov 16, 1942 Nov 26, 1944 Apr 19, 26 (USNM, as "Summit," and as "Summit Garden"); Scholes and Scholes, observations, 1950 Dec 17 (Scholes and Scholes, 1954, as

"Summit Gardens"); MCU, 1955 Nov 7 (USNM, as "Summit"); Willis, observations, 1960-1970 (Willis, 1972:36, as "Summit Gardens"); MARU, 1961 Aug 25 (USNM, as "Summit"); Yunker (from NIID), 1961 Oct 16, 1962 Jan 23 (USNM, as "Summit"); Gale, N. 1963 Jan (GML, as "Summit Gardens"; USNM, as "Summit Gardens"); Olson and Wiese, GL for Carretera Gaillard, 1966 May 8-10, 13, 15, 25-26, Jun 13, 15 (USNM, as "Summit Gardens"); Morton, 1970 Dec 19, 1971 Mar 17, and GL = Carretera Gaillard, near, 1974 Mar 12, observations and studies, 1970-1975 (CAS, as "Summit"; USNM, as "Summit Gardens"; Morton, 1976, as "Summit Gardens"; Morton, 1980, as "Summit Gardens"; Morton, 1982, as "Summit Gardens-Madden Forest"); White, GL for Carretera Gaillard, 1972 Apr 6 (ANSP, as "Summit"); Dyrcz, observations, 1979 Mar-Jun (Dyrcz, 1982, as "Summit Gardens"; Dyrcz, 1983, as "Summit Gardens"); Rodríguez, 1981 (GML, as "Summit"); Collector ?, 1989 Feb 1 (LSUMZ, as "Summit Gardens"); Brawn et al., mist nets, 199? (Brawn et al., 1996, as "Summit"); Gorrichategui, 1992 Aug 15 (MVUP, as "Jardin Botanico Summit"); Braun, M., GL = Carretera Gaillard, near, 1994 Apr 13, May 6 (USNM, as "Summit Gardens"); Collector ?, GL for Carretera Gaillard, 1998 Oct (LSUMZ, as "Summit Golf Club"); Babin, GL for Old Gamboa Road, 1999 Mar 20, 22-25 (LSUMZ, as "Summit Gardens"). (N. Smith, 1975, as "Summit Gardens"; Olson, 1983, as "Summit").

SUMWOOD, CANAL; Bocas del Toro

0909/8209 (USBGN)

A 100-200 m wide marine channel that extends through mangroves between the SW coast of Isla Popa [0911/8207 (USBGN)] and the NE coast of Isla Loma Partida [0909/8210 (USBGN)] (E762). SBE, used to further modify specimens from Isla Popa (q. v.) and from Isla Loma Partida (q. v.), 1991 Feb 24-27, Mar 1-13 (USNM, as "Sumwood Channel"; Olson, 1993a, as "Sumwood Channel").

SUMWOOD CHANNEL; see Sumwood, Canal

SURCURTÍ, RÍO; Darién

0844/7759 (USBGN)

Originates in N Darién province on the Cordillera de San Blas and flows WSW entering the Río Chucunaque [0809/7744 (USBGN)] 70 km NNW of Yaviza [0811/7741 (USBGN)] (MGRP; SMN2B, as "Río Sucubtí"). Baer, GL = Yaviza, Río Chucunaque, mouth of the Río Surcurtí, according to Marsh (1934:153), the expedition made camp just above the river's mouth, 1924 Apr 13 (USNM). For more on Baer see comments under Marsh (1934).

SWAN CAY; Bocas del Toro, Archipiélago de Bocas del Toro

0927/8218 (USBGN)

Sea level, small island with a land area of 0.02 km^2 (2 ha) (Angehr, 2003:94), located in the Caribbean Sea [1500/7500 (USBGN)] 2 km off the N coast of Isla Colón [0924/8217 (USBGN)] (E762; JOG, as "Cayo Swan"). Described by Peters as an islet located between Isla Bastimentos [0918/8208 (USBGN)] and Isla Popa [0911/8207 (USBGN)], which is ca. 30 km SE of its actual location, although he also indicated that the island was not shown on maps available to him (Peters, 1931a:295, as "Swan Key"). The island may also be referred to as Cayo del Cisne or Cayo

Cisne and is considered an Important Bird Area (Angehr, 2003:94). Smith, 1927 Apr 17 (YPM, as "Swan Key"); Benson, 1927 Jun 3 (USNM, as "Swan Key"); Wedel, 1927 Jun 3, 1938 Mar 18, Apr 23-25, May 9, 12 (CMNH, as "Swan Key"; FMNH; MCZ, as "Swan Key, Almirante Bay"; SDSNH; Peters, 1930; Peters, 1931a; and Griscom, 1935b:310, all as "Swan Key"); Wetmore, on the water close to the sheltered S side near a sandy beach with palms and lower vegetation, 1958 Jan 26 (USNM, as "Swan Key").

SWAN KEY; see Swan Cay

TABASARÁ; Chiriquí
0804/8138 (USBGN)
200 m, town in easternmost Chiriquí province on the W banks of the Río Tabasará [0802/8142 (USBGN)] S of the Carretera Interamericana [0832/8242 (USBGN)] ca. 12 km SSE of Tolé [0814/8141 (USBGN)] and ca. 12 km NNW of Puerto Vidal [0803/8136 (USBGN)] (RDP2). Morales, this single specimen appears to have been mislabeled or misread as Morales' other specimen labels indicate that he was working in W Chiriquí on this date, 1968 Jan 30 (GML); Hinds, 1981 Aug 2, 25, Sep 2 (GML); Collector ? [Hinds ?], 1981 Sep 5 (GML).

TABASARÁ, RÍO; Chiriquí/Veraguas (Pacific)
0802/8142 (USBGN)
Originates in NE Chiriquí province in the Serranía de Tabasará [0833/8140 (USBGN)], flows S crossing the Carretera Interamericana [0832/8242 (USBGN)] 11 km ESE of Tolé [0814/8141 (USBGN)] after which it marks the Chiriquí/Veraguas provincial boundary and continues S into the Pacific Ocean. Just above its mouth the river crosses the Carretera Interamericana 6.5 km SE of Quebrada de Piedra [0806/8141 (USBGN)] and 4 km WNW of Puerto Vidal [0803/8136 (USBGN)] (JOG). Wetmore and Perrygo, GL = Quebrada de Piedra, on the road 2 mi (3 km) W of the river in Chiriquí province, trail down a hillside to a creek bordered by mangrove and wet forest that floods at high tide, hill slopes cleared for cultivation, thorny growth, 1953 Jun 6, GL = Puerto Vidal, at the road bridge over the Río Tabasará, 1953 Jun 6 (USNM). (Delgado, 1983).

TABASARÁ, SERRANÍA DE; Chiriquí
0833/8140 (USBGN)
Mountain range within the Cordillera Central [0838/8150 (USBGN)] in E Chiriquí and W Veraguas provinces. The range extends along the continental divide in the Reserva Indígena del Tabasará [0825/8142 (USBGN)] (RDPF, as "Serranía del Tabasará"; MHA).

TABERNILLA; Colón ? (Canal Zone) (Atlantic)
ca. 0908/7949
Former station on the original Panama Railroad [ca. 0922/7954] located on the continuous Río Chagres (see Río Chagres [0919/8000 (USBGN)]) ca. 3 mi (5 km) S of Frijoles [0910/7948 (USBGN)] (IP; PR). According to E866, the site is now submerged in Lago Gatún (Gatun Lake) [0912/7955 (USBGN)] just SE of Isla Barro Colorado [0909/7951 (USBGN)]. This locality is near what is now the Colón/Panamá provincial boundary, and it is estimated to lie within Colón province.

Goldman, 1911 Mar 10-12, Apr 2, 8, 13, 15-18, 20, 29 (ANSP; USNM; Wetmore, 1970; Olson, 1981d; Olson, 1983; Brumfield et al., 2001); Jewel, thicket by a stream, nest in a dead stump in an open field, 1911 Mar 19, Apr 8, 18 (Stone, 1918). (Eisenmann, 1952:12). See also: Caño Quebrado, Río.

TABLAS, LAS; see Las Tablas

TABOGA; Panamá, Isla Taboga (Pacific)
0848/7933 (USBGN)
Sea level, town on the NE coast of Isla Taboga [0847/7933 (USBGN)] (E761B) that was estimated by Wetmore to have ca. 200 houses in 1952 (USNM). "Taboga" on specimen labels most likely refers to the Isla Taboga (q. v.) in general, not the small town.

TABOGA, ISLA; Panamá (W, Pacific)
0847/7933 (USBGN)
307 m, island in Bahía de Panamá [0850/7915 (USBGN)] 20 km S of Panama City [0858/7932 (USBGN)]. The island extends 4 km from NW to SE and varies in width from 700 m to 2.5 km. From its shores the land quickly rises to 400 feet (120 m) with several peaks over 600 feet (180 m) (E671B; JOG). According to Wetmore, in 1952 the central peak (307 m) rose up in forested slopes to exposed rocky soil that was covered in bunch grass and its precipitous W side harbored a colony of pelicans. Several dismantled military installations were located in the SE where 6,000 men had been stationed during WWII (USNM). According to Angehr (2003:248-249), the steep W and S sides of the island, along with Isla Uravá [0847/7932 (USBGN)], are now located in Refugio de Vida Silvestre Taboga (Taboga Wildlife Refuge) [0847/7933 (Angehr, 2003:248)]. Collector ?, date ? (AMNH; ANSP; BMNH; FMNH; USNM; Saunders, 1876, as "Taboga"; Wetmore, 1952:16); Hallinan, 1915 Mar 27, Aug 15, Sep 12, Dec 5 (AMNH; Wetmore, 1952:3, 8, 10); Hicks, 1865 Jan 30-31 (USNM; Wetmore, 1952:3, 15, 16); Heyde and Lux, 1888 Aug-Sep (USNM, as "Taboga," and as "Island Taboga"; Ridgway, 1911b:347, as "Saboga Island" (but see Wetmore 1968: 368); Wetmore, 1952:3, 15, 16); Brown, 1904 Apr 5, 11 (USNM, as "Taboga Island"); Griscom et al., 1924 Feb 17 (Wetmore, 1952:3); Kelsall and Cullingford, 1924 Sep (BMNH, as "Taboga Island"; Douglas and Johnson, 1926, as "Taboga Island"); Miller, 1937 Apr 5 (USNM, as "Taboga Island"); Eisenmann, observations, 1951 Jul 5 (Wetmore, 1952:5); Wetmore and Perrygo, collected throughout the island, along a steep trail on the W side of the central peak with bunch grass and forested slopes, along a trail on the N and W shores through open woodland, extensive papaya, yuca and pineapple plantations, on level land near the sea with steep, wooded pitches above, near a fresh water stream, in the SE near old military installations, in cultivated clearings, in dry open forest that was 30-50 feet (9-15 m) high, 1952 Mar 15-17, 19, 21-23 (USNM; Wetmore, 1952; Huff and Wetmore, 1967, as "Taboga Is."); Wetmore, road and trails in the SE, 1955 Dec 25-26 (USNM), observations at sea and on the mainland near Cerro Cabra [0854/7938 (USBGN)], 1955 Dec 25 (Wetmore, 1965:195); Robins, observations, 1957 Jul 15-26 (Robins, 1958); Pennycuick, observations, 1980 May 21-Jun 6 (Pennycuick, 1983); Montgomery and Martínez, observations, 1981 Jan-

Jul (Montgomery and Martínez, 1984, as "Taboga Island"); Moreno, 1985 Mar 30 (MVUP). (Wicker, 1917; Roubik et al., 1985; Olson, 1997a). See also: Panamá, Bahía de; Restinga; Taboga; Taboga, Morro de; Taboguilla, Isla; and Uravá, Isla.

TABOGA, MORRO DE; Panamá (W, Pacific)
0848/7933 (USBGN)
100 feet (30 m), very small island with a land area of ca. 0.04 km^2 (4 ha) located 200 m off the NE coast of Isla Taboga [0847/7933 (USBGN)] (E761A; E661, as "Morro I."). Described by Wetmore as a wooded islet that is actually part of Isla Taboga as the two are joined by a sandy beach at low tide. Meyer, 1941 Mar 2 (USNM, as "Morro Island"); Wetmore and Perrygo, observations, 1952 Mar 15-23 (Wetmore, 1952, as "El Morro").

TABOGA ISLAND; see Taboga, Isla

TABOGA WILDLIFE REFUGE; see Taboga, Isla

TABOGUILLA, ISLA; Panamá (W, Pacific)
0848/7931 (USBGN)
213 m, island in Bahía de Panamá [0850/7915 (USBGN)] 20 km S of Panama City [0858/7932 (USBGN)] and 3.5 km NE of Isla Taboga [0847/7933 (USBGN)] (q. v.). The island extends 2.5 km from NE to SW and is generally 1 km wide. From its shores the land quickly rises to 200 feet (60 m) with a single peak at 213 m in its center (E671B; JOG). Hallinan, rocky islets offshore covered with trees, cacti and shrubbery, on the water, 1915 Aug 15, Sep 12, Dec 5 (AMNH; Hallinan, 1924, as "Taboguillo Island"; Wetmore, 1952:3, 8-10, 20); Kelsall and Cullingford, 1924 Sep (BMNH, as "Taboguilla Island"; Douglas and Johnson, 1926, as "Taboguilla Island"); Miller, 1937 Mar 30, Apr 1 (USNM); Meyer, GL for Isla Farallón, 1942 Dec 14, 1944 Apr 9 (USNM, as "Toboguilla Island"); Wetmore and Perrygo, landed on the N coast where only 3 families lived in 1952, the area was largely forested with a few cultivated clearings and the high point was covered with dry scrub, to a valley with a small seep of water, in tracts of extensive woodland, up steep slopes, 1952 Mar 18, 20; Wetmore, small beach on the SW shore, steep slope through dense forest to the other side of island, trails inland, in an area known as Papallel (not located) on the E side of the island, 1955 Dec 24 (USNM; Wetmore, 1952); Pennycuick, observations, 1980 May 21-Jun 6 (Pennycuick, 1983); Cardiff, GL = Bahía de Panamá, 2 km E of, 1996 Feb 28 (LSUMZ, as "Isla Tobaquilla"). (Wicker, 1917; Olson, 1980b, as "Taboguilla Island"; Olson, 1997a; Reed, 1988; Angehr, 2003:250-251). See also: Panamá, Bahía de.

TABOGUILLA ISLAND; see Taboguilla, Isla

TABOGUILLO ISLAND; see Taboguilla, Isla

TABORCILLA ISLAND; see Taborcillo, Isla

TABORCILLO, ISLA; Panamá (W, Pacific)
0840/7944 (USBGN)
Sea level, island in the E waters of Bahía Chame [0838/7947 (USBGN)] 5 km NW of Punta Chame [0839/7942 (USBGN)] and 20 km S of La Chorrera [0853/7947 (USBGN)] (JOG). An Important Bird Area that Angehr (2003:230) describes as "a

low sandy crescent-shaped island" containing dry scrub forest on the E and W ends. Ridgely, observations, 1976 Apr 25 (Ridgely and Gwynne, 1989:70, as "Taborcilla Island").

TACARCUNA; see Tacarcuna Village

TACARCUNA; see Tacarcuna, Cerro

TACARCUNA, CERRO; Darién
0811/7717 (USBGN)
1,875 m, peak in the Serranía del Darién [0830/7730 (USBGN)] located 48 km E of El Real de Santa María [0808/7743 (USBGN)]. Several maps indicate that the peak is just W of the continental divide and therefore W of the Colombian border (CARP; MGRP; RDPF), but according to both MHA (1934) and USBGN, the peak is on the divide and the border. Anthony, Richardson, and Ball, at ca. 4,500-5,200 feet (1,350-1,600), according to Wetmore's journal, steep sides limited access to the peak, and along with Fairchild and Handley (1966), Wetmore believed that Anthony and Ball did not arrive at Cerro Tacarcuna in 1915, but worked on "its spur," the nearby Cerro Malí [0807/7714 (FH)] (q. v.). According to Anthony (1916:358), late in March they traveled from Tacarcuna Village (q. v.) to "a still higher camp, about 5,200 feet (1,600 m) above sea level and approximately the highest continuous country attained...on the slopes of Mt. Tacarcuna." In his descriptions of the region Anthony (1916:360-362) states that "the character of the forest remains the same eastward from the mouth of the Tuyra [sic] until an elevation of about 4,500 feet, on the slopes of Mt. Tacarcuna, is reached. The density of the growth decreases with the rise in elevation but at the mountain a different type of forest is found. The growth is neither so heavy nor the trees so large as in the lowland." They collected on both the Colombian (Atlantic drainage) and the Panamanian (Pacific drainage) slopes of the peak, and specimens labeled "W slope" appear to be from Panama, 1915 Mar 28-Apr 18 (AMNH, as "Mount Tacarcuna"; MCZ, as "Mount Tacarcuna"; USNM, as "Mount Tacarcuna"; ZSM, as "Tacarcuna"; Chapman, 1915b, as "Mt. Tacarcuna"; Chapman, 1923, as "Mt. Tacarcuna"; Chapman, 1924b, as "Mt. Tacarcuna"; Chapman, 1926, as "eastern Panama"; Anthony, 1916, as "Mt. Tacarcuna"; Hellmayr, 1921, as "Tacarcuna"; Barbour and Brooks, 1923, as "Mt. Tatarcuna"; Griscom, 1924b, as "Mt. Tacarcuna"; Griscom, 1927a, as "Mt. Tacarcuna," and as "Tacarcuna"; Zimmer, 1939, as "Mt. Tacarcuna"; Zimmer, 1951:13, as "Mt. Tacarcuna"; Wetmore, 1950b; Wetmore, 1967; Parkes, 1954a, as "Mt. Tacarcuna"; Olson, 1981c); Wetmore, at 4,100-4,800 feet (1,200-1,450 m), camp was located on a natural opening 100 yards (90 m) across on a bench above the N fork of the Río Púcuro [0801/7733 (USBGN)], the slope below was grown with a heavy stand of bracken, elsewhere unbroken forest, below 4,700 feet (1,400 m) dense, gallery forest, above 4,700 feet (1,400 m) heavy cloud forest marked by palms and luxuriant growth, 1964 Feb 20, 28-29, Mar 2, at 4,800 feet (1,450 m), on the Colombian border, Feb 28, Mar 3, and as a GL for: Río Púcuro, and Río Tacarcuna, 1964 Feb 23, 25, 28, Mar 1-2 (USNM; Wetmore, 1967); Handley, at 4,100-4,800 feet (1,250-1,450 m), 1964 Feb 29, Mar 3, 5-7, 10-12, Nov 7 (USNM); Hinds and GMLP, at 1,650 m, 1975 Feb 2-12, and as a GL for Río Púcuro, 1975 Jan

26-30 (GML; LSUMZ; USNM; Wetmore, 1967). (Wetmore, 1962; Myers, 1969). See also: Malí, Cerro, and Tacarcuna Village.

TACARCUNA, MOUNT; see Tacarcuna, Cerro

TACARCUNA, RÍO; Darién

ca. 0805/7718

Shown on MRP and on MHA (1934) as a tributary of the Río Capetí [0805/7735 (USBGN)] that originates on the S slopes of Cerro Tacarcuna [0811/7717 (USBGN)]. Wetmore, however, described the river as a tributary of the Río Púcuro [0801/7733 (USBGN)] originating S of Cerro Tacarcuna near Cerro Malí [0807/7714 (FH)] and flowing W, presumably between the Río Púcuro and the Río Topalisa [0801/7728 (USBGN)]. Barrett, specimen labeled "ca. 2 mi from T. Camp," which presumably refers to Tacarcuna Village [0805/7717 (FH)], 1963 Jun 28 (USNM, as "Tacarcuna River"); Wetmore, GL = Cerro Tacarcuna, mouth of the Río Tacarcuna, on the Río Púcuro, 1964 Mar 1 (USNM).

TACARCUNA, SERRANÍA DE; see Darién, Serranía del

TACARCUNA CAMP; see Tacarcuna Village

TACARCUNA CASITA; see Casita

TACARCUNA LAGUNA; see La Laguna

TACARCUNA RIDGE; see Darién, Serranía del

TACARCUNA RIVER; see Tacarcuna, Río

TACARCUNA VILLAGE; Darién

0805/7717 (Fairchild and Handley, 1966)

ca. 2,300 feet (700 m), not shown on our maps, but according to a letter from Pedro Galindo dated 2 Sep 1958 (USNM, as "Tacarcuna Camp"), "the camp is located on the tail end of the massif known as Serranía del Darién, which at this point forms a spur which protrudes between the valleys of the Tapalisa [sic] and the Pucro [sic] rivers." According to Wetmore's journal, the camp was on the Río Tacarcuna [ca. 0805/7718] ca. 3 mi (5 km) above its mouth where the riverbed was rough with many boulders and its fall was considerable. Wetmore (1972:481) also calls it "the ancient village site" on the base of Cerro Tacarcuna. Anthony and Ball, at ca. 2,050-4,400 feet (m), set up camp at 2,600 feet (800 m) at "Tacarcuna," which according to Anthony (1916:359-360) was "an old Indian village (deserted most of the time)." The surrounding region was "very steep and mountainous. The ridges [ran] in long parallel series and [were] terraced off by the intervening ravines in rises of about 1,000 to 2,000 feet (300-600 m). Everywhere the same heavy forest and undergrowth flourish[ed], and [was] so thick that only occasionally [could] one see any distance from the crest of a high ridge." From Tacarcuna the trail reached "the headwaters of the Río Tapalisa [sic]" before reaching Cerro Tacarcuna (q. v.). Presumably specimens labeled simply "Tacarcuna" came from Tacarcuna Village, and those labeled "Mount Tacarcuna" were collected at Cerro Tacarcuna, although these two designations appear to be used somewhat interchangeably in the literature and the dates given by Anthony (1916:376), which are presented here, and the elevations, based on Anthony's (1916) descriptions, may also be used to distinguish between the two localities, 1915 Feb 20-Mar 27 (AMNH, as "Tacarcuna"; MCZ; USNM; Chapman, 1915a; Chapman, 1915b; Chapman, 1923; and Chapman, 1924a, all as "Tacarcuna"; Chapman, 1926, as "eastern Panama"; Chapman, 1929b; Anthony, 1916; Griscom, 1927a; Griscom, 1927b; Griscom, 1928; Griscom, 1929a; Griscom, 1931; Zimmer, 1932:9; Zimmer, 1934:14, 20; Zimmer; Zimmer, 1941b:11; Zimmer, 1950:15; and Eisenmann and Howell, 1962, all as "Tacarcuna"; Adsett and Wege, 1998, as "Old Village Site"); Feinstein, at 3,200 feet (1,000 m), 1959 Jan 19, 29, Feb 1, 6, 9-12, 14, 17, 20-22, 26-28, Mar 6, 9, 12, 20, and as a GL for Casita, 1959 Jan 21-28 (USNM, as "Tacarcuna," and as "Tacarcuna Village Camp"; Wetmore, 1962, as "Old Tacarcuna Village site"); Collector ? [Feinstein ?], GL for Casita, 1959 Jan 24 (USNM, as "Tacarcuna"); Handley, at 3,200 feet (1,000 m), 1959 Feb 24, Mar 1, 19, at 4,100 feet (1,250 m), 1964 Mar 3, and as a GL for: Cerro Malí, and La Laguna, 1959 Mar 10, 15, 18 (USNM, as "Tacarcuna," and as "Tacarcuna Village Camp"); Galindo, Hinds, Mendez, and GMLP, at 2,000 feet (600 m), 1963 Mar 27, Jun 17, 19-30, Jul 1, 4-9, 11-19, and GL = base of Cerro Malí, headwaters of the Río Púcuro, at 950 m, 1963 Jul 4 (GML, as "Tacarcuna," and as "Tacarcuna Village"; LSUMZ; USNM, as "Old Tacarcuna Village Site," as "Tacarcuna," and as "Tacarcuna Village"; Wetmore, 1963b); Wetmore, at 1,900-2,500 feet (600-750 m), collected primarily in the woods along the river and its tributaries, a beach, an open section along the river, 1964 Mar 4-17 (USNM; Olson et al., 1968, as "Tacarcuna"; Adsett and Wege, 1998, as "Old Tacarcuna Site"). (Olson, 1980a; Olson, 1981c; Olson, 1983). See also: Tacarcuna, Cerro.

TACARCUNA VILLAGE CAMP; see Tacarcuna Village

TACOMA; see Tocumen

TACOMA MARSHES; see Tocumen

TAIMATÍ; see Taimatí, Bahía de

TAIMATÍ, BAHÍA DE; Darién

ca. 0808/7815

Not shown on our maps, but must refer to the coast of Ensenada de Garachiné [0808/7820 (USBGN)] (q. v.) near the town Taimatí [0809/7814 (USBGN)] and the mouth of the Río Taimatí [0808/7815 (USBGN)], which are located on MGRP ca. 15 km NE of Garachiné [0804/7822 (USBGN)]. Palomino, GL = Taimatí, 1984 Aug 18 (MVUP).

TAIMATÍ, RÍO; Darién

0808/7815 (USBGN)

Originates in the hills S of La Palma [0825/7809 (USBGN)] and flows NW entering the S coast of Golfo de San Miguel [0822/7817 (USBGN)] 10 mi (16 km) NE of Garachiné [0804/7822 (USBGN)] (MHA; SMN2B; Barbour and Brooks, 1923, map). On MGRP the river flows due W, but enters the bay at the same location. See also: Esnape, Río.

TAIMATÍ, RÍO; see Taimatí, Bahía de

TALAMANCA, CORDILLERA DE; (COSTA RICA)

0930/8340 (USBGN)

Mountain range that extends through E Costa Rica and connects to the Cordillera Central [0838/8150 (USBGN)] in W Panama (MHA). Some collectors have referred to the mountains in western Panamá as the Cordillera de Talamanca. See also: Central, Cordillera.

TAMBO; see Toabré

TAMOGANTÍ; see Las Cañitas

TANK FARM; see Arraijan Tank Farm

TAPALISA; Darién

0759/7726 (Fairchild and Handley, 1966)

Anthony's (1916:358) description of the town is somewhat ambiguous as he initially indicates that "the river [Tuira was] ascended to the limit of canoe navigation at Tapalisa," but later states that Tapalisa is on the Río Púcuro [0801/7733 (USBGN)]. In addition he states that "the character of the country remains the same from San Miguel Bay to Tapalisa, mainly flat, with low, rolling, forested hills," but then goes on to say that "at Tapalisa an elevation of 1,000 feet is reached and here the first line of foothills of the cordillera are met." Fortunately, his map places the town at a location that agrees with Fairchild and Handley's coordinates, just above the mouth of the Río Topalisa [0801/7728 (USBGN)] and according to our maps, the elevation here is ca. 100 m. Richardson, at 400 feet (120 m), 1915 Feb 16-Mar 19 (AMNH; Chapman, 1915a, as "Tapaliza"; Chapman, 1926, as "eastern Panama"; Chapman, 1929b; Anthony, 1916; Hellmayr, 1921, as "Tapaliza"; Hellmayr, 1929:84; Griscom, 1927a; Griscom, 1929a; Griscom, 1935a; Zimmer, 1934:20; Zimmer, 1950:26; Haffer, 1962, as "Tapaliza").

TAPALISA, RÍO; see Topalisa, Río

TAPALIZA; see Tapalisa

TAPIA; see Tapia Número Dos

TAPIA, RÍO; Panamá (E, Pacific)

0902/7924 (USBGN)

Originates NE of Panama City [0858/7932 (USBGN)] and flows S entering an area of grass and coastal mangroves before discharging into Bahía de Panamá [0850/7915 (USBGN)] 18 km E of Panama City (E661; E761A; MGRP). Arbib and Loetscher, observations, "grassy savannah-land, swampy rolling country," the authors do not describe the town, currently called Tapia Número Dos [0903/7924 (USBGN)] (q. v.), but as some entries are listed as simply "Tapia," they presumably worked near the town, 1933 Aug 24, 1934 Aug 29 (Arbib and Loetscher, 1935, as "Tapia," and as "Río Tapia").

TAPIA NÚMERO DOS; Panamá (E, Pacific)

0903/7924 (USBGN)

20 feet (6 m), town 5.5 km E of Juan Díaz [0903/7927 (USBGN)] and 4 km SSE of Pedregal [0904/7926 (USBGN)], located 2 km S of the Carretera Interamericana

[0832/8242 (USBGN)] on the Río Tapia [0902/7924 (USBGN)] just N of the coastal mangroves (E661, as "Tapia Abajo"; E761A, as "Tapia Abaja"). Chapin, Rogers and Mathews, between Juan Díaz and Pacora, "on the road several miles east of Juan Díaz," "the open, marshy country just back of the coast, several miles southeast of Juan Díaz and twelve to fifteen miles (19-24 km) east of Panama City [0858/7932 (USBGN)]," 1923 Jul 21, 24-29, 31, Aug 1-7, 9, 11 (FMNH, as "Tapia"; Griscom, 1935b:347, as "Tapia River Marshes"; Rogers, 1939, as "Juan Díaz," and as "Tapia"); Collector ? [Chapin et al. ?], 1923 Jul 25, Aug 4-5 (FMNH, as "Tapia"); Lance, 1931 Mar 26 (ANSP, as "Tapua"); Benson, at 100 feet (30 m), 1931 Aug 10 (USNM, as "Tapia"); Blanton, 1953 Sep 23 (USNM, as "Tapia Road"). See also: Tapia, Río.

TAPIA ROAD; see Tapia Número Dos

TAPUA; see Tapia Número Dos

TATARCUNA, MOUNT; see Tacarcuna, Cerro

TEJEL, QUEBRADA; Herrera
ca. 0756/8035
Described by Wetmore as a 30-40 foot (9-10 m) wide stream that crossed the road to La Trinidad [0755/8043 (USBGN)] ca. 3 mi (5 km) W of Pesé [0754/8037 (USBGN)] (USNM). Not shown on our maps, but presumed to be a tributary of the Río Parita [0802/8027 (USBGN)]. Wetmore and Perrygo, GL = Pesé, through pastures bordered by tall spreading trees and occasional thickets, 1948 Mar 29 (USNM; Wetmore, 1950a).

TERABLE, RÍO; Panamá (E, Pacific)
0913/7859 (USBGN)
Originates in the hills N of El Llano [0913/7859 (USBGN)] near the Comarca de San Blas/Panamá provincial border and flows S entering the Río Chepo [0900/7906 (USBGN)] just SW of El Llano (SMN2A). Wetmore, GL = El Llano, the clear, rapid stream was 70-75 feet (20-25 m) wide with a stony bed, tall grass in the pastures, scrub and occasional trees along fences and small streams, 1962 Feb 3 (USNM, as "Río Terrable").

TEREBE, RÍO; see Teribe, Río

TEREBE RIVER; see Teribe, Río

TERIBE, RÍO; Bocas del Toro
0922/8232 (USBGN)
Originates on the N and W slopes of Cerro Fabrega [0907/8252 (USBGN)], flows NE then E joining the Río Changuinola [0928/8227 (USBGN)] 9 km S of Changuinola [0926/8231 (USBGN)] (JOG). According to USBGN, the river is called the Río Terebe in Costa Rica, where some of its tributaries originate. Smith, at Zegla (q. v.), 1927 May 21 (Griscom, 1933, as "Zegla, Terebe River"); Handley and Rhymer, 1960 Mar 7 (USNM, as "Boca de Río Teribe"); Handley and Greenwell, 1962 Oct (USNM); SBE, at Zegla (q. v.), and GL = Charagre, at Quebrada Carbon [0920/8233 (USBGN)] (q. v.), 1991 Feb 24, 28 (USNM; Parsons et al., 1993; Brumfield and Braun, 2001; Brumfield et al., 2001).

TERRABA, PUNTA; Bocas del Toro

0926/8220 (USBGN)

Point on the mainland in NW Bocas del Toro province on a peninsula that separates the N waters of Boca del Drago [0925/8220 (USBGN)] from Tirbi Bight [ca. 0926/8222] (MHA, as "Punta Tirbi o Terraba"). See also: Tirbi Bight; and Bocas del Toro, Archipiélago de.

TERRABLE, RÍO; see Terable, Río

TESORERA, RÍO; see Abajo, Río

THATCHER, CARRETERA; Panamá (Canal Zone) (W, Pacific)

0858/7938 (USBGN)

The section of the Carretera Interamericana [0832/8242 (USBGN)] within the former Canal Zone that extends between Arraijan [0857/7939 (USBGN)] and La Boca [0857/7934 (USBGN)] (E661, as "Thatcher Highway"; RM1, as "Thatcher Highway," and as "Road K5"). See also: Arraijan; and West Ferry Ramp.

THATCHER FERRY; see West Ferry Ramp

THATCHER FERRY BRIDGE; see West Ferry Ramp

THATCHER HIGHWAY; see Thatcher, Carretera

THE LAKES; see Las Lagunas

THOMAS BARBOUR TRAIL; see Barro Colorado, Isla

THREE FALLS CREEK; Darién

ca. 0810/7741 ?

Baer was based at Camp Townsend [ca. 0810/7741] in late February, but he also may have been at the Amerindian village described under Camp Creek (Marsh, 1934:128) for a period of time. During late Feb, specimens were also collected on the Río Chico [0811/7741 (USBGN)], and presumably Three Falls Creek was near either Camp Townsend or Camp Creek, Baer, GL = Yaviza, Río Chucunaque, 1924 Feb 23-25 (USNM). For more on Baer see comments under Marsh (1934). Not shown on our maps.

TIARILLO, QUEBRADA; see Pequení

TIBURÓN, CABO; San Blas

0842/7721 (USBGN)

Sea level, small point on the Comarca de San Blas/Colombian border extending ca. 2 mi (3 km) N into the Caribbean Sea [1500/7500 (USBGN)] (SMN2B). See also: Puerto Obaldía; and Urabá, Golfo de.

TIERRA OSCURA; Bocas del Toro

0911/8217 (USBGN)

Sea level, a settlement on the S coast of Laguna de Tierra Oscura [0912/8216 (USBGN)] (Olson, 1993a:101 (map)), although E762 labels the entire region on the W and S coasts of the bay as Tierra Oscura. The settlement is located ca. 17 km S of Bocas del Toro [0920/8215 (USBGN)]. Wedel, GL = Almirante or Bahía Almirante, 1936 Feb 19, 1939 Feb 24 (CMNH, as "Darkland"; FMNH, as "Darkland"); SBE, most specimen labels are unmodified, but one says "16 km S of

Bocas del Toro," 1989 Feb 3-5, Feb 20-Mar 5 (MVUP; USNM; Olson, 1993a; Parsons et al., 1993; Brumfield et al., 2001).

TIERRA OSCURA, LAGUNA DE; Bocas del Toro
0912/8216 (USBGN)
The more western of two well-protected bays between the S coast of Isla San Cristóbal [0915/8216 (USBGN)] and the mainland coast. Laguna de Tierra Oscura extends 2 km from E-W between 2 small peninsulas on the mainland and the bay extends 4 km from N to S (JOG). SBE, 1989 Mar 2 (USNM; Olson, 1993a).

TIGER, CAYOS; Bocas del Toro, Archipiélago de Bocas del Toro
0913/8156 (USBGN)
Group of ca. 6 very small islands located 3 km N of Punta Valiente [0911/8155 (USBGN)] on Península Valiente [0905/8151 (USBGN)] (E762; JOG, as "Cayos Tigre"). According to Angehr (2003:103-104), the 2 largest islets in this Important Bird Area are each approximately 5 ha, "with scattered trees, coarse grass and other dense herbaceous vegetation." Miller and Richardson, GL for Middle Cay, 1936 Mar 12 (UCLA, as "Tiger Rocks"); Wetmore, GL for Tiger Rock, 1958 Feb 25 (USNM, as "Tiger Cays"; Wetmore, 1959a, as "Tiger Cays"); SBE, GL = Península Valiente, at Middle Cay (q. v.), 1990 Apr 24 (MVUP, as "Roca Tigre"; USNM, as "Tiger Rocks"; Olson, 1993a, as "Tiger Rocks").

TIGER CAYS; see Tiger, Cayos

TIGER HILL; Panamá (Canal Zone) (Atlantic)
ca. 0914/7954
28 feet (9 m), former station on the original Panama Railroad [ca. 0922/7954] that was located 2.5 mi (4 km) SSE of Gatún [0916/7955 (USBGN)] (MIP). Hallinan, collected near the former station that he described as being ca. 14 mi (23 km) from the Atlantic entrance of the Panama Canal [0920/7955 (USBGN)], in the jungle, shrubbery, 1909 Jan 24, Feb 20, Mar 21 (AMNH; Hallinan, 1924).

TIGER ROCK; Bocas del Toro, Archipiélago de Bocas del Toro
0912/8155 (USBGN)
Sea level, the easternmost and the largest island in Cayos Tiger [0913/8156 (USBGN)] with a land area of ca. 0.05 km^2 (5 ha) (E762). The western end of the island is "nearly divided by a cleft through which storm waves wash" (Wetmore, 1959a). According to Wetmore, the island rose to 35-40 feet (10 m) and the crests supported coconut palms, other trees and a few creepers. The high crest had much *caña blanca* while the side slopes on the leeward side were covered with an open growth of a woody, broad-leafed plant, Wetmore, GL = Cayos Tiger, Península Valiente [0905/8151 (USBGN)], off Punta Valiente [0911/8155 (USBGN)], on the E side of Tiger Rock climbed up a rough surface of volcanic rock through palms and undergrowth to more level ground, 1958 Feb 25 (USNM; Wetmore, 1959a). (Angehr, 2003:104). See also: Middle Cay.

TIGER ROCKS; see Tiger, Cayos

TIGER TRAIL; Colón (Canal Zone) (W)
ca. 0916/7959

Not named on our maps, but according to Angehr (personal communication), the name refers to a trail in Colón province a few km W of Gatún [0916/7955 (USBGN)]. The N terminus of the trail is located at a central point on Road S1 [ca. 0918/8001] and from here the trail extends ca. 4 km NW-SE between Road S1 and the W banks of the Río Chagres [0919/8000 (USBGN)]. The trail terminates on the NE coast of Lago Gatún (Gatun Lake) [0912/7955 (USBGN)] 1 km W of Represa Gatún (Gatún Dam) [0916/7956 (USBGN)]. According to Engleman et al. (1989:486-7), the trail extends along an old railway with a shady canopy and sparse undergrowth. Engleman, D. and Gardner, observations, near Gatun Dam, 1984 Dec 20, 29 (Ridgely and Gwynne, 1989:371); Seutin, GL = Gatún, 3 km WSW Gatún, 1994 Sep 24 (LSUMZ; MVUP).

TIGRE; Coclé (Atlantic)

ca. 0842/8036

Described by Wetmore as a 10 acre clearing in forested hills, ca. 9 km N of El Copé [0837/8036 (USBGN)] on the Atlantic drainage near the head of the Río Guabal [ca. 0846/8036]. According to Angehr (in litt., Aug 2005), El Tigre appears in this region on map 4042-II (Coclesito) [E762]. JOG incorrectly indicates an elevation of ca. 1,000 m for this region. Wetmore, GL = Río Guabal, head of the Río Guabal, at 475 m, down to the Río Guabal, in high forest that was fairly open, fruiting trees, 1962 Feb 26-Mar 2 (USNM; Olson, 1980a, as "Head of Río Guabal, Tigre"; Olson, 1981d, as "Tigre, head of Río Guabal").

TIGRE, CANAL DEL; Bocas del Toro

0915/8158 (USBGN)

Marine channel that extends between the NW coast of Península Valiente [0905/8151 (USBGN)] and the E coast of Isla Popa [0911/8207 (USBGN)] connecting Laguna de Chiriquí [0903/8200 (USBGN)] and the Caribbean Sea [1500/7500 (USBGN)] (JOG). See also: Bluefield, Laguna de; and Toro, Cayo.

TIGRE, CERRO; see Cerro Tigre Army Reservation

TIGRE, PUERTO EL; see Puerto El Tigre

TIGRE, ROCA; see Tiger, Cayos

TIGRILLO; Panamá (E, Pacific)

ca. 0840/7837

Sea level, described by Wetmore as a small beach immediately S of Playa del Creo [ca. 0841/7837], ca. 3 mi (5 km) S of Chimán [0842/7837 (USBGN)], Wetmore and Perrygo, GL = Chimán, inland along 2 branches of a small stream, 1950 Feb 22 (USNM). Not shown on our maps.

TIRBI BIGHT; Bocas del Toro

ca. 0926/8222

A bay in NW Bocas del Toro province on the Caribbean coast formed on the S and E by the mainland as it curves N to the peninsula that extends to Punta Terraba [0926/8220 (USBGN)], located 16 km E of Changuinola [0926/8231 (USBGN)] and 15 km N of Almirante [0918/8224 (USBGN)] (MHA, as "Ens. Tirvi"). Wetmore, GL = Almirante, from Changuinola Canal [ca. 0919/8221] crossed 150 yards (140 m) N through the tall trees of the wooded swamp, sea grape, and *emajagua*, to the low sand of the coastal beach with much drift wood, lines of salt grass, *Ipomea*, and stands of bushes, followed the beach to the NW for 1 mi (1.5 km), 1958 Feb 20 (USNM).

TISINGAL; Chiriquí

0848/8240 (USBGN)

1,300 m, town 2 km E of Palo Santo [0849/8241 (USBGN)] (JOG). Wetmore, GL = Volcán, at 4,200 feet (1,300 m), Wetmore used the name "Tisingal" for the entire region surrounding Palo Santo and applied the name to both a summit and a valley. Collections made on the summit of Tisingal at 5,500-5,800 feet (1,700-1,750 m) were presumably made on Cerro Pando [0849/8242 (USBGN)] (q. v.) as his 1955 journal entries indicated that Cerro Pando had been renamed Cerro Tisingal. The valley Tisingal is a level stretch along the Río Chiriquí Viejo [0820/8241 (USBGN)] S of the town Tisingal, and appears to be the same general area that was labeled "Río Chiriquí Viejo, at Palo Santo" (q. v.) in 1954 and in 1955. In 1960 the valley had been logged and partly cleared, collected near several small streams, a potato field, some original forest, roads and trails, a brushy pasture and a coffee plantation, 1960 Feb 10-11, 13, 16, 22, GL = Volcán, at 4,600-5,000 feet (1,400-1,500 m), followed a small stream along the logging road from Palo Santo into a deep valley with pastures and tall *rastrojo*, the forested hills had coffee planted in the understory, 1965 Feb 27, Mar 14 (USNM; Olson, 1981c, as "El Volcán, Tisingal").

TOABRÉ; Coclé (Atlantic)

0840/8019 (USBGN)

200 m, town 15 km NNE of Penonomé [0831/8022 (USBGN)] (JOG). The town Tambo [0840/8017 (USBGN)] is 3 km NE of Toabré (MGRP). Tejera, Tambo-Toabré, 1978 Jan 13 (MVUP).

TOABRÉ; Colón (W)

0856/8032 (GA)

Angehr (in litt., Aug 2005), who collaborated with Percís Garcés in 2001-2002, states that "Garcés specimens from Toabré were collected at a campsite we called Quebrada La Bruja, on the S side of the Toabré near the town of La Bruja, and about 2 km SSW of Boca de Toabré. The GPS coordinates for this campsite were 08°55'31" N, 80°31'39" W, elevation 20 m, and the specimens were collected from net lines just east of the camp." Boca de Toabré [0856/8032 (USBGN)] is a town at the mouth of the Río Toabré [0856/8033 (USBGN)] (RDP2), which originates in Coclé province at the merger of several streams and flows NW along the Colón/Coclé provincial boundary discharging into the Río Coclé del Norte [0905/8035 (USBGN)] ca. 50 km NNW of Penonomé [0831/8022 (USBGN)] (MGRP). Garcés, Jiménez, López, Ramos, and Rodríguez, 2001 Jun 16 (MVUP, as "Toabré").

TOABRÉ, BOCA DE; see Toabré

TOABRÉ, RÍO; see Toabré

TOBAQUILLA, ISLA; see Taboguilla, Isla

TOCUMEN; Panamá (E, Pacific)

0905/7923 (USBGN)

ca. 25 m, town 15 km NE of Panama City [0858/7932 (USBGN)] located just N of the Carretera Interamericana [0832/8242 (USBGN)] (JOG). Specimens labeled "Swamp" or "Marshes" were collected at wetlands near the Río Tocumen [0901/7923 (USBGN)]. According to Delgado (1983) and to Engleman et al. (1989:488), much of the Tocumen Marsh was drained and converted to rice fields ca. 1983. According to Angehr, Tocumen Marsh [0903/7920 (Angehr, 2003:256)] includes "freshwater and brackish wetlands along the Pacific coast" and a "small complex of freshwater ponds just south of the airport." Collector ?, date ? (Bangs and Penard, 1921:391, as "Focoume"); André, 1899 Mar 7 (AMNH, as "Tocoumé"; Hartert and Goodson, 1917:418, as "Tocoumé"; Wetmore, 1972:108); Chapin, Rogers and Mathews, one as "near Tapia," 1923 Aug 8 (FMNH); Weber, 1928 Jul 28 (FMNH, as "Tacoma," and as "Tacoma Marshes"; USNM, as "Tacoma," and as "Tacoma Marshes"); Benson, 1931 Aug 21 (USNM; Eisenmann, 1955c); Blanton, 1953 Sep 25, Oct 6-7 (USNM, as "Tocumen Road," and as "Tocumen Swamp"); Curtiss, 1936 Jul 25 (MCZ); Fischer, 5 mi (8 km) E, and 12 mi (19 km) E, 1963 May 25 (USNM); Muñoz, 1961 Sep 12 (GML); Díaz, 1962 Nov 28, 1964 Nov 1, 10, 1965 Mar 14, 21, May, Oct 14 (GML; USNM; Hicks et al., 1967); Morton, 4 mi (6 km) E of airport, 1966 May 13 (FMNH); Emanuel et al., observations, 1978 Jan 14, 21, damp pastureland 1979 Feb 9, marshy pastureland 1980 Feb 10 (Emanuel, 1980; Braun and Wolf, 1987); Flores, 1979 Jul 21 (MVUP); Fong, 1981 Sep 22 (MVUP); Whitney, Rowlett and Barth, observations, 1982 Jan 7 (Braun and Wolf, 1987); Braun, M., 1982 Feb 15 (LSUMZ, as "Tocumen Marsh"); Braun, D., at 20 m, 1982 Aug 20 (LSUMZ, as "Tocumen Marsh"); Quiroz, *Via Tocumen* (Tocumen Highway), 1983 Nov 30 (MVUP); Mendieta, 1985 Mar 20 (MVUP); O'Meallie and Meyer, observations, 1985 Dec 23 (Braun and Wolf, 1987); González, *Estacion de Agronomia*, date ? (GML). (Wetmore, 1953). See also: Tapia Número Dos; and Tocumen, Aeropuerto de.

TOCUMEN, AEROPUERTO DE; Panamá (E, Pacific)

0905/7923 (USBGN)

ca. 25 m, international airport located just S of Tocumen [0905/7923 (USBGN)] (JOG, as "Tocumen Internacional"). Eisenmann and Ambrose, observations, 1961 Apr 30 (Eisenmann, 1962a, as "Tocumen Airport"); Loftin, observations, 1962 Nov (Loftin, 1963a, as "Tocumen Airport"); Harty et al., "open savanna" on the Pan American Hwy "16.2 mi (26 km) E of La Siesta Motel on the Tocomen [sic] Airport Circle," 1963 Jul 6 (ANSP; Harty, 1964).

TOCUMEN, RÍO; Panamá (E, Pacific)

0901/7923 (USBGN)

Originates N of Tocumen [0905/7923 (USBGN)], flows S crossing the Carretera Interamericana [0832/8242 (USBGN)] near town, then broadens and enters an area of grassland 4 km SSE of Tocumen before continuing to the coast where it merges with the Río Tapia [0902/7924 (USBGN)] as it discharges into Bahía de Panamá [0850/7915 (USBGN)] (E761A; MGRP). On E761A, the grassland area is marked "Passable in Dry Season." See also: Tocumen.

TOCUMEN AIRPORT; see Tocumen, Aeropuerto de

TOCOMEN AIRPORT CIRCLE; see Tocumen, Aeropuerto de

TOCUMEN INTERNACIONAL; see Tocumen, Aeropuerto de

TOCUMEN MARSHES; see Tocumen

TOCUMEN ROAD; see Tocumen

TOCUMEN SWAMP; see Tocumen

TOCUOMÉ; see Tocumen

TOLÉ; Chiriquí

0814/8141 (USBGN)

300 m, city in SE Chiriquí province on the Carretera Interamericana [0832/8242 (USBGN)] 15 km W of the Veraguas provincial boundary (JOG). Calderon, Guevara, and Lopez, 1985 Mar 18 (MVUP); Antinori, 1985 Mar 18-19 (MVUP); Talavera, 1985 Mar 20 (MVUP); Lopez, 1985 Mar 24 (MVUP). See also: Tolé, Cordillera de.

TOLÉ, CORDILLERA DE; Chiriquí/Veraguas (Pacific)

ca. 0817/8140

Described by Salvin as a district located between Chiriquí [0824/8219 (USBGN)] and Santiago [0806/8059 (USBGN)] and a locality distinct from either Santiago or Santa Fé [0831/8105 (USBGN)]. He noted that Arcé must have entered "the hot forest of low elevation" as well as reached "a considerable elevation" and a "cool mountain-climate" (Salvin, 1867a:121; Salvin, 1867b:129). Arcé, presumably he collected in the region surrounding Tolé [0814/8141 (USBGN)], but it is unclear if collecting was confined to what is now Chiriquí province, 1866 (BMNH; USNM; Salvin, 1867a; Salvin, 1867b; Salvin and Godman, 1879-1904, vol. 1:305, vol. 2:416; Berlepsch, 1911; Wetmore, 1967). (Peters, 1931:319).

TONOSÍ; Los Santos

0724/8027 (USBGN)

ca. 50 m, city in S Península de Azuero [0740/8035 (USBGN)] 37 km WSW of Los Asientos [0731/8008 (USBGN)] on the Río Tonosí [0721/8022 (USBGN)] (JOG). Wetmore, NE of town along the Las Tablas [0746/8017 (USBGN)] trail, through pastures, trees in bloom, scrub and partly cleared forest, to an area of low ground with brush and scattered trees that floods during the rainy season, to an area of low brush with scattered taller trees and much corozo palm, 1957 Mar 22, 24-25, 28, and as a GL for: El Puerto de Piña, and Río Tonosí, 1957 Mar 23, 26-27 (USNM; Wetmore, 1970; Olson, 1980b, as "Tonosí, Puerto Piña"; Olson, 1981d); Loftin and Olson, observations, 1963 Mar 8 (Loftin and Olson, 1963).

TONOSÍ, RÍO; Los Santos

0721/8022 (USBGN)

Originates in S central Península de Azuero [0740/8035 (USBGN)] and flows SE entering the Pacific Ocean 9 km SE of Tonosí [0724/8027 (USBGN)] (JOG). Delgado (1983) mentions regions of wetlands and mangroves near the mouth that are affected by rice cultivation. Wetmore, GL = Tonosí, 3 mi (5 km) SE of Tonosí along a poorly maintained road that followed the river, almost to the head of tidewaters, but remained above the area of tidal influence, scattered woodland, stretch of broad gravel bars, 1957 Mar 23, 26 (USNM; Wetmore, 1962). See also: El Puerto de Piña.

TONOSÍ VALLEY; see Tonosí, Río

TOPALISA, RÍO; Darién
0801/7728 (USBGN)
Originates on the continental divide NE of Paya [0758/7724 (USBGN)], flows W and from the S enters the Río Púcuro [0801/7733 (USBGN)] 9 km above its mouth, 3 mi (5 km) E of Púcuro [0801/7730 (USBGN)], and 14 km SE of Boca de Cupe [0803/7735 (USBGN)] (MGRP; RDPF; SMN2B, as "Río Tapalisa"). SMN2B mislabels the lower section of the Río Púcuro, below its junction with the Río Topalisa, as the Río Topalisa. Often called the Río Tapalisa. See also: Casita; and Tacarcuna Village.

TORNO ROMPIDO, QUEBRADA; Colón (W)
0908/8012 (USBGN)
Short river that originates W of the Río Membrillar [0911/8011 (USBGN)] in the low hills SE of El Chilar [0909/8012 (USBGN)] and flows NNW joining the Río Indio [0912/8011 (USBGN)] 3 km S of El Chilar (E761B, as "Quebrada Tomás Rompejarro"). Described by Wetmore as a small stream navigable by canoe for only 75-80 yards (75 m), above this it became shallow with occasional duplicate channels, Wetmore and Perrygo, GL = El Chilar, in a broad valley where the forest had been largely cleared for cultivation with a few tracts of tall forest remaining, on a lagoon with brush and grass, the short N branch with a steep and rocky bed, 1952 Mar 7-8, 11 (USNM, as "Quebrada Torno Rompío"; Eisenmann and Howell, 1962, as Quebrada Torno Rompío, Chilar").

TORNO ROMPÍO, QUEBRADA; see Torno Rompido, Quebrada

TORO, CAYO; Bocas del Toro, Archipiélago de Bocas del Toro
0909/8157 (USBGN)
8 m, island with a land area of ca. 0.05 km^2 (5 ha) located in Canal del Tigre [0915/8158 (USBGN)] 1 km due W of Punta de Toro [ca. 0909/8156] (q. v.) (E762). Two very small, unnamed islands are indicated on E762 between Punta de Toro and Cayo Toro, and this group of islands is presumed to be the one referred to by Miller in his field notes as "Toros Rocks" and as "Toros Islands." Described by Miller in 1936 as a group of ca. 6 islets formed by wave dissection just off the coast of Punta Bluefield [0908/8157 (USBGN)]. The islets had very abrupt sides reaching 10-20 feet (3-6 m) and were topped with vegetation, mainly coconut palms. Between the islands water passages varied in width from a few feet to several hundred yards, Miller and Richardson, GL = Laguna de Chiriquí [0903/8200 (USBGN)], Península Valiente [0905/8151 (USBGN)], this locality was included in Miller's field notes, but apparently not included on specimen labels, although some of the specimens indicated in his field notes were not located, 1936 Mar 4-5 (UCLA, as "Toros Ids," and as "Toros Rocks"); SBE, observations, 1990 Apr 24 (USNM).

TORO, CHARCO DEL; Panamá (E, Pacific)
ca. 0844/7833
ca. 50-60 feet (10-20 m), described by Wetmore as a camp on the Río Majé [0840/7835 (USBGN)] ca. 4-5 mi (6-8 km) N of Majé [0840/7835 (USBGN)] at the head of trails leading into the E end of Serranía de Majé [0855/7830 (USBGN)], Wetmore and Perrygo, GL for: Río Chucantí, and Río Majé, 1950 Mar 3, 20-28 (USNM; Wetmore, 1951a; Wetmore, 1959b:21; Wetmore, 1970, as "Charco del Toro, Río Majé"; Payne, 1974; Olson, 1981d, as "Charco del Toro, Río Maje"). Not shown on our maps.

TORO, ISLA, Chiriquí
0809/8028 (USBGN)
Sea level, now a peninsula located in Golfo de Chiriquí [0800/8220 (USBGN)] 5 km SE of Remédios [0814/8151 (USBGN)] opposite Isla La Porcada [0807/8144 (USBGN)] (JOG). On many maps appears as Isla Boca Toro. Batty, specimens at AMNH are dated 1902 Jan 1, Feb 5, but Batty's data were completely fabricated and his specimens labeled "Espartal Island" were actually collected in 1901 near his mainland bases at Boquete [0847/8226 (USBGN)] and Boquerón [0830/8234 (USBGN)] (AMNH; Olson, 2008). Not shown on our maps.

TORO, PUNTA; Colón (Canal Zone) (W)
0922/7957 (USBGN)
Sea level, mainland point on the NW coast of Bahía Limón [0920/7956 (USBGN)] located 5 km WNW of Colón [0922/7954 (USBGN)] (MGRP). Jewel, edge of woods at the head of a small stream, bare beach at the foot of a cliff, 1911 Apr 2, Jul 23, Sep 4, Oct 1, 1912 Jan 28, Mar 3, Aug 4, 23, 1913 Apr 27 (ANSP, as "Toro Pt"; Jewel, 1913, as "Toro Point"; Stone, 1918, as "Toro Point"); Jewel (from Ferguson), bare coral beach, 1911 Oct 1 (Jewel, 1913, as "Toro Point").

TORO, PUNTA DE; Bocas del Toro
ca. 0909/8156
Sea level, point on the NW coast of Península Valiente [0905/8151 (USBGN)], located on the NW coast of the peninsula that forms the S coast of Laguna de Bluefield [0909/8155 (USBGN)]. Cayo Toro [0909/8157 (USBGN)] (q. v.) is 1 km due W of Punta de Toro (E762). The point is very near — if not the same as — Punta Bluefield [0908/8157 (USBGN)], which is not named on E762.

TORO, QUEBRADA DEL; Los Santos
ca. 0732/8000
Described by Wetmore as a creek near the beach at Pedasí [0732/8002 (USBGN)], Wetmore, GL = Pedasí, followed the dry streambed inland to an area of small pools, 1957 Mar 10 (USNM). Not shown on our maps.

TORO POINT; see Toro, Punta

TOROS ISLANDS; see Toro, Cayo

TOROS ROCKS; see Toro, Cayo

TORTADÓ; Darién
ca. 0730/7801

ca. 100 m, described by Wetmore as an area on the Río Jaqué [0731/7810 (USBGN)] near the mouth of the Arroyo Tortadó [ca. 0730/7801] (q. v.) that lies in the transition zone between the humid tropical climate of the interior and the arid tropical climate of the Pacific coast, Wetmore and Perrygo, GL = Río Jaqué, 1947 Apr 18 (USNM). Not shown on our maps.

TORTADÓ, ARROYO; Darién

ca. 0730/7801

Described by Wetmore as a tributary of the Río Jaqué [0731/7810 (USBGN)] that enters from the N above the mouth of the Quebrada Junguradó (not located) and below Moja Culo [ca. 0729/7759]. The streambed of the Río Jaqué becomes rocky and enters a series of rapids above the mouth of the Arroyo Tortadó (USNM). MGRP shows a tributary of the Río Jaqué named the Río Tortadó [0729/7756 (USBGN)], but this river enters the Río Jaqué ca. 7 km upstream from Wetmore's camp at the mouth of the Río Imamadó and therefore is not the stream that Wetmore called the Arroyo Tortadó. Wetmore and Perrygo, GL = Río Jaqué, 1947 Mar 28, Apr 18 (USNM). See also: Tortadó. Not shown on our maps.

TÓRTOLA, ISLA; Panamá (Canal Zone) (W, Pacific)

0852/7934 (USBGN)

100 feet (30 m), very small island with a land area of ca. 0.03 km^2 (3 ha) located in Bahía de Panamá [0850/7915 (USBGN)] 10 km S of Panama City [0858/7932 (USBGN)], 3.5 km S of Punta Bruja [0853/7935 (USBGN)], and 6 km N of Isla Taboga [0847/7933 (USBGN)] (E761A, as "Tortola Island"). Hallinan, 0.5 mi (1 km) offshore, 1915 Mar 27 (AMNH; Hallinan, 1924, as "Tortola Island").

TORTOLA ISLAND; see Tórtola, Isla

TOWNSEND, CAMP; see Camp Townsend

TRANCA, RÍO; Panamá (E, Pacific)

ca. 0905/7908

Originates on the S slope of Cerro Carbunco [0911/7914 (USBGN)] and flows SSW crossing the Carretera Interamericana [0832/8242 (USBGN)] 14 km WSW of Chepo [0910/7906 (USBGN)] before joining the Río Chepo [0900/7906 (USBGN)] 10 km SW of Chepo (E761A). The USBGN coordinates [0905/7913] are not supported by E761A. Wetmore and Perrygo, GL = Chepo, at Cerro Carbunco (q. v.), 1949 Apr 14 (USNM). See also: Carbunco, Cerro.

TRANQUILLA; Panamá (Canal Zone) (E, Atlantic)

ca. 0914/7933

Study area on the SE coast of Lago Alajuela (Madden Lake) [0915/7935 (USBGN)] near the mouths of the Río Chagres [0907/7942 (USBGN)] and the Quebrada Tranquilla [0915/7933 (USBGN)], Tejera, observations, 1992-1993 (Tejera, 1995:9-10).

TRANSISTHMIAN HIGHWAY; see Boyd-Roosevelt, Carretera

TRAPECHE ISLAND; see Trapiche, Isleta

TRAPICHE, ISLA; see Trapiche, Isleta

TRAPICHE, ISLETA; Panamá, Archipiélago de las Perlas
0826/7907 (USBGN)
Sea level, small island located 1 km NW of El Cocal [0825/7906 (USBGN)], which is on Isla Pedro González [0824/7906 (USBGN)] (JOG). Bovallius, described the island as less than 1 km long with a "plateau ca. 30 m above the sea level, gently sloping to the S and covered with a fine cocoa-park," see discussion of Bovallius under Pueblo Nuevo [ca. 0820/8013 ?], 1882 Apr 20-24 (NHRM; Rendahl, 1920, as "Trapiche"); Miller, 1937 Mar 14-15 (USNM, as "Crapeche Island," and as "Trapeche Island"); (MacArthur et al., 1972—species list; see comments in Bibliography).

TREINTICINCO, LA; see Treinticinco, Quebrada

TREINTICINCO, QUEBRADA; Colón (Canal Zone) (W)
0912/8000 (USBGN)
Originates on the NW coast of Lago Gatún (Gatun Lake) [0912/7955 (USBGN)] near La Treinticinco [0911/7958 (USBGN)] and flows NW discharging into Río Providencia [0913/8002 (USBGN)] 4.5 km NW of La Treinticinco. The stream crosses Achiote Road [ca. 0917/8003] 2 km NW of La Treinticinco, which is a town situated at the junction of Achiote Road and Gatún-Escobal Road [ca. 0916/7955] (E8660). Cardiff, Christian, Dittmann, and Lovette, GL = Achiote Road, at 25 m, ca. 2 km W Gatún-Escobal Road, 1997 Mar 19 (LSUMZ).

TRES ALEMENDRAS ISLAND; see Tres Almendras Islands

TRES ALMENDRAS ISLANDS; Colón (Canal Zone) (E)
0912/7951 (USBGN)
USBGN uses the name for a group of islands located just over 1 km NE of Isla Abogado [0912/7952 (USBGN)] (E661, as "Tres Almendras Is."). Wright (1985) used the name for one of the islands with a land area of 7.6 ha located 25 m N of Península Bohío [0912/7950 (USBGN)], censuses, 1979-1981 (Wright, 1985; Wright, 1988, both as "Tres Almendras Island").

TRES DE NOVIEMBRE, CAMINO; see Boquete Trail

TRIBIQUE, RÍO; Veraguas (Pacific)
0800/8119 (USBGN)
Originates in the hills SW of Soná [0801/8119 (USBGN)], flows ENE and enters the Río San Pablo [0751/8110 (USBGN)] just S of Soná (JOG). Described by Wetmore as a swift stream, 40 feet (10 m) wide, with boulders and frequent shallow rapids, Wetmore, GL = Soná, San José [0800/8120 (USBGN)], crossed the river S of San José ca. 2 mi (3 km) above its mouth, forest with creeping bamboo, area of gallery forest, steep slopes planted in corn and rice, 1953 June 12, 15 (USNM).

TRINIDAD; see La Trinidad

TRINIDAD; see Trinidad Arriba

TRINIDAD, BAHÍA; Colón/Panamá (Canal Zone) (W, Atlantic)
0905/8000 (USBGN)
The SW section of Lago Gatún (Gatun Lake) [0912/7955 (USBGN)] that extends

ca. 20 km NNE from the mouth of the Río Cirí [0858/8003 (USBGN)] and the mouth of the Río Trinidad [0901/7958 (USBGN)] to the central portion of the reservoir (JOG).

TRINIDAD, CERRO; Panamá (W, Atlantic, Pacific)
0844/7958 (USBGN)
975 m, peak on the continental divide at the N end of Cordillera del Llorón [0843/7957 (USBGN)] 7 km NNW of Cerro Campana [0841/7955 (USBGN)] (E762; JOG). See also: Capira; and Llorón, Cordillera del.

TRINIDAD, LA; see La Trinidad

TRINIDAD, RÍO; Panamá (Canal Zone) (W, Atlantic)
0901/7958 (USBGN)
Originates NW of Cerro Trinidad [0844/7958 (USBGN)] and flows roughly N entering the SW coast of Lago Gatún (Gatun Lake) [0912/7955 (USBGN)] where the lower course of its original streambed is now submerged. The river originally continued NE through the section of the reservoir now called Bahía Trinidad [0905/8000 (USBGN)] and Bahía Agua Clara [ca. 0911/7957] and entered the continuous Río Chagres (see Río Chagres [0919/8000 (USBGN)]) 6.5 km S of Gatún [0916/7955 (USBGN)] just S of what is currently Islas Las Brujas [0913/7954 (USBGN)] (IP; JOG). Goldman, GL for Agua Clara, 1911 Jan 23, 26 (USNM); Jewel, 8-12 mi (13-19 km) S of Gatún, forest, and as a GL for Agua Clara, 1911 Mar 12, 24, 1912 Mar 19, 24, 31 (ANSP, as "Trinidad R."; CU; Stone, 1918, as "Trinidad River"); Busck, 40 mi (64 km) above mouth, upper, 1911 May 1-10, 1912 Mar 30, Jun 7 (USNM, as "Trinidad River"; Hekstra, 1982, as "Upper Trinidad"); Shiras Expedition, GL for Cuchirvo, 1914 Mar 7-17 (AMNH; Anthony, 1916, as "Maxon Ranch, Río Trinidad"); Collector ? [Jewel ?], caged bird from the wild, date ? (Stone, 1918:251, as "Upper Trinidad"). (Olson, 1981d, as "Río Trinidad, Agua Clara"). See also: Agua Clara Swamp.

TRINIDAD ARRIBA; Panamá (W, Atlantic)
0844/7959 (USBGN)
200 m, town at the W base of Cerro Trinidad [0844/7958 (USBGN)] on the upper reaches of the Río Trinidad [0901/7958 (USBGN)] ca. 12 km WSW of Capira [0845/7953 (USBGN)] and 3 km roughly S of El Aguacate Número Uno [0845/7958 (USBGN)] (JOG; MGRP). Montenegro, GL = El Aguacate Número Uno, 1978 Mar 12, Oct 30 (GML, as "Arriba de Trinidad," and as "Trinidad"). In addition, there is at GML that is labeled Trinidad/Pacora. Presumably Pacora is a locality near Cerro Trinidad that we could not locate, or perhaps it is a misreading of Pacific. Montenegro, date ? (GML, as "Trinidad/Pacora").

TRINIDAD RIVER; see Trinidad, Río

TROPICAL SURVIVAL SCHOOL; see Albrook Air Force Base

TROPIC STAR LODGE; see Piña, Río

TUANCLE, RÍO; Veraguas (Pacific)
0800/8132 (USBGN)
Originates in the hills SW of Soná [0801/8119 (USBGN)], flows N and W into the

Río Bubí [0759/8138 (USBGN)] W of Zapotillo [0800/8131 (USBGN)] (MGRP; MHA, as "Río Tuangle"). Wetmore and Perrygo, GL = Soná, Zapotillo, worked 20 mi (32 km) W of Soná near Zapotillo, crossed to the S side of the swift, 30-60 feet (9-20 m) wide stream and followed a trail with casual foot and horse traffic to a hill covered with a scant remainder of original forest, 1953 May 19, 25, Jun 9 (USNM).

TUANGLE, RÍO; see Tuancle, Río

TUCESA, RÍO; see Tuquesa, Río

TUCUTÍ, RÍO; see Balsas, Río

TUICUSA; see Tuquesa, Río

TUIRA, RÍO; Darién

0821/7803 (USBGN)

Originates on the E slopes of Serranía de Pirre [ca. 0749/7744] and flows E, then S through a valley between the Serranía de Pirre and the Cerro Setetule [0747/7736 (USBGN)]. The river then turns to the NE flowing within 4-12 mi (6-19 km) of the Colombian border until the Río Paya [0757/7730 (USBGN)] enters from the NE, after which the river turns NW and W widening to 4 mi (6 km) as it enters the Golfo de San Miguel [0822/7817 (USBGN)] (MGRP; SMN2B; Woakes, 1900; Anthony, 1916). Shown in Atlas (1988:206) and on RDP2 as a slightly shorter river that originates in the Altos de Quía [0735/7727 (USBGN)]. Salvadori and Festa (1899:1, as "Río Tuyra") indicate that their map labels the river's estuary "Hâvre du Darién," which refers to the Darién Harbor [0823/7806 (USBGN)] at La Palma [0825/7809 (USBGN)] just above the river's mouth (MHA, as "Puerto Darien"; MRP, as "Puerto Darien"). The early literature often refers to the Río Tuyra Valley, Basin or Delta when describing the Darién coast, Golfo de San Miguel [0822/7817 (USBGN)], or river towns such as La Palma, El Real de Santa María [0808/7743 (USBGN)] and Pinogana [0808/7741 (USBGN)] (Salvadori and Festa, 1899, as "Río Tuyra"; Griscom, 1933, as "Tuyra Basin," and as "Tuyra delta"). According to Wetmore, in 1959 the river banks near El Real de Santa María [0808/7743 (USBGN)] were low and covered with grass and *caña blanca*. The river near Boca de Cupe [0803/7735 (USBGN)] ran between steep sided hills with extensive rock exposures, and near the mouth of the Río Paya the river banks had areas of cultivation. Richardson, GL = Boca de Cupé, 1915 Apr 30, May 2, GL = El Real de Santa María, 1914 Dec 24, 29, 1915 Jan 9, 23 (AMNH, as "Río Tuyra"; ANSP, as "Río Tuyra"; Anthony, 1916, as "Río Tuyra"; Chapman, 1917, as "Río Tuyra"; Chapman, 1926, as "eastern Panama"; Hellmayr, 1929:84, as "El Real, Río Tuyra"; Zimmer, 1939, as "El Real, Río Tuyra"; Zimmer, 1950:8, 26, as "El Real, Río Tuyra"; Zimmer, 1955:5, as "El Real, Río Tuira"; Olson, 1981d, as "El Real, Río Tuira," and as "Boca de Cupe, Río Tuira"); Baer, GL = Yaviza, Río Chucunaque, collected on the Río Chucunaque between the Río Canglón [0820/7746 (USBGN)] and the Río Tuira, 1924 Feb 27, and at the junction with the Río Tuira, 1924 Mar 2-3 (USNM) (for more on Baer see comments under Marsh, 1934). ; Wetmore, GL = mouth of the Río Paya, worked from the Paya camp, located at the mouth of the Río Paya, and collected within 15 km of camp on both rivers and inland along their banks. The area was covered with gallery forest and the undergrowth was sometimes open and sometimes dense with thorny vines and areas of dense, 8-10 foot (2-3 m) wild pineapple that limited access. In general the area was rather dry; some trees were in bloom and the crowns of large *cuipo* and mahogany trees towered above the 80-125 foot (25-40 m) forest canopy. In addition he collected in cultivated areas, in old *rastrojo*, along dry creeks, on gravel beaches, on small islands with medium-sized trees, on a sandy beach at Quebrada Cortatita [ca. 0756/7732] (q. v.) and then inland along the stream, on the Río Paya at Playa Cortatita [ca. 0754/7729] (q. v.) and farther upstream to Monte Teodoro [ca. 0754/7727] (q. v.), 1959 Feb 13-26, 28-Mar 17, GL = El Real de Santa María, upriver toward Pinogana [0808/7741 (USBGN)] where the river still had tidal influence, the river banks were lined with shrubs, tall grass and trees farther back, landed at a swampy woodland with pools of water, along a trail through the forest, a stream near the airport, 1964 Jan 9, 11, 26, Feb 16, 18 (USNM; Wetmore, 1962; Haffer, 1962, as "Boca de Paya"; Olson, 1981d, as "mouth of Río Paya, Río Tuira"); GMLP, 1959 Apr 15 (GML); Slattery, GL = El Real de Santa María, 1966 May 31 (GML); Collector ? [Slattery], GL = El Real de Santa María, 1966 May 31 (GML). (Delgado, 1983). See also: Sabana, Punta; San Miguel, Golfo de; Union Choco; and Yaviza.

TULITA, LA; see La Tulita

TUMBA MUERTO; Panamá (E, Pacific)

ca. 0900/7932

Common name for Via Ricardo J. Alfaro, a road that extends through greater Panama City [0858/7932 (USBGN)] W and N of Corregimiento Betania [0900/7932 (USBGN)]. The road passes through Santa María [ca. 0901/7933], which is in N Corregimiento Betania (E762). GMLP, 1976 Jul 14 (GML); Cedeño, Fabrea, Janguez, Magaña, Miranda, Serrano, and Yanguez, GL = Santa María, 1989 Dec 10 (MVUP); Fabrea and Miranda, GL = Santa María, 1990 Mar 21 (MVUP).

TUPISA, RÍO; see El Salto

TUQUESA, RÍO; Darién

0823/7747 (USBGN)

Originates on the continental divide on the S face of Cerro Anachucuna [0836/7735 (USBGN)], flows SW joining the Río Chucunaque [0809/7744 (USBGN)] 18 mi (29 km) NNW of El Real de Santa María [0808/7743 (USBGN)] (SMN2B). Baer, GL = Yaviza, Río Chucunaque, junction of the Río Chucunaque and the Río Tuquesa, and 1 mi (1.5 km) up the Río Tuquesa, 1924 Feb 27, Mar 2-5 (USNM, as "Río Tucesa") (for more on Baer see comments under Marsh, 1934); Johnson, 1924 Mar 7 (USNM, as "Río Tuicusa"); Wetmore, GL for Río Chucunaque, 1959 Mar 20-Apr 3 (FMNH; USNM; Wetmore, 1962); Cavalier, at Quebrada Particheque (not located) near the mouth of the Río Tuquesa, 1967 Sep 28, and 5 km NW of the mouth of the Río Tuquesa, 1967 Sep 29 (MVZ).

TURBI BIGHT; see Tirbi Bight

TUTE, CERRO; Veraguas (Pacific)

0832/8106 (USBGN)

1,575 m, peak located 5 km WNW of Santa Fé [0831/8105 (USBGN)] (MHA, as "C. Tuto"; MRP, as "C. Tuto"). (Wetmore et al., 1984:311). See also: Calovévora, Pico.

TUTO, CERRO; see Tute, Cerro

TUYRA, RÍO; see Tuira, Río

TUYRA BASIN; see Tuira, Río

TUYRA DELTA; see Tuira, Río

U. S. FOREST PRESERVE; see Madden Road Timber Preserve

UCURGANTÍ, RÍO; Darién
0827/7748 (USBGN)
Originates on the continental divide on the W face of Cerro Anachucuna [0836/7735 (USBGN)] and flows SW entering the Río Chucunaque [0809/7744 (USBGN)] 5 mi (8 km) NNW of the Río Tuquesa [0823/7747 (USBGN)] (SMN2B). Wetmore described the mouth as 30-40 feet (9-10 m) wide, Wetmore, GL for Río Chucunaque, 1959 Mar 20-22, 31 (USNM).

ULLAMA, CERRO; see La Ullama

ULLAMA, LA; see La Ullama

ULTIMA, CERRO; Panamá (E, Atlantic, Pacific)
ca. 0911/7923
2,600 feet (790 m), peak on the continental divide 12 km N of Tocumen [0905/7923 (USBGN)] and 3 km NNE of Cerro Azul [0910/7924 (USBGN)] (E761A). Wetmore and Perrygo, GL = Cerro Azul, at 2,000 feet (600 m), walked several miles along the ridge that connects Cerro Azul and Cerro Prominente [0914/7917 (USBGN)], the higher elevations were covered with a dense growth of small, shrubby trees and the lower slopes were covered with tall, green, gallery forest, 1949 Mar 31, Apr 12 (LSUMZ; USNM; Wetmore, 1952:19).

UNÍ; see Uní, Río

UNÍ, RÍO; Panamá (E, Pacific)
ca. 0913/7859
Described by Wetmore in 1962 as the next stream W of the Río Terable [0913/7859 (USBGN)] (USNM). Not shown on our maps, but presumed to originate ca. 5 km NW of El Llano [0913/7859 (USBGN)] near Uní [0914/7901 (USBGN)] and flow SE joining the Río Chepo [0900/7906 (USBGN)] SW of El Llano [0913/7859 (USBGN)]. Wetmore, GL = El Llano, from the Carretera Interamericana [0832/8242 (USBGN)] followed a border of forest along the stream, 1962 Feb 9 (USNM).

UNIÓN CHOCÓ; Darién
ca. 0808/7737
ca. 25 m, town on the Río Tuira [0821/7803 (USBGN)] near the mouth of the Río Yape [0807/7735 (USBGN)] ca. 15 km E of El Real de Santa María [0808/7743 (USBGN)] (RDP2; Seutin, 1998 (map)). Seutin, observations on the Río Tuira

downstream of, 1996 Jul 6 (Seutin, 1998).

UPPER CHANGUINOLA; see Changuinola, Río

UPPER CHANGUINOLA DRAINAGE; see Changuinola, Río

URABÁ, GOLFO DE; Chocó/Antioquia (COLOMBIA)
0825/7653 (USBGN2)
Gulf on the NW coast of Colombia that extends W to Cabo Tiburón [0842/7721 (USBGN)] on the border of Colombia with Comarca de San Blas (MGRP). According to Griscom (1932:303), in 1877 the "Paris Museum received a collection of birds from Dr. Viguier made 'along the borders of the Gulf of Darien'...it is quite possible, however, that some or all of the birds were actually collected on the Caribbean coast of eastern Panama" but cites no source for the quote. Later Griscom (1935b:362) listed *Polioptila schistaceigula* Hartert "in Paris Museum from 'Gulf of Darien' coll. by Viguier (*fide* Hellmayr)," although Hellmayr (1911a, 1934:362) twice listed the species from the *Isthmus* of Darién (q. v.). The original description (Hartert, 1898b) does not include Panama in the range. (Wetmore, 1962:12-13, as "Gulf of Urabá").

URABA, ISLA; see Uravá, Isla

URACILLO, BOCA DE; see Boca de Uracillo

URACILLO, EL; see Boca de Uracillo

URACILLO, RÍO; Coclé (Atlantic)
0858/8011 (USBGN)
Originates in NE Coclé province, flows N and NE joining the Río Indio [0912/8011 (USBGN)] at Boca de Uracillo [0858/8011 (USBGN)] (MHA). Described by Wetmore as a 30-35 foot (9-10 m) wide stream with clear water and rocky shallows, Wetmore and Perrygo, GL = Boca de Uracillo, along the river 0.5-3 mi (1-5 km) above its mouth in open pastures and cultivated areas with tall grass, sawgrass, corn, yuca and sugar cane, over numerous small tributaries, on low hills with forest on their peaks and valleys between them, along the river in a tract of timber and in secondary growth, 2 mi (3 km) above its mouth to a region known as El Silencio [ca. 0857/8011] and to its border with El Águila [0857/8011 (USBGN)], 1952 Feb 23-25, 27-Mar 5 (USNM; Wetmore, 1962).

URAVÁ, ISLA; Panamá (W, Pacific)
0847/7932 (USBGN)
177 m, small island in Bahía de Panamá [0850/7915 (USBGN)] located 23 km S of Panama City [0858/7932 (USBGN)] and separated from the SE coast of Isla Taboga [0847/7933 (USBGN)] (q. v.) by a 200-500 m wide channel. The island extends just over 1 km N-S and is generally 700 m wide. From its shores the land quickly rises to 200 feet (60 m) to two central peaks (E671B; JOG). Described by Bartholomew and Goldstein (1984) as a "steep, volcanic island." According to Angehr (2003:249), the island is uninhabited. Wetmore and Perrygo, from a rocky beach to a saddle between the two central peaks, slopes covered with dry and largely leafless forest with some thorny creepers and a few clearings for papaya and yuca,

at 40-50 feet (10-15 m) on an extensive flat with green forest that was shaded with luxuriant growth, old, large mango trees, 1952 Mar 22; Wetmore, noted clearing since his previous visit in 1952, landed in a small cove on the W shore on a rocky flat, walked E, S and SW along trails through cultivated areas at the base of the steep interior slope that was covered with tall trees and open undergrowth, 1955 Dec 26 (USNM; Wetmore, 1952); Pennycuick, observations, 1980 May 21-Jun 6 (Pennycuick, 1983); Bartholomew and Goldstein, studies, 1983 (Bartholomew and Goldstein, 1984, as "Isla Uraba"). (Olson, 1997a).

URBANIZACIÓN LAS MERCEDES; see El Dorado

URRACÁ, PARQUE; see Bella Vista

URUSECA; Darién

ca. 0807/7744

Described by Wetmore as an area near El Real de Santa María [0808/7743 (USBGN)]. Presumably located WSW of El Real de Santa María along the Río Uruseca [ca. 0807/7744], a short tributary of the Río Pirre [0808/7744 (USBGN)]. Wetmore, GL = El Real de Santa María, through an extensive clearing grown with *Heliconia*, along a trail through heavy woodland, scattered *rastrojo*, yuca and rice fields, and a recent forest clearing, along a bulldozer cut for the Carretera Interamericana [0832/8242 (USBGN)], overgrown fields in a farm 6 mi (10 km) from El Real de Santa María, 1964 Jan 23-25, 27, Mar 21 (USNM). See also: Pirre, Río.

URUSECA, RÍO; Darién

ca. 0807/7744

A short tributary of the Río Pirre [0808/7744 (USBGN)] that discharges 1 km WSW of El Real de Santa María [0808/7743 (USBGN)] (DP). Alvarez, 7 km de El Real de Santa María, 1993 Feb 1 (MVUP, as "Río Uruseca"). See also: Uruseca.

URUTÍ; Darién

ca. 0806/7742

A small village along the W bank of the Quebrada Urutí [ca. 0808/7743], which is a tributary of the Río Tuira [0821/7803 (USBGN)] (SMN2B). The village is 5 km SE of El Real de Santa María [0808/7743 (USBGN)] and 3.5 km S of Pinogana [0808/7741 (USBGN)] (DP). Wetmore, GL = El Real de Santa María, on a gravel road to Pinogana, cattle farm located near the foothills of Cerro Pirre [0756/7743 (USBGN)], cleared pastures, open scrub and coarse grass of a dry swamp, a stream near the airport, 1964 Feb 18 (USNM).

URUTÍ, QUEBRADA; see Urutí

USTUPO; San Blas

0908/7756 (USBGN)

A small island a few hundred meters off the central San Blas coast just NW of Bahía de Masargandí [0907/7754 (USBGN)] (MGRP). USBGN also indicates a second Ustupo, which is also an island, at 0855/7744, although this one was not shown on our maps. Loftin, GL for Río Puprugandi, 1962 Oct 7 (Loftin, 1963a).

UTIVÉ; Panamá (E, Pacific)

0909/7919 (USBGN)

200 feet (60 m), town 9 km NNW of Pacora [0905/7917 (USBGN)] on the Río Utivé [0909/7917 (USBGN)] 3 km NW of its mouth (TM). Wetmore and Perrygo, GL for Río Utivé, 1949 Mar 17, 19, 25 (USNM; Wetmore, 1952:18; Wetmore, 1970; Olson, 1981a, as "Utivé, Río Cabobré"; Olson, 1983).

UTIVÉ, RÍO; Panamá (E, Pacific)
0909/7917 (USBGN)
Originates near Cerro Jefe [0914/7921 (USBGN)] and flows SE joining the Río Cabobré [0906/7917 (USBGN)] 8 km N of Pacora [0905/7917 (USBGN)] (JOG; TM, as "Río Cabobré"). Wetmore referred to this stream as the Río Cabobré but his description and general locator indicate that he worked on the Río Utivé. Described by Wetmore as a rapid stream with clear water, that was 25-30 feet (8-9 m) wide, Wetmore and Perrygo, GL = Utivé, at 300-400 feet (90-120 m), above Utivé, collected in green gallery forest, in forested hills bordering the stream, along a tributary strewn with boulders near recent clear cuts, 1949 Mar 17, 19, 25 (USNM, as "Río Cabobré"; Olson, 1981a, as "Utivé, Río Cabobré"; Wetmore et al., 1984:230, as "Río Colobre").

UVA, ISLA; Veraguas (Pacific)
0748/8146 (USBGN)
167 m, small, irregularly shaped island with a land area of ca. 1.5 km² (150 ha), located in Islas Contreras [0750/8147 (USBGN)] 5 km SSE of Isla Brincanco [0751/8147 (USBGN)] (JOG).

UVITO, EL; see El Uvito

UYAMA, LA; see La Ullama

UYAMA, RÍO; Bocas del Toro
0907/8215 (USBGN)
Originates in the hills W of Punta Róbalo [0902/8215 (USBGN)], flows N, then turns E and SE entering the NW coast of Laguna de Chiriquí [0903/8200 (USBGN)] 8 km N of Punta Róbalo (JOG). Braun, M., Brumfield, and Glenn, GL for Quebrada Garza, 1994 Jan 11-13 (USNM; Brumfield et al., 2001).

VALIENTE, CABO; see Valiente, Punta

VALIENTE, PENÍNSULA; Bocas del Toro
0905/8151 (USBGN)
Peninsula that extends ca. 30 km NW off the coast of Bocas del Toro province between Laguna de Chiriquí [0903/8200 (USBGN)] and Golfo de los Mosquitos [0900/8115 (USBGN)] (JOG and TOG, as "Peninsula Valiente"). Miller and Richardson, GL = Laguna de Chiriquí, the house boat that served as Miller's base was anchored in or near Laguna de Bluefield [0909/8155 (USBGN)] on Península Valiente, Miller and Richardson regularly went ashore and worked primarily along the S coast of the peninsula in eastern Laguna de Chiriquí. However, as they also visited Isla Escudo de Veraguas and labeled those specimens Laguna de Chiriquí, they may have done some collecting on the N coast of the peninsula in W Golfo de los Mosquitos [0900/8115 (USBGN)]. Collected on the S shore of the peninsula along a rude trail through the jungle, on the S side of the peninsula along cattle

trails, near brush and bush and at flowering trees, on the S shore "the terrain is limited to heavy jungle," at a beach on the base of the peninsula, at Cayo Toro [0909/8157 (USBGN)] (q. v.), at Nassau Hill [ca. 0910/8155 ?] (q. v.), and at Laguna de Bluefield [0909/8155 (USBGN)] (q. v.), 1936 Feb 28 (?), 29, Mar 1-2, 4, 5 (?), 7-8, 10-11, 13, 15-16, 18-19, 21, 23, 25-26, Apr 1-2 (UCLA, as "Valiente Peninsula"; MVZ, as "Chiriquí Lagoon"); Wetmore, used as an additional locator for Tiger Rock, 1958 Feb 25 (USNM, as "Peninsula Valiente"; Wetmore, 1959a, as "Valiente Peninsula"); SBE, GL for: Cayo Patterson, Laguna de Bluefield, Muturi, Playa Verde, Punta Alegre, Quebrada Hido, and Sierra Llorona, 1990 Apr 17-29 (MVUP; USNM, as "Valiente Peninsula"; Olson, 1993a, as "Valiente Peninsula"; Parsons et al., 1993, as "Valiente Peninsula"; Brumfield and Braun, 2001, as "Valiente Peninsula"; Brumfield et al., 2001, as "Valiente Peninsula"). (Angehr, 2003:311).

VALIENTE, PUNTA; Bocas del Toro

0911/8155 (USBGN)

Sea level, point on the most northwestern coast of Peninsula Valiente [0905/8151 (USBGN)], located on the W end of the peninsula that forms the N coast of Laguna de Bluefield [0909/8155 (USBGN)] (E762, as "Punta Bluefield"; JOG, as "Punta de Bluefield"; MGRP; MHA). On many maps Punta Valiente is labeled Punta de Bluefield or Punta Bluefield (see Punta Bluefield [0908/8157 (USBGN)]), which is a point 6 km SSW of Punta Valiente on the SW coast of Laguna de Bluefield. Wetmore, used as an additional locator for Tiger Rock, 1958 Feb 25 (USNM, as "Cape Valiente"; Wetmore, 1959a, as "Cabo Valiente"); Bullis, 10 mi (16 km) S, offshore, according to our maps, he collected in the E waters of Laguna de Chiriquí [0903/8200 (USBGN)], 1962 May 30 (USNM, as "Point Valiente"). See also: Bluefield, Laguna de.

VALIENTE POINT; see Valiente, Punta

VALLA DOLID; see Valladolid, Islote

VALLADOLID, ISLOTE; Panamá (W, Pacific)

0843/7936 (USBGN)

Sea level, an island in Bahía de Panamá [0850/7915 (USBGN)] 11 km N of Isla Otoque [0836/7936 (USBGN)] and 4 km SW of Isla Chamá [0844/7935 (USBGN)] (E761A; MHA). Pacific Project, GL = Gulf of Panama, 1969 Mar 3 (USNM, as "Valla Dolid"; Ridgely and Gwynne, 1989:135, as "Valladolid Rock"; Loftin, 1991).

VALLADOLID ROCK; see Valladolid, Islote

VALLE, EL; see El Valle

VALLE DE AGUA; Bocas del Toro

0915/8223 (USBGN)

175 m, settlement 5.5 km SSE of Almirante [0918/8224 (USBGN)] and 2 km W of Punta de Gallinazo [0915/8222 (USBGN)] on a high point midway between the Río Oeste [0916/8224 (USBGN)] and Quebrada Jones [0914/8222 (USBGN)] (E762). Kennard et al., GL = Almirante, 1926 (MCZ, as "Water Valley"; Kennard and Peters, 1928:443-444, as "Water Valley"); Wedel, 1938 Oct 5 (CMNH, as "Water Valley"); Wetmore, GL = Almirante, the forest had been cleared in the lowland

areas and on the gentler slopes, the vegetation was largely cacao with scattered areas of grass and brush and a few remaining large trees, entered the SE shore through mangrove islands and channels just S of Punta de Gallinazo, crossed a walkway to the shore and worked in an area of thickets with tall trees overhead, followed foot trails up the steep slopes through cultivated areas, pastures, *rastrojo*, and a small tract of original forest, climbed a steep slope along a rocky stream with rough boulders in heavy forest, 1958 Jan 16, 20, 22, 27, 31, Feb 4, 8, 17 (UF, as "Water Valley"; USNM, as "Water Valley").

VALLE DE ANTÓN; see El Valle

VALLE DE RISCÓ; Bocas del Toro

ca. 0913/8225

Not shown on our maps but the coordinates indicate a region on our maps (E762; JOG) along a tributary of the Río Riscó [0916/8228 (USBGN)] at 100-200 m, ca. 2 km SE of Cerro Riscó [0914/8225 (USBGN)], and ca. 9 km S of Almirante [0918/8224 (USBGN)]. Weir, at 200 m, 9°13'20"/82°24'40", 2003 Mar 7, 9-10, 12-13 (CU; LSUMZ).

VAN TYNE TRAIL; see Barro Colorado, Isla

VELARIO; see Palo Seco Bosque de Protección

VELÁSQUEZ, RÍO; Panamá (Canal Zone) (W, Pacific)

0858/7935 (USBGN)

Originates on the N slope of Cerro Galera [0856/7937 (USBGN)], flows N, then E and enters mangrove swamps as it discharges into the Panama Canal [0920/7955 (USBGN)] opposite Balboa [0857/7934 (USBGN)] (E761A). Hallinan, jungle, thickets, shrubbery, and trees in the valley of, 1916 Feb 13, May 2, 15 (AMNH; Hallinan, 1924, as "Río Velazquez").

VELAZQUEZ, RÍO; see Velásquez, Río

VELO; Chiriquí

ca. 0849/8229

Described by Blake (1958:502 (map), 503) as a section of the Mönniche properties located in the eastern foothills of Volcán Barú [0848/8233 (USBGN)] N of Finca Lérida [0848/8229 (USBGN)] between the lower reaches of the Río Chiquero [ca. 0849/8229] and the Quebrada El Velo [0849/8228 (USBGN)]. Described by Worth (1939b) as a "narrow dark valley about one mile from the Mönniche's home." He also states (1939a) that Mönniche purchased several tracts of forest with the intent of protecting them and through which he built access trails. Worth includes a mention of Velo as one of these protected forests; "one of the trails…to these [protected] forests descends steeply from a wooded plateau into a humid *oroya* [sic] or gorge, through which flows the Velo, one of the upper tributaries of the Río Caldera." Mönniche, at 5,200-6,700 feet (1,600-2,050 m), 1932-1934, 1936-39 (CAS, as "Velo," and as "Velo River"; FMNH, as "El Velo," and as "Velo"; MCN, as "El Velo," and as "Velo"); Worth, GL = Boquete, specimens are labeled Boquete, although Worth stayed at Finca Lérida (q. v.) and specifically mentions Velo, 1937 (ANSP, as "Boquete"; Worth, 1939a; Worth, 1939b).

VELO, QUEBRADA; see El Velo, Quebrada

VELO, QUEBRADA DE; see El Velo, Quebrada

VELO RIVER; see Velo

VENADO; see Venado, Ensenada

VENADO, ENSENADA; Los Santos

ca. 0725/8011

Described by Wetmore as a small bight behind a rocky, steep-sided hill that became an island at high tide (USNM). Not shown on our maps, but presumed to be one of two small coastal bays on either side of Venado [0725/8011 (USBGN)], which is on the S coast of Los Santo province 21 km WSW of Punta Mala [0728/8000 (USBGN)] (JOG). Angehr (in litt., Aug 2005) indicates that Playa Venado is indicated as a roughly 2.5 km beach in the cove W of the town of Venado on his map 4143-II (E762). He also adds that Wetmore may have used "Ensenada Venado to refer to the smaller cove just in front of the town of Venado, since its entrance is nearly blocked" in the manner that Wetmore described — behind a rocky, steep-sided hill that becomes an island at high tide. Wetmore, GL = Punta Mala, 11 mi (18 km) W of Punta Mala, forest and dry creek, 1957 Feb 25-26 (USNM; Wetmore, 1970). Wetmore (1965:456) also calls this Playa Venado, not to be confused with Venado Beach [0854/7936 (USBGN)] in the former Canal Zone.

VENADO, ISLA; Panamá (Canal Zone) (W, Pacific)

0853/7936 (USBGN)

120 feet (37 m), small island with a land area of ca. 0.13 km^2 (13 ha) located in Bahía de Panamá [0850/7915 (USBGN)] 1.5 km S of Venado Beach [0854/7936 (USBGN)] (E761A, as "Venado Island"). Hallinan, 1915 Mar 24 (AMNH; Hallinan, 1924, as "Venado Island").

VENADO, PLAYA; see Venado, Ensenada

VENADO, QUEBRADA; San Blas

ca. 0840/7729

Described by Wetmore as a river originating SW of Armila [0840/7728 (USBGN)] and flowing N into the Caribbean Sea [1500/7500 (USBGN)] with the Río Pito [0842/7732 (USBGN)] on its W and the Río Armila [0840/7728 (USBGN)] on its E (USNM). Handley, GL = Armila, 1962 Mar 21, 23, 27, Apr 1, 1963 Mar 17, 22-23, 31 (USNM); Wetmore, GL = Armila, camp was at 100 feet (30 m) ca. 4 km above the river mouth in forest near the ridge that divides the valleys of the Río Pito and the Río Armila, collected upstream in heavy forest with trees ranging from 80-130 feet (20-40 m), along a trail leading to the continental divide to an elevation of ca. 600 feet (180 m), and downstream near recent clearings, open fields, an old cacao plantation and other cultivated areas, 1963 Feb 21-Mar 13 (AMNH; LSUMZ; MVUP; USNM; Adsett and Wege, 1998); Feinstein, GL = Armila, 1963 Mar 8 (USNM). Not shown on our maps.

VENADO, RÍO; Panamá (Canal Zone) (W, Pacific)

0854/7936 (USBGN)

Short river that originates on the S slopes of Cerro Galera [0856/7937 (USBGN)] and flows S widening in mangroves swamps as it discharges into Bahía de Panamá [0850/7915 (USBGN)] at Venado Beach [0854/7936 (USBGN)] (E661; E761A). Hallinan, mangrove trees, rocky and sandy beach, muddy shores near the mouth of, 1915 Mar 27 (AMNH; Hallinan, 1924).

VENADO BEACH; Panamá (Canal Zone) (W, Pacific)
0854/7936 (USBGN)
Sea level, beach on the coast of Bahía de Panamá [0850/7915 (USBGN)] 9 km SW of Panama City [0858/7932 (USBGN)] at the mouth of the Río Venado [0854/7936 (USBGN)] (E761A). Wetmore, observations, 1961 Feb 12 (Wetmore, 1965:415). See also: Venado, Río; and Venado, Ensenada.

VENADO ISLAND; see Venado, Isla

VENTORILLO; see Ventorrillo

VENTORRILLO; Coclé (Pacific)
0837/8037 (USBGN)
1,000 m, town located 2.5 km W of El Copé [0837/8036 (USBGN)] (MGRP). Wetmore (1968:296, as "Ventorillo") mentions this locality, although specimens from this region all appear to be labeled Río Guabal [ca. 0846/8036] (q. v.).

VERA CRUZ; see Howard Air Force Base

VERACRUZ; Panamá (W, Pacific)
0853/7938 (USBGN)
Sea level, coastal town located 4.5 km W of Punta Bruja [0853/7935 (USBGN)] and 3 km E of Punta Bique [0853/7939 (USBGN)] (E761A, as "Camaron"; JOG; MGRP). USBGN treats Camerón as an older name for this town. Heyde, 1888 Sep 20 (USNM, as "Cameron"); Lasso de la Vega, 1983 Aug 19 (MVUP, as "Playa Veracruz"); Rodríguez, 1985 Apr 16 (MVUP); Guitton-Mayerma, Lelevier, Maley, and Miller, 7 km W of, 2006 Jan 21-22, 25 (STRI; UAM).

VERACRUZ, PLAYA; see Veracruz

VERAGUA; see Central, Cordillera, and Veraguas, Provincia de

VERAGUA ARCHIPELAGO; Chiriquí/Veraguas

A term of convenience coined by Olson (1997a) to include Isla de Coiba [0728/8147 (USBGN)] and all of the smaller islands of the Pacific coast of Chiriquí and Veraguas between the Península de Azuero [0740/8035 (USBGN)] and the Península Burica [0810/8255 (USBGN)]. Few of the smaller islands have been explored ornithologically, and a large series (AMNH, FMNH) attributed to many of the islands as having been collected in 1902 by J. H. Batty was fraudulently labeled (Olson 1997a) and consisted of specimens collected at Batty's mainland bases at Boquete [0847/8226 (USBGN)] and Boquerón [0830/8234 (USBGN)]. (AMNH, FMNH; LACM, SDSNH; USNM; Olson 1997a; Olson 2008; Olson 2007).

VERAGUA, COCUYOS DE; see Cucuyos de Veragua

VERAGUA, CUCUYOS DE; see Cucuyos de Veragua

VERAGUA, MOUNTAINS OF; see Central, Cordillera

VERAGUAS, PROVINCIA DE; (Atlantic, Pacific)

0810/8115 (USBGN)

Province in central W Panama that extends from Golfo de los Mosquitos [0900/8115 (USBGN)] on the Caribbean coast to Golfo de Montijo [0740/8107 (USBGN)] and the Pacific Ocean on the S coast. Panama's largest island, Isla de Coiba [0728/8147 (USBGN)], is within the provincial limits and is located in the Pacific Ocean just off the SW coast of the provincial mainland. Santiago [0806/8059 (USBGN)], the capital, is located in S central Veraguas province in the lowland regions of the Pacific drainage. The islands of these southern waters and the Pacific slope of the mainland, which reaches just over 1,500 m at the continental divide, have been explored by various ornithologists, but only limited work has been conducted on the Caribbean coast. The W boundary with Bocas del Toro and Chiriquí provinces is formed in part by the continental divide and by the Ríos Calovébora [0848/8113 (USBGN)], Viguí [0811/8134 (USBGN)], and Tabasará [0802/8142 (USBGN)]. The E boundary is shared with Colón and Coclé provinces on the mainland, and with Herrera and Los Santos provinces on Península de Azuero [0740/8035 (USBGN)]. Arcé worked throughout W Panamá on both the Atlantic drainage and the Pacific drainage labeling many specimens simply "Veragua" or "Veraguas." The region "Veragua" extended beyond the limits of modern Veraguas province and included localities within Chiriquí and Coclé provinces, if not others. According to Salvin's map (1870), Veragua referred to the entire region W of the Panama Railroad [ca. 0922/7954] to the Costa Rican border. Within Veraguas province, Arcé's secondary localities of which we are aware included: Calobre, Calovébora, Castillo, Chitra, Cordillera del Chucu, Laguna de la Yeguada, Santa Fé, and Santiago. These localities often did not occur on early maps and therefore conflicting reports of their actual locations appeared in the early literature. For example, Salvin (1870:177) speculated that Arcé collected almost entirely on the Pacific slope, even though the province spanned the continental divide and at least two localities are on the Caribbean versant. Griscom (1935a:199) described Calovébora and Chitra as "two small hamlets in the humid strip at the base of the mountains on the Pacific slope in Veraguas, western Panamá," yet earlier Sclater and Salvin (1867:123) had indicated that both Castillo and Calovevora were on the N slope of Veraguas province. According to our maps, Calovébora and Cordillera del Chucu are on the N slope or the Atlantic drainage, and the remaining localities, including Calobre, Castillo, Chitra, Laguna de la Yeguada, Santa Fé and Santiago, are on the S slope or the Pacific drainage. Furthermore, the difficulties in assigning coordinates to these secondary localities was exacerbated by the fact that Arcé and many early authors were apparently quite liberal with the use of some of the aforementioned names—see Santiago [0806/8059 (USBGN)] and Cordillera del Chucú [ca. 0840/8050] for further detail. In addition to the secondary localities mentioned above, the literature also refers to the district El Mineral de Veraguas [ca. 0830/8100] (q. v.), which included localities on both sides of the continental divide. See also: Central, Cordillera.

VERDE, ISLA; Veraguas (Pacific)

0750/8105 (USBGN)

Sea level, island with a land area of ca. 3 km^2 (300 ha) in NE Golfo de Montijo

[0740/8107 (USBGN)], 11 km S of Puerto Mutis [0755/8103 (USBGN)] and just below the mouth of the Río San Pedro [ca. 0750/8105] (JOG). Wetmore, GL = Puerto Mutis, mouth of the Río San Pedro, the island's center had been cleared and mangrove swamps covered the surrounding lowland areas, landed on a rocky point and covered most of island's cultivated, higher ground, 1953 June 11 (USNM; Olson, 1980b, as "Puerto Mutís, Isla Verde, mouth of Río San Pedro").

VERDE, POTRERO; Chiriquí

ca. 0848/8235

ca. 1,850 m, described by Wetmore as an oval depression bordered by trees that was located between Potrero Seco [ca. 0848/8236] (q. v.) and Potrero Cordillera [ca. 0848/8235] (q. v.) on the upper end of Llanos del Volcán [ca. 0847/8238] near the W base of Volcán Barú [0848/8233 (USBGN)] ca. 2 mi (3 km) from the highway leading N out of Volcán [0846/8238 (USBGN)], Wetmore, GL = Volcán, Volcán Barú, at 6,000 feet (1,850 m), 1955 Mar 13 (USNM).

VERGAYO, PUNTA; see Pastores, Isla

VERGAYO POINT; see Pastores, Isla

VERMEJO, MONTAÑA DEL; Coclé ? (Pacific ?)

ca. 0840/8030 ?

600 m ?, not located on our maps but presumed to be in the mountains on the Pacific slope of the continental divide. Today, a town named Bermejo [0840/8033 (USBGN)] (q. v.) and a river named Río Bermejo [0838/8034 (USBGN)] exist in W Coclé province near the Veraguas/Coclé provincial border, therefore it is possible that Montaña del Vermejo was located in this region. Despite the word "Montaña" in the locality name, some of the specimens attributed to this locality were probably collected at lower elevations, Arcé, 1874, 1875 (BMNH; Salvin and Godman, 1888-1904, vol. 2:144, vol. 3:268).

VIA RICARDO J. ALFARO; see Tumba Muerto

VIBALÁ; see Bibalaz

VIDAL, PUERTO; see Puerto Vidal

VIDAL, RÍO; Veraguas (Pacific)

ca. 0801/8138

Originates in the hills SW of Las Palmas [0808/8127 (USBGN)] and flows SW into Boca de Pajaroncito [0756/8139 (USBGN)] opposite the E shore of Isla de la Corocita [0758/8139 (USBGN)] (MRP, as "Río Virá"; SMN2B). The USBGN coordinates [0757/8146] place the river mouth well off the coast. Wetmore and Perrygo, GL = Puerto Vidal, near Puerto Vidal in muddy, deeply shaded woods that flood during high tide, in forest with heavy undergrowth on banks that rose 4-5 feet (1-2 m) above the high tide mark, in second growth, 1953 May 29-30, mouth of the Río Vidal, at Isla de la Corocita (q. v.), 1953 Jun 8 (USNM; Olson, 1980b, as "Puerto Vidal, Río Vidal, Isla Corocita"). (Delgado, 1983).

VIEJA, PUNTA; see Old Point

VIEJO, CERRO; Chiriquí

0816/8136 (USBGN)

728 m, name of a peak located 8 km NE of Tolé [0814/8141 (USBGN)] and ca. 6 km N of the Carretera Interamericana [0832/8242 (USBGN)] (JOG). A town named Cerro Viejo [0815/8135 (USBGN)] is on the SE slopes of the peak, and the lowlands of the Río Tabasará [0802/8142 (USBGN)] valley are within a few km of the peak (MGRP). This area is in Corregimiento Cerro Viejo [0817/8133 (USBGN)]. There is also a hill named Cerro Viejo [0828/8207 (USBGN)] farther W in Chiriquí province, but because Hinds worked at Tabasará [0804/8138 (USBGN)] in Aug and Sep 1981, it seems likely that his work at Cerro Viejo was also in E Chiriquí province. Hinds, 1981 Apr 30, May 9, 11, 17-19, 22, Jun 8, 11 (GML).

VIEJO, CERRO; Herrera

0737/8048 (USBGN)

3,300 feet (1,000 m), according to Aldrich and Bole (1937, map), the peak is centrally located on Península de Azuero [0740/8035 (USBGN)] in E Veraguas province, but more recent maps indicate that the peak is located in the very S extension of Herrera province. See Península de Azuero [0740/8035 (USBGN)] for further discussion of the provincial borders. See also: Cerro Viejo Camp; and Cavulla Camp.

VIEJO CAMP, CERRO; see Cerro Viejo Camp

VIENTO, CERRO; see Cerro Viento

VILLA, ISLA; Los Santos

0756/8018 (USBGN)

Rocky islet in the Gulf of Panama [0800/7930 (USBGN)] off the E coast of Península de Azuero [0740/8035 (USBGN)] 5 km NE of La Honda [0754/8019 (USBGN)] (JOG). Angehr (2003:207) indicates that the island is also known as Peñon de la Honda Island, which is included along with La Honda within Refugio de Vida Silvestre Peñon de la Honda (Peñon de la Honda Wildlife Refuge). Described in Wetmore's journal as an elongate rock that was 60-80 feet (20 m) high. The summit was covered with low scrub and dropped abruptly on the S shore, but sloped somewhat on the N and NW shores. Wetmore, observations, 1957 Feb 28 (Wetmore, 1965:73; Olson, 1997a). See also: La Honda.

VILLA, RÍO DE LA; see La Villa, Río

VILLA, RÍO LA; see La Villa, Río

VILLA FRANCÉS; see Francés

VILLA GABRIELA; Panamá (E, Pacific)

ca. 0902/7929

50 m, section of NE greater Panama City [0858/7932 (USBGN)] located in Corregimiento Río Abajo [0901/7929 (USBGN)] (E962). Jiménez and Pérez, 2001 Apr 7 (MVUP).

VILLA GRECIA; Panamá (E, Atlantic)

ca. 0906/7934

120 m, town located 14 km E of Gamboa [0907/7942 (USBGN)] on the Carretera

Boyd-Roosevelt [0917/7943 (USBGN)] just N of the continental divide (E762). Rodríguez, 1980 Nov 24 (GML).

VILLA GUADALUPE; Panamá (E, Pacific)
0902/7929 (USBGN)
50 m, section of greater Panama City [0858/7932 (USBGN)] located just NW of La Pulida [0903/7929 (USBGN)] (E962). Martínez, 1980 Feb 22, Apr 15 (MVUP).

VILLA LOBOS; Panamá (E, Pacific)
0905/7926 (USBGN)
Not shown on our maps but USBGN (as "Villalobos") indicates a town ca. 1 km N of Pedregal [0904/7926 (USBGN)]. Valdéz, 1985 Oct 14 (MVUP).

VILLALOBOS; see Villa Lobos

VIQUE, PUNTA; see Bique, Bahía

VIQUE COVE; see Bique, Bahía

VIRÁ, RÍO; see Vidal, Río

VISTAMARES, CERRO; Panamá (E, Atlantic)
ca. 0914/7922
ca. 800 m, "just c. 2 km to the W of Cerro Jefe [0914/7921 (USBGN)]," Observers ?, along the trail on the N side of the mountain, 1992 Dec 26, observations, 1992 Jan and Feb, at 750-800 m on the W end of the mountain, 1995 Dec (Adsett and Wege, 1998). Not shown on our maps.

VISTAMARES, SENDERO; see Jefe, Cerro

VIVEROS, ISLA; Panamá, Archipiélago de las Perlas
0828/7900 (USBGN)
Sea level, irregularly shaped island with a land area of 6.49 km^2 (649 ha) (Wright et al., 1985) located immediately off the NW coast of Isla del Rey [0822/7855 (USBGN)] (JOG). Bovallius, "dense tropical forests," cultivation, see discussion of Bovallius under Pueblo Nuevo [ca. 0820/8013 ?], 1882 Apr 3-11 (NHRM; Rendahl, 1920; Wetmore, 1946b; Wetmore, 1952:31; Wetmore, 1970); (MacArthur et al., 1972—species list; see comments in Bibliography); Wright, Faaborg, and Campbell, mist nets, observations, 1976 Jun-Aug, 1978 Jan-Mar (Wright et al., 1985); Levings et al., observations, 1982-1983 (Levings et al., 1986).

VIVIENDA, ISLA LA; Panamá, Archipiélago de las Perlas
0829/7903 (USBGN)
Sea level, small island located less than 100 m off the SE coast of Isla Bayoneta [0829/7904 (USBGN)] (JOG). Murphy and Correia, Askoy Expedition, 1941 May 23 (AMNH; Murphy, 1944a; Murphy, 1944b, as "La Vivienda Island"; Murphy, 1945, as "La Vivienda Island"). See also: Málaga, Isla.

VOLCÁN; Chiriquí
0846/8238 (USBGN)
1,300 m, town in NW Chiriquí province 12 km WSW of Volcán Barú [0848/8233 (USBGN)] and 12 km SW of Cerro Punta [0852/8235 (USBGN)] (JOG, as "El Hato del Volcán"; RDPF). According to Wetmore in 1954, the town was located

on the lower edge of Llanos del Volcán [ca. 0847/8238] where scattered houses and stores extended 1 mi (1.5 km) W to Nueva California [0847/8239 (USBGN)] (USNM). The older names El Hato del Volcán and El Volcán appear on many maps. Davidson, GL for Barriles, 1929 Dec 14-15, 22, 26-27, 1931 Jan 12, 14-15, 17, 20-22, 24, 26-28, Feb 3-4, 6 (Davidson, 1932, as "Barriles"; Davidson, 1938, as "Barriles, El Volcán"); Mönniche, 1932, 1937, 1940 (FMNH); Worth and Cresson, at 4,100 feet (1,250 m), at least some specimens were collected along the Río Gariché (q. v.), 1937 Jun 17-18, 20-26, 29-Jul 4, 6-11, 18 (ANSP, as "El Volcán"; Worth, 1938a, as "El Volcán"; Worth, 1938b, as "El Volcán"; Worth, 1939a, as "El Volcán"); Enders, at 4,100 feet (1,250 m), 1937 Jun 20, 29, Jul 20 (ANSP, as "El Volcán"); Hartmann, R., at 4,100 feet (1,250 m), 1937 Jun 21, 29, Jul 1, 6 (ANSP, as "El Volcán"); Bishop, 1937 Jun 30 (ANSP, as "El Volcán"); Eisenmann, ca. 4,200 feet (1,300 m), observations, 1945 Aug 1, 1949 Jul 13 (Eisenmann, 1946, as "El Volcán"; Eisenmann, 1962a, as "El Hato"; Eisenmann, 1962c, as "El Hato"); Hartman, at 4,000-4,500 feet (1,200-1,350 m), one specimen at 7,000 feet ? (2,150 m) annotated by Wetmore that Hartman's collectors did not work above 5,000 feet (1,500 m), "outside the village of Volcán," 1948 Feb 27, 1949 Feb 18-19, 21-25, 27, 1951 Feb 19-23, 26-28, Mar 1-2, 12, 16-17, 19-21, 1953 Feb 2, 8-19, Mar 19, 1956 Feb 21, 27, 29, Mar 1-3, 5-9, 11-15, 1958 Feb 14-22, 24, 26, Mar 1, 3, 5, and as a GL for Cerro Pando, 1951 Mar 5-7, 10, GL for Palo Santo, 1958 Feb 13, 15-17, 19-21 (LSUMZ; USNM, as "El Volcán," as "Volcán Chiriquí," and as "Volcán"; Hartman, 1957; Hartman and Brownell, 1961a, as "El Volcán"; Hartman and Brownell, 1961b, as "El Volcán"; Hartman and Lessler, 1963, as "El Volcán"); Wetmore, at 4,500 feet (1,350 m), on the W edge of the savanna, 1954 Mar 25-26. All the secondary localities attributed to Volcán (except for Cerro Picacho) were located within 11 km of town in one of two general areas: E of town on the E end of Llanos del Volcán near the W face of Volcán Barú, or W of town along the Río Chiriquí Viejo, along its tributaries and on the high slopes above the river. GL for: Barriles, Cerro Pando, Las Lagunas, Llanos del Volcán, Palo Santo, Pando, Río Chiriquí Viejo, Río Colorado, and Silla de Cerro Pando, 1954 Feb 24-Mar 13, Mar 21-22, Mar 24-29, GL for: Cerro Picacho, La Lagunita, Las Lagunas, Llanos del Volcán, Palo Santo, Pando, Potrero Cordillera, Potrero Seco, Potrero Verde, Río Chiriquí Viejo, Silla de Cerro Pando, and Volcán Barú, 1955 Jan 31-Feb 7, Feb 9-12, 26, Mar 13-16, in 1960 he noted extensive clearing for pasture land since his 1955 visit to the region, 1960 Feb 24, and as a GL for: Aguacate, Barriles, Cerro Pando, Las Lagunas, Llanos del Volcán, Palo Santo, Quebrada Zumbona, Silla de Cerro Pando, and Tisingal, 1960 Feb 6, 8-13, 15-20, 22-25, Mar 20, 22, in 1965 Wetmore noted additional and extensive clearing of his localities near Volcán since his previous visits, forested or wooded sections were very limited as most land had been converted to pastures, at 4,800 feet (1,450 m), 1965 Mar 19, 10 km NW of Volcán, crossed Silla de Cerro Pando and hunted on the principal tributary of the Río Colorado [0849/8243 (USBGN)], 1965 Mar 23, and as a GL for: Las Lagunas, Palo Santo, Río Chiriquí Viejo, Silla de Cerro Pando, and Tisingal, 1965 Feb 26-28, Mar 2, 7, 9, 14-18, 20, 22-24 (LSUMZ, as "El Volcán"; MVUP, as "El Volcán"; USNM, as "El Volcán"; Huff and Wetmore, 1967, as "El Volcán"; Olson, 1981c, as El Volcán,

Tisingal," as El Volcán, Silla de Cerro Pando," as "El Volcán, Quebrada Zumbona," as "El Volcán, Laguna Grande," as "El Volcán, Lagunas," and as El Volcán, Cerro Picacho"; Olson, 1983, as "El Volcán, Palo Santo"); Willis, at 1,300 m, 1962 Jul (Willis, 1972, as "El Volcán"); Holmes, GL for Las Lagunas, 1964 May (GML, as "El Volcan"); Smith, N., 2.6 mi (4.2 km) SSE, 1964 Jun 20 (AMNH, as "El Hato"); Tyson, observations, 1962 mid-Mar (Loftin and Olson, 1963:192, as "El Volcán"), at 7,500 feet (2,300 m), 7 mi (11 km) NW, 1965 Feb 16 (USNM); Morton, specimens collected 6 mi (10 km) W, 3 mi (5 km) E at Volcán de Chiriquí [= Volcán Barú], and at Finca Florentina (not located), 1966 Apr 24 (FMNH, as "El Hato"); Kleen, 1967 Oct 8 (AMNH); Ridgely, 1968 Mar 13 (AMNH, as "El Volcán"); Hilty, 1968 Apr 25, 27 (LSUMZ); Tejera, Ponce, and Ramos, 1981 Mar 1 (MVUP). (Wetmore, 1952:16, as "El Volcán"; Wetmore, 1967, as "El Volcán"; Wetmore, 1970, as "El Volcán"; Leck, 1972b, as "El Hato"; Parkes, 1976, as "El Volcán"; Olson, 1981c; Olson, 1981d, as "El Volcán"; Willson et al., 1982). See also: Las Lagunas.

VOLCÁN, EL; see Volcán

VOLCÁN, LLANOS DEL; Chiriquí

ca. 0847/8238

ca. 1,300-1,900 m, region just SW of Volcán Barú [0848/8233 (USBGN)] (MHA). Described by Wetmore as a broad, open flat descending from the W base of Volcán Barú SW to Las Lagunas [0846/8241 (USBGN)] that was created by lava flow. Steep-sided creeks near the volcano fanned out over the rocky plain and the area was covered with scanty grass, low weedy growths and brush (USNM). Eisenmann, at ca. 5,000 feet (1,500 m), observations below Bambito [0850/8237 (USBGN)], near, 1948 Jul 9 (Eisenmann, 1962a); Wetmore, GL = Volcán, at 4,200-4,500 feet (1,300-1,350 m), collected near the base of Volcán Barú on small, isolated hills that were covered with forest that was partly open and partly choked with weeds and brush, near fruiting trees and shrubs, in grassland, and along dry creeks and small running creeks, on the road to Las Lagunas at the S end of Llanos del Volcán on open land with a scattering of bushes and heavy grass, and 30-40 foot (9-10 m) trees with dense undergrowth, 1954 Feb 26, 1955 Feb 1, 11, GL = Volcán, at 6,300 feet (1,900 m), on the E end of Llanos del Volcán near Volcán Barú, left the jeep at 6,150 feet (1,870 m) and walked ca. 2 mi (3 km) to 6,400 feet (2,000 m) to steep-sided and deep dry washes, to low and scattered bushes on stony soil with boulders, at 5,500-6,000 feet (1,700-1,850 m) *Agave* was blooming, 1960 Feb 18 (USNM, as "El Llano," as "Llanos," and as "Llanos del Volcán").

VOLCÁN BARÚ PARQUE NACIONAL; see Barú, Volcán

VOLCÁN BARÚ PARQUE NATIONAL; see Barú, Volcán

VOLCÁN DE CHIRIQUÍ; see Barú, Volcán

VOLCÁN LAKES; see Las Lagunas

WATER CAY; see Agua, Cayo

WATER KEY; see Agua, Cayo

WATER VALLEY; see Valle de Agua

WEST BEACH;

0905/8134 (USBGN)

Bocas del Toro, Archipiélago de Bocas del Toro

Presumably the same as or near West Point [USBGN] at the southwestern tip of Isla Escudo de Veraguas (E762); Billo, GL = Isla Escudo, 2007 Apr 12, "West Beach" (USNM).

WESTERN RIVER; see Oeste, Río

WEST FERRY RAMP; Panamá (Canal Zone) (W, Pacific)

ca. 0856/7933

The W ramp for the ferry that crossed the Panama Canal [0920/7955 (USBGN)] between the E terminus of Carretera Thatcher [0858/7938 (USBGN)] and La Boca [0857/7934 (USBGN)]. The ferry was eventually replaced by Thatcher Ferry Bridge [0856/7933 (USBGN)] (E661, as "Ferry"; RM1, as "Thatcher Ferry") which is now called *Puente de las Américas* or Bridge of the Americas [ca. 0856/7933] (q. v.). Blanton, 1953 Sep 27 (USNM). See also: Farfan Beach.

WEST POINT; see West Beach

WILCOX CAMP; Veraguas (Pacific)

ca. 0741/8118

Sea level ?, described by Griscom as a camp located near the mouth of the Río San Lorenzo [0740/8118 (USBGN)] (q. v.) on the SW coast of Veraguas province W of Golfo de Montijo [0740/8107 (USBGN)]. The camp was named after A. R. Wilcox, the president of the Tropical Forest Products Company and was located "at the heart of a primeval forest" (Griscom, 1924a:509 (map, as "Wilcox Lumber Camp"), 517-519). Griscom et al., forest, launch to the coast, 1924 Apr 2-11, 12 ? (AMNH, as "Wilcox Camp, San Lorenzo River"; Griscom, 1924a; Griscom, 1924b, as "Wilcox Camp, San Lorenzo River"; Griscom, 1933; Zimmer, 1941b:14; Zimmer, 1950:26; Olson, 1981a, as "Wilcox Camp, San Lorenzo River [Caribbean Slope]" [sic]).

WILLIAMSTOWN; see Panamá

WINKLOSKI, CASA; see Bambito

YAGUADA, LAGUNA DE LA; see La Yeguada, Laguna de

YAVISA; see Yaviza

YAVIZA; Darién

0811/7741 (USBGN)

ca. 25 m, town in central Darién province on the Carretera Interamericana [0832/8242 (USBGN)]. The town is on the Río Chucunaque [0809/7744 (USBGN)] 6 km ENE of its junction with the Río Tuira [0821/7803 (USBGN)] and 6 km NE of El Real de Santa María [0808/7743 (USBGN)] (MGRP; SMN2B). Baer, an anthropologist on the Marsh-Darien Expedition (see comments under Marsh, 1934), collected several specimens along the Río Tuira, the Río Chucunaque, and their tributaries. The secondary localities, aside from Río Surcurtí, are within 25 km of Yaviza, GL for: Camp Creek, Camp Townsend, Rio Canglón, Río Chico, Río Chucunaque, Río Surcurtí, Río Tuquesa, Río Tuira, and Three Falls Creek, 1924

Feb 11, 14, 18, 22-25, 27, Mar 1-5, 7-8, 27, Apr 13 (USNM); Slattery, 1966 May 16, Sep 19-21, Nov 23, Dec 1, 13, 1967 May 15, and as a GL for Río Chucunaque, 1966 Dec 2 (GML; AMNH); Cavalier and Slattery, a few specimens as 2 km NE of, one as 2.5 km E of, 1967 Sep 13, 15, 17-24, 26-27, Oct 1-2, 8, 10-11, 13, 31, Nov 1-3, 5-8, 12-14, and as a GL for Río Chico, 1967 Oct 4-5, 7 (MVZ); Ridgely, observations near Yaviza along the lower Río Chucunaque and Río Tuira, 1975 Jul (Wetmore et al., 1984:8). See also: El Real de Santa María.

YEGUADA, LA; see La Yeguada

YERBAZALES; Chiriquí
0809/8254 (USBGN)
100 m, town located on Península Burica [0810/8255 (USBGN)] 14 km SSW of Puerto Armuelles [0817/8252 (USBGN)] and 2 km inland from the coast (JOG). Weir, GL = Península Burica, at 20-100 m, 2003 Apr 4-7 (CU; LSUMZ).

ZAINO MADRONIO; see Madroño

ZANJA LIMÓN; see Limón, Zanja

ZAPATILLA, CAYOS; Bocas del Toro, Archipiélago de Bocas del Toro
0916/8203 (USBGN)
12 m, 2 islands, each labeled Cayo Zapatilla on many maps, located 5 km SE of Isla Bastimentos [0918/8208 (USBGN)] in the Caribbean Sea [1500/7500 (USBGN)]. The slightly larger island has a land area of ca. 0.5 km^2 (50 ha) and is ca. 1 km SE of the smaller island (E762, as "Cayo Zapatilla"; MHA). According to Handley's journal (USNM, as "Big Zapatillo Cay"), the larger island was hilly, and the forest was no higher than coconut palm trees. The smaller island (as "Little Zapatillo Cay") had beaches, some coconut palm, but primarily scrub forest. Benson, 1927 Sep 11 (Wetmore, 1968:7, as "Almirante, Sapodilly Cay"); Miller and Richardson, GL = Laguna de Chiriquí [0903/8200 (USBGN)], described by Miller as "a low sand spit covered with trees" with fine white sand beaches, walked around the entire island, 1936 Mar 4 (MVZ (?), as "Chiriqui Lagoon"; UCLA, as "Zapatilla"); SBE, collected on both islands, 1987 Mar 25 (MVUP, as "Cayo La Gran Zapatilla," and as "Cayo Pequeña Zapatilla"; USNM, as "Little Zapatilla Key," and as "Big Zapatilla Key").

ZAPOTILLO; Veraguas (Pacific)
0800/8131 (USBGN)
ca. 100 m, a town on the Río Tuancle [0800/8132 (USBGN)] 22 km W of Soná [0801/8119 (USBGN)] (MGRP). Wetmore and Perrygo, GL = Soná, at Río Tuancle (q. v.), 1953 May 19, 25, June 9; Wetmore, GL for Río Bubí, 1955 Mar 19 (USNM, as "El Zapotillo," and as "Zapotillo"; Wetmore, 1967, as "Zapotillo, 30 km W of Soná"; Eisenmann, 1957).

ZEGLA; Bocas del Toro
ca. 0922/8232
10 m, settlement on the Río Teribe [0922/8232 (USBGN)] 0.5 km W of its mouth and 8.5 km S of Changuinola [0926/8231 (USBGN)] (E762). Kennard et al., GL = Almirante, 1926 (MCZ; Kennard and Peters, 1928:443-444); Smith, some specimens

are also labeled "Terebe River," 1927 May 16-21, 23 (YPM; Griscom, 1933, as "Zegla, Terebe River"; Eisenmann and Howell, 1962, as "Zegla, Terebe River"; Escalante-Pliego and Peterson, 1992, as "Zegla, Terebe River"); Wedel, 1927 Dec 16, 1928 Apr 28-29 (MCZ); SBE, S bank Río Teribe, 1991 Feb 28 (USNM).

ZUMBADOR; see La Zumbadora

ZUMBADORA, LA; see La Zumbadora

ZUMBONA, QUEBRADA; Chiriquí
0848/8241 (USBGN)
Originates as an intermittent stream in the hills N of Palo Santo [0849/8241 (USBGN)] and flows S entering the Río Chiriquí Viejo [0820/8241 (USBGN)] less than 1 km S of Palo Santo (E762). Wetmore, GL = Volcán, at 4,500 feet (1,350 m), along a logging road in a deep valley that was protected from the wind, 1960 Feb 20, 22 (MVUP; USNM; Olson, 1981c, as "El Volcán, Quebrada Zumbona").

Annotated Bibliography of Ornithology in Panamá

In the four volumes of *Birds of the Republic of Panamá*, Wetmore used abbreviated bibliographical citations in the text with the intention of publishing a complete bibliography when the series was finished. The present bibliography incorporates the card file that Wetmore had accumulated but may not include all of the references cited in the series when they do not pertain directly to Panamá. We hope that all of the references cited in the gazetteer appear here, but the bibliography also includes publications that are not cited in the gazetteer. This is not a complete bibliography of the voluminous literature of Panamanian ornithology but it should be reasonably complete for taxonomic, systematic, and biogeographical studies.

Adames, Abdiel J., B. Dutary, Hugo Tejera, E. Adames, and Pedro Galindo. 1993. Relación entre Mosquitos Vectores y Aves Acuáticos en la Transmision Potenciál de dos Arbovirus. [Relationship between Mosquito Vectors and Aquatic Birds in the Potential Transmission of Two Arboviruses]. *Revista Médica de Panamá*, 18(2):106-119, 2 maps, 6 tables. In Spanish with English summary. Work near El Rosario.

Adsett, William J., and David C. Wege. 1998. Natural History of the Little-Known Speckled Antshrike *Xenornis setifrons*. *Cotinga*, 10:24-29, appendix, map, 6 photographs, 2 figures. Mentions specimens and observations, many of which are based on reports from *El Tucán, Journal of the Panama Audubon Society*. Localities mentioned include Cerro Bruja, Río Chico, Cerro Guagaral, Cerro Jefe, La Laguna, Nusagandí, Permé, Obaldía, Old Village Site, Old Tacarcuna Site, Quebrada Venado, and Cerro Vistamares. The appendix contains locality descriptions.

Aid, Charles S., G. Gene Montgomery, and Douglas W. Mock. 1985. Range Extension of the Peruvian Booby to Panama During the 1983 El Nino. *Colonial Waterbirds*, 8(1):67-68. Observations, photographic records and 1 specimen of *Sula variegata* from Isla Boná.

Alba, Alberto Federico. 1946. *Algunas Aves de Panama*. [Some Birds of Panama]. 134 pages, illustrations. Panama: Imprenta Nacional. Descriptions of birds with common and scientific names, some with a photograph or a drawing. Spanish text.

Alderton, Cora C. 1961. The Breeding Cycle of the Yellow-Bellied Seedeater in Panama. *Condor*, 63(5):390-398, 4 tables. Observations of *Sporophila nigricollis* by the author at "South Gamboa" in the Canal Zone, in 1956-1957, and 1959.

Alderton, Cora C. 1963. The Breeding Behavior of the Blue-Black Grassquit. *Condor*, 65(2):154-162, 3 tables. Observations by the author of 14 pairs and 22 nests of *Volatinia jacarina* at Gamboa, in the Canal Zone, 1955-1961

Aldrich, John Warren. 1937. Annotated List of Birds of the Western Slope of the Azuero Peninsula. *Scientific Publications of the Cleveland Museum of Natural History*, 7:27-139, tables. The entire volume is titled *The Birds and Mammals of the Western Slope of the Azuero Peninsula* [Republic of Panama] by J. W. Aldrich, and B. P. Bole. A collection of 509 specimens of 164 species made from February to April 1932, the following being described as new: *Tinamus major brunneiventris*, *Crypturornis soui poliocephalus*, *Ortalis garrula olivacea*, *Chaetura vauxi ochropygia*, *Sittasomus griseicapillus veraguensis*, *Manacus aurantiacus flaviventris*, *Elaenia flavogastra pallididorsalis*, *Habia rubica aurantiicapilla*, and *Atlapetes gutturalis azuerensis*. Presents data for specific CMNH specimens.

Aldrich, John Warren. 1945. Notes on Panamanian Birds. *Auk*, 62(1):111-114. *Ammodramus savannarum obscurus* [= *bimaculatus*], from grasslands near Chepo, and *Lochmias nematura nelsoni* n. subsp. from Mt. Pirre. Both collected by Goldman.

Aldrich, John Warren, and Benjamin Patterson Bole, Jr. 1937. The Birds and Mammals of the Western Slope of the Azuero Peninsula [Republic of Panama]. *Scientific Publications of the Cleveland Museum of Natural History*, 7:1-196, 1 map, 8 plates, tables. Introduction (pages 5-26) with account of topography, collecting stations, ecology, and faunal relationships. Birds are specifically treated in Aldrich (1937).

Aldrich, John Warren, and Allen J. Duvall. 1958. Distribution and Migration of Races of the Mourning Dove. *Condor*, 60(2):108-128, 3 figures, 1 table. Migration of *Zenaidura macroura marginella* and *Z. m. carolinensis* through Panama. Mentions specimens from Divalá, and specimens examined from Antón, El Espino, San Francisco and Soná.

Allen, G. M. 1939. Chapman's 'Life in an Air Castle'. Review of *Life in an Air Castle*, by Frank M. Chapman. *Auk*, 56(1):95-96.

Alvarez del Toro, Miguel. 1952. Contribución al Conocimiento del la Oología y Nidología de las Aves Chiapanecas. [Contributions to the Understanding of the Oology and the Nests of the Birds from Chiapas]. *Ateneo, Organo del Ateneo de Ciencias Y Artes de Chiapas*, 4:11-21, 17 figures. In Spanish.

Alvarez, E. and D. H. Ellis. 1994. Harpy Eagle (*Harpia harpyja*) Nesting in Manipulated Forests. *Journal of Raptor Research*, 28(1):51. Monitored 9 nests in Venezuela and 5 nests in the buffer region of Darién National Park.

Amadon, Dean. 1949. Notes on *Harpyhaliaetus*. *Auk*, 66(1):53-56, 1 table. Suggests that *Urubitornis* J. Verreaux is a synonym to *Harpyhalietus* Lafresnaye, which should include the 2 species *coronatus* and *solitarius*.

Amadon, Dean. 1961. Relationships of the Falconiform Genus *Harpagus*. *Condor*, 63(2):178-179. Concludes that *Harpagus* in not related to *Accipiter*.

Amadon, Dean, and Don R. Eckelberry. 1955. Observations on Mexican Birds. *Condor*, 57(2):65-80, 1 colored frontispiece, 6 figures. Notes from 7 weeks of field work in eastern and southern Mexico.

American Ornithologists' Union. 1957. *Check-list of North American Birds*. 5th edition, xiv + 691 pages. Washington, D.C.: American Ornithologists' Union.

American Ornithologists' Union. 1998. *Check-List of North American Birds: The Species of Birds of North America from the Arctic through Panama, including the West Indies and Hawaiian Islands*. 7th edition, liv + 829 pages. Washington, D.C.: American Ornithologists' Union.

Angehr, George R. 1981. The Role of Interference Competition in the Organization of a Guild of Panamanian Hummingbirds. xiv + 210 pages, 18 figures, 18 tables. Boulder: University of Colorado at Boulder. Observations from 1977-1979 on hummingbird foraging on Barro Colorado Island. Ph.D. Dissertation.

Angehr, George R. 1999. Rapid Long-Distance Colonization of Lake Gatun, Panama, by Snail Kites. *Wilson Bulletin*, 111(2):265-268. Discusses the introduction of the apple snail (*Pomacea lattrei*) to control *Hydrilla* growth, and the subsequent arrival of *Rostrhamus sociabilis*, probably from an established population in Colombia. Observations in Guindilla Cove of Gigante Bay, Lake Gatun.

Angehr, George. 2000. Red Data Bird: Yellow-billed Cotinga. *World Birdwatch*, 22(2):24-25 (June), 3 figures. Observations of *Carpodectes antoniae* at Cerro Batipa. Mentions collections S of David in 1901 and observations on Península Burica.

Angehr, George R. 2003. *Directorio de Áreas Importantes para Aves en Panamá/Directory of Important Bird Areas in Panama.* 342 pages, 4 appendices, 16 colored plates, 5 tables. Panama: Sociedad Audubon de Panama. Overview of BirdLife International's Important Bird Areas Program followed by site descriptions organized by province. All sections of the directory are written in Spanish and in English.

Angehr, George R., and Daniel G. Christian. 2000. Distributional Records from the Highlands of the Serranía de Majé, an Isolated Mountain Range in Eastern Panama. *Bulletin of the British Ornithologists' Club*, 120(3):173-178, appendix, 1 figure. Annotated list of 18 species found in the cloud forest above 1,150 m based on a survey of Cerro Chucantí in 1996. The appendix contains additional species and the map (fig. 1) indicates mountain ranges in central and eastern Panama.

Angehr, George R., Daniel G. Christian, and Karla M. Aparicio. 2004. A Survey of the Serranía de Jungurudó, an isolated Mountain Range in Eastern Panama. *Bulletin of the British Ornithologists' Club*, 124(1):51-62, appendix, 1 table, 1 figure (map). Observations of species that are rare or of limited range in Panama. The appendix contains a list of additional species found above 700 m.

Anthony, H. E. 1916. Panama Mammals Collected in 1914-1915. *Bulletin of the American Museum of Natural History*, 35(10):357-376, 6 figures. Dates and itinerary for work conducted in Darién province in 1914 and 1915 followed by annotated list of 55 species. Darién map (fig. 6) indicates 10 localities, 2 of which are Goldman's (Mt. Pirre and Cana), and the others are Chepigana, El Real, Boca de Cupe, Cituro, Capeti, Tapalisa, Tacarcuna and Mt. Tacarcuna. The text, but not the map, also refers to the Shiras expedition (mammal collections) in the Canal Zone at Gatun, Old Panama, Maxon Ranch (Río Trinidad), and Río Chilibrillo. Also: Río Capetí, Cuchirvo, Río Cupé, and Río Tuyra.

Anthony, H. E. 1923. In the Footsteps of Balboa. *Natural History*, 23:313-324, 11 black and white photographs. Popular account of the 1915 collecting expedition (Richardson, Ball and Anthony) to the Isthmus of Darién (Anthony, 1916). Photographs of local residents and collecting localities, only a few of which are mentioned.

Aparicio, Karla M. 2000. Áreas Importantes para las Aves en Panamá. Panamá Avanza en la Protección de Sus Aves. [Important Areas for the Birds of Panama. Panama Achieves Advances for the Protection of its Birds.] *La Garcilla*, 108:10-11, 1 photograph, 3 tables. In Spanish. Discusses the Programa Nacional de Áreas Importantes para las Aves en Panamá, which was initiated in 1995. See also: Angehr, 2003.

Arbib, Robert S., Jr., and Frederick W. Loetscher, Jr. 1935. Some Notes from the Panama Canal Zone—Summer 1934. *Auk*, 52(3):325-328. Sight records made during August-September 1934 with a few notes from 1933. Brief descriptions of the following localities: Panama Vieja, Río Tapia, Madden Dam, Barro Colorado Island, Pedro Miguel, and Gatun.

Avery, Ralph Emmet. 1915. *The Greatest Engineering Feat in the World at Panama: Authentic and Complete Story of the Building and Operation of the Great Waterway—The Eighth Wonder of the World. The American Triumph in Panama.* 384 pages, illustrations. New York: Leslie Judge Company. Edited by William C. Haskins. A 1913 edition also exists. Information on the French and the American construction efforts on the Panama Canal with photographs of the work in progress.

Backwell, Patricia R. Y., Patrick D. O'Hara, and John H. Christy. 1998. Prey Availability and Selective Foraging in Shorebirds. *Animal Behaviour*, 55(6):1659-1667, 1 figure, 5 tables. Studied 6 of the 13 shorebirds that regularly foraged in their study area, La Playa El Agallito, from Oct 1995 to Jan 1996.

Baird, Spencer Fullerton. 1864-1872. Review of American Birds in the Museum of the Smithsonian

Institution. Part 1. *Smithsonian Miscellaneous Collections*, 181. 2nd edition, vi + 478 pages, tables, diagrams. Part I includes "slender-billed Oscines, with the exception of the Caerebidae." Issued in parts from 1864-1866, reissued in 1872 with the addition of front matter, list of species and an index so that the final volume consisted of vi + 478 pages. 37 species of Passeriformes listed from Panama, with description of *Troglodytes inquietus* n. sp. from the Panama R. R.

Banasch, Ursula, J. Paul Goossen, Alberto Einstein Riez, Clark Casler, and Romeo Dominguez Barradas. 1992. Organochlorine Contaminants in Migrant and Resident Prey of Peregrine Falcons, *Falco peregrinus*, in Panama, Venezuela, and Mexico. *Canadian Field-Naturalist*, 106(4):493-498, 3 tables. Abstract in French. Collected migrant and resident birds in the regions near Chitré in 1984-1985. Tables provide mean levels (ppm) of organochloride residues by species.

Bangs, Outram. 1900. List of Birds Collected by W. W. Brown, Jr., at Loma del Leon, Panama. *Proceedings of the New England Zoological Club*, 2:13-34, tables. List of 752 specimens of 148 species collected in 1900, with the following described as new: *Mionectes oleagineus parcus*, *Myrmelastes ceterus*, and *Saltator lacertosus*.

Bangs, Outram. 1901a. Birds of San Miguel Island, Panama. *Auk*, 18(1):24-32, tables. 42 species collected by W. W. Brown, from late April to mid-May, 1900, with the following described as new: *Melanerpes seductus*, *Phaethornis hyalinus*, *Elaenia sordidata*, and *Rhamphocelus limatus*. Also mentions a few specimens from Loma del Leon and Pedro Gonsales Island.

Bangs, Outram. 1901b. A New Honey Creeper from San Miguel Island, Panama. *Proceedings of the New England Zoological Club*, 2:51-52. *Coereba cerinoclunis*, n. sp.

Bangs, Outram. 1901c. A New *Ortalis* from the Archipielago de las Perlas, Bay of Panama. *Proceedings of the New England Zoological Club*, 2:61-62, tables. *Ortalis struthopus* n. sp. from San Miguel Island. The tables also contain data from specimens collected at Divala, Loma del Leon, and Pedro Gonsales Island.

Bangs, Outram. 1901d. On a Collection of Birds Made by W. W. Brown, Jr., at David and Divala, Chiriqui. *Auk*, 18(4):355-370, tables. Describes 1,183 specimens representing 175 taxa taken from October to December, 1900, with the following described as new: *Odontophorus castigatus*, *Campophilus guatemalensis buxans*, *Mionectes assimilis dyscolus*, *Pipra mentalis ignifera*, *Cercomacra crepera*, and *Automolus exsertus*.

Bangs, Outram. 1901e. Description of a New Woodpecker from Chiriqui. *Proceedings of the New England Zoological Club*, 2:99-100. *Veniliornis neglectus* n. sp. from Divalá.

Bangs, Outram. 1902a. On a Second Collection of Birds Made in Chiriqui, by W. W. Brown, Jr. *Proceedings of the New England Zoological Club*, 3:15-70, tables. Work on Volcan de Chiriquí from January to August, 1901, mainly at Boquete, but ranging to the top of the mountain, with a trip in June to the northern Caribbean slope. Also mentions collections and specimens from Bogaba, David, Dolega, Loma del Leon, Pedregal, Remedios, the Sona River and Sona. Described as new: *Phaethornis guy coruscus*, *Campylopterus hemileucurus mellitus*, *Dendrocopus villosus extimus*, *Myiopagis placens accola*, *Sayornis amnicola*, *Grallaricula vegeta*, *Sclerurus mexicanus pullus*, *Sittasomus levis*, *Catharus gracilirostris accentor*, *Thryophilus modestus elutus*, *Troglodytes browni*, *Henicorhina collina*, *Microcerculus acentetus*, and *Chlorospingus novicius*. *Xiphorhynchus grandis* Cherrie is a nomen nudum.

Bangs, Outram. 1902b. Two New Birds from San Miguel Island, Bay of Panama. *Proceedings of the New England Zoological Club*, 3:71-73, tables. *Formicivora alticincta* and *Vireo insulanus*, n. spp.

Bangs, Outram. 1902c. Description of a New Thrush from Chiriqui. *Proceedings of the New England Zoological Club*, 3:91-92, table. *Merula leuchauchen cnephosa* n. subsp. from Volcán de Chiriquí.

Bangs, Outram. 1903a. A New Race of *Scotothorus veraepacis* from Chiriqui. *Proceedings of the New England Zoological Club*, 3:103-104. *S. v. dumicola* n. subsp. from Divala.

Bangs, Outram. 1903b. A New Wren from San Miguel Island, Bay of Panama. *Proceedings of the New England Zoological Club*, 4:3-4, tables. *Thryophilus galbraithi conditus*, n. subsp. Compared with specimens from Loma del Leon.

Bangs, Outram. 1903c. A New Race of *Vireosylva josephae* from Chiriqui. *Proceedings of the New England Zoological Club*, 4:9-10. *V. j. chiriquensis* n. subsp. from Boquete.

Bangs, Outram. 1905a. Descriptions of Seven New Subspecies of American Birds. *Proceedings of the Biological Society of Washington*, 18:151-156, tables. *Dacnis cayana callaina* n. subsp. from Divala.

Bangs, Outram. 1905b. The Name of the Panama Green Honey Creeper. *Proceedings of the Biological Society of Washington*, 18:186. Indicates that *Chlorophanes spiza exul* ranges from Ecuador to Panama and *C. s. guatamalensis* from Costa Rica to Guatemala. Specimen from Divala.

Bangs, Outram. 1906a. Vertebrata from the Savanna of Panama. I. Introduction. *Bulletin of the Museum of Comparative Zoology*, 46(12):211. Describes the collecting locality, Calidonia, the Savanna of Panama, near Panama City, of W. W. Brown in 1904. The treatment of Aves is by Thayer and Bangs (1906).

Bangs, Outram. 1906b. Notes on Birds from Costa Rica and Chiriqui, with Descriptions of New Forms and New Records for Costa Rica. *Proceedings of the Biological Society of Washington*, 19:101-112, tables. Described as new from Boquete and Volcan de Chiriquí: *Pyrrhura hoffmanni gaudens*, *Xenicopsis variegaticeps idoneus*, and *Cyanolyca blandita*.

Bangs, Outram. 1907. On a Collection of Birds from Western Costa Rica. *Auk*, 24(3):287-312, tables. Descriptions based on a collection of ca. 6,000 skins collected by C. F. Underwood in 1906. *Micrastur interstes*, and *Buarremon costaricensis*, n. spp. *Gymnocichla nudiceps erratilis*, *Synallaxis albescens latitabunda*, *Dendrocolaptes sancti-thomae hesperius*, *Leptopogon pileatus faustus*, and *Cyanerpes lucidus isthmicus*, n. subspp. Twelve new records for Costa Rica. Specimens from Panama and Line of R. R.

Bangs, Outram. 1908. On Certain Costa Rican Birds. *Proceedings of the New England Zoological Club*, 4:23-35, tables. New subspecies from Boquete and Divala: *Myiobius xanthopygus aureatus*, *Troglodytes ochraceus ligea*, and *Myioborus aurantiacus acceptus*. *Phlogothraupis sanguinolenta aprica*, from Costa Rica.

Bangs, Outram. 1930. Types of Birds Now in the Museum of Comparative Zoology. *Bulletin of the Museum of Comparative Zoology*, 70(4):147-426. Lists the holotype or the cotype for 1,241 taxa. Includes a list of authors of new names and a brief history of the MCZ type collection.

Bangs, Outram, and Thomas Barbour. 1922. Birds from Darien. *Bulletin of the Museum of Comparative Zoology*, 65(6):191-229. Collection of nearly 700 birds representing 159 species made by Barbour and W. S. Brooks in 1922 in the Sambu Valley and on the Serranía del Sapo. New record for Panama: *Pilherodias pileatus*. *Phaethornis adolphi nelsoni*, nom. nov. New taxa: *Jacamerops aurea penardi*, *Threnetes ruckeri darienensis*, *Chalybura buffoni micans*, *Phaenostictus macleannani chocoanus*, *Sclerurus mexicanus anomalus*, *Pipra erythrocephala actinosa*, *Craspedoprion olivaceum bardus*, *Oxyruncus brooksi*, *Chlorophanes spiza arguta*, and *Tangara inornata languens*.

Bangs, Outram, and Thomas E. Penard. 1918. Notes on a Collection of Surinam Birds. *Bulletin of the Museum of Comparative Zoology*, 62(2):25-93, tables. Described as new from Panama: *Milvago chimachima cordata*, and *Aramides cajanea latens* (San Miguel Island), *Pitangus lictor panamensis* (Loma del Leon), and *Sporophila minuta centralis* (Panama, near Panama).

Bangs, Outram, and Thomas E. Penard. 1919. Some Critical Notes on Birds. *Bulletin of the Museum of Comparative Zoology*, 63(2):21-40, tables. *Saltator striatipictus speratus* n. subsp. from Saboga

Island. Specimens from Loma del Leon, Panama City, San Miguel Island.

Bangs, Outram, and Thomas E. Penard. 1921. Notes on Some American Birds, Chiefly Neotropical. *Bulletin of the Museum of Comparative Zoology*, 64(4):365-397, tables. *Myiozetetes cayanensis harterti* n. subsp. from Loma del Leon. Panama specimens.

Banks, Richard C. 1978. Nomenclature of the Black-Bellied Whistling Duck. *Auk*, 95(2):348-352. South American and West Indian populations are referred to *Dendrocygna autumnalis autumnalis* with *D. discolor* as a junior synonym, and birds N of Panama are referred to *D. a. fulgens* with *D. a. lucida* as a synonym.

Barajas, Flor C., and Allan R. Phillips. 1994. A *Haplospiza* Finch in Western Mexico; the Lessons of an Enigma. *Bulletin of the British Ornithologists' Club*, 114(1):36-46, 1 figure, 1 table. Includes measurements of specimens from Panama.

Barbour, Thomas, and W. S. Brooks. 1923. The Sapo Mountains and the Sambu Valley. A Biological Reconnaissance in Southeastern Panama. *The Geographical Review*, 13(2):211-222, 8 figures. Details of the author's 1922 trip into Darién, the first zoological expedition to the Sambú Valley. Includes a sketch map and photographs.

Bard, Susanne C., Michaela Hau, Martin Wikelski, and John C. Wingfield. 2002. Vocal Distinctiveness and Response to Conspecific Playback in the Spotted Antbird, a Neotropical Suboscine. *Condor*, 104(2):387-394, 3 figures, 1 table. Recorded songs of *Hylophylax naevioides* at Limbo Plot in 1997.

Barnard, Gladys Cargill. 1954. Notes on the Nesting of the Thick-Billed Euphonia in the Panama Canal Zone. *Condor*, 56(2):98-101. Observations of *Euphonia laniirostris* near Pedro Miguel.

Barnard, Gladys Cargill. 1956. Nesting of the Blue-Black Grassquit in Panama. *Condor*, 58(3):229-231. Observations of 4 nests of *Volatinia jacarina* at Pedro Miguel, with additional notes from Ancon Hill and Gamboa.

Barrantes, Gilbert, and Bette A. Loiselle. 2002. Reproduction, Habitat Use, and Natural History of the Black-and-Yellow Silky-flycatcher (*Phainoptila melanoxantha*), an Endemic Bird of the Western Panama-Costa Rican Highlands. *Ornitologia Neotropical*, 13(2):121-136, 5 figures, 7 tables. Observations at 16 sites in Costa Rica, one of which is near the border.

Bartholomew, G. A., and D. L. Goldstein. 1984. The Energetics of Development in a Very Large Altricial Bird, the Brown Pelican. *Perspectives in Vertebrate Science*, 3:347-357, 8 figures, 4 tables. The volume is titled *Respiration and Metabolism of Embryonic Vertebrates; Satellite Symposium of the 29th International Congress of Physiological Sciences, Sydney Australia, 1983*. Data from embryos of *Pelecanus occidentalis* that were taken from Isla Uraba, then incubated on Taboga Island.

Bartlett, A. D. 1866. Notes on the Breeding of Several Species of Birds in the Society's Gardens During the Year 1865. *Proceedings of the Zoological Society of London*, 1866:76-79, 1 plate.

Batista, Victoria E. and G. Gene Montgomery. 1982. Numbers of Brown Pelicans in the Pearl Island Archipelago, Bay of Panama, in 1981. *Pacific SeaBird Group Bulletin*, 9(2):70. Circumnavigated the islands and counted *Pelecanus occidentalis* 5 times during 1981.

Beason, Robert C. 1976. An Observation of Nocturnal Passerine Migration Through Panama. *Condor*, 78(1):117-118, 1 figure. Observations at Balboa during 1972 Nov 2-3.

Beebe, William. 1942. *Book of Bays*. 1st edition, xviii + 302 pages, 2 appendices, 33 illustrations (map, plates, sketches). New York: Harcourt, Brace, and Company. Chapter 22 includes information on Panama with notes made in 1938 March at Bahía Honda, and off Isla de Coiba. Occasional mention of birds throughout the text.

Beebe, William. 1950. Home Life of the Bat Falcon, *Falco albigularis albigularis* Daudin. *Zoologica*:

Scientific Contributions of the New York Zoological Society, 35(1):69-86. Based on observations in Venezuela.

Beebe, William. 1954. Discovered—The Nest and Egg of the Black-Winged Bell Bird. *Animal Kingdom, Magazine of the New York Zoological Society*, 57(4):115-119, 1 illustration, 3 photographs. Two nests of *Procnias averano* in Trinidad.

Beebe, William, G. Innes Hartley, and Paul G. Howes. 1917. *Tropical Wild Life in British Guiana, Zoological Contributions from the Tropical Research Station of the New York Zoological Society.* Volume I. 504 pages, 143 figures. New York: The New York Zoological Society. Bird lists. Information on nests and eggs of several species.

Beehler, Bruce. 1980. A Comparison of Avian Foraging at Flowering Trees in Panama and New Guinea. *Wilson Bulletin*, 92(4):513-519, 3 tables. Observations at a single flowering tree along the Pipeline Road, Jan 1978.

Belcher, Sir Charles, and G. D. Smooker. 1934. Birds of the Colony of Trinidad and Tobago-Part I. *Ibis*, 13(4):572-595. Description of the islands followed by species accounts. Tinamidae-Falconidae.

Belcher, Sir Charles, and G. D. Smooker. 1935. Birds of the Colony of Trinidad and Tobago-Part II. *Ibis*, 13(5):279-297. Cracidae-Rynchopidae.

Belcher, Sir Charles, and G. D. Smooker. 1936. Birds of the Colony of Trinidad and Tobago-Part III. *Ibis*, 13(6):1-35. Columbidae-Trochilidae.

Belcher, Sir Charles, and G. D. Smooker. 1936. Birds of the Colony of Trinidad and Tobago-Part IV. *Ibis*, 13(6):792-813. Trogonidae-Cotingidae.

Belcher, Sir Charles, and G. D. Smooker. 1937. Birds of the Colony of Trinidad and Tobago-Part V. *Ibis*, 14(1):225-249. Pipridae-Tyrannidae.

Belcher, Sir Charles, and G. D. Smooker. 1937. Birds of the Colony of Trinidad and Tobago-Part VI. *Ibis*, 14(1):504-550. Hirundinidae-Fringillidae.

Beltran, William, and Gustavo Kattan. 2001. First Record of the Slaty-Backed Nightingale-Thrush in the Central Andes of Colombia, with Notes on Its Ecology and Geographical Variation. *Wilson Bulletin*, 113(2):134-139, 1 figure, 1 table. Observational censuses and mist netting of *Catharus fuscater* in Colombia, May 1996-Jul 1999.

Bennett, Charles F., Jr. 1962. The Bayano Cuna Indians: An Ecological Study of Livelihood and Diet. *Annals of the Association of American Geographers*, 52(1):32-50, 2 figures, 9 tables. List of animals hunted includes 11 birds. Maps of the Bayano River Valley and of Icandí [= Icantí] before the formation of Lago Bayano.

Bent, Arthur Cleveland. 1919. Life Histories of North American Diving Birds, Order Pygopodes. *United States National Museum, Bulletin* 107; xiii + 245 pages, 55 black and white plates.

Bent, Arthur Cleveland. 1922. Life Histories of North American Petrels and Pelicans and their allies. Order Tubinares and Order Steganopodes. *United States National Museum, Bulletin* 121: xii + 343, 69 black and white plates.

Bent, Arthur Cleveland. 1925. Life Histories of North American Wild fowl, Order Anseres (Part), Family Anatidae, Ducks, Geese and Swans. *United States National Museum, Bulletin* 126: ix + 250 pages, 46 black and white plates.

Bent, Arthur Cleveland. 1925. Life Histories of North American Wild fowl, Order Anseres (Part), Family Anatidae, Ducks, Geese and Swans. *United States National Museum, Bulletin* 130: x + 376 pages, 60 black and white plates.

Bent, Arthur Cleveland. 1937. Life Histories of North American Birds of Prey (Part 1), Order Falconiformes. *United States National Museum, Bulletin* 167: viii + 409 pages, 102 black and white plates.

Bent, Arthur Cleveland. 1938. Life Histories of North American Birds of Prey (Part 2), Orders Falconiformes and Strigiformes. *United States National Museum, Bulletin* 170: viii + 482, 92 black and white plates.

Bent, Arthur Cleveland. 1940. Life Histories of North American Cuckoos, Goatsuckers, Hummingbirds, and Their Allies. Orders Psittaciformes, Cuculiformes, Trogoniformes, Coraciiformes, Caprimulgiformes, and Micropodiiformes. *United States National Museum, Bulletin* 176: viii + 506, 73 black and white plates.

Berlepsch, Hans Graf von. 1880. Preliminary Descriptions of New Birds from South America, and Remarks on Some Described Species. *Ibis*, 4(4):112-114. *Celeus immaculatus* n. sp. from Agua dulce.

Berlepsch, Hans Graf von. 1883. Descriptions of Six New Species of Birds from Southern and Central America. *Ibis*, 5(1):487-494, 1 colored plate. *Pipra velutina* n. sp. from Veragua.

Berlepsch, Hans Graf von. 1884. Untersuchungen über de Vögel der Umgegend von Bucaramanga in Neu-Granada. [Examination of the Birds from the Vicinity of Bucaramanga, New Grenada]. *Journal für Ornithologie*, 32(3-4):274-320, 1 colored plate. *Damophila [juliae] panamensis*, n. subsp.

Berlepsch, Hans Graf von. 1888. Descriptions of New Species and Subspecies of Birds from the Neotropical Region. *Auk*, 5(4):449-460, tables. *Basileuterus godmani* n. sp. *Eucometis spodocephala stictothorax* n. subsp. both from Veragua.

Berlepsch, Hans Graf von. 1888. Notes on Some Neotropical Birds Belonging to the United States National Museum. *Proceedings of the United States National Museum*, 11:559-566, table. Comments on *Damophila juliae* and *D. panamensis*.

Berlepsch, Hans Graf von. 1902. Beschreibung Zweier Neuer Drosselformen aus Südamerika. [Descriptions of Two New Forms of Thrushes from South America]. *Ornithologische Monatsberichte*, 10(5):69-71. *Catharus fuscater hellmayri* n. subsp. from Veragua.

Berlepsch, Hans Graf von. 1911. Revision der Tanagriden. [Revision of the Tanagers]. *Verhandlungen des V. Internationalen Ornithologen-Kongresses in Berlin, 30. Mai bis 4. Juni 1910:1001-1161*. [Proceedings of the International Ornithological Congress (5th, 1910, Berlin)]. Index of scientific names (p. 1149-1161). According to a footnote in Hartert (1927:38) "the volume, though dated 1911, did not appear before January or February 1912." *Calospiza larvata centralis*, nomen nov. *Calliste larvata* Du Bos (ut supra), *Calliste fanny* autorum nec Lafresnaye, *Calliste francescae* Sclater 1856 (nomen emend.). The type is from Calovevora; also mentioned are: Castillo, David, Loma del Leon, Mina de Chorcha, Stationen Paraiso, Sta. Fé, and Cordillera de Tolé.

Berlioz, J. 1949. Le Polymorphisme chez les Trochilidés. [Polymorphism among the Trochilids]. Pages: 3-6 *in* Ernst Mayr and Ernst Schuz, editors. *Ornithologie als Biologische Wissenschaft, 28 Beiträge als Festschrift zum 60. Geburtstag von Erwin Stresemann.* Heidelburg: Carl Winter, Universitätsverlag.

Bertin, Robert I., and Peggy A. Wilzbach. 1979. Sexual Differences in Feeding Territoriality of the Crowned Woodnymph, *Thalurania colombica*. *Wilson Bulletin*, 91(2):319-321, 1 table. Observations in 1977 near the Limbo Hunt Club.

Binford, Laurence C. 1965. Two Subspecies of Birds from Oaxaca, Mexico. *Occasional Papers of the Museum of Zoology, Louisiana State University*, 30:1-6, 1 colored plate, 1 figure.

Blake, Emmet R. 1949. *Ictinia misisippiensis* collected in Paraguay. *Auk*, 66(1):82.

Blake, Emmet R. 1956. A Collection of Panamanian Nests and Eggs. *Condor*, 58(5):386-388. Made by Tolef B. Mönniche in Chiriquí without mention of specific localities.

Blake, Emmet R. 1958. Birds of Volcán de Chiriquí, Panama. *Fieldiana: Zoology*, 36(5):499-577, 1 map. Detailed account of a collection representing 235 taxa of birds made by Tolef B. Mönniche from 1932 to 1952, in the region near Boquete. Brief descriptions of several localities.

Blake, John G. 1989. Birds of Primary Forest Undergrowth in Western San Blas Panama. *Journal of Field Ornithology*, 60(2):178-189, 3 figures, 4 tables. Mist netting at Cerro Brewster, Cangandi and Nusagandí from Apr-May 1985.

Bonaparte, Charles Lucien. 1855. [Principales Espèces Nouvelles d'Oiseaux qu'il Vient d'Observer dans son Récent Voyage en Éscosse et en Angleterre/On the Principle New Species of Birds Observed on a Recent Voyage to Scotland and England]. *Comptes Rendus Hebdomadaires des Séances de l'Academie des Sciences*, 41:651-661. *Planesticus casius* n. sp. from Panama.

Bond, James. 1950. Results of the Catherwood-Chaplin West Indies Expedition, 1948. Part II. Birds of Cayo Largo (Cuba), San Andrés and Providencia. *Proceedings of the Academy of Natural Sciences of Philadelphia*, 102:43-68.

Bond, James. 1953. Notes on Peruvian *Dendrocolaptidae*, *Conopophagidae* and *Rhinocryptidae*. *Notulae Naturae of the Academy of Natural Sciences of Philadelphia*, 248:1-7. Mention of a large specimen of *Xiphorhynxhus l. lachrymosus* from Garachiné, Darién.

Bond, James. 1956. Check-List of the Birds of the West Indies. 4th edition, ix + 214 pages, 1 map. Philadelphia: Academy of Natural Sciences. The 4th edition was composed and printed in Lancaster at the Wickersham Printing Company. Twenty-five supplements to the check-list were published between 1956 and 1984. Occasional mention of Panama.

Bond, James, and Rodolphe Meyer de Schauensee. 1944. The Birds. C. Birds Collected in Darién, Panama. *The Academy of Natural Sciences of Philadelphia, Monographs*, 6:22-45. Annotated list of 171 species collected on the Vanderbilt expedition mainly from Garachiné and Mt. Sapo, with excellent account of physical features, including the higher slopes of the mountain. The entire monograph is titled *Results of the Fifth George Vanderbilt Expedition (1941)*.

Bonhote, J. Lewis. 1903. On a Collection of Birds from the Northern Islands of the Bahama Group. *Ibis*, 8(3):273-315. Collected 104 taxa in the winter of 1901-1902.

Boucard, Adolphe. 1878. On Birds Collected in Costa Rica. *Proceedings of the Zoological Society of London*, 1878:37-71, 1 plate. Describes ca. 250 taxa collected during 5 months in Costa Rica. Notes on some species found in Panama.

Boucard, Adolphe. 1891. Description of a Supposed New Species of Tanager in Boucard's Museum. *Humming Bird*, 1(7):53. *Ramphocelus chrysopterus* n. sp. from the state of Panama, Columbia.

Boucard, Adolphe. 1892. A Complete List Up to Date of the Humming-Birds Found in Columbia [sic], with Descriptions of Several New Species. *Humming Bird*, 2(9):73-87. *Heliodoxa berlepschi*, from Veragua, and *Phaethornis panamensis*, from Panama and Veragua, n. spp. Mentions Agua Dulce, Volcán de Chiriquí, and Chiriquí, all of which were in Colombia at the time of writing.

Boucard Adolphe. 1892-1895. *Genera of Humming Birds*. xiv + 412 pages. London: Pardy & Son. The printed front pages indicate that the volumes were published in 1893-1895, but handwritten notes on the early sections indicate that they were published in 1892. Describes 539 taxa, with Addenda (with additional taxa), Errata, and an Index (p. 404-412).

Bovallius, Carl Erik Alexander. 1877. *Resa i Central-Amerika 1881-1883*. [*Voyage through Central*

America, 1881-1883]. 316 pages, illustrations. Spanish translation as *Viaje por Centroamerica 1881-1883*. 1977. Translated by Camilo Vijil Tardon. Managua: San José, S. A. Chapters 1-9 describe his trip through Panama with notes on the isthmus, the Perlas Islands, and the western shores of the Bay of Panama. The accuracy of his descriptions are somewhat questionable, see the gazetteer entry under Pueblo Nuevo. The index (p. 313-314) contains scientific names of plants and animals mentioned in the text.

Boyden, Thomas C. 1978. Territorial Defense Against Hummingbirds and Insects by Tropical Hummingbirds. *Condor*, 80(2):216-221. Observations at flowering plants at Morgan's Gardens, 5 km N of Balboa from Jun 23-25 1974 and at a woods trail 2 km W of the Bridge of the Americas on 1974 Jul 3.

Brabourne, Lord, and Charles Chubb. 1912. *A List of the Birds of South America*, Volume 1. xx + 504 pages, 1 map (color). London: R. H. Porter. Also listed as publishers are John Weldon, and Taylor and Francis. Index (p. 445-504). Volume II does not contain additional species descriptions due to the death of Brabourne but contains 38 colored plates prepared by Grönvold and 11 pages describing them. Issued in parts.

Brackbill, Hervey. 1960. Foot-Quivering by Foraging Hermit Thrushes. *Auk*, 77(4):477-478.

Brandreth, Courtney. 1924. Five Studies of Birds of Panama. *Scribner's Magazine*, 76(5):465-469, 5 illustrations. With notes by James P. Chapin of the American Museum of Natural History.

Brandt, Herbert W. 1938. Four Extra-Limital Records. *Auk*, 55(2):287-288. *Belonopterus chilensis cayennensis* collected by Wedel in 1934 at Port Obaldia.

Braun, Michael J., and David E. Wolf. 1987. Recent Records of Vagrant South American Land Birds in Panama. *Bulletin of the British Ornithologists' Club*, 107(3):115-117. *Chrysolampis mosquitus* at El Real. *Coccyzus lansbergi* at Tocumen and at Cana. *C. pumilus* at Tocumen. Photographs by P. Moynihan in the VIREO collection at ANSP.

Brawn, Jeffrey D., T. M. Collins, M. Medina, and E. Bermingham. 1996. Associations Between Physical Isolation and Geographical Variation Within Three Species of Neotropical Birds. *Molecular Ecology*, 5(1):33-46, 2 appendices, 5 figures, 2 tables. Compared mitochondrial DNA polymorphisms and morphological characters between three passerine species from the Archipiélago de las Perlas and the mainland. Collected data from: Bolaños, Chapera, Pacheca, Rey, and Saboga Islands and on the mainland from Atlantic [= Gatun], Gamboa and Summit.

Brawn, Jeffrey D., James R. Karr, and James D. Nichols. 1995. Demography of Birds in a Neotropical Forest: Effects of Allometry, Taxonomy, and Ecology. *Ecology*, 76(1):41-51, 3 figures, 2 tables. Analysis of mark and recapture data obtained from mist netting along the Pipeline Road in Soberanía National Park from 1977-1986.

Brawn, Jeffrey D., James R. Karr, James D. Nichols, and W. Douglas Robinson. 1998. Demography of Forest Birds in Panama: How do Transients Affect Estimates of Survival Rates? *Ostrich*, 69(1 and 2):22. Symposium abstract. Employed new models that adjust for transient individuals and found that estimates for survival rates were not greatly affected.

Briggs, Marjorie A. 1954. Apparent Neoteny in the Saw-Whet Owls of Mexico and Central America. *Proceedings of the Biological Society of Washington*, 67:179-182.

Brodkorb, Pierce. 1936. A New Genus for *Empidonax atriceps* Salvin. *Occasional Papers of the Museum of Zoology, University of Michigan*, 331:1-2, 1 table. *Cnemonax*, gen. nov.

Brodkorb, Pierce. 1940. New Birds from Southern Mexico. *Auk*, 57(4):542-549, 2 tables. Examined specimens from Panama in the description of *Chloroceryle amazona mexicana*, n. subsp. Mentions *C. americana isthmica* in the description of *C. a. vanrossemi*, n. subsp.

Brodkorb, Pierce. 1941. The Pygmy Owl of the District of Soconusco, Chiapas. *Occasional Papers of the Museum of Zoology, University of Michigan*, 450:1-4.

Brodkorb, Pierce. 1943. Birds from the Gulf Lowlands of Southern Mexico. *Miscellaneous Publications, Museum of Zoology, University of Michigan*, 55:1-88, 1 map, 12 tables.

Brodkorb, Pierce. 1948. Taxonomic Notes on the Laughing Falcon. *Auk*, 65(3):406-410, 1 figure. *Herpetotheres c. cachinnans* specimens from Panama.

Brodkorb, Pierce. 1950. Geographical Variation in the Gray Kingbird *Tyrannus dominicensis*. *Auk*, 67(3):333-344, 1 figure, 9 tables. Specimens of *T. dominiscensis sequax* n. sp. from Colon, Perme, and Obaldia.

Bronson, Wilfrid Swancourt. 1931. *Paddlewings*. 106 pages, illustrations. New York: Macmillian. A children's book on the life and adventures of a Galapagos Penguin. The story of a bird's capture and confinement on a ship, its escape off the Perlas Islands, Panama, how it was re-captured by a fisherman, and finally sold in the Panama City market to men from the same ship who took it north to the New York aquarium.

Brown, R. G., and Dodge Engleman. 1986. First Record of Hudsonian Godwit in Panama. *American Birds*, 40(3):429. Sightings of *Limosa haemastica* at Coco Solo.

Brown, R. G. and Lorna Engleman. 1988. First Record of American Avocet From the Republic of Panama. *American Birds*, 42(1):28. Sightings of *Recurvirostra americana* at Aguadulce.

Browning, M. Ralph. 1989. The Type Specimens of Hekstra's Owls. *Proceedings of the Biological Society of Washington*, 102(2):515-519. Data on holotype of *Otus guatemalae centralis*, from Cerro Malí.

Brumfield, Robb T., and Michael J. Braun. 2001. Phylogenetic Relationships in Bearded Manakins (Pipridae: *Manacus*) Indicate that Male Plumage Color is a Misleading Taxonomic Marker. *Condor*, 103:248-258, 3 figures, 2 tables. Abstract in Spanish. Tissues from specimens collected at: Cayo Agua, Cana, Puerto Armuelles, Soberania National Park, Río Teribe,and Valiente Peninsula.

Brumfield, Robb T., Robert W. Jernigan, David B. McDonald, and Michael J. Braun. 2001. Evolutionary Implications of Divergent Clines in an Avian (*Manacus*: Aves) Hybrid Zone. *Evolution*, 55(10):2070-2087, appendix, 6 figures, 7 tables. Examined the hybrid zone between *Manacus vitellinus* and *M. candei* near Chiriquí Lagoon in Bocas del Toro province. The appendix provides specimen data including locality and catalog number.

Brumfield, Robb T., Robert W. Jernigan, David B. McDonald, and Michael J. Braun. 2003. Errata [to Brumfield et al. 2001]. *Evolution*, 57: 2919. Calls belated attention to the fact that over half the specimens used in this study were collected by SBE and not by the authors, as it appeared originally. It fails to mention that the specimens were used without the permission of the collectors.

Brush, Alan H. 1967. Pigmentation in the Scarlet Tanager, *Piranga olivacea*. *Condor*, 69(6):549-559, 3 figures, 2 tables. Pigment extracts from liver and feathers samples of spring and fall specimens.

Buckley, F. C., and P. A. Buckley. 1968. Upside Down Resting by Young Green-Rumped Parrotlets (*Forpus passerinus*). *Condor*, 70(1):89. Observations of a captive flock.

Buckley, P. A., Mercedes S. Foster, Eugene S. Morton, Robert S. Ridgely, and Francine G. Buckley, editors. 1985. Neotropical Ornithology. *Ornithological Monographs* 36: xi + 1041 pages, colored frontispiece, 7 colored plates, figures, tables. Contains separately authored contributions some of which are cited elsewhere herein.

Buehler, Deborah M., Alberto I. Castillo, and George Angehr. 2004. Shorebird counts in the

Upper Bay of Panama highlight the importance of this key site and the need to improve its protection. *Wader Study Group Bulletin*, 105:56-64, 3 figures, 2 tables. Observations of Scolopacidae and Charadriidae at Costa del Este, Panama Viejo, Río Pacora, Río Bayano, and Río Pasiga.

Burleigh, Thomas D. 1960. Geographic Variation in the Western Wood Peewee (*Contopus sordidulus*). *Proceedings of the Biological Society of Washington*, 73:141-146. Specimens from Panama Canal Zone.

Burton, P. J. K. 1973a. Composite Nest of Short-Crested Flycatcher, *Myiarchus ferox*. *Bulletin of the British Ornithologists' Club*, 93:114-115. Eggs and nest in BMNH.

Burton, P. J. K. 1973b. Non-Passerine Bird Weights from Panama and Colombia. *Bulletin of the British Ornithologists' Club*, 93:116-118. Weights of 163 specimens representing 79 species. The specimens were deposited in BMNH and GML. Sex not indicated on all records. Coordinates given for the localities mentioned, which include: C.I.M., El Real, Río Espave (Bayano Valley), La Palma, Jaque, Río Jaque, Río Paya, Boca de Paya (Río Tuira), Pinogana (Río Tuira), and Cerro Pirre.

Buskirk, William H., George V. N. Powell, James F. Wittenberger, Ruth E. Buskirk, and Thomas U. Powell. 1972. Interspecific Bird Flocks in Tropical Highland Panama. *Auk*, 89(3):612-624, 5 figures, 2 tables. Maps and descriptions of 1968-1969 study areas within Florida State University Volcán de Chiriquí Field Station at Nueva Suiza, also called the Cerro Punta study area.

Butler, Robert W., Francisco S. Delgado B., Horacio De La Cueva, Victor Pulido, and Brett K. Sandercock. 1996. Migration Routes of the Western Sandpiper. *Wilson Bulletin*, 108(4):662-672, 2 figures, 1 table. Sightings and recoveries of birds banded at Chitré, as well as sites in Peru, Mexico, British Columbia and AK.

Butler, Robert W., R. I. Guy Morrison, Francisco S. Delgado. 1992. The Distribution of Fish-Eating, Wading, and Raptorial Birds in the Gulf of Panama, October 1991. *Canadian Wildlife Service Progress Notes*, No. 198:1-3, 1 figure, 1 table. Aerial surveys of the Pacific coast, which was divided into 16 sectors between Chitré and Garachiné. Map and text refer to Bahía de Chame, Bahía de Parita, Bahía Vacamonte, Chimán, Chitré, and Punta Chamé. Mentions abundant species and rarities with brief habitat descriptions.

Butler, Robert W., R. I. Guy Morrison, Francisco S. Delgado B., R Kenyon Ross, G. E. John Smith. 1997. Habitat Associations of Coastal Birds in Panama. *Colonial Waterbirds*, 20(3):518-524, 1 figure, 3 tables. Aerial surveys of coastal birds along the entire Pacific and Atlantic coasts of Panama in January 1993.

Butler, Robert W., R. I. Guy Morrison, Francisco S. Delgado B., R Kenyon Ross, G. E. John Smith. 1999. Summary of Butler et al. (1997): Habitat Associations of Coastal Birds in Panama. *Journal of Field Ornithology*, 70(2):289.

Caballero y C., Eduardo, and Esperanza Hidalgo. 1955. Helmintos de la Republica de Panamá. XVI. Descripcion de dos especies de Trematodos *Digeneos* de *Florida caerulea* (L.). [Helminths from Panama. XVI. Description of Two Digenetic Trematod Species Found on a *Florida caerulea*]. *Revista de la Sociedad Mexicana de Historia Natural*, 16(1-4):29-34, 3 figures. Collected from a specimen shot at Panama Viejo, by Dr. Robert G. Grocott in 1954.

Cabanis, Jean. 1874. Protokoll der XLIII. Monats-Sitzung. [Minutes of the 4 May 1874 meeting of Deutsche Ornihologische Gesellschaft]. *Journal für Ornithologie*, 22:348-350. Includes a description of *Chrysotis panamensis*. German text.

Caldwell, Gloria Sullivan. 1980. Underlying Benefits of Foraging Aggression in Egrets. *Ecology*, 61(4):997-998, 1 table. Observations along the Atlantic and Pacific coasts of Panama from

1976-1978.

Caldwell, Gloria Sullivan. 1981. Attraction to Tropical Mixed-Species Heron Flocks: Proximate Mechanism and Consequences. *Behavioral Ecology and Sociobiology*, 8(2):99-103, 1 figure, 2 tables. Observations and experiments with models along the Pacific and Atlantic coasts of Panama, 1976-1978.

Caldwell, Gloria Sullivan. 1986. Predation as a Selective Force on Foraging Herons: Effects of Plumage Color and Flocking. *Auk*, 103(3):494-505, 3 tables. Observations Dec 1975, from Jan 1977-Jan 1978, and from Nov 1982-Nov 1983 on the Atlantic coast from Galeta Island to Ft. Sherman.

Caldwell, Gloria Sullivan, and Roberta Wolff Rubinoff. 1983. Avoidance of Venomous Sea Snakes by Naive Herons and Egrets. *Auk*, 100(1):195-198, 1 figure, 1 table. Study conducted at Naos Marine Laboratory at STRI.

Capparella, Angelo P. 1986. First Record of Yellow-Collared *Chlorophonia* for Middle America. *American Birds*, 40(2):194-195, 2 photographs. Photographs of *Chlorophonia flavirostris* taken at Cerro Pirre, Darién.

Cardiel, José María, Santiago Castroviejo, and Mauricio Velayos. 1997. El Parque Nacional de Coiba: El Medio Físico. Pages: 11-30, 5 figures, 2 tables. In Santiago Castroviejo, editor. *Flora y Fauna del Parque Nacional de Coiba (Panama): Inventario Preliminar*. Madrid: Agencia Espanola de Cooperación Internacional (AECI). Descriptions of the coastline, rivers, climate, geography and history of Isla de Coiba. Spanish with English summary.

Carriker, Melbourne A., Jr. 1910. An Annotated List of the Birds of Costa Rica Including Cocos Island. *Annals of the Carnegie Museum*, 6(2-4):314-915, 1 map. List of 753 taxa with 86 from Río Sicsola [Río Sixaola], at least some of which were taken in Panama. *Crypturellus soui panamensis* n. subsp. from Loma del Leon.

Cassin, John. 1850. Descriptions of New Species of Birds of the Genera Parus, Linn.; Emberiza, Linn.; Carduelis, Briss.; Myiothera, Ill.; and Leuconerpes, Sw., Specimens of Which are in the Collection of the Academy of Natural Sciences of Philadelphia. *Proceedings of the Academy of Natural Sciences of Philadelphia*, 5:103-106, 6 plates. Volume 5 includes contributions for 1850-1851. *Myiothera nudiceps* n. sp. collected by John G. Bell in Panama.

Cassin, John. 1851. Sketch of the Birds Composing the Genera *Vireo*, Vieillot, and *Vireosylvia*, Bonaparte, with a List of the Previously Known and Descriptions of Three New Species. *Proceedings of the Academy of Natural Sciences of Philadelphia*, 5:149-154, 2 plates. *Vireosylvia flavoviridis* n. sp. collected by John G. Bell in Panama.

Cassin, John. 1857. Notes on the North American Species of *Archibuteo* and *Lanius*, and Description of a New Species of Toucan, of the Genus *Selenidera*, Gould. *Proceedings of the Academy of Natural Sciences of Philadelphia*, 9:211-214. *Selenidera spectabilis* n. sp. from Cocuyos de Veragua, New Grenada.

Cassin, John. 1858. Description of a New Tanager from the Isthmus of Darien, and Note on *Selenidera spectabilis*. *Proceedings of the Academy of Natural Sciences of Philadelphia*, 10:177-178. *Calliste lavinia* n. sp. Isthmus of Darién, New Grenada.

Castro, Gonzalo, and J. P. Myers. 1988. Snowy Plover (*Charadrius alexandrinus*) Records from Panama. *American Birds*, 42(3):374. Sightings at Punta Chame in 1987.

Castro, Gonzalo, and J. P. Myers. 1990. Validity of Predictive Equations for Total Body Fat in Sanderlings from Different Nonbreeding Areas. *Condor*, 92(1):205-209, 1 figure, 1 table. Analysis of tissue samples from birds collected at Punta Chame as well as sites in Peru and the U.S.A.

Castro, Gonzalo, J. P. Myers, and Robert E. Ricklefs. 1992. Ecology and Energetics of Sanderlings Migrating to Four Latitudes. *Ecology*, 73(3):833-844, 11 figures, 5 tables. Analysis of tissue samples from birds collected at Punta Chame as well as sites in Peru and the U.S.A.

Castroviejo, Santiago, editor. 1997. *Flora y Fauna del Parque Nacional de Coiba (Panama): Inventario Preliminar.* xii + 534 pages, photographs, figures, tables. Madrid: Agencia Espanola de Cooperación Internacional (AECI). Chapters written by different authors. The chapter on birds (Carlos Ibañez and José Cabot) is largely based on Wetmore's work.

Chapman, Frank M. 1914a. Diagnosis of Apparently New Colombian Birds. *Bulletin of the American Museum of Natural History*, 31(16):139-166, 1 plate (map), tables. Mentions specimens from Chiriqui and Panama.

Chapman, Frank M. 1914b. Diagnosis of Apparently New Colombian Birds, II. *Bulletin of the American Museum of Natural History*, 33(12):167-192, 1 plate (map), tables. Panama specimens mentioned.

Chapman, Frank M. 1915a. Descriptions of Proposed New Birds from Central and South America. *Bulletin of the American Museum of Natural History*, 34(11):363-388, tables. *Odontophorus guianensis panamensis* n. subsp. from Panama R. R. Line. *Atlapetes gutturalis brunnescens* n. subsp. from Boquete. Additional specimens from Divalá, Natá, Tapaliza, and Tacarcuna.

Chapman, Frank M. 1915b. The More Northern Species of the Genus *Scytalopus* Gould. *Auk*, 32(4):406-423, tables. *Scytalopus panamensis* n. sp. Tacarcuna. Also examined specimens from Boquete and Chiriqui.

Chapman, Frank M. 1917. The Distribution of Bird-Life in Colombia, A Contribution to a Biological Survey of South America. *Bulletin of the American Museum of Natural History*, 36:1-729, appendix, 41 plates (some colored), 21 figures, tables. Appendix of collecting localities (Colombia), bibliography (p. 657-660), and index (p. 661-729). Mentions records from the Anthony expedition to Darién that are often the first published reference for a species for Panama. Also mentions numerous localities in Panama including, but probably not limited to: Boca de Cupe, Boquete, Cana, Chiriqui, Chitra, David, El Real, Lion Hill, Panama R. R, Mt. Pirre, and Río Tuyra. Designates David as the type locality for *Asturina ruficauda* Sclater and Salvin (1869).

Chapman, Frank M. 1923. Descriptions of Proposed New Birds from Panama, Venezuela, Ecuador, Peru and Bolivia. *American Museum Novitates*, 67:1-12, tables. *Buarremon atricapillus tacarcunae* n. subsp. from Mt. Tacarcuna. Also examined specimens from Citaro, and Tacarcuna.

Chapman, Frank M. 1924a. Descriptions of New Genera and Species of Tracheophonae from Panama, Ecuador, Peru and Bolivia. *American Museum Novitates*, 123:1-9, 2 figures. *Xenornis setifrons*, n. gen. et sp., from Tacarcuna.

Chapman, Frank M. 1924b. Descriptions of New Birds from Colombia, Ecuador, Peru and Bolivia. *American Museum Novitates*, 143:1-16, tables. *Basileuterus tacarcunae*, n sp., from Mt. Tacarcuna.

Chapman, Frank M. 1926. The Distribution of Bird-Life in Ecuador, a Contribution to a Study of the Origin of Andean Bird-Life. *Bulletin of the American Museum of Natural History*, 55:xiv + 784 pages, 30 plates, 20 figures, tables, lists. Numerous statements of "eastern Panama" in range, which are based on Anthony's expedition to Darién, and are frequently the first published records for the country.

Chapman, Frank M. 1928a. The Homes of a Hummer. A Study in Tropical Bird Life. *Natural History*, 28(3):284-289, 5 illustrations. Barro Colorado Island.

Chapman, Frank M. 1928b. The Nesting Habits of Wagler's Oropendola (*Zarhynchus wagleri*) on Barro Colorado Island. *Bulletin of the American Museum of Natural History*, 58:123-166, 8 black and white plates, 1 figure, tables. Detailed observations.

Chapman, Frank M. 1929a. *My Tropical Air Castle, Nature Studies in Panama.* xv + 417 pages, appendix, 65 plates, 30 illustrations. New York: D. Appleton and Co. Observations on Barro Colorado Island, with an annotated list of 212 species. Chapter 14, "Off the Island", details trips to the Pacora Marshes and to Ancon Hill with a photograph near Pacora of the Savanna of Panama.

Chapman, Frank M. 1929b. Description of New Birds from Mt. Duida, Venezuela. *American Museum Novitates*, 380:1-27, tables. Panama specimens from Boca de Cupe, Cana, Canal Zone (near Panama) [= Panama City], Citaro, El Real, Line P. R. R., Lion Hill, Tacarcuna, and Tapalisa.

Chapman, Frank M. 1931a. The Winter Range of the Chimney Swift (*Chaetura pelagica*). *Auk*, 48(1):119-121. Two specimens of swifts and 31 additional species of North American migrants recorded from Cocoplum.

Chapman, Frank M. 1931b. The Nesting Habits of Wagler's Oropendola on Barro Colorado Island. *Annual Report to the Board of Regents of the Smithsonian Institution, Reprint: From the Smithsonian Report for 1930, 3090*:347-386, 8 black and white plates, tables. Reprint of the 1928 paper with a few additional notes.

Chapman, Frank M. 1933. The Migration of Turkey Buzzards as Observed on Barro Colorado Island, Canal Zone. *Auk*, 50(1):30-34. Records of *Cathartes aura* for Feb-Mar in 1929, 1930, and 1932.

Chapman, Frank M. 1935. The Courtship of Gould's Manakin (*Manacus vitellinus vitellinus*) on Barro Colorado Island, Canal Zone. *Bulletin of the American Museum of Natural History*, 68(7):471-525, 13 figures. (reviewed by Stone, 1936).

Chapman, Frank M. 1937. Swifts at Sea. *Auk*, 54(3):392. Two flocks thought to be *Chaetura pelagica* at sea N of Porto Bello.

Chapman, Frank M. 1938. *Life in an Air Castle, Nature Studies in the Tropics.* xii + 250 pages, appendix, 31 plates, illustrations (by Francis Lee Jaques). New York: D. Appleton-Century Company. Previously copyrighted by *Bird-Lore* (1936) and by *Natural History* (1931-1937). Further observations on Barro Colorado Island, with a revised bird list of 252 species.

Chapman, Frank M. 1940. The Post-Glacial History of *Zonotrichia capensis*. *Bulletin of the American Museum of Natural History*, 77:381-438, 17 figures, tables. Includes an account of *Z. capensis costaricensis* with specimens from Boquete, Chiriqui, Coiba, Cerro Flores, Santa Fé, and Veragua.

Chapman, Frank M. 1941. Mockingbirds in Panama. *Auk*, 58(1):98-99. Discusses accidental introduction vs. normal extension of range of *Mimus gilvus* to the Canal Zone subsequent to 1928. Observations in 1938-1339 at Ancon, Balboa Railway Station, Panama City, and Pedro Miguel.

Cherrie, George, K. 1890. Notes on the Nesting Habits of Several Birds at San Jose, Costa Rica. *Auk*, 7:233-237. Based on observations and collections in Costa Rica.

Cherrie, George, K. 1891a. Description of a New Species of *Ramphocelus* from Costa Rica. *Auk*, 8:62-64. *R. costaricensis* n. sp. from Costa Rica with comparisons with *R. passerinii.*

Cherrie, George, K. 1891b. Description of a Supposed New *Myrmeciza*. *Auk*, 8(2):191-193, tables. *Myrmeciza immaculata occidentalis* n. subsp. from Costa Rica.

Cherrie, George, K. 1891c. A Preliminary List of the Birds of San José, Costa Rica. *Auk*, 8(3):271-279, table. Numbered list of species (1-16, genera *Catharus* through *Dendroica*) with brief descriptions that are continued in subsequent contributions.

Cherrie, George, K. 1892a. A Preliminary List of the Birds of San José, Costa Rica. *Auk*, 9(1):21-

27. Continuation of a numbered list of species (17-57, genera *Dendroica* through *Volatinia*) with brief descriptions.

Cherrie, George, K. 1892b. A Preliminary List of the Birds of San José, Costa Rica. *Auk*, 9(3):247-251. Continuation of a numbered list of species (58-93, genera *Euetheia* through *Milvulus*) with brief descriptions.

Cherrie, George, K. 1892c. A Preliminary List of the Birds of San José, Costa Rica. *Auk*, 9(4):322-329. Final contribution that includes a numbered list of species (94-167, genera *Milvulus* through *Anas*) with brief descriptions.

Cherrie, George, K. 1893. Notes on Two Costa Rican Birds. *Auk*, 10(3):278-280. Additional information on *Ramphocoelus costaricensis* and *Thamnophilus bridgesi*.

Cherrie, George, K. 1895. Exploraciones Zoológicas Efectuadas en el Valle del Río Naranjo en el Año de 1893. *Anales del Instituto Físico-Geográfico Nacional de Costa Rica*, 6:67-73.

Chibas, Eduardo, J. 1898. The Darien Gold-Mining Region in Colombia. *Engineering Magazine*, 16(1):49-65, 15 photographs. Discusses the history of the mines at Cana and the efforts of the Darién Gold Mining Company to relocate the Espiritu Santo Mine operated by the Spaniards. Discusses geology, mining operations, and the author's travel through Darién (beginning in Dec 1896) to determine the feasibility of a railway. The company built a "macadam road" along a route selected by the author.

Child, George I. 1969. A Study of Nonfat Weights in Migrating Swainson's Thrushes (*Hylocichla ustulata*). *Auk*, 86(2):327-338, 5 figures, 5 tables. Measurements of fat, water and nonfat weights of *H. ustulata* collected at 2 sites in Panama (Almirante and San Fransisco) and 5 sites in the U.S.A.

Christian, Daniel G. 2001. Nests and Nesting Behavior of Some Little Known Panamanian Birds. *Ornitologia Neotropical*, 12:327-336, 1 figure. Abstract in Spanish and English. Observations in Peru and in Panamá at Fortuna, Cerro Jefe, Sierra de Jungurudó, Nusagandi, and Río Sambú. The species described are: *Xenornis setifrons*, *Scytalopus chocoensis*, *Phyllomyias zeledoni*, *Sapayoa aenigma*, *Chloropipo holochlora*, *Thryothorus leucopogon*, *Tangara fucosa*, and *Chlorospingus tacarcunae* with comparisons to their closest taxonomic relatives.

Christian, Daniel G., and Dina L. Roberts. 2000. First Description of Nest and Nesting Behavior of the Nightingale Wren. *Wilson Bulletin*, 112(2):284-287. Observations of *Microcerculus marginatus* at Santa Clara and in Parque Nacional Soberanía.

Chubb, Charles. 1919. Notes on Collections of Birds in the British Museum, from Ecuador, Peru, Bolivia, and Argentina. Part 1. Tinamidae—Rallidae. *Ibis*, 11(1):1-55, 1 colored plate, 2 figures, tables. *Odontophorus guianensis panamensis* n. subsp. from Lion Hill.

Clark, Hubert Lyman. 1913. Notes on the Panama Thrush-Warbler. *Auk*, 30(1):11-15. Anatomical descriptions of *Rodinocichla rosea eximia* based on alcoholic specimens with comparisons to other genera. Concludes that it is a specialized tanager.

Cleaves, Howard H. 1944. Panama Ant-Thrush (*Formicarius analis panamensis*). *Wilson Bulletin*, 56(2):plate 2 (Frontispiece). Photograph of nesting bird with legend.

Collar, N. J., and Angela Dunn, editors. 1989. New National Parks. *World Birdwatch*, 11(1):3. Information on Bastimentos National Marine Park and La Amistad International Park.

Collins, Charles T. 1973. The Natal Pterylosis of the Swallow-Tanager. *Bulletin of the British Ornithologists' Club*, 93(4):155-157, 2 tables. Natal pterylosis of *Tersina viridis* based on eight specimens of nestlings and embryos collected in Venezuela.

Coltart, N. B. 1952. Eggs of the Falconinae. *Oologists' Record*, 26(3):42-46. Information on species not previously described by L. R. Wolfe.

Condit, Richard, W. Douglas Robinson, Roberto Ibáñez, Salomón Aguilar, Amelia Sanjur, Raúl Martínez, Robert F. Stallard, Tomas García, George R. Angehr, Lisa Petit, S. Joseph Wright, Tara R. Robinson, and Stanley Heckadon. 2001. The Status of the Panama Canal Watershed and Its Biodiversity at the Beginning of the 21st Century. *Bioscience*, 51(5):389-398, 2 figures, 1 table. Section on the avifauna discusses unique aspects of the canal watershed with mention of a few bird species.

Conover, Boardman. 1944. The Races of the Solitary Sandpiper. *Auk*, 61(4):537-544. Specimens of *Tringa solitaria solitaria* from Chiriquí and Permé.

Conway, Ainslie and Frances Conway. 1943. *The Enchanted Islands*. 254 pages, maps, illustrations. London: Geoffrey Bles. Galapagos Islands.

Conway, William G. 1962. The Weights of 11 Living Eagles and Vultures at the New York Zoological Park. *Auk*, 79(2):274-275. Sex given only when known.

Cooke, May Thacher. 1946. The Winter Range of the Great Blue Herron. *Auk*, 63(2):254. Recoveries of banded birds with one from Minnesota that was recovered in Gatún.

Cooke, Richard G., and Mark Bush. 1989. Out-of-Season Red-Necked Phalaropes in Central Panama. *Journal of Field Ornithology*, 60(1):39-42, 1 photograph. *Phalaropus lobatus* observed at Playa Caleta. Photographs in the VIREO collection at ANSP.

Cooke, Richard G., and Storrs L. Olson. 1984. An Archaeological Record for the White-Faced Whistling-Duck (*Dendrocygna viduata*) in central Panama. *Condor*, 86(4):493-494, 1 figure. Sitio Cerra, SW Coclé province.

Cooper, Daniel S. 1999. Notes on the Birds of Isla Popa, Western Bocas del Toro, Panama. *Cotinga*, 11:23-26, appendix. In English with Spanish summary. The appendix lists the 173 species found on the Traversi-Doggett property, a private reserve located on the SE section of the island.

Cordier, Charles. 1943. The Umbrella Bird Comes to the Zoo. *Animal Kingdom, Bulletin of the New York Zoological Society*, 46(1):1-10, 3 photographs. Account of the capture in Costa Rica of three *Cephalopterus glabricollis* that were brought to the zoo.

Cory, Charles B. 1887. A List of the Birds Taken by Mr. Robert B. Henderson, in the Islands of Old Providence and St. Andrews, Caribbean Sea, during the Winter of 1886-87. *Auk*, 4(3):180-181. A list without descriptions of 25 species from Old Providence and 19 species from St. Andrews. The footnote indicates that "an author's edition of 250 copies of this paper was published May 28, 1887."

Cory, Charles B. 1918. Catalogue of Birds of the Americas and the Adjacent Islands in Field Museum of Natural. Part 2, No. 1. *Field Museum of Natural History Zoological Series* Volume 13. 315 pages, colored frontispiece, 1 illustration. Families covered: Bubonidae, Tytonidae, Psittacidae, Steatornithidae, Alcedinidae, Todidae, Momotidae, Nyctibiidae, Caprimulgidae, Cypselidae, and Trochilidae. Several other parts in volume 13 were completed by Hellmayr.

Cory, Charles B., and Charles E. Hellmayr. 1925. Catalogue of Birds of the Americas and the Adjacent Islands in Field Museum of Natural History. Part 4. *Field Museum of Natural History Zoological Series* Volume 13, iv + 390 pages, 1 colored frontispiece. Families included: Furnariidae and Dendrocolaptidae.

Cox, Charles W. 1973. Hybridization Between Mourning and MacGillivray's Warbler. *Auk*, 90(1):190-191, 1 figure. Survey by the author of *Oporonis philadelphia* and *O. tolmei* in SW Alberta.

Cox, George W. 1960. A Life History of the Mourning Warbler. *Wilson Bulletin*, 72(1):5-28, 1 colored plate, 9 tables. Focuses on the breeding populations of *Oporornis philadelphia* in Minnesote and migrating birds in Illinois, with observations of wintering birds in the Canal Zone from Feb-Apr 1959.

Cox, George W., and Robert E. Ricklefs. 1977. Species Diversity and Ecological Release in Caribbean Land Bird Faunas. *Oikos*, 28(1):113-122, 10 tables. Censuses of 9 habitats in Panama (and other countries) of breeding, non-raptorial landbirds in 1970. Summary in Russian.

Cracraft, Joel. 1967. Comparative Foraging Behavior of *Myiozetetes similis* and *M. granadensis* in Costa Rica. *Wilson Bulletin*, 79(1):115-116, 2 tables. Four sites in Costa Rica, 1965.

Crandall, Lee S. 1948. Notes on the Display of the Three-Wattled Bell-Bird (*Procnias tricarunculata*). *Zoologica: Scientific Contributions of the New York Zoological Society*, 33(2):113-114, 1 plate. Captive birds at the zoo brought from Costa Rica.

Croat, Thomas B. 1997. A Revision of *Philodendron* Subgenus *Philodendron* (Araceae) for Mexico and Central America. *Annals of the Missouri Botanical Garden*, 84(3):311-704. Coordinates were given for specimens of *Philodendron ligulatum* (p. 482-483) that were collected at Pirre Ranger Station in Darién Parque Nacional.

Dalmas, Comte R. de. 1900. [Two New Species of *Calliste*]. *Bulletin of the British Ornithologists' Club*, 11:35-36. Descriptions *C. emiliae* and *C. johannae*. In French.

Darlington, P. J., Jr. 1931. Notes on the Birds of Rio Frio (near Santa Marta), Magdalena, Colombia. *Bulletin of the Museum of Comparative Zoology*, 71(6):349-421, 1 map, 5 lists, tables. Notes on 306 skins representing 237 species collected in 1928-1929. Mentions specimens from Loma del Leon and Isthmus of Panama, Atlantic Side.

Davidson, Mary Ella McLellan. 1932. New Birds from Chiriqui Province, Panama. *Proceedings of the Biological Society of Washington*, 45:167-168. *Automolus xanthippe*, and *Spodiornis barrilesensis*, n. spp., both from Barriles. *Hylophilus viridiflavus pallescens* n. subsp. from Concepcion.

Davidson, Mary Ella McLellan. 1936. A Note on the Collecting of W. W. Brown, Jr., in Chiriqui. *Auk*, 53(1):232-233. Brown's specimens labeled "Caribbean slope of Volcan de Chiriqui" may be from the Boquete Trail in Bocas del Toro.

Davidson, Mary Ella McLellan. 1938. On Some Birds Rare in, or Hitherto Unrecorded from Chiriqui Province, Panama. *Proceedings of the California Academy of Sciences*, 23(17):255-261, 1 table. Notes on 48 species collected at numerous localities during "several winters' work" from 1929-1932. Mentions her localities and those of others.

Davis, David Edward. 1940. Social Nesting Habits of the Smooth-Billed Ani. *Auk*, 57(2):179-218, 1 figure, 7 tables. Field work in Cuba during 2 breeding seasons, 1937-1938.

Davis, David Edward. 1942. The Phylogeny of Social Nesting Habits in the Crotophaginae. *Quarterly Review of Biology*, 17(2):115-134.

Davis, L. Irby. 1962. Acoustic Evidence of Relationship in *Caprimulgus*. *Texas Journal of Science*, 14(1):72-106, 16 figures.

Davis, William James, and Douglas J. Graham. 1991. The Influence of Food on Reproductive Strategies in a Monogamous Kingfisher (*Chloroceryle amazona*). *Auk*, 108(4):780-789, 5 figures, 5 tables. Cascajal River in Portobelo National Park from Dec 1987-May 1988.

Dawn, Walter. 1964. Nest and Eggs of Cabanis Tiger Heron in Chiapas, Mexico. *Auk*, 81(2):230-231, 1 figure. Field work on *Tigrisoma mexicana* in 1962.

De Lattre, Ad., and Jules Bourcier. 1846. Description de Quinze Espèces Nouvelles de Trochilidées, Faisant Partie des Collections Rapportées par M. Ad. De Lattre, dont les Précédentes

Excursions ont déjà Enrichi Plusieurs Branches de l'Histoire Naturelle, et Provenant de l'Intérieur de Pérou, des Républiques de l'Équateur, de la Nouvelle-Grenade et de l'Isthme de Panama. [Description of Fifteen New Species of Trochilid Species, Part of the Collection of M. Ad De Lattre, Introduced by Mr. Ad. De Lattre, of which the Previous Excursions have already Enriched Several Branches of Natural History, and Originating in the Interior of Peru, and in the Republics of Ecuador, New Grenada, and the Isthmus of Panama]. *Revue Zoologique*, 9:305-312. *Trochilus edward* and *T. cuvierii*, n. spp., from the Isthme de Panama. In French.

Deignan, Herbert G. 1933. Notes on Some Canal Zone Birds. *Auk*, 50(1):125. Record of *Mimus gilvus* at Balboa Station. Mentions Balboa and Panama City.

Deignan, Herbert G. 1936. Notes on a Small Collection of Birds from the Republic of Honduras. *Auk*, 53(2):186-193, table. Adds 12 new records for the country with notes on 68 taxa.

Deignan, Herbert G. 1961. Type Specimens of Birds in the United States National Museum. *United States National Museum Bulletin,* 221: x + 718 pages. Includes Cote Ferme as a locality in Panama, which appears to be an error. El Real de Santa María, Río Pirri, and Santa Cruz de Cana.

Delgado B., Francisco S. 1983. Situación Actual de las Zonas Pantanosas de Panamá; Perspectivas de Conservación de la Avifauna Acuática. [Status of Wetland Areas in Panama; Perspectives on the Conservation of Waterbirds]. Pages 112-118, 1 figure (map) *in* Hugh Boyd, editor. *First Western Hemisphere Waterfowl and Waterbird Symposium.* Canada: Minister of Supply and Services. In Spanish with English and French summaries. Map indicates 24 wetlands areas in Panama, the text briefly describes the major areas and indicates aquatic birds found at these areas.

Delgado B., Francisco S. 1985. A New Subspecies of the Painted Parakeet (*Pyrrhura picta*) from Panama. Pages 17-20, 1 colored plate, 1 figure, 1 table *in* P. A. Buckley et al., editors, Neotropical Ornithology. *Ornithological Monographs* 36. *P. picta eisenmanni* n. subsp. is described from 11 specimens taken in humid forests in SW Azuero Peninsula including El Cortezo de Tonosí and Los Tres Cerros (Cerro Hoya). Abstract in Spanish.

Delgado B., Francisco S. 1997. Situación de los Cracidos de Panamá. [Status of Cracids in Panama.] Pages 345-349 *in* Stuart D. Strahl, Silvia Beaujon, Daniel M. Brooks, Alfredo J. Begazo, Gina Sedaghatkish, and Fabio Olmos, editors. *The Cracidae: Their Biology and Conservation.* Surrey, British Columbia: Hancock House Publishers Ltd. In Spanish. Information on natural history, hunting, protection and conservation of Cracids with brief descriptions of *Crax rubra*, *Penelope purpurascens*, *Champaepetes unicolor*, and *Ortalis cinereiceps.*

Delgado B., Francisco S., and Robert W Butler. 1993. Shorebirds in Parita Bay, Panama. *Wader Study Group Bulletin*, 67:50-53, 2 tables, 1 figure. Mist netting near Chitré Sep 1989-Mar 1990 and Oct 1990-Feb 1991.

Dickerman, Robert W., and Dwain W. Warner. 1962. A New Orchard Oriole from Mexico. *Condor*, 64(4):311-314, 2 figures, 1 table.

Dickey, Donald R., and Adriaan Joseph Van Rossem. 1930. The Identity of *Ortyx leucopogon* Lesson. *Condor*, 32(1):72-73. *Colinus leucotis panamensis* n. subsp. from Aguadulce.

Dickey, Donald R., and Adriaan Joseph Van Rossem. 1938. The Birds of El Salvador. *Field Museum of Natural History, Zoological Series* 23: 609 pages, 24 black and white plates, 29 figures.

Douglas, Archibald John Angus, and Philip Henry Johnson. 1926. *The South Seas of To-day, Being an Account of the Cruise of the Yacht St. George to the South Pacific.* xiv + 295 pages, 39 illustrations, 3 charts. London: Cassell and Company, Ltd. Account of a British expedition whose ornithologists, H. J. Kelsall and his assistant Seaman Cullingford, made small collections at Balboa, Corozal, Gamboa, Old Panama, on Coiba, del Rey, Taboga and Taboguilla Islands, and at sea off Azuero Peninsula, east of Mala, south of Panama, S of the Pearl Islands, and in the Gulf of Panama.

Drury, William H., Jr. 1962. Breeding Activities, Especially Nest Building, of the Yellowtail (*Ostinops decumanus*) in Trinidad, West Indies. *Zoologica: Scientific Contributions of the New York Zoological Society*, 47(1):39-58, 4 figures, 1 table.

Dubois, Alphonse. 1875. Description de Quelques Oiseaux Nouveaux. [Descriptions of Some New Birds]. *Bulletins de L'Académie Royale des Sciences, Des Lettres et des Beaux-Arts de Belgique*, 2nd Series, 41(12):797-801. Description of *Icterus xanthornus* Var. á. *Dubusii*, based on a Verreaux specimen labeled "Isthme de Panama," and Var. â. *Marginalis*, with the same data. Localities undoubtedly erroneous. In French.

Dugand, Armando. 1948. Notas Ornitologicas Colombianas, IV. [Notes on Colombian Ornithology, IV]. *Caldasia*, 5(21):157-199, 1 black and white photograph. Describes 178 taxa and 9 new records for Colombia. In Spanish.

Duke, James Alan. 1972. *Isthmian Ethnobotanical Dictionary*. 96 pages, illustrations. Fulton, Maryland.

Duke, James Alan. 1986. *Isthmian Ethnobotanical Dictionary*. 3rd edition. xiv + 205 pages, illustrations. Jodhpur: Scientific Publishers.

Dungel, J., and M. Sebela. 1995. American Dipper *Cinclus mexicanus* in Venezuela. *Bulletin of the British Ornithologists' Club*, 115(3):191-192. The authors quote R. S. Ridgely *in litt* that: "The record of an American Dipper on the river Cana in Darién, eastern Panamá, is not now accepted." Nor, apparently, is their record for Venezuela, as Hilty (2003) makes no mention of it.

Dwight, Jonathan, and Ludlow Griscom. 1942. Description of New Birds from Costa Rica. *American Museum Novitates*, 142:1-5, tables. Panama specimens.

Dyrcz, Andrzej. 1982. Synanthropy in Tropics as Studied in Clay-Colored Robin *Turdus grayi* on Panama Lowland. Pages 27-37, 2 figures, 7 tables *in* Maciej Luniak and Bohdan Pisarski, editors. *Animals in Urban Environment. Proceedings of the Symposium on the Occasion of 60th Anniversary of the Institute Zoology of the Polish Academy of Sciences, Warszawa-Jablonna 22-24 October 1979.* Warsaw: Polish Academy of Sciences. Found 129 nests (83 active) at Morgan's Garden and Summit Gardens and compares habits and predation of birds at the two sites.

Dyrcz, Andrzej. 1983. Breeding Ecology of the Clay-Coloured Robin *Turdus grayi* in Lowland Panama. *Ibis*, 125(3):287-304, 8 figures, 14 tables. Observations at Summit Gardens and Morgans Gardens, Mar-Jun 1979.

Dyrcz, Andrzej. 1984. Breeding Biology of the Mangrove Swallow *Tachycineta albilinea* and the Grey-Breasted Martin *Progne chalybea* at Barro Colorado Island, Panama. *Ibis*, 126(1):59-66, 5 figures, 1 table. N shore of Barro Colorado Island. 1979 May-Jul.

Dyrcz, Andrzej. 1999. Nesting of the Rufescent Tiger-Heron *Tigrisoma lineatum* on Barro Colorado Island (Republic of Panama). *Ornitologia Neotropical*, 10(1):105-106.

Dyrcz, Andrzej. 2000. Nesting of the Mangrove Swallow (*Tachycineta albilinea*) in Nests of the Rusty-Margined (*Myozetetes cayanensis*) and Social (*M. similis*) Flycatchers. *Ornitologia Neotropical*, 11(1):83-84. Observations on Barro Colorado Island.

Dyrcz, Andrzej. 2002. Breeding Ecology of the Social (*Myiozetetes similis*) and Rusty-Margined (*M. cayanensis*) Flycatchers at Barro Colorado Island, Republic of Panama. *Ornitologia Neotropical*, 13(2):143-151, 4 tables.

Eckelberry, Don R. 1957. The Three-Wattled Bellbird. *Condor*, 59(5):273, 1 colored plate. Calls and habits of *Procnias tricarunculata* recorded in Costa Rica.

Edwards, Ernest P. 1967. Nest of the Common Bush-Tanager and the Scaled Antpitta. *Condor*, 69(6):605. Observations on *Chlorospingus ophthalmicus* and *Grallaria guatimalensis* in Mexico.

Edwards, Ernest P., and Robert B. Lea. 1955. Birds of the Monserrate area, Chiapas, Mexico. *Condor*, 57(1):31-54, 8 figures.

Egeler, Oliver, and Tony D. Williams. 2000. Seasonal, Age, and Sex-Related Variation in Fatty-Acid Composition of Depot Fat in Relation to Migration in Western Sandpipers. *Auk*, 117(1):110-119, 3 figures, 4 tables. Birds collected at Playa El Agallito in Dec 1995 and Mar 1996 and at a site in British Columbia.

Eisenmann, Eugene. 1946. Acorn Storing by *Balansphyra formicivora* in Panama. *Auk*, 63(2):250. Observations by the author near El Volcán.

Eisenmann, Eugene. 1948. Surf-Bird and Vermilion Flycatcher in Panama. *Auk*, 65(4):605-606. First records in Panama of *Aphriza virgata*, near Old Panama, and *Pyrocephalus rubinus* at Playa Coronado.

Eisenmann, Eugene. 1950. Some Notes on Panamá Birds Collected by J. H. Batty. *Auk*, 67(3):363-367. Supposed lowland records of highland birds secured in 1901 and 1902, on islands on the Pacific Coast offshore from Chiriqui and Veraguas including: Afuera, Brava, Burica, Cebaco, Gobernador, Insoleta, Jicaron, Parida, and Sevilla. [It has since been determined that Batty fabricated these data (Olson 2008)]. *Turdus assimilis coibensis*, n. subsp. from Coiba Island. Also mentions specimens from Almirante, Bogava, Boqueron, Changuinola, and Divala.

Eisenmann, Eugene. 1951. Northern Birds Summering in Panama. *Wilson Bulletin*, 63(3):181-185. Supposed non-migratory individuals of *Ardea herodias*, *Pandion haliaetus*, shore-birds, gulls, and terns that were seen in 1948-1951 at the following localities: Farfan Beach, Gatun Lake near Barro Colorado Island and near Gamboa, Juan Franco, La Venta Beach, Old Panama, Panama Bay near Balboa, and San Francisco de la Caleta.

Eisenmann, Eugene. 1952. Annotated List of Birds of Barro Colorado Island, Panama Canal Zone. *Smithsonian Miscellaneous Collections*, 117(5):1-62. Summarizes specimen as well as sight records of 306 species.

Eisenmann, Eugene. 1953a. What Bird Lays Black Eggs? *Auk*, 70(3):362-363. References to literature from northern South America, Central America, and the Caribbean.

Eisenmann, Eugene. 1953b. The Nest of the Long-Billed Gnatwren (*Ramphocaenus rufiventris*). *Auk*, 70(3):368-369. The nest was found at Juan Franco, Panama.

Eisenmann, Eugene. 1955a. Status of the Black-Polled, Bay-Breasted and Connecticut Warblers in Middle America. *Auk*, 72(2):206-207. Status of Bay-breasted Warbler in Panama with mention of a specimen from Cocoplum.

Eisenmann, Eugene. 1955b. The Species of Middle American Birds: A List of All Species Recorded from Mexico to Panama, with Suggested English Names, Outlines of Range, and a Distributional Bibliography. *Transactions of the Linnaean Society of New York*, 7: vi + 128 pages. With the collaboration of E. R. Blake and E. L. Chalif on the selection of English names.

Eisenmann, Eugene. 1955c. Cattle Egret, Marbled Godwit, Surfbird and Brown-Chested Martin in Panama. *Auk*, 72(4):426-428. Observations of *Bubulcus ibis, Limosa fedoa, Aphriza virgata*, and *Phaeoprogne tapera* at Chepo, Empire Rifle Range, Fort San Lorenzo, Gatun, Maria Chiquita, Mindi Dairy, Old Panama, Panama City, and specimens from Tocumen collected by Benson.

Eisenmann, Eugene. 1956. Galapagos Penguin in Panama. *Condor*, 58(1):74-75. Captured alive at

Puerto Armuelles and deposited at MCN, Panama.

Eisenmann, Eugene. 1957. Notes on Birds of the Province of Bocas del Toro, Panama. *Condor*, 59(4):247-262. Brief summary of previous work in Bocas del Toro with field observations from 29 June to 4 July 1956 at Almirante Bay, Almirante, Bocas del Toro (village), Changuinola, and San San River. Also mentions specimens and observations at Chagres River, Coco Solo, Fort San Lorenzo, Las Lajas, Pedro Miguel, Volcan Lakes, and Zapotillo. He reports that Carriker collected on the Panama side of the Río Sicsola and mentions that some specimens of *Chalybura melanorrhoa*, *Dysithamnus striaticeps* and *Gymnocichla chiroleuca nudiceps* were "certainly" collected in Panama.

Eisenmann, Eugene. 1958. Mourning Dove Breeding in Costa Rica. *Condor*, 60(6):405. Specimens of *Zenaida macroura* from Natá.

Eisenmann, Eugene. 1959. South American Migrant Swallows of the Genus *Progne* in Panama and Northern South America; With Comments on Their Identification and Molt. *Auk*, 76(4):528-532. Also includes remarks on some swifts. Mentions specimens from Obaldía and "near Colón," which apparently refers to Coco Solito.

Eisenmann, Eugene. 1961. Favorite Foods of Neotropical Birds: Flying Termites and *Cecropia* Catkins. *Auk*, 78(4):636-638. Observations near Gatun Dam, Fort Davis, and Barro Colorado Island.

Eisenmann, Eugene. 1962a. Notes on Nighthawks of the Genus *Chordeiles* in Southern Middle America, with a Description of a New Race of *Chordeiles minor* Breeding in Panama. *American Museum Novitates*, 2094:1-21, 1 table. *Chordeiles minor panamensis* n. subsp. from Cerro Campana. Examined specimens from Balboa, Chepo, Corozal, and Frances. Mentions observations throughout the country.

Eisenmann, Eugene. 1962b. On the Systematic Position of *Rhodinocichla rosea*. *Auk*, 79(4):640-648. Observations in Panama.

Eisenmann, Eugene. 1962c. Notes on Some Neotropical Vireos in Panama. *Condor*, 64(6):505-508. Observations and specimens from the Canal Zone, Cerro Punta, El Hato, Finca Lerida, Isla Coiba, and the Pearl Islands.

Eisenmann, Eugene. 1963a. Mississippi Kite in Argentina; with Comments on Migration and Plumages in the Genus *Ictinia*. *Auk*, 80(1):74-77. Observations by Eisenmann and Wetmore in Panama and on San José Island.

Eisenmann, Eugene. 1963b. Is the Black Vulture Migratory? *Wilson Bulletin*, 75(3):244-249. Observations of *Coragyps atratus* at Chepo, Portobelo, Juan Franco, and Old Panama.

Eisenmann, Eugene. 1966. Comments on Repeats and Recoveries of Migrants in Panama. *Bird-Banding*, 37(4):286. Comments on Loftin et al., 1966 (not 1965 as given in the text).

Eisenmann, Eugene. 1968. Behavior of Orange-Chinned Parakeets in Panama. *Condor*, 70(1):86. Observations on Barro Colorado Island, Sep 1938.

Eisenmann, Eugene. 1969. Wing Formula as a Means of Distinguishing Summer Tanager, *Piranga rubra*, from Hepatic Tanager, *P. flava*. *Bird-Banding*, 40(2):144-145.

Eisenmann, Eugene. 1970a. Avifauna in Panama. Pages 50-57 *in* Helmut K. Buechner and Jimmie H. Buechner, editors. The Avifauna of Northern Latin America: A Symposium Held at the Smithsonian Institution, 13-15 April 1966. *Smithsonian Contributions to Zoology* 26. Deals primarily with habitat preservation and conservation.

Eisenmann, Eugene. 1970b. Letter to the Editor. *Condor*, 72(3):381. Addresses the spelling of the subspecific name "*inexspectata*," a *nomen oblitum* vs. "*inexpectata*" for *Sturnella magna*.

Eisenmann, Eugene, and Thomas R. Howell. 1962. The Taxonomic Status of the Hummingbirds *Chalybura melanorrhoa* and *Chalybura urochrysia*. *Condor*, 64(4):300-310. Concludes that *Chalybura melanorrhoa* should be treated as a race of *C. urochrysia*. Specimens examined from Almirante, Bocca del Toro in Costa Rica [sic], Boquete Trail, Cana, Guaval (Río Calovévora), mouth of the Río Imamado (Río Jaque), Panama, Quebrada Torno Rompío (Chilar), Río Sicsola, Santa Fé, Santiago de Veragua, Tacarcuna, Veragua, Zegla (Terebé River), and from the following, which are not Colón, but San Blas: Permé, Puerto Obaldia, and Ranchón.

Eisenmann, Eugene, and F. Carlos Lehmann V. 1962. A New Species of Swift of the Genus *Cypseloides* from Colombia. *American Museum Novitates*, 2117:1-16, 3 figures, 2 tables. *C. lemosi*, n. sp. from Colombia. Panama specimens of related species mentioned.

Eisenmann, Eugene, and D. L. Serventy. 1962. An Erroneous Panama Record of *Puffinus tenuirostris* and Other Misidentifications of *P. griseus*. *Emu*, 62(3):199-201. A specimen collected by Hallinan near Naos Island is not *P. tenuirostris* but *P. griseus*.

Eisenmann, Eugene, and Lester L. Short. 1982. Systematics of the Avian Genus *Emberizoides* (Emberizidae). *American Museum Novitates*, 2740:1-21, 1 figure, 9 tables. Specimens from Panama.

Ellis, Hazel R. 1952. Nesting Behavior of a Purple-Throated Fruit-Crow. *Wilson Bulletin*, 64(2):98-100. Observations of *Querula purpurata* on Barro Colorado Island.

Emanuel, Victor L. 1980. First Documented Panama Record of Spotted Rail (*Pardirallus maculatus*). *American Birds*, 34(2):214-215. Tocumen, Jan 1978.

Emlen, Stephen T., Natalie J. Demong, and Douglas J. Emlen. 1989. Experimental Induction of Infanticide in Female Wattled Jacanas. *Auk*, 106(1):1-7, 1 figure, 1 table. Chagres River, near Gamboa, Apr-Jun 1987.

Emlen, Stephen T., Peter H. Wrege, and Michael S. Webster. 1998. Cuckoldry as a Cost of Polyandry in the Sex-Role-Reversed Wattled Jacana, *Jacana jacana*. *Proceedings of the Royal Society of London. Series B, Biological Sciences*, 265(1413):2359-2264, 3 figures. Chagres River, 1990-1995.

Engleman, Dodge, Lorna Engleman, and Robert S. Ridgely. 1989. Finding Birds in Panama. Pages 475-500 *in* Robert S. Ridgely, and John A Gwynne Jr. *A Guide to the Birds of Panama with Costa Rica, Nicaragua, and Honduras*. Princeton: Princeton University Press. Describes popular birding localities in central, western and eastern Panama with important travel and birding tips for each region.

Enstrom, David A. 1992. Delayed Plumage Maturation in the Orchard Oriole (*Icterus spurius*): Tests of Winter Adaptation Hypotheses. *Behavioral Ecology and Sociobiology*, 30(1):35-42, 6 figures, 3 tables. Observations near Gamboa at 37 *Erythrina fusca* trees in 1989. Mist netted birds for molt and plumage assessments.

Escalante-Pliego, Patricia, and A. Townsend Peterson. 1992. Geographic Variation and Species Limits in Middle American Wood Nymphs (*Thalurania*). *Wilson Bulletin*, 104(2):205-219, appendix (specimens examined), 1 colored plate, 4 figures, 2 tables. Specimens from 9 localities in Panama.

Fairchild, Graham B., and Charles O. Handley, Jr. 1966. Gazetteer of Collecting Localities in Panama. Pages 9-22, 1 map *in* Rupert L. Wenzel and Vernon J. Tipton, editors. *Ectoparasites of Panama*. Chicago: Field Museum of Natural History. Localities are listed by province and some include brief descriptions.

Festa, Enrico. 1909. *Nel Darien e Nell' Ecuador. Diario di Viaggio di un Naturalista*. [In Darien and in Ecuador. Diary of a Naturalist's Journey]. xvi + 397 pages, 2 maps, 74 illustrations. Torino: Unione Tipographica-Editrice Torinese. Description of the author's trip to Panama and

Ecuador.

Ficken, Millicent S. and Robert W. Ficken. 1966. Behavior of Myrtle Warblers in Captivity. *Bird-Banding*, 37(4):273-279, 3 tables. *Dendroica coronata.*

Fisher, Walter K. 1904. The Home Life of a Buccaneer. *Condor*, 6(3):57-61, 1 frontispiece, 1 photograph. Observations of *Fregata aquila* on Laysan atoll in the Pacific Ocean.

Fleischer, Robert C., and Neal G. Smith. 1992. Giant Cowbird Eggs in the Nests of Two Icterid Hosts: The Use of Morphology and Electrophoretic Variants to Identify Individuals and Species. *Condor*, 94(3):572-578, 2 figures, 4 tables. Collected eggs of *Scaphidura oryzivora* and hosts in 1986 from nests at Arraijan Tank Farm (near Arraijan) and at Rodman Marine Base near Panama City.

Fleischer, Robert C., Cheryl L. Tarr, Eugene S. Morton, Alexandra Sangmeister, and Kim C. Derrickson. 1997. Mating System of the Dusky Antbird, a Tropical Passerine, as assessed by DNA Fingerprinting. *Condor*, 99(2):512-514, 1 figure. *Cercomacra tyrannina* in Parque Nacional Soberanía, Gamboa, 1991-1995.

Fleming, C. A. 1950. Some South Pacific Sea-Bird Logs. *Emu*, 49(3):169-188, 8 figures. Observations made on five voyages between Australasia and the Panama Canal.

Fontaine, Roy. 1980. Observations on the Foraging Association of Double-Toothed Kites and White-Faced Capuchin Monkeys. *Auk*, 97(1):94-98, 1 table. Barro Colorado Island, Feb-Dec 1977. *Harpagus bidentatus.*

Fowler, Henry W. 1944. The Fishes. *The Academy of Natural Sciences, Monographs*, 6:57-530. The entire monograph is titled *Results of the Fifth George Vanderbilt Expedition (1941)* and the section on fishes contains a few notes on ornithological localities. The section on birds was written by J. Bond and R. Meyer de Schauensee.

Fowler, James M., and James B. Cope. 1964. Notes on the Harpy Eagle in British Guiana. *Auk*, 81(3):257-273, 3 figures, 3 tables.

Fraga, Rosendo M. 1989. Interactions Between Nectarivorous Birds and the Flowers of *Aphelandra sinclairiana* in Panama. *Journal of Tropical Ecology*, 5(1):19-26, Observations of bird visits to *A. sinclairiana* on Old Gamboa Road (Road C2F), S of Río Sardinilla in 1987. Captured and banded birds.

Freed, Leonard A. 1986a. Territory Takeover and Sexually Selected Infanticide in Tropical House Wrens. *Behavioral Ecology and Sociobiology*, 19(3):197-206, 3 figures, 2 tables. La Laguna, Panama province, 1982-1984. Reports cases of physical takeover by both males and females of *Troglodytes aedon* and the subsequent death of eggs, nestlings, and fledglings.

Freed, Leonard A. 1986b. Usurpatory and Opportunistic Bigamy in Tropical House Wrens. *Animal Behaviour*, 34(6):1894-1896, 1 table. Reports 6 bigamous pairings of *Troglodytes aedon* at La Laguna, central Panama, from 1982-1985.

Freed, Leonard A. 1987a. Rufous-and-White Wrens Kill House Wren Nestlings During a Food Shortage. *Condor*, 89(1):195-197, 1 figure, 1 table. La Laguna, Panama province, 1982-1985.

Freed, Leonard A. 1987b. The Long-Term Pair Bond of Tropical House Wrens: Advantage or Constraint? *American Naturalist*, 130(4):507-525, 8 figures, 3 tables. *Thryothorus rufalbus* and *Troglodytes aedon* at La Laguna, Panama province, 1982-1986.

Freed, Leonard A. 1988. Forced Fledging: An Investigation of the Lengthy Nestling Period of Tropical House Wrens. *National Geographic Research*, 4(3):395-407, 9 figures (some photographs), 1 table. *Troglodytes aedon* at La Laguna, Panama province, 1985.

Friedmann, Herbert. 1927. A Revision of the Classification of the Cowbirds. *Auk*, 44(4):495-

508. Specimens from Boca de Cupe (Río Tuyra) and Veragua.

Friedmann, Herbert. 1935. A Hawk of the Genus *Leucopternis* New to Science. *Auk*, 52(1):30. *Leucopternis princeps zimmeri*, n. subsp. Specimens of *L. p. princeps* cited from Boquete Trail and Cebaco Island.

Friedmann, Herbert. 1950a. The Forms of the Black Hawk-Eagle. *Smithsonian Miscellaneous Collections*, 111(16):1-4, 1 plate. *Spizaetus tyrannus serus* n. subsp. from Río Indio.

Friedmann, Herbert. 1950b. The Birds of North and Middle America: A Descriptive Catalog of the Higher Groups, Genera, Species and Subspecies of Birds Known to Occur in North America from the Arctic Lands to the Isthmus of Panama, the West Indies and Other Islands of the Caribbean Sea, and the Galapagos Archipelago, Part XI. *United States National Museum, Bulletin* 50 (9): xiii + 793 pages, 51 figures. Includes the families Cathartidae, Accipitridae, Pandionidae, and Falconidae.

Friedmann, Herbert. 1952. Vestigial Claws on the Wings of the Kiskadee Flycatcher, *Pitangus sulphuratus caucensis*. *Auk*, 69(2):200. Based on 13 specimens from Colombia.

Friedmann, Herbert. 1963. Host Relations of the Parasitic Cowbirds. *United States National Museum, Bulletin* 233: ix + 276 pages, tables.

Friedmann, Herbert. 1964. The History of Our Knowledge of Avian Brood Parasitism. *Centaurus, International Magazine of the History of Mathematics, Science and Technology*, 10:282-304.

Friedmann, Herbert, and Foster D. Smith, Jr. 1955. A Further Contribution to the Ornithology of Northeastern Venezuela. *Proceedings of the United States National Museum*, 104:463-524, 4 plates, 5 figures.

Fuertes, Louis Agassiz. 1913. Impressions of the Voices of Tropical Birds. *Bird-Lore*, 15:341-344, 2 illustrations.

Galindo, Pedro, and Eustorgio Méndez. 1965. Banding of Thrushes and Catbirds at Almirante, Panama. Second Year of Observations. *Bird-Banding*, 36(4):233-239, 4 tables. Summaries of captures and recaptures of 4,654 migrant thrushes that were banded Sep 1963-Jun 1964.

Galindo, Pedro, Eustorgio Méndez, and Abdiel J. Adames. 1963. Banding of Migrant Thrushes in Almirante, Panama. *Bird-Banding*, 34(4):202-209, 3 tables. Summaries of captures and recaptures of 2,035 migrant thrushes that were banded from Fall 1962-Spring 1963. Map and descriptions of the study plots.

García-Pelayo y Gross, Ramón. 1993. Gran Diccionario Español-Inglés/English-Spanish Dictionary. Unabridged Edition. New Edition. Paris: Larrouse.

Garella, Napoleon. 1845. Projet d'un Canal de Jonction de l'ocean Pacifique et de l'ocean Atlantique a travers l'Isthme de Panama. viii + 233 pages, 2 maps. Paris: Carilian-Goeury et Vor. Dalmont. Based on surveys made in 1844.

Gifford, Edward Winslow. 1913. Expedition of the California Academy of Sciences to the Galapagos Islands, 1905-1906. No. VIII. The Birds of the Galapagos Islands, with Observations on the Bird of Cocos and Clipperton Islands (Columbiformes to Pelecaniformes). *Proceedings of the California Academy of Sciences*, series 4, 2 (part I): 1-132, 7 plates, 13 tables.

Gjording, Chris N. 1991. *Conditions Not of Their Choosing. The Guaymí Indians and Mining Multinationals in Panama.* xxiv + 409 pages, 4 figures, 9 tables. Washington, D.C.: Smithsonian Institution Press. Includes a map of the Cerro Colorado mining operations.

Godinez, Esteban. 1999. Roadside Vulture Counts in a Central Panama province, Panama. *El Pitirre*, 12(3):98-99, 1 table. Daily counts along the Interamerican Road [= Carretera Interamericana] W of Santiago City.

Godman, Frederick Du Cane. 1907-1910. *A Monograph of the Petrels.* iv + 381 pages, 104 colored plates. London: Witherby and Co. Covers the families of Procellariiformes.

Goeldi, Emil A. 1894. On the Nesting of *Phibalura flavirostris* and *Lochmias nematura. Ibis*, 6(6):484-494, 3 figures. Observations in Brazil.

Goldman, Edward A. 1911. A New Kingfisher from Panama. *Smithsonian Miscellaneous Collections*, 56(27):1-2. *Ceryle americana isthmica* n. subsp. from Río Indio.

Goldman, Edward A. 1920. Mammals of Panama. *Smithsonian Miscellaneous Collections*, 69(5):1-309, 39 plates, 24 figures. Plate 1 is a map of the life zones of Panama included in our map section as LZP. Describes intermittent work from 1910-1912. Assisted by Adan Lizano (1910-1911) and George Scott (1912). Includes itinerary, descriptions of localities with several photographs, and lists of birds found in the life zones.

Goodwin, Derek. 1958. Remarks on the Taxonomy of Some American Doves. *Auk*, 75(3):330-334. Proposes the merger of *Zenaidura, Melopelia,* and *Nesopelia* with *Zenaida* and *Osculatia* with *Geotrygon.*

Goodwin, Derek. 1959. Taxonomic Notes on the American Ground Doves. *Auk*, 76(4):510-516, appendix, 1 figure. Proposes five genera of American ground doves merging *Columbigallina, Eupelia,* and *Oxypelia* with *Columbina,* and *Leptophaps* and *Gymnopelia* with *Metriopelia.*

Gorski, Leon J. 1969. Traill's Flycatchers of the "Fitz-Bew" Songform Wintering in Panama. *Auk*, 86(4):745-747. Observations of *Empidonax traillii* from Dec 1967-Jan 1968. Chagres River, Gamboa, La Jagua, and Pacora River. Habitat descriptions for Gamboa and La Jagua. Absence of the bird at Ft. Sherman, Pan American Highway, Chorrera, Río Hato, Chitré, Barro Colorado Island, Miraflores Lake and the K-6 area.

Gotelli, Nicholas J., and Gary R. Graves. 1990. Body Size and the Occurrence of Avian Species on Land-bridge Islands. *Journal of Biogeography*, 17:315-325, 3 figures, 7 tables. Analyzes data from the Pearl Islands (San José and Rey) and Coiba.

Gould, John. 1838. [New birds in the Society's collection]. *Proceedings of the Zoological Society of London*, for 1837:79-80. Original description of *Ortyx guttata* from Bay of Honduras.

Gould, John. 1843. Nine New Birds Collected During the Recent Voyage of H. M. S. Sulphur. *Proceedings of the Zoological Society of London*, 1843:103-106. *Pipra vitellina* n. sp. from Panama.

Gould, John. 1844. Birds. Pages 39-50, 15 colored plates *in* Richard Brinsley Hinds, editor. *Zoology of the Voyage of H. M. S.* Sulphur, *Under the Command of Captain Sir Edward Belcher, R.N., C.B., F.R.G.S., Etc., During the Years 1836-1842.* London: Smith, Elder and Co. Volume 1 contains mammals, birds, and fish (missing from our copy), and volume 2 contains the mollusks. Apparently, the text was released in 10 parts. The introductory pages to the text indicate that in March of 1839 the ship sailed for Panama from Nicoya [09°55'N], "touching on her way, at the bay of Honda and at Quibo [Veraguas]. On the 26th we finally bid farewell to Panama, and sailed for the Sandwich Islands." Contains a slightly modified re-description of *Pipra vitellina* Gould (p. 41, plate 21) indicating that the collector was Mr. Hinds.

Gould, John. 1850a. Descriptions of New Birds. *Proceedings of the Zoological Society of London*, 1850:91-95, 1 colored plate. *Cephalopterus glabricollis* n. sp. collected by Warszewicz in western Panama, Cordillera of Chiriqué in Veragua.

Gould, John. 1850b. On Six New Species of Humming Birds. *Proceedings of the Zoological Society of London*, 1850:162-164. Descriptions of *Trochilus scintilla, T. chionura, T. venusta, T. coeruleogularis, T. castaneoventris,* and *T. niveoventer* from specimens collected by Warszewicz near David, on the Cordillera of Chiriquí, and on the Volcano of Chiriquí in Veragua. For *T. coeruleogularis* locality is "near David, on the north side of the Cordillera," which presumably should read "south side of the Cordillera." N side would indicate the Boquete Trail, in Bocas del Toro, which seems unlikely as the species has not otherwise been recorded from there.

Gould, John. 1853. Description of a New Species of *Aulacorhamphus*. *Proceedings of the Zoological Society of London*, 1853:45. *A. caeruleogularis* n. sp. collected by Seeman in Veragua, while on board H. M. S. *Herald*. Reprinted in 1855 (see below). See also *The Zoologist*, 1853:3861. Gould's "official" descriptions were often anticipated, either in summary or verbatim, in the more popular journals *The Zoologist* and *The Athenaeum*, in which case they properly should have priority. We have not attempted to sort this out here and merely point out some instances of duplication.

Gould, John. 1855. Description of a New Species of *Aulacorhamphus*. *Aulacorhamphus caeruleogularis*. *Annals and Magazine of Natural History*, Series 2, 15(89):390-391. Reprint of Gould, 1853:45.

Gould, John. 1856. Description of a New Trogon and a New Odontophorus. *Proceedings of the Zoological Society of London*, 1856:107-108. *Trogon aurantiiventris* n. sp. from David and *Odontophorus veraguensis* n. sp. collected in Veragua and at Panama. See also *The Athenaeum*, 1490:620 (17 May 1856).

Gould, John. 1860. Descriptions of Twenty-Two New Species of Humming-Birds. *Proceedings of the Zoological Society of London*, 1860:304-312. *Oreopyra leucaspis* n. sp. collected by Warszewicz at the Volcan of Chiriquí in Costa Rica [= Panama].

Gould, John. 1861a. Descriptions of Two New Species of Humming-bird, Belonging to the Genus *Hypuroptila*. *Proceedings of the Zoological Society of London*, 1861:198-199. *H. urochrysa*, from Panama, and *H. isaurae*, collected by Verreaux, from Bocca del Toro, in Costa Rica [= Panama].

Gould, John. 1861b. *An Introduction to the Trochilidae, or Family of Humming-Birds*. iv + 216 pages. London: Taylor and Francis. Describes 416 species mentioning various early localities in Panama in the habitat descriptions.

Gould, John. 1868. On Some Additional Species of the Genus *Eutoxeres*. *Annals and Magazine of Natural History*, 4(1):455-457. *E. salvini* n. sp. from Veragua.

Gould, John. 1990. *John Gould's Hummingbirds*. A reprinting of Gould's *Monograph of Trochilidae* (1849-1861), to which Gould (1861b) is an introduction, along with the posthumous supplement. The book contains Gould's original 360 plates, 58 supplemental plates, and 74 pages of text titled "Notes to the Plates". London: Wellfleet Press. *Hypuroptila urochrysia* Gould (plate 90) was recorded from "the neighborhood of Panama." This was later changed to western Colombia (Peters, 1945).

Gradwohl, Judy Ann, and Russell Greenberg. 1980. The Formation of Antwren Flocks on Barro Colorado Island, Panama. *Auk*, 97(2):385-395, 3 figures, 2 tables. Observations of *Myrmotherula fulviventris* and *Microrhopias quixensis* from Oct 1977-Mar 1978.

Gradwohl, Judy Ann, and Russell Greenberg. 1982. The Effect of a Single Species of Avian Predator on the Arthropods of Aerial Leaf Litter. *Ecology*, 63(2):581-583, 1 figure, 1 table. Observations of *Myrmotherula fulviventris* on Orchid Island.

Gradwohl, Judy Ann, and Russell Greenberg. 1983. The Breeding Season of Antwrens on Barro Colorado Island. Pages 345-351, 4 tables *in* Egbert Giles Leigh, Jr., Austin Stanley Rand, and Donald M. Windsor, editors. *The Ecology of a Tropical Forest. Seasonal Rhythms and Long-Term Changes*. Oxford: Oxford University Press. Censuses on Barro Colorado Island, 1978-1979.

Gradwohl, Judy Ann, and Russell Greenberg. 1984. Search Behavior of the Checker-Throated Antwren Foraging in Aerial Leaf Litter. *Behavioral Ecology and Sociobiology*, 15(4):281-285, 3 tables. Pairs of *Myrmotherula fulviventris* foraging on Barro Colorado Island, 1979.

Graetz, Erich. 1956. Beobachtungen über das Verhalten einiger Tanagriden, Coerebinen und Icteriden im tropischen Klima. [Observations on the Behavior of some Tanagers, Coeribids and Icterids in Tropical Climates]. *Zoologische Beiträge*, 2:273-278. Notes on several species of captivity. In German.

Grant, P. R. 1966. The Taxonomic Status of the Yellow-Green Vireo, *Vireo flavoviridus* (Cassin), from the Tres Marias Islands, Mexico, with Additional Notes. *Acta Zoologica Academiae Scientiarum Hungaricae*, 12:77-82, 2 tables.

Graves, Gary R., and Nicholas J. Gotelli. 1983. Neotropical Land-bridge Avifaunas: New Approaches to Null Hypotheses in Biogeography. *Oikos*, 41:322-333, 2 appendices, 5 figures, 4 tables. Analyzes data from the Pearl Islands (San José and Rey) and Coiba.

Greenberg, Russell. 1981a. Frugivory in Some Migrant Tropical Forest Wood Warblers. *Biotropica*, 13(3):215-223, 2 figures, 3 tables. Barro Colorado Island, Frijoles, and Gamboa, 1977-1979.

Greenberg, Russell. 1981b. The Abundance and Seasonality of Forest Canopy Birds on Barro Colorado Island, Panama. *Biotropica*, 13(4):241-251, 4 figures, 5 tables. Barro Colorado Island, 1978-1979.

Greenberg, Russell, and Judy Ann Gradwohl. 1980a. Leaf Surface Specializations of Birds and Arthropods in a Panamanian Forest. *Oecologia*, 46(1):115-124, 3 figures, 8 tables. Censuses from 1978-1979 on Barro Colorado Island and along Pipeline Road crossing the Ríos Frijoles, Limbo, Mendoza, and Agua Salud.

Greenberg, Russell, and Judy Ann Gradwohl. 1980b. Observations of Paired Canada Warblers *Wilsonia canadensis* During Migration in Panama. *Ibis*, 122(4):509-512, 2 tables. Observations at Barro Colorado Island during migrations in Oct-Nov 1977 and Apr-May 1979.

Greenberg, Russell, and Judy Ann Gradwohl. 1983. Sexual Roles in the Dot-Winged Antwren (*Microrhopias quixensis*), a Tropical Forest Passerine. *Auk*, 100(4):920-925, 5 tables. Barro Colorado Island, Nov 1978-Sep 1979.

Greenberg, Russell, and Judy Ann Gradwohl. 1985. A Comparative Study of the Social Organization of Antwrens on Barro Colorado Island, Panama. Pages 845-855, 3 figures, 3 tables *in* P. A. Buckley et al., editors. Neotropical Ornithology. *Ornithological Monographs*, 36. Abstract in Spanish. Studies conducted in 1977-1983.

Greenberg, Russell, and Judy Ann Gradwohl. 1986. Constant Density and Stable Territoriality in Some Tropical Insectivorous Birds. *Oecologia*, 69(4):618-625, 3 figures, 4 tables. Barro Colorado Island, censuses, mist netting, and color-banding of birds during the rainy seasons of 1977-1983.

Greenberg, Russell, and Judy Ann Gradwohl. 1997. Territoriality, Adult Survival, and Dispersal in the Checker-Throated Antwren in Panama. *Journal of Avian Biology*, 28(2):103-110, 2 figures. Fourteen years of observations of *Myrmotherula fulviventris* at Barro Colorado Island.

Greenway, James. C., Jr. 1934. *Panyptila cayenensis* (Gmelin) Nesting in a House. *Auk*, 51(1):377-379. Ancón and Pedro Miguel Locks (as "San Miguel Locks"), July 1932.

Greenway, James. C., Jr. 1973. Type Specimens of Birds in the American Museum of Natural History. Part 1. Tinamidae . . . . Rallidae. *Bulletin of the American Museum of Natural History*, 150(3):207-346.

Greenway, James. C., Jr. 1978. Type Specimens of Birds in the American Museum of Natural History. Part 2. Otitidae . . . . Picidae. *Bulletin of the American Museum of Natural History*, 161(1):1-306.

Greenway, James. C., Jr. 1987. Type Specimens of Birds in the American Museum of Natural

History. Part 4. Passeriformes: Tyrannidae . . . . Atrichornithidae. *American Museum Novitates*, 2879:1-63.

Griscom, Ludlow. 1924a. Bird-Hunting Among the Wild Indians of Western Panama. *Natural History*, 24(4):509-519, 1 map, 7 photographs. Account of expedition to Cerro Flores [Chiriquí] in 1924 Mar-Feb. Mentions Cerro Iglesia and Río San Lorenzo. Map details the route from the Canal Zone to Aguadulce and into the interior. More recent publications suggest that Griscom reached Cerro Santiago. See Cerro Flores for discussion.

Griscom, Ludlow. 1924b. Descriptions of New Birds from Panama and Costa Rica. *American Museum Novitates*, 141:1-12. *Trogonurus aurantiiventris flavidior*, *Scytalopus chiriquensis*, *Margarornis rubiginosa boultoni*, *Pseudocolaptes lawrencii panamensis*, *Troglodytes ochraceus remotus*, *Catharus gracilirostris bensoni*, *Planesticus tristis panamensis*, *Phainoptila melanoxantha minor*, *Emberizoides sphenurus floresae*, *Atlapetes gutturalis coloratus*, *Pselliophorus luteoviridis*, and *Chlorospingus pileatus diversus*, described as new from Cerro Flores; *Chlorospingus tacarcunae*, from Mt. Tacarcuna; and *Atalotriccus pilaris wilcoxi*, from La Colorado, Santiago. Additional specimens from Boquete, Calobre, Canal Zone, Chitra, David, eastern Chiriquí, El Villano (Santiago), Mt. Pirre, San Lorenzo River (Wilcox Camp), and Volcan de Chiriquí.

Griscom, Ludlow. 1926. The Ornithological Results of the Mason-Spinden Expedition to Yucatan. Part I.—Introduction; Birds of the Mainland of Eastern Yucatan. *American Museum Novitates*, 235:1-19, 1 table. Annotated list of 167 taxa. *Cochlearius zeledoni panamensis* n. subsp. from Corozal, and specimens from Veraguas and the Canal Zone.

Griscom, Ludlow. 1927a. Undescribed or Little-Known Birds from Panama. *American Museum Novitates*, 280:1-19. *Odontophorus melanotis coloratus*, *Leptotila plumbeiceps malae*, *Premnoplex brunnescens albescens*, *P. b. distinctus*, *Dendroplex picirostris extimus*, *Xiphocolaptes emigrans panamensis*, *Xiphorhynchus guttatus marginatus*, *Campylorhamphus borealis olivaceus*, *Dendrocolaptes picumnus veraguensis*, *Pipromorpha oleaginea lutescens*, *Empidonax flavescens floresae*, *E. f. chitrae*, *Mitrephanes aurantiiventris vividus*, *Heleodytes zonatus panamensis*, *Basileuterus bensoni*, *B. tristriatus chitrensis*, *B. fulvicauda gaffneyi*, *B. f. toddi*, *Diglossa plumbea veraguensis*, *Chlorothraupis carmioli magnirostris*, and *C. c. lutescens*. Described as new primarily from specimens collected by Benson in Coclé and Veraguas, with additional specimens collected by Griscom et al. at Cerro Flores, Batty at Boqueron, Anthony and Ball at Mt. Tacarcuna, and Richardson at Tacarcuna. Localities mentioned are Agua Dulce, Boqueron, Boquete, Calobre, Calovevora, Canal Zone, Cape Garachine, Cape Mala Peninsula, Cerro Montosa, Chiriquí, Chitra, Coiba Island, Guaval, Juan Diaz, Mt. Pirre, Río Calovevora, San Lorenzo River, Santa Fé, Santiago, Tapalisa, and Volcan de Chiriquí.

Griscom, Ludlow. 1927b. An Ornithological Reconnaissance in Eastern Panama in 1927. *American Museum Novitates*, 282:1-10. Mainly records from the Pearl Islands and the coast of Darién province that were collected by Griscom et al., Feb-March 1927. *Saucerrotia edwardi margaritarum*, from Pedro Gonzales Island, and from Cape Garachiné, and *S. e. crosbyi*, *Pachysylvia minor darienensis*, *Compsothlypis pitiayumi nana*, and *Ateleodacnis leucogenys panamensis* described as new. Other localities mentioned are Agua Dulce, Capeti River, Chiman, Colon Harbor, El Rey Island, Juan Diaz, La Chorrera, Río Chiman, Saboga Island, Sambú River, Santiago, Tacarcuna, and Veraguas.

Griscom, Ludlow. 1928. New Birds from Mexico and Panama. *American Museum Novitates*, 293:1-6. Based on specimens from Almirante, Bocas del Toro, collected by Rex R. Benson in 1927. Adds 23 taxa as previously unknown in Panama. *Eupsittula astec extima*, *Chloronerpes simplex aurorae*, and *Hylopezus fulviventris flammulatus* described as new from Almirante. *Chalybura urochrysa incognita*, n. subsp. from Tacarcuna.

Griscom, Ludlow. 1929a. A Collection of Birds from Cana, Darien. *Bulletin of the Museum of*

Comparative Zoology, 69(8):149-190. Describes 193 taxa based on a collection of 618 specimens collected in 1928 by Rex R. Benson at Cana on Mt. Pirre, and at El Real and Cupé, both on the Rio Tuira. *Tinamus major saturatus*, *Odontophorus guianensis chapmani*, *Trogon collaris extimus*, *Eubucco boucieri anomalus*, *Chrysoptilus punctigula lucescens*, *Dechonychura* [sic] *typica darienensis*, *Leptopogon superciliaris troglodytes*, *Pheugopedius spadix xerampelinus*, and *Saltator maximus iungens* described as new from Cana, as well as *Butorides striatus patens*, from Panama City, and *Myrmotherula brachyura ignota*, from Jesusito. Other localities mentioned are Barro Colorado Island, Canal Zone, Chiman, Lion Hill, Mt. Sapo, Pearl Islands, Tacarcuna, Tapalisa, and Volcan de Chiriquí.

Griscom, Ludlow. 1929b. Notes on the Rough-Winged Swallow (*Stelgidopteryx serripennis* (Aud.)) and Its Allies. *Proceedings of the New England Zoological Club*, 11:67-72. *Stelgidopteryx ruficollis decolor* n. subsp. from Divala.

Griscom, Ludlow. 1930. Revisions of Two Central American Birds. *Occasional Papers of the Boston Society of Natural History*, 5:287-292. *Claravis mondetoura pulchra* n. subsp. from Boquete.

Griscom, Ludlow. 1931. Notes on Rare and Little Known Neotropical Pygmy Owls. *Proceedings of the New England Zoological Club*, 12:37-43. *Glaucidium minutissimum rarum* n. subsp. from Permé. Also mentioned are Obaldia and Tacarcuna.

Griscom, Ludlow. 1932a. The Ornithology of the Caribbean Coast of Extreme Eastern Panama. *Bulletin of the Museum of Comparative Zoology*, 72(9):303-372. Collection of 1,396 specimens representing 284 species and subspecies made by H. Wedel from March 1929 to March 1931, at Permé, Obaldia and Ranchón, in eastern Comarca de San Blas. *Ortalis garrula mira*, *Rhynchortyx cinctus hypopius*, *Piaya cayana incincta*, *Pulsatrix perspicillata chapmani*, *Lophostrix cristata wedeli*, *Eutoxeres aquila munda*, *Lepidopyga caeruleogularis confinis*, *Capito maculicoronatus melas*, *Chloropipo holochlora suffusa*, *Thryophilus nigricapillus reditus*, and *T. leucopogon griscescens* described as new, as well as *Dysithamnus puncticeps intensus*, from Mt. Sapo, *Campylorhamphus trochilirostris brevipennis*, from Lion Hill, *Rhynchocyclus brevirostris hellmayri*, from Cana, and *Camptostoma pusillum major*, from San Miguel, El Rey. Additional Panama specimens mentioned. *Thalurania colombica insulicola* n. subsp. is mistakenly ascribed to Panama from San Miguel, El Rey, Pearl Islands, but Wetmore (1968:310) notes that it comes from San Miguel, Sierra Nevada de Santa Marta, Colombia.

Griscom, Ludlow. 1932b. Notes on Imaginary Species of *Ramphocelus*. *Auk*, 49(2):199-203. Reports that Panama is not the type locality for *R. chrysopterus* Boucard or for *R. inexpectatus* Rothschild. Panama, rather than Guatemala,was suggested as the possible origin of *R. uropygialis* Bonaparte. *R. luciani* Lawrence (Lion Hill) = *R. dimidiatus* x *R. icteronotus*, and *R. dunstalli* = *R. dimidiatus* x *R. icteronotus*. See also Olson and Violani, 1995.

Griscom, Ludlow. 1932c. The Distribution of Bird-Life in Guatemala; A Contribution to a Study of the Origin of Central American Bird-Life. *Bulletin of the American Museum of Natural History*, 64:ix + 439 pages, 2 maps, 11 figures.

Griscom, Ludlow. 1933. Notes on the Havemeyer Collection of Central American Birds. *Auk*, 50(3):297-308. Records from the Almirante Region and from Port San Antonio, Río Chepo (mistaken placed in Darién province) based on collections by Austin Paul Smith. *Ramphocelus dimidiatus albirostris* n. subsp. from Divalá. Other localities mentioned are Boqueron, Cana, Coiba Island, David, Lion Hill, Mt. Sapo, Panama City, the Pearl Islands, Permé, Punta de Sabana, Remedios, Tuyra basin, Veraguas, Wilcox Camp, and Zegla (Río Terebe).

Griscom, Ludlow. 1934. The Ornithology of Guerrero, Mexico. *Bulletin of the Museum of Comparative Zoology*, 75(10):367-422, 1 plate, tables. *Sturnella magna subulata* n. subsp. Boquete (Brown). *Amaurospiza concolor australis* n. subsp. Boquete (Benson).

Griscom, Ludlow. 1935a. Critical Notes on Rare Panama Birds. *Occasional Papers of the Boston Society of Natural History*, 8:199-204. *Mecocerculus superciliaris palloris* n. subsp. from Tapalisa. Records of *Vireo carmioli* from Volcan de Chiriquí and *Hylophilus viridiflavus* with comments on the proposed race *pallescens*. Other localities mentioned include Chitra, Concepcion, and El Quiel.

Griscom, Ludlow. 1935b. The Ornithology of the Republic of Panama. *Bulletin of the Museum of Comparative Zoology*, 78(3):261-382. List of 1038 species and subspecies, with brief statements of overall distribution and range in Panama. Also includes a review of ornithological work, important literature, physiography, climate and life-zones. Lists *Polioptila schistaceigula* from Panama based on Viguier's specimen from Gulf of Darién. (Reviewed by Stone, 1935).

Griscom, Ludlow. 1935c. Critical Notes on Central American Birds in the British Museum. *Ibis*, 13(5):541-554. Includes new taxa from Central America.

Griscom, Ludlow. 1937. A Collection of Birds from Omilteme, Guerrero. *Auk*, 54(2):192-199.

Griscom, Ludlow, and James C. Greenway, Jr. 1937. Critical Notes on New Neotropical Birds. *Bulletin of the Museum of Comparative Zoology*, 81(2):417-437. *Micrastur mirandollei extimus* n. subsp., from Permé, and *Caprimulgus rufus minimus* n. subsp., from Panama City.

Griswold, John A., Jr. 1936a. A New Subspecies of *Lurocalis* from Panama. *Proceedings of the New England Zoological Club*, 15:101-103. *L. semitorquatus noctivagus* n. subsp. from Río Pequení, Salamanca Hydrographic Station.

Griswold, John A., Jr. 1936b. Two New Records for Panama. *Auk*, 53(4):457-458. *Dendrocygna b. bicolor* and *Belonopterus chilensis cayennensis* from La Jagua.

Gross, Alfred O. 1927. Barro Colorado Island Biological Station. *Annual Report of the Smithsonian Institution, year ending June 30, 1926*:327-342, 9 black and white plates. Visited the island from Jun-Sep 1925. Includes accounts of nesting birds.

Gross, Alfred O. 1930a. Rainbows on Wings. The Massena Trogons of the Tropics. *Nature Magazine*, 15(4):249-250, 2 figures.

Gross, Alfred O. 1930b. That Bird the Toucan. *Nature Magazine*, 16(5):291-294, 7 photographs.

Gross, Alfred O. 1950. Nesting of the Streaked Flycatcher in Panama. *Wilson Bulletin*, 62(4):183-193, 3 photographs, 1 table. *Myiodynastes maculatus* on Barro Colorado Island.

Gross, Alfred O. 1952. Nesting of Hicks' Seedeater at Barro Colorado Island, Canal Zone. *Auk*, 69(4):433-446, 1 plate, 3 tables. Includes notes on *Sporophila [americana] hicksii* from the Canal Zone and other localities.

Gross, Alfred O. 1964. Nesting of the Black-Tailed Flycatcher on Barro Colorado Island. *Wilson Bulletin*, 76(3):248-256, 2 figures, 2 tables. Fieldwork on *Myiobius atricaudus* in 1925, 1927, and 1949.

Gross, Alfred O. and Josselyn Van Tyne. 1929. The Purple Gallinule (*Ionornis martinicus*) of Barro Colorado Island, Canal Zone. *Auk*, 46(4):431-446, 5 plates, 1 figure, tables.

Guglielmo, Christopher G., Patrick D. O'Hara, and Tony D. Williams. 2002. Extrinsic and Intrinsic Sources of Variation in Plasma Lipid Metabolites of Free-Living Western Sandpipers (*Calidris mauri*). *Auk*, 119(2):437-445, 2 figures, 2 tables. Captured and bled birds wintering in Panama from Dec 1995-Mar 1996 as well as from sites in Canada. Abstract in Spanish.

Guiamarães Sobrinho, José Caetano. 1932. Notas Ornitologicas. Soco-Boi (*Tigrisoma brasiliense*). *Revista do Museu Paulista*, 17(2):917-918. Description of the eggs. In Portuguese.

Gurney, John Henry. 1879. Notes Upon Three American Raptorial Birds Apparently New to Science. *Ibis*, 4(3):171-178, 1 plate, tables. *Micrastur amaurus* n. sp. from the "forest-region of Panama."

Gyldenstolpe, Nils. 1951. The Ornithology of the Rio Purus Region in Western Brazil. *Arkiv for*

Zoologi, Series 2, 2(1):1-320, 1 map, 1 figure.

Haffer, Jürgen. 1959. Notas Sobre Las Aves de la Region de Uraba. [Notes on the Birds of the Uraba Region]. *Lozania, Acta Zoologica Colombiana*, 12:1-49, 3 figures, 1 table. Includes map of the collecting sites. Collections in 1959 along the Gulf of Uraba, Colombia near the eastern Panamá boundary. In Spanish with English summary.

Haffer, Jürgen. 1962. Zum Vorkommen von *Brachygalba salmoni* Sclater & Salvin, mit Beschreibung einer neuen Form. [On the Occurrence of *Brachygalba salmoni* Sclater & Salvin, with a Description of a New Form]. *Journal für Ornithologie*, 103(1):38-46, 4 figures (1 is a map). Mentions: Boca de Paya, Cana, El Real, Río Jesusito, and Tapaliza.

Haffer, Jürgen. 1967a. On Birds from the Northern Choco Region, NW-Colombia. *Veröffentlichungen der Zoologischen Staatssammlungen München*, 11:123-149, appendix, 5 figures. Description of region followed by annotated species list.

Haffer, Jürgen. 1967b. Some Allopatric Species Pairs of Birds in Northwestern Colombia. *Auk*, 84(3):343-365, 6 figures, 4 tables. Discusses three species pairs (*Crax*, *Pipra*, and *Rhynchocychus*) based on his collections in Colombia and specimens from the literature. Mentions: Cana, Río Cupe, Mt. Sapo and Río Jesusito.

Haffer, Jürgen. 1967c. Speciation in Colombian Forest Birds West of the Andes. *American Museum Novitates*, 2294:1-57, 15 figures, 8 tables. Panama is included on the maps with species distributions.

Hallinan, Thomas. 1924. Notes on Some Panama Canal Zone Birds with Special Reference to Their Food. *Auk*, 41:304-326. Description of 159 species that were identified by W. deW. Miller and deposited at AMNH. The descriptions are based on 440 specimens that were collected by the author at more than 30 localities along and near the Panama Canal.

Hardy, John William. 1964. Behavior, Habitat, and Relationships of Jays of the Genus *Cyanolyca*. *Occasional Papers of the C. C. Adams Center for Ecological Studies*, 11:1-14, 9 figures. Fieldwork in Mexico.

Harrowes, David E. 1936. Nesting of *Oncostoma* in the Canal Zone. *Auk*, 53(3):336-337. Description of nest and eggs.

Hartert, Ernst. 1898a. On a Collection of Birds from North-Western Ecuador, Specimens from W. F. H. Rosenberg. *Novitates Zoologicae*, 5:477-505, 2 plates. *Prionorhynchus platyrhynchus minor*, n. subsp. from Panama.

Hartert, Ernst. 1898b. [Descriptions of Six New Birds.] *Bulletin of the British Ornithologists' Club*, 7(50):28-30. Specimens from Rosenberg taken in Ecuador. Does not mention Darién in description of *Polioptila schistaceigula*, only Cachabi, North Ecuador.

Hartert, Ernst. 1900a. Trochilidae. *Das Tierreich*. Lieferung 9: 254 pages, 34 figures. Berlin: R. Friedländer und Sohn. A reivew of all of the genera and species of hummingbirds then known.

Hartert, Ernst. 1900b. Allgemeines und Specielles uber Kolibris (Erläuterungen, Ergänzungen und Zusätze zur 9. Lieferung "Trochilidae" des Tierreiches). [Generalities and Specifics of Hummingbirds (Elucidations, Additions, and Addenda to *Lieferung 9 "Trochilidae"* of *Das Tierreich*)]. *Journal für Ornithologie*, 48(3):350-368. Comments on Hartert 1900a.

Hartert, Ernst. 1901. [Descriptions of Four New Birds.] *Bulletin of the British Ornithologists' Club*, 12:32-33. *Cyclorhis coibae* n. sp. and *Aphantochroa cuvieri saturatior* n. subsp. both from Coiba Island.

Hartert, Ernst. 1902. Some Further Notes on the Birds of Northwest Ecuador. *Novitates Zoologicae*,

9:599-617, 1 colored plate, sketches.

Hartert, Ernst. 1919. Types of Birds in the Tring Collection. B. Types in the General Collection. I. Corvidae-Meliphagidae. *Novitates Zoologicae*, 26:123-178.

Hartert, Ernst. 1922. Types of Birds in the Tring Collection. B. Types in the General Collection. *Novitates Zoologicae*, 29:365-412. Troglodytidae (cont'd) through Trochili. *Aphantochroa cuvieri saturatior* Hartert, collected on Coiba Island by Batty.

Hartert, Ernst. 1927. Types of Birds in the Tring Collection. B. Types in the General Collection. VIII. Columbae-Struthionidae. *Novitates Zoologicae*, 34:1-38. *Leptotila battyi* Rothschild, collected on Coiba Island by Batty.

Hartert, Ernst, and Arthur Goodson. 1917. Notes and Descriptions of South American Birds. *Novitates Zoologicae*, 24:410-419. *Xenops genibarbis ridgwayi* n. subsp. collected by André at Tocoumé. Also mentions Iguaros, Sevilla, Almijas, and Medidor Islands.

Hartman, Frank A. 1954. Cardiac and Pectoral Muscles of Trochilids. *Auk*, 71(4):467-469, 1 table. Heart weight and pectoral weight as a percentage of body weight for 25 different species and subspecies based on 148 specimens collected in Panama.

Hartman, Frank A. 1955. Heart Weight in Birds. *Condor*, 57(4):221-238, 1 figure, 1 table. List of heart weights of 1,340 birds representing 291 species. Birds were collected in the U. S. and in Panama at Juan Mina (Chagres River), La Jagua River, Mount Copete (above Boquete), Palo Santo (near El Volcán village), and Santa Clara (15 mi from the Costa Rican Border).

Hartman, Frank A. 1956. A Nest of the Striped Horned Owl. *Condor*, 58(1):73, 1 figure. Juan Mina Station, 1953. *Rhinoptynx clamator.*

Hartman, Frank A. 1957. Some Additions to Nesting Data on Panamanian Birds. *Condor*, 59(4):269-271, 2 figures. *Amazilia edward niveoventer*, *Lampornis castaneoventris*, *Selasphorus scintilla*, and *Thripadectes rufobrunneus* nests at Volcán and Mount Copete.

Hartman, Frank A., and Katharine A. Brownell. 1961a. Adrenal and Thyroid Weights in Birds. *Auk*, 78(3):397-422, 2 figures, 1 table. Body weights and percentage adrenal and thyroid weights for 249 species of birds in 49 families collected at Río Chagres and at El Volcán.

Hartman, Frank A., and Katharine A. Brownell. 1961b. Lipids In The Locomotor Muscles Of Birds. *Condor*, 63(5):403-409, 1 table. Lipid content of muscle determined in 104 species of birds. All but 3 species were collected in Panama at Juan Mina Field Station or at El Volcán.

Hartman, Frank A., and M. A. Lessler. 1963. Erythrocyte Measurements in Birds. *Auk*, 80(4):467-473, 1 table. Blood smears taken from 124 species of birds taken on various studies at Juan Mina Field Station and at El Volcán.

Harty, Stephen T. 1964. The Discovery of the Black Rail (*Laterallus jamaicensis*) in Panama and the First Breeding Record. *Cassinia*, 48:19-20, 1 plate. Nest and eggs collected east of Tocomen Airport Circle, 1963.

Hau, Michaela, Martin Wikelski, and John C. Wingfield. 1998. A Neotropical Forest Bird Can Measure the Slight Changes in Tropical Photoperiod. *Proceedings of the Royal Society of London. Series B, Biological Sciences*, 265(1391):89-95, 3 figures, 1 table. Spotted Antbirds (*Hylophylax n. naevioides*) captured on Pipeline Road were used in a laboratory study.

Hau, Michaela, Martin Wikelski, and John C. Wingfield. 2000. Visual and Nutritional Food Cues Fine-Tune Timing of Reproduction in a Neotropical Rainforest Bird. *Journal of Experimental Zoology*, 286(5):494-504, 4 figures. *Hylophylax n. naevoides* captured in May 1997 in Soberania National Park were used in laboratory studies.

Haverschmidt, François. 1949. Biographical Notes on the Penard Brothers. *Auk*, 66(1):56-60.

Haverschmidt, François. 1952. Nesting Behavior of the Southern House Wren in Surinam. *Condor*, 54(5):292-295, 2 tables. *Troglodytes* [*aedon*] *musculus.*

Haverschmidt, François. 1954. The Nesting of the Ridgway Tyrannulet in Surinam. *Condor*, 56(3):139-141, 1 figure. *Camptostoma obsoletum.*

Haverschmidt, François. 1955. Beobachtungen an *Tapera naevia* und ihren Wirtsvögeln in Surinam. [Observations of *Tapera naevia* and its Host Birds in Surinam]. *Journal für Ornithologie*, 96(3):337-343, 2 plates.

Haverschmidt, François. 1957. The Nests of *Pitangus lictor* and *Coryphotriccus parvus. Auk*, 74(2):240-242, 2 plates. Surinam.

Haverschmidt, François. 1959. Notes on *Helicolestes hamatus* in Surinam. *Auk*, 76(1):32-36, 2 plates. General biology with notes on a nest.

Haverschmidt, François. 1961. Der Kuckuck *Tapera naevia* und seine Wirte in Surinam. [The Cuckoo *Tapera naevia* and its Hosts in Surinam]. *Journal für Ornithologie*, 102(3):353-359, 2 figures.

Haverschmidt, François. 1962. Notes on the Feeding Habits and Food of Some Hawks of Surinam. *Condor*, 64(2):154-158, 2 figures.

Haverschmidt, François. 1964a. Beobachtungen an *Chondrohierax uncinatus* (Temminck) in Surinam. [Observations of *Chondrohierax uncinatus* (Temminck) in Surinam]. *Journal für Ornithologie*, 105(1):64-66, 1 photograph.

Haverschmidt, François. 1964b. Nesting of the Crane Hawk in Surinam. *Condor*, 66(4):303-305, 1 figure. *Geranospiza caerulescens.*

Haverschmidt, François. 1965. A Nest of the Double-Banded Pigmy Tyrant. *Condor*, 67(6):538-539, 1 figure. *Lophotriccus vitiosus.*

Haverschmidt, François. 1966a. Hybrids of *Thraupis palmarum* and *Thraupis virens. Bulletin of the British Ornithologists' Club*, 86(1):5-6. Museum specimens from South and Central America.

Haverschmidt, François. 1966b. The Eggs of the Giant Cowbird. *Bulletin of the British Ornithologists' Club*, 86(8):144-147. *Scaphidura oryzivora.* Comments on Friedmann, 1963.

Haverschmidt, François. 1967. Additional Notes on the Eggs of the Giant Cowbird. *Bulletin of the British Ornithologists' Club*, 87(8):136-137, 1 photograph. *Scaphidura oryzivora.*

Haverschmidt, François. 1968. *Birds of Surinam.* xxix + 445 pages, 40 colored plates, 30 black-and-white plates, 1 map, illustrations. London: Oliver and Boyd. Illustrated by Paul Barruel.

Haverschmidt, François. 1970a. The Nest of the Red-Breasted Blackbird. *Auk*, 87(1):167. *Leistes militaris.*

Haverschmidt, François. 1970b. Are There Two Species of Shrub Flycatcher (*Sublegatus*)? *Auk*, 87(2):358.

Heath, R. E. 1932. Notes on Some Birds of Barro Colorado Island, Canal Zone. *Ibis*, 13(2):480-486. Annotated list of 37 species observed during a fortnight, winter 1931.

Heck, Heinz. 1963. The Successful Breeding of the King Vultures (*Sarcorhamphus papa* [L.]) in captivity. *Der Zoologische Garten*, 27(6):295-297, 1 figure.

Hekstra, Gerrit P. 1982. Description of Twenty Four New Subspecies of American *Otus* (Aves, Strigidae). *Bulletin Zoologisch Museum Universiteit van Amsterdam*, 9(7):49-63. *Otus guatemalae centralis* n. subsp. from Cerro Malí. Additional specimens from Bocas del Toro, Cana, Colon, La Laguna, and Upper Trinidad.

Hellebrekers, W. Ph. J. 1942. Revision of the Penard Oölogical Collection from Surinam. *Zoologische*

Mededeelingen, 24(1-2):240-275. Descriptions of eggs collected by the Penard brothers in Surinam. Excludes those collected outside of Surinam.

Hellebrekers, W. Ph. J. 1945. Further Notes on the Penard Oölogical Collection from Surinam. *Zoologische Mededeelingen*, 25(1-4):93-100, 1 colored plate. Corrections and additions to the 1942 descriptions.

Hellmayr, Charles E. 1906. [Description of a New Subspecies of Bunting.] *Bulletin of the British Ornithologists' Club*, 19(79):28-29. *Emberizoides macrourus hypochondriacus* n. subsp. from Frances, Volcan de Chiriquí.

Hellmayr, Charles E. 1908. Übersicht der Südamerikanishen Arten der Gattung *Chaetura* (s. str.). [Summary of South American Species of the Genus *Chaetura*]. *Verhandlungen der Ornithologischen Gesellschaft in Bayern*, 8 [for 1907]:144-161, tables. In German. Describes 12 taxa from South and Central America.

Hellmayr, Charles E. 1910a. Notes sur Quelques Oiseaux de l'Amérique Tropicale. [Notes on Some Birds from Tropical America]. *Revue Française d'Ornithologie*, 1(11):161-165. In French. *Calospiza lavinia dalmasi*, nom. nov., from Nicaragua to the Isthmus of Panama. *Calliste l. lavinia* Cassin, Isthmus of Darién, New Grenada, and *C. emiliae* Dalmas, Isthme de Darién, Río Truando [= Colombia].

Hellmayr, Charles E. 1910b. [Descriptions of Two New Races of Neotropical Birds.] *Bulletin of the British Ornithologists' Club*, 25:87-88. *Corapipo leucorrhoa heteroleuca* n. subsp. from Boquete and Frances.

Hellmayr, Charles E. 1911a. Passeres. Fam Regulidae. Part 17, pages 1-18, 1 plate, *in* Philogène Wytsman, editor. *Genera Avium*. Brussels: V. Verteneuil and L. Desmet. Includes measurements of specimens from Lion Hill, Panama R.R., Veragua (Chitra), Chiriqui, and mention of Isthmus of Darien in the range of *Polioptila schistaceigula* Hartert, although the original description (Hartert, 1898b) did not mention the Isthmus of Darien.

Hellmayr, Charles E. 1911b. A Contribution to the Ornithology of Western Colombia. *Proceedings of the Zoological Society of London*, 1911:1084-1213. Based on a collection of 700 specimens. *Calospiza gyroloides bangsi* n. subsp. Boquete.

Hellmayr, Charles E. 1918. Miscellanea Ornithologica III. VIII. Die Formen von *Rhodinocichla rosea* (Less.). *Verhandlungen der Ornithologischen Gesellschaft in Bayern*, 13:302-305. In German. Describes 4 taxa with *R. r. eximia* from Boquete, Bugaba, Calovevora, Chitra, David, Lion Hill, Mina de Chorcha, and Santa Fé.

Hellmayr, Charles E. 1921. [Die Beschreiben 12 Neue Formen aus dem Neotropischen Gebiet/ Descriptions of 12 New Forms from the Neotropical Realm]. *Anzeiger der Ornithologischen Gesellschaft in Bayern*, 1(4):25-32. In German. *Henicorhina leucosticta darienensis* n. subsp. from Tacarcuna and Tapaliza.

Hellmayr, Charles E. 1929. Catalogue of Birds of the Americas and the Adjacent Islands in Field Museum of Natural History. *Field Museum of Natural History Zoological Series*, 13, Part 6. vi + 258 pages. Includes Oxyruncidae, Pipridae, Cotingidae, Rupicolidae, and Phytotomidae. *Schiffornis turdinus panamensis*, n. subsp.

Hellmayr, Charles E. 1934. Catalogue of Birds of the Americas and the Adjacent Islands in Field Museum of Natural History *Field Museum of Natural History Zoological Series*, 13, Part 7. vi + 531 pages. Includes Corvidae, Paridae, Sittidae, Certhiidae, Chamaeidae, Cinclidae, Troglodytidae, Prunellidae, Mimidae, Turdidae, Zeledoniidae, and Sylviidae. *Polioptila schistaceigula* from Isthmus of Darién.

Hellmayr, Charles E. 1936. Catalogue of Birds of the Americas and the Adjacent Islands in Field

Museum of Natural History. *Field Museum of Natural History Zoological Series*, 13, Part 9. vi + 458 pages. Includes Tersinidae and Thraupidae. Río Lara.

Hellmayr, Charles E. 1937. Catalogue of Birds of the Americas and the Adjacent Islands in Field Museum of Natural History. *Field Museum of Natural History Zoological Series*, 13, Part 10. vi + 228 pages. Includes Icteridae.

Hellmayr, Charles E. 1938. Catalogue of Birds of the Americas and the Adjacent Islands in Field Museum of Natural History. *Field Museum of Natural History Zoological Series*, 13, Part 11. vi + 662 pages. Includes Ploceidae, Catamblyrhynchidae, and Fringillidae.

Hellmayr, Charles E., and Boardman Conover. 1942. Catalogue of Birds of the Americas and the Adjacent Islands in Field Museum of Natural History. *Field Museum of Natural History Zoological Series*, 13, Part 1, No. 1. vi + 636 pages. Includes Rheiformes, Tinamiformes, Galliformes, Gruiformes, and Columbiformes.

Hellmayr, Charles E., and Boardman Conover. 1948a. Catalogue of Birds of the Americas and the Adjacent Islands in Field Museum of Natural History. *Field Museum of Natural History Zoological Series*, 13, Part 1, No. 2. viii + 434 pages. Includes Spenisciformes, Gaviiformes, Colymbiformes, Procellariiformes, Pelecaniformes, Ciconiiformes, and Anseriformes.

Hellmayr, Charles E., and Boardman Conover. 1948b. Catalogue of Birds of the Americas and the Adjacent Islands in Field Museum of Natural History. *Field Museum of Natural History Zoological Series*, 13, Part 1, No. 3. vi + 383 pages. Includes Jacanidae, Rostratulidae, Haematopodidae, Charadriidae, Scolopaciidae, Recurvirostridae, Phalaropodidae, Burhinidae, Thinocoridae, Chionididae, Stercorariidae, Laridae, Rynchopidae, and Alcidae.

Hellmayr, Charles E., and Boardman Conover. 1949. Catalogue of Birds of the Americas and the Adjacent Islands in Field Museum of Natural History. *Field Museum of Natural History Zoological Series*, 13, Part 1, No. 4. vi + 358 pages. Includes Cathartidae, Accipitridae, Pandionidae, and Falconidae.

Hernandez-Camacho, Jorge. 1959. La Validez Nomenclatural del Nombre Generico *Pan* Richmond, 1899 (Familia *Capitonidae, Aves*). [The Nomenclatural Validity of the Generic Name *Pan* Richmond, 1899 (Family *Capitonidae, Aves*)]. *Lozania, Acta Zoologica Colombiana*, 11:11-14. In Spanish.

Hespenheide, Henry A. 1975. Selective Predation by Two Swifts and a Swallow in Central America. *Ibis*, 117(1):82-99, appendix, 4 figures, 3 tables. Sampled flying insects and examined the diet of *Chaetura spinicauda*, *C. brachyura,* and *Stelgidopteryx ruficollis* at Contractors Hill and Madden Forest from Aug-Sep 1969 and at three sites in Costa Rica. Habitat descriptions of collecting localities and a four page appendix titled "Characteristics of Flying Insect Communities Relevant to Birds".

Hespenheide, Henry A. 1980. Bird Community Structure in Two Panama Forests: Residents, Migrants, and Seasonality During the Nonbreeding Season. Pages 227-237, 2 figures, 3 tables *in* Allen Keast and Eugene S. Morton, editors. *Migrant Birds in the Neotropics. Ecology, Behavior, Distribution and Conservation.* Washington, D.C.: Smithsonian Institution Press. Censuses at Palo Seco and Pipeline Road during different seasons.

Hewitt, Vivian. 1937. A Description of the Eggs of the Grey Crane-Hawk, Guiana White-Tailed Hawk, Eyebrowed Hawk, and Four Banded Sparrow Hawk. *Oologists' Record*, 17(1):12-14. *Geranospiza caerulescens, Buteo albicaudatus, Accipiter superciliaris*, and *A. bicolor* in Venezuela.

Hicks, David L. 1967. First Record of the White-Eyed Vireo in Panama. *Condor*, 69(1):90. Collected in Almirante by the author.

Hicks, David L., Eustorgio Méndez, and Horace Loftin. 1967. Additions to the Avifauna of

Panama: The Palm Warbler and The Connecticut Warbler. *Condor*, 69(3):319-320. Specimens of *Dendroica p. palmarum* and *Oporornis agilis* collected at Tocumen and Almirante now at USNM.

Hilty, Steven L. 2003. *Birds of Venezuela*. 875 pages, 67 color plates, 4 figures, color photographs, black and white sketches, maps. London: Christopher Helm. A field guide/handbook.

Hobson, Dorothy Madden. 1948. An Annotated List of Birds Seen in the Canal Zone and Panama. *Year Book Indiana Audubon Society*, 26:18-24. Sight records of 90 species, mainly on Barro Colorado Island.

Hofslund, Pershing B. 1959. Fall Migration of Herring Gulls from Knife Island, Minnesota. *Bird-Banding*, 30(2):104-114, 4 maps, 3 figures, 1 table. Banded *Larus argentatus* recovered in the bay at Bocas del Toro, 1958.

Houston, C. Stuart, and Frank Scott. 1992. The Effect of Man-Made Platforms on Osprey Reproduction at Loon Lake, Saskatchewan. *Journal of Raptor Research*, 26(3):152-158, 3 figures, 2 tables. Banded *Pandion haliaetus* recovered in Panama. No specific locality given.

Houston, David C. 1986. Scavenging Efficiency of Turkey Vultures in Tropical Forest. *Condor*, 88(3):318-323, 1 table. Observations at Barro Colorado Island of *Cathartes aura* from Nov-Dec 1982.

Houston, David C. 1987. The Effect of Reduced Mammal Numbers on *Cathartes* Vultures in Neotropical Forests. *Biological Conservation*, 41(2):91-98, 2 figures, 1 table. Barro Colorado Island and Pipeline Road, 1981-1984.

Howe, Henry F. 1981. Dispersal of a Neotropical Nutmeg (*Virola sebifera*) by Birds. *Auk*, 98(1):88-98, 3 figures, 5 tables. Bird activity and fruit sampling at trees on Barro Colorado Island from Sep 1977-Feb 1978.

Howe, Henry F. and Diane De Steven. 1979. Fruit Production, Migrant Bird Visitation, and Seed Dispersal of *Guarea glabra* in Panama. *Oecologia*, 39(2):185-196, 3 figures, 2 tables. Observations on Barro Colorado Island in 1978. Reports on activity of 12 resident and 7 North American migrant species.

Howe, Henry F., and Gayle A. Vande Kerckhove. 1981. Removal of Wild Nutmeg (*Virola surinamensis*) Crops by Birds. *Ecology*, 62(4):1093-1106, 7 figures, 9 tables. Observations of frugivory and seed dispersal at trees on Barro Colorado Island in 1979.

Howell, Thomas R. 1953. Racial and Sexual Difference in Migration in *Sphyrapicus varius*. *Auk*, 70(2):118-126, 2 figures, 1 table. Examined birds from winter range including 12 from Panama.

Howell, Thomas R. 1955. A Southern Hemisphere Migrant in Nicaragua. *Condor*, 57(3):188-189. *Notiochelidon cyanoleuca*.

Howell, Thomas R. 1956. Variation in *Deconychura longicauda* in Central America and Colombia. *Auk*, 73(4):517-528, 2 plates, 1 figure, 1 table. Specimens from Nicaragua to Colombia including the following localities in Panama: Almirante, Bogava, Cana, Divala, Guaval, Lion Hill, and Port Antonio.

Howell, Thomas R. 1957. Birds of a Second Growth Rain Forest Area in Nicaragua. *Condor*, 59(2):73-111, 1 colored plate, 6 figures. Plates and figures by Don Eckelberry. Annotated species list based on field work (1953-1954) and specimen loans. Occasional mention of Panama and a few specific localities.

Howell, Thomas R. 1972. Birds of the Lowland in Pine Savanna of North Eastern Nicaragua. *Condor*, 74(3):316-340, 2 figures, 6 tables.

Huber, Wharton. 1923. Two New Birds from Nicaragua. *Auk*, 40(2):300-302, 1 table.

Huber, Wharton. 1932. Birds Collected in Northeastern Nicaragua in 1922. *Proceedings of the Academy of Natural Sciences of Philadelphia*, 84:205-249. Annotated list with some comparison with specimens from Panama.

Hudson, Peter J. 1984. Observations on Nestlings of the Long-Tailed Hermit *Phaethornis superciliosus*. *Bulletin of the British Ornithologists' Club*, 104(3):97-99, 1 figure, 2 tables. Observations of a nest from Feb 15-Mar 12 1979 at Río Aila, San Blas.

Hudson, Peter J., and A. M. Sugden. 1984. Inflorescence Production by *Brownea rosa-de-monte* (*Leguminosae*) and Feeding Behaviour of the Long-Tailed Hermit *Phaethornis superciliosus*. *Ibis*, 126(3):416-420, 3 figures, 1 table. Observations at Río Aila, San Blas.

Huff, Clay G., and Alexander Wetmore. 1967. Blood Parasites of Birds Collected in Four Successive Years in Panama. *Bulletin of the Wildlife Disease Association*, 3(4):178-181, 2 tables. Blood smears were taken from 183 birds collected by Wetmore in 1952-1955. Table 2 provides specimen data for the 28 positive smears.

Hunt, James H. 1971. A Field Study of the Wrenthrush, *Zeledonia coronata*. *Auk*, 88(1):1-20, 1 colored plate, 7 figures, 2 tables. July 1968 in Costa Rica.

Ihering, Hermann von. 1904. The Biology of the Tyrannidae with Respect to their Systematic Arrangement. *Auk*, 21(3):313-322.

Ihering, Hermann von. 1914. Biologia e Classificação das Cuculidas Brazileiras/Biologie der Brasilianischen Cuculide. [Biology and Classification of the Brazilian Cuckoos]. *Revista do Museu Paulista*, 9:371-410, 2 figures. In Portuguese (p. 371-390) and in German (p. 391-410).

Imhof, Thomas A. 1950. Additional Bird Records for Panamá. *Auk*, 67(2):255-257. Sight records made during military residence from Jan 1942-Feb 1943.

Jacques, Florence Page. 1942. *Birds Across the Sky*. xiv + 240 pages, 25 illustrations. New York: Harper and Brothers. Illustrated by Francis Lee Jacques. Account of a brief visit to Barro Colorado Island on pp. 158-168.

Jewel, Lindsey L. 1913. Some North American Birds in Panama. *Auk*, 30(3):422-429. List of 74 species included in A. O. U. Check-List recorded in 1911 and 1912 in the Canal Zone at: Gatun, Gatun Lake, Gatun River, Mindi, Miraflores, Old French Canal, Río Chagres, and Toro Point.

Jewel, Lindsey L. 1914. Canal Zone Bird Species. *The Canal Record*, 7(23):219-220. List of 251 taxa of Passeriformes compiled by L. L. Jewel, without notations except to indicate migrants.

Johnson, Ned K. and Alan H. Brush. 1972. Analysis of Polymorphism in the Sooty-Capped Bush Tanager. *Systematic Zoology*, 21(3):245-262, 6 figures, 4 tables. Field work in Costa Rica and at the "Camino a Boquete" in 1966.

Johnson, R. A. 1953. Breeding Notes on Two Panamanian Antbirds. *Auk*, 70(4):494-496, 1 table. Nests of *Thamnophilus punctatus atrinucha* and *Microrhopias quixensis virgata*, Barro Colorado Island, 1948.

Johnson, R. A. 1954. The Behavior of Birds Attending Army Ants Raids on Barro Colorado Island, Panama Canal Zone. *Proceedings of the Linnaean Society of New York*, 63-65:41-70, 6 figures, 3 tables, illustrations. Gives a classification of the bird groups attending raids of the army ant *Eciton burchelli*.

Johnson, R. E. 1937. Hunting for the "Poor-me-one": the "Tree-Stump" Giant Nightjar of Trinidad (Nyctibius griseus). *The Illustrated London News*, Saturday, Mar 13:434-436, 6 photographs.

Johnston, Richard F. 1961. The Genera of American Ground Doves. *Auk*, 78(3):372-378, 1 table.

Combines *Columbigallina* with *Columbina*.

Jones, S. E. 1977. Coexistence in Mixed Species Antwren Flocks. *Oikos*, 29(2):366-375, appendix, 2 figures, 1 table. Barro Colorado Island, 1974. In English with Russian summary.

Joseph, Leo. 2000. Beginning an End to 63 Years of Uncertainty: The Neotropical Parakeets Known as *Pyrrhura picta* and *P. leucotis* Comprise More Than Two Species. *Proceedings of the Academy of Natural Sciences of Philadelphia*, 150:279-292, 2 appendices, 3 figures, 4 tables. Examined 257 specimens of *Pyrrhura picta* and *P. leucotis*, and concludes that *P. p. eisenmanni* warrants recognition as a species. Panama localities include Cascajilloso and Las Piraugales, El Cortezo de Tondsi [sic].

Jouanin, Christian. 1950. Catalogue Systématique des Types de Trochilidés du Muséum National d'Histoire Naturelle de Paris. *Bulletin du Muséum National d'Histoire Naturelle*, Series 2, 22 (Supplement II):1-27.

Jouy, P. L. 1894. Notes on Birds of Central Mexico, with Descriptions of Forms Believed to be New. *Proceedings of the United States National Museum*, 16:771-808. Printed in 1894 for the 1893 volume. Lists 112 taxa with *Spinus psaltria croceus* n. subsp. from Panama.

Karr, James R. 1971a. Ecological, Behavioral, and Distributional Notes on Some Central Panama Birds. *Condor*, 73(1):107-111. New records along with information on birds thought to be rare or doubtful in the Canal Zone. Based on work by the author in 1968-1969 and that of others at Bohío Peninsula, Cerro Azul, Cerro Campana, Chimán, Chiva Chiva Road, Limbo Hunt Club, Madden Forest Reserve, Pipeline Road, Río Medio, and Río Mendoza. Locality descriptions with coordinates.

Karr, James R. 1971b. Structure of Avian Communities in Selected Panama and Illinois Habitats. *Ecological Monographs*, 41(3):207-233, 2 appendices, 3 figures, 13 tables. Work in 1968-1969 at La Jagua (grasslands), La Joya (grasslands), Chiva Chiva Road (shrub) and Limbo Hunt Club. Locality descriptions with coordinates.

Karr, James R. 1971c. Wintering Kentucky Warbler (*Oporornis formosus*) and a Warning to Banders. *Bird-Banding*, 42(4):299. Mentions birds captured repeatedly over several years at Navy Pipeline Road and Chiva Chiva Road.

Karr, James R. 1976. Seasonality Resource Availability and Community Diversity in Tropical Bird Communities. *American Naturalist*, 110(976):973-994, 10 figures, 1 table. Study of seven habitats in Panama, 1968-1969. See Karr (1971a, 1971b) for details of the study areas.

Karr, James R. 1977. Ecological Correlates of Rarity in a Tropical Forest Bird Community. *Auk*, 94(2):240-247, 2 tables. Pipeline Road, 1968-1969, 1971, 1973, and 1975.

Karr, James R. 1982a. Avian Extinction on Barro Colorado Island, Panama: A Reassessment. *American Naturalist*, 119(2):220-239, 2 figures, 6 tables. Compares Barro Colorado Island avifauna with that of Limbo Hunt Club avifauna. Mist netting in 1980.

Karr, James R. 1982b. Population Variability and Extinction in the Avifauna of a Tropical Land Bridge Island. *Ecology*, 63(6):1975-1978, 2 tables. Compares avifauna of Limbo Hunt Club to that on Barro Colorado Island.

Karr, James R. 1984. Turnover Rates in Tropical Forest Bird Communities. *National Geographic Society Research Reports*, 16:421-425, 3 tables. Pipeline Road, 1968-1969, 1971, 1973, 1975, 1977.

Karr, James R. 1990. The Avifauna of Barro Colorado Island and the Pipeline Road, Panama. Pages 183-198, 3 figures, 3 tables *in* Alwyn H. Gentry, editor. *Four Neotropical Rainforests*. New Haven: Yale University Press. Barro Colorado Island and Pipeline Road with mention of Río Mendoza, Limbo, Gamboa, and Plantation Road, 1983-1984.

Karr, James R., and Kathryn E. Freemark. 1983. Habitat Selection and Environmental Gradients: Dynamics in the "Stable" Tropics. *Ecology*, 64(6):1481-1494, 7 figures, 3 tables. Mist netting from 1979-1982 at four study plots (Ridge, Limbo, Road, and Valley) along Pipeline Road.

Karr, James R., James D. Nichols, M. Kathleen Klimkiewicz, and Jeffrey D. Brawn. 1990a. Survival Rates of Birds of Tropical and Temperate Forests: Will the Dogma Survive? *American Naturalist*, 136(3):277-291, 1 figure, 4 tables. Estimated survival rates of birds from Panama (25 species) and Maryland (10 species). Parque Nacional Soberanía, 1980-1987.

Karr, James R., Scott K. Robinson, John G. Blake, and Richard O. Bierregaard, Jr. 1990b. Birds of Four Neotropical Forests. Pages 237-269, appendix, 2 figures, 6 tables *in* Alwyn H. Gentry, editor. *Four Neotropical Rainforests*. New Haven: Yale University Press. Barro Colorado Island and Pipeline Road. The appendix (p. 252-269) lists species found in the four countries studied, Costa Rica, Panama, Brazil, and Peru.

Keast, Allen, and Eugene S. Morton, editors. 1980. *Migrant Birds in the Neotropics. Ecology, Behavior, Distribution and Conservation*. 576 pages, figures, tables. Washington, D.C.: Smithsonian Institution Press. Proceedings of symposium held at the Conservation and Research Center, National Zoological Park, Smithsonian Institution, October 27-29, 1977. Various chapters include information on Panama.

Kelso, Leon. 1937. A Costa Rican Race of Jardine's Pygmy Owl. *Auk*, 54(3):304. Panama included in range of *Glaucidium jardinii costaricanum* n. subsp.

Kelso, Leon, and Estelle H. Kelso. 1935. Supplement to a Key to Species of American Owls. *Biological Leaflet*, 5:2 pages [unpaginated]. *Otus clarkii* n. sp. from Calobre.

Kennard, Frederic Hedge, and James Lee Peters. 1927. New Birds from Panama. *Proceedings of the New England Zoological Club*, 10:1-2. *Lanio leucothorax ictus* and *Psilorhinus mexicanus captus*, n. subspp.

Kennard, Frederic Hedge, and James Lee Peters. 1928. A Collection of Birds from the Almirante Bay Region of Panama. *Proceedings of the Boston Society of Natural History*, 38(10):443-465. Annotated list of birds collected by Kennard and J. D. Smith near Almirante from Feb-Mar 1926, including 3 weeks in the interior on the Boquete Trail.

Kielhorn, William V., Kenneth S. Norris, and William E. Evans. 1963. Bathing Behavior of Frigate Birds. *Condor*, 65(3):240-241. Observations of *Fregata magnificens* in Baja California.

Kilham, Lawrence. 1972. Habits of The Crimson-Crested Woodpecker in Panama. *Wilson Bulletin*, 84(1):28-47, 4 figures, 1 table. Observations of *Phloeoceastes melanoleucos* at Madden Forest Reserve, Limbo Hunt Club, Barro Colorado Island, Cardenas Village and Frijoles.

Koga, Tsunenori, Patricia R. Y. Blackwell, John H. Christy, Minoru Murai, and Eiiti Kasuya. 2001. Male-Biased Predation of a Fiddler Crab. *Animal Behaviour*, 62(2):201-207, 3 figures, 2 tables. Studied fiddler crab predation by Great-tailed Grackles (*Quiscalus mexicanus*) at a mudflat on the W side of the Pacific entrance of the Panama Canal. 1998, 2000.

Kreuger, Ragnar. 1963. Details of Three Previously Undescribed South American Raptor Eggs. *Oologists' Record*, 37(1):5-6. Description of eggs of *Morphnus guianensis*, *Busarellus n. nigricollis* and *Spizastur melanoleucus* from British Guiana.

Kreuger, Ragnar. 1968. Some Notes on the Oology of Members of the Family Tyrannidae (Tyrant Flycatchers). *Oologists' Record*, 42(4):70-74.

Kreuger, Ragnar. 1969. Further Notes on the Oology of Members of the Family Tyrannidae (Tyrant-Flycatchers). *Oologists' Record*, 43(2):27-28. *Myiozetetes cayennensis harterti* from Panama.

La Bastille, Anne. 1973. An Ecological Survey of the Proposed Volcano Baru National Park,

Republic of Panama. *IUCN Occasional Paper*, 6:1-77, 3 appendices, 12 figures, 5 tables. List of birds seen during a 1972 survey. Includes a map of the proposed park.

La Bastille, Anne. 1974. 1st Proposed National Park for Panama. *Biological Conservation*, 6(2):102-105, 4 figures. Based on a survey of Volcan Baru (Talamanca Range) with brief descriptions of climate, flora, fauna, human impacts, and recommendations for the proposed park.

Lafresnaye, Frédéric de. 1847. Quelques Oiseaux Nouveaux ou Rares Rapportés par M. Delatre [sic], de Bolivie, de la Nouvelle-Grenade, et de Panama. *Revue Zoologique*, 10:67-79. Remarks on *Pipra vitellina* Gould. *Conophaga naevioïdes*, *Arremon aurantiirostris*, *Coccoborus cyanoïdes*, and *Malacoptila panamensis* n. spp. from Panama.

Lafresnaye, Frédéric de. 1853. Quelques Nouvelles Espèces d'Oiseaux. *Revue et Magasin de Zoologie*, series 2, 5:56-62. *Embernagra striaticeps* n. sp. from Panama.

Land, Hugh C. and William L. Schultz. 1963. A Proposed Subspecies of the Great Potoo, *Nyctibius grandis* (Gmelin). *Auk*, 80(2):195-196, 2 tables. Panama specimen of *N. g. grandis* included in data.

Laughlin, Robert M. 1952. A Nesting of the Double-Toothed Kite in Panama. *Condor*, 54(3):137-139. *Harpagus bidentatus* on Barro Colorado Island.

Lawrence, George Newbold. 1858. Description of a New Species of Humming Bird of the Genus *Mellisuga* Brisson, with a Note on *Trochilus aquila* Bourcier. *Annals of the Lyceum of Natural History of New York*, 6(5)[for 1855]:137-142, 1 colored plate. *Mellisuga albo-coronata* n. sp. from Belén, Veraguas.

Lawrence, George Newbold. 1862a. Descriptions of Three New Species of Humming-Birds of the Genera *Heliomaster*, *Amazilia* and *Mellisuga*. *Annals of the Lyceum of Natural History of New York*, 7(4-9)[for 1860]:107-111 (reprint paged 1-4). *Mellisuga merrittii* n. sp. from El Mineral de Veraguas.

Lawrence, George Newbold. 1862b. Descriptions of New Species of Birds of the Genera *Myiarchus* and *Phlogopsis*. *Annals of the Lyceum of Natural History of New York*, 7(4-9)[for 1860]:284-286. *Myiarchus panamensis* and *Phlogopsis mcleannani*, n. spp., from the Isthmus of Panama.

Lawrence, George Newbold. 1862c. Catalogue of a Collection of Birds, Made in New Grenada, by James McLeannan, Esq., of New York, with Notes and Descriptions of New Species. Part I. *Annals of the Lyceum of Natural History of New York*, 7(10-12)[for 1861]:288-302. A list of 142 taxa collected by John McLeannan "on the Atlantic side of the Isthmus of Panama, along the line of the Panama Railroad from near the coast to about a central point between the two oceans." *Leucopternis semiplumbeus*, *Chlorostibon assimilis*, *Pachyramphus cinnamomeus*, *Tachyphonus cassinii*, *Capito maculicoronatus* and *Corethrura albigularis*, n. spp.

Lawrence, George Newbold. 1862d. Descriptions of Three New Species of Birds. *Annals of the Lyceum of Natural History of New York*, 7(10-12)[for 1861]:303-305. *Grallaria perspicillata* and *Poliptila superciliaris*, n. spp., from the Isthmus of Panama.

Lawrence, George Newbold. 1862e. Catalogue of a Collection of Birds, Made in New Grenada, by James McLeannan, Esq., of New York, with Notes and Descriptions of New Species. Part II. *Annals of the Lyceum of Natural History of New York*, 7(10-12)[for 1861]:315-334. A second collection made "during the past winter" by McLeannan and John R. Galbraith on the Isthmus of Panama, with field notes by the latter, mainly on the Atlantic slope, but a few species are definitely indicated from the Pacific side and in the vicinity of Panama City. *Thryothorus galbraithii*, *T. castaneus*, *Hylophilus pusillus*, *H. aurantiifrons* and *H. viridiflavus*, n. spp.

Lawrence, George Newbold. 1862f. Descriptions of Six New Species of Birds from the Isthmus of Panama. *Ibis*, 4:10-13. Volume 4, no series. *Heleodytes albo-brunneus*, *Pitangus albovittatus*,

Myiozetetes granadensis, *Todirostrum olivaceum*, *Tyranniscus parvus*, and *Tyrannulus brunneicapillus*, n. spp. Collections made by McLeannan and Galbraith on the Panama Railroad. Includes a footnote by Sclater (1862a) proposing the new genus *Oncostoma*.

Lawrence, George Newbold. 1862g. Catalogue of a Collection of Birds, Made in New Granada, by James McLeannan, Esq., of New York, with Notes and Descriptions of New Species. Part III. *Annals of the Lyceum of Natural History of New York*, 7(13-16):461-479. Covers correction of errors in identification in two previous parts, description of new taxa, and notice of further species forwarded by McLeannan (without definite indication of locality, although they likely come from the Panama RR). *Syrinum lineatum*, *Trogon concinnus*, *Dacnis venusta*, *Automolus pallidigularis*, *Dendrocincla olivacea*, *Dendrornis lachrymosus*, *Myrmotherula fulviventris*, *Ramphocaenus semitorquatus*, *Myrmeciza ferruginea*, *Turdus obsoletus*, *Atilla sclateri*, *Empidonax griseigularis*, *Leptopogon flavovirens*, *Psittovius subcoeruleus*, *Pionus coccinicollaris*, and *Chloronerpes callopterus*, n. spp. *Spermophila schistacea*, n. sp. proposed provisionally, although validly.

Lawrence, George Newbold. 1863. Descriptions of Eight New Species of Birds from the Isthmus of Panama. *Ibis*, 5:181-184. *Cotyle uropygialis*, *Dendrornis nana*, *Formicivora virgata*, *Myrmelastes corvinus*, *Myiozetetes marginatus*, *Myiobius atricaudus*, *Platyrhynchus superciliaris* and *Celeus squamatus*, n. spp. Presumably most represent further collections made by McLeannan.

Lawrence, George Newbold. 1864. Descriptions of New Species of Birds of the Families *Caerebidae* [sic], *Tanagridae*, *Icteridae*, and *Scolopacidae*. *Proceedings of the Academy of Natural Sciences of Philadelphia*, 1864:106-108. *Dacnis ultramarina*, *Saltator intermedius*, and *Cassicus vitellinus*, n. spp., from the Isthmus of Panama.

Lawrence, George Newbold. 1865a. Descriptions of Four New Species of Birds from the Isthmus of Panama, New Granada. *Proceedings of the Academy of Natural Sciences of Philadelphia*, 1865, 17(2):106-108. *Tachyphonus rubrifrons*, *Thamnophilus nigricristatus*, and *Geotrygon albiventer*, n. spp. from Line of Panama Railroad, near Lion Hill Station, and *Anthus (Notiocorys) parvus* n. sp. from Savanna near Panama City.

Lawrence, George Newbold. 1867a. Catalogue of a Collection of Birds, Made in New Granada, by James McLeannan, Esq., of New York, with Notes and Descriptions of New Species. Part IV. *Annals of the Lyceum of Natural History of New York*, 8[for 1863]:1-13. List of additional species received from McLeannan "since the publication of Part III." Correction of some identifications in the earlier papers and inclusion of a few on basis of McLeannan's sight records. *Petrochelidon albilinea*, *Cyphorinus* [sic] *lawrencii* "Scl. M. S.," *Pithys bicolor*, *Rhynchocyclus flavo-olivaceus*, and *Lipaugus albogriseus*, n. spp. *Spermophila schistacea* is formally proposed, but was first named in Lawrence, 1862g. Pages 11-12 indicate that McLeannan worked from Colon ("10 miles from the Atlantic") to Panama City ("4 miles from the Pacific"), and specifically mention a stream called the Cabiancho. Wetmore (1968:25) indicates that he could not locate the specimen of *Streptopelia risoria* and that the species probably did not acclimate to Panamá.

Lawrence, George Newbold. 1867b. Descriptions of New Species of Birds of the Families Tanagridae, Dendrocolaptidae, Formicaridae [sic], Tyrannidae, and Trochilidae. *Annals of the Lyceum of Natural History of New York*, 8[for 1865]:126-135. *Disythamnus* [sic] *rufiventris* and *Myrmotherula albigula*, n. spp., from Line of Panama R.R.

Lawrence, George Newbold. 1867c. Descriptions of Six New Species of Birds from Central America. *Annals of the Lyceum of Natural History of New York*, 8[for 1865]:170-173 (reprint paged 171-174). *Spermophila hicksii* n. sp. from Panama, and *S. fortipes* n. sp. from Line of Panama R.R.

Lawrence, George Newbold. 1867d. List of Birds from Near David, Chiriqui, New Granada, Collected for the Smithsonian Institution, Washington, by Mr. Fred. Hicks, with Descriptions of New Species. *Annals of the Lyceum of Natural History of New York*, 8[for 1865]:174-178 (reprint paged 175-179). *Spermophila collaris*, *Elaenia chiriquensis*, and *E. semiflava*, n. spp.

Lawrence, George Newbold. 1867e. Characters of Seven New Species of Birds from Central and South America, with a Note on *Thaumatias chionurus*, Gould. *Annals of the Lyceum of Natural History of New York*, 8[for 1866]:344-350 (reprint paged 1-7). *Geotrygon veraguensis* n. sp. collected in Veragua (presumably Belén or El Mineral) by Merritt.

Lawrence, George Newbold. 1867f. Descriptions of Six New Species of Birds of the Families Hirundinidae, Formicaridae [sic], Tyrannidae, and Trochilidae. *Annals of the Lyceum of Natural History of New York*, 8[for 1866]:400-405. *Thamnophilus leucopygus* and *Empidonax pectoralis*, n. spp., both from Lion Hill Station.

Lawrence, George Newbold. 1867g. Notes on Certain Birds from New Granada, with Descriptions of New Species. *Proceedings of the Academy of Natural Sciences of Philadelphia*, 1867, 19(3):94-95. *Tachyphonus propinquus*, nom. nov. *Phaenicothraupis vinacea* and *Leptoptila cassinii*, n. spp., from Line of the Panama Railroad.

Lawrence, George Newbold. 1868-1869. A Catalogue of the Birds Found in Costa Rica. *Annals of the Lyceum of Natural History of New York*, 9: in two part, 86-141; 141-149. An 1870 reprint contains an appendix with recent additions.

Lawrence, George Newbold. 1874. Descriptions of New Species of Birds from Mexico, Central American and South America, with a Note on *Rallus longirostris*. *Annals of the Lyceum of Natural History of New York*, 10[for 1871]:1-21. *Trogon eximius* n. sp. from the Isthmus of Panama.

Leck, Charles F. 1972a. Seasonal Changes in Feeding Pressures of Fruit Eating and Nectar Eating Birds in Panama. *Condor*, 74(1):54-60, 1 figure, 5 tables. Observations at Barro Colorado Island, 1968-1969.

Leck, Charles F. 1972b. The Impact of Some North American Migrants at Fruiting Trees in Panama. *Auk*, 89(4):842-850, appendix, 2 figures, 4 tables. Observations at El Hato and Cerro Punta, Mar-Apr 1968.

Leck, Charles F., and Steven Hilty. 1968. A Feeding Congregation of Local and Migratory Birds in the Mountains of Panamá. *Bird-Banding*, 39(4):318. Banding near Cerro Punta at the Volcan de Chiriquí Field Station and Nature Center.

Lefebvre, Gaëtan, and Brigitte Poulin. 1996. Seasonal Abundance of Migrant Birds and Food Resources in Panamanian Mangrove Forests. *Wilson Bulletin*, 108(4):748-759, 2 figures, 2 tables. Observations, invertebrate collections, and regurgitation sessions at Juan Díaz and Galeta, 1993-1995. The coordinates given for Juan Díaz [09°00' N, 79°04' W] and Galeta [09°20' N, 79°09' W] appear to be erroneous and presumably should read Juan Díaz [0903/7927 (USBGN)] and [Isla] Galeta [0924/7952 (USBGN)].

Lefebvre, Gaëtan, and Brigitte Poulin. 1997. Bird Communities in Panamanian Black Mangroves: Potential Effects of Physical and Biotic Factors. *Journal of Tropical Ecology*, 13(1):97-113, 4 figures, 3 tables. Compared abundance, foraging behavior, diet and other factors of communities at Juan Diaz (Pacific) and Galeta (Atlantic). Map and habitat descriptions of the study areas, 1993-1995. The coordinates given for Juan Díaz and Galeta appear to be erroneous (see above).

Lefebvre, Gaëtan, and Brigitte Poulin. 2000. Determinants of Avian Diversity in Neotropical Mangrove Forests. Pages 161-179, 9 figures *in* B. Gopal, W. J. Junk, and J. A. Davis, editors. *Biodiversity in Wetlands: Assessment, Function and Conservation*. Volume 1. Leiden: Backhuys Publishers. The coordinates given for Juan Díaz and Galeta appear to be erroneous (see above).

Lehmann V., F. Carlos. 1943. El Genero *Morphnus*. [The Genus *Morphnus*]. *Caldasia*, 2:165-179, 3 plates, 1 table.

Leopold, A. Starker. 1959. *Wildlife of Mexico, the Game Birds and Mammals*. xvi + 568 pages, 2 appendices, 194 figures, 18 tables. Berkeley: University of California Press. Illustrated by Charles W. Schwartz.

Levin, Rachel N. 1996a. Song Behaviour and Reproductive Strategies in a Duetting Wren, *Thryothorus nigricapillus*: I. Removal Experiments. *Animal Behaviour*, 52(6):1093-1106, 6 figures, 1 table. Removal of 5 males and 5 females from 10 pairs of free-living birds in Panamá and Colón provinces.

Levin, Rachel N. 1996b. Song Behaviour and Reproductive Strategies in a Duetting Wren, *Thryothorus nigricapillus*: II. Playback Experiments. *Animal Behaviour*, 52(6):1107-1117, 4 tables. Results of playback experiments on 45 Bay Wrens with territories along small streams in Panamá and Colón provinces.

Levings, Sally C., Stephen D. Garrity, and Linda R. Ashkenas. 1986. Feeding Rates and Prey Selection of Oystercatchers in the Pearl Islands Of Panama. *Biotropica*, 18(1):62-71. Observations and foraging data were collected for *Haematopus palliatus* on the islands of Bajo Boyarena, Bayoneta, Cassaya, Chapera, Chitre, Mogo Mogo, Pedro Gonzales, Saboga, del Rey, Viveros, and several small unnamed islets, 1982-1983.

Leviton, Alan E., Robert H. Gibbs, Jr., Elizabeth Heal, and C. E. Dawson. 1985. Standards in Herpetology and Ichthyology: Part I. Standard Symbolic Codes for Institutional Resource Collections in Herpetology and Ichthyology. *Copeia*, 3:802-832. We have followed these codes, when applicable, with the exception of Universidad de Panama (UP), for which we have used Museo de Vertebratos de Universidad de Panama (MVUP, the acronym used in their database.

Lincoln, Frederick C. 1936. Returns of Banded Birds: Third Paper. (Some Recoveries of Water Birds from Latin America). *Bird-Banding*, 7:139-148. Common Tern banded in Minnesota recovered at La Venta Beach.

Lincoln, Frederick C. 1944. Chimney Swift's Winter Home Discovered. *Auk*, 61(4):604-609, 1 map, 1 table. Summarizes previously published records of *Chaetura pelagica* for Panama.

Loetscher, Frederick W., Jr. 1952. Striped Cuckoo Fed by Rufous-and-White Wren in Panamá. *Condor*, 54(3):169. *Tapera naevia* and *Thryothorus rufalbus* at Boquete, Chiriqui.

Loftin, Horace. 1963a. Notes on Autumn Bird Migrants in Panama. *Caribbean Journal of Science*, 3(1):63-68. Notes on 57 species of North American migrants observed in fall 1961 and 1962 mostly at the Pacific terminus of the Panama Canal and at Almirante. A mimeographed sheet of corrigenda accompanied separates.

Loftin, Horace. 1963b. Some Repeats and Returns of North American Migrants in Panama. *Bird-Banding*, 34(4):219-221, 1 table. Records from Almirante, Bocas del Toro.

Loftin, Horace. 1975. Recaptures and Recoveries of Banded Native Panamanian Birds. *Bird-Banding*, 46(1):19-27. Records of birds banded by Loftin (1962-1964) and Belton (1963-1964) that returned or were recovered two or more years after banding. Each individual is listed with locality data and other available data. Many of the records suggest longevity and locality attachment. Banding localities are briefly described and include Ancon, Gamboa, Pipeline Road, and Curundu in the former Canal Zone as well as Cerro Campana, Cerro Punta, and Chiriquí Viejo River.

Loftin, Horace. 1977. Returns and Recoveries of Banded North American Birds in Panama and the Tropics. *Bird-Banding*, 48(3):253-258. Includes a list of new Panama returns and recoveries as well as a summary of all records of North American migrant returns to tropical winter quarters.

Loftin, Horace. 1991. An Annual Cycle of Pelagic Birds in the Gulf of Panama. *Ornitologia Neotropical*, 2(2):85-94, 4 tables. Seabird observations and counts obtained on ten cruises made in the Gulf of Panama from Aug 1968-Aug 1969. Twenty four species were identified.

Loftin, Horace, and Storrs L. Olson. 1963. Notes on Spring Migrants in Panama (Aves). *Caribbean Journal of Science*, 3(4):191-195. Notes on 48 species of North American migrants observed in the spring of 1962 and 1963 mostly at the Pacific terminus of the Panama Canal and at Almirante.

Loftin, Horace, David T. Rogers, Jr., and David L. Hicks. 1966. Repeats, Returns and Recoveries of North American Migrant Birds Banded in Panama. *Bird-Banding*, 37(1):35-44, 2 tables. Data for 1963-1964 from Almirante and the Canal Zone (Ancon Hill, Curundu, and Gamboa) presented in the two tables. Accounts for 24 species.

Loftin, Horace, George I. Child, and Salvatore Bongiorno. 1967. Returns in 1965-1966 of North American Migrant Birds Banded in Panama. *Bird-Banding*, 38(2):151-152. Omits many records so as not to duplicate information presented in earlier papers. Limited list of 8 species includes records of noteworthy returns to areas previously unrecorded for that species or records with lengthy time intervals between banding and recapture.

Loiselle, Bette A., and William G. Hoppes 1983. Nest Predation in Insular and Mainland Lowland Rainforest in Panama. *Condor*, 85(1):93-95, 1 table. Limbo Hunt Club and Barro Colorado Island, Mar 1981.

Louis Berger Group, Inc., Universidad de Panamá, and Smithsonian Tropical Research Institute. (2003). Recopilación y presentación de datos de recursos ambientales y culturales en la región occidental de la Cuenca del Canal de Panamá : la región occidental de la Cuenca del Canal de Panamá. Unpublished report to the Autoridad del Canal de Panama, Panama City, Panama. Copy in Smithsonian Institution Libraries.

Lowe, Percy R. 1909. Notes on Some Birds Collected During a Cruise in the Caribbean Sea. *Ibis*, 9(3):304-347, 1 colored plate, tables. Islands off the coast of Venezuela.

Lowe, Percy R. 1942. The Anatomy of Gould's Manakin (*Manacus vitellinus*) in Relation to its Display, with Special Reference to an Undescribed Pterylar Tract and the Attachments of the Flexor Carpi Ulnaris and Flexor Sublimis Digitorum Muscles to the Secondary Wing-Feathers. *Ibis*, 14(6):50-83, 10 figures.

Lundy, William E. 1951. "Poor-Me-One." *Natural History*, 60(3):111-113, 1 figure. Description of and mythology behind the call of the Common Potoo *Nyctibius griseus*.

Lundy, William E. 1957. Has the Nest of the King Vulture Finally Been Found? *Animal Kingdom, Magazine of the New York Zoological Society*, 60(4):117-121, 3 photographs. Account of a young *Sarcoramphus papa* found on Barro Colorado Island by Koford.

Maack, G. A. 1872. Report on the Geology and Natural History of the Isthmuses of Choco, of Darien, and of Panama. Pages 155-175 *in* Thomas Oliver Selfridge, editor. *Reports of Exploration and Surveys to Ascertain the Practicability of a Ship-Canal between the Atlantic and Pacific Oceans by Way of the Isthmus of Darien.* Washington, D.C.: Government Printing Office.

Mabey, Sarah E., and Eugene. S. Morton. 1992. Demography and Territorial Behavior of Wintering Kentucky Warblers in Panama. Pages 329-336, 2 figures, 3 tables *in* John M. Hagan, III, and David W. Johnston, editors. *Ecology and Conservation of Neotropical Migrant Landbirds.* Washington, D.C.: Smithsonian Institution Press. Banded and made behavioral observations on *Oporornis formosus* on Barro Colorado Island and along the Pipeline Road in 1988 and 1989.

MacArthur Robert, Jared M. Diamond, and James R. Karr. 1972. Density Compensation in Island Faunas. *Ecology*, 53(2):330-342, 3 figures, 2 tables. Mist netting on Puercos Island. Includes species lists for Puercos and other islands in the Pearl Archipelago supposedly based on Bangs 1901a;Thayer and Bangs 1905; Rendahl 1920; Wetmore 1946b, 1965, 1968; "and observations by MacArthur" in January 1971 (p. 333, table 2). For some islands, such as Cañas, Chitré, Espiritu Santo, Málaga, and Puerco, there was no previous information in the literature, but except for Puerco there is no indication of how much time was spent surveying the islands or any other information, nor is there any indication of whether MacArthur et al. added any new species to islands that had previously been visited by other collectors or observers.

MacArthur Robert, John MacArthur, Duncan MacArthur, and Alan MacArthur. 1973. The Effect of Island Area on Population Densities. *Ecology*, 54(3):657-658, 1 figure, 1 table. Compares mist net catches between Cañas and Puercos Islands.

MacDougall-Shackleton, Elizabeth, and Heidi Harbison. 1998. Singing Behavior of Lekking Green Hermits. *Condor*, 100(1):149-152, 4 figures. Observations of *Phaethornis superciliosus* along Quebrada Aleman Trail near Los Planes, Jan 1997.

Marcus, Mickey J. 1985. Feeding Associations Between Capybaras and Jacanas: A Case of Interspecific Grooming and Possibly Mutualism. *Ibis*, 127(2):240-243. On Guindilla Island in Gatun Lake the author observed *Jacana jacana* removing ectoparasites from a male capybara for more than 10 minutes, Oct 1982.

Marsh, Richard Oglesby. 1934. *White Indians of Darien*. vi + 276 pages, maps on inside covers, 1 frontispiece, 21 illustrations. New York: G. P. Putnam's Sons. Documents Marsh's anthropological expedition to Darién, accompanied by John L. Baer, who collected a few birds. The following undated notes in Wetmore's Panama historical files are worth putting on record. These were hand-written and marked "Personal knowledge of A. W[etmore]."

When Richard O. Marsh asked the Smithsonian Institution to send a representative on his Darién expedition of 1924, through the insistence of Dr. Aleš Hrdlièka, curator of Physical Anthropology in the U. S. National Museum and of Baer himself, John L. Baer, assistant to Dr. Hrdlièka was selected. Baer, short in stature and very corpulent, came to me for advice as to outfit on several occasions before his departure, and finally I recommended strongly against his going, as I did not believe that he was equipped physically because of his obesity and his excessive weight, for strenuous jungle work. He protested vehemently, as according to Marsh he did later with him, and went with the party. The sequel proved that Marsh and I were correct, as though Baer was hard-working and full of energy, when the party had reached a camp near the mouth of the Río Sucubti he was taken seriously ill. Marsh had him transported by bearers across the mountains to Aclá on Bahia Caledonia, not far from Sasardí, and here he died on May 28 and was buried (see plate, opposite p. 186).

While an anthropologist, he was much interested in birds and collected specimens as opportunity permitted. As a result he prepared 46 skins which are now in the U. S. National Museum. The specimens while relatively few have given useful and valuable data. At that period the Chucunaque area was little known, Baer's knowledge of Spanish spelling was somewhat rudimentary, and further it must be remembered that his ornithological work was done under pressure at intervals during his main studies in physical anthropology. The localities on his specimen labels are therefor[e] in part seem confusing. However through the dates and through the running account in Marsh's book I have established the following as his localities for birds:

Marshall, Joe T., Jr., 1943. Additional Information Concerning the Birds of El Salvador. *Condor*, 45(1):21-33, 5 figures. Collections in 1941-1942.

Marshall, Joe T., Jr., 1967. Parallel Variation in North and Middle American Screech-Owls. *Monographs of the Western Foundation of Vertebrate Zoology*, 1:72 pages, 1 colored frontispiece, 31 figures.

Martin, Thomas E., and James R. Karr. 1986. Temporal Dynamics of Neotropical Birds with Special Reference to Frugivores in Second-Growth Woods. *Wilson Bulletin*, 98(1):38-60, 4 figures, 5 tables. Mist netted and banded birds in Jan, Mar 1980 and Jan, Mar 1981 in Parque Nacional Soberanía. Data for 76 species in 1980 and 84 species in 1981. Includes recapture data.

McDade, Lucinda A. 1983. Long-Tailed Hermit Hummingbird Visits to Inflorescence Color Morphs of *Heliconia irrasa*. *Condor*, 85(3):360-364, 2 figures, 3 tables. Pipeline Road with map of study sites. Study indicates that *Phaethornis guy* does not show preference for either color morph (red or yellow).

McDonald, David B., Robert P. Clay, Robb T. Brumfield, and Michael J. Braun. 2001. Sexual Selection on Plumage and Behavior in an Avian Hybrid Zone: Experimental Tests of Male-Male Interactions. *Evolution*, 55(7):1443-1451, 4 figures, 1 table. Experiments with models in lek arenas near the Río San San (*Manacus candei*), on the railroad between Almirante and Changuinola (*M. "cerritus"*), the upper Río Changuinola (*M. candei* and *M. "cerritus"*), and in central Panama near the Chiva Chiva Road and near Gamboa (*M. vitellinus*), 1995-1996.

McHargue, Laurie A. 1977. Nesting of the Turkey and Black Vultures in Panama. *Wilson Bulletin*, 89(2):328-329. Observations in 1976 of *Cathartes aura* and *Coragyps atratus* on a small island (0.5 ha) located in Gatun Lake, 275 m from Barro Colorado Island.

McHargue, Laurie A. 1981. Black Vulture Nesting, Behavior, and Growth. *Auk*, 98(1):182-185, 1 figure, 1 table. Observations of *Coragyps atratus* from 1975-1979 on a small island off Barro Colorado Island.

McKim, Fred. 1947. The Forbidden Land. Reconnaissance of Upper Bayano River, R.P. in 1936. *Etnologiska Studier*, 15:115-186, 17 figures. One of two posthumous works in volume 15 edited by Henry Wassén. Observations of cormorants pp. 137, 151.

Medina Padilla, Gonzalo. 1955. Sobre la Avifauna de "El Paito" y la Nidificacion de *Galbula ruficauda*. [Concerning the Avifauna of "El Paito" and the Nest Building of *Galbula ruficauda*]. *Boletin del Museo de Ciencias Naturales*, 1(3-4):196-200.

Méndez, Eustorgio. 1969. Una Breve Introducción a las Aves de Panama. [A Brief Introduction to the Birds of Panama]. 57 pages, illustrations. Panama: [No publisher given]. Spanish text with introductory sections on birds, bird migration and bird conservation followed by accounts of 20 orders.

Méndez, Eustorgio. 1979. *Las Aves de Caza de Panama* [Game Birds of Panamá]. xvi + 290 pages, 44 illustrations. Panama: Editora Renovación, S.A.

Meyer de Schauensee, Rodolphe. 1948-1952. The Birds of the Republic of Colombia, Their Distribution and Keys for Their Identification. *Caldasia, Boletin del Instituto de Ciencias Naturales de la Universidad Nacional de Colombia-Bogota,* Part I, 1948, 5(22):251-380—introduction includes the geography of Colombia and the localities mentioned, followed by accounts of Tinimidae-Cathartidae. Part II, 1949, 5(23):381-644—Accipitridae-Picidae. Part III, 1950b, 5(24):645-871—Dendrocolaptidae-Tyrannidae. Part IV, 1951, 5(25):873-1112—Alaudidae-Fringillidae. Part V, 1952, 5(26):1115-1214, i-xxxix, 1 colored plate—addenda and corrigenda to the previous lists with additional localities and a special index to the series (pages i-xxxix).

Meyer de Schauensee, Rodolphe. 1950a. Colombian Zoological Survey, Part V. - New Birds from Colombia. *Notulae Naturae of the Philadelphia Academy of Natural Sciences*, 221:1-13, 1 figure, 1 table. Panama specimens from Aqua Clara, Canal Zone, Chiriquí, Darién, Garachiné, Gatún, Mt. Sapo, and Ranchon.

Meyer de Schauensee, Rodolphe. 1957. On Some Avian Types, Principally Gould's in the Collection of the Academy. *Proceedings of the Academy of Natural Sciences of Philadelphia*, 109:123-246.

Meyer de Schauensee, Rodolphe. 1964. *The Birds of Colombia and Adjacent Areas of South and Central America.* xvi + 427 pages, maps, 20 plates, illustrations. Narberth, Pennsylvania: Livingston Publishing Co. Published for the Academy of Natural Sciences. Illustrations by Earl L. Poole and George Miksch Sutton. Mentions Panama and the Pearl Islands in species' range.

Meyer de Schauensee, Rodolphe. 1966. *The Species of Birds of South America and Their Distribution.* xvii + 578 pages, 2 maps. Philadelphia: Livingston Publishing Co. Published for the Academy of Natural Sciences. Prepared with the collaboration of Eugene Eisenmann. Four pages of Addenda and Corrigenda are attached to our volume. Frequent mention of Panama in the extralimital range. Cocoplum also mentioned.

Michener, Martin C. 1964. A Breeding Colony of Agami Herons in Veracruz. *Condor*, 66(1):77-78. *Agamia agami.*

Miller, Alden H. 1960. The Slaty Spinetail. *Condor*, 62(6):413, 1 colored plate. Figure by Don R. Eckelberry of *Synallaxis brachyura* with a nest.

Miller, Alden H. 1963a. Seasonal Activity and Ecology of the Avifauna of an American Equatorial Cloud Forest. *University of California Publications in Zoology*, 66(1):1-78, 2 plates, 3 figures, 3 tables. Species accounts based on fieldwork in Colombia.

Miller, Alden H. 1963b. The Vocal Apparatus of Two South American Owls. *Condor*, 65(5):440-441, 1 figure. Based on specimens from Colombia.

Miller, Loye. 1937. Winter Notes on Some North American Birds in the Tropics. *Condor*, 39(1):16-19. Records from the Canal Zone and Chiriquí Lagoon. Based on a 10 week trip departing from California, passing through the Panama Canal, and ending in Chiriquí Lagoon, Feb-Apr 1936.

Miller, Loye. 1950. *Lifelong Boyhood; Recollections of a Naturalist Afield.* x + 226 pages, figures. Berkeley: University of California Press. A chapter entitled "Cruising Down the Latitudes" describes a voyage from Los Angeles through the Panama Canal to the Chiriqui Lagoon, 1936.

Miller, Waldron DeWitt. 1924. Further Notes on Ptilosis. *Bulletin of the American Museum of Natural History*, 50(5):305-331.

Miller, Waldron DeWitt, and Ludlow Griscom. 1921. Descriptions of Proposed New Birds from Central America, with Notes on Other Little-Known Forms. *American Museum Novitates*, 25:1-13, tables. Discusses status of *Crax panamensis*. Specimens from Boqueron, Canal Zone, Chepigana, and Chiriquí.

Miller, Waldron DeWitt, and Ludlow Griscom. 1925. Further Notes on Central American Birds, with Descriptions of New Forms. *American Museum Novitates*, 184:1-16, tables. Specimens from Chiriquí, Panama, and Veraguas.

Miller, Waldron DeWitt, and Ludlow Griscom. 1925. Notes on Central American Birds, with Descriptions of New Forms. *American Museum Novitates*, 183:1-14, tables. Includes specimens from Panama.

Mindell, David P., and Hal L. Black. 1984. Combined-Effort Hunting by a Pair of Chestnut-Mandibled Toucans. *Wilson Bulletin*, 96(2):319-321. Observations of *Ramphastos swainsonii* in

May 1983 on Barro Colorado Island.

Monroe, Burt L., Jr. 1963a. Three New Subspecies of Birds from Honduras. *Occasional Papers of the Museum of Zoology, Louisiana State University*, 26:1-7.

Monroe, Burt L., Jr. 1963b. Notes on the Avian Genus *Arremonops* with Description of a New Subspecies from Honduras. *Occasional Papers of the Museum of Zoology, Louisiana State University*, 28:1-12, 1 figure, 2 tables. Specimens listed from Alto de Quiel, Bajo Mono, Horqueta, and Lerida.

Monroe, Burt L., Jr., and Thomas R. Howell. 1966. Geographic Variation in Middle American Parrots of the *Amazona ochrocephala* Complex. *Occasional Papers of the Museum of Zoology, Louisiana State University*, 34:1-18, 2 figures, 1 table. Specimens from Aguadulce, Divala, Isla Pedro González, Isla San José, Nata, Panama, and Parita.

Montgomery, G. Gene, and Marta Lucia Martínez. 1984. Timing of Brown Pelican Nesting on Taboga Island in Relation to Upwelling in the Bay of Panama. *Colonial Waterbirds*, 7:10-21, 6 figures, 5 tables. Studied 318 nests of *Pelecanus occidentalis* on Isla Taboga from Jan-Jul 1981.

Montgomery, G. Gene, and Carolina Murcia. 1982. Seasonal Upwelling and Seabird Abundance off the Pacific Coast of Panama. *Pacific Seabird Group Bulletin*, 9(2):69. Marine and other colonial waterbirds on 80 islands in the Pacific waters of Panama. Some comparisons between species in the Bay of Panama and the Gulf of Chiriqui.

Moore, Clara Alma. 1948. Choosing a Bird for Special Study on Barro Colorado Island. *Year Book Indiana Audubon Society*, 26:24-26, 2 plates. Impressions of the birds of Barro Colorado Island.

Moore, Owen R., Bridget J. M. Stutchbury, and James S. Quinn. 1999. Extrapair Mating System of an Asynchronously Breeding Tropical Songbird: The Mangrove Swallow. *Auk*, 116(4):1039-1046, 4 figures, 3 tables. Observations of *Tachycineta albilinea* on Barro Colorado Island in 1996.

Moore, Robert T. 1944. Nesting of the Brown-Capped Leptopogon in Mexico. *Condor*, 46(1):6-8. *Leptopogon amaurocephalus.*

Moore, Robert T. 1950. Northern Record for *Klais guimeti* in Central America. *Auk*, 67(3):387. Mentions the name *Mellisuga merrettii* that was proposed by Lawrence.

Moore, Robert T., and James L. Peters. 1939. The Genus *Otus* of Mexico and Central America. *Auk*, 56(1):38-56. Specimens of *Otus guatemalae vermiculatus*, *O. choliba luctisonus* and *O. clarkii* from Boquete, Calobre, Calovevora, Cana, Chitra, Divala, and San Miguel Island.

Morrison, R. I. Guy, Robert W. Butler, Francisco S. Delgado B., and R. Kenyon Ross. 1998. *Atlas of Nearctic Shorebirds and Other Waterbirds on the Coast of Panama.* 112 pages, 3 appendices, 36 maps, 10 photos, and 34 tables. Ottawa: Canadian Wildlife Service. Aerial surveys of the coasts of Panama conducted in Feb 1988, Oct 1991 and Jan 1993.

Morton, Eugene S. 1971. Food and Migration Habits of the Eastern Kingbird in Panama. *Auk*, 88(4):925-926. *Tyrannus tyrannus* in Panama Canal Zone.

Morton, Eugene S. 1976. Vocal Mimicry in the Thick-Billed Euphonia. *Wilson Bulletin*, 88(3):485-487, 1 figure. Observations at Summit Gardens, Madden Lake Boy Scout Camp, and Gamboa in 1970, 1971, and 1974. Reports *Euphonia laniirostris* using "alarm" notes of other species.

Morton, Eugene S. 1979. Effective Pollination of *Erythrina fusca* by the Orchard Oriole (*Icterus spurius*): Coevolved Behavioral Manipulation? *Annals of the Missouri Botanical Garden*, 66(3):482-489, 1 figure, 2 tables. Studies in the Canal Zone suggest that the flower color of *E. fusca* in Panama matches that of the male orchard oriole only once it has been opened and drained of nectar as an aggressive signal that sends the male to other trees thereby assisting with cross pollination. Some comparisons with other regions that lack orchard orioles. Pedro Miguel, Jan, Mar 1974, and Mar 1975.

Morton, Eugene S. 1980. Adaptations to Seasonal Changes by Migrant Land Birds in the Panama Canal Zone. Pages 437-453, 3 figures, 5 tables *in* Allen Keast and Eugene S. Morton, editors. *Migrant Birds in the Neotropics. Ecology, Behavior, Distribution and Conservation.* Washington, D.C.: Smithsonian Institution Press. Results of research conducted from 1973-1975 at 13 study areas: Achiote Road, Barro Colorado Island, Chiva Chiva Road, Contractor's Hill and Empire Range, Curundu, Fort Kobbe (Far Fan Beach), Fort Sherman, Galeta Island, K-6 Road, Madden Forest Preserve, Madden Lake Boy Scout Camp, Pipeline Road, and Summit Gardens (GH12 Road).

Morton, Eugene S. 1982. Adaptations for Tropical Survival in Wintering North American Birds. *National Geographic Society Research Reports*, 14:491-498, 1 figure. Map of study sites. Observations and specimen collections from Fort Kobbe, Fort Kobbe mangrove island, Far Fan, Balboa, Morgan's Gardens, Summit Gardens-Madden Forest, Gamboa-Pipeline Road, Madden Lake, Gatun, Fort Sherman, and Galeta mangroves. More details are presented in Morton (1980).

Morton, Eugene S., and Kim C. Derrickson. 1996. Song Ranging by the Dusky Antbird, *Cercomacra tyrannina*: Ranging Without Song Learning. *Behavioral Ecology and Sociobiology*, 39(3):195-201, 5 figures. Playback of degraded and undegraded songs to 14 pairs near Gamboa, 1991.

Morton, Eugene S., and Susan M. Farabaugh. 1979. Infanticide and Other Adaptations of the Nestling Striped Cuckoo *Tapera naevia*. *Ibis*, 121(2):212-213, 1 plate. Observations in the Panama Canal found that 4 of 11 *Thryothorus rufalbus* nests were parasitized by *T. naevia*. Adaptations of the cuckoo included hooked bill, call mimicry, and plumage displays that concealed the bird's actual size.

Morton, Eugene S., and Bridget J. M. Stutchbury. 2000. Demography and Reproductive Success in the Dusky Antbird, a Sedentary Tropical Passerine. *Journal of Field Ornithology*, 71(3):493-500, 2 figures, 1 table. Observations and banding of *Cercomacra tyrannina* in Parque Nacional Soberanía during the dry seasons of 1991-1998.

Moynihan, Martin. 1962a. The Organization and Probable Evolution of Some Mixed Species Flocks of Neotropical Birds. *Smithsonian Miscellaneous Collections*, 143(7):1-140, 3 diagrams, 45 tables. Based on observations made in 1957-1960 in the Canal Zone and at Volcan de Chiriquí.

Moynihan, Martin. 1962b. Display Patterns of Tropical American "Nine-Primaried" Songbirds. II. Some Species of *Ramphocelus*. *Auk*, 79(4):655-686, 6 figures. Observations at Barro Colorado Island and on captive birds.

Moynihan, Martin. 1963. Display Patterns of Tropical American "Nine-Primaried" Songbirds. III. The Green-Backed Sparrow. *Auk*, 80(2):116-144, 6 figures. Observations of *Arremonops conirostris* on Barro Colorado Island and near Frijoles.

Murphy, Robert Cushman. 1936. *Oceanic Birds of South America: A Study of Species of the Related Coast and Seas, Including the American Quadrant of Antarctica Based Upon the Brewster-Sanford Collection in the American Museum of Natural History*. 2 volumes. xx + 1245 pages, 16 colored plates (unnumbered), 72 plates (numbered), 80 figures. New York: MacMillan Company (for American Museum of Natural History). Many of the illustrations are from paintings by Francis L. Jaques. Also includes photographs, maps and drawings.

Murphy, Robert Cushman. 1938a. Dark Skies—A Scientific Voyage in a 38-Foot Long Boat Along the Rainy Pacific Shores of Colombia, the Least Known Continental Seacoast in the World. *Natural History*, 41(3):165-178, 231, 1 map, 21 photographs. Voyage began at Balboa and passed along the coast of Darién.

Murphy, Robert Cushman. 1938b. The Wandering Albatross in the Bay of Panama. *Condor*, 40(3):126, 1 figure. First record of *Diomedea exulans* for the area.

Murphy, Robert Cushman. 1944a. To the Chocó in the Schooner "Askoy." *Natural History*, 53(5):200-208, 9 maps, 14 photographs, 1 sketch. General information about the trip. Does not mention specific localities, although the map indicates the course and stations of the schooner. The map includes coordinates, but the stations are simply numbered, not given specific names.

Murphy, Robert Cushman. 1944b. Among the Pearl Islands. The Second Installment in the Story of the "Askoy" Expedition. *Natural History*, 53(6):274-281, 1 map, 13 photographs. Descriptions of several of the islands.

Murphy, Robert Cushman. 1944c. To the Continental Shelf: The Third Installment in the Story of the "Askoy" Expedition. *Natural History*, 53(7):303-309, 14 photographs, 2 tables. Cruise across Bahía de Panamá from Bahía Piña to Punta Mala, spring 1941, with observations of seabirds including Galápagos Albatross *Diomedea* [*Phoebastria*] *irrorata.*

Murphy, Robert Cushman. 1945. Notes on Pearl Islands Ornithology. *Auk*, 62(1):114-117. Collections and observations by Cushman and Correia from the schooner *Askoy* in Feb and May 1941, on Pacheca, Saboga, La Vivienda, and El Rey Islands.

Murphy, Robert Cushman. 1956. The Vertebrates of SCOPE, November 7-December 16, Birds. Pages 121-140 *in* Robert W. Holmes, editor. *Physical, Chemical, and Biological Observations in the Eastern Tropical Pacific, November-December 1956.* La Jolla: Scripps Institute of Oceanography. Also published as *Fish and Wildlife Service Special Science Report*, Fisheries No. 279:101-110. Observations on birds were made in Panamanian waters in 1956 from Nov 25 to Dec 1.

Myers, Charles W. 1969. The Ecological Geography of Cloud Forest in Panama. *American Museum Novitates*, 2396:1-52, 19 figures. Includes information regarding, and a map depicting, the following 12 regions: Cerro Sapo; the Jaqué-Imamadó Divide; Serranía de Pirre; Cerros Tacarcuna and Malí; Camp Summit (Serranía del Darién); Serranía de Cañazas; Cerro Jefe; Cerro Bruja; Cerro Campana; Cerro Hoya; Cerro Santiago; and Cerro Pando.

Myers, Charles W. 2003. Rare Snakes—Five New Species from Eastern Panama: Reviews of Northern *Atractus* and Southern *Geophis* (Colubridae: Dipsadinae). *American Museum Novitates*, 3391:1-47, 2 maps, 20 figures, 3 tables. Coordinates for Mortí Hydro, which is also indicated on the map.

Myers, Charles W., and William E. Duellman. 1982. A New Species of *Hyla* from Cerro Colorado, and other Tree Frog Records and Geographical Notes from Western Panama. *American Museum Novitates*, 2752:1-32, 25 figures, 2 tables. Coordinates for Cerro Colorado and notes on Cerro Santiago based on aerial surveys of the continental divide.

Navarro S., Adolfo G., A. Townsend Peterson, Esteban Lopez-Medrano, and Hesiquio Benitez-Diaz. 2001. Species Limits in Mesoamerican *Aulacorhynchus* Toucanets. *Wilson Bulletin*, 113(4):363-372, 5 figures. Examined and measured 305 specimens of *Aulacorhynchus* toucanets from Central and South America. Examined specimens of *A. cognatus* and *A. caeruleogularis* from eastern and western Panama, respectively, without mention of specific locality data or individual specimen data.

Nehrkorn, Adolph. 1899. *Katalog der Eiersammlung; Nebst Beschreibungen der Aussereuropaischen Eier.* [*Catalogue of the Egg Collection; Together with Descriptions of Eggs from Outside of Europe*]. viii + 256 pages, 4 colored plates. Braunschweig: Harald Bruhn (Verlagsbuchhandlung für Naturwissenschaft und Medicin). In German.

Nehrkorn, Adolph. 1901. *Eugenes spectabilis chiriquensis* Nehrk. n. subsp. *Ornithologische Monatsberichte*, 9(9):132-133. From Chiriquí. In German.

Nelson, E. W. 1904. A Revision of the North American Mainland Species of *Myiarchus*. *Proceedings of the Biological Society of Washington*, 17:21-50. *Myiarchus lawrencei bangsi* n. subsp. from Boquete.

Nelson, E. W. 1911a. Description of a New Genus and Species of Hummingbird from Panama. *Smithsonian Miscellaneous Collections*, 56(21):1-2. *Goldmania violiceps*, n. gen. et sp., from Chepo.

Nelson, E. W. 1911b. Two New Subspecies of Birds from Panama. *Smithsonian Miscellaneous Collections*, 56(22):1. *Hylomanes momotula obscurus* n. subsp. from Cerro Brujo [= Cerro Bruja], and *Pachysylvia ochraceiceps brevipennis* n. subsp. from Lion Hill.

Nelson, E. W. 1912a. Descriptions of Two New Species of Nun Birds from Panama. *Smithsonian Miscellaneous Collections*, 56(37):1-2. *Monasa fidelis* and *M. similis*, n. spp., from Cerro Azul.

Nelson, E. W. 1912b. Descriptions of New Genera, Species and Subspecies of Birds from Panama, Colombia and Ecuador. *Smithsonian Miscellaneous Collections*, 60(3):1-25. The following new taxa described from Panama from Goldman's collections at Cana, Marraganti, and Mt. Pirre: *Geotrygon goldmani*, *Chloronerpes chrysochlorus aurosus*, *Aulacorhamphus coeruleigularis cognatus*, *Momotus conexus reconditus*, *Electron platyrhynchus suboles*, *Goethalsia bella*, *Eriocnemus floccus*, *Phaethornis adolphei fraterculus*, *Thamnistes anabatinus coronatus*, *Dysithamnus mentalis suffusus*, *Herpsilochmus rufimarginatus exiguus*, *Grallaricula flavirostris brevis*, *Margarornis bellulus*, *Mitrephanes eminulus*, *Praedo audax*, *Caryothraustes canadensis simulans*, *Tangara fucosus* [sic], *Hylospingus inornatus*, *Chrysothlypis chrysomelas ocularis*, *Vireolanius eximius mutabilis*, *Basileuterus melanogenys ignotus*, *Troglodytes festinus*, *Myadestes coloratus*, and *Catharus fuscater mirabilis*. *Goethalsia*, *Praedo*, and *Hylospingus* are new genera. *Basileuterus melanogenys eximius*, n subsp., from Boquete.

Nelson, E. W. 1913a. Two New Subspecies of Birds from the Slopes of Mount Pirri, Eastern Panama. *Smithsonian Miscellaneous Collections*, 60(21):1-2. *Capito maculicoronatus pirrensis* n. subsp. from Cana, and *Pseudotriccus pelzelni berlepschi* n. subsp. from Mt. Pirre.

Nelson, E. W. 1913b. A New Subspecies of Nun Bird from Panama. *Proceedings of the Biological Society of Washington*, 26:67-68. *Monasa pallescens minor* n. subsp. from Marraganti.

Neudorf, Diane L., and Paul J. Blanchfield. 1995. The Slate-Colored Seedeater (*Sporophila schistacea*): A Bamboo Specialist? *Ornitologia Neotropical*, 5(2):129-132, 1 figure. Surveys and mist netting along Pipeline Road in Feb 1994.

Noble, G. K. 1916. The Resident Birds of Guadeloupe. *Bulletin of the Museum of Comparative Zoology*, 60(10):359-396.

Nolan, Val, Jr. 1962. The Swaying Display of the Red-Eyed and Other Vireos. *Condor*, 64(4):273-276. *Vireo olivaceus.*

Norris, Joseph Parker, Jr. 1926. A Catalogue of Sets of Accipitres' Eggs in the Collection of Joseph Parker Norris, Jr., Philadelphia, Pa., U.S.A. *Oologists' Record*, 6(2):25-41.

Norris, Joseph Parker, Jr. 1927. Nesting of the Harpy Eagle, *Thrasaëtus harpyia* in Brazil. *Oologists' Record*, 7(2):25-26.

Nottebom, Fernando, and Robert K. Selander. 1972. Vocal Dialects and Gene Frequencies in the Chingolo Sparrow (*Zonotrichia capensis*). *Condor*, 74(2):137-143, 4 figures, 3 tables.

Oates, Eugene W. 1901. *Catalogue of the Collection of Birds' Eggs in the British Museum (Natural History). Vol. I. Ratitae, Carinatae (Tinamiformes-Lariformes)*. xxiii + 252 pages, 18 colored plates. London: Trustees of the British Museum (Natural History).

Oates, Eugene W. 1902. *Catalogue of the Collection of Birds' Eggs in the British Museum (Natural History). Vol. II. Carinatae (Charadriiformes-Strigiformes)*. xx + 400 pages, 15 colored plates. London: Trustees of the British Museum (Natural History).

Oates, Eugene W., and Savile G. Reid. 1903. *Catalogue of the Collection of Birds' Eggs in the British*

Museum (Natural History). Vol. III. Carinatae (Psittaciformes-Passeriformes). xxiii + 349 pages, 10 colored plates. London: Trustees of the British Museum (Natural History).

Oates, Eugene W., and Savile G. Reid. 1905. *Catalogue of the Collection of Birds' Eggs in the British Museum (Natural History). Vol. IV. (Passeriformes continued).* xviii + 352, 14 colored plates. London: Trustees of the British Museum (Natural History).

Oberholser, Harry Church. 1904. A Monograph of the Genus *Dendrocincla* Gray. *Proceedings of the Academy of Natural Sciences of Philadelphia*, 56(2):447-463, 6 tables. *D. homochroa acedesta*, n subsp., from Chiriquí. Specimens from Boquete, Chiriquí, El Banco, Panama Railroad, Panama, and Veragua.

Oberholser, Harry Church. 1912. A Revision of the Subspecies of the Green Heron (*Butorides virescens* [Linnaeus]). *Proceedings of the United States National Museum*, 42:529-577, tables. *B. v. margaritophilus* n. subsp. from San Miguel Island.

Oberholser, Harry Church. 1913. A Revision of the Forms of the Great Blue Heron (*Ardea herodias* Linnaeus). *Proceedings of the United States National Museum*, 43:531-559, tables.

Oberholser, Harry Church. 1914. A Monograph of the Genus Chordeiles Swainson, Type of a New Family of Goatsuckers. *United States National Museum Bulletin* 86: vii + 123 pages, 6 black and white plates, tables. Records from Panama. In our copy Wetmore changed the label of plate 2 from *Antrostomus carolinensis* to *Nyctidromus.*

Oberholser, Harry Church. 1932. Description of a New *Odontophorus* from Costa Rica. *Proceedings of the Biological Society of Washington*, 45:39-42.

Odum, Howard T. 1947. The Possible Effect of Cloud Cover on Bird Migration in Central America. *Auk*, 64(2):316-317, 1 figure, 1 table. Data on cloudiness at Balboa Heights, Canal Zone.

Oehser, Paul. 1980. In memoriam: Alexander Wetmore. *Auk*, 97(3): 608-615. [This contains a direct quote (p. 615) attributed to Olson. The first sentence of this quote is indeed from Olson (1976: xi), the second sentence is gibberish that must have been invented by Oehser, and the third part is badly misquoted from Ripley's Appreciation in the same volume (p. viii).]

Ogilvie-Grant, W. R. 1893. *Catalogue of the Birds in the British Museum.* Volume 22. xvi + 585 pages, 8 colored plates. London: Trustees of the British Museum (Natural History). *Crax panamensis* n. sp. from Panama and Lion Hill.

Ogilvie-Grant, W.R. 1912. *Catalogue of the Collection of Birds' Eggs in the British Museum (Natural History). Vol. V. Carinatae (Passeriformes completed).* xxiii + 547 pages, 22 colored plates. London: Trustees of the British Museum (Natural History).

Olson, Storrs L. 1965. Notes on Ant-Tanagers in Panama. *Bird-Banding*, 36(2):113. Distraction display in *Habia fuscicauda* at Albrook Field. Immature plumaged *H. rubica* with enlarged testes collected at Curundu.

Olson, Storrs L. 1966. A Partly Chestnut Specimen of Variable Seedeater. *Wilson Bulletin*, 78(1):127. *Sporophila aurita* on the Navy Pipeline Road near Gamboa.

Olson, Storrs L. 1968. Possible Breeding of Some Seedeaters in Subadult Plumage. *Florida Naturalist*, 41(1):20. Observations on *Sporophila aurita*, *S. minuta*, *Oryzoborus funereus* from Curundu and Miraflores Locks in the former Canal Zone.

Olson, Storrs L. 1976. Alexander Wetmore and the study of fossil birds. Publications in avian paleontology by Alexander Wetmore. Index to fossil avian taxa described by Alexander Wetmore. Pages xi-xxvi, 1 figure. *In* Storrs L. Olson, editor. Collected papers in avian paleontology honoring the 90th birthday of Alexander Wetmore. *Smithsonian Contributions to Paleobiology*, 27.

Olson, Storrs L. 1980a. Revision of the Tawny-Faced Antwren, *Microbates cinereiventris* (Aves: Passeriformes). *Proceedings of the Biological Society of Washington*, 93(1):68-74, 1 figure. Includes systematics and distribution of subspecies in Panama.

Olson, Storrs L. 1980b. Geographic Variation in the Yellow Warblers (*Dendroica petechia*: Parulidae) of the Pacific Coast of Middle and South America. *Proceedings of the Biological Society of Washington*, 93(2):473-480, 1 figure. Mainly a revision of the populations found in Panama. *Dendroica petechia aithocorys* n. subsp. from La Honda, *D. p. iguanae* n. subsp. from Isla Iguana, and *D. p. jubaris* n. subsp from Colombia, with range extending to E Darién.

Olson, Storrs L. 1980c. The Subspecies of Grasshopper Sparrow (*Ammodramus savannarum*) in Panamá (Aves: Emberizinae). *Proceedings of the Biological Society of Washington*, 93(3):757-759. *A. s. pratensis* a rare migrant, *A. s. bimaculatus* widely disjunct in Chiriqui and eastern Panamá Province, and *A. s. beatriceae* n. subsp. from El Coco, Coclé.

Olson, Storrs L. 1981a. A Revision of the Subspecies of *Sporophila* ("*Oryzoborus*") *angolensis* (Aves: Emberizinae). *Proceedings of the Biological Society of Washington*, 94(1):43-51, 1 figure, 1 table. Merges the genus *Oryzoborus* with *Sporophila* and the species *funerea* with *angolensis*. Discussion of races in Panama. *S. a. ochrogyne* n. subsp. from Utivé, Río Cabobré [= Río Utivé].

Olson, Storrs L. 1981b. A Revision of the Northern Forms of *Euphonia xanthogaster* (Aves: Thraupidae). *Proceedings of the Biological Society of Washington*, 94(1):101-106, 1 figure. Birds from Darién referred to as *E. x. oressinoma* n. subsp. from Andes of Colombia.

Olson, Storrs L. 1981c. Systematic Notes on Certain Oscines from Panama and Adjacent Areas. *Proceedings of the Biological Society of Washington*, 94(2):363-373. *Vireo gilvus dissors* from Darién. Distribution of races of *Chlorospingus ophthalmicus*, *Chrysothlypis chrysomelas*, *Icterus chrysater* and *I. mesomelas* in Panama. *Zonotrichia capensis orestera* Wetmore from Cerro Campana synonymized with *Z. c. costaricensis*.

Olson, Storrs L. 1981d. The Nature of the Variability in the Variable Seedeater in Panama (*Sporophila americana*: Emberizinae). *Proceedings of the Biological Society of Washington*, 94(2):380-390, 1 figure. Revives *S. a. hicksii* Lawrence 1865 (revised type-locality Buenaventura, Colombia) for birds of the Pacific coast from Azuero Peninsula into Colombia, with *S. a. chocoana* as a synonym. Resurrects *S. a. hoffmani* Cabanis 1861 for birds of the Pacific coast in western Panama and Costa Rica. The name I. "*aurita*" is applied to hybrid *corvina* x *hicksii* in the Canal area. Distribution of *corvina*, *hicksii*, *hoffmani*, and intergrades indicated on a map.

Olson, Storrs L. 1983. The Distribution of the Races of *Arremon aurantiirostris* (Emberizinae) in Panama. *Bulletin of the British Ornithologists' Club*, 103(3):84-86. Lectotype designated for *A. aurantiirostris* Lafresnaye and characters and distribution of the races in Panama redefined.

Olson, Storrs L. 1993a. Contributions to Avian Biogeography from the Archipelago and Lowlands of Bocas del Toro, Panama. *Auk*, 110(1):100-108, 3 figures. New distributional records for the province resulting from the SBE.

Olson, Storrs L. 1993b. Intergradation Between the Bush-Tanagers *Chlorospingus punctulatus* and *C. opthalmicus* in Western Panama (Aves: Thraupidae). *Auk*, 110(1):148-150, 1 figure.

Olson, Storrs L. 1996. History and ornithological journals of the *Tanager* Expedition of 1923 to the Northwestern Hawaiian Islands, Johnston and Wake Islands. *Atoll Research Bulletin*, 433: 210 pages, 49 figures, 5 tables.

Olson, Storrs L. 1997a. Avian Biogeography in the Islands of the Pacific Coast of Western Panama. Pages 69-82, 1 figure, 3 tables *in* Robert W. Dickerman, compiler. *The Era of Allan R. Phillips: A Festschrift*. Albuquerque: Horizon Communications. Species lists and locality descriptions for the following 13 islands of the newly coined Veragua Archipelago: Afuerita, Bolaños,

Brincanco, Canal de Afuera, Cébaco, Coiba, Farallón del Chirú, Frailes del Sur, Gobernadora, Iguana, Parida, Ranchería, and Villa. Based on specimen data and information from Alexander Wetmore's specimen catalogs and journals.

Olson, Storrs L. 1997b. In memorium: Beatrice Wetmore, 1910-1997. *Auk*, 114 (4): 751-753.

Olson, Storrs L. 2005. Refutation of the historical evidence for a Hispaniolan macaw (Aves: Psittacidae: *Ara*). *Caribbean Journal of Science*, 49 (2): 319-323. Mentions Oviedo's observations on parrots of Tierra Firme in the Darién region in the early 16th century.

Olson, Storrs L. 2007. A new subspecies of *Sporophila angolensis* (Aves: Emberizidae) from Isla de Coiba and other additions to the avifauna of the Veragua Archipelago, Panama. *Proceedings of the Biological Society of Washington*.120(4): 377-381. *Sporophila angolensis fractor*, new subspecies, characterized by a larger bill than *S. a. ochrogyne*, is described from Isla de Coiba. The Isla de Coiba subspecies *Contopus cinereus aithalodes* is reported from Isla Jicarón and *Elaenia flavogaster* is new for Isla Brincanco.

Olson, Storrs L. 2007. A Carriker trilogy: chapters in a saga of neotropical ornithology. [Review of 3 books]. *Auk*, 124(1): 357-361.

Olson, Storrs L. 2008. Falsified Data Associated with Specimens of Birds, Mammals, and Insects from the Veragua Archipelago, Panama, Collected by J. H. Batty. *American Museum Novitates*, 0000. Provides detailed documentation showing that all of the specimens labeled by Batty as coming from the smaller islands of the Veragua Archipelago are fraudulent and came from mainland localities. Although some of Batty's specimens from Isla de Coiba are genuine, others are also fraudulent and came from the mainland.

Olson, Storrs L., and Kurt E. Blum. 1968. Avian Dispersal of Plants in Panama. *Ecology*, 49(3):565-566, 1 table. Seeds germinated from various portions of the digestive tract and feces of birds collected in Panama in 1966.

Olson, Storrs L., and Eugene Eisenmann. 1966. Notes on Breeding of the Common Nighthawk in Panama. *Auk*, 83(3):469-470. Female collected with two eggs on outskirts of Panama City.

Olson, Storrs L., Horace Loftin, and John [Joachim] Wiese. 1967. Observations on the Behavior of Black and Turkey Vultures at Traps and in Captivity. *Bird-Banding*, 38(1):75-76. Includes observations of *Coragyps atratus* and *Cathartes aura* at Pedro Miguel Dump and at Panama City and a report of competition between vultures and and Brown Pelicans *Pelecanus occidentalis* for offal.

Olson, Storrs L., George V. N. Powell, and Eugene Eisenmann. 1968. Distributional Records from Cerro Campana, Panama, with Notes on a Nesting of the Quail-Dove, *Geotrygon lawrencii*. *Condor*, 70(2):179-180. Discussion of physiography and zoogeographic significance of Cerro Campana with notes on *Claravis mondetoura*, *Eutoxeres aquila*, and *Trogon aurantiiventris* in addition to the *Geotrygon*.

Olson, Storrs L., and Carlo Violani. 1995. Some Unusual Hybrids of *Ramphocelus*, With Remarks on Evolution in the Genus (Aves: Thraupinae). *Bollettino Museo Regionale di Scienze Naturali – Torino*, 13(1):297-312, 8 figures (photographs). *R. festae* from western Panamá is a hybrid between *R. passerinii* and *R. dimidiatus* rather than an atypical *R. passerinii*. *R. inexpectatus*, probably from Central Panama, is a backcross between *R. icteronotus* and a hybrid *R. icteronotus/dimidiatus*.

Oniki, Yoshika. 1972. Some Temperatures of Panamanian Birds. *Condor*, 74(2):209-215, 3 tables. Cloacal temperatures and body weights of birds netted at Barro Colorado Island (38 spp.), Limbo Hunt Camp (11 spp.), and Cerro Punta (25 spp.), along with data from a few captive birds.

Otis, Fessenden N. 1859. Tropical Journeyings. *Harper's New Monthly Magazine*, 18(104):145-169, illustrations. Describes the building of the Panama Railroad and gives detailed descriptions of his trip along the route. Mentions M'Clellan [sic] at Frijoles.

Oustalet, Jean-Frédéric Émile. 1883. [Description of *Dacnis viguieri* n. sp.]. Page 246, plate 15a, figure 3 *in* Osbert Salvin and Frederick Ducane Godman, 1879-1904, vol.1. Although often attributed to Salvin and Godman, the description and all information concerning the unique type in the Paris Museum was written entirely by Oustalet, who cited by Salvin and Godman as the author of the name *Dacnis viguieri.*

Oviedo y Valdés, G. Fernández de. 1851. *Historia General y Natural de las Indias.* Vol. 1. Madrid: Real Academia de la Historia. Mentions parrots and probably other birds of Tierra Firme in the Darién region in the early 16th century.

Parkes, Kenneth C. 1954a. A Revision of the Neotropical Finch *Atlapetes brunneinucha. Condor,* 56(3):129-138, 2 figures, 1 table. *A. b. elsae* n. subsp. from Boquete, Cerro Campana, and Mt. Chiriquí, and *A. b. frontalis* n. subsp. from Mt. Pirre, and Mt. Tacarcuna.

Parkes, Kenneth C. 1954b. The Generic Name of the Rice Grackle. *Condor,* 56(6):229. *Scaphidura* Swainson, 1837, replaces *Psomocolax* Peters, 1929, for the Giant Cowbird, which becomes *Scaphidura oryzivora.*

Parkes, Kenneth C. 1955. Systematic Notes on North American Birds. 1. The Herons and Ibises (*Ciconiiformes*). *Annals of the Carnegie Museum,* 33:287-293.

Parkes, Kenneth C. 1967. Prealternate Molt in the Summer Tanager. *Wilson Bulletin,* 79(4):456-458. *Piranga rubra.*

Parkes, Kenneth C. 1969. The Red-Throated Ant-Tanager (*Habia fuscicauda*) in Panama and Colombia. *Proceedings of the Biological Society of Washington,* 82:233-242. *Habia fuscicauda willisi* n. subsp. from Boca del Río Indio.

Parkes, Kenneth C. 1976. Geographic Variation in the Common Tody-Flycatcher (*Todirostrum cinereum*) with Special Reference to Middle America. *Nemouria,* (18):1-9. *T. c. wetmorei* n. subsp. from San Félix.

Parkes, Kenneth C, and Emmit R. Blake. 1965. Taxonomy and Nomenclature of the Bronzed Cowbird. *Fieldiana: Zoology,* 44(22):207-216. *Molothrus aeneus.*

Parsons, Thomas J., Storrs L. Olson, and Michael J. Braun. 1993. Unidirectional Spread of Secondary Sexual Plumage Traits Across an Avian Hybrid Zone. *Science,* 260(5114):1643-1646, 2 figures, 2 tables. Examines the collar (white, golden, intermediate) of male *Manacus* at the following 10 localities: Guápiles (CR), Río Sixaola, Río Teribe, Río Changuinola, Río Oeste, Quebrada Pastores, Tierra Oscura, Chiriquí Grande, Valiente Peninsula, and the Panama Canal.

Payne, Robert B. 1974. Species Limits and Variation of the New World Green Herons *Butorides virescens* and Striated Herons *B. striatus. Bulletin of the British Ornithologists' Club,* 94(2):81-88, 2 figures, 1 table. Examined 837 New World and 500 Old World *Butorides* specimens from various museums. Discussion of neck color in Panama specifically mentioning USNM specimens collected by Wetmore from Almirante, Charco del Toro, La Jagua, Pesé, Río Indio (Colón/Panama/Coclé), and by Goldman from Río Indio (Canal Zone).

Paynter, Raymond A., and Miguel Alvarez del Toro. 1957. Blue and White Swallow in México. *Condor,* 59(4):268. *Notiochelidon cyanoleuca.*

Peck, Morton E. 1910. The Effect of Natural Enemies on the Nesting Habits of Some British Honduras Birds. *Condor,* 12(2):53-60, 4 figures, 2 tables.

Pelzeln, August von. 1871. *Zur Ornithologie Brasiliens. Resultate von Johann Natterers Reisen in den Jahren 1817 bis 1835.* 462 pages. Wien: A. Pichler's Witwe and Sohn. Originally issued in 4 parts from 1868 to 1870.

Penard, Frederick Paul, and Arthur Philip Penard. 1908. *De Vogels van Guyana (Surinam, Cayenne en Demerara).* xliii + 587 pages, numerous figures. Paramaribo: Penard and Penard. Published in two parts.

Pennycuick, Colin J. 1983. Thermal Soaring Compared in Three Dissimilar Tropical Bird Species, *Fregata magnificens*, *Pelecanus occidentalis* and *Coragyps atratus*. *Journal of Experimental Biology*, 102:307-325, 10 figures, 6 tables. Observations from Flamenco Island. Mentions Perico, Taboga, Taboguilla, and Uravá Islands.

Peters, James Lee. 1927. A New Manakin from Panama. *Proceedings of the New England Zoological Club*, 10:9-10. *Manacus cerritus* n. sp. from Almirante.

Peters, James Lee. 1929. An Ornithological Survey in the Caribbean Lowlands of Honduras. *Bulletin of the Museum of Comparative Zoology*, 69(12):397-478. *Glyphorynchus spirurus sublestus* n. subsp. from Changuinola, and *Tanagra gouldi praetermissa* n. subsp. from Western River.

Peters, James Lee. 1930. Two Undescribed Races of *Phaëthon aethereus*. *Occasional Papers of the Boston Society of Natural History*, 5:261-262. *P. a. mesonauta* n. subsp. from Swan Key.

Peters, James Lee. 1931a. Additional Notes on the Birds of the Almirante Bay Region of Panama. *Bulletin of the Museum of Comparative Zoology*, 71(5):293-345. Summary account of the ornithology of this area based on a collection made by Wedel (1926-1929), along with those of Kennard and Smith (1926), Benson (1927), and a few specimens collected by H. S. Blair, of the United Fruit Co. Descriptions of many of Wedel's collecting localities, although a few of those mentioned are omitted. *Leptotila plumbeiceps notius* n. subsp. from Almirante.

Peters, James Lee. 1931b. *Check-List of Birds of the World.* Volume 1 [1st Edition]. xviii + 345 pages. Cambridge: Harvard University Press.

Peters, James Lee. 1934a. *Check-List of Birds of the World.* Volume 2. xvii + 401 pages. Megapodiidae to Alcidae. Cambridge: Harvard University Press.

Peters, James Lee. 1934b. The Classification of Some American Pigeons. *Condor*, 36(5):213-215.

Peters, James Lee. 1935. The Range of *Amazona oratrix*. *Auk*, 52(3):449-450. Includes *Amazona ochrocephala panamensis* with *A. oratrix* and *A. ochrocephala* being conspecific.

Peters, James Lee. 1937. *Check-List of Birds of the World.* Volume 3. xiii + 311 pages. Cambridge: Harvard University Press. Pteroclididae to Psittacidae.

Peters, James Lee. 1938. Systematic Position of the Genus *Ciccaba* Wagler. *Auk*, 55(2):179-186, tables. Notes on external ear openings with measurements of several species of related owls.

Peters, James Lee. 1940. *Check-List of Birds of the World.* Volume 4. xii + 291 pages. Cambridge: Harvard University Press. Musophagidae to Hemiprocnidae.

Peters, James Lee. 1943. First Supplement to the List of Types of Birds Now in the Museum of Comparative Zoölogy. *Bulletin of the Museum of Comparative Zoology*, 92:(2):53-118.

Peters, James Lee. 1945. *Check-List of Birds of the World.* Volume 5. xii + 306 pages. Cambridge: Harvard University Press. Trochilidae to Bucerotidae.

Peters, James Lee. 1948. *Check-List of Birds of the World.* Volume 6. xii + 259 pages. Cambridge: Harvard University Press. Galbulidae to Picidae.

Peters, James Lee. 1951. *Check-List of Birds of the World.* Volume 7. x + 318 pages. Cambridge: Harvard University Press. Eurylaimidae to Rhinocryptidae.

Peters, James Lee, and Ludlow Griscom. 1929. The Central American Races of *Rhypornis magnirostris*. *Proceedings of the New England Zoological Club*, 11:43-48. *R. m. arguta* n. subsp. from Almirante, and *R. m. alia* n. subsp. from San Miguel, El Rey. Veragua designated as type locality of *R. m. ruficauda* Sclater and Salvin.

Petit, Lisa J., Daniel R. Petit, Daniel G. Christian, and Hugh D. W. Powell. 1999. Bird Communities of Natural and Modified Habitats in Panama. *Ecography*, 22(3):292-304, 5 figures, 1 table. Surveyed migratory and resident birds from Dec 1993-Feb 1994 in 11 different habitat types ranging from extensive tracts of undisturbed forest to areas of intensive agriculture (grazed and agricultural land). Central Panama from Cerro Azul and Cerro Jefe west to Santa Fé including study sites near Arraijan and Río Hato.

Phelps, William H., and William H. Phelps, Jr. 1951. Four New Venezuelan Birds. *Proceedings of the Biological Society of Washington*, 64:65-72, 1 map.

Phillips, Allan R. 1961. Notas Sistematicas Sobre Aves Mexicanas. I. [Systematic Notes on Mexican Birds. I.] *Anales del Instituto de Biologia*, 32:333-381, 1 figure. In Spanish with English summary.

Phillips, Allan R. 1969. An Ornithological Comedy of Errors: *Catharus occidentalis* and *C. frantzii*. *Auk*, 86(4):605-623, 2 figures, 1 table. *C. f. wetmorei* n. subsp. from Boquete.

Phillips, Allan R. 1986. *The Known Birds of North and Middle America. Distributions and Variation, Migrations, Changes, Hybrids, etc.* Part I. Hirundinidae to Mimidae; Certhiidae. With the cooperation of Robert W. Dickerman, Amadeo M. Rea, and J. Dan Webster. 1xi + 259 pages, 4 appendices, colored frontispiece, 1 colored plate, figures, tables. Denver: Allan R. Phillips (Denver Museum of Natural History).

Phillips, Allan R. 1991. *The Known Birds of North and Middle America. Distributions and Variation, Migrations, Changes, Hybrids, etc.* Part II. Bombycillidae; Sylviidae to Sturnidae; Vireonidae. With the cooperation of Daniel D. Gibson, Kenneth C. Parkes, Mario A. Ramos, and Amadeo M. Rea. liii + 249 pages, colored frontispiece, 4 colored plates, figures. Denver: Allan R. Phillips.

Phillips, Allan R., and Robert W. Dickerman. 1965. A New Subspecies of *Icterus prosthemelas* from Panama and Costa Rica. *Wilson Bulletin*, 77(3):298-299. *I. p. praecox* n. subsp. from Almirante.

Phillips, John C. 1922. *A Natural History of the Ducks.* Volume 1: Plectropterinae, Dendrocygninae, Anatinae (in part). xii + 264 pages, 18 colored plates, 27 maps. Boston: Houghton Mifflin Company. Plates are based on drawings by Frank W. Benson, Allan Brook, Louis Agassiz Fuertes, and others. The complete work is four volumes.

Pinto, Oliverio M. de Oliveira. 1938. *Catalogo das Aves do Brasil, e Lista dos Exemplares Que as Representam no Museu Paulista.* Parte 1. Aves não Passeriformes, e Passeriformes não Oscines, excluida a Fam. Tyrannidae e Seguintes. [Catalog of the Birds of Brazil, and a List of the Specimens of Them Represented in the Museu Paulista. Part 1. Non Passeriformes, and Passeriformes non-Oscines, excluding the Fam. Tyrannidae and Onward]. xviii + 566 pages. São Paulo: Museo Paulista. Index of common and scientific names.

Pinto, Oliverio M. de Oliveira. 1944. *Catalogo das Aves do Brasil e Lista dos Exemplares Existentes na Coleção do Departamento de Zoologia.* Parte 2. Ordem Passeriformes (continuação): Superfamília Tyrannoidea e Subordem Passeres. [Catalog of the Birds of Brazil, and a List of the Existing Specimens in the Collection of the Department of Zoology. Part 2. Order Passeriformes (continued): Superfamily Tyrannoidea and Suborder Passeres)]. xii + 700 pages, 1 colored frontispiece, 4 colored plates, 10 black and white plates. São Paulo: Departamento de Zoologia, Secretaria da Agricultura, Indústria e Comércio. Index of common and scientific names.

Pinto, Oliverio M. de Oliveira. 1947. Contribuição à Ornitologia do Baixo Amaxonas. Estudo Crítico de uma Coleção de Aves do Estado do Pará. [Contributions to the Ornithology to the

Lower Amazonas. A Critical Study of a Collection of Birds from the State of Pará]. *Arquivos de Zoologia do Estado de São Paulo*, 5(6):311-482, 4 black and white plates, maps, tables. Index of scientific names (p. 481-482). Describes a collection of birds obtained by Lass from 1932-1939. Locality descriptions. In Portuguese.

Pinto, Oliverio M. de Oliveira. 1966. *Estudo Crítico e Catálogo Remissivo das Aves do Território Federal de Roraima.* [Critical Study and Submitted Catalog of the Birds of the Federal Territory of Roraima]. 176 pages. Manaus: Presidencia de Republica, Conselho Nacional de Pesquisa, Instituto Nacional de Pesquisas da Amazonia. Also published as *Instituto Nacional de Pesquisas Amazonia, Cadernos da Amazonia*, no. 8.

Pitelka, Frank A. 1950. Geographic Variation and the Species Problem in the Shore-Bird Genus *Limnodromus*. *University of California Publications in Zoology*, 50(1):1-108, 10 plates, 9 figures, 30 tables. Specimen of *Limnodromus* from Perme (p. 44, 49, 79).

Pitelka, Frank A. 1951. Race Names in the Central American Jay, *Cyanolyca argentigula*. *Journal of the Washington Academy of Sciences*, 41(3):113-115, 1 table. *Cyanolyca a. blandita* Bangs is a synonym of *C. a. argentigula* found in the Talamanca mountain range from Costa Rica to the Volcan de Chiriqui. A more northern form *C. a. albior* is described as new.

Pittier, Henry. 1912. Little-Known Parts of Panama. *The National Geographic Magazine*, 23(7):627-662, 1 map, 30 photographs.

Plótnick, Rubén. 1956. Posición Sistemática del Género *Heterospizias*. [The Systematic Position of the Genus *Heterospizias*]. *El Hornero*, 10(2):136-139, 18 figures.

Poulin, Brigitte, and Gaëtan Lefebvre. 1995. Additional Information on the Use of Tartar Emetic in Determining the Diet of Tropical Birds. *Condor*, 97(4):897-902, 1 table. Tested the effects of antimony potassium tartrate on over 2,500 birds (137 species) collected from Aug 1993-Nov 1994 near Gamboa in Soberanía National Park, Galeta, and Juan Díaz. The coordinates given appear to be erroneous and presumably should read Gamboa [0907/7942 (USBGN)], [Isla] Galeta [0924/7952 (USBGN)], and Juan Díaz [0903/7927 (USBGN)].

Poulin, Brigitte, and Gaëtan Lefebvre. 1996. Dietary Relationships of Migrant and Resident Birds from a Humid Forest in Central Panama. *Auk*, 113(2):277-287, 3 figures, 4 tables. Regurgitation samples collected from birds netted along Pipeline Road (between the Ríos Frijoles and Frijolito) indicate little overlap between the diets of migrants and of residents, 1993-1994. The coordinates given for Soberanía National Park [09°10' N, 79°07' W] appear to be erroneous and presumably should read Soberanía National Park [ca. 0907/7940] near Gamboa [0907/7942 (USBGN)].

Poulin, Brigitte, S. Joseph Wright, Gaëtan Lefebvre, and Osvaldo Calderón. 1999. Interspecific Synchrony and Asynchrony in the Fruiting Phenologies of Congeneric Bird-Dispersed Plants in Panama. *Journal of Tropical Ecology*, 15(2):213-227, 6 figures, 3 tables. Censuses on Barro Colorado Island, 1991-1993, and mist netting in Soberanía National Park, 1993-1994. The coordinates for Barro Colorado Island are correct but those for Soberanía National Park appear to be erroneous (see above).

Poulin, Brigitte, Gaëtan Lefebvre, Roberto Ibáñez, César Jaramillo, Carlos Hernández, and A. Stanley Rand. 2001. Avian Predation upon Lizards and Frogs in a Neotropical Forest Understorey. *Journal of Tropical Ecology*, 17(1):21-40, 7 figures, 1 table. Diet analysis via regurgitation samples of 91 species of birds in Soberania National Park, 1993-1994. The coordinates for Soberanía National Park appear to be erroneous (see above).

Pujals, Jaime J., John W. Wall, and David S. Wilcove. 1977. First Record of the Pearl Kite in Panama. *American Birds*, 31:1099-1100, photograph. Sighting of *Gampsonyx swainsonii* along

K-2 Road near Empire (firing) Range, Jun 1977.

Rabenold, Kerry N. 1980. The Black-Throated Green Warbler in Panama: Geographic and Seasonal Comparison of Foraging. Pages 297-307, 6 figures, 3 tables *in* Allen Keast and Eugene S. Morton, editors. *Migrant Birds in the Neotropics. Ecology, Behavior, Distribution and Conservation.* Washington, D.C.: Smithsonian Institution Press. Observations of *Dendroica virens* in Chiriquí province at Nueva Suiza, and along the Boquete Trail above Cerro Punta, Jan 1977. Foraging was more generalized in birds wintering in Panama than in birds observed in their breeding range, Maine and N. Carolina.

Raitt, Ralph J., and John William Hardy. 1970. Relationships Between Two Partly Sympatric Species of Thrushes (*Catharus*) in Mexico. *Auk*, 87(1):20-57, 15 figures, 3 tables. Field work in 1965 on *C. occidentalis* and *C frantzii.*

Rand, A. L., and M. A. Traylor, Jr. 1953. The Systematic Position of the Genera *Ramphocaenus* and *Microbates. Auk*, 70(3):334-337, 1 table. Relationship to African *Macrosphenus kempi* suggested.

Reed, Jonathan R. 1988. Inca Terns in the Bay of Panama During the 1982-1983 El Nino Event. *American Birds*, 42(2):172-173, 1 photograph. Observations and photographs of *Larosterna inca* and a swallow-tailed gull (*Creagrus furcatus*) in the Bay of Panama, between Islas Chepillo and Taboguilla, on San Jose Rock, and on Islas Pacheca and Pachequilla.

Remsen, J. V., Jr., F. Gary Stiles, and Peter E. Scott. 1986. Frequency of Arthropods in Stomachs of Tropical Hummingbirds. *Auk*, 103(2):436-441, appendix, 2 tables. Argues that tropical hummingbirds regularly include arthropods in their diets. Based on data collected by LSUMZ personnel from 1980-1985 and by Stiles et al. from 1971-1985 in Darién province and other localities in South and Central America.

Rendahl, Hialmar. 1919. Notes on a Collection of Birds from Panama, Costa Rica and Nicaragua. *Arkiv för Zoologi*, 12(8):1-36. Annotated list of collection made by Carl Bovallius in 1882, including 28 species from mainland Panama, where he made several stops from Feb-Jun.

Rendahl, Hialmar. 1920. A List of the Birds of the Pearl Islands, Bay of Panama. *Arkiv för Zoologi*, 13(4):1-56, 1 map. Based on a collection made by Carl Bovallius from Mar-May 1882, with additional species added from the literature.

Restrepo, Vicente. 1952. *Estudio Sobre Las Minas de Oro y Plata de Colombia.* [*Study of the Gold and Silver Mines of Colombia*]. 295 pages. Bogotá: Banco de la Repúbica. Contains a chapter on Panama and Darien. Published as volume 7 in the series *Arquivo de la Economía Nacional.* The original volume was published in 1888 and an English translation was prepared in 1886 by C.W. Fisher.

Richardson, Frank. 1942. Adaptive Modifications for Tree-Trunk Foraging in Birds. *University of California Publications in Zoology*, 46(4):317-368, 1 colored plate, 1 black and white plate, 16 figures, 10 tables.

Richmond, Charles W. 1893. Notes on a Collection of Birds from Eastern Nicaragua and the Rio Frio, Costa Rica, with a Description of a Supposed New Trogon. *Proceedings of the United States National Museum*, 16:479-532, tables. Use of "*Odontophorus rubigenis* Lawr." in such a way as to validate the name from this point. Panama specimens cited without specific locality.

Richmond, Charles W. 1899. New Name for the Genus *Tetragonops. Auk*, 16(1):77. *Pan* suggested as replacement for what is now *Semnornis.*

Richmond, Charles W. 1900. Some Necessary Changes in Nomenclature. *Auk*, 17(2):178-179. *Semnornis* to replace *Pan.*

Ricklefs, Robert E. 1971. Foraging Behavior of Mangrove Swallows at Barro Colorado Island.

Auk, 88(3):635-651, 5 figures, 8 tables. Observations at the nest of a pair of mangrove swallows (*Iridoprocne albilinea*) on Barro Colorado Island, 1968.

Ricklefs, Robert E. 1976. Growth Rates of Birds in the Humid New World Tropics. *Ibis*, 118(2):179-207, appendix, 11 figures, 11 tables. Measured growth (weight gain) from published studies and unpublished data of 40 species of lowland, neotropical birds from Panama, Trinidad, Surinam, and Colombia. Concludes that neotropical passerines (30 spp.) grow more slowly than temperate passerines and discusses various hypotheses.

Ricklefs, Robert E. 1977. Reactions of Some Panamanian Birds to Human Intrusion at the Nest. *Condor*, 79(3):376-379, 3 tables. Collected data on nests and nestlings of 17 passerine species and categorized the birds' reactions to the human observer.

Ricklefs, Robert E. 1980. 'Watch-Dog' Behaviour Observed at the Nest of a Cooperative Breeding Bird, the Rufous-Margined Flycatcher *Myiozetetes cayanensis*. *Ibis*, 122(1):116-118, 2 tables. Barro Colorado Island, 1968.

Ridgely, Robert S. 1989. Costa Rica, Panama, Trinidad, Brazil. *American Birds*, 43(4):1208-1210, 1 sketch, 1 photograph. Christmas counts for the region.

Ridgely, Robert S., and John A. Gwynne. 1989. *A Guide to the Birds of Panama with Costa Rica, Nicaragua, and Honduras*. 2nd edition, xvi + 534 pages, 48 colored plates, illustrations, tables. Princeton: Princeton University Press. Includes a section on finding birds that contains locality descriptions written by Engleman et al. (1989).

Ridgway, Robert. 1887. *A Manual of North American Birds*. xi + 631 pages, 124 plates. Philadelphia: J. B. Lippincott Company. *Piranga leucoptera latifasciata* n. subsp. from Costa Rica and Veragua.

Ridgway, Robert. 1888. Descriptions of Some New Species and Subspecies of Birds from Middle America. *Proceedings of the United States National Museum*, 10:505-510. *Dendrornis lawrencei* n. sp. from Panama.

Ridgway, Robert. 1901. New Birds of the Families Tanagridae and Icteridae. *Proceedings of the Washington Academy of Sciences*, 3:149-155. *Calospiza florida arcaei* n. subsp. from Veragua and *Ramphocelus dimidiatus isthmicus* n. subsp. from Frijole Station.

Ridgway, Robert. 1902a. Descriptions of Three New Birds of the Families Mniotiltidae and Corvidae. *Auk*, 19(1):69-70. *Compsothlypis pitiayumi speciosa* n. subsp. from Boquete.

Ridgway, Robert. 1902b. The Birds of North and Middle America: *United States National Museum Bulletin*, 50: Part II. xx + 834 pages, 22 plates. . Includes families: Tanagridae, Icteridae, Coerebidae, and Mniotiltidae. *Rhodinicichla rosea eximia* n. subsp. Isthmus of Panama and north to southern Costa Rica.

Ridgway, Robert. 1903a. Descriptions of New Genera, Species and Subspecies of American Birds. *Proceedings of the Biological Society of Washington*, 16:105-112. *Vireolanius pulchellus viridiceps* n. subsp. from Panama.

Ridgway, Robert. 1903b. Diagnosis of Nine New Forms of American Birds. *Proceedings of the Biological Society of Washington*, 16:167-170. *Cistothorus polyglottus lucidus* n. subsp. from Boquete.

Ridgway, Robert. 1906. Descriptions of Some New Forms of Oligomyodian Birds. *Proceedings of the Biological Society of Washington*, 19:115-120. *Scotothorus furvus* n. subsp. from Boquete de Chitra and *Lathria unirufa clara* n. subsp. from "(Lion Hill station ?)".

Ridgway, Robert. 1908. Diagnoses of Some New Forms of Neotropical Birds. *Proceedings of the Biological Society of Washington*, 21:191-196. *Myrmeciza boucardi panamensis* n. subsp. from Line of Panama Railroad, and *Formicarius moniliger panamensis* n. subsp. from "(Lion Hill station ?)".

Ridgway, Robert. 1909. New Genera, Species, and Subspecies of Formicariidae, Furnariidae, and

Dendrocolaptidae. *Proceedings of the Biological Society of Washington*, 22:69-74. *Synallaxis albescens hypoleuca* n. subsp. from Natá, and *Xiphorhynchus punctigula insolitus* n. subsp. from Cascajal.

Ridgway, Robert. 1911a. Diagnoses of Some New Forms of Picidae. *Proceedings of the Biological Society of Washington*, 24:31-36. *Veniliornis kirkii darienensis* n. subsp. from El Real and *Picumnus olivaceus panamensis* n. subsp. from Lion Hill Station.

Ridgway, Robert. 1911b. The Birds of North and Middle America. *United States National Museum Bulletin,* 50: Part V. xxiii + 859 pages, 33 plates. Includes the families: Pteroptochidae, Formicariidae, Furnariidae, Dendrocolaptidae, Trochilidae, Micropopidae, and Trogonidae. Wetmore (1952:19) indicates that Saboga Island (p. 347) is a typographical error for Taboga Island.

Ridgway, Robert. 1912. Descriptions of Some New Species and Subspecies of Birds from Tropical America. *Proceedings of the Biological Society of Washington*, 25:87-92. *Nyctibeus griseus panamensis* n. subsp. from Natá.

Ridgway, Robert. 1915. A New Pigeon from Chiriqui, Panama. *Proceedings of the Biological Society of Washington*, 28:139-140. *Oenoenas chiriquensis* n. sp. from Chiriqui.

Ripley, S. Dillon. 1945. The Barbets. *Auk*, 62(4):542-563, 2 figures. Includes *Capito maculicoronatus* and races.

Ripley, S. Dillon. 1952. The Thrushes. *Postilla*, 13:1-48, 1 figure, 1 table.

Ripley, S. Dillon, and James A. Steed. 1987. Alexander Wetmore. *National Academy of Sciences Biographical Memoirs*, 56: 597-626.

Robbins, Mark B., Theodore A. Parker III, and Susan E. Allen. 1985. The Avifauna of Cerro Pirre, Darien, Eastern Panama. Pages 198-232, 2 figures, 1 table, appendix *in* P. A. Buckley et al., editors. Neotropical Ornithology. *Ornithological Monographs* 36. s. Abstract in Spanish. History and description of the region followed by accounts for 25 species of little known birds. First account of the Purple-Throated Woodstar (*Calliphlox mitchellii*) in Central America. The appendix (p. 209-232) contains data from specimens collected by LSUMZ personnel in 1982.

Robbins, Mark B., and F. Gary Stiles. 1999. A New Species of Pygmy-Owl (Strigidae: *Glaucidium*) from the Pacific Slope of the Northern Andes. *Auk*, 116(2):305-315, 2 figures, 2 tables. Includes specimens of *G. costaricanum* from western Panama.

Robins, C. Richard. 1958. Observations on Oceanic Birds in the Gulf of Panama. *Condor*, 60(5):300-302. Sight records in the Gulf of Panama made from a fishing boat near the Pearl Islands and Piñas Bay, 1957 Jul 15-26.

Robinson, Tara Rodden, William Douglas Robinson, and E. Ceinwen Edwards. 2000. Breeding Ecology and Nest-Site Selection of Song Wrens in Central Panama. *Auk*, 117(2):345-354, 2 figures, 3 tables. *Cyphorhinus phaeocephalus* at Río Limbo in Soberanía National Park from 1995-1998.

Robinson, William Douglas. 1999. Long-Term Changes in the Avifauna of Barro Colorado Island, Panama, a Tropical Forest Isolate. *Conservation Biology*, 13(1):85-97.

Robinson, William Douglas. 2001. Changes in Abundance of Birds in a Neotropical Forest Fragment Over 25 Years: A Review. *Animal Biodiversity and Conservation*, 24(2):51-65, 1 figure, 4 tables. In English with abstract and table explanations also in Spanish. Examines species abundance on Barro Colorado Island using Robinson (1999) and Willis (1980).

Robinson, William Douglas, Jeffrey D. Brawn, John Terborgh, and Scott K. Robinson. 1998. Structure of a Lowland Panama Bird Community: Effects of Spatial Scale and Biogeographic History. *Ostrich*, 69(1 and 2):134. Symposium abstract. Comparisons of Limbo Plot to

Amazonian communities.

Robinson, William Douglas, George R. Angehr, Tara R. Robinson, Lisa J. Petit, Danil R. Petit, and Jeffrey D. Brawn. 2004. Distirbution of Bird Diversity in a Vulnerable Neotropical Landscape. *Conservation Biology*, 18(2): 510-518, 3 figures, 2 tables. Review of the importance of forested tracts in the former Canal Zone for conservation of lowland birds in Panama.

Robinson, William Douglas, Jeffrey D. Brawn, and Scott K. Robinson. 2000a. Forest Bird Community Structure in Central Panama: Influence of Spatial Scale and Biogeography. *Ecological Monographs*, 70(2):209-235, 6 figures, 5 tables. Soberanía National Park and Limbo Plot, 1994-1995.

Robinson, William Douglas, Tara Rodden Robinson, Scott K. Robinson, and Jeffrey D. Brawn. 2000b. Nesting Success of Understory Forest Birds in Central Panama. *Journal of Avian Biology*, 31(2):151-164, 2 appendices, 3 figures, 4 tables. Observations in 1996-1997, focusing on 10 spp. at Limbo Plot, Gamboa Woods, Howard Air Force Base, and Cocolí.

Robinson, William Douglas, and Tara Rodden Robinson. 2001. Observations of Predation Events at Bird Nests in Central Panama. *Journal of Field Ornithology*, 72(1):43-48. Describes predation events by mammals, birds, snakes, and army ants at sites near Gamboa in Soberanía National Park, 1997-1999.

Rogers, Charles H. 1937. Chimney Swifts in Eastern Panama. *Auk*, 54(3):392. Collected at Puerto Obaldia, San Blas.

Rogers, Charles H. 1939. The Swifts of Panama. *Auk*, 56(1):81-83. List of 12 taxa with new records.

Rogers, David T. Jr. 1965. Fat Levels and Estimated Flight-Ranges of Some Autumn Migratory Birds Killed in Panama During a Nocturnal Rainstorm. *Bird-Banding*, 36:115-116, 1 table. Data from birds killed at Almirante, on Oct 13, 1963.

Rogers, David T. Jr. 1968. Energy Expenditures of Caged Post Migrants in Panama. *Bird-Banding*, 39(1):60-61, 1 table. Reports the weight losses of "several fat postmigrants" that were captured in fall 1963 and caged for 24 hours in western Panama.

Rogers, David T. Jr., and Eugene P. Odum. 1966. A Study of Autumnal Postmigrant Weights and Vernal Fattening of North American Migrants in the Tropics. *Wilson Bulletin*, 78(4):415-433, 6 figures, 7 tables. Recorded weights of birds netted at Almirante by Gorgas Memorial Labs Personnel in fall 1963, as well as at sites in British Honduras, Florida and Louisiana. Based on estimates of fat free weights some individual post migrants in Panama had exhausted their migratory fat reserves and apparently began losing muscle mass. Includes weights from recaptures.

Rogers, Gerald. 1945. Costa Rican Sparrow Near the Canal Zone. *Auk*, 62(4):641. *Zonotrichia capensis costaricensis* on Cerro Chame.

Roper, James J. 2000. Experimental Analysis of Nest-Sites and Nest Predation for a Neotropical Bird: Stuck Between a Rock and a Hard Place. *Ararajuba*, 8(2):85-91, 2 figures, 3 tables. Conducted experiments along the Pipeline Road from 1991-1994. The effects of nest location and nest characteristics were examined by placing a single quail (*Coturnix coturnix*) egg in slaty antshrike (*Thamnophilus punctatus*) nests and recording predation.

Roper, James J., and Rachel R. Goldstein. 1997. A Test of the Skutch Hypothesis: Does Activity at Nests Increase Nest Predation Risk? *Journal of Avian Biology*, 28(2):111-116, 1 figure, 1 table. Observations of *Thamnophilus* [*punctatus*] *atrinucha* in Parque Nacional Soberanía do not support the hypothesis that increased activity at the nest increases predation.

Rothschild, Walter. 1895. A New Species of Tanager. *Novitates Zoologicae*, 2(4):481. *Ramphocoelus dunstalli*, n. sp. "By the make of the skin it evidently came from Central America, and probably Panama." See Griscom, 1932b.

Rothschild, Walter. 1901. [Description of *Leptotila battyi* n. sp.]. *Bulletin of the British Ornithologists' Club*, 12:33-34. From Coiba Island.

Roubik, David W. 1982. The Ecological Impact of Nectar-Robbing Bees and Pollinating Hummingbirds on a Tropical Shrub. *Ecology*, 63(2):354-360, 1 figure, 7 tables. Observations at Pipeline Road and Cerro Jefe from Nov-Dec 1979.

Roubik, David W., N. Michelle Holbrook, and German Parra V. 1985. Roles of Nectar Robbers in Reproduction of the Tropical Treelet *Quassia amara* (Simaroubaceae). *Oecologia*, 66(2):161-167, 2 figures, 2 tables. Observations in 1980 and 1981 on Barro Colorado Island, Scheelea Island, Pipeline Road and Taboga Island.

Routledge, Katherine Scoresby. 1920. *The Mystery of Easter Island, The Story of an Expedition.* xxiv + 404 pages. London: Hazell, Watson, and Viney. A brief stop at Coiba Island in early March, 1916, with mention of two small settlements on Bahia Damas (p. 352-358).

Rowley, J. Stuart. 1962. Nesting of the Birds of Morelos, Mexico. *Condor*, 64(4):253-272, 13 figures.

Rowley, J. Stuart. 1966. Breeding Records of Birds of the Sierra Madre del Sur, Oaxaca, Mexico. *Proceedings of the Western Foundation of Vertebrate Zoology*, 1(3):107-204, colored frontispiece, 48 figures, 9 tables.

Rowley, J. Stuart, and Robert T. Orr. 1964. The Status of Frantzius' Nightingale Thrush. *Auk*, 81(3):308-314, 2 figures. Nests of *Catharus frantzii* from Mexico.

Rowley, J. Stuart, and Robert T. Orr. 1965. Nesting and Feeding Habits of the White-Collared Swift. *Condor*, 67(6):449-456, 3 figures. *Streptoprocne zonaris* in Mexico.

Ruegg, Kristen C., and Smith, Thomas B. 2002. Not as the Crow Flies: A Historical Explanation for Circuitous Migration in Swainson's Thrush (*Catharus ustulatus*). *Proceedings of the Royal Society B, Biological Sciences*, 269(1498):1375-1381, 2 figures, 3 tables.

Ruschi, Augusto. 1949. Ninhos e Ovos dos Trochilideos. [Nests and Eggs of Hummingbirds]. *Boletim do Museu de Biologia Prof. Mello-Leitao, Series Biologia.* Published in 3 parts in numbers 4-6 as follows: 4:1-69, 17 photographs; 5:1-55, 18 photographs; and 6:1-53, 14 photographs.

Salvadori, Tomaso. 1891. *Catalogue of the Birds in the British Museum.* Volume 20. xvii + 658 pages, 18 plates. London: Trustees of the British Museum (Natural History). *Chrysotis virenticeps* n. sp. from Bugaba, *C. inornata* n. sp. from Veragua, and *C. salvini* n. sp. from Lion Hill Station. These are now in the genus *Amazona.*

Salvadori, Tomaso. 1893. *Catalogue of the Birds in the British Museum.* Volume 21. xvii + 676 pages, 15 plates. London: Trustees of the British Museum (Natural History). *Columba crissalis* n. sp. from Boquete de Chitra.

Salvadori, Tomaso. 1895. *Catalogue of the Birds in the British Museum.* Volume 27. xv + 636 pages, 19 plates. London: Trustees of the British Museum (Natural History). *Tinamus castaneiceps* n. sp. from Volcan de Chiriquí.

Salvadori, Tomaso. 1896. Descrizione di una Nuova Specie del Genere *Rhamphocoelus* di Chiriqui. [Description of a New Species of the Genus *Rhamphocoelus* from Chiriqui]. *Bollettino dei Musei di Zoologia ed Anatomia Comparata della R. Università di Torino*, 11(249):1-2. *R. festae*, n. sp. In Italian.

Salvadori, Tomaso, and Enrico Festa. 1899. Viaggio del Dott. E. Festa nel Darien e Regioni Vicine.

Ucelli. [Voyage of Doctor E. Festa to Darien and the Surrounding Region]. *Bollettino dei Musei di Zoologia ed Anatomia Comparata della R. Università di Torino*, 14(339):1-13. Report on a collection of 202 specimens made from Punta de Sabana to Chepigana, Darién, with additional records from Chiriqui and Veraguas based on specimens purchased from Arcé. In Italian.

Salvadori, Tomaso, and Enrico Festa. 1900. Viaggio del Dr. Enrico Festa nell'Ecuador. XXII. Uccelli. Parte Terza—Trochili-Tinami. [Voyage of Dr. Enrico Festa to Ecuador. XXII. Birds. Part Three—Trochili-Tinami]. *Bollettino dei Musei di Zoologia ed Anatomia Comparata della R. Università di Torino*, 15(368):1-54. Record of *Aramus scolopaceus* from Laguna della Pita, Darién.

Salvin, Osbert. 1863. Description of Thirteen New Species of Birds Discovered in Central America by Frederick Godman and Osbert Salvin. *Proceedings of the Zoological Society of London*, 1863:186-192, 2 plates. *Petrochelidon littoria* mentioned from Panama.

Salvin, Osbert. 1865. The Sea-Birds and Waders of the Pacific Coast of Guatemala. *Ibis*, new series (1):187-199.

Salvin, Osbert. 1866. Descriptions of Eight New Species of Birds from Veragua. *Proceedings of the Zoological Society of London*, 1866:67-76, 2 colored plates. *Catharus griseiceps*, *Microcerculus luscinia*, *Euphonia rufivertex*, *Buarremon mesoxanthus*, *Philydor fuscipennis*, *Dysithamnus puncticeps*, *Formicarius rufipectus*, and *Trogon clathratus*, n. spp. Collected by Enrique Arcé near Santiago and Santa Fé, Veraguas.

Salvin, Osbert. 1867a. On the Genus *Cinclus*. *Ibis*, new series (3):109-122, 1 plate. *Cinclus ardesiacus*, n. sp.

Salvin, Osbert. 1867b. On Some Collections of Birds from Veragua [Part 1]. *Proceedings of the Zoological Society of London*, 1867:129-161, 1 plate. List of 220 species from Santa Fe, near Santiago, and Cordillera de Tole, collected by Enrique Arcé. Also mentions species found at David by Bridges and Hicks along with other published records from Veraguas and Chiriqui. *Dorifera veraguensis*, *Odontophorus leucolaemus*, and *Chamaepetes unicolor*, n. spp.

Salvin, Osbert. 1869. Notes on Mr. Lawrence's List of Costa-Rica Birds. *Ibis*, new series (5):310-319.

Salvin, Osbert. 1870a. Additional Notes on Mr. Lawrence's List of Costa-Rica Birds. *Ibis*, new series (6):107-116.

Salvin, Osbert. 1870b. On Some Collections of Birds from Veragua. Part II *Proceedings of the Zoological Society of London*, 1870:175-219, 1 plate (map). Report on further collections by Arcé from Calovevora, Calobre, Chitra, Boqueti de Chitra, Castillo, Laguna del Castillo, and Cordillera del Chucu, with the inclusion of the species from the earlier collections already reported, making a total of 432 species. *Thryothorus semibadius*, *Tachyphonus nitidissimus*, *Empidonax atriceps*, *Chiromachaeris aurantiaca*, *Antrostomus saturatus*, *Chaetura fumosa*, *Lophornis adorabilis*, *Selasphorus torridus*, *S. ardens*, *Chloronerpes simplex*, *Melanerpes chrysauchen*, n. spp.

Salvin, Osbert. 1871. On the *Psittacidae* of Central America. *Ibis*, 3(1):86-100, 1 plate. *Conurus finschi* n. sp. from Bugaba.

Salvin, Osbert. 1872. Remarks on the Mniotiltine Genus *Geothlypis*. *Ibis*, 3(2):147-152. *Geothlypis chiriquensis* n. sp. from Volcan de Chiriquí.

Salvin, Osbert. 1874. A Visit to the Principal Museums of the United States, and Notes on Some of the Birds Contained Therein. *Ibis*, 3(4):305-329, 2 plates. Critical remarks mainly on Neotropical birds in the collections of the Smithsonian Institution, George N. Lawrence, American Museum of Natural History, Academy of Natural Sciences of Philadelphia, Boston Society of Natural History, Vassar College, S. Cabot, and T. K. Merritt. *Geotrygon lawrencii* n. sp. from Calobre.

Salvin, Osbert. 1876. On Two Additional Species of Central American *Odontophorinae*. *Ibis*, 3(6):379-381. *Odontophorus cinctus* n. sp. from Veragua (district around Calobre).

Salvin, Osbert. 1878. Descriptions of Two New Species of Birds from Central America, and One from South America. *Ibis*, 4(2):445-449. *Odontophorus spodiostethus* n. sp. from Veragua.

Salvin, Osbert. 1882. *A Catalogue of the Collection of Birds Formed by the Late Hugh Edwin Strickland.* xvi + 652 pages. London: C. J. Clay, M. A., and Son. Record of *Tigrisoma cabanisi* (p. 520) collected by Kellett and Wood in Panama Bay. See Archipiélago de las Perlas or Sharpe (1906) for a discussion of the data from Kellett and Wood.

Salvin, Osbert. 1883. A List of the Birds Collected by Captain A. H. Markham on the West Coast of America. *Proceedings of the Zoological Society of London*, 1883:419-432. Includes ten species from Panama taken in January, 1882.

Salvin, Osbert. 1891. Descriptions of New Species of Upupae and Trochili in the Collection of the British Museum. *Annals and Magazine of Natural History*, 6(7):374-378. *Polyerata decora* n. sp. from Chiriquí, Volcan de Chiriquí.

Salvin, Osbert. 1896. Tubinares (Petrels and Albatrosses). Pages: ix-xi + 340-455, 7 colored plates. In *Catalogue of the Birds in the British Museum*, Volume 25: Catalogue of the Gaviae and Tubinares. London: Trustees of the British Museum (Natural History).

Salvin, Osbert, and Frederick Ducane Godman. 1879-1904. *Biologia Centrali-Americana.* Aves. Volume 1 (text). xliv + 512 pages. London: Taylor and Francis. The following new taxa were proposed from Panama: *Thryothorus hyperythrus*, *Dacnis viguieri* Oustalet, *Lanio melanopygius*.

Salvin, Osbert, and Frederick Ducane Godman. 1888-1904. *Biologia Centrali-Americana.* Aves. Volume 2 (text). iv + 598 pages. London: Taylor and Francis. The following new taxa were proposed from Panama: *Myiobius fulvigularis*, *Automolus fumosus*, *Thamnophilus atrinucha*, *Melanerpes wagleri*, *Bucco fulvidus*.

Salvin, Osbert, and Frederick Ducane Godman. 1897-1904. *Biologia Centrali-Americana.* Aves. Volume 3 (text). iv + 510 pages. London: Taylor and Francis.

Salvin, Osbert, and Frederick Ducane Godman. 1879-1904. *Biologia Centrali-Americana.* Aves. Volume 4 (plates). vii + 79 colored plates. London: Taylor and Francis.

Saunders, Howard. 1876. On the Sterninae, or Terns, with Descriptions of Three New Species. *Proceedings of the Zoological Society of London*, 1876:638-672, 1 plate, 5 figures. *Sterna dougalli*(pp. 652-653) in BMNH "registered as obtained at Taboga," which the author presumes is Panama.

Saunders, Howard. 1896. Gaviae (Terns, Gulls, and Skuas). Pages: vii + 339, 1 colored plate. In *Catalogue of the Birds in the British Museum*, Volume 25: Catalogue of the Gaviae and Tubinares. London: Trustees of the British Museum (Natural History).

Schaefer, Ernst. 1953. Contribution to the Life History of the Swallow-Tanager. *Auk*, 70(4):403-460, 8 plates, figures, tables. *Tersina viridis.*

Schemske, Douglas W., and Nicholas Brokaw. 1981. Treefalls and the Distribution of Understory Birds in a Tropical Forest. *Ecology*, 62(4):938-945, 4 figures, 1 table. Mist netting at Limbo Hunt Club from May 1977-Jan 1978.

Schneider, David C. 1985. Predation on the Urchin *Echinometra lucunter* by Migratory Shorebirds on a Tropical Reef Flat. *Journal of Experimental Marine Biology and Ecology*, 92(1):19-27, 1 figure, 4 tables. Observed the foraging of 9 species of birds in 1978 on Galeta Island. The longitude [09°24'18"/79°57'48"] given for Galeta Island appears to be erroneous and presumably should read 79°52', not 79°57'.

Schneider, David, and Elizabeth P. Mallory. 1982. Spring Migration of Shorebirds in Panama.

Condor, 84(3):344-345, 1 figure. Counts from Jan-May 1978 at Punta Mala, Farfan, and Mangrove (all near Panama City) found no evidence of a spring build-up.

Scholes, Robert T., and Kathryn T. Scholes. 1954. Notes from Panama and the Canal Zone. *Condor*, 56(3):166-167. Sight records for *Aythya collaris*, *Buteo jamaicensis*, *Larus pipixcan*, *Sphyrapicus varius*, *Bombycilla cedrorum*, *Sicalis luteola*, and *S. flaveola.*

Schomburgk, Richard. 1848. *Reisen in Britisch-Guiana in den Jahren 1840-1844. Dritter Theil. Versuch einer Fauna und Flora von Britisch-Guiana.* [*Travels in British Guyana during the Years 1840-1844. Part Three. The Fauna and Flora of British Guyana*]. x + 533-1260 pages. Leipzig: Verlagsbuchhandlung von J. J. Weber. In German. Birds are treated by J. Cabanis on pp. 662-765.

Schönwetter, Max. 1960-1967. *Handbuch der Oologie*, Band 1: Nonpasseres. [*Handbook of Oologie.* Volume 1. Nonpasseres]. viii + 928 pages, 10 colored plates, tables. Berlin: Akademie-Verlag.

Schreiber, Elizabeth A., and Ralph W. Schreiber. 1988. Insights Into Seabird Ecology from a Global 'Natural Experiment.' *National Geographic Research*, 5(1):64-81, 6 figures, 2 tables. Effects of El Niño-Southern Oscillation on bird populations in the Pacific. Neal Smith sight record of Peruvian boobies (*Sula variegata*) in the Gulf of Panama.

Sclater, Philip Lutley. 1854. Characters of Some New or Imperfectly Described Species of Tanagers. *Proceedings of the Zoological Society of London*, 1854:95-98, 2 plates. Specimens from Nova Grenada and Chiriqué in Veragua.

Sclater, Philip Lutley. 1855a. Remarks on the Arrangement of the Jacamars (Galbulidae), with Description of Some New Species. *Proceedings of the Zoological Society of London*, 1855:13-16, 1 plate. *Galbula melanogenia* included from Veragua.

Sclater, Philip Lutley. 1855b. On Some New or Little Known Species of Birds in the Derby Museum at Liverpool. *Proceedings of the Zoological Society of London*, 1855:74-77, 4 plates. *Cyphorinus* [*sic*] *albigularis*, n. sp. Mentions *Ramphocaenus rufiventris* from Panama.

Sclater, Philip Lutley. 1856a. Synopsis Avium Tanagrinarum. A Descriptive Catalogue of the Known Species of Tanager. Part I. *Proceedings of the Zoological Society of London*, 1856:64-94. Records for *Saltator magnoides* and *Arremon aurantiirostris.*

Sclater, Philip Lutley. 1856b. Synopsis Avium Tanagrinarum. A Descriptive Catalogue of the Known Species of Tanagers. Part II. *Proceedings of the Zoological Society of London*, 1856b:108-132. Records for *Pyranga aestiva*, *P. ludoviciana*, *Ramphocelus dimidiatus* and *R. passerinii.*

Sclater, Philip Lutley. 1856c. List of Mammals and Birds Collected by Mr. Bridges in the Vicinity of the Town of David in the Province of Chiriqui in the State of Panama. *Proceedings of the Zoological Society of London*, 1856:138-143. List of 46 species of birds collected January to mid-March 1856. *Thamnophilus bridgesi* and *Geotrygon chiriquensis*, n. spp. The name *Calliste franciscae* Sclater dates from this publication, although it is clearly only an emendation.

Sclater, Philip Lutley. 1856d. Synopsis Avium Tanagrinarum. A Descriptive Catalogue of the Known Species of Tanagers. Part III. *Proceedings of the Zoological Society of London*, 1856:230-281, 1 table. Reference to *Tanagra diaconus*, *Calliste gyroloides* and *Calliste franciscae* from Panama.

Sclater, Philip Lutley. 1856e. On the Species of the American Genus *Parra*. *Proceedings of the Zoological Society of London*, 1856:282-283. Includes *Parra* [= *Jacana*] *hypomelaena* from Chiriqui.

Sclater, Philip Lutley. 1857a. Notes on the Birds in the Museum of the Academy of Natural Sciences of Philadelphia, and Other Collections in the United States of America. *Proceedings of the Zoological Society of London*, 1857:1-8. Discussion of *Cyphorinus albigularis.*

Sclater, Philip Lutley. 1857b. Review of the Species of the South American Subfamily Tityrinae.

Proceedings of the Zoological Society of London, 1857:67-80, 2 sketches, 1 table. *Tityra personata* from Chiriqui.

Sclater, Philip Lutley. 1858a. Synopsis of the American Ant-Birds (*Formicariidae*). Part I. Containing the Thamnophilinae. *Proceedings of the Zoological Society of London*, 1858:202-224, 2 plates. *Thamnophilus melanurus*, *T. bridges,i* and *T. doliatus* from Panama.

Sclater, Philip Lutley. 1858b. Synopsis of the American Ant-Birds (*Formicariidae*). Part II. Containing the Formicivorinae or Ant-wrens. *Proceedings of the Zoological Society of London*, 1858:232-254, 2 plates. *Hypocnemis naevioides* from Panama.

Sclater, Philip Lutley. 1858c. Synopsis of the American Ant-Birds (*Formicariidae*). Part III. Containing the Third Subfamily Formicariinae, or Ant-Thrushes. *Proceedings of the Zoological Society of London*, 1858:272-289, 1 plate. *Gymnocichla nudiceps* from Panama.

Sclater, Philip Lutley. 1858d. On Two Species of Ant-Birds in the Collection of the Derby Museum, at Liverpool. *Proceedings of the Zoological Society of London*, 1858 (=1859):540-541. *Myrmeciza exsul* n. sp. from Panama and Nicaragua, restricted "to near Gatun, Canal Zone" (Wetmore 1962:7).

Sclater, Philip Lutley. 1861a. Characters of Some New Species of American Passeres. *Proceedings of the Zoological Society of London*, 1861:127-131. *Saltator isthmicus*, n. sp.

Sclater, Philip Lutley. 1861b. On a New Species of Bird of the Genus *Lipaugus* of Boié. *Proceedings of the Zoological Society of London*, 1861:209-212. Discussion of *L. unirufus* (Cassin, 1860) from the Isthmus of Darién.

Sclater, Philip Lutley. 1861-1862. *Catalogue of a Collection of American Birds Belonging to Philip Lutley Sclater.* xvi + 368 pages, 20 colored plates. London: N. Trubner and Co.

Sclater, Philip Lutley. 1862a. [Footnote regarding a new genus, *Oncostoma*] in Lawrence, G. N. *Ibis*, 4:12. Proposed for *Todirostrum cinereigulare* with *T. olivaceum* Lawrence included.

Sclater, Philip Lutley. 1862b. Additional Notes on the American Barbets. *Ibis*, 4:1-2, 1 colored plate. Treats *Capito maculicoronatus* (from the Isthmo of Panama) and *Tetragonops ramphastinus.*

Sclater, Philip Lutley. 1863a. Synopsis of the Known Species of *Dacnis*. *Ibis*, 5:311-317, tables, 1 plate. Includes *Dacnis venusta* from the Isthmus of Panama.

Sclater, Philip Lutley. 1863b. List of Recent Additions to the Genus *Calliste*. *Ibis*, 5:450-452, 1 plate. Includes *Calliste lavinia* from the Isthmus of Darién.

Sclater, Philip Lutley. 1864a. On the Species of the American Genus *Coccyzus*. *Proceedings of the Zoological Society of London*, 1864:119-122. *Coccyzus seniculus* from Chiriqui and *C. erythrophthalmus* from Panama.

Sclater, Philip Lutley. 1864b. Characters of Three New American Parrots. *Proceedings of the Zoological Society of London*, 1864:297-298, 1 plate. *Chrysotis viridigenalis* from the Isthmus of Panama.

Sclater, Philip Lutley. 1865. On a New Species of the Genus *Basileuterus* of Cabanis, with a Synopsis of the Known Species of the Genus. *Proceedings of the Zoological Society of London*, 1865:282-286, 2 plates. Includes *B. uropygialis* from Isthmus Panama.

Sclater, Philip Lutley. 1866a. On a New American Cuckoo of the Genus *Neomorphus*. *Proceedings of the Zoological Society of London*, 1866:59-60, 1 plate. *Neomorphus salvini*, n. sp.

Sclater, Philip Lutley. 1866b. On a New Species of the Genus *Accipiter* from New Granada. *Proceedings of the Zoological Society of London*, 1866:302-304. Mention of *Accipiter tinus* [= *A. superciliosus*] from Veragua.

Sclater, Philip Lutley. 1868. Description of Some New or Little Known Species of Formicarians.

Proceedings of the Zoological Society of London, 1868:571-575, 1 plate, 1 figure. Includes a list of new species, some of which are from Panama, since publication of his *Catalogue of American Birds* (1862).

Sclater, Philip Lutley. 1870a. Further Notes on the Cuckoos of the Genus *Coccyzus*. *Proceedings of the Zoological Society of London*, 1870:165-169. Specimens from Panama and Chiriqui.

Sclater, Philip Lutley. 1870b. Description of Three Apparently New Species of Tyrant-Birds of the Genus *Elainea*, with Remarks on Other Known Species. *Proceedings of the Zoological Society of London*, 1870:831-835, 3 figures. *Elainea pudica* n. sp. with one specimen from Chiriqui.

Sclater, Philip Lutley. 1879-1882. *A Monograph of the Jacamars and Puff-Birds, or Families Galbulidae and Bucconidae.* liv + 171 pages, 55 plates. London: R. H. Porter. Issued in 7 parts.

Sclater, Phillip Lutley, and Osbert Salvin. 1864. Notes on a Collection of Birds from the Isthmus of Panama. *Proceedings of the Zoological Society of London*, 1864:342-373, 1 plate. List of 272 species based mainly on birds obtained by James McLeannan along the Panama Railroad. Detailed description of Lion Hill based on Salvin's visit to the area. *Cassiculus microrhynchus*, *Myrmeciza immaculata*, *Camptostoma flaviventre*, *Conurus ocularis*, n. spp.

Sclater, Phillip Lutley, and Osbert Salvin. 1866-1869. *Exotic Ornithology. Containing Figures and Descriptions of New or Rare Species of American Birds.* 192 pages, 100 colored plates. London: Bernard Quaritch.

Sclater, Phillip Lutley, and Osbert Salvin. 1868a. Descriptions of New Species of Birds of the Families Dendrocolaptidae, Strigidae, and Columbidae. *Proceedings of the Zoological Society of London*, 1868:53-60, 1 plate, 2 figures. *Dendrocincla ruficeps* n. sp. from Isthmus Panamensis.

Sclater, Phillip Lutley, and Osbert Salvin. 1868b. Descriptions of New or Little Known American Birds of the Families Fringillidae, Oxyrhamphidae, Bucconidae and Strigidae. *Proceedings of the Zoological Society of London*, 1868:322-329, 1 plate, 2 figures. *Oxyrhamphus frater n. sp.* from Veragua.

Sclater, Phillip Lutley, and Osbert Salvin. 1868c. Descriptions of Four New Species of Birds from Veragua. *Proceedings of the Zoological Society of London*, 1868:388-390. *Pyranga testacea*, *Chlorospingus hypophaeus*, *Leptotriccus superciliaris*, and *Eupherusa egregia*, n. spp.

Sclater, Phillip Lutley, and Osbert Salvin. 1868d. Synopsis of the American Rails (*Rallidae*). *Proceedings of the Zoological Society of London*, 1868:442-470, 1 plate, 11 figures, 1 table. Includes records from Panama.

Sclater, Phillip Lutley, and Osbert Salvin. 1869a. Notes on the Species of the Genus *Asturina*. *Proceedings of the Zoological Society of London*, 1869:129-134. *A. ruficauda* n. sp. with specimens from David and Veragua. See Chapman (1917).

Sclater, Phillip Lutley, and Osbert Salvin. 1869b. Descriptions of Six New Species of American Birds of the Families Tanagridae, Dendrocolaptidae, Formicariidae, Tyrannidae and Scolopacidae. *Proceedings of the Zoological Society of London*, 1869:416-420, 1 plate. *Grallaria princeps* n. sp. from Veragua.

Sclater, Phillip Lutley, and Osbert Salvin. 1869c. Descriptions of Three New Species of Tanagers from Veragua. *Proceedings of the Zoological Society of London*, 1969:439-440, 2 plates. *Buthraupis arcaei*, *Tachyphonus chrysomelas*, and *Chlorospingus punctulatus*, n. sp.

Sclater, Phillip Lutley, and Osbert Salvin. 1870a. On Birds Collected by Mr. George M. Whitely on the Coast of Honduras. *Proceedings of the Zoological Society of London*, 1870:835-839. Records from Panama and Veragua.

Sclater, Phillip Lutley, and Osbert Salvin. 1870b. Descriptions of Five New Species of Birds from the United States of Columbia. *Proceedings of the Zoological Society of London*, 1870:840-844, 1 plate. *Trogon chionurus* n. sp. from Panama and Veragua.

Sclater, Phillip Lutley, and Osbert Salvin. 1873. *Nomenclator Avium Neotropicalium, Sive Avium Quae in Regione Neotropica Hucusque Repertae Sunt Nomina Systematice Disposita Adjecta Sua Cuique Speciei Patria Accedunt Generum et Specierum Novarum Diagnoses.* viii + 163 pages. London: Published by the Authors. *Leptoptila rufinucha* n. sp. from Veragua.

Sclater, Phillip Lutley, and Osbert Salvin. 1876. A Revision of the Neotropical Anatidae. *Proceedings of the Zoological Society of London*, 1876:358-412, 1 plate, 1 figure, 1 table. Specimens from Panama, Lion Hill and Veragua.

Sclater, Phillip Lutley, and Osbert Salvin. 1879. On the Birds Collected by the Late Mr. T. K. Salmon in the State of Antioquia, United States of Columbia. *Proceedings of the Zoological Society of London*, 1879:486-550, 4 plates, map. Observations on nesting of some species found in Panama.

Seemann, Berthold. 1853. *Narrative of the Voyage of H.M.S.* Herald *During the Years 1845-51, Under the Command of Captain Henry Kellett. R.N., C.B.; Being a Circumnavigation of the Globe, and Three Cruises to the Arctic Regions in Search of Sir John Franklin.* Volume 1: xvi + 322 pages, colored frontispiece, map. Volume 2: vii + 302 pages, colored frontispiece. London: Reeve and Co. Seemann was the ship's naturalist. The account covers the itinerary of Capt. Kellett of the *Herald*, Lieut. Wood of the *Pandora*, and Seemann in their visits to Panama from 1846-1849.

Selfridge, Thomas Oliver. 1874. *Reports of Explorations and Surveys to Ascertain the Practicality of a Ship Canal Between the Atlantic and Pacific Oceans by Way of the Isthmus of Darien.* 268 pages, maps, plates, tables. Washington, D.C.: Government Printing Office. The volume was also issued as Miscellaneous Document No. 113, House of Representatives, 42nd Congress, 3rd Session.

Seutin, Gilles. 1998. Two Bird Species New for Panama and Central America: White-Whiskered Hermit *Phaethornis yaruqui* and White-Winged Swallow *Tachycineta albiventer. Cotinga*, 9:22-23, 1 figure, 1 drawing. Sight records along with those of several other species at Manené, on the Tuira River, and in Darién National Park, in 1996. The coordinates given for Darién National Park Station [08°52' N, 77°51' W] appear to be erroneous and presumably should read ca. 0800/7744.

Seutin, Gilles, Jeffrey Brawn, Robert E. Ricklefs, and Eldredge Bermingham. 1993. Genetic Divergence Among Populations of a Tropical Passerine, the Streaked Saltator (*Saltator albicollis*). *Auk*, 110(1):117-126, 4 figures, 3 tables. Compared mitochondrial DNA from blood and muscle tissues taken in the Pearl Islands, mainland Panama, Peru, and the Lesser Antilles. Found little genetic differentiation (0.0034) between *S. a. isthmicus* and *S. a. speratus*, but high levels (0.035 and 0.063) between Panama vs. Peru and Lesser Antilles vs. Panama/Peru, respectively.

Seutin, Gilles, and Mark Letzer. 1995. The Short-Tailed Nighthawk is a Tree Nester. *Journal of Field Ornithology*, 66(1):30-36, 3 figures. Describes a nest of *Lurocalis semitorquatus* found in Gamboa in 1993 with discussion of other nests in the suborder Caprimulgi.

Sexton, Owen J. 1975. Black Vultures Feeding on Iguana Eggs in Panama. *American Midland Naturalist*, 93(2):463-468, 2 figures. Observed *Coragyps atratus* and nesting iguanas on Slothia Island, just off the coast of Barro Colorado Island in 1973.

Sharpe, R. Bowdler. 1882. *Catalogue of the Birds in the British Museum.* Volume 6. xii + 420 pages, 18 colored plates. London: Trustees of the British Museum (National History). *Thryothorus melanogaster* n. sp. from Bugaba.

Sharpe, R. Bowdler. 1885. *Catalogue of the Birds in the British Museum.* Volume 10. xiii + 682 pages,

13 colored plates. London: Trustees of the British Museum (Natural History). *Basileuterus [leucopygius] veraguensis* n. subsp. from Paraiso Station.

Sharpe, R. Bowdler. 1894. *Catalogue of the Birds in the British Museum*, Volume 23. xiii + 353 pages, 9 plates. London: Trustees of the British Museum (Natural History).

Sharpe, R. Bowdler. 1898. Plataleae (Ibises and Spoonbills) and Herodiones (Herons and Storks). Pages: ix-x, xiii-xv, 1-328, 9 colored plates. In *Catalogue of the Birds in the British Museum*, Volume 26. London: Trustees of the British Museum (Natural History).

Sharpe, R. Bowdler. 1906. Birds. Pp. 79-515 in *The History of the Collections Contained in the Natural History Departments of the British Museum*. London: Trustees of the British Museum (Natural History). Species accounts, chronological listing of major acquisitions, alphabetical listing of collectors/donors.

Sheffler, W. J., and Adriaan Joseph Van Rossem. 1944. Nesting of the Laughing Falcon. *Auk*, 61(1):140-142. *Herpetotheres cacchinnans* in Mexico.

Shelley, G. E. 1891. Families Indicatoridae, Capitonidae, Cuculidae, and Musophagidae. Pages: vii-viii, x-xii, 1-123, 209-484, 8 colored plates. In *Catalogue of the Birds in the British Museum*, Volume 19. London: Trustees of the British Museum (Natural History). *Capito salvini* n. sp. from Panama and Costa Rica.

Shelvocke, George. 1726. *A Voyage Round the World by Way of the Great South Sea, Performed in the Years 1719, 20, 21, 22 in the* Speedwell *of London*. xxxii + 468 pages, map, plates. London: J. Senex. Account of Isla Coiba based on visits on January 13, 1720 and early May, 1721. mentions (p. 360) a "great variety of birds." Isla Rancheria is called Quivetta.

Shiras, George, III. 1915. Nature's Transformation at Panama. Remarkable Changes in Faunal and Physical Conditions in the Gatun Lake Region. *The National Geographic Magazine*, 28(2):159-194, 2 maps, 32 photographs. Description of Gatun Lake soon after it was filled. Maps show the Canal Zone before and after the formation of the lake.

Short, Lester L. 1968. Sympatry of Red-Breasted Meadowlarks in Argentina, and the Taxonomy of Meadowlarks (Aves: *Leistes*, *Pezites*, and *Sturnella*.) *American Museum Novitates*, 2349:1-30, 11 figures, 1 table.

Shurcliff, Sidney N. 1930. *Jungle Islands, the* "Illyria" *in the South Seas. The Record of the Crane Pacific Expedition, Field Museum of Natural History, Chicago, Illinois, with a Scientific Appendix by Karl Patterson Schmidt*. xv + 298 pages, colored frontispiece, map, plates, illustrations. New York: G. P. Putnam's Sons. While in Panama City in Dec 1928 for ship repairs, a party traveled to Barro Colorado Island. Mention a few birds on pages 50 and 52.

Sibley, Charles G., and Lester L. Short, Jr. 1964. Hybridization in the Orioles of the Great Plains. *Condor*, 66(2):130-150, 3 figures, 12 tables. *Icterus galbula* and *I. bullockii*.

Sick, Helmut. 1958. Geselligkeit, Schornstein-Benutzung und Überwinterung beim Brasilianischen Stachelschwanzsegler *Chaetura andrei* [Sociability, Chimney Use, and Overwintering in the Brazilian Spine-tailed Swift *Chaetura andrei*]. *Die Vogelwarte*, 19(4):248-253, 3 figures.

Sick, Helmut. 1959. Notes on the Biology of Two Brazilian swifts, *Chaetura andrei* and *Chaetura cinereiventris*. *Auk*, 76(4):471-477, 1 plate. Reference to *C. a. meridionalis* in Panama.

Sick, Helmut. 1962. Escravismo em Aves Brasileiras. [Brood Parasitism in Brazilian Birds]. *Arquivos do Museu Nacional*, 52:185-192, 2 figures. In Portuguese with an English summary.

Sick, Helmut. 1967. Courtship Behavior in the Manakins (Pipridae): A Review. *Living Bird*, 6:5-22, 2 colored plates, 6 figures.

Sieving, Kathryn E. 1990. Pheasant Cuckoo Foraging Behavior, with Notes on Habits and Possible

Social Organization in Panama. *Journal of Field Ornithology*, 61(1):41-46, 1 figure. Observations of *Dromococcyx phasianellus* in Soberanía National Park.

Sieving, Kathryn E. 1992. Nest Predation and Differential Insular Extinction Among Selected Forest Birds of Central Panama. *Ecology*, 73(6):2310-2328, 8 figures, 6 tables. Two study plots on Barro Colorado Island and two plots along the Pipeline Road.

Skinner, Kenneth L. 1925. The Eggs of the Hirundinidae. *Oologists' Record*, 5(2):36-41.

Skutch, Alexander F. 1930. The Habits and Nesting Activities of the Northern Tody Flycatcher. *Auk*, 47(3):313-322, 1 black and white plate. *Todirostrum cinereum* near Almirante, Bocas del Toro.

Skutch, Alexander F. 1934. A Nesting of the Slaty Antshrike (*Thamnophilus punctatus*) on Barro Colorado Island. *Auk*, 51(1):8-16, 1 black and white plate. Nesting birds observed in December 1930 and January 1931.

Skutch, Alexander F. 1937. Life-History of the Black-Chinned Jacamar. *Auk*, 54(2):135-146, 2 black and white plates. *Galbula [ruficauda] melanogenia* in Honduras.

Skutch, Alexander F. 1940a. Social and Sleeping Habits of Central American Wrens. *Auk*, 57(3):293-312, 1 black and white plate, 1 figure. A few notes from Panama included.

Skutch, Alexander F. 1940b. Some Aspects of Central American Bird-Life. I. Family Life in a Non-Migrating Bird Population. *Scientific Monthly*, 51(11):409-418.

Skutch, Alexander F. 1940c. Some Aspects of Central American Bird-Life. II. Plumage, Reproduction, and Song. *Scientific Monthly*, 51(12):500-511, 8 photographs.

Skutch, Alexander F. 1943. The Family Life of Central American Woodpeckers. *Scientific Monthly*, 56(4):358-364.

Skutch, Alexander F. 1944a. Life History of the Blue-Throated Toucanet. *Wilson Bulletin*, 56(3):133-151. *Aulacorhynchus caeruleogularis.*

Skutch, Alexander F. 1944b. Life History of the Quetzal. *Condor*, 46(5):213-235, 4 figures. Reprinted in Smithsonian Annual Report for 1946. *Pharomachrus mocinno.*

Skutch, Alexander F. 1945a. Incubation and Nestling Periods of Central American Birds. *Auk*, 62(1):8-37. Includes observations from western Panama, Almirante, and the Canal Zone.

Skutch, Alexander F. 1945b. On the Habits and Nest of the Ant-Thrush *Formicarius analis*. *Wilson Bulletin*, 57(2):122-128. Barro Colorado Island.

Skutch, Alexander F. 1945c. The Migration of Swainson's and Broad-Winged Hawks Through Costa Rica. *Northwest Science*, 19(4):80-89. *Buteo swainsoni* and *B. platypterus.*

Skutch, Alexander F. 1946a. Life Histories of Two Panamanian Antbirds. *Condor*, 48(1):16-28. Barro Colorado Island. *Hylophylax naevioides* and *Myrmotherula axillaris.*

Skutch, Alexander F. 1946b. Life History of the Costa Rican Tityra. *Auk*, 63(3):327-362. *Tityra semifasciata.*

Skutch, Alexander F. 1947a. A Nest of the Sun-Bittern in Costa Rica. *Wilson Bulletin*, 59(1):38. *Eurypyga helias.*

Skutch, Alexander F. 1947b. Life History of the Turquoise-Browed Motmot. *Auk*, 64(2):201-217, 1 black and white plate, 1 figure. *Eumomota superciliosa*, with notes on denudation of tail in *Electron platyrhynchum minor.*

Skutch, Alexander F. 1948a. Life History Notes on Puff-Birds. *Wilson Bulletin*, 60(2):81-97. Bucconidae. Barro Colorado Island.

Skutch, Alexander F. 1948b. Life History of the Golden-Naped Woodpecker. *Auk*, 65(2):225-260. Mentions *Melanerpes chrysauchen* in western Panama.

Skutch, Alexander F. 1949a. Life History of the Yellow-Thighed Manakin. *Auk*, 66(1):1-24, 1 plate. Observations of *Pipra mentalis* on Barro Colorado Island and in Costa Rica.

Skutch, Alexander F. 1949b. Do Tropical Birds Rear as Many Young as They Can Nourish? *Ibis*, 91(3):430-455. Comments by David Lack follow on pages 455-458.

Skutch, Alexander F. 1951. Life History of Longuemare's Hermit Hummingbird. *Ibis*, 93(2):180-195.

Skutch, Alexander F. 1952a. Life History of the Chestnut-Tailed *Automolus*. *Condor*, 54(2)93-100. Observations on *Automolus ochrolaemus* made in Costa Rica.

Skutch, Alexander F. 1952b. Life History of the Blue and White Swallow. *Auk*, 69(4):392-406. *Notiochelidon cyanoleuca.*

Skutch, Alexander F. 1953a. How the Male Bird Discovers the Nestlings. *Ibis*, 95(1):1-37; 95(3):505-542, appendix, 2 tables.

Skutch, Alexander F. 1953b. Life History of the Southern House Wren. *Condor*, 55(3)121-149, 1 table. Detailed account of life history of *Troglodytes* [*aedon*] *musculus* based in part on studies made in Panama.

Skutch, Alexander F. 1954a. Life Histories of Central American Birds I. Families Fringillidae, Thraupidae, Icteridae, Parulidae, and Coerebidae. *Pacific Coast Avifauna*, 31: 448 pages, 1 colored frontispiece, 68 figures, 13 tables. Illustrations by Donald R. Eckelberry.

Skutch, Alexander F. 1954b. Life History of the White-Winged Becard. *Auk*, 71(2):113-129, 1 black and white plate, 1 table. *Pachyramphus polychopterus.*

Skutch, Alexander F. 1954c. Life History of the Tropical Kingbird. *Proceedings of the Linnaean Society of New York*, 63-65:21-38. Observations of *Tyrannus melancholichus* in Central America.

Skutch, Alexander F. 1956a. The Bird's Nest as a Dormitory. *Animal Kingdom, Magazine of the New York Zoological Society,* 59(2):50-55, 4 photographs. Relates mainly to Central American species.

Skutch, Alexander F. 1956b. Life History of the Ruddy Ground Dove. *Condor*, 58(3):188-205. Based in part on observations of *Columbina talpacoti* in Panama.

Skutch, Alexander F. 1957a. The Friendly Bicolored Antbird. *Animal Kingdom, Magazine of the New York Zoological Society*, 60(3):75-79, 1 illustration. Nest of *Gymnopithys leucaspis* in Costa Rica.

Skutch, Alexander F. 1957b. Life History of the Amazon Kingfisher. *Condor*, 59(4):217-229, 4 figures. *Chloroceryle amazona.*

Skutch, Alexander F. 1958a. Life History of the White-Whiskered Soft-Wing *Malacoptila panamensis*. *Ibis*, 100(2):209-231.

Skutch, Alexander F. 1958b. Roosting and Nesting of Aracari toucans. *Condor*, 60(4):201-219, 1 colored plate, 1 figure. *Pteroglossus.*

Skutch, Alexander F. 1960. Life Histories of Central American Birds II. Families Vireonidae, Sylviidae, Turdidae, Troglodytidae, Paridae, Corvidae, Hirundinidae, and Tyrannidae. *Pacific Coast Avifauna*, 34: 593 pages, 1 colored frontispiece, 100 figures, 8 tables. Illustrations by Donald R. Eckelberry.

Skutch, Alexander F. 1961a. Life History of the White-Crested Coquette Hummingbird. *Wilson Bulletin*, 73(1):5-10, 1 colored plate. *Lophornis adorabilis.*

Skutch, Alexander F. 1961b. The Purple-Capped Fairy Hummingbird. *Audubon Magazine*, 63(1):8-

9, 13, 2 illustrations by Don R. Eckelberry. *Heliothrix barroti.*

Skutch, Alexander F. 1962a. Life Histories of Honeycreepers. *Condor*, 64(2):92-116, 1 table. Coerebidae.

Skutch, Alexander F. 1962b. On the Habits of the Queo, *Rhodinocichla rosea*. *Auk*, 79(4):633-639.

Skutch, Alexander F. 1963a. Life History of the Little Tinamou. *Condor*, 65(3):224-231, 1 figure. *Crypturellus soui.*

Skutch, Alexander F. 1963b. Habits of the Chestnut-Winged Chachalaca. *Wilson Bulletin*, 75(3):262-269. *Ortalis garrula.*

Skutch, Alexander F. 1963c. Life History of the Rufous-Tailed Jacamar *Galbula ruficauda* in Costa Rica. *Ibis*, 105(3):354-368.

Skutch, Alexander F. 1964a. Life Histories of Hermit Hummingbirds. *Auk*, 81(1):5-25, 1 figure. *Phaethornis.*

Skutch, Alexander F. 1964b. Life History of the Scaly-Breasted Hummingbird. *Condor*, 66(3):186-198. *Phaeochroa cuvieri.*

Skutch, Alexander F. 1964c. Life Histories of Central American Pigeons. *Wilson Bulletin*, 76(3):211-247. Columbidae.

Skutch, Alexander F. 1964d. Life History of the Blue-Diademed Motmot *Momotus momota*. *Ibis*, 106(3):321-332, 1 table.

Skutch, Alexander F. 1965a. Life History Notes on Two Tropical American Kites. *Condor*, 67(3):235-246. *Elanoides forficatus* and *Harpagus bidentatus.*

Skutch, Alexander F. 1965b. Life History of the Long-Tailed Silky Flycatcher, with Notes on Related Species. *Auk*, 82(3):375-426, 8 figures, 4 tables. *Ptilogonys caudatus* and *Phainoptlia melanoxantha.*

Skutch, Alexander F. 1966a. A Breeding Bird Census and Nesting Success in Central America. *Ibis*, 108(1):1-16, 6 tables.

Skutch, Alexander F. 1966b. Life History Notes on Three Tropical American Cuckoos. *Wilson Bulletin*, 78(2):139-165. *Piaya cayana, Morococcyx erythropygus,* and *Crotophaga ani.*

Skutch, Alexander F. 1966c. Western Kingbird and Inca Dove in Costa Rica. *Auk*, 83(4):669. *Tyrannus melancholichus* and *Scardafella inca.*

Skutch, Alexander F. 1967. Life Histories of Central American Highland Birds. *Publications of the Nuttall Ornithological Club*, 7:i-vi, 1-213, illustrations.

Skutch, Alexander F. 1968. The Nesting of Some Venezuelan Birds. *Condor*, 70(1):66-82, 2 figures.

Skutch, Alexander F. 1969a. Notes on the Possible Migration and the Nesting of the Black Vulture in Central America. *Auk*, 86(4):726-731. Observations of *Coragyps atratus* by the author while in Costa Rica in 1964 with mention of Eisenmann's and Wetmore's published observations in Panama.

Skutch, Alexander F. 1969b. Life Histories of Central American Birds III. Families Cotingidae, Pipridae, Formicariidae, Furnariidae, Dendrocolaptidae, and Picidae. *Pacific Coast Avifauna*, 35: 580 pages, 1 frontispiece, appendix, 28 figures, 8 tables. Illustrations by Donald R. Eckelberry.

Skutch, Alexander F. 1971a. Life History of the Bright-Rumped Attila, *Attila spadiceus*. *Ibis*, 113(3):316-322.

Skutch, Alexander F. 1971b. *A Naturalist in Costa Rica.* x + 378 pages, 2 appendices, maps on inside cover, illustrated. Gainesville: University of Florida Press. Appendix 1 includes a list of birds in found in the study area.

Slater, P. J. B.. 1997. Singing in the Rain Forest: The Duets of Bay Wrens. *Trends in Ecology and Evolution*, 12(6):207-208. *Thryothorus nigricapillus.*

Slud, Paul. 1964. The Birds of Costa Rica. Distribution and Ecology. *Bulletin of the American Museum of Natural History*, 128:1-430, 1 appendix, 1 figure, 2 plates. Detailed map.

Smith, Hubert L. 1936. The Cage Birds of Panama. *Aviculture*, volume 6, 3(1):1-4. Account of species found in bird market with a few remarks on wild species.

Smith, James N. M. 1977. Feeding Rates, Search Paths and Surveillance for Predators in Great-Tailed Grackle Flocks. *Canadian Journal of Zoology*, 55(6):891-898, 4 figures, 4 tables. Gatun, 1972-1973. *Quiscalus mexicanus.*

Smith, Neal Griffith. 1971. Biovarianism in an Insular Population of a Neotropical Bird. *American Midland Naturalist*, 86(1):238-241, 1 figure. Compares the reproductive tracts of female *Thamnophilus doliatus* sampled at four sites on the Pacific coast and on the following Pearl Islands: Bayoneta, Casaya, Chapera, Contadora, Pedro Gonzales, Rey, and San José. Smith's later retraction of records from Bayoneta, Casaya, Chapera, (seeWright et al. 1985) is of more than biogeographical importance, as he indicated that he obtained and examined the ovaries of 7 specimens from these 3 islands where the species does not occur. Such a prevarication casts doubt over the entire publication.

Smith, Neal Griffith. 1975. "Spshing Noise": Biological Significance of Its Attraction and Nonattraction by Birds. *Proceedings of the National Academy of Sciences of the United States of America*, 72(4):1411-1414, 7 tables. Playback experiments at Summit Gardens in 1973 and at Nueva Suiza in 1974.

Smith, Neal Griffith. 1980. Hawk and Vulture Migrations in the Neotropics. Pages 51-65, 10 figures, 1 table *in* Allen Keast and Eugene S. Morton, editors. *Migrant Birds in the Neotropics. Ecology, Behavior, Distribution and Conservation.* Washington, D.C.: Smithsonian Institution Press. Counts and photographs of *Buteo platypterus*, *B. swainsoni*, and *Cathartes aura* near Panama City, 1969-1977.

Smith, Neal Griffith. 1982. Lesser Black-Backed Gull in Panama. *American Birds*, 36(3):336-337, 1 photograph. Observations of *Larus fuscus graellsii* at beaches in Panama City 1979-1982 and at Ft. Amador in 1980.

Smith, W. John. 1966. Communication and Relationship in the Genus *Tyrannus*. *Publications of the Nuttall Ornithological Club*, 6: ix + 250 pages, figures, tables.

Smith, W. John. 1971. Behavioral Characteristics of Serpophaginine Tyrannids. *Condor*, 73(3):259-286, 12 figures. Based on studies in South and Central America including observations along the Río Chiriquí Viejo below Cerro Punta, Panama, in 1962.

Snow, David W. 1956. The Dance of Manakins. *Animal Kingdom, Magazine of the New York Zoological Society*, 59(3):86-91, 5 figures, 1 photograph. Observations in Trinidad. Pipridae.

Snow, David W. 1961. The Displays of the Manakins *Pipra pipra* and *Tyranneutes virescens*. *Ibis*, 103a(1):110-113, 1 figure.

Snow, David W. 1962. A Field Study of the Black and White Manakin, *Manacus manacus*, in Trinidad. *Zoologica: Scientific Contributions of the New York Zoological Society*, 47(2):65-104, 3 appendices, 21 figures, 8 tables. Trinidad.

Sousa, Octavio E., and Carlton M. Herman. 1982. Blood Parasites of Birds from Chiriqui and Panama Provinces in the Republic Of Panama. *Journal of Wildlife Diseases*, 18(2):205-221, 1 figure, 4 tables. 3,715 birds representing 281 species from La Fortuna, Cuestra de Piedra, Juan Mina, and Bayano [= Isla Majé].

Spear, Larry B., and David G. Ainley. 1999. Seabirds of the Panama Bight. *Waterbirds*, 22(2):175-198, appendix, 10 figures, 7 tables. Data from surveys conducted on 10 cruises from 1984-1991 were used to determine distribution and abundance of seabirds off the coasts of Panama, Colombia, and Ecuador.

Stiles, F. Gary. 1983. Systematics of the Southern Forms of *Selasphorus* (Trochilidae). *Auk*, 100(2):311-325, 6 figures, 2 tables. Based on examination of museum specimens and field observations in Costa Rica the author recognizes *Selasphorus flammula flammula* Salvin 1864, *S. f. torridus* Salvin 1870, *S. f. simoni* Carriker 1910, *S. scintilla* Gould 1850, *S. ardens* Salvin 1870. Suggests that *S. scintilla* and *S. ardens* are closely related and may compromise a superspecies and that *S. "underwoodii"* is a hybrid between *S. f. flammula* and *S. scintilla.*

Stiles, F. Gary. 1983. The Taxonomy of *Microcerculus* Wrens (Troglodytidae) in Central America. *Wilson Bulletin*, 95(2):169-183, colored frontispiece, 1 figure, 3 tables. Based on examination of museum specimens and song identification in the field (Costa Rica), the author recognizes *M. philomela* Salvin and *M. luscinia* Salvin.

Stiles, F. Gary. 1995. Distribucion y Variacion en el Ermitaño Carinegro (*Phaethornis anthophilus*) in Colombia. [Distribution and Variation of the Black-faced [=Pale-Bellied] Hermit (*Phaethornis anthophilus*) in Colombia]. *Caldasia*, 18(86):119-129, 1 figure, 3 tables. In Spanish with an English abstract. Recognized the following subspecies in Colombia: *P. a. anthophilus* Bourcier for birds near the boundaryof Panama and *P. a. fuscicapillus* Cory for birds near the boundary of Venezuela.

Stiles, F. Gary, and Rafeal G. Campos. 1983. Identification and Occurrence of Blackpoll Warblers in Southern Middle America. *Condor*, 85(2):254-255. *Dendroica striata* captured during banding operations in Costa Rica. Other records include a bird banded in Bocas del Toro.

Stone, Witmer. 1899. A Study of the Type Specimens of Birds in the Collection of the Academy of Natural Sciences of Philadelphia, with a Brief History of the Collection. *Proceedings of the Academy of Natural Sciences of Philadelphia*, 1899:5-62.

Stone, Witmer. 1918. Birds of the Panama Canal Zone, with Special Reference to a Collection Made by Mr. Lindsey L. Jewel. *Proceedings of the Academy of Natural Sciences of Philadelphia*, 1918:239-280. Report on 236 species collected mainly at Gatun, but also at Agua Clara, Chagres River, Colón, Fort Lorenzo, Gatun Lake, Gatun River, Mindi, Miraflores, Mt. Hope, Pedro Miguel, Río Indis, Río Siri, Río Trinidad, Tabernilla, and Toro Point. Most of the collection was sent to ANSP, but a few specimens went to AMNH, and at least one to USNM. The report contains many of the collector's field notes, a history of ornithological collections in the Canal Zone, as well as species known to the area, but not encountered by Jewel, bringing the list to 432 taxa.

Stone, Witmer. 1935. Griscom's 'Ornithology of Panama'. *Auk*, 52(3):338. Review of Griscom (1935b).

Stone, Witmer. 1936. Chapman on the Courtship of Gould's Manakin. *Auk*, 53(1):111-112. Review of Chapman (1935).

Storer, Robert W. 1989. Geographic Variation in the Yellow-Rumped Tanager (Aves: Thraupinae). *Proceedings of the Biological Society of Washington*, 102(3):553-554, 1 table. Measurements and color comparisons of 143 female and immature male specimens collected in Panama, Colombia, and Ecuador. Reconizes *Ramphocelus flammigerus variens* Lafresnaye for birds in Panama and parts of Colombia, and *R. f. icteronotus* Bonaparte for birds in Ecuador. Birds from SW Colombia are intergrades.

Strahl, Stuart D., Silvia Beaujon, Daniel M. Brooks, Alfredo J. Begazo, Gina Sedaghatkish, and Fabio Olmos, editors. *The Cracidae: Their Biology and Conservation.* xvii + 506 pages, appendix, figures, tables. Most chapters are in Spanish with a few in Portuguese or English.

Strauch, Joseph G., Jr. 1975a. A Specimen of the Yellow-Headed Blackbird from the Panama Canal Zone. *Condor*, 77(3):355. First record of *Xanthocephalus xanthocephalus* S of Mexico; specimen collected at Gatun Dam in Nov 1972 for UMMZ.

Strauch, Joseph G., Jr. 1975b. Observations at a Nest of the Black-And-White Hawk Eagle. *Condor*, 77(4):512. The author observed nest construction and abandonment at a site 9 km E of Chepo. *Spizastur melanoleucus.*

Strauch, Joseph G., Jr. 1977. Further Bird Weights from Panama. *Bulletin of the British Ornithologists' Club*, 97(2):61-65. Weights and other data for 504 birds of 207 species collected for UMMZ in 1971-1973 in the Canal Zone (including Pipeline Road) and at Cerro Jefe, Cerro Campana, Cerro Punta, and Changuinola.

Strong, R. M. 1952. A Peculiar Pigmentation. *Auk*, 69(2):199-200, 1 figure. In throat feathers of male Purple-throated Fruit-Crow *Querula purpurata.*

Sturgis, Bertha Bement. 1928. *Field Book of Birds of the Panama Canal Zone, a Description of All the Species of the Birds of the Panama Canal Zone, with Notes on Their Habits, Call Notes and Songs, for the Purposes of Identifying Them.* xxxix + 466 pages, map, 6 color plates, 18 black and white plates, 107 figures. New York: G. P. Putnam's Sons.

Stutchbury, Bridget J. M., Paul R. Martin, and Eugene S. Morton. 1996. Nesting Behavior of the Slate-Colored Seedeater (*Sporophila schistacea*) in Panama. *Ornitologia Neotropical*, 7(1):63-65. Observations on Pipeline Road, Feb 1995.

Stutchbury, Bridget J. M., Eugene S. Morton, and Walter H. Piper. 1998. Extra-Pair Mating System of a Synchronously Breeding Tropical Songbird. *Journal of Avian Biology*, 29(1):72-78, 3 figures. Data annually (1994-1996) collected near Gamboa for 20-40 nests of *Turdus grayii* showed that 53% of females produced extra-pair young (EPY) and that 38% of nestlings were EPY.

Sullivan, John O. 1965. "Flightlessness" in the Dipper. *Condor*, 67(6):535-536. *Cinclus mexicanus* flightless in late summer.

Sundevall, Carl J. 1870. Översigt af Fogelslägtet *Dendroeca. Öfversigt af Kongl. Vetenskaps-Akademiens Förhandlingar*, 26(6)[1869]:605-618. *Dendroeca petechia aequatorialis* n. subsp. from "Guayaquil," but see Peters (1931).

Sutton, George Miksch. 1951. *Mexican Birds, First Impressions Based Upon an Ornithological Expedition at Tamaulipas, Nuevo León, and Coahuila, with an Appendix Describing Briefly all Mexican Birds.* xv + 282 pages, 16 colored plates, 65 illustrations by the author. Norman: University of Oklahoma Press.

Sutton, George Miksch. 1954. Blackish Crane-Hawk. *Wilson Bulletin*, 66(4):237-242, 1 colored plate. *Geranospiza caerulescens.*

Sutton, George Miksch, and Olin Sewall Pettingil, Jr. 1942. A Nest of the Brown Jay. *Wilson Bulletin*, 54(3):213-214, 1 black and white photograph. *Psilorhinus morio.*

Swann, H. Kirke. 1921-1922. *A Synopsis of the Accipitres.* 2nd edition, viii + 233 pages. London: Wheldon and Wesley. Issued in four parts. *Urubitinga anthracina bangsi* n. subsp. from San Miguel Island.

Swann, H. Kirke. 1924-1945. *A Monograph of the Birds of Prey (Order Accipitres).* Issued in 16 parts to make up 2 Volumes. Volume I: Cathartidae, Aegypiidae, Sagittariidae, Falconidae (Polyborinae to Buteoninae). 1924-1930. Parts 1-9, lxviii + XI + 1-487 pages, colored plates. Volume II: Falconidae (continued) (Gypaetinae to Pandioninae). 1931-1945. Parts 10-16, 1-538 pages, colored and monochrome plates. London: Wheldon and Wesley. The volumes were edited by Alexander Wetmore.

Swarth, Harry S. 1931. The Avifauna of the Galapagos Islands. *Occasional Papers of the California Academy of Science*, 18: 299 pages, 1 map, 57 figures, tables. Uncertain record for *Creagrus furcatus* in Panama (p. 65-66).

Taczanowski, Ladislas. 1884. *Ornithologie du Pérou*. Volume 2. 566 pages, figures. Rennes: Oberthur. *Lophotriccus squamicristatus luteiventris* n. sp. from Chiriquí. Text in French. Table of contents (p. 559-566) includes supplements and additions to volume 1.

Tanner, James T. 1942. *The Ivory-Billed Woodpecker*. xii + 111 pages, appendix, 20 black and white plates, 22 figures. New York: Dover Publications. Reprint of Research Report Number 1 of the National Audubon Society.

Tejera, Víctor Hugo. 1995. Inventario Biológico del Canal de Panamá. Estudio Ornitológico. [Biological Inventory of the Panama Canal. An Ornithological Study]. Pages 5-107, appendix (with 17 tables), 6 figures *in* V. H. Hugo, El Inventario Biológico del Canal de Panamá. II. El Estudio Ornitológico, Herpetológico y Mastozoológico. *Scientia (Panamá) Revista de Investigación de la Universidad de Panamá*, Enero de 1995, Número Especial 2: 281 pages. Surveys at six locations within the Panama Canal drainage including: South Entrance, Gaillard Cut, North Entrance, Tranquilla, Pequeni and Ciri. Locality descriptions with coordinates. ISSN 0258-9702.

Tejera, Víctor Hugo. 2001. Contribución al Conocimiento de la Aves de la Reserva de Fortuna, Chiriquí, República de Panamá. [Contributions to the Knowledge of the Birds of Fortuna Reserve, Chiriquí, Republic of Panama]. *Tecnociencia*, 3(1):31-67, 1 figure, 7 tables. In Spanish with English summary. Lists birds seen during a survey of the region in 1984 at sites between Los Planes and the continental divide. Includes a map.

Tejera, Víctor Hugo, and Rafael R. Rincón. 1982. *Columbina passerina* [Linne, 1758] en Panama. *Natura (Revista Cientifica y Humanistica, Centro Regional Universitario de David)*, 3(1):18-21, 3 figures, 1 table. In Spanish. First record of *C. passerina* in Panama. Specimens collected near the mouth of the Río La Villa.

Tejera, Víctor Hugo, and Artemis V. de Tejera. 2001. Pericos, Casangas, Loros, Guacamayos y Afines. [Parrots, Parakeets, Macaws, Cockatoos, and their Allies]. *Tecnociencia*, 3(1):19-30, 2 illustrations. In Spanish with English summary. Status and distribution in Panama with nests found at the University of Panama.

Tejera, Víctor Hugo, Ana M. Jimenez, and Ricardo J. Pérez. 2001. Observaciones Relacionadas con los Nacimientos y los Polluelos Exitosos de *Columbina talpacoti* (Temminck, 1811) durante un Año, en la Universidad de Panamá. [Observations Related to Success of Hatchlings and Fledglings of *Columbina talpacoti* (Temminck, 1811) During a Year, at the University of Panama]. *Scientia*, 16(1):1-15, 1 map, 6 figures, 3 tables. In Spanish with English summary. Campus map and description.

Thayer, John E., and Outram Bangs. 1905. The Mammals and Birds of the Pearl Islands, Bay of Panama. *Bulletin of the Museum of Comparative Zoology*, 46(8)137-160. List of 92 species taken by W. W. Brown, Jr. from Feb-Apr 1904, on San Miguel, Saboga and Pacheca Islands. *Myiobius naevius furfurosus* n. subsp. from Saboga Island, and *Tanagra cana dilucida* n. subsp. from San Miguel Island.

Thayer, John E., and Outram Bangs. 1906. Vertebrata from the Savanna of Panama. III. Aves.

Bulletin of the Museum of Comparative Zoology, 46(12):213-224, tables. List of 86 taxa collected by W. W. Brown, Jr., in May 1904, in the Savanna of Panama while based at Calidonia (now a part of Panama City). Mentions a few specimens from other localities. *Momotus conexus* n. sp., *Tyrannulus reguloides panamensis* and *Tiaris olivacea dissita*, n. subspp.

Thorstrom, Russell, Richard Watson, Aaron Baker, Serena Ayers, and David L. Anderson. 2002. Preliminary Ground and Aerial Surveys for Orange-Breasted Falcons in Central America. *Journal of Raptor Research*, 36(1):39-44 1 figure, 1 table. Ground and aerial surveys with coordinates given for sightings of *Falco deiroleucos* in Darién province [see Sierra de Jungurudó] and in Coclé province [see Cordillera Central] in Apr 2000. Suggests that the S. American population is isolated from the Guatemala/Belize population.

Todd, W. E. Clyde. 1912. Descriptions of Seventeen New Neotropical Birds. *Annals of the Carnegie Museum*, 8(2):198-214, tables. *Piaya rutila panamensis* n. subsp. from Loma del Leon.

Todd, W. E. Clyde. 1913. A Revision of the Genus *Chaemepelia*. *Annals of the Carnegie Museum*, 8(3-4):507-603, tables. *C. rufipennis nesophila* provisionally proposed from Pearl Islands.

Todd, W. E. Clyde. 1917. New Genera, Species, and Subspecies of South American Birds. *Proceedings of the Biological Society of Washington*, 30:127-130. *Tachyphonus luctuosus panamensis* n. subsp. from Gatun.

Todd, W. E. Clyde. 1929. A Review of the Vireonine Genus *Pachysylvia*. *Proceedings of the Biological Society of Washington*, 42:181-206. *Pachysylvia ochraceiceps nelsoni*, nom. nov., from Lion Hill.

Todd, W. E. Clyde. 1942. A List of the Hummingbirds in the Collection of the Carnegie Museum. *Annals of the Carnegie Museum*, 29:271-370, 3 figures. Occasional Panama specimens some from the Río Sicsola.

Todd, W. E. Clyde. 1948. A New Booby and a New Ibis from South America. *Proceedings of the Biological Society of Washington*, 61:49-50. *Sula nebouxii excisa* and *Theristicus caudatus hyperorius* n. subspp.

Todd, W. E. Clyde, and M. A. Carriker, Jr. 1922. The Birds of the Santa Marta Region of Colombia: A Study of Altitudinal Distribution. *Annals of the Carnegie Museum*, 14:i-viii, 1-611, 1 map, 9 plates, 7 figures. Includes a forward by W. J. Holland, a supplement, and an index.

Tordoff, Harrison B. 1953. Irazu Junco - A Primitive Member of the Genus. *Wilson Bulletin*, 65(1):51. *Junco vulcani* considered primitive but northern in origin.

Townsend, Charles H. 1890. Scientific Results of Explorations by the U.S. Fish Commission Steamer *Albatross*. No. XIV. Birds from the Coasts of Western North America and Adjacent Islands, Collected in 1888-89, with Descriptions of New Species. *Proceedings of the United States National Museum*, 13(799):131-142. Second known specimen of *Halocyptena microsoma* from Panama Bay.

Trainer, Jill M. 1987. Behavioral Associations of Song Types During Aggressive Interactions Among Male Yellow-Rumped Caciques. *Condor*, 89(4):731-738, 4 figures, 2 tables. Observations, song recordings and banding of *Cacicus cela vitellinus* during the breeding seasons of 1981-1983 on Howard Air Force Base.

Trainer, Jill M. 1988. Singing Organization During Aggressive Interactions Among Male Yellow-Rumped Caciques. *Condor*, 90(3):681-688, 5 tables. *Cacicus cela* on Fort Howard Air Force Base, 1982-1983.

Trainer, Jill M. 1989. Cultural Evolution in Song Dialects of Yellow-Rumped Caciques in Panama. *Ethology*, 80(1-4):190-204, 8 figures, 3 tables. Banded male *Cacicus cela* and recorded songs during 3 breeding seasons (1981-1983) at colonies in Cardenas, Tank Farm, Rodman, Farfan,

Howard, Lookout, Antenna, Vera Cruz and Kobbe Beach (all within 10 km of the Pacific entrance of the Canal). Includes a map. Concludes that most new songs arise from gradual change within the colony.

Trainer, Jill M., and Tom C. Will. 1984. Avian Methods of Feeding on *Bursera simaruba* (Burseraceae) Fruits in Panama. *Auk*, 101(1):193-195. Observed 26 species near Kobbe Beach in Feb 1982.

Travis, J., and Robert E. Ricklefs. 1983. A Morphological Comparison of Island and Mainland Assemblages of Neotropical Birds. *Oikos*, 41(3):434-441, 7 tables. Summary in Russian. Examined museum specimens and conducted censuses in Panama, Trinidad, Jamaica, St. Lucia and St. Kitts.

Traylor, Melvin A. 1968. Winter Molt in the Acadian Flycatcher, *Empidonax virescens*. *Auk*, 85(4):691. Includes specimens from Panama, Colombia and Ecuador.

Trimble, Ruth. 1943. Birds Collected During Two Cruises of the "*Vagabondia*" to the West Coast of South America. *Annals of the Carnegie Museum*, 29:409-441. Lists birds collected by A. C. Twomey on 13 April 1939 near Capira and Parita.

Tristam, H. B. 1889. *Catalogue of a Collection of Birds Belonging to H. B. Tristam, D. D., LL. D., F. R. S.* xvi + 278 pages. Durham: Printed at the "Advertiser" Office, 48 Saddler Street. Lists occasional specimens from Panama, some of which were gifts from O. Salvin.

Underdown, C. Eliot. 1932. A Study of *Chlorospingus ophthalmicus* (Dubus) and its Allies. *Ibis*, 13(2):638-655. Includes *C. o. novicius* from Chiriqui.

United States Board on Geographic Names. . 1990. *Gazetteer of Panama. Names Approved by the United States Board on Geographic Names.* 3rd edition. xvii + 509 pages, maps. Defense Mapping Agency, Washington, D.C. Used throughout the text with the acronym USBGN.

United States Board on Geographic Names. 1988. *Gazetteer of Colombia. Names Approved by the United States Board on Geographic Names.* 3rd edition. xviii + 859 pages, maps. Defense Mapping Agency, Washington, D.C. Used in the text with the acronym USBGN2.

Van Rossem, Adriaan Joseph. 1934a. A New Race of *Piranga bidentata* from Central America. *Transactions of the San Diego Society of Natural History*, 7(32):367-368. *P. b. citrea* n. subsp. from Boquete.

Van Rossem, Adriaan Joseph. 1934b. Notes on the Races of *Claravis mondetoura*. *Transactions of the San Diego Society of Natural History*, 8(3):5-8.

Van Rossem, Adriaan Joseph. 1935. A New Name for the Large-Billed Hawk of Western Costa Rica and Panama. *Condor*, 37(4):215. *Buteo magnirostris petulans* nom. nov. for *B. m. ruficauda* (Sclater and Salvin).

Van Rossem, Adriaan Joseph. 1939. Some New Races of Birds from Mexico. *Annals and Magazine of Natural History*, 11(4):439-443. Panama included in ranges.

Van Tyne, Josselyn. 1926. The Nest of *Automolus ochrolaemus pallidigularis* Lawrence. *Auk*, 43(4):546. On Barro Colorado Island.

Van Tyne, Josselyn. 1929. The Life History of the Toucan *Ramphastos brevicarinatus*. *Miscellaneous Publications of the Museum of Zoology of the University of Michigan*, 19:1-43, 8 plates.

Van Tyne, Josselyn. 1935. The Birds of Northern Petén, Guatemala. *University of Michigan Museum of Zoology Miscellaneous Publications*, 27:1-46, 1 map, 2 black and white plates, tables.

Van Tyne, Josselyn. 1937. South American Pied-Billed Grebe in the Canal Zone. *Auk*, 54(3):379. *Podilymbus podiceps* at Barro Colorado Island.

Van Tyne, Josselyn. 1944. The Nest of the Antbird *Gymnopithys bicolor bicolor*. *Occasional Papers of the Museum of Zoology*, University of Michigan, 491:1-5, 1 plate. On Barro Colorado Island.

Van Tyne, Josselyn. 1950. Bird Notes from Barro Colorado Island, Canal Zone. *Occasional Papers of the Museum of Zoology*, 525:1-12, 4 black and white plates. Summary of field observations on several species.

Vanderbilt, George. 1944. Introduction and Itinerary. Pp. 1-6 in Results of the Fifth George Vanderbilt Expedition (1941). *Academy of Natural Sciences of Philadelphia Monographs*, 6. Overview of the expedition. Birds are treated in J. Bond and R. Meyer de Schauensee (1944).

Vaurie, Charles. 1971. Notes Systematiques sur des Furnariides Rares des Genres *Philydor* et *Xenerpestes*, et Parallelisme de la Forme du Bec au "Type Xenops." [Systematic Notes on Rare Furnariids of the Genera *Philydor* and *Xenerpestes*, and Parallelism of the Beak Form in "Type Xenops"]. *L'Oiseau et la Revue Francaise d'Ornithologie*, 41(2-3):117-126, 2 figures.

Verreaux, Jules, and Édouard Verreaux. 1853. Description d'Oiseaux Nouveaux. [Descriptions of New Birds]. *Revue et Magasin de Zoologie*, 2(5):193-197. *Casmarhynchus tricaruncalatus* n. sp. from Bocos del Toro. In French.

Vleck, Carol Masters 1981. Hummingbird Incubation: Female Attentiveness and Egg Temperature. *Oecologia*, 51(2):199-205, 3 figures, 3 tables. Includes observations of Purple-crowned Fairy (*Heliothryx barroti*) nests on Barro Colorado Island.

Voous, K. H. 1957. The Birds of Aruba, Curacao, and Bonaire. *Studies of the Fauna of Curacao and Other Caribbean Islands*, 7(29):1-260, 13 plates, 1 figure, 1 table.

Wafer, Lionel. 1699. *A New Voyage and Description of the Isthmus of America, Giving an Account of the Author's Abode There, the Form and Make of the Country, the Coasts, Hills, River, etc., Woods, Soil, Weather, etc. Trees, Fruit, Beasts, Birds, Fish, etc. The Indian Inhabitants, Their Features, Complexion, etc. Their Manners, Customs, Employments, Marriages, Feasts, Hunting, Computation, Language, etc. With Remarkable Occurrences in the South Sea, and Elsewhere.* 224 pages, 4 plates. London: Printed for J. Knapton. A second edition, published in 1704, and a third in 1729 in vol. 3 of Dampier's *Voyages*, had as a supplement "An additional Account of several beasts, birds, fishes, reptiles, etc. and particularly many trees, shrubs, and herbs, with their names, use Vertues, etc. as has been observed in those parts. Communicated by a member of the Royal Society." The list of birds in this later section is compiled from species known from tropical America in general and has no direct reference to Panama.

Wagner, Helmuth O. 1951. Beobachtungen an Motmots. [Observations of Motmots]. *Natur und Volk*, 81(10):245-250, 4 figures. Studies in Mexico. In German.

Wagner, Moritz. 1870. Ueber die hydrographischen Verhältnisse und das Vorkommen der Susswasserfische in den Staaten Panama und Ecuador. Ein Beitrag zur Zoogeographie Amerikas. [On the Hygrographic Conditions and the Occurrence of Freshwater Fish in Panama and Ecuador. A Contribution to American Zoogeography.] *Abhhandlungen Mathematisch-Physikalischen Classe, der Königlich Bayerischen Akademie der Wissenschaften*, 10(1):65-113. Wetmore (1965:p.?) evidently erred in citing the year as 1866.

Wagstaffe, Reginald. 1978. *Type Specimens of Birds in the Merseyside County Museums (Formerly City of Liverpool Museums)*. 33 pages, 2 black and white plates. Liverpool: Merseyside County Museums.

Wallace, George J. 1965. Studies on Neotropical Thrushes in Colombia. *Publications of the Museum, Michigan State University, Biology Series*, 3(1):1-47, 2 tables.

Warkentin, Ian G., and Eugene S. Morton. 1995. Roosting Behavior of Prothonotary Warblers in the Non-Breeding Season. *Wilson Bulletin*, 107(2):374-376. Observations of *Protonotaria citrea* in Costa Rica and in Panama on Far Fan Road, 1973-1974, 1991.

Warren, Rachel L. M. 1966. *Type-Specimens of Birds in the British Museum (Natural History)*. Volume 1. *Non-Passerines*. ix + 320 pages. London: Trustees of the British Museum (Natural History).

Watson, David M., and A. Townsend Peterson. 1999. Determinants of Diversity in a Naturally Fragmented Landscape: Humid Montane Forest Avifaunas of Mesoamerica. *Ecography*, 22(5):582-589, 2 figures, 1 table. Examined 31 montane patches of humid forest from Mexico to Panama. Analyzed species lists and 7 patch characteristics that were measured from maps.

Weathers, Wesley W. 1997. Energetics and Thermoregulation by Small Passerines of the Humid, Lowland Tropics. *Auk*, 114(3):341-353, 6 figures, 2 tables. *Sporophila aurita* and *S. minuta* captured in 1984 for laboratory experiments in Panama and the U. S.

Webster, J. Dan. 1968. A Revision of the Tufted Flycatchers of the Genus *Mitrephanes*. *Auk*, 85(2):287-303, 5 tables.

Wege, David C. 1996. Threatened Birds of the Darien Highlands, Panama: A Reassessment. *Bird Conservation International*, 6(2):175-179, 1 map. Discusses birds above 1,000 m at Cerros de Quía, and above 1,200 m from the Pirre ridge and the Tacarcuna ridge including four species that qualify as globally threatened for occupying areas of 100 km^2 or less.

West, Steve. 1975. Gray Hawk Feeding on Worm Lizard. *Condor*, 77(3):354. Observed *Buteo nitidus* feeding on *Amphisbaena fuliginosa*. Remains of the lizard were collected from Gamboa Pipeline Road on Sep 1973 for identification and preservation at GML.

Westwater, R. L. 1988. Bird Memories of Almirante, Panama. *Sea-Swallow*, 37:8-10. Military captain recounts birds seen as his vessel passed a day at Almirante.

Wetmore, Alexander. 1917. On Certain Secondary Sexual Characters in the Male Ruddy Duck, *Erismatura jamaicensis* (Gmelin). *Proceedings of the United States National Museum*, 52:479-482, 1 figure.

Wetmore, Alexander. 1923. List of Sea Birds Collected by Mr. Chas. Fagan. *Condor*, 25(5), 170-171. *Puffinus subalaris*, S of Balboa and *Sterna anaethetus nelsoni*, abeam Cape Mala.

Wetmore, Alexander. 1926. Observation on the Birds of Argentina, Paraguay, Uruguay, and Chile. *United States National Museum Bulletin* 133: iv + 448 pages, 20 black and white plates, tables. Washington, D.C.:

Wetmore, Alexander. 1943. The Birds of Southern Veracruz, Mexico. *Proceedings of the United States National Museum*, 93(3164):215-340, 3 black and white plates, 1 map.

Wetmore, Alexander. 1944. A Collection of Birds from Northern Guanacaste, Costa Rica. *Proceedings of the United States National Museum*, 95(3179):25-80, 4 black and white plates.

Wetmore, Alexander. 1945. A Review of the Forms of the Brown Pelican. *Auk*, 62(4):577-586. *Pelecanus occidentalis carolinensis* listed for Caribbean (Brujas Point) and Pacific Coasts (El Rey, San José, Pearl Islands and Panama Bay) of Panama, although Brujas Point is on the Pacific not the Caribbean coast.

Wetmore, Alexander. 1946a. New Forms of Birds from Panama and Colombia. *Proceedings of the Biological Society of Washington*, 59:49-54. *Nyctanassa violacea caliginis*, *Aramides cajanea morrisoni*, *Elaenia flavogaster silvicultrix*, n. subspp., from Isla San José.

Wetmore, Alexander. 1946b. The Birds of San José and Pedro González Islands, Republic of Panamá. *Smithsonian Miscellaneous Collections*, 106(1):1-60, 4 black and white plates. Habitat, ornithological history, and annotated list of species incorporating previous literature and collections made by Wetmore and Morrison in 1944.

Wetmore, Alexander. 1950a. An Additional Form of Pepper-Shrike from Western Panamá. *Proceedings of the Biological Society of Washington*, 63:61-62. *Cyclarhis gujanensis perrygoi* n. subsp.

from Ciénaga Macana, near El Rincón.

Wetmore, Alexander. 1950b. Additional Forms of Birds from the Republics of Panama and Colombia. *Proceedings of the Biological Society of Washington*, 63:171-174. *Geotrygon goldmani oreas* n. subsp from Quebrada Cauchero, and *Cyclarhis gujanensis flavens* n. subsp. from Punta Madroño [= Punta Madroña].

Wetmore, Alexander. 1950c. The Identity of the American Vulture Described as *Cathartes burrovianus* by Cassin. *Journal of the Washington Academy of Science*, 40(12):415-417. Mentions Ciénaga Macana, near El Rincon.

Wetmore, Alexander. 1951a. Additional Forms of Birds from Colombia and Panamá. *Smithsonian Miscellaneous Collections*, 117(2):1-11 First records for Panama for one species and three subspecies of birds. *Chaetura spinicauda aetherodroma* n. subsp. from Cerro Carbunco, *Premnoplex brunnescens mnionophilus* and *Zonotrichia capensis orestera* n. subspp. from Cerro Campana.

Wetmore, Alexander. 1951b. Four Additional Species for Panama. *Auk*, 68(4):525-526. First records for *Ardea cocoi*, *Plegadis falcinellus*, *Theristicus caudatus*, and *Sarkidiornis sylvicola*.

Wetmore, Alexander. 1952. The Birds of the Islands of Taboga, Taboguilla, and Uravá, Panamá. *Smithsonian Miscellaneous Collections*, 121(2):1-32, 3 black and white plates. Habitats, annotated list of species based mainly on Wetmore's collections in 1952. *Amazilia edward ludibunda* n. subsp from Quebrada Chitabé [= Q. Chitrabé], *A. e. collata* n. subsp. from Río Antón, *Elaenia flavogaster cristula*, and *Saltator albicollis melicus* n. subspp. from Taboga Island. Frequent mention of Pearl Islands and Panama Bay as well as other localities.

Wetmore, Alexander. 1953. Further Additions to the Birds of Panamá and Colombia. *Smithsonian Miscellaneous Collections*, 122(8):1-12. *Nonnula frontalis stulta* n. subsp. from El Uracillo, *Phaeomyias murina eremonoma* n. subsp. from the Río Santa María, *Phyllomyias griseiceps quantulus* n. subsp. from Cana, and *Sicalis luteola eisenmanni* n. subsp. from the savannahs of Coclé.

Wetmore, Alexander. 1956a. Additional Forms of Birds from Panama and Colombia. *Proceedings of the Biological Society of Washington*, 69:123-126. *Zenaidura macroura turturilla* n. subsp. from El Espino.

Wetmore, Alexander. 1956b. The Muscovy Duck in the Pleistocene of Panamá *Wilson Bulletin*, 68(4):327. Bone of *Cairina moschata* collected by Gazin at El Hatillo in 1951

Wetmore, Alexander. 1957. The Birds of Isla Coiba, Panama. *Smithsonian Miscellaneous Collections*, 134(9):1-105, 4 plates, 15 figures. Habitats, ornithological history, and annotated list of species based mainly on Wetmore's collections from 6 January to 6 February 1956. Wetmore's secondary localities on the island are not frequently mentioned in the text, although they are indicated on most specimen labels. Descriptions of the following 16 new subspecies in addition to the 4 previously described from the island: *Centurus rubricapillus subfusculus*, *Cranioleuca vulpina dissita*, *Thamnophilus doliatus eremnus*, *Contopus cinereus aithalodes*, *Camptostoma obsoletum orphnum*, *Leptopogon amaurocephalus idius*, *Troglodytes aedon carychrous*, *Polioptila plumbea cinericia*, *Hylophilus flavipes xuthus*, *Parula pitiayumi cirrha*, *Basileuterus delattrii actuosus*, *Thraupis virens cumatilis*, *Ramphocelus dimidiatus arestus*, *Saltator albicollis scotinus*, *Tiaris olivacea ravida*, *Arremonops conirostris viridicata*.

Wetmore, Alexander. 1958. Extralimital Records for the Eastern Kingbird, Tree Swallow, and Blackpoll Warbler. *Auk*, 75(4):467-468. *Tyrannus tyrannus*, *Tachycineta bicolor*, and *Dendroica striata*.

Wetmore, Alexander. 1959a. Description of a Race of the Shearwater *Puffinus lherminieri* from Panama. *Proceedings of the Biological Society of Washington*, 72:19-22. *P. l. loyemilleri* n. subsp. from Tiger Rock, Bocas del Toro.

Wetmore, Alexander. 1959b. The Birds of Isla Escudo de Veraguas, Panamá. *Smithsonian Miscellaneous*

Collections, 139(2):1-27, 1 plate, 3 figures. Habitats and annotated list of species based mainly on Wetmore's collections in 1958 Feb 28-Mar 2 with the following new endemic subspecies: *Manacus vitellinus amitinus*, *Thryothorus nigricapillus odicus*, *Thraupis virens caesita.*

Wetmore, Alexander. 1960. A Classification for the Birds of the World. *Smithsonian Miscellaneous Collections*, 139(11):1-37.

Wetmore, Alexander. 1962. Systematic Notes Concerned with the Avifauna of Panamá. *Smithsonian Miscellaneous Collections*, 145(1):1-14. Remarks on variation in *Coragyps atratus*. *Colinus cristatus mariae* n. subsp. from Alanje, *Myrmeciza exsul niglarus* n. subsp. from the Río Chiman, and *Tangara icterocephala oresbia* n. subsp. from Cerro Campana. New records for Panama: *Tigrisoma salmoni*, from the Río Uracillo and other localities, *Helicolestes hamatus*, from the Río Tuira, *Steatornis caripensis*, from Old Tacarcuna Village, and *Chaetura b. brachyura*, from Corozal.

Wetmore, Alexander. 1963a. An Additional Race of the Pileated Tinamou from Panamá. *Proceedings of the Biological Society of Washington*, 76:173-176. *Crypturellus soui capnodes* n. subsp. from Almirante.

Wetmore, Alexander. 1963b. Additions to Records of Birds Known from the Republic of Panamá. *Smithsonian Miscellaneous Collections*, 145(6):1-11. *Amazilia handleyi*, from Isla Escudo de Veraguas n. sp., *Odontophorus dialeucos* n. sp. from Cerro Mali, *Elaenia caniceps absita* n. subsp. from Tacarcuna Village, and *Tangara inornata rava* n. subsp. from Almirante.

Wetmore, Alexander. 1965. The Birds of the Republic of Panamá, Part 1. Tinamidae (Tinamous) to Rynchopidae (Skimmers). *Smithsonian Miscellaneous Collections*, 150(1):iv + 483 pages, 73 figures (figure 1 = colored frontispiece).

Wetmore, Alexander. 1967. Further Systematic Notes on the Avifauna of Panamá. *Proceedings of the Biological Society of Washington*, 80:229-242. Descriptions of the following new subspecies: *Neocrex columbianus* [sic] *ripleyi* from Achiote, *Phaeochroa cuvierii furvescens* from Zapotillo, *Haplophaedia aureliae galindoi* from Cerro Mali, *Hylocharis eliciae earina* from Quebrada Chucanti, *Lampornis castaneoventris homogenes* from Chitra, and *Trogon collaris heothinus* from Río Pucro. First records for Panamá of *Myiobus villosus villosus* from Cerro Tacarcuna and *Myiodynastes chrysocephalus minor* from Cerro Pirre, Cerro Tacarcuna.

Wetmore, Alexander. 1968. The Birds of the Republic of Panamá, Part 2: Columbidae (Pigeons) to Picidae (Woodpeckers). *Smithsonian Miscellaneous Collections*, 150(2):v + 605 pages, colored frontispiece, 75 figures. Includes Corrigenda to Part 1.

Wetmore, Alexander. 1970. Descriptions of Additional Forms of Birds from Panamá and Colombia. *Proceedings of the Biological Society of Washington*, 82:767-776. Descriptions of the following new subspecies: *Glyphorhynchus spirurus pallidulus* from Charco del Toro, *Xenops rutilans incomptus* from Cana and Cerro Pirre, *Thamnophilus doliatus nesiotes* from Río Cacique, Isla del Rey, and *Oryzoborus crassirostris loftini* from Almirante.

Wetmore, Alexander. 1972. The Birds of the Republic of Panamá, Part 3: Passeriformes: Dendrocolaptidae (Woodcreepers) to Oxyruncidae (Sharpbills). *Smithsonian Miscellaneous Collections*, 150(3):iv + 631 pages, colored frontispiece, 48 figures. *Schiffornis turdinus acrolophites* n. subsp. from Cerro Mali. See comments under Wright et al. (1985)

Wetmore, Alexander, and Pedro Galindo. 1972. Additions to the Birds Recorded in Panama. *Proceedings of the Biological Society of Washington*, 85(25):309-311. Collections by Galindo of the Chocó Tinamou *Crypturellus kerriae* and the Purple Honeycreeper *Cyanerpes caeruleus chocoanus* near the Río Mono on Cerro Quiá, 1970-1971. Observations by Wetmore in Apr 1947 on the Río Imamadó and at Los Peñitas.

Wetmore, Alexander, Roger F. Pasquier, and Storrs L. Olson. 1984. The Birds of the Republic of Panamá, Part 4: *Hirundinidae* (Swallows) to *Fringillidae* (Finches). *Smithsonian Miscellaneous*

Collections, 150(4):vi + 670 pages, appendix, colored frontispiece, 49 figures.

Whitney, Bret M., and Gary H. Rosenberg. 1993. Behavior, Vocalizations and Possible Relationships of *Xenornis setifrons* (Formicariidae), a Little-Known Choco Endemic. *Condor*, 95(1):227-231, 2 figures. Observations and recordings in Nusagandí, Feb 1987 and Jan 1992.

Wicker, Cyrus French. 1917. The Pearl Islands of the Pacific. *Pan-American Magazine*, 24(5):225-234, map, 16 photographs. Personal account of a trip to the islands with mention of Taboga and Taboguilla. While anchored at Saboga made visits to other islands in the group.

Wiedenfeld, David A. 1991. Tropical Bird Species Numbers in Second-Growth vs. Primary Forest Habitats at Large Scales. *Ornitologia Neotropical*, 2(1):40-43, 5 figures. Compares species lists for Barro Colorado Island with those from sites in Amazonia.

Wiedenfeld, David A. 1992. Foraging in Temperate and Tropical Breeding and Wintering Male Yellow Warblers. Pages 321-328, 1 figure, 5 tables *in* John M. Hagan, III, and David W. Johnston, editors. *Ecology and Conservation of Neotropical Migrant Landbirds*. Washington, D.C.: Smithsonian Institution Press. Observations of "yellow" and "mangrove" warblers (*Dendroica petechia*) at sites in the U. S., Ontario, Mexico and Panama with brief descriptions of the study sites. Collected foraging data at Isla Galeta, Playa El Agallito, and the mouth of the Río Juan Díaz in 1986-1988.

Wikelski, Martin, Michaela Hau, and John C. Wingfield. 1999. Social Instability Increases Plasma Testosterone in a Year-Round Territorial Neotropical Bird. *Proceedings of the Royal Society Series B, Biological Sciences*, 266(1419):551-556, 3 figures. Male Spotted antbirds along Pipeline Road in Soberania National Park.

Wikelski, Martin, Michaela Hau, and John C. Wingfield. 2000. Seasonality of Reproduction In A Neotropical Rain Forest Bird. *Ecology*, 81(9):2458-2472, 6 figures, 1 table. Spotted Antbirds *Hylophylax n. naevioides* studied on Pipeline Road in Soberania National Park.

Wilczynski, Walter, Michael J. Ryan, and Eliot A. Brenowitz. 1989. The Display of the Blue-Black Grassquit: The Acoustic Advantage of Getting High. *Ethology*, 80(1-4):218-222, 2 figures. Conducted acoustic transmission experiments on *Volatinia jacarina* at a grassy, abandoned air field (Chagres Field) near Gamboa. Determined that songs broadcast above the grass level vs. within the grass at the same height suffered less attenuation.

Wiley, R. Haven. 1971. Cooperative Roles in Mixed Flocks of Antwrens (Formicariidae). *Auk*, 88(4):881-892, appendix, 3 figures.

Willis, Edwin O'Neil. 1966a. The Role of Migrant Birds at Swarms of Army Ants. *Living Bird*, 5:187-231, 1 colored plate, 3 black and white plates, 2 figures, 17 tables. Drawings by Arthur Singer. Observations in Panama at Barro Colorado Island, Bohio Peninsula, Buenavista Point, Forest Reserve, and Río Agua Salud.

Willis, Edwin O'Neil. 1966b. Interspecific Competition and the Foraging Behavior of Plain-Brown Woodcreepers. *Ecology*, 47(4):667-672, 4 figures, 3 tables. Compares the foraging of *Dendroica fuliginosa* in the presence and in the absence of the Ocellated Antthrush *Phaenostictus mcleannani*. Field work at Barro Colorado Island, and in Trinidad and British Guiana, 1961-1962.

Willis, Edwin O'Neil. 1967. The Behavior of Bicolored Antbirds. *University of California Publications in Zoology*, 79:1-127, 3 appendices, 3 plates, 21 figures, 9 tables. Observations at of *Gymnopithys leucaspis* at Cerro Campana, Escobal Road, Bohio Peninsula, Buenavista Point, and Madden Forest Reserve. Appendix 1 includes locality descriptions with coordinates, habitat descriptions and study dates.

Willis, Edwin O'Neil. 1968. Studies of the Behavior of Lunulated and Salvin's Antbirds. *Condor*,

70(2):128-148, 9 figures, 5 tables. *Gymnopithys lunulata* and *G. salvini* in South America.

Willis, Edwin O'Neil. 1972. Taxonomy, Ecology, and Behavior of the Sooty Ant-Tanager (*Habia gutturalis*) and Other Ant-Tanagers (Aves). *American Museum Novitates*, 2480:1-38, 7 figures, 1 table. Habitat and behavior of *H. fuscicauda* in Panamá based on observations at numerous localities. List of study areas includes dates and coordinates.

Willis, Edwin O'Neil. 1972. The Behavior of Plain-Brown Woodcreepers, *Dendrocincla fuliginosa. Wilson Bulletin*, 84:377-420, 12 figures, 4 tables. Mainly on Barro Colorado Island, 1961-1971, with a few measurements from specimens at AMNH and MCZ.

Willis, Edwin O'Neil. 1972. The Behavior of Spotted Antbirds. *Ornithological Monographs*, 10:1-162. *Hylophylax naevioides.*

Willis, Edwin O'Neil. 1973. The Behavior of Ocellated Antbirds. *Smithsonian Contributions to Zoology*, 144:1-57, 25 figures, 9 tables. *Phaenostictus mcleannani* mainly on Barro Colorado Island, 1960-1971.

Willis, Edwin O'Neil. 1974. Populations and Local Extinctions of Birds On Barro Colorado Island Panama. *Ecological Monographs*, 44(2):153-169, 5 figures, 12 tables, 2 appendices. Appendix II lists breeding birds that have disappeared from Barro Colorado Island.

Willis, Edwin O'Neil. 1980. Ecological Role of Migratory and Resident Birds on Barro Colorado Island, Panama. Pages 205-225, 2 figures, 9 tables, appendix *in* Allen Keast and Eugene S. Morton, editors. *Migrant Birds In the Neotropics. Ecology, Behavior, Distribution and Conservation.* Washington, D.C.: Smithsonian Institution Press.

Willis, Edwin O'Neil. 1982. *Notharchus* Puffbirds (Aves, Bucconidae) as Army Ant Followers. *Ciencia e Cultura*, 34(6):777-782, 3 tables. Abstract in Portuguese. Includes Barro Colorado Island, 1960-1981.

Willis, Edwin O'Neil. 1983a. *Piaya* Cuckoos (Aves, Cuculidae) as Army Ant Followers. *Revista Brasileira de Biologia*, 43(1):29-32, 2 tables. Includes Barro Colorado Island, and Madden Forest Reserve, 1960-1981. Summary in Portuguese.

Willis, Edwin O'Neil. 1983b. Trans-Andean *Xiphorhynchus* (Aves, Dendrocolaptidae) as Army Ant Followers. *Revista Brasileira de Biologia*, 43(2):125-131, 2 tables. Includes Barro Colorado Island, 1960-1981. Summary in Portuguese.

Willis, Edwin O'Neil. 1983c. Longevities of Some Panamanian Forest Birds, with Note of Low Survivorship in Old Spotted Antbirds (*Hylophylax naevioides*). *Journal of Field Ornithology*, 54(4):413-414, 1 table. Observations in 1977 and 1981 of ant-following birds banded on Barro Colorado Island from 1960 to 1970.

Willis, Edwin O'Neil. 1984. Manakins (Aves, Pipridae) as Army Ant Followers. *Ciencia e Cultura*, 36(5):817-823, 2 tables. Abstract in Portuguese. Includes Barro Colorado Island.

Willis, Edwin O'Neil. 1985. Behavior and Systematics Status of Gray-Headed Tanagers (*Trichothraupis penicillata*, Emberizidae). *Naturalia*, 10:113-145, 13 figures, 12 tables. Abstract in Portuguese. Banded and observed birds at Belém (Brazil) and on Barro Colorado Island, 1961-1971, 1977, 1981.

Willis, Edwin O'Neil. 1992. Comportamento e Ecologia do Arapaçu-Barrado *Dendrocolaptes certhia* (Aves, Dendrocolaptidae). [Behavior and Ecology of Barred Woodcreepers, *Dendrocolaptes certhia* (Aves, Dendrocolaptidae)]. *Boletim do Museu Paraense Emilio Goeldi Serie Zoologia*, 8(1):151-216. In Portuguese with English abstract. Captured and marked a few birds at each of the three study areas including Barro Colorado Island (1960-1969). Examined 1177 specimens from museums in the U.S. and Europe.

Willis, Edwin O'Neil, and Yoshika Oniki. 1972. Ecology and Nesting Behavior of the Chestnut-backed Antbird (*Myrmeciza exsul*). *Condor*, 74(1):87-98, 8 figures, 1 table.

Willis, Edwin O'Neil, and Eugene Eisenmann. 1979. A Revised List of Birds of Barro Colorado Island, Panama. *Smithsonian Contributions to Zoology*, 291, 1-31, 1 figure. Annotated list of 366 species with introductory remarks on the island.

Willis, Edwin O'Neil, and Yoshika Oniki. 1992. As Aves e as Formigas de Correição. [Birds and Army Ants]. *Boletim do Museu Paraense Emilio Goeldi Serie Zoologia*, 8(1):123-150, 1 table. In Portuguese with English abstract. Summary of research from the Americas and Africa. Mentions Panama and Barro Colorado Island.

Willson, Mary F., Elizabeth A. Porter, and Richard S. Condit. 1982. Avian Frugivore Activity in Relation to Forest Light Gaps. *Caribbean Journal of Science*, 18(1-4):1-6, 2 tables, 1 appendix. Investigation in Illinois with additional data from Barro Colorado Island, Fipeline Road, El Volcán, Cerro Punta, and Rancho Ron (Darién). We could not locate the last.

Woakes, Ernest R. 1900. Modern Gold-Mining in the Darien. Notes on the Reopening of the Espiritu Santo Mine at Cana. *Transactions of the American Institute of Mines and Metallurgical Engineers*, 29:249-280, 1 map, 7 figures. Details of the mining operation including a history of the Spanish operations, photographs of the region, and an excellent map of the region although the longitude does not agree with recent maps.

Wolfe, L. R. 1936. The Eggs of the Eyebrowed Sparrow-Hawk, Red-Thighed Hawk and Brazilian Laughing Hawk. *Oologists' Record*, 16(4):83-84. *Accipiter superciliosus, Harpagus diodon,* and *Herpetotheres cachinnans.*

Wolfe, L. R. 1938. Eggs of the Falconiformes. *Oologists' Record*, 18(1):2-10; 18(2):25-35, 1 plate; 18(3):49-60; 18(4):74-87. Based on the author's collection.

Wolfe, L. R. 1939. Eggs of the Falconiformes. *Oologists' Record*, 19(1):7-8. Supplementary notes to Wolfe 1938.

Wolfe, L. R. 1951a. Eggs of the Falconiformes. Part I. *Oologists' Record*, 25(2):17-22. Descriptions of new specimens obtained by the author since his 1938-1939 publications.

Wolfe, L. R. 1951b. Eggs of the Falconiformes. Part II. *Oologists' Record*, 25(3):36-42.

Wolfe, L. R. 1951c. Eggs of the Falconiformes. Part III. *Oologists' Record*, 25(4):49-54.

Wolfe, L. R. 1954. Nesting of the Laughing Falcon. *Condor*, 56(3):161-162, 1 figure. *Herpetotheres cacchinnans.*

Worth, C. Brooke. 1938a. Nesting of Salvin's Barbet. *Auk*, 55(3):535-536. Two nests of *Eubucco bourcieri salvini* at 4,100 feet near El Volcan, Chiriqui.

Worth, C. Brooke. 1938b. Nesting of the Pepper-Shrike. *Auk*, 55(3):539-540. *Cyclarhis flaviventris subflavescens* at El Volcan, Chiriqui.

Worth, C. Brooke. 1939a. Nesting of Some Panamanian Birds. *Auk*, 56(3):306-310. Worth and Cressin made observations and collections of *Thripadectes rufobrunneus* at Finca Lerida (Boquete). *Turdus assimilis cnephosa*, *Chlorospingus ophthalmicus novicus* and *Cyanocompsa cyanoides caerulescens* observed near the Río Garaché [= Gariché] near Volcán.

Worth, C. Brooke. 1939b. Cedar Waxwings and Other Birds in Panama. *Bird-Lore*, 41(5):279-284, 2 figures. Illustrated by Conrad Roland. Observations on *Bombycilla cedrorum* and other birds made during three weeks stay at Mönniche's plantation, Finca Lerida, Chiriquí. Specimens at ANSP labeled Boquete.

Worth, C. Brooke. 1942. Notes on Hummingbirds at Chiriqui, Panama. *Auk*, 59(3):364-368.

Songs, flight and nesting, mainly of *Phaethornis guy coruscus*, *Campylopterus hemileucurus* and *Glaucis hirsuta aenea*.

Worthington, A. 1983. Population Sizes and Breeding Rhythms of Two Species of Manakins in Relation to Food Supply. Pages 213-225, 1 figure, 3 tables *in* Egbert Giles Leigh, Jr., Austin Stanley Rand, and Donald M. Windsor, editors. *The Ecology of a Tropical Forest. Seasonal Rhythms and Long-Term Changes*. Oxford: Oxford University Press. *Manacus vitellinus* and *Pipra mentalis*, Isla Orchid in Gatun Lake, 150 m N of Barro Colorado Island, 1978-1979.

Wright, Stewart Joseph. 1979. Competition Between Insectivorous Lizards and Birds In Central Panama. *American Zoologist*, 19(4):1145-1156, 6 figures, 5 tables. Map of the study sites including: Abogado Island, Barro Colorado Island, Gatun Dam, Juan Gallegos Island, Lion Hill Island, Orchid Island, Pantera Island, and Pipeline Road, 1976-1978.

Wright, Stewart Joseph. 1985. How Isolation Affects Rates of Turnover of Species On Islands. *Oikos*, 44(2):331-340, 3 appendices, 5 figures. Censuses (1977-1981) on Abogado, Lion Hill, Orchid, Pantera, Puma, and Tres Alemendras islands. Map of the study area. Appendix 1A lists the resident species for each island. List of common forest trees.

Wright, Stewart Joseph. 1988. Patterns of Abundance and the Form of the Species-Area Relation. *American Naturalist*, 131(3):401-411, 4 figures. Censuses on Barro Colorado, Juan Gallegos, Lion Hill, Pantera, Puma, and Tres Almendras Islands in Oct 1980 and Feb-Mar 1981. No maps or species lists.

Wright, Stewart Joseph, John Faaborg, and Claudia J. Campbell. 1985. Birds Form Tightly Structured Communities in the Pearl Archipelago, Panama. Pages 798-812, 5 figures (includes a map), 3 tables *in* P. A. Buckley et al., editors. Neotropical Ornithology. *Ornithological Monographs* 36.Washington, D.C. Abstract in Spanish. Species lists for 14 islands based on censuses and published literature. The authors indicate (footnotes 7, 8, p. 807) that "N. G. Smith (pers. comm.) now retracts records [of *Thamnophilus doliatus*] for Casaya, Chapera, and Bayoneta (Smith 1971)," and that a record of *Formicivora grisea* for Saboga (Wetmore, 1972:183-184) could not be verified.

Wyse, Lucien Napoléon Bonaparte. 1886. *Le Canal de Panama: l'Isthme Americain, Explorations; Comparaison des Tracés Étudiés, Négociations; État des Travaux*. [The Panama Canal. The American Isthmus, Explorations; Comparison of Contour Studies, Negotiations, State of the Work]. 399 pages, 2 maps, 90 wood engravings. Paris: Librairie Hachette et Cie. Account of two expeditions 1876-1877, and 1877-1878. Viguier was the surgeon and naturalist on the first expedition when the party reached Panama in Nov 1876. It appears that the group traveled into Darién via the Río Tuira reaching Chepigana on Dec 11. They worked throughout the region until the rains interfered with the surveys in April. Frequent casual mention of birds.

Young, Allen M. 1971. Roosting of a Spotted Antbird (Formicariidae: *Hylophylax naevioides*) in Costa Rica. *Condor*, 73(3):367-368, 1 table.

Young, Orrey P. 1982. Aggressive Interaction Between Howler Monkeys and Turkey Vultures: The Need to Thermoregulate Behaviorally. *Biotropica*, 14(3):228-231. *Cathartes aura* on Barro Colorado Island, 1974.

Yudins, Richard, and Clyde S. Stephens. 1989. Banana Railroads. Pages 23-32 *in* Clyde S. Stephens., *Bananeros in Central America, True Stories of the Tropics*. Fort Myers, Florida: Press Printing Company. Narratives concerning the United Fruit Company.

Zaret, Thomas M. and R. T. Paine. 1973. Species Introduction in a Tropical Lake. A Newly Introduced Piscivore Can Produce Population Changes in a Wide Range of Trophic Levels. *Science*, 182(4111):449-455, 5 figures, 2 tables. Explores changes in Gatun Lake following the introduction of peacock bass *Cichla ocellaris*. Mentions a few species of birds.

Zimmer, John T. 1929. The Birds of the Neotropical Genus *Deconychura*. *Field Museum of Natural History, Zoological Series*, 17(1):1-18. Includes specimens from Panama.

Zimmer, John T. 1932. Studies of Peruvian Birds. VIII. The Formicarian Genera *Cymbilaimus*, *Thamnistes*, *Terenura*, *Percnostola*, *Formicarius*, *Chamaeza*, and *Rhegmatorhina*. *American Museum Novitates*, 584:1-20. Includes specimens from Panama.

Zimmer, John T. 1934. Studies of Peruvian Birds. XIII. The Genera *Dendrexetastes*, *Campylorampbus* and *Dendrocincla*. *American Museum Novitates*, 728:1-20. Includes specimens from Panama.

Zimmer, John T. 1938. Notes on Migrations of South American Birds. *Auk*, 55(3):405-410.

Zimmer, John T. 1939. Studies of Peruvian Birds. XXXI. Notes on the Genera *Myiotriccus*, *Pyrrhomyias*, *Myiophobus*, *Onychorhynchus*, *Platyrinchus*, *Cnipodectes*, *Sayornis* and *Nuttallornis*. *American Museum Novitates*, 1043:1-15, 1 figure. *Cnipodectes subbrunneus panamensis* n. subsp. from Barro Colorado Island. Includes specimens from Panama.

Zimmer, John T. 1940. Studies of Peruvian Birds. XXXV. The Genera *Phylloscartes*, *Euscarthmus*, *Pseudocolopteryx*, *Tachuris*, *Spizitornis*, *Yanacea*, *Uromyias*, *Stigmatura*, *Serpophaga*, and *Mecocerculus*. *American Museum Novitates*, 1095:1-19. Includes specimens from Panama.

Zimmer, John T. 1941a. Studies of Peruvian Birds. XXXVII. The Genera *Sublegatus*, *Phaeomyias*, *Camptostoma*, *Xanthomyias*, *Phyllomyias*, and *Tyranniscus*. *American Museum Novitates*, 1109:1-25. Includes specimens from Panama.

Zimmer, John T. 1941b. Studies of Peruvian Birds. XXXVIII. The Genera *Oreotriccus*, *Tyrannulus*, *Acrochordopus*, *Ornithion*, *Leptopogon*, *Mionectes*, *Pipromorpha*, and *Pyrocephalus*. *American Museum Novitates*, 1126:1-25. Includes specimens from Panama. Aspecimen credited to Panama (p. 11) from "East slope of Tacarcuna," is probably from the Colombian side of the range.

Zimmer, John T. 1944. Two New Subspecies of *Catharus aurantiirostris*. *Auk*, 61(3):404-408. *C. griseiceps* merged with *C. aurantiirostris*. Includes specimens from Panama.

Zimmer, John T. 1945. A New Swift from Central and South America. *Auk*, 62(4):586-592. *Cyspeloides cryptus* recorded from Port Obaldia, Panama.

Zimmer, John T. 1947. *Sublegatus arenarum* - A Correction. *Auk*, 64(3):456. Correction to Zimmer (1941b). Specific name should be *arenarum*, not *glaber*.

Zimmer, John T. 1950. Studies of Peruvian Birds. No. 55. The Hummingbird Genera *Doryphera*, *Glaucis*, *Threnetes*, and *Phaethornis*. *American Museum Novitates*,1449:1-51, 1 table. Includes specimens from Panama.

Zimmer, John T. 1951. Studies of Peruvian Birds. No. 60. The Genera *Heliodoxa*, *Phlogophilus*, *Urosticte*, *Polyplancta*, *Adelomyia*, *Coeligena*, *Ensifera*, *Oreotrochilus*, and *Topaza*. *American Museum Novitates*, 1513:1-45. Includes specimens from Panama.

Zimmer, John T. 1955. Studies of Peruvian Birds. No. 66. The Swallows (Hirundinidae). *American Museum Novitates*, 1723:1-35. Includes specimens from Panama.

Zimmerman, Dale A. 1957. Display of the Least Grebe. *Auk*, 74(3):390. Observations of *Podiceps dominicus* in Mexico.

Zimmerman, John L., and James V. Morrison. 1972. Vernal Testes Development In Tropical Wintering Dickcissels. *Wilson Bulletin*, 84(4):475-481, 1 figure. Collected *Spiza americana* in Panama, the Canal Zone, and Illinois in 1961 and measured testes length to calculate testicular growth rate.

Deborah Siegel is a biologist and a teacher. She received her B.S. (Biochemistry) from the University of Wisconsin-Madison and her M.S. (Fishery Resources) from the University of Idaho. She has worked in Africa, Europe, and the Americas, and began her career as a Peace Corps Fisheries Volunteer in Sierra Leone. She is currently teaching biology at the International School of Helsinki.

Storrs L. Olson was born in Chicago in 1944 and moved to Tallahassee, Florida, with his family in 1950. In high school he was mentored by Horace Loftin and spent many enjoyable days with him in the field trapping and banding shorebirds and raptors. He got his first taste of the Neotropics in 1963 when he stayed with the Loftin family in Panama while completing a semester at Canal Zone Junior College. He returned to Panama in 1966 as part of a program investigating the effects of botulinal toxins on vultures. While in Panama he collected birds, including a number of important new specimen records that brought him to the attention of Alexander Wetmore, whom he first met on a trip to Washington in 1967. After receiving Bachelor and Master's degrees from Florida State University, Olson moved to the Washington area in 1968 and held various positions in the Smithsonian Institution until moving to Baltimore to attend Johns Hopkins University from which he received his doctorate in 1972. In August 1971, he moved into the National Museum of Natural History on a predoctoral fellowship and never left, attaining the post of Curator of Birds in 1975. He has published nearly 400 titles on fossil and living birds of all ages and many different parts of the world, including co-authorship of the final volume of Wetmore's Birds of the Republic of Panamá. Olson participated in the Smithsonian Bocas Expeditions from 1987 to 1991, during which specimens of vertebrates were collected on all of the main islands of the archipelago of Bocas del Toro, Panama.

Longitude West from Greenwi
C A R I B
COSTA RICA
Carreta Pt.
BOCA DEL DRAGO
COLUMBUS ISLAND
BOCA DEL TORO
BOCAS DEL TORO
PROVISION ISLAND
Cape Valiente
MOSQUITO GULF
Bluefields Pt.
CHIRIQUI LAGOON
Chiriqui Grande
Plantation Bay
St. Christopher Bay
Boppen Bluff
BOCAS DEL TORO
SERRANIA DE TABASARA
Cerro Colorado
CHIRIQUI
SERRANIA DE CARONES
GULF DULCE
Bugaba
Boqueron
DAVID
Padregal
Divala
Bartolome
Gualaca
Chorcha
Horconcitos
San Felix
Remedios
SEVILLA I.
DAVID BAY
PARIDA I.
BURICA PEN.
Burica Pt.
Pueblo Nuevo
Las Palmas
Santa Fe
San Francisco
SANTIAGO
Sona
Rio de Jesus
Montijo
VERAGUAS
Calobre
Nata
Aguadulce
Santa Maria
Parita
Chitre
Muertos Pt.
CONTRERAS ISLANDS
Bahia Honda
AFUERA I.
RANCHERIA I.
MONTIJO BAY
Lorenzo Bay
GOBERNADOR I.
CEBACO ISLAND
Hermosa Bay
COIBA ISLAND
DAMAS BAY
JICARON I.
JICARITA I.
Mariato Point
Las Minas
AZUERO PENINSULA
Temperate Zone
Upper Tropical Zone
Lower Tropical Zone
Dotted and undotted purple areas indicate humid and arid divisions, respectively, of Lower Tropical Zone.